GRAVITY'S SHADOW

GRAVITY'S SHADOW

SHADOW

The Search for Gravitational Waves

Harry Collins

The University of Chicago Press
Chicago & London

Harry Collins is distinguished research professor of sociology at Cardiff University, where he also directs the Centre for the Study of Knowledge, Expertise, and Science. Professor Collins has published over one hundred papers in the peer-reviewed literature, and he has edited or coedited four books, including *The One Culture? A Conversation about Science* (with Jay Labinger; University of Chicago Press, 2001). In addition, he has written or cowritten six books, most recently *The Shape of Actions: What Humans and Machines Can Do* (with Martin Kusch, 1998).

The University of Chicago Press, Chicago 60637
The University of Chicago Press, Ltd., London
© 2004 by The University of Chicago
All rights reserved. Published 2004
Printed in the United States of America

13 12 11 10 09 08 07 06 05 04   1 2 3 4 5

ISBN: 0-226-11377-9 (cloth)
ISBN: 0-226-11378-7 (paper)

Library of Congress Cataloging-in-Publication Data

Collins, Harry.
  Gravity's shadow : the search for gravitational waves / Harry Collins.
    p. cm.
  Includes bibliographical references and index.
  ISBN 0-226-11377-9 (hardcover : alk. paper)—ISBN 0-226-11378-7
(pbk. : alk. paper)
    1. Gravitational waves—Research.   2. Gravitational waves—Measurement.
I. Title.
QC179.C65   2004
539.7'54—dc22

                                            2003023823

To Joe
and Lily

CONTENTS

PREFACE

For more than 40 years, scientists have been trying to detect gravitational waves. At the turn of the millennium, their search is
being conducted with devices costing hundreds of millions of dollars that are operated by teams from the United States, Britain,
France, Germany, Italy, Japan, and Australia. When the waves are
found, the discovery will demonstrate the truth of Einstein's theories, enable us to look into the heart of black holes, and found a
new gravitational astronomy. Sometime in the not-too-distant future, it will be agreed that gravitational waves have been detected.
I started watching the gravitational scientists about 30 years ago,
and when this exciting moment comes for them, I hope I'll still be
watching. But in some respects gravitational wave detection has
already had more than its fair share of excitement. The scientists
must hope that there is less of this kind of excitement to come.

This book introduces the science of gravitational waves and
examines the way scientific knowledge is made. The traumas of
the field's past are what make it an especially good case study of
the methods of science. The search for gravitational waves began in

the late 1960s and was pioneered by Joseph Weber.[1] By the early 1970s, Weber believed he had seen the waves, but he could not turn his claims into accepted scientific truth. Right up to the present day, gravitational wave scientists have been making positive claims, but they have been unable to convince the wider community; most of these claims, including Weber's, are nearly forgotten. Among other things, I try to bring the claims back to life so that I can explain how forgetting works within the scientific community. I have been asking the scientists questions about these things and recording their replies since 1972.

There is a danger that social analysis of science will become a private conversation within a narrow disciplinary group. The revolution in our understanding of the nature of the sciences, which began in the 1970s, has been an enormous success within the social science and humanities disciplines, but it will disappear as one more academic fashion unless the social studies of science makes contact with the natural scientists. For this reason among others, in writing this book I have had the community of gravitational wave scientists very much in mind. Nearly every sentence has been written with one of my respondents metaphorically looking over my shoulder. I try never to take refuge in forms of words, or unspoken understandings, or philosophical flights of fancy, that can belong only to an exclusive community.

In this book, then, I try to bring to nonspecialists the esoteric world of the social studies of science; I also try to bring to nonscientists the esoteric world of physics. The model of physical science presented here is, however, "irreverent." This is how it works: The prevailing model of science—what we might call the canonical model—establishes the idea of science by contrasting it with the mundane. Science is seen as divine and has replaced religion in more ways than one. But I describe science as itself essentially mundane—that's where the irreverence comes in. To those whose way of thinking is informed by the canonical model—and that includes science's fiercest critics, who blame it for its failure to live up to the model, as well as its most zealous supporters, who think the canonical model describes it correctly—this irreverence sounds like blasphemy. But almost nothing in this book is intended as criticism, and there are very few passages of actions described that I believe should have been done differently. Even when there were times when I wished things had been different, my views on these matters, as I try to point out, are worth little or nothing.

1. Joe Weber died on September 30, 2000, at the age of 81.

The problem is not that science does not live up to the canonical model but that the canonical model is wrong. Nothing that humans do is divine; all is mundane. What we see described here is the mundane, but it is, nevertheless, an example of the very best that humans can do. There is one more irony: to do the mundane as well as we can do it, we must act as though we are divine. Here, as in so much of life, the "is" of how things are done does not provide a guide to the "ought" of how we should try to do them. This book could be misread, as so much social studies of science has been misread (sometimes by its authors), as though the "ought" followed from the "is"; in such a case it would be a dangerous book. On the contrary, I want it to be a book that safeguards science against its critics by setting more reasonable *expectations* for what science can do, even while maintaining our *aspirations* for how scientists do it. The search for gravitational waves is a heroic and exemplary passage of human activity. It counts among the things that make it worthwhile to be human.

ACKNOWLEDGMENTS

I have been working on this project for so long that I owe an acknowledgment to almost everyone I know in the academic world, not least those who commented on those sections of the book that were first published as papers. The extended work itself, however, was first put into some kind of coherent shape in May of 2002, when I held a fellowship at the Max Planck Institute for the History of Science in Berlin. Otto Sibum and his colleagues invited and suffered my obsessions for this most enjoyable and exceptionally fruitful period. Jed Buchwald initiated my stay at California Institute of Technology as Andrew W. Mellon Visiting Professor of History from January to March 2003; here I found a cultured and intellectually stimulating environment where I could bring the work close to publishable standard—not to mention avoid the worst of a British winter. An unexpected bonus was Jed's technical skill and his cornucopia of modern gadgets; with these we knocked into shape the photographs illustrating this book.

I asked the scientists in my study to give me a lot; they responded generously. They gave me access to inner recesses of their research facilities in exotic places, something normally granted only after a long and demanding scientific apprenticeship; they gave me access to their private and provisional thoughts, something normally never granted to outsiders, because exposure of these things is antagonistic to the very project of science; and they made me a member of their communities—oftentimes even a welcome member, sharing their food and sustaining me personally as well as intellectually. Over the last ten years I have spent more time with gravitational wave physicists than with any other academic community outside my home university, and I am a better person for it; it has been wonderful to live a good life within a society driven by intense intellectual curiosity, a constant reminder of how lucky I am to have stumbled into an academic career.

Of course, it hasn't all been plain sailing: convincing some of the skeptical scientists that my project—so different from what they must have expected at the outset—had its own integrity has been almost as fascinating and demanding as the project itself. Luckily, a number of respondents have been sympathetic from the start and have helped me through the hard times. It would be invidious to draw up a list of names of those who have been especially helpful because I would never get it right, but I ought to mention Rich Isaacson, who guided me into the National Science Foundation (NSF), who gave me strong support during one particularly unfortunate misunderstanding about confidentiality, and who was always there for a good conversation about the science and the politics of the field. Above all I must thank my new friend, the author of the standard text on contemporary gravitational wave detection, Peter Saulson. Peter is one of nature's gentlemen. He always supported me when I felt like giving up, continually discussing the sociological themes of my work as well as resolving all the scientific conundrums I could throw at him.

The manuscript of the first draft of this book was 1,000 closely printed pages—a pile about six inches high. When you first write such a thing you really do not know whether it is any good or if it is all a massive folly. Martin Kusch was the first to give me a sanity check, and after just one week of reading the draft pronounced it good. Peter Saulson then gave the whole thing a line-by-line reading, picking up all manner of subtle points of science and sociology and, donating his time generously to discussion, helped me avoid a number of errors. Historian of science Simon Werrett also undertook the task of reading the whole manuscript and gave me a view from the professional community that resolved a number of doubts, particularly over length. Dan Kennefick read the whole thing as well, reassuring me about the interest of the story. Don Mackenzie (as I was to discover later) read the entire manuscript for the University of Chicago Press and provided a set of comments overwhelming in their generosity along with a series of astute criticisms. Another reader for Chicago, who would remain anonymous, wrote a report laden with superlatives and some criticisms which caused me to reshape the ending of the book. One section of the book about which I was left unsure was rapidly read for me by Trevor Pinch, who reinforced my initially good opinion of it in very positive terms.

In the summer of 2002, the massive manuscript hit the desks of a number of scientists, and at one time or another sections of it were projected at most of the scientists quoted in its pages. Those who read sections nearly always responded with grace, sometimes even when what I said about them could be read as less than complimentary. In the main I gave the manuscript to these scientists so that they could do an "interested reading": that is, make sure I had not said anything that was unfair to them or their projects. The proportion of disinterested comment to interested comment was higher than I had any right to expect, however. Notable in this respect was Gary Sanders, who started his review at the beginning, making many useful comments which had nothing to do with his special interest and much to do with an enthusiastic engagement with the social nature of science. Ron Drever, too, began at the beginning and offered comments quite different from those I had expected along with a series of important technical corrections. Robbie Vogt and Kip Thorne read sections that could be construed as critical of their actions and responded, once more, with more generosity of spirit than I had any right to expect even within a community driven primarily by academic values. None of this is to say that any of these scientists endorse or support any of the specific claims made in this book. I would hate it to be the case that scientists who are generous with their time should feel co-opted into my project against their will; all arguments, judgments, mistakes, and infelicities are solely my responsibility.

In the early years of the project I was strongly supported by my colleagues at the University of Bath. In more recent years, Cardiff University has given me every support, and the Cardiff group has patiently listened to my "tales from the field," read my draft papers, and skillfully maintained the impression that yet more sociological theses about gravitational wave detection were both fascinating and intellectually worthwhile. Many of my colleagues from the science studies community have provided friendship and hospitality on my travels as well as reading drafts of manuscript sections and of papers. I hope they will forgive me for not mentioning them individually, but I am sure I would miss someone from such a long list; for that I could not forgive myself.

On the advice of my publishers, the book as printed is somewhat shorter than the original manuscript. One way it has been condensed was through the removal or abridgment of certain lengthy quotations or extracts from documents—a wrench because, though this revision increases the readability of the book, it makes it less of a historical resource. I hope that in years to come, as gravitational astronomy grows ever more important, the words of the scientists of the time will come to seem more interesting. Such passages as have been removed from the book are, however, still accessible on a Web site, which is referenced wherever appropriate with the word *WEBQUOTE*. The site itself can be accessed at www.cardiff.ac.uk/socsi/gravwave/webquote, and in due course its quotations may also find their way onto the Web site of the American Institute of Physics.

In this research I have been given very good (nonfinancial) support by the NSF, which allowed me possibly unprecedented access to its archives for a non-US citizen. Reading the NSF papers has been extremely useful for confirming what I have learned elsewhere and for understanding the scope of some of the arguments. I used the archives under the terms of a signed legal agreement, which, quite properly, allows me to quote from them only rarely and attribute them hardly ever.

The project also has been supported by a number of grants from the UK Economic and Social Research Council (ESRC), which in the early days was the SSRC (Social Science Research Council). It all started in 1971 as part of an SSRC-funded Ph.D. dissertation, "Further Exploration of the Sociology of Scientific Phenomena," followed by an SSRC grant in 1975. This grant was for $893, within which budget I managed to drive across America twice, investigating three different fields of science. Then in 1995–96, ESRC R000235603 funded "The Life after Death of Scientific Ideas: Gravity Waves and Networks"; from 1996 to 2001, ESRC R000236826 funded "Physics in Transition"; and from 2002 to 2006, all being well, ESRC R000239414 will continue to fund "Founding a New Astronomy."

I must also thank my friends and family, who have tolerated the sometimes overlong accounts of my travels that I have inflicted on them from abroad via email; this was my way of making a lonely hotel room feel like home.

The late Susan Abrams, who was my editor in the case of books I published previously with the University of Chicago Press, was a constant source of encouragement whenever I discussed the new project with her. It was to her that I first dispatched the draft manuscript, but unfortunately she was too ill to read it. Susan will be deeply missed.

COMMON ACRONYMS IN GRAVITATIONAL WAVE RESEARCH

ACIGA	Australian Consortium for Interferometric Gravitational Astronomy
AIGO	Australian International Gravitational Observatory
ARS	Australian Research Council
CERN	European Organisation for Nuclear Research
CNRS	Centre Nationale de la Recherche Scientifique
EGO	European Gravitational Observatory
GWADW	Gravitational Wave Advanced Detector Workshop
GWDAW	Gravitational Wave Data Analysis Workshop
GWIC	Gravitational-Wave International Committee
IGEC	International Gravitational Event Collaboration
INFN	Istituto Nazionale di Fisica Nucleare
LIGO	Laser Interferometer Gravitational-Wave Observatory
LSC	LIGO Scientific Collaboration
NAS	National Academy of Sciences
NASA	National Aeronautics and Space Administration
NSB	National Science Board
NSF	National Science Foundation

TWO KINDS OF SPACE-TIME

Basically, we don't know what caused this large peak. It conceivably could have been due to a burst of very intense gravitational radiation from an explosive type of astronomical event, but more likely it was a malfunction of the equipment.

GRAVITATIONAL WAVES

Gravitational wave detection is about seeing the biggest things that ever happen—the collisions, explosions, and quakings of stars and black holes—by measuring the smallest changes that have ever been measured: the effect of gravitational waves on experimental apparatus mounted on the surface of the earth. So far scientists have spent about 40 years talking about whether there was anything to detect and whether it has yet been detected. The epigraph records the very first of such speculations, but it stands for nearly all of them.

One claim that is different was recognized by the awarding of the 1993 Nobel Prize in physics to two astronomers, Russel Hulse and Joseph Taylor, for an indirect confirmation of the existence of gravitational waves.[1] Here, however, we are concerned with direct detection. The majority view among scientists is that there will be agreement that gravitational waves have been directly detected by the time the 40 years of searching has turned into half a century; in other words, we expect to be sure we have seen gravitational waves in the first decade of this millennium or shortly thereafter.

According to the theory of relativity, we are constantly bathed in gravitational radiation but quite insensible to its passage. We are penetrated by many other invisible forces and entities; for example, each hour roughly a thousand million million neutrinos born in the heart of the Sun pass through our bodies, leaving no mark. Scientists did not succeed in detecting these neutrinos until 1956, yet they are much easier to see than gravitational waves.

To see gravitational waves, we have to look at their effect on the distance between things. Very roughly, what we have to do is measure the distance between two objects and wait for it to change when a gravitational wave passes by. It will not change by much.

On March 13, 1991, members of the US House of Representatives learned about the size of gravitational waves from Dr. Tony Tyson. He told them,

> Imagine this distance: travel around the world 100 billion times (a total of 2,400 trillion miles, or one million times the distance to Neptune). Take two points separated by this total distance. Then a *strong* gravitational wave will briefly change that distance by less than the thickness of a human hair. We have perhaps less than a few tenths of a second to perform this measurement. And we don't know if this infinitesimal event will come next month, next year, or perhaps in thirty years.[2]

Dr. Tyson was hoping to dissuade Congress from spending about $200 million on a program to build two huge new gravitational wave detectors, of a type known as interferometers. The scientists wanted to build one in Washington State and the second in Louisiana. Tyson and his colleagues were briefly successful in reducing political support for the program, but it was

---

1. Hulse and Taylor observed the very slow degradation of the orbit of a pair of binary stars over many years and showed that it was consistent with the emission of gravitational radiation according to Einstein's theory. For an engaging and informative account of this episode, see Bartusiak 2000.

2. Testimony of Dr. J. Anthony Tyson before the Subcommittee on Science of the Committee on Science, Space and Technology, United States House of Representatives, on March 13, 1991 (p. 4). Accessed in an academic library in Washington, DC.

funded a year later and the instruments are now becoming operational. Slightly smaller interferometers are also starting up in Italy, Germany, and Japan. These huge and expensive detectors are just the latest stage in the passage of scientific research that is entering its fifth decade.

As we have seen, some scientists believe gravitational waves have already been detected using earlier generations of simpler and cheaper detectors. Our big questions are the following: Why were these early claims disbelieved? Will the findings of the modern generation of detectors be more believable, and if so, why? The answers to the second question will not be known for a few years.

GRAVITATIONAL RADIATION AND BELIEFS IN GRAVITATIONAL RADIATION

We have two things to talk about: gravitational radiation—what I call "gravity's shadow"—and belief in gravitational radiation—gravity's shadow's shadow, as it were. Let us start with gravitational radiation. Tony Tyson explained that to see gravitational waves, we must see a change roughly equal to the thickness of a human hair in the distance from Earth to the nearest star. But since the measurement instruments we will use here on Earth are necessarily built on a terrestrial rather than an astronomical scale, we will have to measure changes in a much smaller distance. The distances we have to play with are between a couple of meters for the smaller devices and four kilometers for the largest ones. The latest detectors may be "huge" on a terrestrial scale, but what is huge so far as the earth is concerned is tiny when we consider the scale of the heavens.

To measure in such huge/tiny devices the equivalent of the thickness of a human hair in the distance to a star means watching for changes of the order of a fraction of the diameter of an atomic nucleus. It is quite hard to get a sense of just how small that is, so appendix 1 to this introductory chapter is devoted just to this question.

The big things we are looking for are happening all the time. For example, stars are exploding or colliding constantly, but it doesn't make news because most stars are such a long way away we know nothing about them. In 1987, the explosion of a star did make news, but only because it was just visible from Earth—very close as supernovae go, but it was still 169,000 light-years distant. A visible explosion of a star is a rare event; the last known occurrence was hundreds of years ago.

When these very big things happen, a lot of mass gets moved around very fast. Lumps of matter the size of suns start to get pushed around at near the speed of light. Just before two stars collide they might spend a few moments of their lives swirling around each other at thousands of revolutions per second; when a star explodes it shoots out much of the material it is made of about as fast as it is possible for anything to go.

The theory of relativity tells us that under extreme circumstances such as this, the constituents of reality lose their identity. As things approach the speed of light, mass forgets its constancy and starts to increase while time seems to slow to a crawl. Matter and energy forget what they are, so what was once a little bit of mass can become a huge amount of energy and vice versa. In the vicinity of sun-sized lumps of stuff cavorting at near the speed of light, the very fabric of space-time is shaken. This shaking of space-time is a gravitational wave, and the vibration spreads through the universe like ripples on a pond.

Why, then, given that they are produced by such massive events, and that they involve such huge amounts of mass turning into vast outpourings of energy, is it so difficult to detect gravitational waves? It is because the surface of space's pond is so stiff that the ripples are minute—like the "ripples" caused by the passage of sound in a solid piece of steel, but much smaller.[3] And such vibrations as there are weaken as they spread evenly and in all directions through the huge distances of space, so that when they reach us there is almost nothing left to sense.

To repeat, in space the cavortings and collisions of countless suns cause them each to give up a good proportion of their mass and turn it into energy. A good proportion of this energy becomes gravitational waves— tiny vibrations in space-time. Earth is bathed in these gravitational waves, but they are so weak that nothing of them can be felt or seen except with instruments of unprecedented sensitivity—and even then hardly ever. Although this is the story of the building of those instruments, it is still more the story of the seeing of those waves.

SEEING'S SPACE-TIME

It is easy enough to understand what a story about building instruments is like, but how can there be a story about the seeing of things? How can there be a historical dimension to seeing?

Reflect on what you know for sure about the things that are the business of science. The answer is almost nothing. We are all the same in this

---

3. As nicely explained in Blair and McNamara 1997.

respect; we all know almost nothing. How can I be so confident about writing this when you, reader, might be anyone, perhaps even a gravitational wave scientist? It is because even if you are among the best and most brilliant scientists in the world, you know for sure almost nothing more than the most scientifically ignorant of us when by "knowing for sure" we mean knowing to the standards of scientific proof: direct and repeated witnessing. Areas of expertise are like crevasses: deep and narrow. Even the best and most brilliant scientists have directly witnessed real proof in only that tiny part of the natural world in which they are specialists—and it is not so clear what "directly witnessed" means even for them, since "witnessing," and this is becoming more and more noticeable nowadays, is merely the conclusion of a very long chain of inferences.

As for the rest of the natural world, scientists know about things in the same way as we know about things: from hearsay. And even if you are one of the scientists I describe in these pages—one of the gravitational wave specialists—you know most of what you know about even gravitational waves from hearsay; that sounds odd, but think about it! Nearly all the science you know you learned from the printed page, the lecture theater, or other scientists' talk and actions. Even the results you know by so-called direct witnessing are tiny corks bobbing on a huge sea of trust—trust in the results of earlier experiments, trust in the colleagues who work with you, trust in the meters and the materials which make up your apparatus, and trust in the computers that analyze the experiment.

We do not normally think of these things as trust because one only needs to "trust" in a self-conscious way, when there is reason to distrust. Thus, one does not say that a child "trusts" its parents, because it never occurs to us that trust is needed.[4] But given the relative power of parents and children, one immediately sees that the relationship works only because of what the child can take for granted. Likewise with science. Scientists do not notice that almost everything they know depends on trust and hearsay, or "socialization," as the sociologist might say, because there is rarely an opportunity to notice it. Usually we become aware of the fabric of trust only when it is torn. When a scientist cheats, we suddenly find we have been swimming in the sea of trust all along. And think how easy it would be to fool everyone, including the scientists, into thinking that a gravitational wave had been found. One could coordinate some disturbances

---

4. An extended analysis of the role of trust in science is Shapin 1994. This has become the standard reference; if it has a fault, it is that it does not clearly disentangle the active and passive senses in which others are trusted.

at each detector site; one could plant a little subroutine in a computer or two; one could bribe or threaten a few colleagues and editors. These things happen in sport and finance all the time. And sometimes they happen in science. Indeed, gravitational wave scientists have belatedly discovered that security is an issue now that the big detectors are coming "on air." In 2001, new firewalls were installed in the American program's computers to keep out hackers who might enjoy a prank. But, of course, most of the key scientists work inside these firewalls—they just have to be trusted.

It is because most of what we know we know from trusted hearsay that there can be a sociology of science and a sociology of gravitational waves; these sociologies explore the way we come to trust some bits of hearsay and not trust other bits of hearsay. More than half this book is about gravitational wave hearsay and how we come to believe it or not. The study of how hearsay comes to be believed by the majority is what is meant by the study of "the social construction of" this or that.

It is because the world is made of hearsay that there cannot only be a historical dimension to seeing, but at least two spatial dimensions in addition. That is how it is that "seeing" has its own space-time.

I see a table, a cup and saucer, a tree, the Sun: my senses apprehend them immediately; nothing stands between me and the object. That is simple seeing.[5] In advanced science it is different: when scientists see things, there is a tenuous chain of instruments and inferences between the senses and the sensed. Peer down a microscope into a drop of pond water and you see strange shapes. What tells you that you are seeing *through* the microscopic at magnified life-forms in motion, rather than, say, *into* the microscope at some kind of kaleidoscope-like effect? It is training and persuasion; it is what you have been told by trusted teachers about what the microscope reveals. This, in short, is the beginning of scientific socialization. Would someone from a culture with little science and technology see miniature life-forms when looking into a microscope?

In most modern experiments, the chain of inference is far longer than in the case of the microscope. What the contemporary physicist or astronomer "sees" is usually a string of numbers generated by a computer. Those numbers have to be "seen" as some entity or other, and it has to be accepted that every link in the chain of instruments from the far end of the sensing device to us has been working properly. Included in this chain are all the transformations that happen to the vestigial signal as it passes

---

5. I use the term *simple seeing* as a commonsense category. We know seeing is not simple at all—as I will argue below.

through the apparatus, as it is changed and amplified from one form to another, finally being spat out as a string of digital symbols, which themselves are transformed by statistical analyses. The fainter the object being sensed, the greater the transformations it must undergo before it can be "seen." In the case of gravitational waves, the signals are as weak as anything that has ever been measured, and their journey is unprecedentedly long and hazardous; every link in this long, long, chain of inference provides an occasion for doubt or dispute about the veracity of the apprehension. To "see" in a science like gravitational wave detection is to agree, at some point, to cease to doubt the chain of inference—to see, then, is to *agree* to see.

Dispute among scientists amplifies seeing's space-time and makes it easier for us to apprehend. As with the theory of relativity, under extreme circumstances, seeing's space-time dimensions blur. Among the extreme circumstances that can beset scientific seeing are faintness of signal, a high degree of novelty, and competing social pressures to see things one way rather than another. Even in ordinary life we come across such circumstances when scientific disputes enter law courts and the like.

The trick is to understand the relationship between this complicated, disputable, "stormy-weather" kind of seeing and the ordinary, everyday, immediate-apprehension-of-the-senses, calm-weather, cup-and-saucer kind of seeing. Most people have the relationship the wrong way around. They think that cup-and-saucer, calm-weather seeing is the basic kind of seeing, and they spend their time marveling at the cleverness of scientists who seem to make stormy-weather seeing so secure and certain. For example, philosophers of science might try to show that scientific seeing is just a very elaborate form of ordinary seeing, with every step in the inferential chain safeguarded by indisputable tests or robust mathematical and statistical analyses. Their effort, in other words, is directed to showing that though the inferential chain in scientific seeing is long, each tiny step is either an immediate apprehension or an indisputably justifiable argument. But stormy-weather seeing is the norm, and calm-weather, cup-and-saucer seeing is the special case. Calm-weather seeing is not beset by doubts and uncertainties only because we do not notice the inferential chains that are found even in ordinary seeing: perhaps when I think I am seeing the cup and saucer I am experiencing a hallucination; perhaps when I am seeing the Sun in the sky it is a dream; perhaps when I am not seeing a tree it is because I am suffering from a migraine that has removed part of my field of vision; perhaps I am in a virtual reality environment; perhaps I am watching a piece of stage magic that is "all done with smoke and mirrors."

In ordinary seeing, this kind of doubt is hardly ever raised.[6] But just as the calm Newtonian world is a local and bounded region in the turmoil of an Einsteinian universe, so calm-weather seeing is a local and bounded special case of stormy-weather seeing. To know how the universe works, one must look to the tumultuous regions of space-time, and it is the same with finding out how seeing works.

Nowadays, detecting gravitational waves is an exemplary case of stormy-weather seeing because first, gravitational waves are so faint, and second, it is a field of science born in dispute; here the hidden dimensions of seeing's space-time are made visible. The time dimension is revealed because some scientists say they have been seeing gravitational waves for the last 30 years, though most others disagree; resolving the true moment of clear seeing to everyone's satisfaction will take time. Most scientists who are in the field think that the resolution will happen when the new, more powerful instruments come on line. If they are right, then seeing gravitational waves will have taken about half a century—not an unusual seeing time in physics.[7]

The two spatial dimensions of seeing gravity waves are revealed in the way disagreements about what has been seen, and with what degree of certainty, are distributed in "social space." On one dimension of social space are laid out the locations occupied by scientists with divergent opinions about whether gravity waves have been seen. The disagreeing groups of scientists each have their own detectors and can support their divergent views because they are located in their own scientific communities—their own locations in social space—which sometimes coincide with locations in geographical space. Thus, in chapter 22 we will see a divergence of seeing between a group of Italian physicists and a group of American physicists.

The second dimension of social space is the chain of hearsay. All this stormy-weather seeing starts with the very best scientists: the specialists, that small number who live closest to the start of the inferential chain of proof. The spread of hearsay starts only after this small group of specialists agree what the chain of scientific inference in their instrument should be taken to mean. If we wanted to measure hearsay as a dimension, we would need units of "social distance": the distance of the seers from the start of the chain—the sites of theory and experiment. We can represent this dimension of social space with figure Intro.1, which I will refer to as the target diagram.

6. Harold Garfinkel's "breaching experiments" (1967) revealed how angry people get when everyday taken-for-granted reality is questioned.

7. See some of the cases discussed in *The Golem* series (Collins and Pinch 1993/1998).

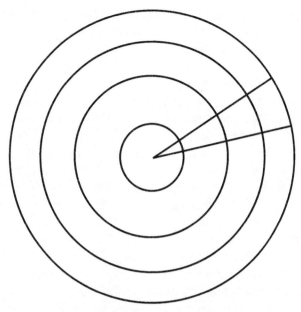

Fig. Intro.1 The "target diagram" showing rings of the scientific community more or less distant from the core.

In the center of the target are those scientists who build and operate the experiments designed to detect some faint phenomenon, along with the theorists who calculate what should be seen, or work out how the apparatus ought to operate and what it ought to "see." This is the *core* of the science. When there is scientific controversy, this central area of the science has been referred to as the core set. The term reflects the fact that the relations between scientists involved in such a controversy can be strained—they may never so much as meet each other; thus, to refer to them as a "group" would be to suggest more social solidarity than there is. I will use the term *core scientists* as the generic term for all those making real contributions at the center of the science, whether they are involved in controversy or not. The term *core group* will be reserved for a more solidaristic group of scientists pushing forward a program together without divisions caused by scientific disagreement. We will revisit these terms in chapter 19.

In the second ring of the diagram are members of the scientific community who work in different areas and therefore know of what has happened only through what they read or hear. We could divide this ring further into those who share a discipline, such as physics or astronomy, with scientists in the center and those who do not—say, biologists. But there is no need to complicate the matter at the outset. The third ring is composed of the policy makers, politicians, and funders—those who come into contact with science through their administrative dealings but do not actually do any science themselves; their knowledge of what happens in the center is again second- or thirdhand, but it is likely to be filtered differently than the knowledge of those in the second ring. (Once more, a single division between this ring and the others oversimplifies while providing the general idea.) Outside these rings are the general public and those who serve them by filtering, popularizing, and perhaps even constituting the science they consume: the journalists and broadcasters. This target diagram will be a recurring motif as I tell the story of both the development of gravitational wave knowledge and the organizational transformations in the science of gravitational wave detection.

Here I am talking about knowledge. Epistemology, the theory of knowledge, is the philosophical study of the grounds—the justification—for knowledge. But what I am describing is more a mapping of the *location* of knowledge through time and space. This more descriptive work is, perhaps, more aptly called "epistemography" in reference to cartography, the drawing of maps.[8] Each space on the diagram is a region of social space that can be divided into segments by drawing radial spokes. These segments can be thought of as representing the first dimension of social space described above: a difference between groups within the same "ring" of the target who may or may not be found in distinct geographical locations. Perhaps the inner circle contains a segment who believe gravitational waves have already been seen, and another segment who believe they have not. The second ring is divided into segments influenced differently by the segments in the center, and so forth. These, then, are the two dimensions of social space: the first dimension divides the target into segments, like those of an orange; the second dimension divides it into concentric rings. My job is to describe these regions and the way news, certainty, and "seeing" move around the target, from center to periphery and, perhaps, back again.

Let us start immediately by describing one weird feature of seeing's second social space dimension: those in the outer rings do not seem to feel disadvantaged by the secondhand quality of their information; far from it. For example, I am part of the community that has, for the first time, seen the dark side of the Moon. I am certain that the dark side of the Moon contains no buildings but is dusty and cratered like the side we can see

8. The word *epistemography* was invented for this purpose by the historian of science, Peter Dear (2002).

from Earth. But I have not really seen the dark side of the Moon; I have seen some photographs and television pictures which purport to represent it. There is no particular problem here, because the chain of inference from the back of the Moon to me is relatively short, and apart from those groups who consider the whole story of moon exploration to be a fake organized by NASA in the Arizona desert, there is not much dispute about it; this is not a case of turbulent seeing. Thus, my feeling that I have seen the dark side of the Moon is not so different from my feeling that I have seen a cup and saucer—though it is a useful comparison, because the chain of inference is a little more obvious in the case of the Moon; it does include rockets, cameras, radio transmissions, and so forth.

That we do not feel terribly disadvantaged by the indirect quality of the seeing in the case of the dark side of the Moon is quite striking, but the really odd thing about the second spatial dimension of seeing reveals itself, as might be expected, when the seeing gets stormy. In heavily disputed areas, people who are far removed in social space from the instruments of seeing are often more certain about what has been seen than those who actually peered through the instruments. A big part of the sociology of seeing concerns the way that those distant from the instruments of seeing come to learn about what has been seen, and the way that they form their often very strong opinions.[9]

One element in the explanation of this strange phenomenon is that, as a general rule, as you move from the scientific core group the message gets simpler and more straightforward. Those in the core group are aware of every argument and every doubt as it unfolds, whereas those a little more distant have their views formed from more-digested sources. In the nature of things, digested sources must simplify. The medium of transmission has, as it were, too narrow a "bandwidth" to encapsulate the sum total of the core's activities. Other things being equal, then, "knowledge waves," at least weak knowledge waves, behave in the opposite way from gravitational waves. Gravitational waves weaken as they spread; knowledge waves get stronger.[10]

9. Peter Saulson pointed out to me that Einstein complained that everyone believes an experiment except the person who conducted it (but also that no one believes a theory except the theorist).

10. This is the general rule, originally formulated as "distance lends enchantment" (Collins 1985/1992), which has exceptions. MacKenzie (1998) argues that science administrators are also presented with the uncertainties of science, so that there can be another high point of uncertainty further out in the target. The reverse of this point—that the activities of the core remain private—is a very important principle in the history and sociology of scientific knowledge. It enables one to establish the difference between an "experimental manipulation," which is private, a "result" announced

Because weak knowledge waves get stronger as they spread, it is hard for most of us to remember how faint they were at their source. In a modern science there are often only a very small number of scientists "looking through" the instrument and handling the strings of numbers that emerge. The study of seeing is about how this indirectness, this faintness, and all this scope for disagreement get turned into the kind of widespread certainty that allows us to say the equivalent of "We have seen the dark side of the Moon" and one day will allow us to say, without fear of contradiction, "Gravitational waves have been directly detected."

If the two spatial dimensions of seeing are represented in the way I have described, as a segmented "target" with concentric circles, we may imagine the time dimension as coming out of the page, turning the target diagram into an extended cylinder, which we can call the "time cylinder." We can imagine a series of cross sections cut through the time cylinder, each section representing a stage in the development of the science. The pattern inscribed on each of these sections would be a little different; as we move up the time cylinder we would see changing patterns of alliance, beliefs, and certainties.

To summarize again, this is our picture: stars cavort and generate massive amounts of energy, shaking space-time. But space-time is so stiff that the gravitational ripples are very weak, and they weaken still further as they spread. Earth, so far as we know the only planet populated by sentient beings, sits bathed in gravitational waves. Groups of humans get together to build instruments that will be sensitive to these almost indescribably weak gravitational ripples. Picture gravitational waves spreading from stars and impinging on those tiny areas on the surface of Earth that are the locations of gravitational wave detectors. Inside those instruments, assuming they respond, a long chain of events and processes unfolds, designed to enhance the effect of the impact of the gravitational waves. The effect is transformed by physical and electronic means into a string of numbers, which is in turn analyzed by computers using statistical techniques. It eventually achieves a form and magnitude that can be perceived by human beings, who report

---

in public, and a "demonstration," which is a display of well-established results, perhaps using techniques to enhance the effect that would be considered cheating in an experiment. (See, e.g., Gooding 1985, Shapin and Schaffer 1987, Collins 1988.) Notoriously, historians and sociologists of scientific knowledge invade the privacy of the laboratory and threaten these distinctions. Holton's famous study (1978) of Millikan's oil-drop experiment and many other historical and sociological analyses of famous experiments exhibit the problems. This very book may cause discomfort because of the way it exposes what would normally be the private thoughts and conversations of scientists, some of whom I now count as my friends, not yet honed and refined for public consumption. I am sorry that my job sometimes makes me appear to trespass upon the conventions of friendship.

it to other humans in the vicinity. A second set of ripples then begins to spread out through social space—a space whose medium is human interaction. The ripples strengthen as they spread, but like any wave motion, they have peaks and troughs that interact; some groups of humans become certain that the gravitational ripples have been seen, while other groups become certain that they have not. Eventually, all the humans come to agree, let us say, that gravitational waves have been detected.

To put this another way, in talking about experiments I have said that there is a long chain of inference from the effect to the conclusion and that the length of this chain of inference, though it is always there, is especially noticeable in modern science, which looks for small effects. What is rarely remarked upon is that this chain of inference stretches beyond the individual scientists "looking down the microscope" or its equivalent, and continues through social space-time until we reach social agreement. For some reason, perhaps because they are themselves part of it, the existence of the final stretch of the chain of inference is less visible to people than the earlier links.

Gravity's Shadow is concerned with both sets of ripples, those that spread through physical space-time and touch the detectors, and those that spread from the detectors through social space-time. The Laser Interferometer Gravitational-Wave Observatory (LIGO), the large American project, takes such a set of ripples as its logo. Similarly, this book's logo is two sets of ripples, as shown in figure Intro.2. The first set, lightly drawn, represents

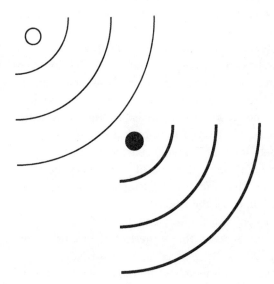

Fig. Intro.2 The logo: ripples in space-time and social space-time.

gravitational waves emitted from some cosmic catastrophe (the hollow circle) and traveling through space-time. These waves hit the detector (the black disk) and then their influence spreads through social space-time, this being represented by the set of heavily drawn ripples.

RELATIVISM

Here is an odd thing: I have described a causal sequence starting with cavorting stars and ending with agreement among humans. Yet everything we know of stars and their cavortings we know only because of the ripples in social space-time. If there were no ripples in social space-time, there would be no cavorting stars in our universe (just as there are no cavorting stars in the universe of those who have no modern science, nor none of their gods or witches in ours). Thus one could argue that the causal sequence runs the other way, not from stars to human apprehension but from human agreement to the stars (as most of us believe it does for gods, witches, and the latest fashions). Everything you have just read about gravitational waves and their impact is, as anticipated at the start of this chapter, based on trust, hearsay, and socialization. All I have said about Newton, and Einstein, and mass, and energy, is itself the product of social agreement, now settled down in the flat, calm weather of that social location called modern physics.

In terms of our logo, we know nothing about the hollow circle or the waves it emits until these waves hit a detector on the surface of Earth—the solid black disk—that is noncontentious. But, equally, we know nothing about the solid black disk until the second set of ripples have made their effect felt on another detector—the scientific community. If we were to superimpose an arrow of causality on our logo, it would normally run from top left to bottom right. But accepting the argument of these paragraphs, it could also run the other way, from bottom right to top left. The first direction is the realist way; the second, the relativist, or "social constructivist," way. Fortunately, just as the laws of the physical universe stay the same whether we run things forward from the Big Bang until now, or backward, from now to the Big Crunch, the laws of seeing look the same whichever way the arrow runs. The difference is metaphysical or, for some of a fundamentalist persuasion, quasi theological.

I will return briefly to this point at the very end, but as far as the bulk of the book is concerned, that is pretty much the end of the discussion of realism versus relativism as a philosophical topic. Relativism will surface again and again, however, as a principle of methodology. A complete understanding of the dark ripples can be prejudiced all too easily by working as though there is too much causal determination of their pattern coming from the top left. For this reason, at the very heart of the analysis is an approach called methodological relativism. This approach, which treats the causal arrow as running from bottom right to top left without making any claim about which way it really runs, is the approach used wherever I socially analyze the science (though not where I simply accept the established consensus). Methodological relativism is little more than the scientific prescription to investigate one cause at a time by holding everything else constant. In this case, where the science is contentious, we hold the science constant (treat it as a not–causally contributing variable), and concentrate on the social variables.[11] This does not mean that scientific arguments do not have to be explained, and *Gravity's Shadow* should be unobjectionable to the most adamant realists unless they think the story of a science can be told without discussion, and a great deal of discussion, of the social.

For those who remain entirely sure that the world must always be described in terms of the realist, top-left-to-bottom-right direction, it may be worth pointing out one more thing about nature and society. Let me talk in a realist register about the first sighting of gravitational waves. For argument's sake, imagine this first sighting is of a burst, a couple of milliseconds in length, emitted by a supernova. Thus in the half-century history of gravitational wave detection, nature would have spoken for only that instant. Nature would have spoken for only about 0.00000000013% of this half century, and it would have been people thinking, building, calculating, hypothesizing, interpreting, spending money, writing, organizing, leading, and persuading that took up 99.99999999987% of the time. And remember, hardly anyone would actually have "seen" the event that lasted even the milliseconds. So whether you are a realist or a relativist, there is an awful lot for the sociologist to deal with.[12] Every now and again throughout the book I will point out how much there is for the sociologist to do even if an adamant philosophical realism is maintained.[13]

---

11. For further discussion, see chapter 42 and 43 of this book and various chapters in Labinger and Collins 2001.

12. This thought follows from a remark by Peter Ladkin.

13. I endorse realism as an attitude both for scientists at their work and for sociologists at theirs. Thus, I write this book in an attitude of "social realism"—that is to say, I treat the social as real. (In Durkheim's words, I "treat social facts as things.") The difficulty is that for me to treat the social facts of science as real, I need to treat the natural facts in a different way. I believe all of us should be capable of stepping outside ourselves from time to time so as to understand that there are other ways of talking about our taken-for-granted worlds.

One other way of thinking about realism and relativism is to ask what is in the world. Can a science support only one truth or many? What is clear with only a moment's thought is that a science can support more than one truth for quite long periods of time; scientific controversies can easily last for 50 years or so, during which two or more views of the world are supported by competing groups of scientists.[14] The determined realist will say that only one of such views is truly scientifically viable, and insofar as one view survives and the other(s) die, it is hard to prove otherwise (the philosophical agnostic can counter that because only one view survives long term does not necessarily prove anything more about that view than its political or cultural robustness, as with competing styles of art). The "rationality debate" which began in the 1960s discussed these issues. If there is more than one rationality, there can be more than one science.[15] But, without settling the rationality debate, the periods of many decades during which alternative scientific views do survive again provide a wide territory for the sociologist to work in. In the middle section of the book, I will describe some of the ways that our social institutions handle claims outside the consensual view and in this way provide a sense of the social meaning of competing realities in our historical epoch. It can reveal our "way of being" in this, the hardest of the sciences.

THE PROJECT AND THE BOOK

What I have described is the idea of the project of which this book is a part. The sociological project will not have been completed until there is wide agreement that gravitational waves have been seen directly and the ripples have reached their final destination in social space-time: taken-for-granted reality or, literally, common sense. Only then will we be able to make the big comparison between what it is to fail to see gravity waves and what it is to succeed. What is completed herein sets the scene for the comparison. I describe the (near) snuffing out of the several ripples in social space-time which might have indicated the first direct detections of gravitational waves but failed to reach their destination in the land of the taken for granted. The stories of the snuffings-out are described at length in parts I and III.

14. See, for example, chapter 2 of *The Golem* (Collins and Pinch 1993/1998).

15. For the rationality debate, see Wilson 1970, Winch 1958, Wittgenstein 1953. For sociological implications, see, for example, Bloor 1973, Collins 1985/1992, and Kuhn 1962.

Part II is a short and fairly straightforward account of the development of two new technologies, cryogenic bar detectors and interferometers. Part III describes the period of uneasy coexistence of these two technologies and the way in which interferometry triumphed over bar technology. It shows how two cultures can exist side by side, but not for long if the one is powerful enough.

Part IV describes the growth of interferometry and, concentrating on the transition of the American project from a small to a big science, the stresses, strains, and dramas associated with the change. Part V develops two general themes, the process of pooling data from competing international groups and the curious process of setting upper limits on the flux of a phenomenon as opposed to making positive detections. This takes us to the present state of interferometry, which has recently issued its first upper-limits claims.

Part VI draws conclusions, but the most important of those are in chapter 42, which develops a new theme, the relationship between experts and nonexperts, as exemplified by this study itself. Implications for the relationship between scientists and the public are drawn out.

OTHER THEMES

A word of caution: there is a lot of history in *Gravity's Shadow;* some might wonder why some of the parts are so brutally long. Indeed, I have given part I the title "À La Recherche des Ondes Perdues" in wry acknowledgment that it does go on a bit. But, like Proust, I have to try and recapture something that is gone, and that is harder than recapturing something that is still around.

Nevertheless, though it covers a lot of historical ground, the book is not primarily intended to answer to the selection criteria or professional priorities of the historian. The principles of selection of the material are first, sociological interest; second, just plain interest; and third, a sense of duty to the history. Most of the episodes drawn from the history of gravitational wave detection are set out with things other than history in mind. They include the construction of truth and falsity; the triumph of one technology over another; the necessity for choice without complete justification; the pruning of the potentially ever-ramifying branches of scientific and technological possibility if science is ever to move forward; the growth of science from small to big; conflicting styles and cultures of science; and the tension between the world seen as an exact, calculable, plannable sort of place, just waiting for us to get the sums right, and the world seen as dark and amorphous, bits of which, from time to time, we are lucky enough to catch in our speculatively thrown nets of understanding—if we throw them with sufficient skill.

NOTE ABOUT PART I FOR ANALYSTS OF SCIENCE

The main arguments of part I of the book will be familiar to those who know my published work on the rise and fall of Joe Weber's claims to have detected gravitational waves. Since the section comprises 12 chapters, whereas the material has been described before at no greater length than two or three academic papers, it is clearly not just a repetition of published work. It is the analysis that will be familiar, but here the details and sequence of the work, the people involved, and much more of the nature and texture of the arguments are included. There are, however, no new analytic ideas in that part of the book, just more data, more on the theory, and a new metaphor.

Readers who skip this section because they are already familiar with the bones of the argument will miss out the following: In chapter 2 I describe Joel Sinsky's calibration experiment, which I have never mentioned before. I would have discussed this work elsewhere if I had thought about it more carefully. Chapter 3 is not really part of the Weber story but contains some of the *science* of gravitational waves which will be needed in later parts of the book. Chapter 5 offers the new metaphor—a reservoir held back by a dam of orthodoxy—for understanding the way a heterodox science survives or fails to survive; this, too, will be referred to in later parts of the book. Chapter 12 contains some new bibliographic material. And, of course, the whole section contains much more historical detail.

PART I

À LA RECHERCHE DES ONDES PERDUES

CHAPTER 1

THE START OF A NEW SCIENCE

I began with Tony Tyson's evidence before Congress. Tyson failed to stop or delay for very long the funding of the interferometer project, and the use of huge interferometers to search for gravitational waves is now beginning. These interferometers are the subject of the later sections of the book. Two earlier generations of detectors preceded the interferometers, and there might have been another generation had it not been stillborn. The first generation of detectors, whose heyday was from the late 1960s to the mid-1970s, were solid metal cylinders known as bars. Roughly a couple of meters long and a couple of feet in diameter, and weighing a ton or so, they were encased in vacuum tanks and insulated from every other disturbance as much as possible. Because they were meant to vibrate in response to a passing gravitational wave, they came to be known as room-temperature resonant bars. Since the year 2000 only one room-temperature bar, at most, has been recording data; this is run by an Italian group; the other remaining room-temperature bars died in the year 2000 at the same time as Joe Weber, who pioneered the technology in 1960s.

The second generation of detectors, which dominated the science from roughly the mid-1970s to the mid-1990s, were a development of the room-temperature bars, but they were cooled to the temperature of liquid helium or below; these devices are known as cryogenic bars. In recent years, up to five of them have been operating: one in the United States, one in Australia, and three run by Italian teams, of which one is located at CERN (European Organisation for Nuclear Research) in Geneva. The stillborn generation of devices, which may yet revive, were a further development of the cryogenic technology and used spherical resonators instead of cylinders. They, too, would have been cooled to the lowest temperature possible, but each would have weighed about 30 tons rather than the couple of tons of the bars. The current generation of detectors, increasingly dominant since the mid-1990s, are the interferometers.

This part of the book is about the room-temperature bars, devices that opened up the science of gravitational wave detection at a time when it seemed that such a science belonged only in a world of fantasy.

EARLY DAYS

The person who began the whole business of looking for gravitational radiation was Joseph Weber. Weber, pronounced "Webb-er," is always referred to as Joe Weber. Most of the text of this book, where it discusses Weber, was first written in the present tense; for most of my professional life Weber has been an "is," not a "was." Changing the tense has been painful. From 1948 until his death September 31, 2000, at the age of 81, Weber remained an active physicist and was at various times professor of electrical engineering, professor of physics, or both at once at the University of Maryland, located in the suburbs of Washington, DC. From 1973 to 1989 he was also a visiting professor at the University of California at Irvine. He continued to maintain offices at both institutions until his death.

Weber was born in Paterson, New Jersey, on May 17, 1919, the son of Jewish immigrants from Lithuania. During the Great Depression he worked as a golf caddy for a dollar a day, increasing his earnings tenfold after teaching himself to fix radios. He received the highest score in a public examination and was invited to become a cadet at the US Naval Academy at Annapolis, where he graduated with a bachelor of science degree in 1940. He found himself in active naval service during the Second World War and soon demonstrated his technical expertise, becoming a skilled radar operator and an excellent navigator. One ship on which he served, the aircraft carrier USS *Lexington*, was sunk during the Battle of the Coral Sea, but Weber survived and became the commander of a submarine chaser, the USS *SC-690*. As he put it, this was an unusual role for a Jewish boy from a poor background.

After the war, Weber continued to study physics, obtaining a Ph.D. from the Catholic University of America in 1951. He also studied physics at the Institute for Advanced Study in Princeton, New Jersey, being influenced by, among others, Robert Oppenheimer and John Wheeler; he was also strongly encouraged by Freeman Dyson. He spent time at the institute in 1955–56, 1962–63, and 1969–70. Among Weber's accomplishments is the development in the 1950s of ideas that prefigured the maser, itself the forerunner of the laser; *laser* stands for "light amplification by stimulated emission of radiation," whereas *maser* stands for the same thing, except *light* is replaced with *microwave*. Perhaps he should have shared in the Nobel Prize for that work, or at least been given more recognition; he certainly thought so, as did a number of others. A wall of the engineering building at the University of Maryland displays a photograph of Weber with a certificate of appreciation for his pioneering maser work.

When I last saw him, in 1996, Weber, then 78, was still putting in a full workday in his shambles of an office in the University of Maryland's Physics Department. Moreover, he told me that he continued to take a three-mile run at four in the morning to keep fit. Widowed after the death of his first wife in 1971, Weber subsequently married astronomer Virginia Trimble, 24 years his junior, his occasional coauthor, and later very well known in her own right. He told me with a smile that when he married her he was famous and she was not, and now their roles were reversed.

Up to his death, Weber continued to put in applications to the funding agencies, not only for projects on gravity waves but for research on the detection of neutrinos and for a revolutionary type of laser. Up to the early 1990s, some agencies were still responding to him in small ways.

To say that Joe Weber was famous or not famous is to oversimplify. Everyone in the world of physics knows who he was, but his fame largely turned to notoriety between the 1970s and the 1990s. I suspect that now that he is dead, the notoriety will revert to renown, and he will regain his place in history. The peak of Weber's reputation came about 1969 and lasted for five years or so. He became famous when he announced that he had detected gravitational waves. His reputation began to change, however, when others said it was not gravitational waves he was seeing but a problem with the apparatus of or the statistics produced by the experiment. These others—the large majority of physicists in the relevant communities—said that what he claimed to have seen was theoretically impossible. They could calculate the maximum strength of the gravitational radiation that should be passing through Earth and they could calculate the sensitivity of Weber's detector, and the calculations did not match.

To the end of his life, Joe Weber insisted that it was gravitational waves that he saw; he continued to insist that he had seen them 30 years later as he drew to the close of his career. Thus, his fame changed to notoriety because he would not admit he was wrong. Worse, he would not desist from telling people, in the most forthright manner, that he was right and they were wrong. In his later years Weber was very bitter about what he saw as the injustices that had befallen him, and he had a sharp edge to his tongue; as a result he drove some erstwhile colleagues to despair.

A NEW FORCE DETECTED?

If the physics community has a "house journal," it is probably *Physics Today*, which is published by The American Institute of Physics. The April 1968 is-sue contained an article by Weber as its lead story.[1] The front cover featured a schematic rendering of two masses joined by a spring—which is how We-ber liked to represent the theory of the gravitational wave detector that he had invented. Weber reported an experiment conducted in his Maryland laboratory by his graduate student, Joel Sinsky. It was said to show that vibrations in one aluminum cylinder could be transmitted to another by the changing gravitational attraction between the two bars.

If we were to look very closely at the end of a vibrating bar, we would see it moving in and out—that's what vibrating is. The end might move about one-tenth of a millimeter—a movement almost invisible to the naked eye. But these tiny movements would cause the gravitational field associ-ated with the mass of the bar to change. Any object near the end of the bar would experience slightly less gravitational pull during that part of the cycle of vibration when the end of the bar was one-tenth of a millimeter further away and slightly more gravitational pull when it was one-tenth of a millimeter nearer.[2] Thus, a vibrating bar gives rise to an oscillating

---

1. Weber 1968a.

2. Readers of chapter 3 should note that if the vibrating mass was a sphere rather than a cylinder, and if its vibrations were symmetrical about all the axes, so that no change in shape was involved, no change in gravitational field would result from such a vibration, as the center of mass would remain fixed. To know the gravitational field of a sphere, you need know only the mass and the position of the center of mass; to know the gravitational field of a cylinder, you also need to know the dimensions of the cylinder.

gravitational field. Weber was claiming that the tiny changes in gravitational field that resulted from the vibration of the one bar were being sensed by the other bar, causing it to vibrate as well. The vibrations in the second bar would be much smaller than those in the first, however; Weber explained that responses in the second bar on the order of 10^{-16} meters were being detected in this experiment. That is to say, Weber was claiming that he could see changes in the length of a two-ton aluminum alloy bar that were somewhat less than the diameter of an atomic nucleus. (See appendix Intro.1 for further elaboration of what it means to say this.)

The second bar detected the minute vibrations via piezotransducers glued to its surface. These are crystals that produce electrical signals when they are squeezed or stretched, thereby linking electricity with physical force. (The spark produced by modern devices for lighting gas stoves is made this way.) Hence the tiny stretchings in the aluminum bar putatively produced by the passage of gravitational waves were to be detected by the electrical signals coming from the tiny stretchings of piezocrystals glued to the surface of the bar and made visible by amplifiers of unprecedented sensitivity.

This experiment is worth describing again to make clear just how ambitious it was. The relationship between force and electricity exhibited by piezocrystals can be reversed. If you apply an electrical potential to them, they will compress or expand according to whether the potential is positive or negative. If you apply an alternating potential to them, the crystals will vibrate. In the Sinsky experiment, two bars with piezocrystals glued to their surfaces were placed near each other. The "driving bar," eight inches in diameter and five feet long, was caused to vibrate by energizing the piezocrystals on its surface with an alternating electrical potential. These vibrations were sensed by the second bar—the detector—which was 22 inches in diameter and five feet in length, and weighed approximately 1.5 tons. The means of transmission of the vibrations from one bar to the other was gravitational attraction.

This experiment was extraordinary, for the gravitational field associated with an object that is less than the size of a planet is hard to sense. It was a triumph of experimental science when, in 1798, Cavendish measured the gravitational attraction between two massive lead balls, because the attractive force between them comprised only one 500-millionth of their weight. Now what Sinsky and Weber were doing was detecting the minute *changes* in the gravitational field caused by the vibration of such a mass—that is, the minute changes in the field caused by the changes in the location of one end of a massive aluminum bar as it "rang." It is not as though the bar was either there or not there, as in the case of the lead balls used by Cavendish; the bar stayed in place and just changed its shape slightly, and that phenomenon was recognizable because it meant that the part of the bar nearest to the detector moved fractionally nearer to and further from the detector with each cycle of vibration.

Incidentally, this force is a change in a gravitational field having the same frequency as the gravitational radiation that Weber's bar was designed to detect—around 1660 Hertz (cycles per second). The force will change with time in the same way as a wave, though this changing gravitational force is not a gravitational wave; gravitational waves cause changes that are transverse to their direction of travel, but this kind of wave, consisting of changes in the force of gravitational attraction, causes changes which are parallel to the direction from which they are coming. Thus, we are *not* talking about gravitational radiation in this experiment but changes in ordinary gravitational attraction. This force is no different in principle from the force that causes the tides—the changing attraction of the Moon for movable objects on Earth as the Moon changes its position in the sky. Gravitational radiation is, as I will explain in chapter 3, something far more subtle. Nevertheless, Sinsky's experiment was meant to show the sensitivity of Weber's apparatus to any potential gravitational waves.[3]

Compared with the strains induced in the second bar, the driven bar's vibrations were very large. The second bar, it was calculated, was seeing strains on the order of 10^{-16} in response to strains in the first bar of 10^{-4} or so (which is about one-tenth of a millimeter). In other words, the secondary vibrations were about 100,000 million times smaller than the first (the first strain $\times 10^{-11}$). But that result, of course, is exactly what was wanted if this experiment were to demonstrate that the second bar could indeed detect the tiny effect of gravitational waves.

Even this calibration experiment was an exciting business. The first bar had to be caused to vibrate so hard that the piezocrystals often fell off, and it heated itself though its own vibrations. The vibrating bar was also enclosed in a vacuum chamber to try to get rid of any nongravitational sources of disturbance transmitting themselves to the first bar. If vibrations were transmitted through the air, the floor of the laboratory, the housings

---

3. Whether an oscillating gravitational field such as was generated in the Sinsky experiment has the same effect on a gravitational wave detector as a gravitational wave was one of the things Weber was to come to argue about with his critics (see chapter 10 below).

of the bars, or some kind of coupled electrical or magnetic forces, then nothing would be proved. Below, I will discuss how Weber and Sinsky tried to show that their experiment was sound and that none of these other couplings were taking place.

Taking this experiment at face value, Joe Weber had proved that his detector was very sensitive to tiny oscillating gravitational forces. Still more important, however, in the *Physics Today* article he reported that after this "calibration" experiment was finished, the driving cylinder was converted to a second detector and mounted about a mile (1.5 kilometers) from the first. He claimed to have seen about ten coincident energy pulses between the two detectors which were not correlated with vibrations in the ground or with electromagnetic disturbances. These appeared to be interesting signals from an outside source.

What did Weber's claims mean, and why were they important enough to feature on the cover of *Physics Today*? The first half of the claim—the calibration experiment—represents a triumph of engineering, and if it is correct, it makes the second claim interesting. The second claim intimates that the coincident energy pulses could have been caused by something very small indeed—something that, given a generous interpretation, could just conceivably have been gravitational radiation.

But what is gravitational radiation? Are we sure it exists and that shaking, spiraling, and exploding stars really do emit it? Scientists took a long time to reach a consensus about gravitational waves; the flavor of the theoretical debate which preceded Weber's work is indicated in chapter 3. Einstein, who originally proposed their existence, changed his mind in the 1930s and then changed it back again. Though by the 1960s the consensus was broad, as late as 1962 Richard Feynman wrote to his wife that the continuing argument about whether detectable waves were emitted that he encountered at a conference was bad for his blood pressure.[4] As it turned out, published arguments about whether the waves could have an effect on gravitational wave detectors continued into the 1990s. For example, in 1996, the abstract of an electronically circulated paper by Luis Bel read as follows:

> We present a new approach to the theory of static deformations of elastic test bodies in general relativity [resonant detectors].... We argue on the basis of this new approach that weak gravitational plane waves do not couple to

4. Kennefick 1999, p. 91.

[affect] elastic bodies and therefore the latter, whatever their shape, are not suitable antennas to detect them.[5]

By this time, however, a dozen or so gravitational wave detectors had been built, making the continuing argument more of a curiosity than a serious concern. Above all, it was Joe Weber who settled the argument, first by working out how to build a gravitational wave detector and then by building one. When we say, "Scientists have reached a decision," what we mean—or at least what we ought to mean—is that they have decided to invest their scientific energies in one way rather than another. What Joe Weber did was to cause scientists to act as though gravitational waves did exist. He built an apparatus to detect them and caused others to build apparatuses, too. By the time a few people had built apparatuses to detect gravitational waves, there was little room for doubt that the waves existed in principle (*pace* Luis Bel), though there was still plenty of doubt to come about whether they had been detected in practice.

THEORY AND EXPERIMENT

The theoretical debate concerning gravitational waves is a microcosm of the whole story of this book. One finds scientists disagreeing, reaching a majority view, then getting on with their scientific lives while a minority continue to dispute the matter. Technical arguments are rarely enough to convince everyone in a scientific controversy, and the effective resolution emerges from the way most scientists decide they should act from day to day. In theoretical debates the "incoming signal" is produced by what theorists write on paper, but thereafter the waves spread out through the academic and lay communities in the same patchy way as happens in the case of experimentally derived signals. The cylinder extending through time with varyingly figured cross sections applies as much to theory change as to the acceptance of experimental data.[6]

---

5. Bel 1996, p. 1. Other theoretical contributions which would significantly change the way the interaction between gravitational waves and detectors is visualized include Cooperstock 1992 and Cooperstock, Faraoni, and Perry 1995. These claims, which would render certain classes of detector in principle incapable of seeing the waves, though published, have been largely ignored by the community. Kundu (1990) claimed that no gravitational waves can emerge from the galaxy in which they were emitted, so this would render the most likely and numerous sources undetectable (though this claim was later withdrawn [Price, Pullin, and Kundu 1993]). Another claim, by Christodoulou (1991), which has been accepted by the community (Thorne 1992a), makes a difference to the interaction, but not one that disables the detectors.

6. See Kennefick 2000 for a more complete analysis of this process in respect of theory.

Together, theory and experiment make up the time cylinder for a whole field of science. The news as it travels out through the layers of the community is mediated by what the theoreticians have said or what they are said to have said. And the significance of what the theoreticians are said to have said is mediated by what the experimentalists are said to have seen. Sometimes theory will dominate the news, and sometimes experiment will dominate. We will see the way the relationship between theory and experiment changes as the story of gravitational waves unfolds. Up to about 1960, it was all theory—no one had thought of trying an experiment—but for a decade and a half, from the late 1960s onwards, things changed, thanks to Joe Weber. He inspired, or irritated, other scientists enough to get them to do something about it, and experiment began to dominate. Thanks to Joe Weber's influence, experiment became the major effector of the way theorists, and general relativists as a whole, thought about the world of gravitational waves right through to the middle 1980s, when theory began to dominate once more.

AN IMPOSSIBLE EXPERIMENT

Weber actually had begun thinking about gravitational waves more than a decade earlier.[7] When he was at Princeton he became involved with the debate about the reality and detectability of gravitational waves. With John Wheeler, the famous interpreter of quantum theory, he published a paper in 1957 arguing the positive case. In a paper published in *Physical Review*, dated, perhaps auspiciously, January 1, 1960, Weber published another paper, which began to calculate possibilities and put forward basic ideas for a detector. Actually, these ideas were first enunciated at a 1959 conference in Royaumont, near Paris, and in an essay written for the Gravity Research Foundation, which won the $1,000 annual prize for 1959.[8] The statement

---

7. The following brief sketch of the early years of gravity wave detection is by no means the kind of history that would be written by the biographer that Weber deserves. Weber is a fascinating instance of the postwar generation of physicists who came from poverty to do extraordinary things. Unfortunately, it is generally only those who are eventually accounted a success whose lives are studied in detail, and this distorts the history of science.

8. The Royaumont conference proceedings paper was not published until 1962 (Weber 1962), but the essence of both it and the prize-winning essay were published in 1961 in Weber's book. The Gravity Research Foundation essay was entitled "Gravitational Waves" (Weber 1959). The foundation was founded by Roger Babson to encourage research into gravity. Babson's motivations are nicely set out in a strange essay entitled "Gravity—Our Enemy Number One," published by the foundation; the full text can be found in appendix II.1. The demise of the GRF has been variously reported over the years, though it continues to exist. It continues its work of awarding a prize for an essay on

is repeated that if two masses separated by a spring are affected by a gravitational wave, the two masses will move in respect of each other, and "Therefore energy may now be extracted from the wave" (fig. 1.1a).[9] In practice, as the three latter papers go on to explain, at this stage the receiver was intended to be a massive, freestanding, piezoelectric crystal, rather than a metal bar with piezocrystals glued onto it. The crystal, in spite of its size, would produce only a tiny electrical impulse when affected by a gravitational wave; an extraordinarily sensitive amplifier would be needed to produce anything visible (fig. 1.1b).

At about this time, Weber was also working out the cross section—that is, sensitivity—of his proposed devices to gravitational waves. His way of working things out seems to have been confirmed by Remo Ruffini and John Wheeler at Princeton, and became known as the classic cross section.[10] As we will see in part IV, 25 years later Weber was to change his mind about this calculation and find a way to work it out which gave radically different results.

Even at this very early stage, where experimentation was just beginning, Weber was anticipating the key developments upon which the new science of gravitational waves would depend. Any signal would be concealed in the noise produced by the random vibrations of the atoms of the crystal and other random fluctuations, but Weber suggested that the waves might still make their presence felt because their intensity would change as the earth rotated during the day. This would mean that sometimes the crystal would be favorably oriented to any constant source of radiation and sometimes it would not, and the variation in the output of the crystal in the course of 24 hours would indicate the effect of the waves: "If radiation is incident from some given direction it may be observed from the diurnal change in amplifier noise output."[11]

As regards direction, Weber talks of the Sun as a possible source, and the Royaumont paper makes clear what he means. There is a short extract of responses to questions at the end of the paper, and Weber is represented

---

gravity under the leadership of George M. Rideout Jr. It was this Rideout who gave me permission to reproduce the essay in appendix II.1. The current address of the foundation is PO Box 81389, Wellesley Hills, MA 02481-0004, USA.

9. Figure 1.1 consists of close reproductions of the originals drawn by me.

10. In 1993 and 1995, Weber told me that he worked out the result, and Ruffini and Wheeler subsequently confirmed it. In 1997, Ruffini implied it was he and Wheeler who did it first: "In the case of Weber, of course, we have been interacting since the early days. When Wheeler and I computed the cross section of gravitational waves in detail—it had not been computed before us." In conversation, however, physicists generally attribute the classic cross section to Weber.

11. Weber 1960, p. 311.

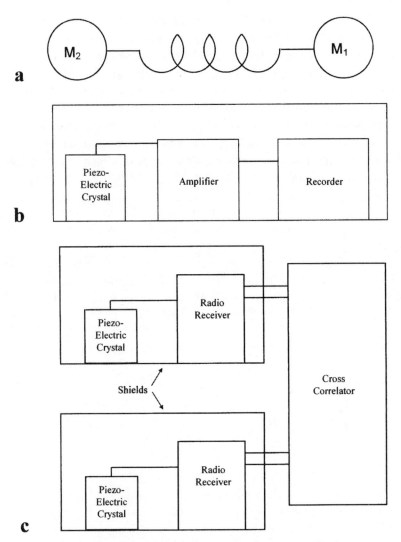

Fig. 1.1 Schematic: gravitational radiation detectors from Weber's early papers and book.

as explaining in French to famous relativist Felix Pirani that protuberances or disturbances in the solar atmosphere (which would involve movements of great masses of gas) would give rise to the radiation.

In a second sketch presented in all three early papers, Weber depicts the comparison of the output of two crystals affected by gravitational waves (fig. 1.1b). The idea is that simultaneous effects on the two crystals would indicate an external source, putatively gravitational waves, and their being simultaneous would distinguish them from noise. This would mean that one would not need to distinguish them from noise by reference to the way they changed as the detector changed its orientation due to the rotation of Earth: "The arrangement of Fig. 3 [fig. 2.1c in this book] should not require rotation. If radiation is incident it will cause correlated outputs. All sources of internal fluctuation will be uncorrelated" (p. 311).

To this day, looking for correlations between two detectors is still a crucial part of the methodology of detection of gravitational waves. All three early papers go on to speculate about the possibility of generating detectable gravitational waves in the laboratory, concluding that the chances are slim.

In 1961, Weber published his book, *General Relativity and Gravitational Waves*. Weber explains that this was a book in a series commissioned by the publishers, and that John Wheeler, who had been first approached, put forward Weber's name as a suitable author. In the book Weber further developed his ideas about detectability and detectors, illustrating them with the same three diagrams. By now the experiments were under way.

CHAPTER 2

FROM IDEA TO EXPERIMENT

The Team

When I first interviewed Joe Weber in 1972, he told me that he had begun experimental work on the detection of gravitational waves in 1958 or 1959. By 1960 he had begun to gather a small team of helpers. The *Physical Review* paper published in that year records the beginning of experimentation.

> *Note added in proof.*—Experimental work along these lines has begun recently. It is being carried out by Dr. David M. Zipoy and Mr. Robert L. Forward, in collaboration with the author.... Excitation of resonant acoustic vibrations in a block of metal ... is being considered along with the arrangements of Figs. 2 and 3 [fig. 2.1b, c in this book]. (Weber 1960, p. 311)

Weber's team had increased in size by the time of the article in *Physics Today* eight years later: "One detector is an aluminum cylinder, with a mass of approximately 1400kg, developed by David M. Zipoy, Robert L. Forward, Richard Imlay, Joel A. Sinsky and me."[1]

Unmentioned in this list are Daryl Gretz, a laboratory technician, and Jerome Larson, of the University of Maryland's electrical engineering department, who also put much effort into the experiments.[2] Richard Imlay was an undergraduate student and worked with the team from 1958 to 1962, after which he left to work in particle physics. When I interviewed him in 1997, Imlay considered himself to have played only a minor part in the development work, working directly for Zipoy rather than Weber. He said that he had helped out in the calibration experiment by doing a calculation concerning the transmission of gravitational energy from the vibrating bar to the detector, and had helped Zipoy with the design of the seismic isolation for the bars (see below).[3]

Both Zipoy and Sinsky had left the field of gravitational research by the end of the 1960s, but both played a significant role in its development during that decade. Robert Forward played a major role in designing and building the first detector in Weber's lab, independently built a series of detectors after he left Weber's lab, and also constructed the first ever laser-interferometer gravitational wave detector.[4]

Forward, in fact, was interested in gravity before he came across Weber and his ideas. He first heard of Weber when he, too, submitted an essay to the Gravity Research Foundation, in the same year (1959) that Weber won the prize. The foundation's first prize amounted to $1,000, which was enough to encourage quite a few now-famous individuals to submit their work.[5] Forward saw Weber's essay, which, he says, was the first paper demonstrating rigorously that gravitational radiation would be emitted in appropriate circumstances, would carry energy, and could be detected. He had completed his undergraduate degree at the University of Maryland, but was working at UCLA while employed by Hughes Aircraft in California. Because of family connections, Forward had already decided to return to Maryland to pursue his Ph.D. in solid-state physics, but he visited Weber

1. Weber 1968c, p. 37.

2. Larson's photograph appears in the first *Science News* article to discuss the experiments (Thomsen 1968), and there is a journal article as well: Weber and Larson 1966.

3. Imlay, though he left gravitational wave physics in 1962, remained in touch with the field, because, by coincidence, he found a job at Louisiana State University, the site of the most successful, and only remaining, cryogenic bar detector in the United States. Imlay remains in contact with the LSU team's work, though he does not participate in it.

4. Forward, who was also a well-known writer of "technical" science fiction, died on September 21, 2002.

5. Stephen Hawking has won the prize three times.

and, having completed his doctoral coursework, transferred to Weber's program. Zipoy had also joined the team by then.

As explained, the first idea for a gravitational "antenna" was a single massive piece of piezoelectric crystal.[6] The team thought that the whole vibrating mass would have to be made of piezoelectric substance if it was to have a chance of being sufficiently sensitive, and they tried to buy a chunk of crystal, as Forward put it, "as big as a desk."

> [1972] [T]he more we looked at it, the more we investigated it, the more we realized that really what we wanted was some mass to interact, and it didn't have to be a crystal, and the crystal could just be a small part of the mass.[7]

After moving to the idea of using a large vibrating lump of metal, the team had initially tried to detect its vibrations using a variable capacitor (a technique used by the well-known Russian experimentalist Vladimir Braginsky). Forward reported, however, that Zipoy, who was using piezoelectric crystals to monitor Forward's experiment, discovered that they were getting more signal from the crystals than from the variable capacitor, so they switched to crystals as the principal means of extracting energy from the bar. Forward remarked,

> [1972] [T]he thing that is most amazing about my experience in research is the stupid, idiotic mistakes in engineering, in mathematics, and electronics that you make because you are working on such a difficult problem. There is just example after example of where, after we've gone through the thing, we've made a very stupid engineering decision because we're so involved in the physics. That was one of them.

It seems that it was Forward who finally discovered that aluminum alloy was as good for a gravitational antenna as any other reasonably inexpensive material, having experimented with tungsten and other metals that turned out to be quite unsuitable. Single (non-piezoelectric) crystals of sapphire or

6. The term *antenna* seems to have been used by Joe Weber to describe a gravitational wave detector so as to signal the continuity between gravitational radiation and electromagnetic radiation. Weber, remember, was an electrical engineer. It was said that some of his early theoretical papers, written in the language of electrical engineering, were initially incomprehensible to physicists, and it was only after some hard thinking that they were able to appreciate the integrity of the underlying arguments.

7. Throughout the book, the year in which the interview was recorded is given in square brackets if not otherwise mentioned in the text. More accurate dates are not given so as to make identification of the source of unattributed quotations less easy (it would make it possible to group quotations by speaker).

some such would be better still, but large lumps of sapphire were not practicable in those days. For the money they had to spend, aluminum would give the best result.[8]

Forward continued to work on the project with the Maryland team until mid-1962, when he decided to return to his work at Hughes in Malibu, California.[9] First, however, he took 48 hours' worth of data from the single detector. This was the very first data run from a gravitational wave detector. Forward's research notebooks show that this data-gathering run began on Saturday, May 12, 1962, at 5:30 a.m. and finished about 6:30 a.m. on the following Monday morning.[10] The notebooks also show, however, that for unknown reasons the chart recorder went off scale from about 6:30 a.m. on Saturday morning until the early hours of Sunday.

Figure 2.1 shows a plot taken from Forward's Ph.D. dissertation covering this run.[11] Electrical transients, identified by their sharp rise times, had already been removed from the data, but the off-scale anomaly can be clearly seen.

In his discussion, Forward is cautious, recommending that the small peak at the far left of the plot be ignored as an anomaly, while of the large peak he says, as quoted in the epigraph to the introduction, "Basically, we don't know what caused this large peak. It conceivably could have been due to a burst of very intense gravitational radiation from an explosive type of astronomical event, but more likely it was a malfunction of the equipment" (p. 142).

The dissertation then explores various possible sources of noise, and the tone of the judgments is worth recording.

> [T]he onset of the noise was very sudden and was strong enough to drive the recorder off scale almost as fast as the system time constraints would let it. This very same type of thing had happened once before . . . on 8–9 May, 1962, and at that time it was thought that something was wrong with the apparatus and all the components were thoroughly checked out and various substitutions and changes made in an attempt to find the cause of the noise. However, nothing could be done to affect it in any way and the cause was never ascertained. . . . While waiting for the system to return to normal behavior, the surrounding buildings were checked for evidence of any new or

8. Nowadays it is the material of preference for all the resonant bar groups except the Australians—who use/d niobium alloy.
9. Where I interviewed him in 1972.
10. The notebooks are now in the archives of the Smithsonian Institution.
11. Forward 1965.

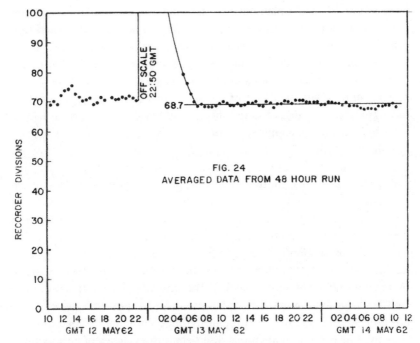

Fig. 2.1 The off-scale anomaly in Forward's early data-gathering run.

unusual experiments or equipment that could be the source of the disturbance, but as would be expected on Saturday night on the campus, there was no evidence of anyone else working except myself. Also, having spent many nights with the equipment, I knew most of the experiments and housekeeping equipment that could cause disturbances and a detector response of this magnitude was out of the ordinary....

The two system sources that could be thought of were a cryogenic instability [the amplifier, but not the whole bar, was cooled with liquid helium] or a swinging of the bar in its harness. (pp. 142–44)

The cryogenic instability cannot be ruled out, according to Forward, but the circumstances that might lead to it are shown to be very unlikely. The swinging of the bar might be caused by an earthquake, but a check of records showed no earthquake activity at that time in the whole of the United States. Electromagnetic storms were also ruled out.

In spite of the inability to pin down a reasonable source for the off-scale peak, only the second 24 hours of data are taken into account in the detailed analysis. The very small fluctuations found in the second 24 hours of data are used to set an upper limit on the amount of gravitational radiation that could have been affecting the bar, while it is nowhere suggested that residual noise was actually caused by gravitational waves.

This first 48 hours of data and its analysis are, in terms of the history of gravitational radiation research, trivial, yet they encapsulate the dilemmas not only of this field but of any experimental research designed to find signals at the edge of detectability. The device emits a trace which, it has to be assumed, comprises a mixture of noise and, with luck, signal. But which is which? On the face of it, the off-scale peak could represent one of the most momentous discoveries of twentieth-century physics, suggesting that such huge sources of gravitational energy were being emitted that either the general theory of relativity contains serious mistakes, or that cosmologists have completely failed to understand the constitution of the universe, or that extraordinarily violent cosmological events are taking place in a close region of the sky while remaining invisible to any other detection apparatus including the human eye. Yet none of these possibilities are entertained except in passing. The experimenter's judgment of the likely behavior of his apparatus, and his sense of the probable and credible, force him to conclude that nothing of significance should be read into the anomaly; this is not a momentous discovery, just evidence of an unknown source of noise.

What kind of decision is this? It is a "rational" decision—whatever that means. Certainly any other decision, such as to announce the discovery of some new force or new phenomenon, would not, at this stage, count as rational, and would make the claimant a laughingstock. And yet it is not a decision forced upon the experimenter by any incontrovertible or obvious logic, or by any irrefutable or even undeniably persuasive data. It is a matter of judgment—and this judgment is as much about others' potential judgments and responses to one kind of claim rather than another as it is about anything else.[12] The history of gravitational radiation research is a history of judgments of this sort. This was an uncontested judgment, whereas, as we shall see, for at least 25 years, the history of gravity waves has been a history of contested judgments. The future of gravitational radiation will be about making outcomes that are uncontested.

Forward, upon his return to Hughes Laboratories, built a series of bar detectors so that he could compare their output with one another and with Weber's bar. These were smaller bars than Weber's, because he was short of space; for example, one was set up in a closet in the bedroom of

---

12. See also Lynch, Livingstone, and Garfinkel 1983.

his home in Oxnard. He reported that he built three detectors because he found another one of the unexplained peaks that sent the chart recorder off scale, but that after building the three he never found another such event.

In the meantime, Weber had built a second detector, and his pair of much more massive and sensitive devices were detecting nothing similar, so Forward had to conclude that the off-scale effects he had seen were of no significance. In the meantime, Forward decided to build a laser interferometer detector, having been given the idea either by Weber or by a friend of his, Phillip Chapman. We will return to this later.[13]

David Zipoy had heard of the gravitational work going on in Maryland while a graduate student at Cornell. He wrote to Weber expressing his interest, was invited to an interview, and was given a job. Zipoy worked on the project for about eight years, beginning in 1959. At the time he arrived, the design was still embryonic. As he put it: [1997] "It would probably be a resonant bar, but maybe, if we thought of something else, it would be that." The Weber team explored various possibilities, including laser interferometers. At that time, however, lasers were known to be very difficult to stabilize, so they eventually settled on the bar design.

Zipoy and Weber found out that the crucial feature of the design of all parts of the apparatus was the "Q" factor, a measure of how long a vibration will continue in the whole system once it has started "ringing." They considered other features of the design, such as the material that the bar should be made of and the number of piezocrystals that should be glued to it. Like the others, Zipoy explained that originally they had intended to use a complete crystal as the antenna and then discovered that crystals glued to a mass made of something else would do. They found out that tungsten was useless, not ringing but going "clunk" when it was hit with a hammer; and they found out that "a good old hard alloy of aluminum was as good as you get."

Zipoy described the feeling of working in the Maryland laboratory as follows: "It was a nice atmosphere because everyone was working together as a group—a team effort, I guess—everyone was doing—working on some aspect of the thing and very amicable—everyone was enjoying themselves. Just a very nice time."

It has to be said that respondents' accounts of what it was to work with Weber vary quite markedly. On the one hand, we have this sense of a

---

13. Chapman was then working at MIT, and it seems likely that he in turn had got the idea from Rainer Weiss, later to become one of the pioneers of large-scale laser interferometry (see chapter 11).

solidaristic team working together on a joint pioneering project; the idea of Weber as a witty, amusing, and brilliant leader is reinforced by other reminiscences, which we will encounter later. On the other hand, some respondents said that Weber was difficult to work with, especially in the period after his first wife died: [1972] "One of the things that you're going to find if you haven't found it already is that for a number of years Professor Weber was very difficult to work with—very difficult." The respondent who gave me this warning went on to list the graduate students who had worked with Weber and subsequently left the field of gravitational radiation research, and it certainly is true that Weber trained no cohort of graduate student supporters. Those who became his loyal supporters in the field of gravity waves did not work closely with him in the laboratory. Of course, as the same respondent went on to point out, sustained outside criticism (discussed throughout this book) does not help, especially when tragedy strikes in one's personal life. By all accounts (and my own experience of him matches these accounts), Weber became more easygoing after he remarried. Yet there was tension in the laboratory even in later years. We have two Joe Webers, then: the Weber who was difficult to work with in a long-term relationship, and a witty, lovable Weber, whom we will encounter frequently in upcoming chapters.

Returning to Zipoy's more rosy account of the times:

> [1997] You'd go to a meeting of general relativity types and essentially everyone was theoretical except Weber and Dicke, and that was about it—Dicke and his students, and Weber, and that was it. And I remember at one time, Peter Bergman, who was very famous—a very theoretical type—mentioned to me at one meeting—he was standing next to me for some reason—and he just turned to me and said he can't believe now that people are actually doing experiments—testing general relativity—he said he was just delighted, fantastically happy, he said. And that was the attitude. Everybody was upbeat.... The field was—as far as I'm concerned, starting in the late '50s was when the field [experimental general relativity] took off. And I'm convinced it was due to Dicke and Weber. The fact that there were actually people doing experiments, trying to test the theory. OK.
>
> [There were people who] said it was not something that they would spend their life doing, or would want to do, because the payoff was essentially zero. They thought there was absolutely no way of getting a positive answer out of this thing, and so it was that sort of thing. They wouldn't want to do it themselves, and they don't understand why you would want to do it, but, you know, "if you want to do it, do it!"

I don't think there were people disparaging it because it was a long shot. The attitude was nice because there wasn't competition. As soon as you get competition in there, people tend to denigrate other people just because they're competing with them . . . you can't be good unless somebody else is bad.

Zipoy said he was not confident that they would find anything: [1997] "'Cause I couldn't think of any sources, certainly at that time, that was before, sort of, anything. . . . I was just intrigued by the idea that he'd even consider building such a thing and build it to the state of the art, and, you know, might get lucky—who knows—wouldn't be the first time."

Richard Imlay, the undergraduate student who worked with Zipoy, put the philosophy as follows: [1997] "I think when you're looking in a new area, experimentalists have to take the attitude that you don't really know 'til you look, to some extent."

The sense of pioneering is also well captured in Zipoy's description of the work.

> [1997] Everything was pushing technology at that time. The Dewar [vacuum flask] we were going to get had to take the inductor we were going to get. . . . at the time just finding Niobium wire—everything was new—Niobium wire was new stuff. And we finally ended up with a coil that was about this high and about that big around [indicating two or three feet high and two or three inches wide][14] of practically solid Niobium after it was all wound, and then we had to get that, of course, down into a Dewar. Dewars at that time normally had a little fill tube about half an inch in diameter. So we had to get a Dewar designed that would take this big coil going down.
>
> And the other cute thing was the amplifier. At that time transistors were pretty new stuff—OK—and transistors themselves—I'm not even sure if they would work at low temperature. . . . The old-fashioned junction transistors—they might not even work down there—but they were so lousy at that time that we really didn't consider it. And we finally ended up with a vacuum-tube amplifier at liquid helium temperature, down with the coil . . . and a little circuit sat on top of the coil in a little vacuum can that was all stuffed down in the helium . . .
>
> And then there was worries about just noises in the room getting at both the bar and the amplifier system, so the bar of course was in its little vacuum and the Dewar was in—this never did really work very well, I don't know why—we built a big plywood box to put the Dewar in, about six feet

---

14. Do not rely on these measurements, but the original can be found in the Smithsonian.

high and three feet square, and in order to make it an acoustic shield I took
my handy power saw and cut almost through the box. Three-quarter-inch
plywood and I cut through about five-eighths of an inch, almost to the edge,
so that the box itself was just a lot of loosely connected squares and it wasn't
clear that ever did anything [laughter]. We put microphones inside and
blared at it, and it wasn't clear that all this trouble of cutting the box up
into little pieces really helped at all.

But anyway, that really never seemed to cause us any problems. You
really had to take a loudspeaker and really blast away at that Dewar in order
to get any response out of it, so apparently that filtering was quite good.

For the first few years, the detection apparatus was set up in a laboratory belonging to the engineering department: [1997] "The mechanical engineers
were sitting ten feet away breaking concrete slabs [laughter]. . . . Fortunately,
that only happened occasionally, and we would stay away when that happened. But there was a nice little pit for some reason in the middle of the
building, and we set up down there. That was probably for the first two or
three years."

The first serious bar was a two-ton cylinder of aluminum, five feet long
and two feet in diameter. It was Zipoy and the undergraduate student,
Richard Imlay, who designed and built the first acoustic filters—piles of
rubber and iron sheets that filtered out vibrations coming through the
ground sufficiently well to avoid the 10^{-16} strains caused by Sinsky's bar
being swamped by other noises: [1997] "Then the other thing was how to
isolate it from the surroundings, and I went into the design of the acoustic
stack to mount it on and the sort of wire suspension system used."

Zipoy said that he does not remember where the idea for the acoustic
stacks originally came from, but he designed them. He went from the optical analogue—using layers with the equivalent of lots of different indexes
of refraction, resulting in lots of reflections at the interfaces. "You can get a
tuned system": [1997] "The testing was for what the spongy layer should be;
things like cork and felt and things like that and some corrugated rubber.
We got a bunch of stuff like that. We ended up with just the rubber slabs
[between iron slabs]."

After a couple of years, the Weber team decided to move the apparatus
to a quieter building. A disused missile silo was considered, but eventually
a purpose-built structure was constructed near the University of Maryland
golf course. This building, which was partly belowground, had to be reinforced more than once to stop the walls caving in from the buildup of
ground forces, slowing the work.

Sinsky's calibration experiment was begun once the apparatus had been moved. Zipoy described it thus:

> [1997] That was interesting. He had a fun time with that one, building the resonant bar—that one was a little shorter than five feet and I guess it was eight inches in diameter, in its own little vacuum tank.
>
> And that was tuned by heating. When you are vibrating the thing to get as strong a signal as you can get, the thing would just heat up by internal friction and that would change the resonant frequency. And so we made some tests on a little bar—a little one-inch-diameter bar... to see roughly what the heating was—to see what we were working with, but eventually we wound up with buying a five-foot bar, the same length as the instrument, and putting it in the vacuum system, cranking it up and turning it on and letting it run for a few hours to let it heat up until it equilibrated and measuring the frequency shift—and it was a lot. And then calculating how much we had to cut off the bar to make it shift to the right frequency. And that was a traumatic time for old Joel [Sinsky]—[loud laughter]. But he got it right. And then the thing was tuned by heating—the tuning was done by heating the bar up and cooling it down to tune the frequency to be same as the resonant bar.

Zipoy left the project before the first results were reported.[15] He explained that he had become tired of making the endless adjustments that had to be made to eliminate noise. [1997] "I guess it was mainly because I didn't really want to spend the rest of my life tweaking. I just saw it as endless tweaking. I guess I just got bored by the whole thing."

It is very hard to eliminate noise from resonant-bar gravity wave detectors for two reasons. First, every adjustment requires that the vacuum chamber be opened and then evacuated once more after it has been closed again. Evacuating a large vessel to a low vacuum is time consuming. Second, it is hard to work out where the noise is coming from, because one cannot see its characteristics; this is because the bar is designed to store energy in its vibrations, thereby smoothing out and hiding the identity of the noise. Zipoy put it this way:

> [1997] It took a long time to sort-of beat down the noise—and that was years, to get the noise level down.... A problem with these high-"Q" systems

15. At the time I interviewed him in 1997, he lived in retirement in Punta Gorda, Florida; he was then 65 years of age.

[systems that vibrate for a long time] is that it's very hard to know when you've got noise and when you've got stray signal. Because it takes so long to do anything. If you have a "Q" of about a million, which is what we had, the ring down is like half a minute or so, and that means you have to wait many of those—minutes—for any transients to die out—so if anything gets disturbed it takes minutes for that to go away. And the thing is if you look at the noise output of this thing on an oscilloscope, it looks like a signal generator—it's a sine wave—a beautiful sine wave. And the only thing you notice about it is the amplitude changes slowly. [That is, the noise is turned by the system into a smooth curve.] . . . And that was a problem, it just took forever. You set the thing up and you'd let it go for days to measure a noise level. It just was a long, tedious process.

Zipoy, when I interviewed him in the spring of 1997, knew nothing of developments in gravitational radiation research much beyond the time he left the field. For example, neither he nor Sinsky knew of the existence of the cryogenic bars program. It is remarkable that these scientists should have left the field in such a thoroughgoing way, but it was a phenomenon I was to encounter several times. Scientists assure me that it surprises them little, since when one leaves a field, one's interest in it has usually been exhausted.[16]

SINSKY AND THE CALIBRATION EXPERIMENT

Joel Sinsky, according to his curriculum vitae, began his doctoral work at the University of Maryland in 1959, but he did not join Weber's team until 1961. Supported by a grant from the US Air Force Office of Scientific Research, Sinsky's calibration experiment, which was the core of his Ph.D.

16. After leaving Weber's team in early 1969, Sinsky joined the Navy and came to be in charge of logistics for the overall US cold-war submarine detection effort. When I spoke to him in 1997, he knew practically nothing of the current state of gravitational radiation research, having ceased to follow developments in the field soon after he left it. On the other hand, Sinsky did have contact with Weber when Weber became interested in new methods of neutrino detection (which we will encounter in part III). Sinsky, bear in mind, was in the submarine detection business and could well understand the potential of a simple neutrino detector for locating the reactors of Soviet nuclear submarines. He told me: "He [Weber] and I had contact about him getting funding from the Navy, either to continue the gravity wave work, or to do the neutrino work. The Navy had some interest in that. . . . So, I was trying to interest the Office of Naval Research to pick up some of his work, and in that regard he and I got together a number of times, and I went with him down to the Office of Naval Research to make presentations. So through the years that I worked for the Navy, I used to see him from time to time." Sinsky retired from the Navy in 1994, seemingly unaware that Weber had actually succeeded in obtaining that funding.

dissertation, did not begin to gather reliable data until the end of 1966. His degree was awarded in 1967, six years after he began the research. During and after his doctoral studies, Sinsky was a member of the team developing the gravity wave detector. He helped to install the equipment in the new building put up specially for the experiment on the University of Maryland's golf course, a mile or so from the main campus. This building was started in the early 1960s, and his contribution to the supervision of its construction caused a considerable delay to his doctoral research.[17] Sinsky also ordered the equipment for the subsequent experiments.

> [1997] I ordered all the bars. And I remember sweating out the orders because the bars had to be faultless, and they had to be cut exactly right, and then I also ordered all the vacuum chambers [which were of the same kind as the fuel storage tanks underneath automobile gas stations] and the suspension systems. I didn't design them—Weber had designed them.... When I started work, Bob Forward and he had already started doing that.

Weber, Forward, Zipoy, and Imlay had in fact begun the design work with the specifications for the 22-inch-diameter resonant bar that was installed in the Electrical Engineering Laboratory—the bar that was to be the receiver for Sinsky's calibration experiment. Sinsky told me that Zipoy chose five feet for the length of the bar, because a frequency of 1660 Hertz multiplied by 2 times pi gives a result of 10,000 (actually, 10,430), which would make calculations easy. Bob Forward told me it was chosen because he, Forward, knew he would have to move the bars around, and wanted the dimensions to be such that he could reach from one end to the other.

Sinsky had a graduate student's appreciation of the personnel in the laboratory: being just at the beginning of his apprenticeship, he was seeing things for the first time. Furthermore, he left the team as soon as the experiment was over, departing the world of universities for government service and maintaining no contact with the fast-changing professional world of gravitational wave physics.[18] As a result, Sinsky's memories may be colored by fewer layers of reinterpretation than those of scientists who knew Weber in his later years; Sinsky's experiences may be a specially good way to get at the texture of the experimental physics of the time. The nature of his

---

17. Sinsky also helped to transport one of the bars for the subsequent coincidence experiment to the Argonne National Laboratory outside Chicago.

18. Though he did maintain some contact with Joe Weber, and this should be borne in mind.

calibration experiment also introduces the main theme, which unites physics and sociology throughout the whole period: the establishment of the existence of a signal by separating it from noise. Third, Sinsky's experiment attempted to calibrate the first gravitational wave detector, another vital theme of this story. Finally, the history of the Sinsky experiment is very little known—it virtually disappeared from professional discourse fairly soon after it was completed and published. That is interesting in itself in that it indicates some of the qualities of social space-time. For all these reasons I will concentrate on Sinsky and his experiment for a while longer. In fact, this next section is going to be one of the most detailed descriptions of an experiment in the whole book—and ironically so, since the experiment is of almost no importance in the history of gravitational wave detection or even the arguments within this history. But bear with it—with this experiment we start to get a feel for the texture of experimental science and what it is to have six or more years of intense labor described in a few lines and then forgotten about. The length and detail of the description is part of the point.

I interviewed the now-retired Dr. Sinsky at his home in Baltimore in November 1997, and we were able to look through his laboratory notebooks and various other documents.[19] Sinsky joined Weber's team in 1961 and began working under the supervision of Zipoy. He found him to be an unsympathetic taskmaster, however, and transferred to Weber's supervision. Toward the end of the experiment, Sinsky found himself working in the laboratory throughout the night, as he had to move the resonant bar to a new position every four hours. At this time, Sinsky's newborn tended to wake the family every few hours, and he saw the two sources of disturbance to his sleep as similar. He confessed that at the end he was desperate to get away from the experiment, because it seemed as though it would never end; it was consuming his life. He had entered the laboratory as an idealist, wanting to do the hardest experiment he could find, but he left it wanting to live a normal life with his family. As Zipoy was quoted above, "that was a traumatic time for old Joel." In spite of this, Sinsky described Weber as follows:

> [1997] He was never hard on me in terms of the work. We had a difficult experiment, but I always felt that he was good to me, he was fair to me, he was kind to me. He worked harder than I did—longer hours—even though I used to sleep in the lab at night so I could move that thing every four hours.

19. I later deposited the notebooks at the Smithsonian Institution.

But still he was there early in the morning and he used to work nights and days and weekends. He never stopped. And I always felt that he was very supportive, and I was very loyal to him when I was working for him.... I couldn't have asked for a more supportive guy.

As Zipoy mentioned, even though the driven bar was in a vacuum chamber, the vibrations heated it enough to change its dimensions significantly and thus affect its frequency. This meant that they had to cut the bar a little short, so that it would vibrate too fast if left at room temperature. The heating caused by the vibration, however, expanded the bar and made it vibrate at a slightly lower frequency. Finally, a further slight increase in temperature accomplished with the use of an electrically powered heater would expand the bar a little more until its dimensions were just right to accomplish ringing at the same frequency as the detector bar.

When pulses of energy are injected at just the right frequency into something that vibrates, the energy and the size of the vibration build up and up, just like the effect of correctly timed pushes on a child's swing. (For this reason, soldiers break step when marching over a bridge; should the frequency of their in-phase marching match the "natural frequency" of the bridge, they could accidentally destroy it.) The ability of a bar to "integrate" the energy of any vibration that hits it with a frequency near its natural frequency is crucial. It is this feature that makes it conceivable that the impact of gravitational waves could cause such a bar to vibrate, even though the integrated energy would still only cause vibrations smaller than the diameter of an atomic nucleus. The integration applies equally to the calibration experiment and to the putative effect of gravitational waves, which is why these detectors are called resonant bars. Sinsky's first laboratory notebook entry appeared in July 1962. The notebook records the technique for heating the transmitting bar in order to tune it as having been invented in October 1962.

The essence of the experiment is as follows: there is a signal-generating bar and a signal-receiving bar (the detector); sometimes I will refer to these as the "driving bar" and the "driven bar." The gravitational relationship between driving and driven bars has to be measured. The trouble is that even though a signal can be detected in the driven bar when the driving bar is running, it may not be caused by gravitational forces. The driving bar can emit energy at the right frequency in the form of vibrations in the air (sound), vibrations in the fabric of the laboratory, electromagnetic vibrations traveling though space (radio frequency noise), and electrical signals traveling through the wiring of the laboratory and even through the ground. At the outset of the experiment, it is likely that any of these other "couplings" will have a greater effect than the gravitational coupling between the two bars; therefore, before one has much hope of seeing the gravitational coupling, one must eliminate these "spurious couplings." If this "noise" is sufficiently reduced, there is a chance that a signal will emerge that will indicate a gravitational coupling. This still does not prove that it is gravity that has given rise to the data—it could still be that one of these other couplings is still present but at a very low level which mimics the gravitational effect. These two stages of noise elimination are typical of all the experiments I will describe in this book: step 1 is to drive down the noise as far as possible; step 2 is to find ways of being certain that the remaining signal is signal not noise.

We now move forward eighteen months. The team tested the detection apparatus for acoustic coupling. Just like a loudspeaker, the driving bar made a sound, and the team needed to find the potential effect of that sound on the receiving bar so that they could be sure it was eliminated. The generator was run in the open air, where the researchers noted that not only did it have an effect on the receiving bar, it had an effect on the receiving bar's *amplifier*—the device that was used to enhance the signals emitted from the piezocrystals glued to the second bar's surface.

This is a typical example of practical physics. One has a notional model of the way the experiment is meant to work, but when a device is first put together, the model is nearly always confounded by mundane effects. In this case, one could take away the receiving mass and the piezocrystals, and a large effect would still have been registered simply because the sound made by the vibrating bar was affecting the electronic circuits! The experimenters decided that in this case the sound of the vibrating bar was affecting the amplifier's vacuum tubes (valves); when the vacuum tubes were caused to vibrate, this resulted in slight differences in their electrical qualities and happened at a frequency mimicking acoustic coupling between the bars. As soon as one sees this, one can see that any electrical circuit is also a physical entity, and anything that causes it to vibrate—and thus change the dimensions of its components—could cause a spurious signal. In this case, the Weber team found that by enclosing the amplifier in a sound-insulating box, they could almost completely eliminate this effect.

Having a reasonably insulated amplifier, the experimenters reconnected it to the receiving bar and noted the increase in output. As they changed the sound by tuning the generator (with the heater, which changed its length) they could see the receiver's output increase as the sound swept through the natural frequency of the receiving bar. This, of course, was largely acoustic coupling, not gravitational coupling. The sound was vibrating the air, which was causing the two tons of aluminum to "ring." They had to eliminate this much larger effect if they wanted to see the effect of gravity.

The team could reduce the acoustic coupling by putting the receiver into a vacuum chamber and taking out most of the air. But, though reduced, the effect was still marked, even at a high vacuum. How did they know they were still seeing an acoustic effect rather than a gravitational effect? First, the effect was too large to fit with what they expected from gravitational coupling. Second, the size of the effect did not change as would be expected as the distance between driver and driven bar was changed. Gravitational forces diminish rapidly with distance, but here the vibrations in the driven bar did not reduce fast enough as the distance between the bars was increased. Thus the coupling between the bars seemed to be acoustic—transmitted through the air and the vacuum chamber, or through the walls and floor of the laboratory.

The researchers noted that in spite of the driving bar being set within a vacuum chamber, the vacuum chamber itself, and the cart and rails on which the vacuum chamber were supported, were all vibrating at the same frequency as the bar. It seemed likely that vibrations were being transmitted to the bar via the wire support that held the bar within the vacuum chamber. They also noted that it was not only the driving bar that was making a noise: the amplifier that was used to drive the piezoelectric crystals that in turn drive the driving bar was itself emitting an audible hum at the driving frequency.

The amplifier noise was eliminated by placing the device in a steel chest with four-inch-thick walls, which would also block off its electromagnetic emissions—another potential source of unwanted linkage between the driving and driven side of the experiment.

Electromagnetic waves traveling through the laboratory were only one source of potential electrical coupling between the two sides of the experiment. Another was the laboratory wiring. Experienced experimenters know that stray signals can pass along the mains wiring that links all the mains-powered apparatus in a laboratory (and links it to other mains-powered apparatuses in other laboratories, for that matter). This source of potential coupling was eliminated by powering the receiver amplifiers and chart recorder with batteries, thereby isolating the receiver from mains power.

The large vacuum pumps for the receiver vacuum chamber were electrically isolated from the detector itself and the smaller final-stage "diffusion

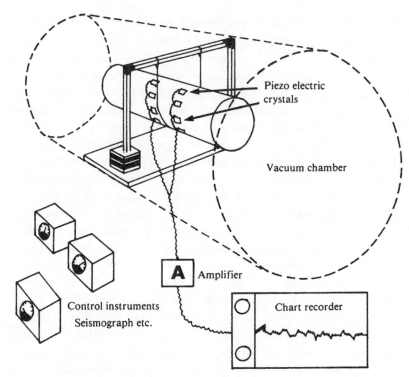

Piezo electric crystals

Vacuum chamber

A Amplifier

Control instruments
Seismograph etc.

Chart recorder

Fig. 2.2 Schematic diagram of the early Weber-bar detector.

pump" was switched off during data taking. Thus the team made sure that neither vacuum pump was able to act as a conduit for electromagnetic noise transmitted as "radio" signals or for mains wiring.

Another way of linking the two sides of the apparatus was through electrical signals passing through the earth. The detection system and the generator's circuits were therefore run ungrounded (unearthed) to eliminate communication via earthborne electrical signals. All other electrical pathways were isolated with metal housings of one sort or another: solid sheaths, lead foil, copper mesh, or combinations of both. Sinsky's dissertation remarks of the final design, "This configuration of grounds and shielding ultimately proves to be the one from which no [electromagnetic] coupling of any kind was detected."[20]

The next step was to try to eliminate the sounds coming from the driving bar. The Weber team enclosed it in its own vacuum tank, with the tank itself being insulated from its support trolley by an acoustic stack of

20. Sinsky 1967, p. 108.

alternating layers of steel and rubber such as was already used for the detector vacuum tank.[21] The whole vacuum tank and acoustic stack assembly was placed within a plywood box lined with acoustic tiles. Further, the metal trolley was replaced with a wooden version that did not require rails. In spite of these precautions, a sound from the generator could still be heard and could not be eliminated even after the experimenters filled in the space between vacuum chamber and the acoustic box.

Having failed to eliminate completely the sounds emitted by the driving side of the experiment, the team decided to make the receiving side insensitive to sound. To do this, they set up loudspeakers that could make sounds far louder than the driving bar could ever make, with a view to making the receiver insensitive to even these loud noises. But as Sinsky wrote in his notebook, the team was "still well below the sensitivity required for grav radiation detection."

Slowly the effect of noise transmitted through air was understood and eliminated with appropriate arrangements of insulation. Sinsky's dissertation sums up some of this progress.

> The electrical fitting at the top of the detector vacuum tank was seated on a shielded rubber O-ring to dampen vibrations between the vacuum tank and the coax [coaxial cable]. The preamp and tuning capacitors on top of the helium dewar were enclosed in a wooden box lined with acoustic tile. Finally, a sonic shield of wood and acoustic tile was placed between the generator and detector and the generator was enclosed in an 8 foot × 8 foot × 12 foot room whose walls were lined with acoustic tile. These latter changes, particularly the improvements in the detector, ultimately reduced the acoustic interaction between the two systems to acceptable values.[22]

By August 1965, Sinksy's notebook records the first indications of an interaction between the two bars that appears to be something other than acoustical or electromagnetic noise, but the experiment was still plagued by some kind of acoustical leak. Thus, the October 30, 1965, entry reads, "As of this date Dr. Weber has, for the last four weeks, tried intensive tests to discover the source of the acoustic pickup in the detection system as well as to eliminate it."

---

21. The driving bar was on a *trolley* so that it could be moved from one place to another.

22. Sinsky 1967, p. 37. Because dates are not recorded in the Ph.D. dissertation, the sequence of events recorded there may not correspond to the notebook entries.

Eventually, the source of the pickup was tracked to a coaxial cable linking the bar and the amplifier, so this assembly was modified in various ways. A notebook entry records the continuing troubles.

> *November 5 1965:* Braid inner conductor did not reduce noise, nor did wrapping coax [coaxial cable] in felt. Dr. Weber now feels that the noise is due to teflon spacers in coax and is having them replaced with polyethylene. 3 old crystals were chipped from generator and one new one was epoxied on with Eastman 910. All of the old crystals chipped off showed that the Hysol epoxy used to bond them bonded to the bar and not strongly to the crystal. A layer of epoxy was left on the bar and the undersides of the crystals were clear.

Sinsky's dissertation explains how unwanted effects could be separated from gravitational effects by various combinations of tuning. For example, the receiving circuit might be detuned from the bar and the changes in output noted as other things remained constant. Or the generating bar might be detuned from the receiving bar and the effects noted. These tests were tried with the bars in all their different relative positions. It was found that the elimination of radio-frequency noise (that is, electromagnetic noise transmitted through space) in one position did not guarantee its elimination in another, so it was necessary to run the full range of tests in every position. Similarly, the effects of acoustic coupling could be examined and eliminated by using the loudspeaker as a very strong source of acoustic noise. Work of this kind delayed the completion of Sinsky's dissertation to well beyond what he expected at the outset of the project.

Nevertheless, it seems that the last sources of noise were finally eliminated late in 1966, when data could, at last, begin to be gathered. Sinsky tells a funny story about the culmination of this phase of the work. He began to feel desperate as the completion date of his Ph.D. grew increasingly remote. [1997] "My problem was that I hooked up with an experiment that was incredibly hard. It involved moving an observatory and building a building. I mean, most guys come in—and you know—"I'll take this apparatus, turn the crank to another frequency, and get a degree.""

Of his friends who had embarked on their doctoral work at the same time as he, most had long completed their dissertations; Sinsky, longing for a return to a normal family life, was sleeping with the apparatus. Sinsky, an Orthodox Jew, eventually asked a religious friend for advice. The friend said he should visit the famous Lubavicher rabbi—Rabbi Schneerson—in Crown Heights, Brooklyn. (Rabbi Schneerson is now dead, but many Orthodox Jews took him to be the new Messiah.) In those days, Schneerson would spend two whole nights every week aiding petitioners, so Sinsky joined the queue and eventually got to see him in the early hours of the morning.

Sinsky explained the trouble he was having in about ten sentences. An initial surprise was Schneerson's first question. He asked if it could be a problem of electromagnetic noise and whether his supervisor knew anything about these things (the rabbi, it turned out, had been trained in science at the Sorbonne). Then, having been reassured that the supervisor, Joe Weber, was himself a professor of electrical engineering, Schneerson offered more in the way of what Sinsky was expecting.

> [1997] He said, "Fine." Then he said, you should take a charity box and put it in the laboratory. And every time you take a break from trying to get this experiment to work, put a penny in the charity box—the amount of money isn't important, it's just the act of putting the money in the charity box that's important. So if you take a break for lunch, put a penny in; if you take a break for dinner, put a penny in; if you take a break to go home to sleep, you put a penny in.

When the box was full, said Schneerson, the experiment would work.

Sinsky did as the rabbi suggested, and the box was full at the beginning of Chanukah 1966. Then the experiment finally started to work properly, just as Schneerson had predicted.

Sinsky found Joe Weber and told him the good news—after five years they finally had all the bugs eliminated and could start taking reportable data. He also told Weber about the Lubavicher rabbi's contribution. Weber's response was to grin. "Can we hire him?" he asked.

Though increasingly cranky and bad-tempered in his later years, Weber retained that underlying twinkle and sense of humor that was much more marked in his younger days. Another anecdote from Sinsky reinforces the image. Sinsky told me that, as an observant Jew, he would occasionally upbraid Weber for being less than modest; modesty was the appropriate demeanor for a Jew. Perhaps after a long night's struggle to solve a problem, Weber would make some remark that, as Sinsky put it, "exhibited a little bit of arrogance." Responding to Sinsky's criticism, Weber once asked, "Isn't it true that the Bible says Moses was the most humble man in creation?" "Yes," replied Sinsky. "Well," said Weber, "that position's taken."[23]

23. For Weber's willingness to extract the humor from a theoretical physics calculation, see the contribution by Blair in his 1991 edited book.

The Calibration Findings Proper

Now that the apparatus seemed free of bugs and the unintended transmission of energy through unwanted channels, the calibration experiment proper could begin. This consisted of two parts. The first was to build a theoretical model of the expected gravitational coupling between the driving bar and the receiving bar which would show what ought to happen as the spacing between the two bars was changed. The two bars were initially placed end to end and the distance between them increased and decreased according to a prearranged schedule. Also, the driving bar could be shifted sideways. Calculations showed how the relative positions of the driving bar should affect the strength of the gravitational signal—it should diminish in a well-understood fashion as the driving bar was moved away and as it was moved sideways. The second part of the experiment was to move the driving bar into various positions and observe the changes in energy in the detector's output. The experiment would be successful if the observed changes in energy matched the calculated changes.

In the first run, Sinsky kept the bars on the same axis while the generating bar was moved a distance of five centimeters every two hours. For example, from 10 a.m. to just before 12 noon, the spacing was five centimeters; from then to around 1:30 p.m., the bar was at a distance of ten centimeters. Subsequently, it was moved out to fifteen centimeters, then twenty centimeters, and then 25 centimeters before ultimately being returned to a five-centimeter separation.[24] As can be seen in figure 2.3, the average output of the detector decreased when the spacing was increased to 25 centimeters, then increased steadily as the distance was reduced to the original five centimeters. Disappointingly, the final level was higher than the original starting point. But this was but one run of the experiment, and Sinsky's dissertation describes others that were more convincing.[25]

These chart recordings strongly suggest that distance affects energy transfer, but they do not prove that the effect was gravitational; there may have been other ways for the driving bar to influence the receiving bar. Nevertheless, it should be possible to separate the different effects by modeling them. Different forces would bring about different patterns of change as distance between the bars changed. What had to be done was fit the data

24. I read these figures from copies of the actual chart recordings supplied to me by Joel Sinsky. The actual separation of the end faces of the bars must have been greater by a fixed amount: because the bars were in vacuum chambers, and the generator was in turn enclosed in an acoustic box, the nearest the ends of the bars could come to each other was 18 cm.

25. Much more of the chart recording is held at the Smithsonian Institution.

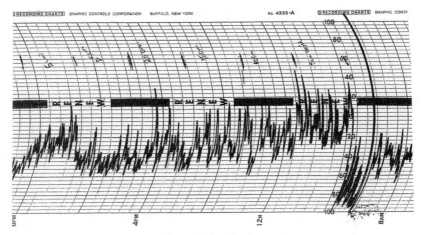

Fig. 2.3 Section of chart-recorder output from Sinsky's calibration experiment.

with a model for change due to gravitational forces. A good fit between model and data would argue against the possibility that the bars were communicating via some other force that followed a different law of physics relating separation and strength. Figure 2.4 shows a graph from page 49 of Sinsky's thesis which compares theoretical prediction with observations for coaxial separations. In this figure, the solid line is the result of the theoretical calculation, and the crosses are actual measurements. Calculating the averages of the observed results and comparing these with theoretically generated figures, Sinsky found that the two figures agree to within 10% in every case, perhaps all that could be expected from an experiment as difficult as this.

What Became of the Sinsky Calibration?

Experimenters have at their disposal various means of convincing others that their results mean what they say they mean. Fitting data output to a theoretical model is one such means. But none of these means can ever be decisive in a quasi-logical sense. A determined critic might say of the Sinsky data that the fit between model and data is not good enough. But what is a good-enough fit? The fit will never be exact, because even if at first glance the fit looks good, more detailed examination can always reveal mismatches at ever-finer levels. What counts as a good-enough fit is, then, a matter of convention. Notoriously, these conventions vary from science to science, but they also vary within sciences. As we will see in due course, the

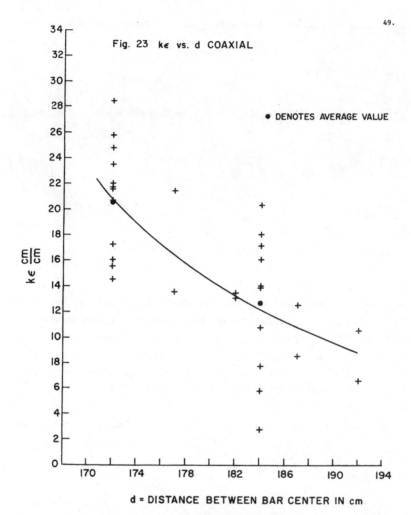

Fig. 2.4 Graph from Sinsky's doctoral dissertation summarizing calibration results.

conventions of high-energy physics are not the same as those that developed during the first two decades of gravitational radiation research.[26]

26. Looking back from today, one can see ways in which the experiment might have been improved. For example, it would be good to set the fit of the data with the gravitational model against the fit with the model for other potentially confounding forces and then compare the statistics for each fit. But this is the finding that actually entered the literature after the due process of peer review, so it must have passed muster according to the conventions of the time. For a brilliant discussion of the meaning of "fit" between theory and data, see Kuhn 1961.

But even if a critic is convinced that the fit between theory and model is good enough, he/she must still be convinced that the experiment was competently carried out. In this case the most obvious weak point is the shielding for other possible coupling forces. As we have seen, the detector, because that is what it is designed for, will respond with exquisite sensitivity to anything that penetrates the isolation of either bar or amplifier at the right frequency. But, as we have also seen, located within the same laboratory space is a bar not only emitting a changing gravitational field at the appropriate frequency, but emitting vibrations through its supporting wires; through any residual air within the vacuum chamber; and, if the vibrations can penetrate the shieldings and acoustic pilings, through the floor, the walls, and the air of the laboratory. At the same time, the bar is driven by an electrical amplifier, so that there are electromagnetic signals at the principal resonant frequency of the bar potentially being emitted into the space between the generator and detector, and electrical connections via the main electricity circuits, which provide power for both amplifiers, chart recorders, and so forth, and the ground. We have seen the steps that were taken to eliminate these effects.

The Weber team's long years of effort prior to Chanukah 1966 were devoted to eliminating all these possibilities for spurious communication, but the critic still has to be convinced that they have been reduced to such an extent that the minute gravitational coupling is a more credible explanation for the results shown on the chart recordings than any of the other forces. To be convinced, the experimenter must accept the integrity and the competence of all the experimenters, because these years of research can only be described in the journals in the most general terms. In the case of this experiment, it is very hard to tell that all these problems have been eliminated. For what it is worth, I did not believe the results of these experiments until I spoke to Joel Sinsky and learned about the care that had been taken.

When Sinsky finished his experiment, he considered that he had done a great thing toward the development of the new science of gravitational wave detection. He had sixty copies of his dissertation printed, expecting, as he said to me, that it would become a best-seller. He continued to work with the Weber group for another couple of years, helping with the first coincidence experiment. He set down in a ring binder, entitled *Gravity Wave Detector Designer's Handbook,* what he had learned about building such devices. It includes details of where to order the parts, how to get them machined, and how to deal with the phone company that was arranging the link between detectors. The original correspondence and orders, complete with cost estimates, for bars, vacuum tanks, piezoelectric crystals, and so forth, are to be found in this text. Sinsky expected it to be in high demand, too.

So what became of the Sinsky results? How did they travel through social space-time? The material from the thesis was extracted and published in two professional papers. *Physical Review Letters* carried a two-page paper by Sinsky and Weber in May 1967, while *The Physical Review* published a seven-page paper by Sinsky in March 1968. The longer paper concludes as follows:

> The gravitational induction field communications experiment described in this paper represents the first laboratory-achieved generation and detection of dynamic Newtonian fields in the kilocycle frequency range. Although the theory of this experiment is well known, its execution is difficult. The experiment fulfils its primary motivation, which is to calibrate a detector of gravitational radiation. The laboratory results conform very closely to theoretical predictions. (p. 1151)

The popular science journal *Science News* reports on page 409 of its April 1968 issue,

> Dr Joel Sinsky performed an experiment to test their [the bars'] sensitivity. He wanted to see whether vibrations in one of them would induce vibrations in the other by gravitational interaction. He found that this did indeed happen at centre-to-centre distances of just under two yards.

And as we have seen, *Physics Today,* in publishing Weber's six-page comprehensive survey of the state of gravitational wave theory and detection, included the following:

> To verify the sensitivity of the detector and to test a new type of gravitational generator Sinsky carried out a high-frequency Cavendish experiment. A second cylinder was suspended in a vacuum chamber. Electrically driven piezoelectric crystals bonded to its surface built up large, resonant, mechanical oscillations. The dynamic gravitational field of the generator was observed at the detector as a function of longitudinal and lateral displacements. Agreement of theory and experiment confirmed that strains of a few parts in 10^{16} were being observed.... This communication represents a major advance in the technology of gravitation.[27]

27. Weber 1968c, p. 37.

But this was the high point of the Sinsky experiment. In the argument that followed about whether Weber had detected gravitational waves, and the embedded argument about whose apparatus was the most sensitive, Sinsky's experiment was forgotten. When I spoke with Sinsky in 1997, he told me that he still had about 55 copies of the thesis left, and the *Handbook* had experienced no demand at all. In fact, it was I who took the original notebooks, *Handbook,* paper chart-recordings, copies of the dissertation, and various other documents from Sinsky's basement in his house in Pikesville near Baltimore and deposited them with the Smithsonian Institution.

The generations of printed output describing the Sinsky experiment show the now well-understood filtering down of the complexities of laboratory practice into unadorned and certain-looking conclusions. The *Handbook* contains original letters, part numbers, and descriptions of conversations with outside contributors such as the phone company; the laboratory notebooks include dates, stumblings, and meanderings; the doctoral dissertation discards the temporal sequence found in the notebooks and replaces it with an order that better reflects the scientific logic which emerged as the experiment was completed, with untidy elements being relegated to appendices. Relatively raw data may be found in the dissertation alongside diagrams that summarize the data. In the papers published in the academic journals, the years of work, the false trials, and the difficulties that were encountered and overcome are compressed. In the *Physical Review Letters* paper, only a couple of sentences bear on these things: "Extreme precautions were therefore required to avoid acoustic and electromagnetic leakage" (p. 795), and "It was discovered that a structure induced by weight stresses on Teflon made it slightly piezoelectric, and this was a major source of acoustic coupling through transmission line supports" (p. 796).[28] The quotations shown in this chapter comprise the entire discussion found in *Science News* and *Physics Today;* the conclusions are worth repeating.

> [Sinsky] wanted to see whether vibrations in one of the [bars] would induce vibrations in the other by gravitational interaction. He found that this did indeed happen at centre-to-centre distances of just under two yards. (*Science News* (93):409)

> Agreement of theory and experiment confirmed that strains of a few parts in 10^{16} were being observed. (*Physics Today*)

---

28. Though the *Physical Review* paper contains more of this than one might expect.

Sociologists have argued that this very process of writing and rewriting, describing and redescribing is what gives scientific findings their factlike status. Ludwik Fleck, who was a sociologist of scientific knowledge before the term was invented, wrote in the 1930s that

> [c]haracteristic of the popular presentation is the omission both of detail and especially of controversial opinions; this produces an artificial simplification.... [and] the apodictic valuation simply to accept or reject a certain point of view. Simplified, lucid, and apodictic science—these are the most important characteristics of exoteric knowledge. *In place of the specific constraint of thought by any proof, which can be found only with great effort, a vivid picture is created through simplification and valuation.*

I myself have said that "[d]istance lends enchantment": The more distant in social space or time is the locus of creation of knowledge, the more certain it is. This is because to create a certainty, the skill and fallible effort that goes into making an experiment work has to be hidden. When the human activity that is experimentation is seen clearly, then one can also clearly see what could be going wrong. Latour and Woolgar talk about the way that an idea becomes real as it moves from the seat of its creation and the modalities and qualifications with which it was originally presented are removed from the sentences that describe it in the literature.[29]

The important thing to notice is that distance and literary transformation account for certainty but not for the content of that certainty. If Sinsky's results were generally believed, the quality of belief would be related to the distance from the experiment and the modes of expression within the literature: distance would lend enchantment. But if the results came to be disbelieved, the quality of *disbelief* would be related to distance and mode of expression in the same way. Thus, while exposure to the details of Sinsky's experiment may make one cautious in accepting its conclusions because one learns of the many things that could have gone wrong, exposure to the details also makes it impossible to dismiss the experiment out of hand, because one can see that the experimenters were aware of the problems and expended immense amounts of time and effort to eliminate them; moreover, one can see how many things went right. In a case where

---

29. The Fleck quotation is from Fleck 1935/1979, pp. 112–13; emphasis in the original. For distance lends enchantment, see Collins 1985/1992. For literary transformation, see Latour and Woolgar 1979.

distance renders all this care invisible, distance leads to enchantment with clear and certain disbelief.[30]

Perhaps Sinsky's own description of the events described above is the most eloquent way of helping us recapture the experimental ingenuity that went into eliminating noise.

> [1997] Every day I would come in, we would try to experiment. And, you know, it would take a few days to determine whether it was working or not, because we'd have to get enough statistics to see if it was reading the thing the way it should read it [If the distance relationships between detector and generator were correct for gravitational coupling]. And if it wasn't, then we'd try and isolate something even more than we'd isolated it before. We'd either put more acoustic stacks under the receiver, or we'd put acoustic stacks under the generator. We'd put floor vibration detectors all around the thing to see if it was coupling through the floor. I'd add more isolation material around the bar. What I was doing, I was driving the bar with this huge amplifier, so it was easy to couple through electromagnetic radiation. So we put the amplifier in a Mosler safe. It was one of these safes that you buy to put your valuables in, but it was a huge safe, with maybe four-inch-thick steel walls [laughter], and we put the amplifier in there. It was incredible.[31]
>
> And each day Joe would think of some new way to measure the spurious coupling and drive it down, and I long gave up after about a year of this that I could possibly think of any other way to improve the isolation. And the thing that amazed me about Weber, and I'll go to the grave thinking that he was the greatest experimentalist that I have ever seen, was because every day he would come in with a new idea on how to reduce the coupling more. Where there might be some leakage and how he might be able to take care of it . . .
>
> I remember coming in every day and saying, "Dr. Weber, I can't think of anything else I can do." And he'd say, "Well, why don't we try this." . . .
>
> That we were able to detect this gravitational interaction—that I can be sure of—because we got all the appropriate falloffs [appropriate reductions in strength of coupling as the driver was moved away from the driven bar].

30. The work that treats the establishment of scientific knowledge as a matter of literary transformation misses the point that positive or negative certainty can be the outcome; and that to explain which it is, more than the literature that needs to be looked at.

31. A propos the language of science, it is interesting that this readymade commercial safe is referred to as "a chest" in the Ph.D. dissertation. Reading the thesis, it seems as though a purpose-built chest was designed to house the amplifier and that the walls were specified to be four inches thick as a result of design calculations. The true source of the "chest" specifications become clear only in conversation.

But the Sinsky calibration turned out to have little importance in the long debate about the existence of gravitational waves. Today one can hear physicists suggest that the strain sensitivity of Weber's bar may have been no greater than 10^{-13}, that is, 1000 times less than Sinsky's calibration suggested. And when, in the 1970s, the fashion for electrostatic calibration became dominant, the Sinsky calibration was never mentioned as a serious alternative candidate. This was in spite of the fact that it was a gravitational calibration, which might have been considered nearer to a *gravitational wave* calibration than an electrostatic calibration.[32]

Of this I will say more in due course, but first let us consider why it was so easy to ignore this result.

Sinsky's calibration technique came with one huge disadvantage: it was, as our description makes clear, very difficult. Therefore it could never be a competitor where many experimenters had to compare their bars rapidly; the electrostatic pulse technique—something that will be described in chapter 10—was far too easy in comparison. But this answers only half the question. Even though others were not going to copy the technique, it does not follow that the results of those painstaking years should be treated as unreliable. But perhaps the result of a complex experiment is more likely to be treated as unreliable when the credibility of an alternative simple technique is at stake.

There are more subtle forces at work, too. The experiment was easy to ignore because scientists had asymmetrical access to its details. No one had special access to Sinsky's account of his hard work and caution—especially as Sinsky had left the field and Weber's credibility was in question anyway. But everyone in the field was being continually presented with the negative face of the experiment, because everyone was continually being presented with, continually discussing, and continually worrying about the difficulties of isolating a massive and acutely sensitive detector; detectors of this sort were exactly what everyone was designing. They would know, then, that to isolate a massive detector that was acutely sensitive to vibrations at

---

32. But consider the following from Tyson 1973a, which is, so far as I can find, the only thing that could count as an exception:

[The] electrostatic technique is absolutely equivalent to the near-field gravitational calibration techniques [citation to Sinsky's dissertation] (A. C. Cavandish [sic] calibration), to the extent that the physics of both is known to sufficient accuracy. So far there is no means of calibrating a gravitation wave antenna using a far field of known intensity, and until we have such calibration we must rely on applying known forces to gravitational wave antennae. The electrostatic technique is simply much more convenient than the A. C. Cavandish [sic] near-field gravitational technique and has the advantage that it can be used often during the course of measurements. (p. 77)

1660 Hertz from another mass being driven by massive amplifiers to vibrate at 1660 Hertz, and was placed at a distance of only a few centimeters from it, was a still more difficult task. Exactly how difficult would be especially evident: from 1970 onwards, the search for gravitational waves required every scientist to be acutely aware of the sources of nongravitational coupling between two such masses, precisely because for a coincident excitation to be a candidate to be a gravity wave, all such nongravitational couplings had to be eliminated. The very first step in eliminating nongravitational coupling was to separate the detectors by thousands of miles, not by a few centimeters. To have a hope of eliminating common sources of disturbance, thousands of miles were needed, not tens of centimeters. Thus, the first thing that would strike anyone in the field who thought about the Sinsky calibration was the manifest number of nongravitational couplings that could link two vibrating masses lying side by side. *Indirectly,* everyone was exposed to the downside of the details of the Sinsky experiment, even if they knew nothing of it directly except via some sanitized version published in the academic journals.

Sinsky's calibration depended on energy changes that are lasting and correlated with distance, whereas the bursts that might be candidates for gravitational radiation sources are much harder to separate from noise. For example, there was no way that an electrical storm, traffic noise, or an earthquake could upset the experiment. Nevertheless, the big separations between detectors needed to cancel these subtle effects were also intended to eliminate local effects, such as coincident signals on the mains electrical supply. Consequently, detectors close to each other must seem a horrendous proposition for adequate isolation. Given no access to the actual precautions taken, I suggest that for those who encountered the Sinsky experiment, distance led to radical disenchantment.

CHAPTER 3

WHAT ARE GRAVITATIONAL WAVES?

ARE THERE ANY GRAVITATIONAL WAVES?

Einstein may not have been the first to discuss the idea of gravitational waves, but, in the second decade of the twentieth century, he was the first to incorporate them properly into a theory.[1] It is a tricky subject, however, and in 1936 Einstein coauthored another paper in which, contrary to his earlier work, he proved—at least in the first draft of the paper—that they did not exist.[2] At a conference in the United States as late as 1957, the existence

---

1. What I write in this section is heavily dependent on the paper by Dan Kennefick (1999). Kennefick, a physicist who is himself a minor contributor to the theoretical analysis of gravitational waves, writes on the history and sociology of the *theory* of gravitational radiation.

2. The paper is by Einstein and Rosen, and was eventually published in 1937. According to Kennefick (private communication), Einstein submitted this paper to *Physical Review*, which returned it with referee's comments asking for reconsideration. Einstein, apparently unused to American refereeing conventions, withdrew the paper in a huff. When it was eventually published, in the *Journal of the Franklin Institute* (Einstein and Rosen, 1937), gravitational waves had reappeared, though still with some reservations. It turns out that the *Physical Review*'s referee had been right.

and detectability of gravitational waves was hotly debated, with Herman Bondi and among those who argued both that the waves were emitted and that they were, in principle, detectable—that is, they had a sensible effect on the things through which they passed.

Let us try to give some sense of how it is possible to argue about such things. First, we must see why we need something like a gravitational wave: Earth continues in its roughly circular orbit around the Sun because of the Sun's gravitational pull, the size of which depends on the Sun's mass. Now imagine that the Sun starts to lose mass; suppose, for example, there is an internal explosion which has the effect of shooting out two large lumps of the Sun in opposite directions at right angles to the plane of Earth's orbit. The bulk of the Sun will stay in the same place, but Earth's orbit will be affected. Because the Sun will now be a bit lighter, Earth will be less strongly attracted to it, and its orbit will grow a little bigger. The question is, how long does it take Earth to "realize" that the Sun is no longer as massive as it was? Does it start to embark upon its new course immediately, or does it take a period for Earth "to get to know" that something has happened to the Sun?

Given that, according to Einstein's theory, nothing can travel faster than light, one would think that Earth would not know that the Sun was losing mass for at least eight minutes—the time it would take for light to travel from Sun to Earth; the Sun, as it were, would have to send a message to Earth, and that message could not travel faster than the speed of light. We need, then, to think of something like a wave, a gravitational wave, that transmits the information that the shape of space-time is changing. So, one way to think about gravitational radiation is as the messenger that carries information about changes in the gravitational fields that attract one thing to another.

A more graphic way to think about this is in terms of the geometry of space-time. A popular way to represent the shape of space-time is as a rubber sheet (fig. 3.1). Heavy objects like the Sun make funnel-like depressions in the rubber sheet. The size of Earth's orbit around the Sun is a matter of its speed—this will determine the point on the wall of the funnel where it can circle without either falling into the Sun or flying upward and outward. If the Sun were to suddenly lose a bit of weight, then it would make less of a dent in the rubber; the funnel would become a little shallower, the walls a little less steep, and Earth, still traveling at the same speed, would find a new stable orbit a little further out. It is easy to see that if the Sun, which is depressing the rubber sheet, is suddenly raised up a little (by becoming lighter), the outer parts of the funnel will not be affected until the ripple

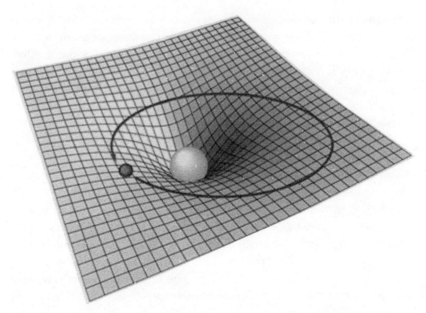

Fig. 3.1 Space-time as a rubber sheet, with Earth circling the Sun; simulation by Patrick Brady, Department of Physics, University of Wisconsin-Milwaukee. Courtesy of Patrick Brady.

has passed through the rubber. A gravitational wave is like that ripple in the rubber.[3]

Another way of thinking about a gravitational wave is that it is like the sound emitted by some eventuality. If I drop my book on the desk, there is a crash. The sound comprises vibrations in the air, which my ear can detect. The vibrations carry energy, and my ear extracts a small proportion of that energy, enabling me to hear the sound. But the amount of energy carried by the sound waves is not directly related to the amount of force involved in the fall of the book onto the desk. First of all, it is likely to be only a very small proportion of all the energy exchanges associated with the event. Second, the amount of energy emitted by the fall of the book onto the desk, though it will depend partly on the weight of the book and the force of the fall, will also depend on the materials of the book and desk and the exact way the book falls. In the same way, the energy contained in a gravitational wave can be very small compared with the magnitude of the events that gave rise to it, and the amount of energy emitted will depend on the exact way those events are patterned.

3. Though most of the waves we will discuss later—the kind that scientists are actually trying to detect—are not single, once-and-for-all, changes but oscillations in the rubber.

The magnitude of a gravitational wave emitted by a spinning rod is set out by Weber in his early book; he imagined the emissions generated by a rod spun to breaking point. The Australia-based gravitational physicist David Blair imagines the use of a nuclear submarine as the spinning rod. We spin the submarine at its center until it is going so fast that it is about to break apart from centrifugal force. It will be emitting gravitational radiation with an energy equal to that used by an ant climbing up a wall divided by 1, followed by seventeen zeros (ant's energy output $\times 10^{-17}$).[4]

Do "Stable Gravitational Systems" Emit Gravitational Waves?

But when are gravitational waves emitted? When Earth is quietly circling the Sun in its normal orbit, we do not seem to need gravitational waves at all. What, under the Newtonian model, was a force—the force of gravity—is just the shape of space-time under the Einsteinian model, in the normal way Earth is just circling around the path in space-time that is easiest for it to follow. The same thing applies to an artificial satellite circling Earth. If you were in such a satellite, with no windows, you would have no sense of movement (as we have no sense of the movement of Earth around the Sun unless we look at the sky). Nothing seems to be going on that involves expending energy; there are no engines firing, and no accelerations that can be felt. Why should such a system need information- and energy-carrying waves? Nothing is happening to either party or to the orbit that they need to be informed about! It is just business as usual.[5]

Furthermore, if there were energy being expended in such waves, it would have to be coming from somewhere. A crucial principle at work here is the conservation of energy, a principle that is accepted by every scientist. (I should say, "the conservation of mass plus energy," since we now know that one can be turned into the other.) The only potential sources of the gravitational wave energy would be the decay of the orbit of the bodies themselves or a reduction in their mass; the satellite, whether moon or space capsule, would have to be slowly falling toward Earth, getting closer

---

4. See Weber 1961, p. 140 and Blair and McNamara 1997.

5. To clear up one possible source of confusion, it is not quite true that the Moon circling Earth does not use energy, because the tides, and other distortions affecting the shape of the two bodies, do involve the expenditure of energy. Furthermore, both the Moon and artificial satellites are plowing through the residue of the atmosphere and the dust that fills even empty space. One or both these effects mean that the Earth-Moon system is slowing down and that artificial satellites eventually fall to Earth. To understand the principles of this argument, however, one has to imagine completely rigid bodies with no tides, floating in a completely empty space. The question is whether such systems would emit energy and slow down.

and closer as it circled, or getting lighter. That seems another good reason for not believing that such an undisturbed system would emit energy-
carrying gravitational waves.[6] In a conference held in 1957, some scientists were arguing that masses that were being accelerated would emit the
waves, while those that were just going where gravity took them, such as
the examples just given, would not. The flavor of the debate taking place
only 40 years ago can be seen from a small news report in the popular
scientific journal *New Scientist*. On June 25, 1959, it reported the following
in its "Trends and Discoveries" page:

> Do objects moving under the influence of gravitation alone give out
> gravitational radiation? We have reported the conclusion of W. B.
> Bonnor—that machine-driven systems give out gravity waves (*The New Scientist*,
> Vol 5, p 708). Now, L. Infeld of Warsaw (who worked with A. Einstein on this
> problem) has produced an argument for supposing that gravitational
> radiation does not emerge from a purely gravitational system—or, at any rate,
> that it cannot be regarded as something absolute. The reason is that it can be
> reduced to zero simply by the mathematical device of choosing the right
> system of co-ordinates (measuring frame) for describing the system. (p. 1393)

Though I have talked of the Moon circling Earth or some such, the same
considerations apply to enormously massive objects circling each other, so
long as their movements are solely under the influence of attractive gravitational forces. For example, two massive stars circling close to each other
just before they spiral in to each other are also following their "natural"
paths, and they, too, would not be expected to emit gravitational waves.

Moving forward to the modern consensus position, it is now believed
that systems such as these *will* emit energy-bearing waves. A consistent way
to think about which objects will emit gravitational waves and which will
not does seem to be the idea of a message. Although, when you are sitting
on the Moon you are receiving no message to tell you that you are being
pulled toward Earth, there are messages traveling into space. Thus, because
a second Earth satellite beyond the Moon would feel the pull of the Moon
as it approached, and notice the pull diminishing as the Moon receded,
there must be messages being sent. The amount of gravitational radiation
emitted by such a system is tiny, however. It grows only when the masses
are huge and the speeds are high—a substantial proportion of the speed of

---

6. At some stages in the historical debate, the idea of non-energy-carrying gravitational waves
was canvassed.

light. The gravitational waves emitted by the Moon as it circles Earth do cause it to lose energy and fall inward toward Earth, its orbit speeding up as it does so. But the change due to this cause is so slight that nothing will be noticed during the life of the universe. On the other hand, as two binary neutron stars draw together at the end of the life of their mutual orbit, the final inspiral should produce a characteristic "chirp" of gravitational radiation of enormous power.

What Sort of Things Emit Gravitational Waves?

Not every violent occurrence in the universe emits gravitational radiation. As I have suggested, to think about whether some change will emit gravitational waves, we need to ask whether it has a gravitational message to pass on. If there is no message, there are no gravitational waves. From this argument we came to see that the Earth-Moon system must emit gravitational waves. Note, however, that if we take away the Moon, and just let Earth spin on its own axis, we would feel nothing (assuming the surface of Earth is smooth), because nothing gravitational changes as Earth spins. That is the key: something gravitationally asymmetric must be going on if gravitational waves are to be emitted.

Oddly enough, this means that an exploding star will not necessarily emit gravitational waves either. If the star explodes symmetrically, everything shooting out in all directions from a central point, or even everything bouncing in and out together, an observer outside the system will feel no gravitational effect. The way to calculate the gravitational effect of a spherical mass is to assume that all the mass is at the center of gravity; this means the gravitational force is the same for a small, spherical neutron star and a huge, spherical ball of gas so long as they have the same mass. It also means that the center of mass, the mass, and the gravitational force remain the same if the star explodes symmetrically, so there is no gravitational message to be transmitted and hence no gravitational wave.

In the exploding Sun example, the term *gravitational wave* was used to refer to a sudden, permanent change. Under these circumstances, we find a "step" rippling through the rubber sheet of space-time. A more enduring undulating wave would be produced by a rotating system, such as two stars closely circling each other. The strength of a gravitational wave is related to such a system's speed of rotation and the size of the masses. When a star explodes, it can produce a wave of this sort, because there is usually a rapidly spinning core left behind. If the core is asymmetrical, it will give rise to a wave; but if it is symmetrical, there will be no wave.

Fig. 3.2 Stretchings and squeezings of space-time as a gravitational wave passes.

Are Gravitational Waves Detectable in Principle?

So gravitational waves are emitted by certain systems according to the consensus view, but can they be detected, and why is this a puzzle? I have described a gravitational wave as a ripple in space-time. Should an oscillating gravitational wave pass through the page on which these words are printed, it would first be slightly stretched in the sideways direction and slightly squeezed in the vertical direction, and then squeezed sideways and stretched vertically, returning to the original state once the wave had passed (fig. 3.2). From the original state, through one stretch and one squeeze in each direction, and back to the original state, mark the passing of one wave. These squeezings and stretchings would go on for as long as waves pass.

Imagine that the page changes shape by about a centimeter each way. (I am talking principles!) Would you see the change? The reason that you could be confused into thinking that you would see nothing is that if it is space-time that is changing shape, it suggests that exactly the same proportional change is going to happen to the chair you are sitting in, the whole room, the whole house, your hands, your eyes, and every atom in your hands, eyes, and brain. And, it seems, the same change would happen to any ruler that you put on the page to measure it, and to every atom in that ruler. To see the effect, some relationship between things has to alter.

Again, the modern consensus is that the effect will be seen. The idea of a ripple in space-time is, perhaps, a little misleading, because gravitational waves produce a force just like any other, and to change the dimensions of a rigid object, they have to work against the electrical forces that hold solid matter together.[7] The more rigid the material object that is subject to the

7. For deep explanations of how a gravitational wave detector works and the resolution of some of the apparent paradoxes, see Saulson 1997a, 1997b. For a very neat explanation of why transverse gravitational waves are emitted by moving masses, and many other facets of the science and technology, see Saulson 1998.

passage of the wave, the less it will change, or the more work that will have to be done to change it by a certain amount. The fact that work is being done means that energy is being used, and this can be measured. Weber himself seems to have been one of the first to show that this kind of effect was in principle detectable in an experiment. He showed this in the late 1950s, and, of course, he then spent a decade building the corresponding experimental apparatus.

By the beginning of the 1960s, scientists such as Richard Feynman, John Wheeler, Herman Bondi, and Felix Pirani had concluded that gravitational waves were emitted by systems quietly circling each other and that the point had been firmly established. They believed that gravitational waves took energy from purely gravitational systems, and that the energy could be sensed by the right kind of receiver. They also believed there was no point in arguing the point further; it was now time to turn to calculating the size of the effects.

CHAPTER 4

THE FIRST PUBLISHED RESULTS

Now let us turn to the first experiments to detect gravitational waves. Our attention will pass more and more from Weber's laboratory to the ripples in social space-time. This must begin as a retrospective account, as my field studies did not begin until 1972. Before this I have a view of only those ripples that propagated through the medium of the printed word. To understand the period properly, we would need to know about the professional conferences, the news conferences, the visits of scientists to Weber's laboratory, the telephone calls, and the talks in corridors and cafes. As it is, we have only the bones of science, not the flesh; the bones of publications tend to be systematically different from the flesh in ways that I will discus in chapter 42.

The earliest and most important of Weber's claims to have seen indications of an external signal on his detectors were published in the journal *Physical Review Letters*. This is a highly prestigious journal that publishes short articles relatively speedily; it is the place to publish results that are considered important enough to be brought to the attention of the world of physicists with a minimum of delay—or so it was until electronic diffusion of results

Table 4.1 Events seen by Weber's detectors as published in 1967

Time	Date
0924	September 21, 1965
2342	August 5, 1966
1015	August 7, 1966
1645	November 22, 1966
0130	December 2, 1966
0720	December 17, 1966
0140	December 20, 1966
1730	January 20, 1967
2309	January 22, 1967
1320	February 17, 1967

became the norm. Between 1967 and 1970, five such papers were published under Weber's sole authorship.[1]

THE 1967 PUBLICATION

1967 was the year of publication of the first suggestive results in the main-stream journals. The March 27 issue of *Physical Review Letters* carried an article entitled simply "Gravitational Radiation," which contained a list of events recorded by the bar detectors in Maryland. In this paper, which the journal had received on February 4, Weber briefly mentions Sinsky's calibration and dwells for some time on the instruments used in the team's attempt to control for uninteresting sources of disturbance to the detector. He explains that they have two seismometers working at different frequencies to monitor vibrations in the ground; they have tiltmeters and gravimeters, which can detect vertical movements in the ground; they record any fluctuations in the voltage of the power supply; and they control the temperature of the environment. Nevertheless, disturbances to the detector had been seen which were not seen by these other instruments. The dates and times (Greenwich Time) of ten such events are recorded on page 499 of the article, as depicted in table 4.1:

Weber explains that during the course of these observations, a second detector (which must be the driving bar first used in the Sinsky experiment and now turned into a detector) has been mounted three kilometers from

1. Some respondents complained to me on behalf of Bob Forward and the other Maryland collaborators that they were not given enough credit in terms of coauthorship of the early papers. None of the collaborators themselves made this complaint, however. (But I did not ask them the question until long after Weber had been marginalized and the claims discredited.)

the larger bar; both bars have the same resonant frequency of 1660 Hertz, but the second is much smaller and is instrumented very differently. A third bar having a lower frequency response has also been brought into operation. Weber says that the last three events on the above table were recorded simultaneously on both of the 1661-Hertz bars, but the lower-frequency bar did not detect them.

According to Weber, the energy flux of gravitational radiation is "so large that observable astrophysical effects would be expected. Since none have been reported, an origin in gravitational radiation appears very unlikely" (p. 500). That is to say, if the events he records were really gravitational waves, one would have expected to see stars exploding or some such. "[P]erhaps some seismic events are not being observed by the gravimeter-seismometer-tilt installation," he continues. "This possibility is being explored further" (ibid.).

What Is Happening in the Laboratory?

Now let us see what is happening here. Sinsky's experiment has shown to the team's satisfaction that the detectors can see tiny forces such as would result in a change in length of the bars on the order of 10^{-16} meters. They are now starting to try to use the detector to see effects coming not from the driving bar but from sources outside the experiment. The work required to do the calibration has not only convinced them that the detector is sensitive, but also that they have made it insensitive to a whole range of sources of noise, acoustic, electrical, and seismic, and the effect of temperature changes on the delicate instruments. The bar(s) are hung from wires, supported by "seismic stacks" inside vacuum chambers. To make them sensitive and relatively noise-free, the amplifiers are cooled to the temperature of liquid gases, so this keeps them relatively stable. The vacuum chambers, being made of metal, isolate the bars from radio frequency noise and keep the temperature of the bars stable; the seismic stacks—piles of rubber and lead plates which have been designed by various members of the team (see chapter 1)—isolate the wires that support the bars from vibrations in the ground.

But we have seen that outside forces can still "leak" in through wires that enter and exit the device (notably the coaxial cable that was discussed in chapter 1), via vibrations of any kind affecting the electronic components, and so forth—one can never eliminate these noises entirely. For this reason, one has to move beyond the first stage—of suppression of noise—to the second stage—separation of noise from "signal." The first step in this process is to measure every external disturbance that can be measured.

Thus the detector is surrounded by instruments to measure movements in the ground, monitor temperature, and monitor the electromagnetic and electrical environment. Anything that disturbs these instruments, which I will refer to collectively as the "environmental monitors," at the same time as the bar is discounted. The "events" of interest are those that affect the bar but do not affect the monitors. The list of ten events comprises this remainder.

The last three events are even more interesting. The calibration experiment now being completed, the team has started to use the driving bar in a different way, by converting it to a second detector. The two detectors are mounted three kilometers apart from each other. The last three events register on both detectors but not on the environmental monitors. This suggests that whatever disturbed the bars was of a kind that could affect them both simultaneously, pointing to some nonlocal and therefore especially interesting source.

Weber is here staking a claim just in case things work out, but he is also making it clear that the forces involved do not square with what is known of gravitational waves. Theorists had already worked out that gravitational forces are tiny and, even though the bars are sensitive to strains of 10^{-16}, the events seem to have been caused by something far too powerful to be gravitational waves as we understand them. Thus Weber's article reports not that he has found gravitational waves but that he is searching for a problem.

THE 1968 PUBLICATION

Fourteen months later, on April 4, 1968, *Physical Review Letters* received an article from Weber entitled "Gravitational-Wave-Detector Events" and published it June 3. In it, Weber describes more coincidences between what appear to be the same small and large detectors discussed earlier (though their separation is now given as about two kilometers). A *coincidence* is here defined as a simultaneous threshold crossing where *simultaneous* means within 0.2 seconds. Four coincident events recorded in "the past 3 months of operation" are set out in table 4.2, with a number of columns recording the power of the pulses and the performance of the detectors. Included in the information is the date and time of the coincident events and a calculation of the frequency with which an event of such a size could be expected to occur by chance if it were due entirely to the random fluctuations of the detectors. This calculation is based on the (much greater) number of times each individual detector crosses energy thresholds of similar size

Table 4.2 Events recorded by Weber as published in 1968

Date and Greenwich mean time		Frequency of random coincidences
February 7 [1968]	2101	Once in 8000 yr
March 13	1150	Once in 40 yr
March 29	0732	Once in 300 d[ays]
March 29	0358	Once in 150 d

without the other becoming excited. Such individual threshold crossings are taken to indicate background noise, thereby enabling the likelihood of the occurrence of combinations of noise that might mimic the putative events to be calculated.

Weber's penultimate paragraph, before the acknowledgments, reads as follows:

> Conclusion.–The extremely low probability of random coincidences enables us to rule out a purely statistical origin. The separated detectors are responding, on rare occasions, to a common excitation which might be gravitational radiation. (p. 1308)

In the table, Weber has provided a column labeled "Frequency of Random Coincidences." What this shows is the expected period between accidental coincidences at these signal strengths. He is arguing that the energy associated with these signals is so high that coincidences between signals of this magnitude could occur only very rarely. Let us explore this idea a little further.

Three Key Developments in the Science of Gravitational Wave Detection

Three important steps have been taken since Weber's first publication. First, what counts as an *event* has been defined. An "event" occurs when the recorded trace of the energy state of the detector crosses a certain threshold. But what should the threshold be? The choice is not forced by any scientific principle. This is vitally important, because the nature of this choice has haunted gravitational wave science from these very first days right up to the present. The future promises more debate on essentially the same issue, even though we have 40 years of development behind us and have grown from a project involving three or four people in one laboratory costing a few thousand dollars to a worldwide enterprise involving hundreds of

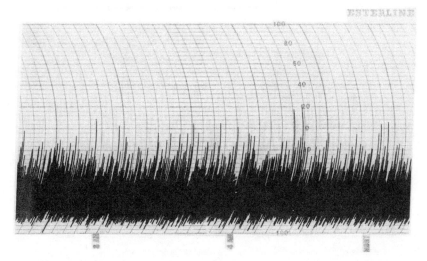

Fig. 4.1 Section of chart-recorder output from Weber's detector at the University of Maryland golf course.

people and half a billion dollars. Let us look closely at the impact of the choice of threshold.

The state of the bar is monitored by a chart recorder. Figure 4.1 shows a section of such a recording taken from the early days, in this case from one of Weber's bars at the University of Maryland golf course.[2] Figure 4.2 is a simplified version of such a trace, overlaid with three different potential choices of threshold. As can be seen, threshold A is crossed twice, threshold B is crossed 6 times, and threshold C is crossed sixteen times. To choose the threshold, then, is to choose the number of "events" seen by the detector— two, six, or sixteen in this case—though, of course, some or all of these "events" will be noise. Thus, within a certain range one cannot say that a detector has "seen" something or that it has not "seen" something—to a surprisingly large extent, the choice about whether the detector has seen a potential event lies with the observer. We will find this problem coming up with especial force in chapter 22.

The second advance is in the notion of "coincidence." A coincidence occurs between two detectors when both cross a threshold simultaneously. But simultaneity is a not a simple term. What counts as simultaneous depends on the accuracy of the clocks, the devices used to record their

2. In this case, the chart recorder was running very slowly, and this short section covers about six hours. If the chart recorder runs faster, then what appears as solid black in this figure would be spread out and resolved, as it is in figure 3.3 and the schematic representations of figures 5.2 and 9.1.

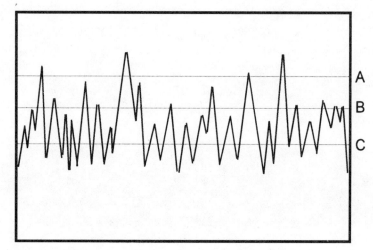

Fig. 4.2 The effect of the choice of threshold on the number of "events" seen.

readings, and the time it takes gravitational waves to travel between two places. In this case, simultaneity has been defined as within a time window of 0.2 seconds.

Again, the choice is to some extent with the investigator—there is nothing in the science to define it exactly. The wider the time interval that is allowed to count as "simultaneous," the more "coincident" threshold crossings there will be. Two-tenths of a second seems rather a long time period to count as "simultaneous," but the point is that the bars have to build up the signal into a noticeable ringing before their output has a chance to cross a threshold, and this might take more or less time depending on the initial state of the bar. It is the resonating of the bars that makes it not unreasonable to count a window of 0.2 seconds as simultaneous. The advantages and disadvantages of resonance as a means of integrating or amplifying the energy in a signal will haunt the history of the bars.

The third advance is the recognition that some coincidences will occur not because the bars have been excited by an external signal but because both of the bars just happen to be excited by "noise" at the same time. Before one can know how many "interesting" coincidences there are, one has to subtract the number of "accidental" coincidences that were likely to have been seen anyway even if there were no nonlocal forces at work.

At first sight, a gravitational wave detector seems to be something like a telescope. It may be very hard to build, but once it is finished and tested, one might think there would not be much doubt about whether it was detecting gravitational waves or not detecting them. But now we see that "observation" is not a passive activity at all—we can see how far it is from our commonsense model of "cup-and-saucer seeing." Observation involves choice and chance: there is a choice about thresholds, a choice about the meaning of *coincidence,* and statistical calculation about what might happen by chance if there are no gravitational waves to be detected. When the statistical calculation has been completed, there is another choice to be made: the sums will tell us only the *likelihood* that in any one time period a certain number of accidental coincidences will occur. Suppose that when we subtract this number from the actual number of coincidences we are left with a positive remainder. We now have a choice about how big that positive number has to be. How big must it be in comparison with the calculated number of accidentals to make it interesting? How big does it have to be to make it credible that we have detected something that looks like the effect of a common external influence—perhaps gravitational waves?

THE THRESHOLD DILEMMA

Here is one of the crucial tensions at the heart of gravitational wave detection science: the greater the number of threshold crossings in any one bar, the greater will be the number of accidental coincidences between two bars. The number is not hard to calculate: if one knows the average number of threshold crossings there are in each bar over a period of time, one can work out how likely it is that two of them will occur within the same 0.2-second window; hence one can work out the number of accidentals. The bigger the number of expected accidentals, the more likely that any excess of coincidences above this number was a result of chance. That is, the bigger the number of threshold crossings, and the bigger the number of accidental coincidences, the more likely is it that an excess of coincidences in any one time period is itself due to chance—one might be looking just when noise in one or both detectors was unusually high. This argues for caution and the setting of a high threshold. With a high threshold there will be few crossings and few accidentals, and any coincidences are likely to signal interesting events.

The trouble with the cautious approach and the high threshold is that we know from theorists' calculations that gravitational waves are likely to be very weak; if we set our threshold too high, we can make certain that we will see nothing. Only low thresholds will capture weak gravitational waves, and it is weak gravitational waves that we are looking for; to put it another way, lowering the threshold enables the detector to pick up signals from a further away. But as the threshold is lowered you will certainly see more accidental events, while it is far from certain that you will see real events. Thus, the combination of choices that I have described can be aggregated into one bigger choice between two strategies. One strategy is to set the threshold as low as is reasonable, looking out as far as possible so as to maximize the chance of capturing any gravitational wave signals that just might be there while increasing the risk of finding spurious results. The other strategy sets the threshold high enough to reduce the chance of finding spurious signals but makes it less probable that any weak gravitational signal that is likely to be passing by will be seen. In other words, one has to trade off the risks of the two well-known statistical errors—missing signals that are there and finding signals that are not there—false negatives and false positives. False positives are also known as type I errors, and false negatives as type II errors.

Unfortunately, when the tradeoff is thought about more carefully, the false-positive route looks less attractive. Lowering the threshold may buy very little improvement in terms of eliminating a false negative while increasing the risk of a false positive disproportionately. If this is true, it is a bad scientific trade to lower the threshold even though it buys some greater chance of success.

The problem is that on the one hand, the typical pattern of noise which emerges from an instrument tends to increase very sharply below a certain level, as can be seen in figure 4.1. If the threshold is set just above that level, there will be few crossings, but if the threshold is set just below that level, there will be many crossings due to noise. Lowering the threshold still further will mean that there is so much noise that it would be impossible to extract a signal anyway. The point is that everyone agrees that it is right to lower the threshold as far as the top of the sudden increase in noise, and everyone agrees there is no point in going below the level where the signal would be completely swamped. This means the choice for the "threshold lowerer" is quite constrained—just a narrow horizontal band within the top of the "forest" of noise.

To make it worthwhile to lower the threshold just this small distance, one has to assume that the signal that one is seeking is just visible within that narrow horizontal band that has been opened up and not above it. What the consequences are in terms of astrophysics can be illustrated with figure 4.3.

Figure 4.3 is a kind of map of the heavens with Earth at the center. Each set of small concentric circles is a splash of gravitational waves, lasting for a second or so. The figure shows how these splashes might look if

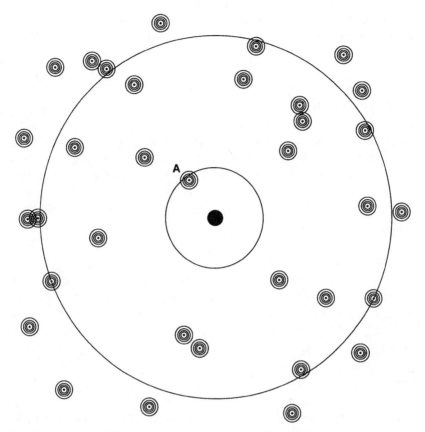

Fig. 4.3 Splashes of gravitational radiation cumulated over time, with Earth at center and rings representing sensitivity of detectors.

accumulated over a period of a few years.[3] Imagine, for the moment, that the inner circle represents the sensitivity of a bar with the threshold X. To make it worthwhile to lower the threshold just a little implies that there are a large number of sources (not shown here) not scattered over the heavens but just outside the inner circle waiting to be discovered by a detector having a slightly greater reach.

Astrophysical scenarios can be imagined that would produce such a distribution, but the history of gravitational wave detection has not taught us

3. I have simply located splashes over the sky in an arbitrary manner, but it would be reasonable to expect the distribution to follow the rough pattern of the distribution of stars in the sky, something I have not tried to reproduce. Bear in mind also that increases in sensitivity sweep a larger sphere rather than a larger disk, giving an eightfold increase in volume of space for a twofold increase in sensitivity rather than the fourfold increase shown here. This figure will be used for a slightly different purpose in a later chapter; hence it has features not discussed here.

to expect to see them. In the past, whenever detectors that were associated with claimed positive results have been made more sensitive, the signal has not suddenly become clear and powerful; rather, it has tended to disappear into the edge of the noise again. There is nothing wrong with this happening in logic or astrophysics, but it does not seem to be a good way to bet. It is these kinds of considerations that sometimes infuriate scientists who think thresholds must be treated conservatively.

There is yet another complexity, however, that appeals less to those who want to minimize the risk of false positives. Suppose you set the threshold so high that you see only one "real" signal in a year of observation. Will anyone believe in the reality of that lonely signal? Surely the test of a science is that it can repeat its observations! Ironically, then, setting a high threshold, even though it decreases the likelihood that the signal is due to chance, increases the likelihood that no one will accept it as genuine. These considerations taken together amount to what I call the threshold dilemma.

Over and over again in the above paragraphs, I have used the word *choice*. *Choice* is a word drawn from the language of individual free will. The word is being used here to point to moments when the "logic of science" does not dictate a decision. Nevertheless, a word that expresses the fact that the choices I have discussed are not so free as they look would have been better, for the choices are made under the constraints of social space-time. Social space-time is a medium with form and texture. It will not support just any choice—it sets limits to the choices that an individual can make. Weber could not have claimed that he knew his bars were detecting gravitational waves because, say, a fortune teller told him it was his fate and his astrologer confirmed that this was his moment; the social space-time of Western science will not flex in the way needed to generate a ripple from such a splash. The texture of social space-time also determines how far a ripple that does get started will spread. Some choices will be transmitted freely and easily, others will be extinguished at or near their source. Weber was faced with a social space-time made stiff with theories which suggested that an apparatus like his was insufficiently sensitive to detect the waves from heavenly sources. He had to shake the fabric of social space-time hard and long to have any chance of generating a lasting ripple. Thus, when we think about the ripples emitted by Weber's detectors, or, to give the analogy a rest, the range and credibility of his choices, we have to think of an individual in a time, place, and social context that provided him with both the outer envelope of the range of things he could choose and the extent to which those choices would be convincing to others. As the years pass, we will see the context of the science transformed, and the envelope of reasonable choices changing its shape as different groups of scientists come to dominate the field. We will see the force of theory become modified as one experimental regime gives rise to another; we will see the culture of high-energy physics interact with the cultures of other physical sciences—"culture" being another way of talking about the local textures of social space-time; we will see the way that politics and finance contribute to the texture of the medium.

THE 1969 PUBLICATION

A year later, the June 16, 1969, issue of *Physical Review Letters*[4] carried a still more confident article by Weber, presenting new data, of which he concluded,

> This is good evidence that gravitational radiation has been discovered.
> (p. 1324)

By now Weber and his team had two new detectors, one of which was located 600 miles away in suburban Chicago, at the Argonne laboratories. (Sinsky, we saw, had helped to move it there.) In all, then, they now had a total of four detectors. The two new ones were linked by a telephone line so that coincident threshold crossings could be recorded automatically. In the 1969 paper, simultaneity is defined as within 0.44 seconds, but there is a new and complex discussion of the delay in effect caused by the use of electronic components cooled to liquid helium temperatures (cryogenic electronics). It seems that the signal of a bar using a cryogenic amplifier takes eleven seconds to reach a maximum. The first large bar developed by the Weber team used cryogenic electronics. It was used in the coincidences mentioned in the 1968 paper, although the long time taken to generate a signal was not mentioned. In any case, the two new bars use room-temperature electronics in which there is no problem with a delayed signal.

One might think that if one has four detectors "on air," then a genuine event would be signaled only by the excitation of all four detectors. This is not the case, however. Whether a detector output will cross a threshold of energy depends on the state it is in immediately before it receives a boost from a putative gravitational wave. If its internal noise is such that

---

4. Received April 29; this article also speculated that what was being seen was the triggered release of energy stored within the bar, so that a small input of energy would produce a large output. This claim would be directly addressed, tested, and confronted in Drever et al. 1973.

its energy is going down at that point or if random factors make it unreceptive for some other reason, it may not record the signal. Therefore, in an
experiment like this, where the size of the signal is small compared with
the noise that is found in the bars even when they are not being excited by
an outside force, not all incoming signals will have an effect; the bars, in
other words, are *inefficient* at detecting signals—they will see only a few of
those that come in. The more detectors you have, the more chance there
is that one or more of them will be in an unreceptive state when a signal
comes by. Therefore, the chance of a two-detector coincidence is much better than the chance of three-detector coincidence, which is in turn much
better than the chance of a four-detector coincidence even when the signals
are caused by an external exciter such as gravitational waves. Conversely,
if a three- or four-detector coincidence does occur, the chance of that being due to a chance alignment of noise in all detectors is very small. Thus,
multiple coincidences boost your confidence that you are seeing something
real, whereas absence of effect on some of your detectors does not show
that no signal was present.

Weber provided a table of coincidences discovered between December 30, 1968, and March 21, 1969. There are nine two-detector coincidences,
five three-detector coincidences, and three four-detector coincidences. The
likelihood of these occurring by chance is given at approximately once every 70 million years in the case of one of the events; periods of tens of
thousands of years for four of the events; and periods of hundreds of years
to hundreds of days for the remainder. Weber concludes (p. 1322) that he is
quite certain that all the coincidences cannot be accidental. As he explains
earlier (p. 1321), "We may conclude that such coincidences are due to gravitational radiation if we are certain that other effects such as seismic and
electromagnetic disturbances are not exciting the detectors."

Forward's Letter

Weber and his team were confident that they were doing exciting physics.
Bob Forward, by then back at Hughes Research Laboratories in Malibu,
California, wrote a letter dated December 10, 1969, to the community of ex-
perimenters who had contacted him, indicating their interest in the work.

> A number of you, inspired by the recent publications of Prof. Joseph Weber at
> the University of Maryland, concerning the possible detection of gravitational
> radiation, have written letters to me requesting information on the
> techniques for constructing gravitational wave antennas. Since it is essential

that we learn more about the events that Prof. Weber has detected, and the best way would be to have independent groups study them, I would like to take this opportunity to make the same offer to you that I have made to those who have contacted me.

If you think you can obtain the funds necessary for the purchase of the equipment, I will be willing to take time from my projects to supply technical advice and assistance during the critical phases of your work...[5]

This letter went to Professor W. D. Allen at the Rutherford High Energy lab in Britain; Dr. H. Billing at the Max Planck labs in Munich; Professor V. B. Braginsky at Moscow State University; Dr. P. K. Chapman at Avco-Everett, Massachussetts; Dr. Ron Drever at Glasgow University; Dr. Bill Hamilton at Louisiana State University;[6] Dr. J. Levine at JILA (formerly known as the Joint Institute for Laboratory Astrophysics), University of Colorado at Boulder; Dr. K. Maischberger at European Space Research Organisation laboratories in Frascati, near Rome; Prof. G. Papini at the University of Saskatchewan; Dr. G. Pizzella in Rome; Dr. J. A. Tyson at Bell Labs, Holmdel, New Jersey; Professor Weber himself; Prof. R. Weiss at MIT; and Dr. H. S. Zapolsky at the NSF, Washington, DC. That is a remarkably large subset of individuals who were to figure strongly in the decades of research on gravitational waves to come. I do not know how many of those named took up the offer of a visit, but Forward's letter does indicate a level of confidence that it is important to understand at this historical distance.

THE FIRST 1970 PUBLICATION

Weber next switched to a new way of presenting data. Instead of a list containing times and dates suggesting the discovery of specific astronomical events, he began to provide tables showing the number of expected coincidences (that is, noise-induced, accidental coincidences) at two or three signal strengths, and the number of coincidences actually found. In all cases the number of coincidences found exceeds the expected number of noise-induced coincidences, but the enormously long times for an individual event to occur by chance are no longer given. The data now look just a little more "tame." He explains the change in data-presentation policy in his

5. Given to me by Bob Forward.
6. Hamilton must have been in the process of moving from Stanford to Louisiana State University.

next publication in *Physical Review Letters* (received September 8, 1969),[7] the first of two published in 1970. He says, "I agree with respected colleagues that earlier conclusions should have been based on the kind of tables given here." In the acknowledgments, Weber states that he has "enjoyed stimulating discussions with L. Alvarez, F. Crawford, and T. Tyson." From my later discussions with Weber and others, it seems that high-energy physicist Luis Alvarez was principally responsible for changing Weber's mind on data analysis and presentation.

In this first 1970 paper, Weber also explains that the number of coincidences recorded varies according to what is taken as the definition of simultaneity. We have seen this change from 0.2 seconds, to 0.44, to 0.35, and this kind of variation opens the door to data manipulation. If one can choose whichever "resolving time" maximizes the signal, the statistical confidence in the results is reduced. Weber therefore gives some results for alternative resolving times and demonstrates that there are an excess of coincidences in either case.

The paper presents 330 days of coincidence data gathered during the period January 1 to November 30, 1969. This is old data from an earlier period reanalyzed along with new data from about March onward. Weber picks four arbitrary strength (threshold) levels for signal. The strength can be expressed in terms of the number of times one might expect thresholds to be crossed by chance. As we have seen, if the strength is very low, then it is easy for noise to imitate a signal and make the detector cross an energy threshold; if the strength is high, it is much more unlikely to happen.

Weber's table has four categories of strength of event given by numbers of threshold crossings by an individual detector. The table is as follows (table 4.3; I have changed the labels for clarity):

Table 4.3 Events from Weber's first 1970 paper

Class (expected crossings by one detector)	Expected accidental coincidences	Observed coincidences
<10	0.18	7
<40	2.8	24
<80	11	90
<100	18	115

Since this table reports 330 days' worth of data, at this time we are looking at about one weak-signal coincidence every three days.

---

7. Weber 1970a.

In this paper, Weber introduces the first stage of what will become one of the most important (and, to some anyway, convincing) methods of data analysis for gravitational radiation detectors—the introduction of a time delay in the coincidence analysis.

> It is more convincing if statistical arguments can be supported by some experimental procedure to measure the rate of accidental coincidences.... A second coincidence detector was set up with time delay of two seconds in one circuit channel, in a manner which did not alter the concurrent experiment with no time delays. (p. 278)

The argument is that if an artificial delay is introduced between the signals from two detectors, no *apparent* coincidences between the two can be ascribed to forces that affect the detectors simultaneously. Therefore, a delayed comparison provides a direct experimental method for measuring the number of pseudocoincidences resulting from noise alone. Instead of merely calculating the accidental coincidence rate, one can now measure it directly.

Weber then provides a table covering twenty days of observations, comparing the statistically expected number of coincidences due to noise alone, the observed number of coincidences in the channels with delay, and the observed number of coincidences with no delay. For three strength classes of signal, the results are as follows:

Table 4.4 "Events" with and without delay

Class	Expected accidental coincidences	Observed with delay	Observed no delay
<100	1.2	0	11
<120	1.7	3	15
<??[8]	3.2	3	18

8. For reasons which are not explained, this number, which is given as 6000, seems to refer to some less-straightforward combination of both detectors. We can see what is going on, however, if we concentrate on the first two rows of the table alone. The data given here are taken from only twenty days of observation. Such a short period does not allow time for many coincidences to have occurred, as compared with earlier papers, which recorded events over a much longer time. This is probably why the strongest coincidences in the Class column of this table are coincidences between relatively weak signals. (From the discussion surrounding figure 5.2, above, we saw that as we lower the threshold we get more and more signals; in other words, there are many more weak signals than strong signals. The figure of 100 crossings per day per single detector suggests weak signals. Signals that give only ten threshold crossings per day, as given in the previous table, are much stronger. There can be many of these listed in the previous papers because there has been a much longer observation period for them to be built up in number.)

Quoting Weber's words once more,

> The marked reduction in coincidence rate for the coincidences in the delay
> channels is evidence that the gravitational-radiation detectors are being
> excited by a common source with propagation time between detectors
> substantially less than 2 sec. It is also evidence that the expected accidental
> coincidences are understood. (p. 278)

It is hard to disagree with this conclusion: the match between the second
and third columns in the table, while not exact, is certainly as good as
might be expected with such low numbers, while the difference with the
final column is very marked indeed. Provided that the thresholds in column
1 and the twenty-day time period have not been specially picked so as to
produce a good result, the table is very convincing.

> Conclusion.–The time-delay and radio-receiver experiments[9] support the
> earlier claim that gravitational radiation is being observed. (p. 279)

THE SECOND 1970 PUBLICATION

Weber had submitted a second paper to *Physical Review Letters* on the same
date as the one just described. It was published in the July 20, 1970, issue.[10] Entitled "Anisotropy and Polarization in the Gravitational-Radiation
Experiments," it provides the last piece in the jigsaw puzzle of data analysis
for those prepared to be convinced that gravitational radiation had been
detected.

What it showed was that the coincidences recorded on the detectors
appeared to be caused by an effect emanating from the direction of the
center of the galaxy. An isotropic effect is one that is uniform in all directions; an "anistropy" means there is a directional effect. That there is a
preferred direction is indicated by a periodicity in the signal.

If the resonant bar is responding to gravitational waves, it will be more
sensitive to waves coming from some directions than from others. The
transverse gravitational waves will stretch and squeeze the bar most when

---

9. The paper also contains a report of a radio receiver being set up to look for pulses of
electromagnetic radiation that might be affecting both detectors simultaneously; it found none.

10. Weber 1970b.

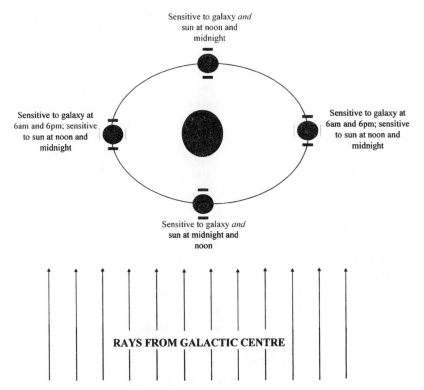

Fig. 4.4 Why peak sensitivity of bar detectors is correlated with sidereal time.

it is at right angles to the direction from which they are coming; they should affect the bar hardly at all when one of its ends is pointing toward the source. The relationship of a bar on the surface of Earth to the Sun and the galaxy is represented in figure 4.4.

Because the bar is fixed on the surface of Earth and aligned east–west, the direction in which it points is affected as Earth rotates. Once every 24 hours, someone looking toward Earth from the direction of the center of the galaxy would see the bar sideways on—in its most sensitive position. Six hours later, they would see the bar end-on, its least sensitive orientation. Another six hours later, the bar would be hidden behind Earth—hidden from observation via the medium of light. But it would still be sideways on to observers viewing it from the center of the galaxy even if they could not see it. Because Earth is almost completely transparent to gravitational waves (they are so hard to detect precisely because nearly all matter is almost transparent to them), the bar would be just as sensitive to waves coming from the center of the galaxy as it was twelve hours earlier, when it was in full view; it would, as it were, still be in full view to gravitational waves. Another six hours after this, the observer would again see the bar end-on as it reappeared around the edge of the rotating Earth, and it would again be in an insensitive state to gravitational waves. Thus there must be two equally sensitive periods for an east–west oriented bar during the course of 24 hours; in other words, there should be a "periodicity" of twelve hours.

A bar located in Maryland will be sideways on to the Sun at roughly noon and midnight, Maryland time, every day. Let us pick a date when the Sun lies between the center of the galaxy and Earth. On that date, when you look at the Sun you are also looking at the center of the galaxy. Also on that date, when the bar is sideways on to the Sun, it will also be sideways on to the center of the galaxy. But Earth orbits the Sun once a year. Three months later, if you look at the Sun from Earth you won't be looking at the center of the galaxy, you will be looking out into space with the center of the galaxy off to one side. This means that at this time, when the Sun is overhead in Maryland, the end of the bar is pointing at the center of the galaxy while in its *least* sensitive state. At this time, the bar is in its sensitive state with respect to the center of the galaxy at 6 p.m. and 6 a.m. Maryland time. Three months later again, if you look at the Sun from Earth you will have your back to the center of the galaxy. But because Earth is transparent to gravitational waves, the sensitive times will again be noon and midnight at Maryland. And so on.

Thus as the Maryland year passes, if the source is the center of the galaxy, the peaks and troughs in the twelve-hour periodicity will drift around the clock. If the peaks are initially at noon and midnight, they will be at 6 p.m. and 6 a.m. three months later, back to noon and midnight three months after this, back to 6 a.m. and 6 p.m. three months further on, and back to the starting point as the year comes to an end. The difference between the orientation of the bar in respect to the Sun and in respect to the galaxy defines two kinds of time: solar time, which has to do with the Sun; and "sidereal time," which has to do with the galaxy. These times drift in and out of phase during the year. Gravitational waves are more likely to reveal a regular periodicity in sidereal time than solar time, because the galaxy is the most likely source of the waves.

In his second 1970 paper, Weber shows two plots of coincidences gathered during a six-month period (there are 311 altogether). Both plots divide the day into six four-hour bins and put each coincidence in whichever bin is appropriate. Using sidereal time, we find that the bins contain the following number of coincidences:

36 71 38 48 73 45

As one can see, there are two high peaks twelve hours apart.

If the same data are plotted against solar time, the time experienced in Maryland, the same total number of coincidences are distributed as follows:

53 50 43 57 45 63

Represented graphically, the figures look as follows:

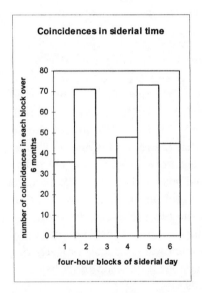 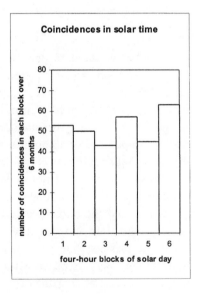

Fig. 4.5 Weber's claimed sidereal correlation.

The periodicity disappears in the second set of figures and the second graph because it is smeared over the clock face as one would expect it to be if the source is the galactic center. Looking at galactic time with a Maryland clock produces a smear; looking at it with a galactic clock produces a clear periodicity.

If you want to see the twelve-hour periodicity, then, you have to use sidereal time. To repeat, if the twelve-hour periodicity is genuine, the direction of the source of the effect must be the galactic center (or something in the exact opposite direction, which seems unlikely), not the Sun or Earth or anything to do with the solar system. This is a very convincing argument; it means that the coincidences are nonlocal in two senses: not only are they not associated with the laboratory, they are not associated with the solar system. This also strongly suggests that the source of the effect is gravitational radiation.

Weber, according to the convention he has adopted, finishes with a formal conclusion.

> Conclusion.—The large (exceeding 6 standard deviations) sidereal anisotropy is evidence that the gravitational-radiation-detector coincidences are due to a source or sources outside the solar system. The location of the peaks suggests that the source is the 10^{10} solar masses at the galactic center. (p. 184)

Weber says that the sidereal anisotropy exceeds six standard deviations.[11] The number of standard deviations expresses the confidence that something other than random fluctuations is giving rise to differences in the data. Different sciences accept different conventions for how many standard deviations are needed in order for an effect to be counted as real. In the social sciences, two standard deviations is counted as providing a warrant to publish a finding; it means that the chances of the result arising out of random effects is 1 in 20 or less. In the laboratory sciences, the number of standard deviations demanded tends to be higher. Six standard deviations is a very high level of confidence; it means that there is almost no chance that the difference between the two arrays of figures, the one on Maryland time and the other on sidereal time, could have arisen by chance. It does not show that the experiment was conducted properly, nor that the data were extracted without bias, but it does show that bad luck was almost certainly not the cause of this result.

The cascade of claims we have seen in the papers discussed so far might be an exemplary case of how to establish a scientific discovery and generate a ripple in social space-time that cannot be suppressed, but it took place against a background of skepticism. It was a very difficult experiment that seemed to be producing impossible results. We have already seen that

---

11. I will use Weber's calculations of statistical significance (calculations that passed the scrutiny of the peer reviewers of the time) without considering whether these calculations would pass today's scrutineers.

Weber had been forced by his critics to begin presenting his results in different ways.

There is now a three-year silence before more of the central positive results are announced. Weber published other papers in this period covering related matters. He built a new detector in the form of a disc rather than a cylinder; this, it is claimed, settles an argument between different theories of gravity.

Weber did not publish anything more referring to his major claim—that he was seeing coincidences and that these represent gravitational waves—until 1973. By this time he had become much more defensive. Given that the results seem to be impossible, he had been assailed on many fronts in the intervening time. And he had made some mistakes and been forced to retreat here and there. For example, problems emerged concerning the sidereal anisotropy.

SIDEREAL ANXIETY

A conference in Israel held in July 1969 (the proceedings of which were published in 1971) was the place where Weber first reported on the way the rotation of Earth affected his sightings.[12] He said in the subsequently published paper, "These data justify the conclusion that gravitational radiation has been discovered." This statement was the strongest emerging from these early days. Later Weber was to report that the strongest claim he ever made was that he had discovered "evidence for" the existence of gravitational waves, but the claim made in Israel was more direct.[13]

The published proceedings of the Israeli conference reveal a number of graphs, which show a high peak at 24-hour intervals and smaller peaks at other intervals. There is a *low* point twelve hours away from the high peak. The text, on the other hand, seems to suggest that there should be a peak every twelve hours.

> The 12 hour effect is not as pronounced as [the equations] predict. This may
> be the result of systematic errors in the experiment or the random character
> of emission from the source. For a source which emits roughly every

12. Weber 1971a.

13. The "evidence for" remark was made in various interviews with the author and also Weber 1992a, p. 230: "A background of signals was observed between 1968 and 1974. The claim made for these was 'evidence for.'"

24 hours there would be no 12-hour effect and for random emission times fluctuations might attenuate the 12 hour effect. Since the intensity is determined statistically...it is a measure of the signal to noise ratio rather than signal intensity alone. Therefore the 12 hour effect might also be attenuated by inadequate shielding of the hours of the day. (p. 320)

A "note added in proof" states that

[i]mproved methods have given data showing anisotropy with two peaks, one in the direction of the galactic center and the second one twelve hours away.

The note refers to a paper published in *Physical Review Letters* in 1970 where data for the 12-hr periodicity are given.[14]

This is a pretty serious mistake. Astonishingly, Weber seems not to have been clear about the transparency of Earth at the time of his presentation in Israel. Even the note added in proof is phrased rather oddly. It talks of a peak in the direction of the galactic center and another twelve hours away, whereas the second peak should also be spoken of as "in the direction of the galactic center." The absence of an equal-sized peak twelve hours after the 24-hour peak in the graphs found in the Israeli conference proceedings paper is, if anything, evidence against the findings representing gravitational waves. The excuses given for the absence of a twelve-hour peak are weak and self-serving: Why should a source emit every 24 hours? Why should the noise in the apparatus mask the effect only every other twelve hours? The subsequent appearance of a strong twelve-hour periodicity after its initial absence suggests that the analyst might be bending the data to fit the model.

I have dwelt on the Israeli conference proceedings because the ambivalence between twelve- and 24-hour periodicity as presented by Weber at that time is remembered by a number of respondents. As late as 1998, a respondent explained that it was the initial source of his doubt about the Weber findings and that it had remained important to him ever since. On the other hand, it was only a subset of respondents who mentioned this problem to me—most seemed to think of it as the kind of mistake that anyone might make in the early days of a difficult experiment. But assaults were coming from all sides; lots of people wanted to suppress the splash that Weber had made.

14. Weber 1970b.

CHAPTER 5

THE RESERVOIR OF DOUBT

Theory: How Much Gravitational Radiation Is There?

As the 1960s wore on, even before the first publications emerged, Weber had begun to tell other scientists, informally and in seminars, what he was finding. At first, as we have seen, his confidence was low. In November 1967, just after he published his first findings, he wrote to a scientific colleague and friend.

> We tested it [the first cylinder detector] by doing an induction field gravity communications experiment with center to center distance of two meters. Strains of a few times 10^{-16} were gravitationally induced and detected so it is quite certain that the apparatus measures the Reimann Tensor and operates as theory predicts. As a generator we employed a second cylinder driven electromagnetically.
>
> When this experiment was completed, the second cylinder was converted to a detector and mounted on a concrete pier roughly two miles away.

97

The isolation of the detectors is good but far from perfect. They see no automobiles or people, but violent local earth motion or lightning striking the building is seen. We have several seismometers and two earth tilt meters in the main building to assist in identification of events.

We see no diurnal effects whatsoever. The main detector sees roughly one of two sharp spikes a week, not coincident with obvious seismic activity. The separated detectors see roughly one coincidence every one or two months, not correlated with seismic activity. These coincidences are surely causally connected. The signals can only be judged to be shorter than the relaxation time of thirty seconds. I cannot be sure they are not seismic since it is probably impossible to monitor all conceivable earth motions....

I am very cautiously optimistic, having reported what we see, in the literature. My guess is that the probability that one of these events was truly gravitational is about 1/50.[1]

Contrary to everyone's expectations, including, perhaps, his own, Weber was finding more and more solid evidence for the existence of gravitational waves. In 1972, describing why he started the experiments, he said to me,

[1972] It's just about inconceivable that there would never be anything seen at any level of sensitivity... or at least, so it seemed to me.... If there was some truth in the theory, then you'd expect the radiation would be there, and if you just kept looking and improving the sensitivity, you'd eventually detect it. No one had ever looked, so I thought I would start and build the best machines I could, and if I got a negative result, then leave it to someone else to do better.

The results he was finding were hard to believe because they seemed to fly in the face of much of what was already known. To restate some of what we have already learned:

- Gravitational waves are very weak.
- Once they are emitted, they are expected to spread uniformly through space.
- Gravitational wave receivers are very insensitive and extract only a tiny proportion of the energy from a wave.

1. Letter given to me by respondent.

Putting these three factors together, and assuming the source is at a certain distance, we can calculate how much energy must have been emitted to give rise to a signal in the detector.

In Weber's case, the answers all looked crazy. If the source of the gravitational waves was near the center of our own galaxy—and if we imagine the source to be more distant, that only makes the problem worse—the amount of energy being emitted was so huge that hundreds or thousands of stars' worth of mass would have to be continually turning into energy emitted as gravitational waves, and the galaxy would be vanishing before our eyes. Indeed, stars would be disappearing at such a rate that the galaxy's gravitational pull upon its own constituents would be noticeably diminishing and what was left of it would be noticeably expanding, soon to fly apart and spread itself through space. In any case, no one had any idea of what kind of processes could give rise to such huge conversions of mass. On the face of it, then, Weber's results were in conflict with the very theory he had set out to corroborate: general relativity. General relativity, it was now widely agreed, predicted the existence of energetic gravitational waves, but general relativity, taken with what we knew of the constituents of the universe, could not, it appeared, be squared with the existence of such vast amounts of the radiation.[2]

So much for gravitational radiation having been detected, one might think, but the relationship between an empirical finding and a theory is not straightforward. A theory and a finding are linked by a series of presumptions, some simple, some complex. The theory and the finding do not bear directly upon each other—they bear upon the whole set of presumptions that support them. At the beginning of the previous paragraph, three of these presumptions are set out, and there are many more.

When the complexity of the relationship between theories and findings was first elaborated, philosophers of science tended to think of experimental

---

2. Actually, according to Rai Weiss (and see Saulson 1998), special relativity requires the existence of gravitational waves, too. Thus in 1972 he explained to me:

And this is a myth that runs through the whole bloody field. That somehow whoever discovers gravitational radiation will have proved Einstein right. That's a bunch of crap. Gravitational radiation has to exist just on the basis of special relativity alone. It doesn't require Einstein's [general] theory. The specific form it has requires Einstein's theory if you believe in the tensor-field kind of gravitational radiation, but the fact that there might be some sort of information process in the gravitational field that travels with the velocity of light is sort of within the fabric of general physics, leave out the general theory of relativity. If it wasn't that way, you'd have a lot of trouble with some things that already hold quite dear and hold very firmly. So that gravitational radiation may be of a quite different character than the Einstein [type], but there must be something that communicates in a gravitational field with near the velocity of light or the velocity of light.

findings as the fixed part of science; the philosophers' problem was to explain how theories were built upon the foundations of data that experiments provided. A key insight, as I have just said, was that a data point did not bear directly on the theory it was meant to test, but only on the theory plus a set of assumptions.[3] More recently, historians and sociologists have shown that the relationship between theory and data is less one-sided. Data may be as disputable as theories; a theory with strong adherents can reduce the credibility of a finding that conflicts with it, just as much as an unexpected finding can bring a theory into question.

The metaphor that is often used to describe the interweaving of theory and data is the network. Theories and data are thought of as held together by a kind of fabric of mutually supporting parts. But here we are interested in the life of one particular "node" in the network, Joe Weber's claim to have detected gravitational waves. I have already talked about the ripples of Weber's findings propagating or not propagating through social space-time, but here I want to concentrate not on the whole of the society into which his findings were introduced, with its different layers as represented in the target diagram, but only on the arguments within the bull's-eye of the diagram—what I have called the core set of scientists. In the core are three main nodes. The first is the theory of relativity, the second is Weber's findings, and the third is the complex of theories and findings about how waves spread, how astronomical sources are distributed in the sky, how detectors extract energy from the waves, and so forth. Each of the nodes would themselves look like complex networks if we "zoomed in" on them, and in the case of the third node we do not even need to zoom; but let us think of it as one node for a moment. The triangle of nodes thus selected from the network was rendered unstable by Weber's finding huge fluxes of gravitational waves. One of the three nodes had to give.

Relativity is a theory with a long history and extensive support in the whole network.[4] If relativity were to be taken away, then almost the whole network of physical findings and theories would become a tangle of loose strands. Ironically, though the first justification for Weber's experiment was to test the theory of relativity, it could never be more than an asymmetrical test because relativity was so strong. Weber might have been able to confirm

---

3. This idea is most closely associated with Pierre Duhem (1981) and Willard Quine (1953). More recently, Hesse (1974) used the idea, and Imre Lakatos (1970, 1976) builds upon it. I claim little originality in this chapter, just a useful metaphor.

4. As explained in note 2, gravitational waves are required by special relativity, but it seems that general relativity would be required to work out the likely flux and produce the conflict with cosmological observations.

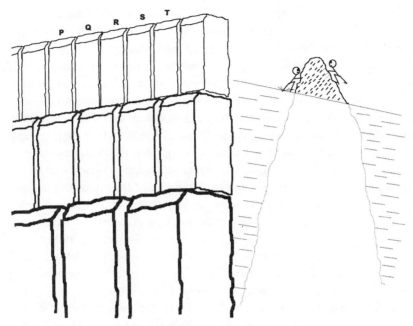

Fig. 5.1 The dam of scientific assumptions and the reservoir of doubt.

relativity, but he could never prove it wrong, however badly the measured flux matched with the theory.[5] And no one wanted to disconfirm relativity, not even the experimenters. Weber certainly did not want to disconfirm it; for him Einstein and his ideas stood for all that was best in physics. And of the others who supported Weber's views, none wanted to present them as opposed to relativity. Only those who wanted to be rid of Weber's results and the like were interested in setting the matter up as a straightforward choice between Weber and general relativity.[6] They wanted to set the matter up in this way because there could be only one winner in such an unequal battle. But there was another choice. The third node was complex, which meant it could be modified here and there without destroying the whole network.

Let us try a more graphic metaphor to describe the relationship (fig. 5.1). Think of a conical island in a reservoir that is steadily filling behind a dam.

5. Incidentally, Weber's findings, even if valid, could not legislate between general relativity and its near rivals, such as the Brans-Dicke theory.

6. For a misunderstanding of this point, see Franklin 1994. Franklin states that the confirmation of general relativity by Hulse and Taylor disconfirmed Weber's findings "once and for all." He seems unaware of the complex relationship between a theory and finding and the fact that Weber's findings were never taken, by any of their supporters, to bear on the theory of relativity in a negative way.

As the waters rise the area of the island gets smaller and smaller. At the beginning there are two people on the island representing the theory of relativity and Joe Weber's findings respectively. When the reservoir is full, there will be room for only one person on the island, and we know it isn't going to be Weber's findings—the theory will always be on higher ground.

The water in the reservoir is the growing unhappiness of physicists with the tension in the network—it is the water of doubt—and if it continues to rise, it will drown one of the inhabitants of the island. The dam is the third node. We can imagine the dam being built of identifiable stones, each consisting of an assumption or theory belonging to astrophysics or the theory of the detector. Stone P might be the idea of the uniform spreading and weakening of gravitational waves as they emerge from a source. Stone Q might be the very reasonable assumption that gravitational waves from a source are distributed across a broad band of frequencies rather than being concentrated, by chance, in just that frequency that Weber could best detect. Stone R is the assumption that the sources of the waves are distant—at least as far away as the center of our galaxy—not so close-by (e.g., some mysterious process in the heart of the Sun) that their energy has not had time to disperse and weaken before it reaches us. Stone S is the assumption that the waves are still being emitted more or less continually rather than consisting of one huge burst that happened about the time that Weber was "on air" (operational) in the late 1960s and early 1970s. Stone T might be the standard calculation of the sensitivity of gravitational wave detectors, which makes them relatively insensitive. I will describe the other stones that go into the dam as the description unfolds.

The point is that both Weber and general relativity can survive if we can take stones out of the dam. We are looking at a dynamic system in which the waters of doubt rise as the scientists initially add stones to the dam, and fall again as stones are removed. The relationship between the different components of the argument is not fixed once and for all, in the way that the relationship between the elements in a mathematical proof are fixed (once the proof is completed). Instead we have a process unfolding over time—some scientists removing stones, others adding them or replacing discarded ones. Thus stone P is removed if someone suggests that gravitational waves do not spread uniformly but might be focused toward Earth. This lets water out and lowers the level of the reservoir. Stone Q can be removed by claiming that the waves are concentrated in just the frequency band that coincides with the frequency of Weber's detector. Stone R can be removed by claiming that the source is very close even though invisible. Stone S can be removed by saying that what Weber is seeing is a unique historical event. And stone T can be removed by saying that we have done our calculations about the sensitivity of the detector incorrectly, and really it is much more sensitive than we thought. Only if all the stones are held in place will there be a stark choice. The theoretical physicist Kip Thorne puts the matter rather nicely, using a different metaphor. He talks of "wriggle room." As theoretical and other arguments gain detail and consensus the wriggle room for a new finding becomes less and less.[7]

The weight and firmness of each stone is itself a product of the historical and scientific context in which that particular scientific argument is embedded, so, as one might anticipate, we can imagine each stone as being set in its own web of arguments which confer extra weight. Bigger and heavier stones are foundational; they are found at the bottom of the dam and are hard to shift, whereas light stones, at the top of the dam and well away from the foundations of physics, are relatively easy to extract. The metaphor suggests quite nicely that if even one of the really heavy foundational stones is removed, most of the water of scientific consensus will run out.

CONSTITUTIVE AND CONTINGENT FORUMS

What I have described here is the kind of argument that happens in science's "constitutive forum."[8] The constitutive forum includes laboratories, refereed papers, and calm conference debates—the places where scientists put their ideas together in a visible way so as to constitute and certify new knowledge, and occasionally to dispute one another's claims. Usually, only a certain type of idea or argument is allowed in this forum; theories and findings might be disputed, but one is not allowed to say that their proponents are cheats, fools, or incompetents—one has to provide formal arguments, counterfindings, alternative data analyses, and other kinds of recognizably technical debating points. This kind of interchange belongs to the special discourse of science; it is what scientists feel comfortable exposing to the wider world. It is what gives rise to the idea of science as esoteric yet democratic and universalistic.[9]

7. Kip Thorne, private conversation.

8. Collins and Pinch 1979.

9. The sociologist Robert Merton (e.g., Merton 1957), built a social theory of science based on this kind of constitutive forum activity. Merton's theory is an excellent prescription for good scientific debate, but it is a flawed description.

There is another forum, the contingent forum, in which is found a kind of argument that does and indeed *must* play a role in causing scientists to come to one view or another, but which rarely appears in the constitutive forum. Scientist X may well conclude that he will never believe what scientist Y says, because he thinks Y is a liar, self-deluded, a fool, unprofessional, insufficiently careful about keeping his or her own desires separate from the data analysis, from a small country, from a weak university, obscure, or whatever. These factors play a role in causing scientists to decide who and what to believe; they have to play such a role because, as we will see, technical argumentation alone cannot exhaust the ground for debate; the need for trust in others' integrity, and in their theoretical or experimental abilities, is necessary if the constitutive forum is to exist.

The need for trust in science, and the role of distrust in science, is rarely exposed in public. It is nearly always impossible to state publicly, "I do not believe the results of Professor X, because he has been wrong too many times in the past," or "because I don't give any weight to the views of people from Podunk University," or "because Professor X is dishonest," or "because Professor X is incompetent." In the journals and similar outlets, one has to show that Professor X has analyzed the data incorrectly, reveal that a specific mistake can account for the results, describe a similar experiment that you yourself have done that reaches a different conclusion, or some such. We will see all these kinds of reasons for belief coming into play in the debate that surrounded Joseph Weber's claims, but what the "reservoir of doubt" metaphor captures especially well is the more limited nature of legitimate scientific dispute.

THEORY: BUILDING AND BREACHING THE DAM OF DISBELIEF

Given the dynamics of the flow of water into and out of the reservoir of doubt, we should not be surprised that not all theoreticians felt threatened by Weber's findings.[10] Depending on temperament, they approached Weber's findings in one of two different ways. Those whom I will call the speculators were delighted with his results, because the findings opened up a whole new world to explore; for example, forgetting, ignoring, or remaining unaware of the difficulties posed by the mounting waters, they could start to build models of the kind of cosmic scenarios that would

10. For an "upbeat" assessment of the possibilities, see Press and Thorne 1972. Press and Thorne think that Weber is probably right, but even if he isn't, strong sources will have been seen by 1980.

emit such large gravity wave signals. Freeman Dyson, a respected theorist who was much involved with Weber in the early days, gave a sense of the atmosphere of the times in a conversation with me.

> [1999] If you took his results at face value, you couldn't really believe there were that many sources [of gravitational waves]. But, of course, there was always an uncertainty there. There was the well-known analogy with radio-astronomy, where progress was held back for twenty years because nobody believed there were sources. So it wasn't worth looking, since you knew there was nothing there. And of course it turned out that there were millions of sources. So the same thing could have happened with the gravity waves. So you couldn't say dogmatically that it was absurd.
>
> ...I certainly wasn't [certain that Weber's findings were wrong]. Because I had this analogy with radio-astronomy in mind. There was a famous calculation—I forget who it was who did that—that you calculate the radio power of the Sun, and that's 10^{-20} of the optical power, and so you look at all the other stars in the sky, and you multiply by 10^{-20} and it's clearly unobservable [laughter]. That was certainly well known to both Joe Weber and me. There could have been all kinds of sources that nobody dreamed of. At that time, certainly, nothing was clear.

Speculators could also try to lower the water by removing a stone or two, for example; by thinking up mechanisms that would focus gravitational waves so that they might be beamed toward Earth from the center of the galaxy rather than spreading uniformly.

Those whom I will call the skeptics, on the other hand, were disinclined to believe the result of the experiment and did not think it merited the theoretical troubles it would bring in its wake if accepted. Skeptics might try to show how impossible it was to accept Weber's results by suggesting a sharp choice between the findings and relativity. Speculators would accept Weber's claims and try to reconcile them with what was already known, while the skeptics would attack him.

There were a number of candidates for possible sources so long as the overall energy problem was ignored. In 1963, long before Weber had found any results, and therefore before the absolute energy problem had raised its head, Freeman Dyson had speculated that neutron stars that were in close orbit around one another could be strong emitters of gravitational waves as they reached the end of their dance and spiraled in to collide. This mechanism, as we will see, was eventually to emerge as the favorite for the first assured detection of gravitational waves by the generation of detectors that would come "on air" in the first decade of the new millennium. By the late 1980s, the rate of emission of gravitational waves from such a source had been calculated carefully. The result showed that, other things being equal, Weber was not likely to be seeing them with his apparatus. In a later section I will argue that for various reasons, from the 1980s and into the early 1990s, theory began to dominate experiment, crushing any tentative empirical results, while in this early period, Weber's findings were to some extent leading theory.

Dyson's speculation was published in a book about interstellar communication.[11] Dyson was based at the Princeton Institute for Advanced Study, where Weber had spent time in the 1950s and '60s, so it is not surprising that Weber's experiment was on his mind. Dyson's paper is mainly about the way binary stars could be used to accelerate objects that passed near to them, using them as a means of speeding up spacecraft and so forth for interstellar travel. For this purpose he needed to think about dense, fast-moving, binary pairs and binary neutron stars, which were then still a subject of speculation among physicists. Dyson presents the emission of gravitational radiation by these binaries as a nuisance, because it would cause them to inspiral rapidly and lose their effectiveness as spaceship accelerators.

> [T]he whole of the gravitational energy is radiated away in a violent pulse of radiation lasting less than 2 sec.... [T]he loss of energy by gravitational radiation will bring the two stars closer with ever-increasing speed, until in the last second of their lives they plunge together and release a gravitational flash at a frequency of about 200 cycles and of unimaginable intensity.
>
> [Such a pulse] should be detectable with Weber's existing equipment [here he quotes *Phys Rev* 117:306, (1960)] at a distance of the order of 100 Mparsecs. So the death cry of a binary neutron star could be heard on earth, if it happened once in 10 million galaxies. It would seem worthwhile to maintain a watch for events of this kind, using Weber's equipment or some suitable modification of it.
>
> Clearly the immense loss of energy by gravitational radiation is an obstacle to the efficient user of neutron stars as gravitational machines...
> (p. 119)

The physicist Remo Ruffini was also at Princeton when he wrote an article in 1970 that suggested another possible source. In the popular science

---

11. Dyson 1963.

journal *Science News*, he was reported as suggesting a mechanism whereby spinning black holes would emit large fluxes of gravitational radiation in the process of ingesting new matter.[12] Ruffini's mechanism could have accounted, at least in part, for the frequency of Weber's reported observations.

Another favorite candidate for emissions of gravitational waves were exploding stars, or supernovas. But when a star explodes, it will not produce gravitational waves unless something asymmetrical happens. If the star explodes in a completely symmetrical way, there is no gravitational message to anything outside the system and no associated gravitational radiation (see chapter 3). As time has gone on, however, the work of theoreticians has suggested that supernovas might not emit intense waves as frequently as had been thought at first. Their degree of asymmetry is unknown, and developments in the theory of the way supernovas evolve have not been promising.

Though these ideas could, perhaps, provide mechanisms capable of generating individual observable pulses of gravitational waves, they could not deal with the consequences of the high overall flux. The skeptical view, which took the overall flux as a refutation of Weber's findings, was taken by, among others, British astrophysicist Martin Rees along with his colleagues Dennis Sciama and George B. Field.

Rees and his colleagues reported that a mass loss of more than about 70 solar masses per year at the center of the galaxy would be bound to have other visible effects on the galaxy that had not been seen.[13] Weber's reported findings corresponded to a mass loss of around 1000 solar masses per year if they were interpreted in a straightforward way. Nevertheless, the paper in *Physical Review Letters* concluded: "Since the high rate of mass loss indicated by Weber's experiments is not ruled out by direct astronomical considerations discussed here, it would clearly be desirable for these experiments to be repeated by other workers" (p. 1515). In other words, though the calculations done by Rees and his colleagues made Weber's results unpalatable, they were not yet ready to reject them outright in the medium of print.

In a conference in January 1972, however, at the Royal Astronomical Society, Martin Rees appears to have built on these results to mount the very forthright claim that Weber must be wrong if general relativity was right. Rees's claim was that the galaxy would be expanding noticeably as

---

12. Thomsen 1970; Ruffini was later to become the editor of the Italian physics journal *Il Nuovo Cimento*, and he will reappear in our story in this role.

13. Sciama, Field and Rees 1969; Field, Rees and Sciama 1969.

its mass was consumed and converted to gravitational waves at the rate required for Weber's findings to be true. If that much mass were being consumed, there would be too little gravitational attraction left to hold the galaxy together at its current size.

A still more uncompromising position was taken by Peter Kafka of the Max Planck Institute in Munich, who was also part of the German team building a room-temperature resonant bar to check Weber's results. In an essay offered to the Gravity Research Foundation in 1972, Kafka extrapolates Weber's findings, revealing that if uniformity was assumed along the various dimensions, including the inefficiency of Weber's bars, *three million* solar masses per year from the galactic center were required to be converted to gravitational energy.[14]

This kind of argument did not go unchallenged. Charles Misner, a theoretician and a colleague of Weber's at the University of Maryland, wrote the following regarding the status of Weber's results:

> In part, simply because of the fundamental significance of the detection of gravitational waves, but in part also because of the lack of more appealing astrophysical proposals for a source for this radiation, Weber's observations are not yet considered definitive. I am not, however, aware that there remains outstanding any proposal to explain the observations as artifacts arising from any cause other than unconscious observer bias. This last possible source of error is being eliminated by increasing automation of the experiment. Since, further, several independent attempts to verify Weber's observations are underway, the observations could soon become undebatable. In this paper I presume that the gravitational-wave flux at the earth is that indicated by Weber's experiments, and suggest directions in which more satisfactory theories of the source can be sought.[15]

Misner clearly was not taking the doubts seriously at this stage. He is an example of what I have called a speculator, and was, therefore, effectively a defender of Weber.

> If Weber's gravitational-wave observations are interpreted in terms of a source at the Galactic center, both the intensity and the frequency of the waves are more reasonable if the source is assumed to emit in a synchotron mode (narrow angles, high harmonics). (ibid.)

14. Kafka 1972.
15. Misner 1972, p. 994.

Misner continues,

> Weber has estimated that a straightforward interpretation of his observations
> would involve a source at the center of the galaxy radiating isotropically
> [equally in all directions] an average power...[equivalent to the total
> conversion of 1,000 solar masses per year] in the form of gravity waves.
>
> ...In the following Letter it is shown that sources can, in principle, be
> manufactured which emit gravitational synchotron [focused] radiation. The
> remainder of this note indicates why the observational evidence leads to an
> assumption that the source is emitting gravitational synchotron radiation,
> and then indicates some directions in which one might hope to create a
> theory of the source mechanism. (ibid.)

Misner, as we see, was trying to remove stone P by inventing a focusing mechanism. He was also offering to explain the source.

Rees's erstwhile coauthor, Dennis Sciama, also took issue with the pessimism of Rees and expressed himself in an article published in *New Scientist* in February 1972. Sciama agrees that the slight expansion of the galaxy that had been observed was not compatible with this rate of loss of stars, but only with an upper limit of 70 solar masses per year. But he asks,

> ...[I]s it the case of either Weber or the existing laws of physics, especially
> general relativity, being incorrect?...It seems to me that one should give up
> the accepted laws of physics only after a severe struggle has shown that they
> cannot account for any new observations which threaten them—otherwise
> the discipline of science would rapidly degenerate into chaos.[16]

Sciama goes on to give examples of previous anomalous observations that had taken years to explain. For example, he points out that it took 50 years for superconductivity to be explained, but during the whole of this time no one questioned the accepted laws.

Sciama suggests that we might be living in a short period of especially strong gravitational radiation, but his favorite hypothesis for reconciling Weber with relativity, and thus lowering the water in the reservoir, was a focusing effect.

> It follows that either Weber (and the rest of us) are very lucky to be living
> during a relatively short intense spell of gravitational activity or the sources
> do not radiate isotropically [uniformly in all directions]. (p. 374)

16. Sciama 1972, p. 373.

Citing Charles Misner and various others, he suggests that a particular sort of massive black hole rotating in the center of the galaxy could cause objects near its equator to emit gravitational radiation in the form of a narrow beam which would illuminate Earth.[17] This would reduce the required mass loss by a factor of up to 1000, making Weber's findings compatible with general relativity and the slight expansion of the galaxy which had been observed. He adds,

> I should stress that a satisfactory theory of this beaming still remains to be worked out. At the moment it is no more than a gleam in the eyes of a few optimistic relativists. They feel that if Weber turns out to be right all may not be lost for general relativity. (ibid.)[18]

THE FOUNDATION STONES

The sheer flexibility of this kind of theorization should not come as a complete surprise. After all, according to the current best wisdom, we see only 10% of the matter of the universe, while 90% is hidden. Who knows what is going on in the hidden 90%, and who knows whether we understand the universe at all? One way to make the skeptical waters recede is to try to pull out the stones that make up the foundations of the dam. But this, as I have said, is a dangerous game. The dam has been carefully built up for centuries; though it is never finished, it represents our accumulated knowledge. If we remove a foundation stone, the whole dam collapses and all the water rushes out. Then, unless the dam is rebuilt quickly—essentially Kuhn's model of scientific revolution—the reservoir of acceptable knowledge will sink, and new theories and findings will grow without control—weeds on the fertile lake bed. If there is no control, there is no science.

This is the kind of fear that besets those who worry about the threat of something radical, such as the paranormal. For example, if you believe that mind can affect matter, then a physicist's desires could affect the readings of a meter, and this might make any meter reading compatible with any theory. How, then, could physics be done?[19] When the first negative results

---

17. The others cited are Donald Lynden-Bell, Jim Bardeen, and Stephen Hawking.

18. Kafka had ruled out the focusing effect because of the polarization of the waves Weber observed. He also said, "It is unsatisfying to appeal to accidents" ("in space, in spectrum or in time") [1972, p. 6], by which he meant that he was unprepared to consider that Weber was observing a short run or otherwise special phenomenon.

19. I should note that the serious parapsychologists would say that these fearsome consequences can be avoided so long as the strange effects happen only rarely and then weakly.

of other gravitational wave teams began to be broadcast, leaving Weber as the only person finding positive results, I heard it said that Weber paid a visit to J. B. Rhine, the founder of parapsychology, to discuss the possibility that it was his mind that was affecting the results; but I could never pin this down beyond a rumor. Of course, for a physicist to admit, or even consider, the influence of the paranormal on an experiment would be to tug at one of the largest foundation stones of the dam.[20]

It may be worth remembering that at this moment in the twentieth century, the foundations of the dam were looser than they had been for some time. We were, so to speak, still living in the 1960s. The student protests had equated law and order with fascism; drug intoxication was thought by many to be a valid or even a superior state of a mind; Carlos Castaneda and the adventures of Don Juan seemed no less fantastical than the Vietnam War;[21] under some interpretations, Thomas Kuhn's book, *The Structure of Scientific Revolutions*, had made even science into a cultural playground; and Paul Feyerabend had come up with the easily overinterpreted phrase "anything goes" to describe scientific method.

The feverish quality of the times seems to have affected physics, too, and there were some odd stories around; if Weber's thoughts did turn to psychokinesis, he was not alone. In California and elsewhere, psychokinetic effects were all the rage.[22] This was the time of the notorious Stanford Research Institute experiments on the paranormal, in which Uri Geller, the Israeli psychokineticist and/or showman, was the principal star.[23] Another well-known psychic, Ingo Swann, was one of the subjects in the institute's

20. Other physical scientists who have played with this foundation stone have had their credibility destroyed.

21. A professor of psychology at the University of California at Davis, Charlie Tart, wrote a much cited paper, published in *Science* (Tart 1972), in which he discussed the idea of sciences that depended on the state of mind of the scientist. The paper referred in its title to "State Specific Sciences." Noting that Kuhn's idea of paradigms implied the boundedness and self-containedness of sets of ideas, Tart concludes that there were effective "paradigms" which lived only within certain drug-induced and other states of mind. He thought that various paranormal effects might belong in this category. Carlos Castaneda wrote a series of best-selling, supposedly "anthropological" accounts of his adventures with Don Juan that saw him engaged in numerous physics-defying feats with his master. These were taken very seriously by some academics, and only much later did it become clear that the tales depended more on Castaneda's inventive genius than on his fieldwork.

22. Weber was anything but a hippy. Whenever I saw him he was wearing a suit and tie, and his bearing, at least at the start of any conversation, was always stiff and proper. He never addressed me as "Harry" in any of the letters he sent—it was always Professor Collins. In our longer conversations we did discuss personal matters such as his family, the circumstances of his second marriage to Virginia Trimble, of which he was very proud, and sometimes his feelings of despair about his research. He did, it was said, have a penchant for wearing formal leather shoes without socks, but I was told this was because his hero, Albert Einstein, had done the same (I cannot vouch for this).

23. Collins and Pinch 1982.

militarily sponsored "remote viewing" experiments. Among his other exploits, Swann entered the Stanford University laboratory of William Fairbank, who was then building the first ever cryogenic-resonant-bar gravitational wave detector. Though Fairbank was not present at the incident, his assistants reported that Swann, apparently by mind power alone, had produced a huge kick on what should have been a thoroughly shielded magnetometer. Fairbank was noncommittal about the effect—because "one can't make it happen at will"—but spoke seriously about the incident; nowadays such a thing would never even be discussed by a physicist who wanted to retain the respect of his colleagues. He said to me that "the whole thing is just preposterous from the point of view of physics," but he added that he would be prepared to let Ingo Swann have a go at his gravitational wave detector (once it was completed) if he wanted.

Even if the story of Weber and J. B. Rhine is untrue, later in the book we will see that Weber would tug at stones at the middle level of the dam in years to come when he devised his new theory of the sensitivity of the detector. In the meantime, we simply note the unknowns in the universe, but note also that citing these unknowns as a way of releasing "skeptical water" is a dangerous game because of the risk to the lake. One of the arts of physics, as with any other science, is not to drown everyone on every island by an excess of conservatism while not allowing the waters of doubt to fall so low that weeds flourish uncontrolled.

WEBER'S RESPONSES

Weber could make two kinds of responses to the skeptics. He could mount a head-on attack, working still harder to make his experimental results convincing. By forcing critics to accept the positive result and finding ways to account for it, he would effectively bludgeon the skeptics into becoming speculators. Alternatively, along with the speculators, he could question the assumptions on which the disbelief was based and try to make the dam less secure—a flank attack, as it were. Weber was to spend much of the next decades endeavoring to make progress both frontally and on the flank.

Weber understood the complex relationship between his findings and the theory. In his article in *Scientific American*, he describes the conclusions of Sciama and Rees, saying that they had concluded that a mass loss of 200 solar masses was just about consistent with what we could see of the expansion of the galaxy, but if his "unequivocal" experimental facts represented a source at the center of the galaxy, along with the correct estimates of his bars' sensitivity and efficiency, he must be seeing the conversion of 1000 solar masses per year.[24] He realized, in other words, that his findings were at odds with the theoretical consensus generated even by some of those who were sympathetic toward him. He concluded nevertheless,

> The origin of the observed gravitational radiation has not been determined, only the direction of its arrival. It is conceivable that the source might be an unusual object such as a pulsating neutron star very much closer than the galactic center. It is also conceivable that the mass at the galactic center is acting as a giant lens, focusing gravitational radiation from an earlier epoch of the universe.... The relatively large intensity apparently being observed may be telling us when time began. (p. 29)

In the same way, when I asked him in late 1972 if he was seeing too much radiation to be compatible with cosmological theories, he replied,

> [1972] No, that's one of the most widely misunderstood things.... [T]he experiment that was done says there's a certain number of events, and it says that you get more when the antennas face the direction of the center of our galaxy. Now, the center of our galaxy is an obvious source, because there's a huge concentration of energy in a small region. Now, that evidence doesn't prove that the galactic center is the source—all it says is there's an ... effect associated with that direction.
>
> The energy estimates are always made by making certain assumptions that are pretty well untested by experiment. And some can never be tested by experiment.... The assumptions that are made are that the source is the galactic center—that it's really that far. However, the source could be in that direction but much closer, and there would be no energy problem. Another assumption that is made is that the galactic center is isotropic—radiates equally well in all directions. Yet another assumption that is made is that the radiation extends over a certain bandwidth. Now the experiments always observe over a narrow bandwidth, and you have to assume what is being seen over the narrow bandwidth extends over [beyond] the narrow bandwidth.
>
> These are very reasonable assumptions; I can't be critical of the people who make these assumptions. However, they are assumptions—they aren't

24. Weber 1971b: "If we assume that the experiment is observing gravitational radiation from the galactic center and that the analysis of the sensitivity of the detector is correct, each event corresponds to an amount of energy being radiated that equals roughly a fifth of the mass of the sun.... Allowing for the small detection efficiency, it corresponds to an energy of perhaps 1,000 solar masses per year or more" (p. 29).

things that I've ever insisted on. . . . I know . . . that there was a meeting in England and at that meeting well-known British astrophysicists made the statement that either Einstein is wrong or Weber is wrong—that's ridiculous. The experimental fact is that there's a bias in the direction of the galactic center, there's a certain energy flux in a narrow bandwidth. And that astrophysicist was saying that just because he's not clever enough to understand my data, it has to be wrong.

The 1970 paper in which Weber first revealed the sidereal anisotropy also contained flanking attacks. These were expressed in a very interesting way.

> If we assume the validity of existing theory, that the radiation is isotropic and that there are no focusing effects, what is being observed may only amount to roughly 10^{-2} solar masses per year. This is a very small energy loss which can be converted to a very large one if certain further assumptions are made. The narrow detector bandwidth, $\sim$0.03 Hz, suggests a flux of roughly 0.3 solar masses per Hz bandwidth. We may guess that perhaps we are only seeing 10% of what is emitted and that the power spectrum extends over several hundred Hz. Numbers like 1000 solar masses per year then come into view. Lack of isotropy and focusing effects might very substantially reduce this.[25]

Let us translate this passage into more readily accessible language. Weber first puts the onus on the reader to "assume" the validity of a certain aspect of existing theory. What we are asked to assume is stone P in the theory dam—namely that the radiation spreads uniformly through space without any focusing. Cleverly, at least in terms of the rhetoric of the paragraph, Weber appears terribly reasonable at this point, "even if I am gracious enough to assume my critics are right you still need only 10^{-2} solar masses to account for my results." That is all in the first sentence.

When we get to the second sentence, we discover that to get to the point of requiring more solar masses to account for the result, we have to make still more "assumptions" than have been granted already. That the radiation does not favor the narrow bandwidth to which Weber's bar is most sensitive—is said to be "a further assumption" (stone Q in the dam)—and that his bar is seeing 10% of the waves that hit it is treated in the same way, as yet a further concession (this is stone T in our dam). As we

25. Weber 1970b, p. 183.

can see, the onus is put on the reader of the paper to make these yet further "assumptions" in order to establish that there is anything untoward about the findings. While most physicists would have simply accepted these assumptions as part of the taken-for-granted background of their subject, Weber is suggesting that they are far from the natural way to look at the world. In the final sentence we are reminded that the concession made in the first sentence might not be valid.

The paper goes on to examine stone T a little further.

> Expressions for the antenna cross-section require very reasonable assumptions and approximations that have not been tested by experiment. It is also possible that the antenna is operating in a more sensitive mode than ordinarily assumed. Frozen-in metastable configurations within each detector might decay to equilibrium as a result of collective excitation by gravitational radiation, releasing far more energy than implied by the gravitational-radiation flux. (pp. 183–84)

It is interesting that even at this early stage Weber was speculating that the cross section of the bar might have been wrongly calculated, an idea that would become salient in the mid-1980s.

As a final flanking attack, reflecting on discussions he must have held with Misner, he writes: "I am also studying the possibility that cosmological gravitational radiation is being observed, focused by the galactic center" (p. 184).

CHAPTER 6

THE FIRST EXPERIMENTS BY OTHERS

THE NEXT EIGHTEEN MONTHS

And so the next eighteen months were played out, and "the several independent attempts to verify Weber's observations" mentioned by Charles Misner (1972) began to report results. These findings did not, however, make Weber's findings "undebatable," as Misner hoped they would. There were some positive results reported, but these did not come out of experiments that copied Weber's bar. An Israeli scientist called Sadeh and an American called Tuman from California's Stanislaus State University both claimed to be able to see the effects of gravitational waves by measuring vibrations in Earth. Earth is a mass like any other and has the advantage of great size and therefore a very large "cross section" for gravitational waves. That is, if there are gravitational waves capable of having an effect on a two-ton aluminum bar, they will, in principle, have a proportionately greater effect on the six-thousand-million-million-million-ton Earth. Compared with the vibrations caused by earthquakes and similar effects, however, the gravitational wave effects will still be very small, and the question is whether they can be separated from the rest of the seismic noise. Weber himself had pointed out the possibility of searching for gravitational waves in this way, and was to be involved in placing a seismometer for this very purpose on the surface of the seismically much quieter Moon.[1] In the meantime, his success had encouraged efforts and reports such as the Earth vibration findings, though their credibility was short-lived.[2]

The earliest publication in a journal to claim negative findings using a bar detector did not appear until 1972, but it was already widely known that other bar groups were having difficulty confirming Weber's results. It was at this stage that I conducted a first round of interviews with the experimental groups in Britain and the United States.

THE 1972 INTERVIEWS

When I conducted my interviews in 1972,[3] the first detectors built by teams independently of Weber had been "on air" for a short time. People were talking about results produced by Vladimir Braginsky in Moscow, by Tony Tyson at Bell Laboratories in Holmdel, New Jersey, and by Ron Drever at Glasgow University. Opinions as to the status of Weber's results, and even

1. This device did not work, but it was no fault of Weber's, as I understand it. I am told it was a problem in the manufacture of the gravimeter.

2. Other scientists also looked for gravitationally induced vibrations in Earth. Wiggins and Press (1969) reported that they could find nothing. Judah Levine at the University of Colorado at Boulder also conducted a negative search. But here is a quotation from a very well-known and respected scientist (not Weber) writing a letter in 1972 to another scientist doing work on seismic detection of gravitational waves: "An Israeli physicist called Sadeh claims to have found a seismic signal at precisely twice the frequency of the pulsar CP 1133. His instrument is a simple vertical seismometer and he claims an amplitude for the resonant signal of about 10–11 cm. Of course this is enormously larger than any theory predicts. Still the theory is uncertain and he may conceivably be right."

3. As a graduate student I was working on a very tight budget, so I bought an old car and drove between interviews, from East Coast to West, covering 5000 miles in all. I was covering four fields of science at the time, of which gravitational radiation was only one; hence the extensive itinerary including a trip to Quebec to talk to laser scientists. As luck would have it, my field trip began with me in a heightened state of sensitivity. I was nearly killed in an eighty-mile-per-hour car crash on the motorway near Heathrow a week before I set off for the US. After I landed in New York, I was queuing, exhausted, for the follow-up flight to Philadelphia when the police in the terminal suddenly drew their guns and shouted at everyone to take cover. As it turned out, the man in front of me in the line was carrying a concealed knife. When death is around, you see the world with enhanced clarity. In retrospect the car crash, the exhaustion, the guns, and the knife were a good start to my field trip—though I cannot recommend them as a routine method.

Some of the interview extracts in this chapter have been published at least three times before, while a smaller number are being seen in public for the first time. I apologize to those who are already familiar with the published material, but it seemed inappropriate to write a whole book on gravitational radiation without including these quotations.

opinions about what these groups had produced, varied widely. One scientist, whose apparatus was not yet running, said to me, "Personally, I have the impression that the critics are now swinging the other way and thinking there might be something in it. Perhaps, because it is nearer the point of confirmation, it is sensible to hedge your bets." On the other hand, another scientist, also only at the very beginning of an experimental program, said, "I may be wrong, but at this moment in time I'll bet you a hundred to one that Weber is full of it." It is quite interesting to see the differing reasons that scientists gave for their belief or disbelief in Weber's results, because they are another good indicator of how indecisive the results of an experiment are when taken on their own.

As we have seen, Weber first began to make his experiment convincing by looking for coincident signals on two more detectors separated by large distances. Some scientists found this convincing.

> [1972] "X" wrote to him specifically asking about quadruple and triple coincidences because this to me is the chief criterion. The chances of three detectors or four detectors going off together is very remote.

Nevertheless, others were less impressed, believing that the coincident effects could be produced by an artifact.

> [1972] [F]rom talking it turns out that the bar in Maryland and the bar in Argonne didn't have independent electronics at all.... There was some very important common contents to both signals. I said ... no wonder you see coincidences. So all in all I wrote the whole thing off again.

The time delay experiment again eliminated this kind of worry for some scientists but not for others. Several respondents made remarks such as "the time delay experiment is very convincing," but not everyone found it so.

Again, the correlation with sidereal time was the outstanding fact requiring explanation for some scientists.

> [1972] [B]ut then you've got the next line of evidence, that it's correlated with sidereal time, not solar time. This seems to me a major puzzle for the skeptics.

> [1972] I couldn't care less about the delay line experiment. You could invent other mechanisms which would cause the coincidences to go away.... The sidereal correlation to me is the only thing of that whole bunch of stuff that

makes me stand up and worry about it.... If that sidereal correlation disappears, you can take the whole Weber experiment and stuff it someplace.

By this time (though I have not yet mentioned it because the results were not yet published), Weber had started to analyze his data with less human intervention. The first reports involved "eyeballing" the output of pen-and-ink chart records of the vibrations of the bar, and some scientists suspected that this could too easily lead to unconscious bias in the results. Weber's response, explained in a paper he published in 1972, was to move to more automated data analysis.[4] This was the crucial move for two of my respondents.

> The thing that finally convinced a lot of us...was when he reported that a computer had analyzed his data and found the same thing.

> The most convincing thing is that he has put it in a computer....

But another said,

> You know he's claimed to have people write computer programs for him "hands off." I don't know what that means.... One thing that me and a lot of people are unhappy about, is the way he's analyzed the data, and the fact that he's done it in a computer doesn't make that much difference....

Whatever their degree of confidence, Weber's stream of innovative findings had brought other scientists to the point at which they felt they had to build their own experiments and run them. What were they saying they had found in 1972?

The group in Glasgow, then led by Ron Drever, had some data but were not ready to commit themselves. It had been reported to me by another group that the Glasgow team had seen a few peaks, but they were not saying they were gravity waves. When I put this to Drever, he said,

> [B]efore you can say you've got any real evidence, you have to have your apparatus running for quite a time, particularly in this field now, where there's been so much controversy about Weber's experiment, nobody will stick their neck out unless they're certain.... so I think everybody is going to

4. Weber 1972.

be very careful. They won't say anything very much until they're certain about the result, and that is bound to take a long time.

... [B]ut we do now have equipment which is operating and maybe detecting gravity waves, but I could not possibly say if they were gravity waves.

Tony Tyson at Bell Laboratories also had been running his detector for a while and had found nothing he was prepared to report in a positive way. He told me he had been observing with two detectors, one located in New Jersey and the other at the University of Rochester, in northern New York State, the home university of his Ph.D. supervisor, David Douglass; the two devices were separated by 300 miles. Tyson said he had found a few coincidences, but they were "not distributed interestingly in sidereal time"; in a year's data, "you do get just a little bump correlated with sidereal time, but I am not willing to interpret that as anything." Tyson said he thought that these detectors were only one-sixtieth as sensitive as Weber's, but he had more recently built a much larger one that he believed was ten times the sensitivity of Weber's. This had been on air for a couple of months without detecting anything significant. He said that Douglass would soon have a similar device in operation at Rochester and that they would then be able to run coincidence experiments between them.

At this time the most outspoken group was that of Vladimir Braginsky in Moscow, who had published a negative paper by the time of my US interviews.[5] Unfortunately, I did not visit Braginsky's group, but about this time he had been confronted by Weber, who advised him to withdraw his negative claims; he refused. In 1995, Braginsky told me the following:

[1995] A few weeks before the article appeared, some guy informed Joe Weber about this. Joe Weber sent a letter to me saying that being very concerned about my reputation, he strongly recommend me to withdraw this paper from *JETP Letters*. I said to him, I will not, because I have not seen anything at the level of sensitivity you claim to reach.

THE DAM OF DISBELIEF AND INTERPRETATIVE FLEXIBILITY

Sensitivity Considerations

After the first detectors built by other teams came on air, the experimenters eventually reached the conclusion that their first published results could

5. Braginsky et al. 1972.

not support Weber's claims. But this is not to say that they found nothing that could be interpreted as a gravitational wave signal. As we will see, the Glasgow group found a signal that they say might have been a gravitational wave; the Munich group found a section of data consistent with Weber's findings, which Weber insisted supported his results, but which they eventually concluded was not reliable. Also, the Bell Labs group, so Weber claimed, found a four-standard-deviation result when their data output over a certain time was compared with his. Weber said he was unable to publish this data, because there was an agreement in existence with Bell Labs which prevented it.[6] Thus, there were various phenomena that could have been interpreted positively but which the scientists decided were unreliable.

We can see that an experimental result is not just an experimental result—it has to be interpreted. And we can see in what has gone before, and in what will come after, that there is a lot of interpretative flexibility between data and result. Likewise, there is a great deal of interpretative flexibility when it comes to working out the meaning of a whole series of results.

We can speculate that these teams reached their conclusion—that any potentially positive data in their detectors' output should not be counted as confirming—in part because of the discussions they were having with one another at conferences and the like. Thus, if they had gone to conferences and found that all or most of the other detector teams were finding clear, positive results, a potentially positive reading would very likely have been interpreted more positively. But given that nearly everyone else was finding negative or unclear data, it would not seem sensible to say anything very positive about a patch of "scruff" in an otherwise signal-free data stream.

There is nothing sinister in the above claim. We can make it with confidence only because the signals emerging from most new experiments of this kind are found deep within the noise and are, therefore, ambiguous. Exactly what the experimenter says about them is bound to be a matter of interpretation, and the interpretation will be formed in part by the developing "agreed view" on which every individual experimenter draws and to which every individual experimenter contributes.[7]

---

6. I have seen the letters that passed between Tyson and Weber and can confirm the general drift of this claim, though not the details in respect of motives or the importance of the data as a confirmation of Weber.

7. I will set this claim out more formally in due course. For a participant observer description of the way "taken-for-granted" scientific reality affects the way ambiguous data are interpreted in parapsychology, see Collins and Pinch 1982.

What I have just described is the way the *contingent* forum works. In a scientific paper—part of the *constitutive* forum—one cannot make it explicit that the conclusion draws on the agreed view that is being aired around the conference circuit; one has to present formal arguments. I will discuss formal arguments in the next chapter. The confrontation with Weber's findings that scientists would eventually make in the constitutive forum would depend on comparisons of the detectors' sensitivity. To confront Weber, one's detector had to be more sensitive. Therefore, the authors of papers had to calculate, or otherwise determine, the sensitivity of their own detectors and either accept Weber's claims about his detectors' sensitivity or recalculate his sensitivity for themselves. One trouble with the Sinsky-style calibration was that it was too difficult to repeat; waiting for such a calibration to be completed would mean putting off any confrontation for years and losing one's place at the frontier of research. A second problem was that its extreme complexity made it easy to refuse to accept its results. This put the onus on other kinds of calibration (to which we will return in due course), or *calculations* of the sensitivity of the competing bars. But calculations of sensitivity are not straightforward, or there would be no need for calibration. Their complexity is revealed in a couple of papers published in 1973 and 1974 by Richard Garwin and James Levine, Weber's most salient and effective critics.

These papers express a reassuring feature of natural science: a scientific disagreement is not fought out with complete ruthlessness. In these papers, which, overall, are critical of Weber, we find Garwin and Levine nonetheless defending him against some of his other critics. Thus, the 1973 paper agreed with Weber's analysis of the sensitivity of his amplifier and refuted the more pessimistic estimate of Tyson.[8] Similarly, in spite of being part of the highly skeptical team, Levine found fault with the sensitivity calculations of another group who were criticizing Weber. He said of the experiment done at the European Space Research Organisation laboratories in Frascati, "We believe they have greatly over-stated the sensitivity of their system... by neglecting the effects of the filters used to reduce broadbandwidth noise."[9] The implication was that neither the results collected by Tyson nor the Frascati experiment confronted Weber findings in the way that the authors had claimed. Once more, the mistakes pointed out by Garwin and

---

8. Levine and Garwin 1973, p. 178; Tyson 1973a, p. 81.

9. Levine 1974, p. 280. Note that those involved in this Frascati experiment, Bramanti, Maischberger, and Parkinson (1973), did not work on the later collaborations between Frascati and Weber.

Levine were mistakes in calculating the relative sensitivities of the competing bars.

But the problems of making comparative calculations of sensitivity form merely the tip of the iceberg of the problem of testing an experimental result by replicating the experiment. When we test an experimental result by replication, we try to *do the same experiment again.* This innocent-looking phrase hides immense complexities. If an experimenter wants to confirm someone else's result, then, up to a point, it is better for the confirming apparatus to appear to differ from the original;[10] that way the confirming apparatus will appear less likely to reproduce some artifact that mimics the effect being sought. One of my respondents put it this way:

> [1972] [S]o if you do a carbon copy, you might find something, too, [but] you
> build the same pitfalls into it right off... if the people who are building a
> Chinese copy find the same thing, it's not so wondrous that they are
> susceptible to the same, for instance, magnetic disturbances, as Weber is.
> That's why I don't think building an exact copy makes much sense.

On the other hand, when disconfirmation is at stake, as in this case, any differences between the initial experiment and those that try to disconfirm it weaken the disconfirmation, because the features of the original experiment that are changed might be crucial. An added difficulty is that in a new area, where we are still groping toward an understanding of the phenomena, we are not completely sure what counts as similarity and what counts as difference.

Every experiment is different from every other. Let us consider the resonant bar design of the gravitational wave detector. If we expect to disconfirm the original findings, should we use the same material for the bars, bought from the same manufacturer? Should the bar be cast from the same batch of metal? Should we buy the piezoelectric crystals from the same manufacturer as the original ones? Should we glue them to the bar using the same adhesive as before, bought from the same manufacturer? Should the amplifiers be identical in every visible respect, or is an amplifier built to certain input and output specifications "the same" as another amplifier built to "the same" specifications? Should we be making sure that the length and diameter of every wire is the same? Should we be making sure that the color of the insulation on the wires is the same? Clearly, somewhere one has to stop asking such questions and use a commonsense

10. Collins 1985/1992, chap. 2.

notion of "the same." The trouble is that in frontier science, tomorrow's common sense is still being formed.[11]

In an earlier study (Collins 1985/1992), I discovered that scientists trying to build a relatively noncontentious kind of device—a certain kind of laser—went wrong when they used their common sense. Their normal laboratory practice led them to ignore the exact length of the wires linking components in the electrical circuit of the laser. Yet as time went on, it turned out that one of the electrical leads needed to be shorter than about eight inches for the laser to work; this crucial dimension never appeared on any circuit diagram or in any paper, because scientists were quite unaware of its importance. For this reason, before the importance of the lead length was understood, two scientists might build two lasers that appeared to be identical to them and to any other member of the laser-building community, yet one would work and one would not.[12]

In the early days of gravitational radiation detection, the science was still more exploratory. What, then, should one look at when deciding whether two detectors counted as "the same"; what counted as crucial differences where disconfirmation was concerned? In the interviews with bar-detector scientists I conducted in 1972, physicists' varying replies brought out the difficulty very clearly. Not surprisingly, two scientists did not see any significant differences between detectors.

(1) [1972] You can pick up a good textbook, and it will tell you how to build a gravity wave detector.... At least based on the theory that we have now. Looking at someone else's apparatus is a waste of time anyway. Basically, it's all nineteenth-century technology and could all have been done a hundred years ago except for some odds and ends. The theory is no different from electromagnetic radiation.

(2) [1972] The thing that really puzzles me is that apart from the split bar antenna [the distinctive British version of the device], everybody else is just doing carbon copies. That's the really disappointing thing. Nobody's really doing research, they're just being copycats. I thought the scientific community was hotter than that.

11. To my astonishment, on reading this passage Gary Sanders told me that in 1988 or 1989 he had been in a laboratory near Tokyo when a Russian physicist, examining a Japanese group's apparatus, declared their results on Tritum Beta decay to be invalid because they had used wires with red insulation! Apparently the red dye contains traces of radioactive uranium, which can confound the measurements.

12. Collins 1985/1992.

A third scientist, on the other hand, thought that there could be unknown but significant differences between the competing bars.

> (3) [1972] [I]t's very difficult to make a carbon copy. You can make a near one, but if it turns out that what's critical is the way he glued his transducers, and he forgets to tell you that the technician always puts a copy of *Physical Review* on top of them for weight, well, it could make all the difference.

And a fourth scientist thought that differences would be crucial.

> (4) [1972] Inevitably in an experiment like this, there are going to be a lot of negative results when people first go on the air because the effect is that small, any small difference in the apparatus can make a big difference in the observations.... I mean, when you build an experiment, there are lots of things about experiments that are not communicated in articles and so on. There are so-called standard techniques, but those techniques, it may be necessary to do them in a certain way.

For these kinds of reasons, even if the community could agree on the right way to calculate the sensitivity of a bar detector, the calculation would not take into account these unknowns.

To use a single term to make the point, we might think of some experimenters as more skillful than others in the way they put their apparatus together. The phenomenon of "golden hands" is well known in experimental sciences—it is little different from our everyday experience of those who are handy and those who are not. Just as only one or two violin makers can produce an instrument that sounds superb, so only one or two experimenters can make an instrument that is superb in terms of its sensitivity, even if the *calculated* sensitivity of other apparatuses rivals theirs. As a respondent put it,

> [1972] Even if [someone] tried to make a Chinese copy of Weber's, it might not succeed. Suppose I went to Weber and I asked him about his detector and I came back and I tried to build a detector just like Weber's. It might not be just like Weber's. If Weber built it, it probably would be like it.[13]

---

13. Though my earlier studies of laser builders showed that even this might not be so. Thus, I studied a laser builder, trying to make an identical copy of a device he had already made and failing. Fortunately, in this case he had a clear criterion of success and failure and could go on making adjustments until he got it right (Collins and Harrison 1975).

From this it might follow that a gravitational wave detector that seemed to be more sensitive in terms of the calculations might be less sensitive in operation. But no one would know this for certain!

The Experimenter's Regress and Attempts to Break Out of It

In the field of gravitational wave detection in 1972, we can see the working-out of what has been called the experimenter's regress. Experimenters argue about whose apparatus is the best. If they were arguing about whose violin was the best, they would have it played by a virtuoso and they would listen, or have an expert listen for them. If they were talking of lasers, they would soon know which of two identical-looking lasers was the best—only the one with the short lead would "lase" (though it might well not occur to them that such an insignificant item as a lead was the crucial variable even after the comparison had been made).[14] But what if they were trying to decide whose gravitational wave detector is the best? In a gravitational wave detector, what is the equivalent of making a beautiful sound or of lasing? At first sight the equivalent must be "detecting gravitational waves," but not if the theorists are right and there is little in the way of gravitational waves to detect. The normal test we use to decide whether an experimental apparatus is working is to try it and see if it works, but here we are not sure what *working* means. We will not know what *working* means in the case of gravitational wave detectors until we know how much gravitational radiation there is to detect and what form it takes. Then we will know that a working gravitational radiation detector ought to see just so many waves of just such-and-such a form and intensity. But the only way we can find out what to expect is to build the experimental apparatus and look for gravitational radiation. Thus, the experimenter's regress as it manifests itself in gravitational wave detection goes like this: To know whether you have built a good gravitational wave detector, you should try it and see if it works properly. But to know what "works properly" means you have to know what it should see. But to know what it should see, you have to know what gravitational waves look like. But to know what gravitational waves look like, you have to build a good gravitational wave detector and look at them. But to know whether you have built a good gravitational wave detector, you should try it and see if it works properly. And so on!

---

14. Likewise with violins; it is not clear just why Stradivarius violins are so good, though there is a recent theory that it has to do with the composition of the varnish.

Much of what happens in the kind of experimental controversy in which Weber's claims became enmeshed consists of arguments about who has the best apparatus. (Bear in mind that hardly anyone was even trying to build their apparatus in the same way that Weber had built his.) There are at least three kinds of arguments for building something different from the original experiment when one tries to test the findings. The "scientific" justification, as I have already explained, is to avoid building into your own apparatus the faults of the original; should the original experiment be looking at an artifact because of some flaw in design, one would not want to reproduce the flaw. A second reason is that scientists each have their own special area of expertise, and they like to use their skills creatively: why build a carbon copy of someone else's apparatus when something different, even superior, could be constructed?

> [1972] [I]n some respects the least creative and the least interesting thing to do is just build a carbon copy of somebody's apparatus. From the outside you could look in and say that it is very desirable that Drever, Tyson [or ourselves], or somebody build an exact carbon copy of Weber's apparatus, although I don't think you go into this field without trying to understand how the apparatus works, and most people in this field are an expert in some part of it. They are either an electronics expert or they are an astrophysicist or they are a low-temperature expert or there is some reason why they think they know about one part of the experiment. With regard to that part of the experiment in which they are an expert, they usually come to the conclusion that they can do it better than Weber. So that when Tyson and Miller build their apparatus, Miller is one of the world's top experts in electronics, they are going to put their own electronics in it; they are not going to try and copy Weber's.
>
> So what you copy is the part that you are not an expert on, and the part you are an expert on you build something that you think is better. That may or may not be true [that it is better].
>
> Braginsky, for instance, is using a capacitor and gave two hours of lectures on his very special way of doing things. But the actual detector may very well be less sensitive than Weber's. This is a problem.

Another scientist said that he had decided to do experimental work on the problem a very long time ago, but only if he could think of some way of doing it better than, or at least differently from, Weber. He did not think it was worth just copying it.

[1972] I'd usually want to do an experiment which is just my own idea. That's
much more exciting than to do something that somebody else has thought
of. And even if somebody had told me a way of doing better, I don't know if
I'd have done it, unless I felt I'd made some contribution to the idea myself.

The third reason for wanting to build something very different from Weber's
original experiment is a matter of calculating the potential rewards.

[1972] Mostly people just don't believe [Weber], and that's why not many
people are going in for it. And those that are going in for it are doing it as
something peripheral mostly. Many of them don't want to go in for it
because they know if they see something, it's a Nobel Prize for Weber, and if
they don't find anything—well—big deal! See, there's nothing to win in this
thing unless you really think there's something in the whole field.

A reason that Weber gave me for why no one had yet confirmed his
result in 1972 followed from the above motivations. He thought that those
who were supposedly testing his results were not trying hard enough to
repeat exactly what he had done.

[1972] Well, I think it is very unfortunate, because I did these experiments
and I published all relevant information on the technology, and it seemed to
me that one other person should repeat my experiments with my technology,
and then, having done it as well as I could do it, they should do it better.…
It is an international disgrace that the experiment hasn't been repeated by
anyone with that sensitivity.

Trying to decide who has the best apparatus regresses to an argument
about the *criteria for deciding* who has the best apparatus. If any of these
matters were obvious, the circle of the experimenter's regress would be
broken. Different members and communities each try to break the regress
in their own ways.

The theorists try to re-establish what counts as "working" by calculating
the form and intensity of gravitational waves; calculating the sensitivity of
the apparatuses; and reviewing the data analysis processes to see if they are
statistically sound. Thus, if the theorists could convince the experimental
community that they had worked all this out correctly, the argument would
be over. But, as we have seen, the theorists rarely come up with anything
so definite. There is too much space to argue; there are too many loose
stones in the dam. The theorists make their contribution, but it is not an

outright solution. To illustrate, I asked an experimentalist, "Who is ahead in the race to confirm or disconfirm Weber?" His reply once more revealed the looseness of fit of the theoretical restraints.

> [1972] [Y]ou can't really say who is ahead yet, because you don't know yet what the nature of the gravitational wave pulses are, and that affects who detects them first. If in fact they turn out to be short pulses coming at the rate Weber said they might be coming at, then a crude detector like the one we have built might pick them up quickly. On the other hand, if there aren't so many and they are longer pulses, it might have to wait until there was a more sensitive one built.... We've got an operating system, but it doesn't necessarily mean that we're ahead.

Another powerful way of trying to break out of the experimenter's regress is to calibrate the various experiments. When James Levine wrote a paper claiming that the Frascati group had done their sensitivity calculation incorrectly, he concluded as follows:

> [T]he above observations emphasize the need for a direct calibration, by using the authors' electrostatic end-plate, to produce impulsive mechanical excitations similar to those being sought, with known energies. In this way all uncertainties in the final interpretation of the results can be eliminated.[15]

As one might anticipate, an experiment designed to break out of the experimenter's regress is going to be subject in turn to its own experimenter's regress; we will return to this in due course.

What I have mentioned so far are what might be called scientific ways of breaking out of the regress. That is to say, they involve calculation, theorization, or measurement—the things that are discussed in the constitutive forum of science. In addition, in the face of their inevitable indecisiveness, scientists looked for other information that they could use to help decide the issue. Inevitably, then, they tried to assess the integrity and competence of the *experimenters* as well as the *experiments*. These kinds of assessments are not normally thought of as belonging to science proper. They are found in the contingent forum. The rules for making such assessments cannot be found in scientific textbooks or standard treatises on the philosophy or methodology of science; science, at least when it comes to *justifying* its findings, is supposed to have no need of personalities. Yet, unless one was

---

15. Levine 1974, p. 282.

oneself in the laboratory at the time that an experiment was being carried out, one has to rely on reports, and one is immediately thrown back on the reliability of the reporter.

Thus, one needs a report on the competence of the experimenter. Science, in other words, like every social activity, is shot through with the need for trust.[16] Trust itself has its criteria, and again we will see that scientists differ in what they think makes someone trustworthy, either as an experimenter or as a reporter. I have discussed the bad impression made by Weber's confused conference presentation on the sidereal correlation. There also can be more-direct experience, as in the following remark:

> [1972] He published a paper in mid-'69 where he reported setting up the other bar a thousand miles [km] away, and he had seventeen events including four triples and three quadruples. Three quadruples out of seventeen is pretty high. That's well above chance rate. But that's what he says. But when you see his actual sheet [showing me some of Weber's chart-recorder output], and it's here you see where the personal uncertainty comes in. Look here! Here's what he marks as a coincidence, but there's hardly any signal at all, so when he says "quadruple" you don't know exactly what he's looking at. Most of these are hardly associated with any significant values at all.

And there can be secondhand experience, which one scientist explained as follows:

> [1972] And now you want to know why I don't believe Weber. Well, it has to do with inside information more than anything he's published. Our paths crossed one time earlier, before he ever started on gravitational radiation ... [description of a much earlier experiment] ... It turned out that Weber had left out some very important effects. He wasn't being very careful about it.
> ... [A] friend of mine "X" went to work for him. "X" came back and told me all kinds of terrible stories.

As we see, attempts to find ways out of the experimenter's regress range from astrophysical theorizing to the common-sense reasoning we use to decide whether someone is trustworthy. In breaking out of the regress, the boundary between what is normally recognized as scientific reasoning and what is normally recognized as everyday reasoning is of little significance.

16. Shapin and Schaffer 1987; Shapin 1994.

The different kinds of warrants for belief may appear in different places and may be expressed in different written or spoken styles, sometimes publicly and sometimes privately. Each, however, contributes to decision making in controversial science; each makes its contribution to the amplification or suppression of the ripples in social space-time. It is hard to see how it could be otherwise outside the textbook.

In my 1972 interviews, I found that scientists used all these means of assessing the value of gravitational wave experiments done not only by Weber but also their other colleagues. In each of the following sets of extracts, three scientists at different establishments (each group of three is different) are reporting on one of four experiments. The experiments they refer to include both bar detectors and earth vibration detectors.

EXPERIMENT W

Scientist (a): [1972] [T]hat's why the "W" thing, though it's very complicated, has certain attributes so that if they see something, it's a little more believable.... They've really put some thought into it ...

Scientist (b): [1972] They hope to get very high sensitivity, but I don't believe them, frankly. There are more subtle ways round it than brute force ...

Scientist (c): [1972] I think that the group at ... W ... are just out of their minds.

EXPERIMENT X

Scientist (i): [1972] [H]e is at a very small place ... [but] ... I have looked at his data, and he certainly has some interesting data.

Scientist (ii): [1972] I am not really impressed with his experimental capabilities, so I would question anything he has done more than I would question other people's.

Scientist (iii): [1972] That experiment is a bunch of shit!

EXPERIMENT Y

Scientist (1): [1972] Y's results do seem quite impressive. They are sort of very businesslike and look quite authoritative ...

Scientist (2): [1972] My best estimate of his sensitivity, and he and I are good friends ... is ... [low] ... and he has just got no chance [of detecting gravity waves].

Scientist (3): [1972] If you do as Y has done and you just give your figures to some ... girls and ask them to work that out, well, you don't know anything. You don't know whether those girls were talking to their boyfriends at the time.

EXPERIMENT Z

Scientist (I): [1972] Z's experiment is quite interesting, and shouldn't be ruled out just because the ... group can't repeat it.

Scientist (II): [1972] I am very unimpressed with the "Z" affair.

Scientist (III): [1972] Then there's "Z." Now the "Z" thing is an out and out fraud!

At that time, with so much uncertainty, with so many new experiments coming on line, with many hoping to be the first or the second to see gravitational waves while others were determined to disprove Weber's findings, intrigue was in the air. The editor of *Physical Review*, Sam Goudsmit, had circulated a "samizdat" explaining what he thought was wrong with Weber's statistical analysis, and particularly what the problems were with the sidereal correlation (of which more later).[17] I was also told that Goudsmit had tried to prevent at least one of Weber's papers from being published in the journal, but the deputy editor had let it through when Goudsmit was away on leave.

Physicists seemed to have been meeting clandestinely and magnifying any rumor that came their way. One rumor had it that Ron Drever's detector was ten times more sensitive than Weber's and that he had been running his apparatus long enough to declare that there was no effect. Another scientist told me that Drever had seen a few pulses that he considered possible gravitational radiation. But a third physicist had telephoned Drever from the United States and learned that the "order of magnitude" improvement claimed by some for Drever's detector was denied by Drever himself. As we have seen, Drever was being very cautious at this time. As another scientist explained,

> [1972] There was some excitement at the Varenna Conference Summer School that we were at. When we first arrived, it was understood that Braginsky had had his detector running for 30 days and not seen any events, though he had his sensitivity as great as Weber's. Then Weber and Braginsky had many hours of talks, and the official opinion when we left was that Braginsky's apparatus was not as sensitive as Weber's. Of course, unofficially I don't know if Braginsky was that satisfied.

Braginsky, as we now know, was not satisfied.

17. The copy I have in my possession, entitled "Critique of Experiments on Gravitational Radiation," is dated 1970 and is marked "not for distribution." At the time of my 1972 interviews, many scientists were talking about the circulating samizdat, but I did not come into possession of a copy until some years later.

I asked most of the respondents about their visits to one another's laboratories and their social contacts and telephone calls. A common theme was that Weber was unforthcoming with his data and the details of his experiments and statistics. On the other hand, a more sympathetic scientist said that this was understandable, as experience had taught him that raw data should never be given away to anyone unless they could be trusted to do their very best with it.

> [1972] It is a very difficult thing in a controversial field, because somebody can take your data, claim they have looked at it, and claim they didn't see anything. And that's what makes it a very difficult thing.

More than one physicist told me that they had garnered "inside information" about the goings-on in Weber's laboratory. In one case, this led to a big increase in skepticism. In another, a scientist explained to me,

> [1972] Some of us undertook to do some skullduggery and investigate his experiment, in his absence very often. I really can't say what I found, [but] let's put it this way: if I had found he was all wet, I wouldn't still be in the field.

Scientists seemed to differ in their willingness to talk to others. A history of working together as supervisor and student led to easy relations for some. Other scientists were reticent about making contacts, fearing that they would be wasting others' time unless they had something definite to say, and wasting their own time if the other party did not have interesting data to talk about.

> [1972] It's not a nice thing to do. It embarrasses me a little to visit somebody else's lab unless one knows them well. But, you know, when I'm abroad or something I think twice about just going into somebody's lab. It takes up their time and, you know, sometimes people just aren't very helpful.

Still others, again, preferred to feel they were working independently on their own ideas. Ron Drever explained to me that he could not afford to travel much—it would cost £30.00 to visit anyone in England from Glasgow—and when I asked him if he ever telephoned Weber or any of the other gravitational wave groups, he explained, "[T]he University wouldn't allow it. It would be too expensive."[18]

To summarize, the "nonscientific" reasons that my respondents gave for their acceptance of the gravitational wave results of others included the following:

- faith in experimental capabilities and honesty, based on a previous working partnership
- personality and intelligence of experimenters
- reputation of running a huge lab
- whether or not the scientist worked in industry or academia
- previous history of failures
- "inside" information
- style and presentation of results
- psychological approach to experiment
- size and prestige of university of origin
- integration into various scientific networks
- nationality

One highly critical scientist told me, referring in this case to his disbelief in Weber's results,

> [1972] You see, all this has very little to do with science. In the end we're going to get down to his experiment, and you'll find that I can't pick it apart as carefully as I'd like.

---

18. I can verify that in British universities in the early 1970s, an overseas phone call was considered a great luxury. Even now, only senior professors can make overseas calls without special permission or a grant to cover the cost.

CHAPTER 7

JOE WEBER'S FINDINGS BEGIN TO BE REJECTED IN THE CONSTITUTIVE FORUM

The 1972 interviews reveal some of the reasons that scientists give for believing, or not believing, the results of other scientists but which they would not normally utter in the constitutive forum. That is, these are things they would not want to present as "science" and therefore would not publish in the professional journals.[1] A small portion of this material can be found unevenly presented in sets of conference proceedings, for these stand somewhere between the contingent and constitutive forums. One of the best ways of understanding the spread of ideas through space-time is by attending conferences; unfortunately, the published proceedings are not so useful, except under unusual circumstances.[2] For

---

1. Collins and Pinch (1979) show that in low-status sciences, this boundary is often crossed. For example, parapsychologists are often said by their opponents to be incompetent or untrustworthy, and these claims are uttered in the constitutive forum.

2. Alan Franklin (1994), having discovered one particularly fulsome set of proceedings, has argued that published works are an adequate and exhaustive resource in the history and sociology of science.

example, the editors' introduction to the proceedings of a conference held in Warsaw in 1973 includes the following:

> There was a general consensus that the Symposium should concentrate on the discussion of present and future experiments to detect gravitational waves of cosmic origins, and it was clear to all of us that a Symposium such as this could never have taken place in 1973 without the pioneering work of J. Weber.
>
> A spirited discussion among the authors of several experiments on the detection of gravitational radiation was one of the highlights of the meeting.[3]

By 1973, one can imagine that the discussion was lively indeed, but all one finds of it in the proceedings is a paper by J. A. Tyson, who claims to have a detector more sensitive than Weber's that failed to see any coincidences, but agrees that his and Weber's experiments were sensitive to different types of signal. Weber himself, in common with many of the other speakers, presents only a single-paragraph, purely descriptive summary of what he said at the conference; there is no mention of controversy. If one had only this set of proceedings to go on, one would have no idea what the editors were talking about. Nevertheless, as the waves propagated out in social space-time, the conference circuit of the time must have been an important medium, albeit one largely lost to us except through secondhand reports. The waves, however, were by now beginning to have a noticeable effect in the professional journals, too. What was being published in the constitutive forum was also beginning to turn against Weber.

THE FIRST NEGATIVE EXPERIMENTAL PUBLICATIONS

Moscow State University—Braginsky

First into print, in August 1972, was the team of Vladimir Braginsky from Moscow State University.[4] Braginsky was to remain closely associated with

---

3. Dewitt-Morette 1974, p. xi. A comment by Peter Kafka also expressing the flavor of the times on the conference circuit can be found at WEBQUOTE under "Kafka on the Conference Circuit." www.cf.ac.uk/socsi/gravwave/webquote.

4. Braginsky et al. 1982. Date order of publication is not necessarily date order of the first announcement of some findings; because announcements at conferences, or private letters or circulated pre-prints tend to precede journal publication the priority of the initial announcement of findings is not necessarily preserved when the publications appear.

the search for gravitational waves, later developing a strong partnership with the California Institute of Technology (Caltech) interferometry group, though he remained based in Russia; he will reappear frequently throughout this story. In his paper, Braginsky reports that his group found no statistically significant excess coincidences in their experiment. He concludes with this somewhat less than transparent statement (rather typical of Braginsky's indirect style):

> The decrease in the number of coincidences following the introduction of delay in one of the channels, and the anisotropy of the distribution of the coincidences in sidereal time, are significant arguments in favor of the correlated bursts in Weber's experiments. The absence of coinciding bursts above [a statistically significant level of 3 standard deviations] in our experimental scheme does not contradict the astrophysical estimates. (p. 111)

To translate, the Braginsky researchers are saying that though Weber's delayed signal comparisons and results favoring the galactic center make it look as though his positive results are sound, they had found what one would expect on the basis of extrapolated known theory (the astrophysical estimates)—that is, nothing. Thus their statement is much stronger than it first seems. Braginsky was from the outset adamant that Weber was wrong, expressing this view forcefully and unambiguously at conferences.[5]

Braginsky was less cautious than many of those who would report in the following year. 1973 saw the publication in the physics journals of papers by teams from IBM, Bell Laboratories, European Space Research Organisation laboratories in Frascati, near Rome, and Glasgow University.[6]

IBM—Garwin and Levine

The IBM team of Richard Garwin and James Levine ran a single bar that was smaller than Weber's, arguing that if Weber had really seen as much energy as he claimed, they should have seen something, too—about six non-noise events in their nine-day run, while, in fact, they detected nothing.

---

5. In 1998, Braginsky told me that Weber had informed him early on that he should change his mind about his claim about lack of coincidences if he did not want to make himself look foolish, but he had refused.

6. Levine and Garwin 1973, Garwin and Levine 1973; Tyson 1973a, 1973b; Bramanti, Maischberger, and Parkinson 1973; Drever et al. 1973.

Their results, published as two sequential papers in *Physical Review Letters*, depended on their doing their own calculation of Weber's sensitivity. They conclude,

> Our experiment suggests that (a) Weber's events of 1969–1970 were not produced by gravity waves, (b) an intense source or sources of gravity waves active in 1969–1970 is less active in 1973, or (c) the duration of these gravity waves is [longer than] 24 msec.[7]

They also said that "a more severe confrontation must await the operation of our larger bar" (p. 179), which they were then constructing.

Garwin and Levine left the gravitational wave field after 1975, although Garwin will reappear in this text as an assessor of the US interferometry project.

Bell Labs—Tyson

Tony Tyson from Bell Laboratories also reported the results from a single bar, "which has sufficiently improved sensitivity over Weber's apparatus to allow comparison with his two-detector coincidence results." Again, his conclusions depended on comparisons of sensitivity. According to his calculations for his much bigger bar, he should have seen at least 450 events, but he, too, saw nothing of any significance. Tyson concludes,

> We do not speculate here on what Weber is observing, but if seems unlikely that these events are gravitational radiation (splash, slowly swept, or otherwise).[8]

Apart from the further results he reports regarding room-temperature resonant bars, we will not meet Tyson again for another twenty years. In the early nineties he was prominent in opposing the building of the Laser Interferometer Gravitational-Wave Observatory (LIGO), and his statement to Congress can be found at the beginning of chapter 1.

---

7. Garwin and Levine 1973, p. 180.
8. Tyson 1973b, p. 329.

Frascati—Bramanti, Maischberger, and Parkinson

D. Bramanti, K. Maischberger, and D. Parkinson were at that time based at the European Space Research Organisation laboratories in Frascati. They also ran only one detector. They conclude as follows:

> Assuming pulsed gravitational wave sources [less than 0.1 seconds in length], we conclude, from our measurements, that Weber's results can be correct only if his instrumentation is able to detect jumps smaller than 0.1kT [a small number].[9]

They did see a number of energy jumps in their detector, but were able to explain all of them in terms of known errors, power failures, "creep" in the detector system, and so forth.

Glasgow—Drever, Hough, Bland, Lesnoff (and Aplin)

The Glasgow team, whose names appeared on the key paper, were Ron Drever, James Hough, R. Bland, and G. Lesnoff. Drever, who will figure large in our story, was to leave Glasgow in the early 1980s and become, for a while, the leader of the California Institute of Technology team involved in the development of large interferometers. Later he would be a member of the three-person team who initially ran the LIGO project. Hough was to remain in Glasgow and become joint leader of the German-British laser interferometer team that was to make major contributions to the science of interferometric detection. Though his tale is not as dramatic as Drever's, he is a very important figure in the continuing story of gravitational-wave detection.

The Glasgow team used a new design of detector in which the bar was split in two and piezoelectric crystals were sandwiched between the two halves.[10] This so-called split-bar design was invented by Peter Aplin of Bristol University and shown to have numerous advantages in theory. Aplin was later to invent a long-bar gravitational-wave detector and build a prototype, but he eventually realized that this idea was based on an incorrect theory and dropped out of gravitational wave field by the early

---

9. Bramanti, Maischberger, and Parkinson 1973, p. 699.

10. The Frascati group also experimented with a split bar (Bramanti and Maischberger 1972), but they were to conclude as they explained in their 1973 paper (Bramanti, Maischberger, and Parkinson 1973) that their own design, which contained embedded piezocrystals, rather than crystals glued to the surface of the bar, was nearly as good in practice.

1980s.[11] Nevertheless, Aplin is still credited with the very clever invention of the split bar, and many scientists have expressed their surprise to me that he did not get more credit for it. It seems that though the principles were first grasped by Aplin, the complete theory was worked out by Stephen Hawking working with Gary Gibbons. Many scientists are surprised that Aplin was not given co-authorship of the paper that Gibbons and Hawking published in 1971.

Returning to the Glasgow experiments, Drever's team report the following:

> We conclude that it is unlikely that the signals reported by Weber in 1970 were due to pulses of gravitational radiation of duration less than a few milliseconds, assuming that the source has not changed significantly since then. But our present observations do not exclude the possibility that Weber may have detected bursts of gravitational radiation of much longer duration.[12]

The Glasgow team compared their runs only with Weber's results for the period May 20 to November 20, 1970, as they did not think they knew enough about the sensitivity of Weber's apparatus at other times. It is also interesting to see the Glasgow team's comments on the previous results that I have discussed.

> An unsuccessful search for gravitational radiation was reported about a year ago by Braginsky et al., with detectors of comparable sensitivity to those of Weber. Very recently, while the present paper was being prepared, null searches with single detectors have been reported by Tyson, using a large and sensitive detector having, however, a pass band which did not overlap with any of those of Weber's experiments, and by Garwin and Levine with a smaller detector of lower sensitivity. (p. 343)

One can see that the Glasgow group were not as convinced of the power of the experiments of Tyson and Garwin and Levine as were the authors themselves, while they make no comment on Braginsky.

The Glasgow paper also contains a positive claim, albeit in respect of only one gravitational wave pulse. The team found on their output one coincidence that satisfied their criteria for a real effect.

---

11. Aplin (1976) provides an account of his own contributions.
12. Drever et al. 1973, p. 344.

This is our best candidate for a true gravitational radiation event. It registered equivalent energies of 0.64 and 0.61kT [quite high and quite similar] on detectors 1 and 2, respectively, and was recorded on September 5, 1972, at 13 h 07 min 29 s UT. (p. 342)

... At present we know of no experimental reason to reject the hypothesis that this signal was caused by gravitational radiation. (p. 343)[13]

Twenty-five years later, Jim Hough was to express to me his regret that this published provisional claim did not lead to anything.

Thus, by 1973 we have five negative publications, though the latter four contain reservations which make them less than full-blooded confrontations with Weber. The showdown was not to happen for another year.[14]

13. In this paper they also discuss Weber's "triggered metastable states" idea and find it does not work.

14. For the sake of completeness, I must mention several other groups. First, a Japanese team with a rather different kind of resonant detector consisting of a large square plate with slots and holes, and searching at the 145-Hertz frequency of the Crab pulsar, found no positive results (Hirakawa and Nariha 1975); understandably, the Hirakawa group's results were never mentioned to me in interviews as significantly confronting Weber's work. Mentioned to me once was the work of a group in Paris. They set up an experiment, but no report of it or reference to it will be found in the literature except for a few lines in an abstract in the published report of the Warsaw conference (Bonazzola, Chevreton, and Thierry-Mieg 1974). In Paris, starting in 1972, Silvano Bonazzola built three bars in the same tank, each weighing about a ton. He did this, he explained to me, with a view to disproving Weber's claims (the interview was conducted in Tirrennia, near Pisa, in September, 1998). He conducted a coincidence search with bars in Frascati and Munich and found nothing. He gave up in 1974 and moved on to other, more theoretical things. Members of the community knew about his results, but if they were cited in the literature, it was not in the mainstream. The Bonazzola results do not seem to have any sociological significance. The same applies to the work of other groups in Regina Canada who began a program using quartz crystals as the resonator, and groups in Guangzhou and Peking using, at least in 1972, room-temperature bars. The results of these groups were never mentioned to me (Stayer and Papini 1982, R. Hu et al. 1982, and E. Hu et al. 1986).

CHAPTER 8

JOE WEBER FIGHTS BACK

NATURE, 1972

In November 1972, Weber's fight back against his critics began to appear in the journals. *Nature* published an analysis of his data recorded between October 1970 and February 1971 which seemed to satisfy one complaint of his critics, in that his analysis had been done by computer. This paper represents his first major publication since 1970 to address the main matter of his contested findings.

Two features of this paper are especially important. First, as indicated, Weber stresses that the analysis was done by computer without the intervention of human judgment to determine what constitutes a satisfactory threshold crossing. He describes the computer program in considerable detail to turn aside critics who claim that his method of having an assistant inspect charts by eye, even though the assistant is working "blind," leaves the result open to doubt. At this distance in time, it is very hard to imagine that anyone would use the human eye as the principal discriminator in an experiment as delicate as this one, but in 1970 computers were expensive and of limited capacity, and the skills to use one in an application such as this were not so widely available. Further-more, it was not at all clear back then that the computer was the better tool for making a discrimination. Bob Forward once explained this to me in the context of the problem of detecting a submarine using a sonar scanner. He pointed out that the form of the pattern in a sequence of sonar traces could not be predicted in advance, thereby making it impossible to be sure that a computer could be programmed to detect it.[1] Nevertheless, Weber had been persuaded that his results had to be analyzed by computer if they were to be convincing.

One may wonder why it took Weber so long to begin to analyze his data using the techniques then thought proper in at least some branches of physics. This might have had to do with his training as a radar operator and commander of a submarine chaser. In these roles the job is to tune the apparatus in order to find any vestigial signal. While the price of not finding a vestigial signal can be death, finding a false signal costs only time and effort. In science, of course, it is different. Failing to find a vestigial signal leaves the experimenter with nothing, positive or negative, whereas finding a signal that is not there carries a high cost in time and credibility. All Weber's instincts and training were directed toward finding the vestigial signal, and much of his experimental practice could be said to follow the same guiding principle. His complaint was that others did not try hard enough to find that (hidden-submarine-like) signal!

The second innovation in the *Nature* paper—and this was to have a last-ing impact on the field—was the introduction of the delay histogram, a very attractive graphic device for expressing the significance of a coincidence re-sult. In his 1970 paper, Weber gives his results in the form of a table with columns. One column represents the *calculated* number of spurious "events" that would be expected to appear due to coincidental peaks of noise in the two detectors caused by chance alone, and another column signifies the number of such chance coincidences *measured* by introducing a time delay in one data stream as compared with the other: where there was a time delay, coincident peaks could not be caused by coincident outside causes; therefore such apparent events could be caused only by chance. The co-incidences caused by chance were called accidentals. The numbers in the *measured* column representing "accidentals" roughly agree with the num-bers in the *calculated* column representing accidentals, while the scientific

---

1. The problem remains to this day; computers are still very bad at recognizing patterns against a background of noise and are likely to be so for the foreseeable future. To understand this, one only has to ask which would be better at interpreting a poorly recorded word in a taped conversation: a human or a computer? The answer is going to be "the human" for a very long time to come.

data are in the third column. The third column shows measured coincidence with no time delay and lists numbers of events far in excess of those in the previous two columns.

In 1970, Weber had decided to delay one data stream in respect of the other by two seconds. But why two seconds? The delay could be anything at all so long as it was large compared with the ringing time of the bar.[2] Any delay from two seconds to two hours or more could serve exactly the same purpose—the choice was arbitrary so long as neither of the two bars could have drifted too far, in terms of sensitivity, from the state at which real-time coincidences were measured during the period of the delay. Furthermore, since the data from both bars were recorded, the noise *measurement* could be made at any time after the data had been gathered. Thus the tapes could be run and rerun with different delays inserted between the two data streams. This gave rise to the idea of the delay histogram, which showed the result of introducing a whole series of such delayed comparisons with delays of different values. Figure 8.1 is a schematic explanation of how a delay histogram works.

In figure 8.1, one stream of data is labeled "A," and the dotted line shows a threshold with six peaks crossing it. Data stream B, as it happens, also has six peaks crossing the threshold while the little circles mark the four of them that coincide in time with threshold crossings in A. What counts as a "coincidence" is a matter of the width of the vertical stripes which represent time periods. B is shown advanced by one and then two seconds and then delayed by one and then two seconds. Coincidences between A and B in the case of the four time offsets are also indicated by small circles. The data are assembled on the delay histogram at the foot of the figure. This has only five bins, but such histograms can have as many delays as one cares to calculate; they usually have many more. This "toy" histogram shows a "zero-delay excess" over the mean background noise (the mean height of the four non-zero-delay bins) of 2.

Figure 8.2 is a rough reproduction of the delay histogram labeled "Fig. 1" in Weber's 1972 *Nature* paper. The horizontal dimension shows delays of various lengths, from one second up to 40 seconds, with negative delays to the left and positive "delays" to the right.

Weber explains that something is wrong with this figure. He says that the computer program was not accurate for delays greater than twenty seconds, and that the outer bins underrepresent the number of accidental

2. The crucial importance of the delay being large compared with the ringing time of the bar will become clear later.

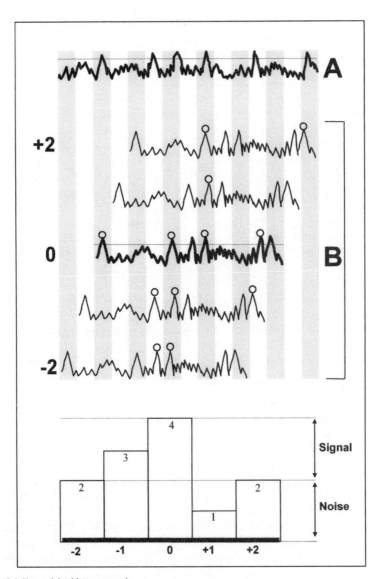

Fig. 8.1 How a delay histogram works.

coincidences at these delays. This is just as well, since there is no reason why the number of "accidentals" should become fewer as the delay interval increases. The rough level of accidentals should vary around a constant no matter how far we move away from the central bin. The really convincing thing, if one is prepared to be convinced, is the height of the zero-delay

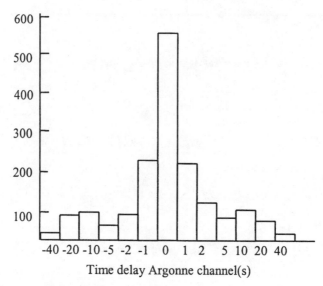

Fig. 8.2 Delay histogram from Weber's 1972 paper (fig. 1).

bin above the delayed bins. Weber says of the results represented on this diagram,

> These data, untouched by human hands, leave no doubt whatsoever that the gravitational radiation detectors, separated by 1,000 km, are being excited by a common source.[3]

Quite properly, Weber points out, "Before we can conclude from Fig. 1 that the source is gravitational radiation, it is essential to rule out other presently known interactions including seismic, electromagnetic, and cosmic ray effects." He then goes on to explain the checks made and the reasoning against the coincidences being caused by common excitation from seismic, electromagnetic, and cosmic ray sources.

What worried some scientists about this paper was that the evidence for the sidereal anisotropy was no longer there—the center of the galaxy was no longer the preferred direction of the source. Weber explains,

> Small effects in the local environment can affect the anisotropy in a significant way. To average these properly at least six months of data are required. The magnetic tapes considered here did not have times

3. Weber 1972, p. 29.

continuously written. Gaps and recorder failures left insufficient data for a
study of the anisotropy. (p. 30)

He concludes,

> The present data are free of human observer bias. No assumptions are made
> concerning the duration of expected pulses or their shapes....It is
> considered established beyond reasonable doubt that the gravitational
> radiation detectors at ends of a 1,000 km baseline are being excited by a
> common source as a result of interactions which are neither seismic,
> electromagnetic nor those of charged cosmic rays. (p. 30)

PHYSICAL REVIEW LETTERS, SEPTEMBER 1973

A little less than a year later, in September 1973, Weber, with five collaborators, published "New Gravitational Radiation Experiments" in *Physical
Review Letters*.[4] In this paper, three more delay histograms are presented,
covering data gathered in 1973. It had been suggested that the telephone
that linked the Argonne and Maryland sites might be a source of spurious
coincidences, but Weber argues that the electrical capacity of the line is too
small to produce an effect. He also says that the telephone line was disconnected for a later series of runs, and the coincidences remained (though no
data are given). He concludes that the current experiment is yielding about
seven real coincidences per day. Figure 8.3 is an approximate reproduction
of one of the figures on page 780 of the *Physical Review Letters* paper.

The other two delay histograms in the paper are broadly similar. It
should be noted that these histograms differ from that of the *Nature* paper
in several ways. First, a horizontal line is drawn showing the number of
measured "accidentals," the mean height of the nonzero-delay bins (414 in
the diagram reproduced here). There is, of course, variation in the height
of these delayed bins—just as one would expect as a result of the play
of chance. Correspondingly, the central zero-delay bin is shown with an
"error bar"—the vertical line—which indicates the unreliability in the excess
height of the central bin.[5] It is also worth noting that the vertical scale has
been exaggerated in effect, since the bottom of the bins lies at about 350

---

4. Weber et al. 1973. (One of the authors is Virginia Trimble, Weber's second wife.)
5. Though at what statistical significance I am not quite sure—I would guess a one-standard-
deviation error is shown.

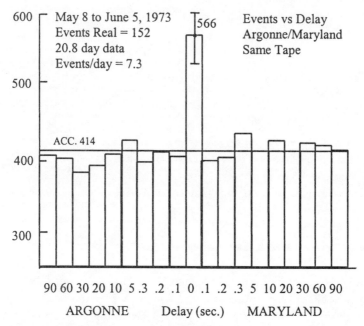

Fig. 8.3 Delay histogram from Weber's 1973 paper (fig. 1C).

accidentals; to get a true visual impression of the height of the central bin above the background noise, one should imagine the bottom of the figure extended downward so that the noise bins are more than three times as high. All that said, the way that the central bin stands proud is very convincing. A glance at the diagram produces a strong conviction that some external source is having its effect on the experimental apparatus. And we are able to see that the height of the central bin represents about 566 − 414 = 152 "real" coincidences over 20.8 days, which is indeed a real coincidence rate of 7.3 per day.

There are two other differences between this histogram and the one published in *Nature*. The odd tailing-off of the noise bins is no longer there—they are roughly of the same height however far away from the center one moves. More problematic, as we will see, is that the noise bins drop to the average noise level as soon as one moves from the center bin. If one looks at the figure from the *Nature* paper—figure 8.1—one sees that the bins adjacent to the zero-delay bin, those at .1 second delay, stand proudly above the other noise bins, representing what one might call "shoulders" on the zero-delay bin. In the *Physical Review Letters* paper, there are no shoulders—even at a delay of .1 second the delayed bins are at the average level. This was going to cause Weber trouble.

KAFKA AND TYSON

Behind the scenes, visible to very few scientists, Weber was arguing about the meaning of two other experiments. He often spoke favorably of the Munich-Frascati experiment, because it tried to copy his own design as nearly as possible. Thus, toward the conclusion of his address at a conference in Tel Aviv in June 1974, Weber said, "On the whole our results are in fair agreement with the observations of the Munich-Frascati group." In December 1974 he would claim in print, "It is interesting that the one other group that has observed a statistically significant number of coincidences (Munich-Frascati), employed two detectors having the same mass as the Maryland antennae and similar instrumentation."[6]

In the last remark, Weber is referring to a coincidence experiment that found some apparently positive results. He was not alone in reading a positive conclusion into the work of these groups. Thus in a paper published in September 1974 and translated from the Russian, the translator, M. E. Mayer, added a note to the version published in English: "*Translator's note.* The Munich-Frascati experiment has recently (March 724, 1974) reported coincidences at a level of 3 standard deviations (H. Billing, P. Kafka, K. Maischberger, F. Meyer, and W. Winkler, Munich Preprint, June 1974.)"[7]

These results were relayed to Weber at a landmark conference: the Fifth Cambridge Conference on Relativity (known as CCR-5) at MIT on June 10, 1974. In 1976, this is how the matter was described to me by Billing and Winkler, the German half of the experiment (Winkler's remarks in curly brackets):

> [1976] At this time we had a short period, of about a fortnight, when we had significant results in the sense of Weber, but just for this fortnight. And I spoke to Weber before the conference and he was very very happy to see these results. And I told him we do not believe that this is significant. {You see this is always a problem of statistics. You cannot put the finger on some event and say, "There was a gravitational wave." What we do is always look around in the noise, and you always have fluctuations, either in the positive or negative direction. And for this period we had a positive peak at the zero time delay with a peak of 3.6 sigma.}
>
> And in this lecture I gave I said that this could be a hint that Weber's results are right, but it could be just accident. {And as this was just at the

6. The quotations are from Shaviv and Rosen 1975, p. 254, and Weber 1974, pp. 11–13.
7. Braginsky et al. 1974, p. 84.

beginning, we did not have very much data, we had some, but not so many, and we did not believe that this was due to gravitational waves.}

The experiment itself was carried out by Maischberger in Frascati using one bar, constructed independently of the second bar in Munich which had been built by Billing and Winkler.[8] But it was Peter Kafka who was given the chief responsibility for analyzing the data. Kafka had already written a devastating theoretical critique of the possible existence of gravitational waves of the magnitude apparently seen by Weber (see chapter 5); he had already suggested that Weber's correlations could be explained away as due to correlations caused by the telephone line that linked the apparatuses.[9] Weber realized that for Kafka to find results so blatantly contradicting his already published theoretical findings would be embarrassing, so he reached the conclusion that Kafka would put most of his effort into explaining the findings away.[10] Weber was very disappointed that Kafka would not allow the data to be used to support his claims. Indeed, in Tel Aviv, Kafka responded to Weber by projecting a slide onto a viewing screen and declaring,

> ... Joe Weber thinks [this picture shows] we have discovered something too. This is for 16 days out the 150 [of total observation time]. There is a 3.6 [SD] peak at zero time delay, but you must not be too much impressed by that. It is one out 13 pieces for which the evaluation was done, and I looked at least at 7 pairs of thresholds. Taking into account selection we can estimate the probability to find such a peak accidentally to be of the order of 1%. Hence, before I say definitely that we have not found any evidence for Weber pulses, I shall study carefully the fluctuations in time.[11]

Kafka would report the following in March 1975:

> The surplus near zero time delay, which appeared for a period of two weeks in March 1974, seems to have been accidental.... The "accident" was, that we

8. It was only after the two separate groups had nearly completed their apparatuses that they discovered each other and realized that they both had the same design philosophy: to build a bar that was very close to Weber's original design so as to avoid the charge that their failure to see gravitational waves had to do with differences in design. The two groups did then share design details with each other and agreed on a joint coincidence run.

9. Kafka 1973.

10. For a study of the different ways in which effort can be expended in explaining apparently unbelievable results, see Collins and Pinch 1982, esp. pp. 114–21.

11. Shaviv and Rosen 1975, p. 265.

had inspected a small selection of pairs of thresholds for a short period of observation (defined by the length of a tape), and that up to this time, the largest deviation printed out was near zero delay. Now, after inspection of the full results (as stored on magnetic tape), this feature does no longer seem unlikely. We always stressed that we did not believe to have discovered real events! Nevertheless, some authors and translators seem to have misunderstood us. [He refers to Weber's comments at Tel Aviv and to the paper published by Braginsky et al. in 1974 which contained the translator's note quoted above.][12]

In 1976, I asked Billing and Winkler, the experimenters, if they could re-member the very first reaction of the team to the discovery of this brief 3.6-standard-deviation peak. One of them replied, with a laugh,

> [1976] I would say there was a positive reaction by the people who had done the experiment, and there was a negative reaction by the [other members of the same group].

Weber also believed that Tony Tyson at Bell Laboratories had results that confirmed his findings but was refusing to publish them. He thought that positive results were to be found in coincidence runs between the Maryland and the Bell Laboratories detectors. But this collaboration had been set up under a confidentiality agreement, and Weber was not prepared to break it unless he could first persuade Tyson to agree. Tyson would not agree. Weber was now making his case to senior members of his department. On December 3, 1974, he wrote a letter to colleagues that set out his claims of a positive result in the BTL (Bell Telephone Laboratories)-Maryland data. He says, "One 4 day period had a 5 standard deviation zero delay excess. The best result for the entire 16 days is better than 4 standard deviations." He agrees that the remaining two months of data show no positive result but argues that this could easily be due to poor temperature control in the BTL antenna. The data seem to have been collected in April and May 1974.

At the end of 1974, it was agreed that a delegation of Maryland physi-cists (including Weber and senior members of his department) would visit Bell Labs to discuss the matter; this visit took place on December 5.

---

12. Kafka 1975, p. 109. A rather similar account is given in Billing et al. 1975. The Braginsky paper referred to is Braginsky et al. 1974.

A subsequent letter to Weber from members of the delegation, dated December 9, states,[13]

> Dr. Tyson agreed during the meeting to study his own tapes for those periods in which a significant zero delay excess was indicated at Maryland. Following the large meeting, Dr. Tyson took the Maryland contingent to his office for about a half hour. While there, a study of his log indicated that a recorder malfunction resulted in loss of the first eight days of data. Other problems with Tyson's equipment, maintenance, and calibration schedules apparently resulted in the unavailability of all other data covering the fifty days of Weber's analysis except for a few days.

Weber told me more than once that Bell Laboratories, a telecommunications company, was embarrassed about failures of communication that were its responsibility and that this was one of the reasons it did not want to talk about this run of data.

The Maryland delegation was clearly disappointed that no data were available from Tyson's laboratory for the period in question, but they did not give Weber what he wanted. One senior member of the department did write back on December 17, explaining that he thought Weber's analysis of the BTL-Maryland data was convincing on the surface, but that such results would have to be confirmed by someone else before they could gain credibility. He wrote to Weber, "The careful documentation which you should now provide, together with a similar effort on the part of others should allow for definitive comparisons with an objective resolution of the question."

Weber was to continue the debate, and my collection of letters includes letters from him to Bell Telephone Laboratories senior personnel up to late 1975. For instance, Weber wrote on November 3, 1975, to Tyson, enclosing data on the coincidences and concluding, "While these data were taken 'for our edification' the results are of wide interest and deserve publication. We are proposing publication, without making claims, together with you as co-author." Tyson was not to take up the offer.

I include these discussions of the Munich-Frascati results and the BTL-Maryland results not to show that Weber had been "cheated" but to gainsay any impression that the data of his critics never gave any positive indications of the existence of gravitational waves. And we should not forget the one positive indication by the Glasgow group, not to mention the stream

---

13. Since Weber was present at Bell Labs, this must have been written "for the record."

of positive results to be produced in due course by Weber's allies. My idea is to try to show that the output of the experiments was not as uniform as might be thought when we look back on them with hindsight, nor is the criticism against Weber so uniformly damning as it now seems.[14] For example, over the course of this debate, Tyson made a mistake about the sensitivity of Weber's amplifiers, discovered a completely irrelevant geophysical source for Weber's signals, and refused to agree even to publication of some putatively positive results. This is likely to suggest something of a bias to the person who is on the receiving end. Weber certainly did feel that certain groups, including those of Kafka and Tyson, were so fearful of ridicule that they were determined to find a way of discounting any data that would back up his claims. As he told me,

> [1975] I can't think of ever setting up a complex experiment that worked well when it was first turned on, and right now if Douglass were to publish a paper confirming all our previous results, he would find himself in mortal combat with Garwin. . . .
>
> You really have to have motivation to do a complex experiment and go through the agony of getting everything right. Tyson learned in 1972 that if he were critical of us, he was swamped with invitations to give talks, and his personal situation improved if he was critical of us. Kafka was critical of us long before the Munich-Frascati experiment was put together—I just don't see how you can expect significant changes in that situation.[15]

THE BEGINNING OF THE END

In December 1974, Weber submitted a fourteen-page paper to *Physical Review D* that was meant to summarize and reinforce his claims. It was not to be published until August 1976,[16] indicating a very long refereeing process. The trouble, we may surmise, had to do with the next round of criticisms that were about to explode. We will return to this paper and its reception in chapter 11, after looking at the response to his 1972 and 1973 publications.

---

14. See, for example, Franklin (1994), who writes, "I have found no reports of positive results with a Weber bar detector by anyone other than Weber and his collaborators."

15. For more comments on this interview, see the section entitled "Joe Weber Is Dead, and My Life as a Sociologist is Harder" in chapter 44.

16. This is Lee et al. 1976.

CHAPTER 9

THE CONSENSUS IS FORMED

As we have seen, in 1972 about a dozen groups were engaged in active experimentation directed at confirming or rejecting Weber's findings. Over the next three years, everything changed. By 1975, when I conducted a second set of interviews, Weber's work was about to enter a limbolike state, during which it would be ignored by almost the entire mainstream gravitational wave community—the ripples in social space-time simply stopped spreading.[1] Though a few teams were still finishing up their work in a ritualistic kind of way, by 1975 no one was putting serious effort even into rejecting Weber's experimental claims. This is not to say the gravitational wave field was quiet—far from it—but scientists were now looking for something very different—something a million or more times weaker that would fit the astrophysical theories. Even Weber faced problems over financial support for his own experiment. By 1975, then, scientists had decided to their satisfaction that Weber had not detected gravitational waves, and seven groups were now building, or considering the design of, antennae of far greater

1. As we will see, it would briefly emerge from this limbo ten years on.

sensitivity. To use the phrase of one of my respondents, 1975 belonged to "the post-Weber era."

The "postness" of the post-Weber area was signaled in the constitutive forum not by papers rejecting his findings outright, but by an absence of papers discussing his work at all. In science there are, one might say, three stages in the evolution of what will one day become a rejected radical claim. At stage 1 the claim is ignored: the struggles of the unborn embryo are hidden in an integument of silence, and no ripples emanate through social space-time. Weber forced his way out of the chrysalis with his series of ingenious additions and modifications to his data, especially the correlation of peaks with sidereal time and the use of the delay histogram to present results. Stage 2 is active rejection, the stage we find ourselves in at this point of the book; here the adolescent creature is assailed from all sides. In the third and most terrible final stage, the claim is once more ignored: by the end of this part of the book, we will see that Weber's main findings had reached this point. To change the metaphor, by the end of 1975 or so, they were floating coldly, lifelessly, and invisibly in space, where "no one can hear you scream."[2]

RESULTS OF OTHERS' EXPERIMENTS (AND CALCULATIONS OF SENSITIVITY)

But how did we get from the Weber era to the post-Weber era? It almost goes without saying that the almost uniformly negative results of other laboratories were important—a great weight of empirical evidence was mounting against Weber. But complete certainty cannot emerge from experimental work in novel areas because of the experimenter's regress. Because Weber had given more time, effort, and dedication to his experiment than anyone else, he could still claim that his was the best. This was understood by both Weber and some of the others in the debate. Thus, one respondent reported,

> [1975] [A]bout that time [1972], Weber had visited us and he made the comment, and I think the comment was apt, that "it's going to be a very hard time in the gravity wave business," because he felt that he had worked for ten or twelve years to get signals, and it's so much easier to turn on an experiment and if you don't see them, you don't look to find out why you don't see them, you just publish a paper. It's important, and it just says, "I don't see them." So he felt that things were going to fall to a low ebb...

2. I associate this phrase with Ridley Scott's film *Alien*.

Another experimenter, who had worked with Weber and was sympathetic to him, commented,

> [1975] [A major difference between Weber and the others is that Weber] spends hours and hours of time per day per week per month, living with the apparatus. When you are working with and trying to get the most out of things, you will find that, [for instance], a tube that you've selected, say one out of a hundred, only stays as a good noise tube for a month if you are lucky, but a week's more like it. Something happens, some little grain falls off the cathode, and now you have a spot that's noisy, and the procedures for finding this are long and tedious. Meanwhile, your system to the outside, looks just the same.
>
> So lots of times you can have a system running, and you think it's working fine, and it's not. One of the things that Weber gives his system that none of the others do is dedication—personal dedication—as an electrical engineer, which most of the other guys are not...
>
> Weber's an electrical engineer, and a physicist, and if it turns out that he's seeing gravity waves, and the others just missed it, that's the answer, that they weren't really dedicated experimenters.... Living with the apparatus is something that I found is really important. It's sort of like getting to know a person—you can, after a while, tell when your wife is feeling out of sorts even though she doesn't know it.[3]

We can contrast these comments with the remark of one of Weber's critics, who admitted, [1975] "It's one of these fields in which people do it with their left hand," going on to explain that it was not an area on which graduate students could found a career. This meant that scientists with graduate students had to keep other projects going as their main contribution to science so that they could have something more sensible for their protégés to work on.

If we do look at the experiments, we find that all but one of those with negative outcomes were criticized by one or more of Weber's *critics*— yet these are people whose biases might be expected to lead them to favor other experiments that yielded negative results.[4] Thus, in the course of

---

3. Weber's own comment along these lines can be found at WEBQUOTE under "Weber on Dedication."

4. Except that even scientists who agreed about the absence of gravitational waves would prefer it if they themselves came to be seen as the ones to have conducted the uniquely decisive negative experiment. This works against their overlooking deficiencies in their rivals' work even when they agree with their rivals' results.

my 1975 interviews, one respondent spoke of the experiment conducted by Vladimir Braginsky, the earliest of Weber's forthright critics.

> [1975] I'm not too sure that too many people were very strong on Braginsky's negative results, mainly because he had only five or six days of data— something like that—not a lot more. . . . He said that if he had seen events—if the events that Weber claimed had been there—he would have seen them.

Another experiment, that of the University of Rochester group run by Dave Douglass, was widely thought to be troubled by "creep" in the metal of the bar which was producing so much spurious noise that it could not do a clean search for any signals that might be there.

The experiment of J. Anthony Tyson at Bell Laboratories was criticized because, like the Rochester experiment, it ran at too low a frequency to be comparable with Weber's experiment and therefore could not make a head-on confrontation with Weber's results. In any case, as we have seen, Tyson had been criticized in print for making claims about the sensitivity of his apparatus which turned out, in most scientists' view, to be unsupportable, so his credibility had been at least a little damaged.[5]

Ron Drever's experiment at Glasgow University was thought to be clever but small, and of such a radically new design—the split bar—that it could not be directly compared with Weber's. The IBM experiment, run by Garwin and Levine, was described to me as a "toy."

Only one experiment remained immune to criticism by Weber's critics, and this was the bar built in Munich by the team associated with Peter Kafka, running in coincidence with a bar in Frascati and designed to be as near as possible to a carbon copy of the original Weber design. Indeed, such criticisms of this bar as there were pointed to its being unimaginative in its slavish copying of Weber's apparatus. Weber's criticism of it, however, turned on differences in the signal-processing algorithm. All Weber's competitors believed he had made a mistake in his choice of signal-processing algorithm, but Weber stuck doggedly to his view. He insisted that the maximum net sensitivity was to be obtained by a nonlinear, or energy, algorithm (the algorithm relates to the circuitry and computer program that processes the raw signal). Weber's critics insisted that a linear, or amplitude, algorithm was the best, and they uniformly made use of it.

---

5. This criticism, as we have seen, was made by Garwin and Levine (1973, p. 178), Weber's most vituperative critics.

When we turn to the other things that were causing critics to lose faith in Weber's work, we find that the loss of the sidereal correlation is salient. This feature simply disappeared from his data, and we can see from remarks recorded above that at least one or two of his critics but not all would count this loss as fatal. Another reason for Weber's loss of credibility was the apparent superiority of other devices in response to calibration pulses; this I will treat at greater length in chapter 10. Yet another serious problem was that Weber had not managed to increase the signal-to-noise ratio of his results even though his apparatus was becoming more sensitive. In fact, considering that his apparatus was undergoing continual improvement, the net signal seemed to be decreasing. This was not felt to be typical of emerging scientific work.

> [1976] *First respondent:* Weber increased his sensitivity, too, and his results are always just on the limit of his sensitivity, and this makes one a little bit suspicious . . . if you are always on the limit, you don't believe this is a real thing. And we are rather sure when Weber had some real events, we must have seen it with our experiment very strongly coming out. . . .
>
> *Second respondent:* If you reduce the noise with two apparatuses with different noise, with one at the limit of significance, then the other one should find very significant results. And that is the main reason why we don't believe in Weber.

Nonetheless, these mounting problems did not have to be fatal, especially as several of his critics were unimpressed with them. Thus, Braginsky aside, criticism of Weber in the constitutive forum was still qualified. The difference between one experiment and another let enough water leak from the dam to allow Weber to keep a foothold on the island of credibility. Physicists' minds had been opened to new possibilities over the years of excitement about Weber's findings, and to make an unambivalent declaration that there was nothing there was still to risk being on wrong end of the one of the great scientific discoveries of the age. But things were to get worse.

COMPUTER ERROR AND FOUR-HOUR ERROR

Weber, unfortunately, made mistakes still worse than the confusion over the sidereal periodicity (24 hours versus twelve hours). The first of these was a technical error that exaggerated the count of coincidences due to a flaw with the way the computer counted signals at the margins of time intervals. Weber acknowledged that this fault was responsible for the majority of the reported signals in one of his runs. After some delay, he changed the program and, as he put it, corrected some "additional errors," with the net result that the same run still produced a statistically significant result. This kind of move does not inspire confidence, because it looks like post-hoc statistical manipulation.

Suppose you buy a raffle ticket and win. If the odds are very long, as in something like the National Lottery, the prize is very valuable. If you win an ordinary raffle, which has lower odds, the prize is less valuable. You can always lower the odds by buying more tickets, but the ratio of the value of the prize to the cost of entry is reduced. If you win a raffle by buying all the tickets, nothing will have been gained.

The statistics used in science are rather like this. Finding a significant signal in your data is like winning a prize. The higher the odds against winning, the more valuable the prize in terms of the information it provides about the nature of things. Therefore, to know how much your prize is worth, you have to know the odds against winning. The assumption on which a calculation of statistical significance is made is that the data is looked at only once (only one raffle ticket is purchased). Thus, if you look at the data and get a two-standard-deviation level of significance, this means that, if the data extraction was fair, the result would have occurred due to a chance selection of the data only one time in twenty. This, as we have said, is a result worthy of esteem in some sciences but not in others, but the same reasoning applies in principle for a three-standard-deviation result and a seven-standard-deviation result. But suppose you are hoping for a two-standard-deviation result and fail to get it. You might then take the data and do another slightly different kind of calculation on it. But this is like buying a second raffle ticket—it reduces the odds against your winning and reduces the value of the result. The trouble is that the outcome of the statistical calculation will look the same—a two-standard-deviation result, because no one knows you have bought the extra ticket—but the value of the result is less than it seems.

That is why scientists are extremely suspicious if they find out that statistical calculations are being made in many different ways on the same piece of data. Indeed, if you allow the result you want to guide the way you arrange the data, you can favor yourself even more strongly; this is like looking in the barrel for the right number before you take your raffle ticket. But, once more, the statistical outcome will still look the same to someone who doesn't know you have been delving about in the sawdust barrel. Rearranging data until a result emerges is known as statistical massage. It is rarely done deliberately, but it can happen quite unconsciously during the proper process of examining data repeatedly for some hidden signal.

As one can see, a strange feature of statistics is that what you see is not what you get; it is only the tip of the iceberg. A published result, in a domain where statistics are applicable, typically includes the statistical significance of the result. What it does not include is the *history* of that result—all the things that were tried out before the result emerged. And this is good, because that history would probably be long and tedious. Yet to understand the true statistical significance, you need the history.

In making his computer error, Weber had given scientists grounds for thinking that they could not trust his statistics without knowing the history of each result. He seemed to have shown that he was capable of unconsciously massaging his statistics, which could have generated a seemingly statistically significant signal from streams of noise. There were plenty of opportunities to enhance the statistical significance of the output artificially because of the number of choices available to the analyst. For example, since the threshold must be chosen, there is the opportunity to try different thresholds until one occurs that produces a good excess of zero-delay coincidences. This is like searching the barrel of raffle tickets for the winning number, so it invalidates the statistics. The issue is nicely put in a letter written to Weber by another experimentalist with whom he was disputing the analysis of a tape of data.

> [1974] I do not deny that a different analysis could produce a zero delay excess on this tape (or any tape).... Given that the criterion for a coincidence, the thresholds and the bin width are now all free variables, I am surprised that the zero delay excess was not made larger than you now report it to be.

Another scientist wrote the following to me:

> [2001] Weber's statistical methods were too flexible, allowing trial-and-error adjustments of thresholds and timing offsets to maximise coincident detections.
>
> ... during a private meeting in Maryland, [Weber made] the observation that his computer programmer was not very good [and] his original program could not find coincidences and had to be rewritten many times ... until they finally worked.... Joe showed me a printout of the FORTRAN program. It was

several inches thick, because it included a vast number of subroutines which had been tried in a[n]... attempt to find coincidences.[6]

Worse still, Weber conducted an accidental experiment on himself that proved for all to see that he *could* get something very close to a signal out of pure noise. He ran a coincidence test between his bar and the one set up by David Douglass at the University of Rochester.[7] The method was for Douglass to send his data tapes to Weber, who would then compare them with his own. To perform a coincidence run with someone else's apparatus is more convincing than doing it with two bars controlled by oneself alone; a lot of the artifacts and possibilities for bias are eliminated. One might say that data from experiments run by two scientists using two apparatuses are separated by a "moral space" as well as a physical space. If coincident signals are shown to be present in data streams from two such profoundly independent detectors, the rhetorical force is strong.

With Douglass's agreement, Weber did such a comparison and did indeed find a positive result at a level of 2.6 standard deviations (meaning that if the data were fairly produced and selected, such a result could be due to chance only about one time in 100). In the social sciences, 2.6 standard deviations would be counted as an excellent level of statistical significance, and that Weber broadcast the outcome on the conference circuit suggests that he felt it meant something in physics, too. Unfortunately, it then turned out that Weber had made a mistake about the clock settings. There was a small error of 1.2 seconds and a large error of four hours due to the different time standard used in his and Douglass's laboratory. Thus Weber had generated a 2.6-standard-deviation zero-delay signal out of data streams that actually had a delay of four hours and 1.2 seconds: he had generated signal out of what should have been pure noise!

In making this mistake, Weber gave a terrible impression to all those in the core set who were trying to work out which way to jump. But note that there was nothing in the *logic* of the situation, in which we will include the force of experiment, that proved that Weber had not seen gravitational waves; he had simply made some terrible errors regarding *one* of his many claims. Only because the logic of the situation cannot be decisive, for all the reasons that we have been discussing, could these mistakes have seemed, to some, to be so important. It was their way of finally reaching a conclusion

6. I have edited out evaluative phrases and adjectives so as to leave the bones of the description.

7. When I first published this story, these names were withheld, but the identity of certain of the participants has been revealed by others (e.g., Franklin 1994), and in any case the incidents took place so long ago that the time to preserve anonymity now seems to have passed.

for themselves, and as we will see, the force of this conclusion becomes more inescapable the more the details of it are known. But it is a conclusion about the man being careless in once instance—the conclusion about the phenomena has to be inferred from the conclusion about the man.

But even this remark has to be qualified; some scientists thought these errors were just unfortunate mistakes. They were not willing to make the errors known beyond the small circle of insiders in the core set, and they felt that the errors were of such a personal nature that it would be improper to mention them anywhere near the constitutive forum. One scientist, however, had decided that enough was enough. He wanted to make sure that the ripples in space-time died immediately before more people were misled and more taxpayers' money was spent.

Crystallizing the Evidence

Richard Garwin is an extremely well known and powerful scientist and an advisor to many high-powered committees. He held the prestigious post of IBM Fellow at the Thomas J. Watson Research Center, roughly an hour's drive north of New York City, and collaborated with James G. Levine on the gravitational wave work. He and Levine had built one of the smallest antennae, though they argued that it was at least as sensitive as Weber's because of its sophisticated design. Nevertheless, most of the other scientists I spoke with dismissed the Garwin team's experimental findings because their bar was so small. But Garwin's impact was high in spite of this skepticism because of the way he presented his results. As one scientist put it,

> [1975] [A]s far as the scientific community in general is concerned, it's probably Garwin's publication that generally clinched the attitude. But in fact the experiment they did was trivial—it was a tiny thing.... But the thing was, the way they wrote it up.... Everybody else was awfully tentative about it.... It was all a bit hesitant.... And then Garwin comes along with this toy. But it's the way he writes it up, you see.

Another scientist said,

> [1975] Garwin had considerably less sensitivity, so I would have thought he would have made less impact than anyone; but he talked louder than anyone, and he did a very nice job of analyzing his data.

And a third:

> [1975] [Garwin's paper] was very clever because its analysis was actually very
> convincing to other people, and that was the first time that anybody had
> worked out in a simple way just what the thermal noise from the bar should
> be.... It was done in a very clear manner, and they sort of convinced
> everybody.

The first negative results challenging Weber's findings had been reported with careful exploration of all the logical possibilities such as would
be consonant with a lack of complete and publishable certainty that Weber's results were entirely spurious. Following closely came the outspoken
second experimental report of the Garwin team, with their careful data
analysis and the claim that their own results were "in substantial conflict with those reported by Weber."[8] Then, as one respondent put it, "that
started the avalanche, and after that nobody saw anything."

The picture that emerges with regard to the importance of experimental results is that the series of experiments made strong and confident
disagreement with Weber's results openly publishable, but this confidence
came only after what one might call a "critical mass" of experimental reports had built up. This mass was "triggered" by the way Garwin presented
things; he, as it were, made the discourse of outright rejection a possibility.[9]

Garwin believed from the beginning that Weber was mistaken and acted
accordingly. But he was less closed minded than the preceding paragraphs
would suggest from a quick reading. Garwin's experimental strategy took
account of the possibility that he might turn out to be wrong and find
indications of high fluxes of gravity waves—even though his initial intention
was to disabuse those who still thought Weber might be right. Garwin
explained his experimental strategy to me as follows:

> [1975] [W]hat we could have done in the beginning was simply to have
> analyzed Weber's performance and to have shown in principle that he
> couldn't have detected the gravity waves that he said he was detecting.... We
> could have argued from the abstract that he couldn't have been detecting
> them even under ideal circumstances. But we felt that we wouldn't have any

8. Levine and Garwin 1974, p. 794.
9. Braginsky was equally forthright in his rejection of Weber, but insofar as we can be sure of
historical facts in this case, it was Garwin who actually succeed in turning the tide of the discourse.

credibility if we did that...and that the only way we could get standing was to have a result of our own.

After completing the work and publishing the report on their "tiny" antenna, the Garwin group built a second antenna of greater size and sensitivity but small enough to utilize the same peripheral equipment (vacuum chamber, etc.). I was interested in their reasons for going ahead with this if they believed that their first antenna, though small, was large enough to do the job of legitimating their disproof of Weber's results. Garwin himself answered simply in terms of maximum utilization of available equipment. The new experiment cost next to nothing and pushed down the upper boundary of possible gravity waves still further. However, another of the group said,

> [1975] [W]ell, we knew what was going to happen. We knew that Weber was building a bigger one, and we just felt that we hadn't been convincing enough with our small antenna. We just had to get a step ahead of Weber and increase our sensitivity, too.
>
> ... At that point it was not doing physics any longer. It's not clear that it was ever physics, but it certainly wasn't by then. If we were looking for gravity waves, we would have adopted an entirely different approach [e.g., an experiment of sufficient sensitivity to find the theoretically predicted radiation]...there's just no point in building a detector of the [type]...that Weber has. You're just not going to detect anything [with such a detector— you know that both on theoretical grounds and from knowing how Weber handles his data], and so there is no point in building one, other than the fact that there's someone out there publishing results in *Physical Review Letters*.... It was pretty clear that [another named group] were never going to come out with a firm conclusion... so we just went ahead and did it... we knew perfectly well what was going on, and it was just a question of getting a firm enough result so that we could publish in a reputable journal, and try to end it that way.

The last phrase in the above quotation is particularly significant. Garwin's group had circulated to other scientists and to Weber himself a paper by Irving Langmuir;[10] it was quoted to me also. This paper deals with several cases of "pathological science"—"the science of things that aren't so." Garwin believed that Weber's work was typical of this genre; he tried to persuade Weber and others of the similarities. Most of the cases cited by

---

10. Langmuir 1989 had been circulating since 1969 as a General Electric report.

Langmuir took many years to settle. Consequently, as a member of the Garwin group put it, "We just wanted to see if it was possible to stop it immediately without having it drag on for twenty years."

Garwin and Levine were worried because they knew that Weber's work was incorrect, but they could see that this was not widely understood. Indeed, the facts were quite the opposite. To quote a member of the group,

> [1975] Furthermore, Weber was pushing very hard. He was giving an endless number of talks. . . . We had some graduate students—I forget which university they were from—came around to look at the apparatus. . . . They were of the very firm opinion that gravity waves had been detected and were an established fact of life, and we just felt something had to be done to stop it. . . . It was just getting out of hand. If we had written an ordinary paper that just said we had a look and we didn't find, it would have just sunk without trace.

In sum, though they were prepared for all eventualities, Garwin and his group set out to kill Weber's findings in the shortest possible time, and they pursued the aim in an unusually vigorous manner. In the first place, they did their experiment so as to put themselves in a position to criticize Weber's findings. They probably would not have bothered to carry out any experimental work if it hadn't been that they "looked at what some other people were planning to do and decided that there wasn't anybody who was going to make this confrontation."

Thus, Garwin acted as though he did not think that the simple presentation of results with only a low-key commentary would be sufficient to destroy the credibility of Weber's results. In other words, he acted as one might expect a scientist to act who realized that simple evidence and arguments aren't enough to settle unambiguously the existential status of a phenomenon.

The Garwin Letters

Something of the flavor of this episode in the controversy can be experienced from a long exchange of letters between Weber, Garwin, and others. Over the years I have collected a large part of the correspondence between Weber and Garwin between February 1973 and April 1975, the period when Weber's credibility dropped to almost zero.

From the Garwin correspondence I can tell that he, probably together with Levine, visited Weber's laboratory on December 18, 1972. The first letter I have is dated February 12, 1973, and is from Weber to Garwin. It concerns methods of calibration. Following the very complex experiment of Joel Sinsky, Weber had used a "noise generator" to calibrate his bars. As I understand it, he injected white noise of known amplitude into the bar via the electronic system. He could then watch the effect on the level of random coincidences recorded by the bars and thus work out their sensitivity. But he had been persuaded to use another calibration method favored by all the other bar groups. This was to inject discrete "kicks" into the bar via an electrostatic end plate. One of Weber's worries about this method was that the electronics for providing the electrostatic "kicks" might feed directly into the amplifier, giving a spuriously high level of sensitivity. His letter of February 1973, however, records the completion of such an experiment by his student Gustaf Rydbeck. He says that the results were as expected.

> This is consistent with the noise generator measurements for his set-up and is consistent with the high frequency Cavendish experiment result [Sinsky's]. By now we can say that the sensitivity of our installations has been studied by three different methods which give consistent results—extremely far from your estimates of our equipment! [Which were lower.]

On March 30, 1973, Weber again wrote to Garwin, sending him some reprints of his article in *Nature* and again stressing the consistency of his calibration results. He refers in this letter to an article in *Physics Today* by Jonathan Logan, in which he had been criticized.[11] In the article, Logan reports that "Garwin estimates that the events observed a few times per day by Weber . . . would produce excitations in the Yorktown Heights [Garwin's lab] apparatus . . . which they would expect to observe. . . . So far, data for only a few days' running have been reduced, and they show no excitations of such strength." In rebuttal, Weber wrote to Garwin,

> We observed coincidences for two years before claiming anything, and I hope you will be able to repeat our experiments—with a two detector system— before adding comments like those in Logan's article, to the literature.

11. This article (Logan 1973) is a first-class summary of the state of play in gravitational radiation research at that date. The overall tone of the article is negative, but Logan is careful to include most of the reservations that prevent the evaluation from being decisive. The article could be read as a well-documented illustration of the nature of the "dam of disbelief." Weber, in his letter to Garwin, is objecting to the estimates of the relative sensitivity of his bar compared with the other antennas that Logan describes. As we have seen, in the case of the relative estimates of his and Tyson's apparatus, he won the argument. Logan turned his attention away from gravitational waves after writing this excellent review (personal communication).

On May 10, Garwin replied to the two letters, copying his reply to Jonathan Logan. He says, "Thank you very much for your letter of February 12th, 1973, which I really did not know how to answer except by sending you preprints of our papers, which I enclose."[12] Garwin explains his difficulty with Weber's claims about noise and detection efficiency and urges him, once more, to calibrate his bars using a different signal-processing algorithm.

Weber's reply, also copied to Logan, was dated May 18, 1973. In it he contests various details of Garwin's understanding of his claims. He also remarks, "Tyson now informs me that his estimates of our sensitivity, presented at the New York Texas [General Relativity] meeting, are incorrect." He returns to the Logan article, implying that Garwin no longer believes his estimates of the apparent high strength of Weber's signals: "If you do not believe it then a letter by either you or Logan seems appropriate.... Needless to say, incorrect reports about our work are very damaging, even when quietly retracted."

Garwin replied on the day he received the letter (May 22), once again copying his response to Logan. While Weber's letter was just over a page in length, Garwin's response is three and a half pages. The argument covers much ground in much detail—there are seven numbered sections dealing with substantive issue. For example, Weber has been claiming that the disappearance of certain features of the signal could be due to decay of the source. Garwin tries to show that this is an unlikely, if not impossible, occurrence, and that if Weber really believes it to be the case, he should state it categorically. Garwin returns to the calibration issue and argues that the IBM group have been extremely careful to make sure there is no direct leakage of calibrator signals into the electronics. The eighth and final numbered paragraph reads,

> We hope we have not incorrectly reported your work, but we have certainly not retracted any of our comments, quietly or noisily.

Weber responded on June 13, again with a copy to Logan. He enclosed some reprints and restates his position, saying that it is reinforced by new findings and new procedures. He complains that if Garwin had asked more pertinent questions during his visit to Maryland, Weber could have supplied answers that would have cleared up confusion about earlier reported results. He concludes,

12. These were Levine and Garwin 1973; Garwin and Levine 1973.

> Anyway, I don't really care whether you publish or not, and the current
> experiment—with much easier algorithm [for signal processing]—with a
> computer programmer who is still on the job and a more easily understood
> program—is of more interest to almost everyone.

Somewhat surprisingly, he signs this letter "Cordially yours" rather that
the usual "Thank you, yours truly."

The next item in my file concerns the papers that Garwin and Levine
were about to publish in *Physical Review Letters* (see chapter 7). The editor had
sent a referee's report to the authors which recommends that the papers
not be published. It would seem almost certain that this report is from
Weber. On June 21, Garwin and Levine responded to the two and a half
pages of criticisms, and the papers, as we know, were to be published.

On July 2, 1973, Garwin replied to Weber's letter of June 13, reminding
him of some of the interchanges that took place during the December visit.
He agrees that the new experiments will be of more interest than continued
reanalysis of old data, but urges Weber to do more calibration.

A letter from Weber must have followed, but I do not have it. In Garwin's
letter of October 11 he starts, "Of course I have read with great interest your
'New Gravitational Radiation Experiments' published in Physical Review
Letters."

Garwin then asks a series of numbered questions. The first is about Weber's choice of threshold for reporting results. As I understand it, Garwin is
pointing out to Weber that as his threshold gets lower—that is, the number
of accidental coincidences increases—the number of zero-delay coincidences
increases faster; consequently, his statistical confidence gets higher. He asks
Weber why he chose a relatively high threshold to publish rather than the
much lower threshold that would produce better statistics [but an absurd
number of zero-delay coincidences: 800 events over four days]. Garwin also
suggests that Weber might like to visit the IBM laboratory and that he
would invite Douglass and Tyson, too (he copied the letter to these).

Weber answered on October 22. He says that in his 1973 article, he
included only findings that had "been independently verified by an on-line
computer coincidence experiment." He says that much higher event rates
can occur because if the bar is already in an energetic state, a further
injection of energy can cause it to hover around a threshold and produce
many crossings from only one external signal. This is an effect that will
come to be called proliferation and will feature again in the subsequent
debate. He says he did not discuss this proliferation in the 1973 paper,
because "all the facts were not known." He concludes,

We do find that the zero delay excess and level of confidence depend on the crossing rate. At this time we are deeply involved in review of our computer programs and study of new ways of data processing which appear to significantly improve the level of confidence. We find that some relatively minor changes in the data processing may lead to negative results. We are also exchanging tapes, and other information with other groups. The present phase of our work will require about another month to complete and I would prefer not to have any meeting until we can make some definite statements about our new procedures.

On December 12, Weber followed up the above letter with another. The second paragraph reads, fatefully, as follows:

At Warsaw we reported on an analysis of separate tapes of the Maryland and Rochester detectors. There is no telephone line. We found the largest bin at 1.2 seconds delay. However, the Rochester clock was found to have a 1.2 second error and the largest bin was therefore, zero delay. The difference between the 1.2 second bin and the mean of the accidentals exceeded three standard deviations of the accidentals and the level of confidence in the amount of excess was 2.6 standard deviations.

On December 20, Garwin wrote asking for a copy of one of Weber's data tapes (this request had also been made earlier, in passing) and offering to exchange one of his own.

On February 7, 1974, Garwin again wrote to Weber (sending copies to Douglass and to Goudsmit, I believe). The letter was sent from University of California at Santa Barbara, where Garwin was spending some time, and is addressed to University of California at Irvine, where Weber was working for a while. I reproduce the letter in full:

Dear Joe,

I was glad to have the opportunity to talk with you by telephone yesterday. I am sorry not to have been able to convince you that it would be in the best interests of physics and of gravitational-radiation science for you to publish the Phys. Rev. Letters a brief Erratum stating simply that:

"Professor David Douglass has demonstrated to us the existence of an error in the computer program used to produce the delay histogram of Fig. 1 of the reference article. When this error is corrected the data presented as Fig. 1b (for example)—a zero-delay total of 566 events and an average accidental total of 414 events in 20.8 days—became instead 430/414?? New data, to be presented soon, show that the appropriate algorithms give a high confidence in a zero-delay excess of about seven events per day when the antennas are working well."

I believe it is important for you and science to publish such a note. I think it unconscionable to have data stand in the scientific literature that is known to be seriously in error, and it is also wrong to have it simply rendered "inoperative" by later algorithms without plainly showing what results that published data yield with minimal charge [change?] of algorithms—i.e. eliminating a single known error.

Were I Editor of Phys. Rev. Lett., I would feel to some extent unhappy that the data of your September 17, 1973 paper turned out to be wrong, but they were submitted and published in good faith. I would, however, be very concerned that you have not seen fit to publish an Erratum in the long period since September, when I believe you confirmed the existence of this programming error. I believe that such an Erratum must be published, and its publication must precede the publication of any new data on your detection of gravitational radiation. In what way does physics benefit from you not publishing this erratum?

I enclose a copy of Langmuir's lecture "Pathological Science." Whatever the reality, it seems to me that your results over the years appear to fit more-and-more the common characteristics of Langmuir's examples of "pathological science." I am sure that you want to avoid such appearances, and I urge you to start by frankly publishing the Erratum suggested above, rather than by publishing new results from new data.

> Sincerely yours
> Richard L. Garwin

It seems likely that this letter crossed in the mail with one from Weber dated February 8, 1974, which also refers to a telephone conversation between himself and Garwin. In this letter, Weber, as was his style, refers to a series of reasons for believing in his data and a series of experiments he intends to undertake. There is no discussion of the proposed Erratum, merely the claim that reanalysis with a correct algorithm produced result "in reasonable agreement with the September letter" (i.e., *in Physical Review Letters*). Weber concludes,

> We will continue to improve our experiments to the point where
> coincidences can be demonstrated with other much less sensitive detectors.
> At that point, and only then in my opinion, can accurate statements be made
> about relative sensitivities.

In early March, Douglass sent a document dated 1974 to Garwin and Levine, noting that it might amuse them. The document was part of a progress report and was entitled "Analysis of the Claim of a Positive Effect between Gravitational Antennae Situated at Maryland and Rochester." The document begins with a "Stated Claim by J. Weber (Warsaw Meeting, Tucson Meeting, etc.)."

The "Stated Claim" concerns the 78 coincidences between the Maryland and Rochester detectors, which showed a 2.6-standard-deviation significance level. These we have already seen mentioned earlier in one of Weber's letters to Garwin. The claim and the observations leading up to it are discussed in detail. Page 5, however, is the fatal page:

> In order to look for these events it is necessary to know what time is written on the Maryland tape. All computer output that we received indicated that the tapes were written using local time which would be Eastern Daylight Savings Time [EDST]. Telephone calls to Maryland confirmed this. So in the search for these 78 events 4 hours (GMT = EDST + 4 hrs) was added to the Maryland tape times. The events could not be found. On a hunch, the card adding 4 hours was pulled from the program and all of the events were found both in time and in amplitude. This means that the Maryland analysis was done assuming that the times written on the Roch. and Maryland tapes were the same. Since this is evidently not true, then the positive effect reported by Weber is in fact for a delayed coincidence of 4 hours + 1.2 sec.
> Conclusion
> We conclude that there is no evidence for gravitational wave coincidences between the Rochester and Maryland antennae as reported by Weber.

The IBM team responded with a letter to Douglass dated March 5, 1974, and copied to Goudsmit and Tyson. Garwin says, "It was quite interesting to see that Weber and his group can find coincidences between essentially random numbers!"

It appears that there must have been some earlier correspondence or telephone calls prior to this letter in view of Garwin's opening sentiment:

> In regard to your decision not to write to Sam Goudsmit, I would like to suggest that your responsibility to "Science" and also to PRL [*Physical Review Letters*] is somewhat greater than your responsibility to Weber. Verbal agreements notwithstanding, Weber has:

1. Felt free to report positive results between your antenna and his, using your data, and:

2. Has discussed in public the computer error, in a form which makes it sound like some minor mistake. This is of course rather misleading, and I think that the fact that he has discussed it relieves you of any obligation to remain quiet. After all, he is effectively rebutting the error, while not permitting you to reply.

Now, I would like to say why I think it is very important to get this out in the open and in print. Although Weber may lose part or all of his NSF funding, he will almost certainly get his Argonne 4K antenna into operation and in coincidence with a room temperature antenna in Maryland. [Garwin is referring to plans for a liquid helium temperature detector in Chicago, which was never finished.] If he gets enough money, he will have a 4K antenna at Maryland as well. In either case, he will almost certainly find signals—he always does, even when they are known not to be there, as you well know. Since this property of his data analysis is unknown publicly, the coincidences will again be accepted by many as evidence for g-waves.... He will certainly have you and Tony [Tyson] by the short hairs, as you will be at 300K, improved transducers notwithstanding. Of course, as usual, his signals will be smaller than before, but this didn't bother him, or many other people, when it happened with his 1973 improved system. He might just claim that the strong source of 1969–1970, which became weaker in 1973, has continued to weaken!

Thus, it seems to me that the only way to forestall these problems is to demonstrate that Weber and/or his group is quite capable of:

1. Publishing data obtained with a defective computer program and later refusing to publish an erratum.

2. Finding and publishing signals given any two tapes containing random numbers.

Like it or not, I think you are elected! Although all of us know Weber's statistical methods, only you have more or less documented proof. I think you must use it. I have asked Sam Goudsmit to send you another letter, and will send him a copy of this as well.

I'm sorry if this letter has a somewhat righteous tone. I wouldn't be happy receiving one like it if I were in your position! But I do hope you will consider some of my arguments seriously and act according to your conscience.

In my collection of letters, I now find indications that Weber was becoming worried about his continuing support from the National Science

Foundation in the light of criticisms his work was now suffering. On May 23 and June 5 respectively, two of his computer programmers wrote to Marcel Bardon, then NSF program director for physics, explaining the accuracy of their work and its independence from Weber.

The Force of the Argument

Reading the exchange above, it is almost impossible not to be convinced that Garwin and Levine had revealed, once and for all, that Weber had generated his findings by unknowing statistical massage. But why is the exchange so convincing? Barring mass hallucination or some such, what the exchange reveals for certain is that the 2.6-standard-deviation Rochester-Maryland result should not have been there. Even Weber does not question the fact of the timing mistake. This result was, then, generated by a fortuitous post-hoc choice of statistical parameters, such as threshold, or by a piece of extremely bad luck. (That is, since a 2.6-standard-deviation result will emerge about one time in 100 out of pure statistical fluke, this may have been one of those one-in-a-hundred occasions; this was, effectively, what Weber would go on to claim.) The other thing we learn is that Weber was not ready to go public immediately and withdraw his incorrect claim from the scientific literature. I suggest—and it is worth the reader thinking about his or her own reaction to the account—that this second thing that we learn is even more damning than the first.

What are Garwin and Levine doing? Essentially, they are urging Weber to reinterpret his own findings in the light of Irving Langmuir's "Pathological Science" paper, a paper that fulfils a useful role in helping scientists to deal with heterodox claims where signal remains hard to extract from noise. The Langmuir document, which has for most of its life circulated as an increasingly faint and grubby photocopy of an unpublished lecture, has frequently been thrust at me in my encounters with scientists. The paper deals with a series of cases of the science of "things that aren't so" and generates a set of their common characteristics. The trouble with the Langmuir approach is that it does not deal with the problem of when to give up. If all scientists setting out to make a new and difficult discovery had Langmuir in the forefront of their minds, they would never persevere in the face of difficulty or criticism. Langmuir says "give up" if someone cannot find a way of extracting the signal cleanly from the noise in the experiment, but he doesn't say how long to go on for. (Someone should write an anti-Langmuir paper, featuring cases of scientists making their discoveries only after pursuing a recalcitrant phenomenon well beyond the normal limits of good sense.) So one of the things that Garwin and Levine were doing was trying to persuade Weber that he had gone on long enough and that his findings were now best interpreted as Langmuir-like rather than anti-Langmuir-like. They were, if you like, engaged in an attempt to help Weber reconstruct his reality.

And they were also engaged in a more widespread exercise in "the social construction of reality." They were urging that Weber's mistakes, and his unwillingness to admit them, be made public. This would encourage the wider community to reshape its view of Weber and his work just as your view, reader, has been reshaped by reading the exchange of letters.

It is easier to understand this activity if it is contrasted with what was being done by Garwin and Levine in their experimental reports in the *Physical Review Letters*. There they made some observations, made the clearest analysis yet of signal and noise in a resonant bar detector, and as a result of their measurements and analysis, convinced at least some people that there were no Weber-type signals to be seen. In principle, these findings could have stood by themselves as a "disproof" of Weber's results even if Weber had made no mistakes and even if he had acted with good grace in admitting any mistakes he did make. But, as we have seen, Garwin and Levine set out with a broader project in mind, and they acted as though the results of their experiment, taken on their own, would be insufficient to do the job. Garwin and Levine clearly understood the meaning of the phrase "social construction of reality" much better than many of science's self-appointed spokespersons.

The Fifth Cambridge Conference

Garwin, as we have seen, was concerned that the ripples in social space-time emanating from Weber's experiment would continue to propagate without a further intervention. He did not want to see the continuation of what he was sure was an episode of "pathological science." He decided, therefore, that irrespective of whether he and Levine could persuade Douglass to take further action, he would expose Weber's mistakes to a wider audience. The occasion was the Fifth Cambridge Conference on Relativity, or CCR-5, which has been mentioned in chapter 8. Garwin prepared a paper for the conference entitled "The Evidence for Detection of Kilohertz Gravitational Radiation," and it revealed all.[13]

13. Garwin 1974a.

In his paper, Garwin points out that Weber's 1973 results, collected with an improved apparatus, suggested a less powerful source without the sidereal anisotropy found in 1970. He concludes,

> [T]hese results therefore do not satisfy the simplest requirement to be regarded as a physical fact, and I propose that they now be dropped from all consideration as evidence for gravitational radiation.[14]

Garwin then turns to the 1973 data and describes the computer error, the four-hour error, and the devastating critique of the 1973 "shoulderless" histogram, which I will describe later in this chapter. Garwin says,

> We thus conclude that the Maryland group has published no credible evidence at all for their claim of detection of gravitational radiation.

At this conference, Douglass was scheduled to speak before Garwin. He told me that he had prepared a presentation that did not include any mention of the mistakes that he had discovered in Weber's data analysis. He said, however, that Garwin preempted him and that he had to change the talk he was going to give.

> [1975] As regards the Cambridge conference, Garwin forced my hand. I went to the Cambridge conference not intending to mention the computer error unless Weber made a misstatement. . . . But when I got there, Garwin presented me with a copy of his remarks already written up, and since I was heading off the session . . . I didn't get any lunch that day, putting in to what I was going to say what happened, in what I felt was an accurate way without being emotional . . . that was the first public announcement.

Douglass's presentation included hastily handwritten "viewgraphs" revealing the facts that Garwin had described in the previously prepared manuscript.[15] Thus Garwin was able to say when he presented his session, "You have heard at this meeting from Professor Douglass that the University of Maryland group misunderstood the time origin of one of these

14. This seems a rather harsh judgment, as there are many observations of unique events in physics. Weber would insist that the existence of an event rate at time A does not imply the same event rate at time B in the case of an astronomical source.

15. I have two copies of the Garwin talk, both typed, which are almost identical. One bears the legend "To be presented orally at . . ."; the other, "presented at . . ."

antennae..."[16] Tony Tyson, who was also at the meeting, said of this event and what was to follow,

> [1975] I felt that was a very inflammatory issue. It was clearly a case where Weber had tripped himself up because of his data analysis, and I felt that it spoke for itself and that those few people who knew about it were enough. But Garwin did not feel that way, and he went after Weber ... and I just stood on the sidelines covering my eyes, because I'm not really interested in that kind of thing, because that's not science.

The Cambridge conference revelations were merely a prelude to Garwin's revealing Weber's mistake to a wider audience still.[17]

The Letters to *Physics Today*

On July 9, 1974, Garwin wrote a letter to the editor of *Physics Today* for publication in that journal—effectively the "house journal" for American physicists if not physicists worldwide. In a cover letter bearing the same date, he explains that he is copying the letter to Weber, so that *Physics Today* can publish Weber's reply should he wish to make one; and that

---

16. Alan Franklin has attempted to write a history of gravitational wave detection that turns entirely on what he calls "rational" actions. As part of his argument he denies the role of Garwin in making Weber's errors public. Franklin, however, used only published sources, such as the conference proceedings discussed here. It is easy to see how Franklin was misled. This illustrates the danger of using published sources alone as a resource in the history of science.

17. In my subsequent fieldwork, many scientists expressed strong reservations about Garwin's actions in revealing the computer errors and the manner of his attack on Weber's findings. Garwin, it seems, had broken the rules of the "club." Here is a European view:

> [1976] All the people who work together on these experiments know each other, and [someone] told us of this wrong time [at CCR-5] and I laughed and thought, "It can happen." And then came Garwin and brought this into the open in the full session and I did not find this right. This was certainly unnecessary. There was a very big mistake—that can happen. Of course this was told—[everyone told each other] but no one laughed.
> {*Collins:* Everyone knew it?} Everyone knew it, but he brought it into the open session.

Speaking for myself, as a sociologist of science who has suffered more or less continuous attack from scientists throughout my career—without resenting it except when my work is carelessly or deliberately misunderstood or misrepresented—I do not see anything to take exception to in what Garwin did; but then, I am not a member of that particular scientific club but rather the taxpayers' club that pays for scientific mistakes.

Note that I am not saying that Garwin's attacks were scientifically sound (or unsound)—that is not my business. I am saying only that if Garwin thought they were scientifically sound (which I am sure he did), then he had every right to choose to go after Weber in the way he did. It is interesting that Tyson, who felt embarrassed by Garwin's attacks, would be on the receiving end of still worse criticism from members of the "club" when he gave evidence against the LIGO project to Congress.

[b]ecause this kind of clarification is not really original research, it seems more appropriate to use this means of publication rather than, for instance, Physical Review Letters.

Garwin wrote a cover letter to Weber on the same day, saying, "I hope that you will join me in this enterprise by providing him [the editor of *Physics Today*] with a speedy reply."

Blind copies of the letter were sent to Douglass, Goudsmit, Logan, and Tyson. The letter was eventually to be published, with only minor corrections, in the December issue.[18] The letter begins, "Your readers are probably unaware of the present situation, which I summarized in a paper presented at the 'Fifth Cambridge Conference on Relativity.'"

Garwin then asks, "Can anything other than gravitational radiation have produced his delayed-coincidence plots?" and "Could gravity waves themselves have produced his delayed-coincidence plots?" Garwin attacks Weber's *Physical Review Letters* publication of September 1973. Answering the first question, he discusses the computer error discovered by Douglass and concludes, "Thus not only some phenomenon besides gravity waves *could*, but in fact *did* cause the zero-delay excess coincidence rate." On the second point, Garwin shows a computer simulation, developed by his colleague James Levine, of the way a gravitational wave detector ought to respond to signals given its relatively poor time resolution due to its resonance. He shows that the delay histogram plots should have what I have referred to in chapter 8 as shoulders. The delayed bins immediately adjacent to the zero-delay bin should show some effect above background noise due the "smearing out" in the time dimension of the response of a resonant detector to a signal. Garwin asserts, "Therefore I claim that it has been demonstrated that the coincidence data of reference 4 *did not* result from gravity waves, and furthermore *could not* have resulted from gravity waves."

Garwin then suggests, "Two other facts were discussed at CCR-5 that should be more widely known." He goes on to describe Weber's announcement of a 2.6-standard-deviation "coincidence" effect between the Maryland and Rochester antennae and points out that these coincidences had been generated out of pure noise, given that the detectors were actually running not in coincidence but with a four-hour delay. Then he says that "in view of the fact that Weber at CCR-5 explained that when the Maryland Group

18. The published version takes up the whole of page 9 and a little of page 11 of the journal. Quoted passages from the letter are all to be found on page 9 of the published version unless otherwise indicated.

failed to find a positive coincidence excess 'we try harder,'" his colleague Levine had simulated a method of producing delay-histogram signals from two streams of noise by an appropriate data selection procedure; and as a result, he had produced a "'six standard deviation' zero-delay excess"–the figure resulting from the simulation (p. 11), showing a prominent central bin with no shoulders. Garwin concludes,

> This "experiment" demonstrates in a simple manner the extreme importance of publishing details of the selection of data in the processing algorithm that might be used by the Maryland Group in any future publications. (p. 11)

Garwin's conclusion originally included the following: "In the meantime your readers should be aware of the lack of *any* published evidence from the Maryland Group regarding the existence of gravitational radiation." But this sentence was removed from the published version of the letter.

My collection of correspondence shows that Weber responded by writing, on August 29, to Ralph Gomory, vice president and director of research at IBM. In the two-page letter, Weber, as he often did, restates all his findings and the evidence supporting them. He says that Garwin has been promulgating the incorrect conclusions of Douglass and that his positive results were due to computer error, reiterating that his 2.6-standard-deviation claim was not presented anywhere as a positive result. Weber adds that he believes that he can rebut Garwin's criticism as submitted to *Physics Today*, but that

> [n]onetheless I feel that publication of Garwin's letter is not in my best interests, not in the best interests of IBM, not in the best interests of Garwin or physics in general. Therefore, I am suggesting that you study these documents [as enclosed], and if you agree, then perhaps you would be willing to talk to Garwin about withdrawing his letter. More important would be your providing funds (less than $20,000) to Garwin to construct a detector like ours. I must say that such an instrument, in his hands, would resolve all present controversies, and that the Bell Laboratories-Rochester efforts have in fact been poor.

Gomory responded on September 6, saying that "nothing in Dr. Garwin's letter, which expresses technical judgments about scientific issues, would justify my intervention." He says that Levine and Garwin do not intend to build a bigger detector and that they "plan to terminate their efforts in gravity wave detection in view of the apparent confirmation of their results by the experiments now in progress in Europe."

Garwin wrote for his own files a note (more than two pages long) dated September 10 that he copied to Gomory. He replied to Weber point by point. On the issue of a bigger detector, he wrote,

> As for building a detector as big as Weber's, if I had thought that it was necessary, I am sure that our budget could have stood the strain. Our present detector has half the mass of his and is more sensitive. Unfortunately, there is no way to resolve the controversy unless Weber pays attention to the results of others and also to the necessities of scientific publication.

Weber wrote again to Gomory on September 11.

> Garwin made incorrect statements about our work at the Los Alamos Laboratory recently, and at a number of other places. In each instance, I followed with letters from scientists at the Stanford Linear Accelerator Center and the M.I.T. supporting our published results. The one magnetic tape which was widely circulated does have 8 events per day. Garwin insists, without having looked at it, that the event rate is one per day.
>
> These visits, followed by my letters, and a Physics Today exchange, are not good for anyone. Some statesmanship would be a positive contribution.

Gomory wrote to Weber on September 18. The draft I have was copied to Garwin with this note: "Dr Gomory asks that you please look over the attached. The original will not be mailed until we have your agreement." Garwin's written response was "I agree completely." Gomory wrote,

> I certainly understand that this type of controversy is very painful for all the people involved. However, such controversy is not new to the history of science. Eventually the disputed points are resolved. I think this process is speeded up when both sides are willing to discuss their views and the relevant facts as publicly as possible. Therefore, I will not attempt to prevent Dr Garwin from publishing. I would rather urge you to publish and articulate your position.
>
> I hope these questions will all be resolved in a sensible way and in the near future.

On September 20, Garwin wrote to Weber (copying to Gomory), referring to Weber's last letter to Gomory. He says he was misquoted and

mentions the availability of a tape of his talk at Los Alamos. He says he
did not claim that there was one event per day on the disputed tape, as
he knows nothing of it except what was said at the June 10 MIT conference, where reports from different laboratories "varied enormously." He
adds, "Furthermore, that statement is inconsistent with my entire position,
namely that the Maryland group has published no credible evidence for having detected gravitational radiation." And, "I am not willing to withdraw
that letter, and I would like to see your reply either in that same issue or
later."

This letter may have crossed with one from Weber to Garwin dated
September 19, again urging on Garwin that he was disseminating incorrect information and concluding, "I hope you will see your way clear to
withdrawing your Physics Today letter, and permitting us to get back to
doing Physics."

On September 26, 1974, Garwin again wrote to Weber, saying, "I called
Dave Douglass today to see whether he was aware of the 'far worse errors'
which you say he made, and he tells me that he has no idea what you are
talking about. I am in the same position." Garwin explains that he is sure
that Weber is grateful to Douglass for pointing out the computer error and
asking for details for the subsequent corrections.

> If you will review your correspondence with me, you will note that in a letter
> you volunteered to me that you had reported at Warsaw a 2.6 standard
> deviation zero-delay excess for coincidences between Maryland and Rochester.
> How can you say that you did not claim this is a "positive result"? I am sure
> that some of your other analyses, with other thresholds, gave no zero-delay
> excess, and yet these were not reported.

He reiterates: "I will certainly not withdraw my Physics Today letter. I think
it is long overdue."

This two-and-a-half-page letter continues with the arguments about
technical details, proliferations, and so forth; suggests that Weber might be
more fulsome in citing criticisms of his work; and once more looks forward
to seeing the response to the *Physics Today* letter.

Weber had submitted his reply to *Physics Today* on September 24. It
was published, with only minor editorial changes, as "Weber Replies" on
pages 11 and 13 of the December issue of *Physics Today,* immediately following Garwin's letter. Weber says that "[t]he reader who takes time to study

Richard Garwin's letter and my reply will want to know what is really going on."[19] He continues,

> Richard Garwin is a respected scientist associated with the respected IBM Research Laboratory. He attempted to check our two-detector coincidence experiment by a search for sudden changes in output of one detector and obtained negative results. The real issue here is whether physicists can have a higher degree of confidence in the Garwin negative result than the result published by us.
>
> The IBM detector... had a mass one tenth as great as our smallest operational detectors. Garwin apparently overlooked the great importance of temperature control or automatic tracking of his cylinder with a reference oscillator.
>
> It is most unfortunate that for these reasons this IBM detector and a second one with a mass of 480kg appear to be the least sensitive of nine operating installations.
>
> Other physicists have joined me in wondering why there wasn't a very small fractional increase in cost to make the IBM installation the one with the largest mass and highest sensitivity in the world. If this had been done, the IBM effort might have terminated the present controversy.
>
> It is interesting that the one other group that has observed a statistically significant number of coincidences (Munich-Frascati), employed two detectors having the same mass as the Maryland antennas and similar instrumentation.

The letter then goes on to discuss the computer errors, explaining that the errors had been acknowledged and that subsequent reanalysis of the computer error tape showed that: "other laboratories processed the tape incorrectly making larger errors than ours and these incorrect results were widely disseminated by Garwin."

Weber also discusses the four-hour error without mentioning it explicitly, saying,

> [We] have not claimed a positive result in open-literature publications or at meetings. Review of these data, applying all known corrections, leads us to conclude that there is a zero-delay excess, which is in fact larger than reported in Warsaw in 1973 [the 2.6 standard deviation].

---

19. All quotations are from the published version.

> Computing errors have been an important factor in the politics, but not
> in the physics of our experiment. (p. 13)

Garwin immediately (December 20) wrote a reply to *Physics Today* that would be published in the issue of November 1975, along with a response from Weber.

Between submission and publication of the first round of letters in *Physics Today*, Weber had been talking about his side of the argument and had circulated a technical report entitled "Gravitational Radiation Experiments in 1973–1974."[20] This was also submitted as Weber's contribution to the Tel Aviv conference, but he gave a much shorter address at the event. Given that this report bears the same title as the major review paper published in *Physical Review D* in 1976, and that the date of first reception of that paper is December 1974, it is likely that this report is a first draft of the eventually published paper. A footnote on page 6 of the report declares,

> Douglass discovered a program error and incorrect values of the unpublished
> list of coincidences. Without further processing this tape, he reached the
> incorrect conclusion that the zero delay excess was one per day. This
> incorrect information was widely disseminated by him and Dr R. L. Garwin
> of the IBM Thomas J. Watson Research Laboratory. After all corrections are
> applied the zero delay excess is 8 per day. Subsequently, Douglass reported a
> zero delay excess of 6 per day for that tape.
>
> Douglass has also reported computing errors in the data reported at the
> Warsaw Copernicus Symposium in September, 1973 involving detectors at
> widely separated frequencies. A 2.6 standard deviation zero delay excess was
> reported for a six day period with no claims for it being a positive or
> negative result. Analysis of other data make it clear that on a time scale at
> least exceeding 4 days there is observed zero delay excess with a level of
> confidence associated with more than 6 standard deviations . . . (p. 6)

The report deals with pulse proliferation, the difference between algorithms, Tyson's claims, and various experimental results before concluding, as in the Tel Aviv address, "On the whole, our results are in fair agreement with the observations of the Munich-Frascati group" (p. 11).

On October 11, Douglass wrote an angry three-and-a-half-page letter to Weber, which he copied to Garwin and Levine. In it he complains that Weber had been misrepresenting the events as they had taken place. For

20. Gretz, Lee, and Weber 1974.

example, Douglass, not unreasonably, charges that Weber keeps saying that the Rochester group had made a "much worse error" in analyzing the co-incidence data which contained the initial computer error. He explains that Weber had not found an "error" but just a different way of analyzing the tape. He also complains about Weber, saying that he had had helped Douglass solve technical problems and rebuilt his apparatus for him, when the Rochester group had merely accepted, in a collegial way, the offer of a preamplifier from the Maryland group at a time when theirs had dete-riorated and a replacement could not otherwise be speedily obtained. The last section of the letter reads as follows:

> V. What if all present disputes about computer errors and their interpretation could be resolved.

> Let us assume that we could all agree that your analysis is free from error and we agree that you have found a significant zero delay excess. Have you discovered gravitational radiation? The answer would be—not proven. The reason is that in the present discussion over your experiments and methods, a fundamental criticism of your experiment has been put to one side. That criticism is that you record the outputs of both antennas on a single tape. This I consider to be a dangerous thing to do. You have not yet proven that this procedure is free from some local effect having to do with the Maryland tape recorder or instrumentation.

> In closing let me state that I regret that our recent communications and correspondence has been side-tracked somewhat to a personal discussion of who said what, away from the central question of whether there is gravitational radiation.

Levine and Garwin also submitted a paper to *Physical Review Letters* just before the *Physics Today* submission; it was received by *PRL* on June 24, 1974, and published on September 23.[21]

The *PRL* paper is very forthright. It records experimental results—which are much more appropriately presented in such an outlet as opposed to *Physics Today*—and claims that these results are in "substantial conflict with the detections reported by Weber."[22] Levine and Garwin do, however, report that they found one large unexplained pulse, but since there was no cor-relation with the detector at Rochester, they feel justified in discounting

21. Levine and Garwin 1974.
22. Levine and Garwin 1974, p. 794.

it. They also say that they are forced to estimate the sensitivity of Weber's antenna, because figures for his detection efficiency have not been published.

> We are thus forced to estimate these quantities, while noting that such information is easily obtained by the experimenter and is normally provided in the publication of a detection experiment. (p. 796)[23]

Levine and Garwin consider whether the difference between their result and Weber's might stem from fluctuations in the source or difference in the nature of the signal that the two apparatuses could optimally detect, but they eliminate these by citing the "no shoulders" argument. Surprisingly, given that we are looking in the "constitutive forum" they go on to remark,

> Indeed, the data of [Weber's 1973 publication] were processed with a faulty computer program.... This error was shown to account for essentially all of the zero-delay excess events on a four-day tape of the data...
> We thank D. H. Douglass for many useful discussions. (p. 797)

It seems that they did not send a copy of this *PRL* submission to Weber, who wrote to the editor of *Physical Review Letters* on October 28, 1974, complaining that the journal had published "incorrect and very misleading information about our experiments." The burden of the complaint was that the computer error was said, in the article, to account for all the zero-delay excess, but that it had been reanalyzed and an excess of eight per day had been found. Weber wrote,

> I am astonished that this paper could appear in Physical Review Letters without the editors even having my comments about it. My first view of the Garwin Levine paper occurred when I saw the journal.
> Our experiments have repeatedly and incorrectly been criticized. The result is loss of financial support and destruction of an important research program.

He had written to IBM on October 16 as follows:

---

23. That is, the result of electrostatic calibration tests.

Dear Dick and James,

... Your recent Physical Review Letter changes the picture. You have published statements about our work which are incorrect and in disagreement with documented studies by other groups.

Our more recently evaluated data may convince you, along with the documents mentioned in earlier correspondence.

During the next few weeks I hope to visit IBM, if I would be welcome, and bring all these data for your examination.

In early 1975 there was an exchange of letters between Garwin and Howard Laster, Weber's department head at Maryland. Garwin included a full copy of all the correspondence between him and Weber. Laster's replies to Garwin are not hostile. In particular, he expresses thanks for Garwin's sending a copy of the "Langmuir talk."

What of the letters to be published in the November 1975 letters section in *Physics Today*? It will be recalled that Weber, in his reply to Garwin published in that journal in December 1974, had complained that the IBM detector was the least sensitive of all the detectors, while the Munich-Frascati devices, which were most like his, had found a zero-delay excess. He went on to express puzzlement that the smallest of all the detectors in the world should be taken to cast serious doubt on the findings of his own much larger detector, and asked why IBM did not "beef up" their own detector and settle the controversy.

Garwin's reply to *Physics Today* explained that the size of the IBM detector had nothing to do with the issue at hand: "Weber brings my own experiments into discussion, but these have nothing at all to do with the point of my letter."[24] Garwin says, on the other hand, that

> Joseph Weber's reply to my letter ... in no way addresses my claim "that the Maryland group has published no credible evidence at all for their claim of detection of gravitational radiation." He states "My first-hand knowledge is based entirely on other data including real-time counting and pen-and-ink records," but this evidence has not been published. Q.E.D. (p. 13)

In his Response, Weber says, "It is regrettable that Garwin continues to publish incorrect information about Maryland experiments after

24. Garwin 1975, p. 13.

submission of his earlier letter in PHYSICS TODAY."[25] He reiterates that the computer error, as set out again by Levine and Garwin in *Physical Review Letters,* has been corrected and that those tapes show a 5.7-standard-deviation effect.

Next, Weber returns to the experiment, arguing that his own "energy algorithm" is more sensitive than Garwin's "amplitude algorithm."

> Expecting increased sensitivity of his amplitude algorithm led Garwin to use a much smaller mass than that used at Maryland. Furthermore he abandoned the well established technique of a two-detector coincidence experiment.
>
> Garwin's results were negative. Maryland experiment using the larger mass and both algorithms simultaneously show, for a recent $2^1/_2$ month period, a larger event rate for the energy algorithm. An explanation suggested by our brilliant student Gustaf Rydbeck is that the pulses are longer than Garwin thought and that they frequently sweep through the detector bandwidth. (p. 15)

Weber explains, "Garwin's use of too small a mass, together with an algorithm tailored to signals having different spectral character than most of the ones we observe, account for his negative result" (p. 99).

These exchanges in *Physics Today* seem to give us a good idea of the "texture" of the debate in these crucial couple of years—the period when Weber's results moved from being intriguing to unlikely to "wrong." On November 3, 1975, Weber wrote to Joseph A. Burton of Bell Laboratories, complaining about Bell's and IBM's treatment of him and claiming that

> [t]he end effect of this BTL-IBM activity has been to reduce our level of federal funding from $200,000 to $43,000 per year. Our research program has been seriously curtailed and may be destroyed.

The *Physics Today* letters contain a very nice illustration of a central feature of the whole debate. Weber always fell back on experiment, saying, in effect, "If you build apparatus that is like ours, you will see what we see; only then should you try to explain what is wrong with the observations." Not unreasonably, he values his own experiment above others; it is, perforce, the only experiment he knows intimately—the only one he can be sure has been done with integrity. And Weber generally values experiment

25. Weber 1975, p. 13.

above theory. Garwin's response is to minimize the importance of experiment and substitute theory. There are three kinds of theories at stake.

First is cosmological and astrophysical theory, which tells us that Weber must be wrong because he sees too much gravitational radiation. Weber is never directly involved in this argument, leaving the astrophysicists to argue about whether we know enough about the universe to be sure that he is wrong. A few years later (see chapters 19–21), he is to try to show how his findings can, in any case, be reconciled with the conservative astrophysical convention.

Second is the theory of data processing, where Garwin and Levine win indisputable victories by indicating that Weber's data have no shoulders where there ought to be shoulders; demonstrating that his theory of proliferation cannot be supported by the data; and using Weber's computer errors to back up their case that the zero-delay excess is the outcome of statistical massage; on almost any interpretation, Weber loses this phase of the argument.

What do I mean by "on almost any interpretation," and by an "indisputable victory," given that my overall point is that every victory in science rests on a body of taken-for-granted assumptions, each of which can be questioned? I mean that the assumptions on which this victory was based were so ubiquitous—for example, they would be a foundational part of the world of even a Joe Weber—that to attempt to remove the corresponding stones from deep in the dam (chapter 5) would be an act of self-destruction. Therefore, the victory in these battles had to be ceded for there to be any continuing chance of victory in the war. It might be done quickly, explicitly, and gracefully or slowly and gracelessly, in the hope that no one would notice. Weber made the mistake of choosing the second option. He had damaged himself enormously by making the computer error and the four-hour error, and he then damaged himself further by the ill-judged way he handled these mistakes. If he had thanked Douglass fulsomely—especially given that Douglass himself was not interested in making the debacle public—and continued his work while being manifestly contrite, other physicists, nearly all of whom make mistakes from time to time, would almost certainly have sympathized. Weber could have enhanced his reputation for integrity rather than damaged it. By resisting so hard he exposed himself to more and more attacks from Garwin.

Finally, there are the midrange theories of mainstream physics, which link the IBM lightweight detector to Weber's detector via a nexus suggesting that the output of the former does stand in direct contradiction to the output of the latter even though the latter uses a much heavier mass.

Garwin and Levine argue that the IBM detector is *the same as* the Maryland detector in the sense necessary for one to test the findings of the other. It is this last stone in the dam that Weber is intent on levering out; he refuses to accept this theoretical linkage, arguing instead that the experiments should be done his way irrespective of the linking theory. He says the two experiments are *different,* not the same, and this enables him to claim that his experiment is *better.*

Weber does not believe that an experiment can be described in a formal way such that two such formulaic descriptions can be directly compared, away from the laboratory bench, and reveal which one is the most sensitive. He thinks there is more to experiment than is captured in the formal description. To put this another way, Weber values his data above the theoretical arguments that would render the experiments comparable. He also provides an argument—the peculiar spectrum of the signals—that shows one way in which the experiments, even when considered in terms of their formal descriptions, should not be considered the same and why his data should be valued above conventional theory.[26] If Weber is right about the peculiar signals, then the difference between his apparatus/algorithm, and Garwin's apparatus/algorithm, has resulted in a *discovery* rather than a refutation of Weber's claims. While Garwin is saying that the difference in output of his and Weber's apparatus shows that Weber is seeing noise, according to Weber this difference signifies a signal of great importance. But the sheer weight and flourish of Garwin's arguments would lead to closure in favor of Garwin, not Weber.

I have indicated how the experimenter's regress was resolved in this case. The growing weight of negative reports, all of which were indecisive in themselves, were crystallized, as it were, by Garwin and Levine. Henceforward, only experiments yielding negative results were included in the envelope of serious contributions to the gravitational waves debate. Once the IBM team had made its contribution to the transformation in socially acceptable opinion, there simply were no high fluxes of gravity waves. Henceforward, all experiments that produced positive results like Weber's must, by that very fact, be flawed.

26. In later work, Weber would reveal that he believed the difference in transducer design was the key to the difference between his apparatus and all the others, rendering all previous theoretically deduced arguments for their similarity redundant in his eyes.

CHAPTER 10

AN ATTEMPT TO BREAK THE REGRESS: THE
CALIBRATION OF EXPERIMENTS

Though the demise of Joe Weber's gravitational wave claims has been largely explained, it is worth reexamining one other classic attempt to break out of the experimenter's regress—calibration. Calibration is especially interesting because it attempts to use an experiment rather than a formal description to measure the efficacy of another experiment. As I mentioned earlier, one must anticipate that the measuring experiment will be subject to the regress even as it is used to try to break the regress in the measured experiment. Now we can see how this works out in practice.

The calibration of instruments is a familiar procedure. Imagine a prototype voltmeter has been constructed. It consists of a needle which swings across a scale, but as yet the scale is blank. To calibrate the instrument, known voltages are applied to the terminals, and the corresponding positions at which the needle comes to rest are recorded. Thus, marks corresponding to known voltages can be inscribed on the scale. Henceforward the meter may be used to measure unknown voltages; the unknown voltage is applied to the terminals, and the mark against which the needle comes to rest gives its value.

The assumption built into this procedure is that the unknown voltage acts upon the meter in the same way as the standard voltages that were applied to calibrate it. This is so slight an assumption as hardly to be worthy of the name. After all, a voltage is a voltage is a voltage! Nevertheless, it would be correct to say that during the calibration of a voltmeter, standardized voltages are used as a surrogate for as-yet-unmeasured signals. In more controversial science, the assumptions underlying the process of calibration are of greater moment and visibility.

The very first experiment to calibrate a gravitational wave detector has already been described at length; this was Joel Sinsky's experiment. But the medium of calibration—the surrogate force—was not gravitational waves but plain old Newtonian gravitational pull, made to alternate at a frequency that was the same as the natural frequency of the bar detector. It was a tiny force, likely to induce in the receiving bar changes in length of about the same order as would be induced by a gravitational wave—approximately 10^{-16} meters. I have already explained why the Sinsky method did not become a standard: it was very difficult and time consuming to carry out successfully, and it was very hard to eliminate flaws.

In subsequent work, Joe Weber used another method called a noise generator to calibrate his instrument. He was the only person ever to use this method, and I confess that I do not have a full grasp of what it was. In any case, it did not figure largely in any debate.

By the time Weber found himself involved in arguments with his detractors, a different calibration method entirely had become the norm. This involved giving the bar a little "nudge" by inserting a pulse into an electrostatic end plate mounted close to the end of the bar. One could know exactly how much of a nudge had been given, and one could use one's various signal processing algorithms to try to see the nudge. This method of calibration is now standard practice among the bar community. Such false signals are inserted every now and again into the bars in order to gauge their state of sensitivity on a continuing basis. Weber, however, resisted the use of the method even though it was becoming the standard. One of my respondents put it this way:

> [1975] We had calibrated our own antennae in a unique way which depended in no way on calculation. So we knew what our sensitivity was, and at that time we could only calculate what Weber's sensitivity was. So you're right in saying that the relative sensitivity was something that was on the one hand calculated and on the other hand known to absolute accuracy.... Soon

thereafter, we did get an opportunity to go down and calibrate Weber's antenna, and we found...our calculations were correct.

As this respondent suggests, the outcome of the tardy electrostatic calibration of Weber's apparatus was seen by most to be a vindication of the critics' calculations. It was felt to be a decisive demonstration that the sensitivity of the critics' antennae was at least as great as Weber's. In particular, an argument concerning the correct way to process incoming signals seemed to be settled. As we have seen, Weber insisted that the maximum net sensitivity was to be obtained by a nonlinear, or energy, algorithm, while his critics insisted that a linear, or amplitude, algorithm was the best. As one respondent explained,

> [1975] For a signal with a sine wave underlying...it turns out that a system which is linear can be shown theoretically and quite soundly to be the best system for detecting things. But Weber's always used the nonlinear system, and so his initial claim was that it was just clearly superior because he finds gravity waves with it, whereas people with the linear systems don't. Despite the fact that you can prove rigorously that it's not so.
>
> Well, Weber was pushed very hard on this, and he finally implemented both systems...and he hooked up to the same detector both a linear system and a nonlinear system...and what he found was that he did indeed find *gravity waves* more often with his system. However, finally, after much pushing, he put calibrators on—things that could simulate gravity waves—and it turned out that the linear system was about twenty times better at finding the calibrator signal. [my emphasis]

But using this method of calibration moves us a long way from gravity and further still from gravitational radiation, and Weber exploited the differences in his arguments. As we have seen, he argued that the electrostatic end-plate method was dangerous, because it required the introduction of extra electronic circuitry into the neighborhood of the bar. He argued—and he had good reason to argue, given his early experiences in trying to insulate the bar from external forces—that the circuitry could be directly exciting the bar, giving the impression that it was far more sensitive to the nudges than it really was.

Second, he argued that the nudges did not resemble gravitational waves in their shape. This argument is nicely brought out in an extract from a 1975 interview:

> *Collins:* In reading your 1974 publication, I understand that you did a calibration experiment using both algorithms and that you got a better result with the linear algorithm?
>
> *Weber:* No, that's not right. The linear algorithm used by other people is unquestionably superior for short pulses—let me make that absolutely clear. There are certain arguments given for use of the linear algorithm. These arguments are applicable to short pulses, and in my opinion they are correct arguments. And the fact that the linear algorithm is not in fact more sensitive is giving us information about the character of the pulse. It means that the character of the pulses does not fit the assumptions which went into that method of analysis . . . so far, we think of several kinds of signals that would give results somewhat similar to the ones we see.

This is the move that I described at the end of chapter 9, in which a difference in outcome of the experiments is interpreted as a discovery rather than a refutation.

Weber's critics interpreted this move less positively. One remarked,

> [1975] What he did was to change the nature of the signals. He said, "Well, the signals must not be of the form which we've been assuming. They've got to be something else now." Some strange waveform of which he failed to give a single example. "And so my algorithm is now best again." In fact that resolved a lot of difficulties for him. He was wondering why we didn't see his signals. And he said, "Now I know why. The signals are of a weird form."

Another respondent remarked on the failure of Weber's algorithm in the calibrator test.

> [Y]ou have this incredible conflict that when you look for gravity waves, the other system seems to do a better job—that's a perfect example of a negative experiment done by the author. It demonstrates that there's nothing there.

To go straight to the end of the story, Weber's interpretation of the calibration results was greeted with skepticism. He did manage to invent hypothetical signals compatible with the calibration test; they had a pulse profile such that they would be more easily detected by his antenna, using his algorithm, than by his critics' methods. However, the existence of such signals was thought unlikely by most scientists. According to one respondent, signals having such a profile were "pathological and uninteresting."

In other words, it would be difficult to think of cosmological scenarios that would give rise to signals with such strange and exact signatures. In the current state of astronomy, Weber's hypothesized signal shapes were too implausible to be considered seriously. Thus, Weber's explanation of the reasons electrostatic calibration was unsuitable proved unconvincing, thereby leaving it open for electrostatic calibration to play a large part in diminishing the credibility of his overall claims.

In retrospect, Weber would have served his case better had he maintained his refusal to use electrostatic calibration—not just because the results proved unfavorable but because of the assumptions taken on board by the act of calibration and the restrictions of interpretation imposed as a result. In agreeing to calibrate his bar electrostatically, and in choosing his arguments when the test went against him, Weber set beyond debate certain assumptions about the relationship between the surrogate and the real signal. He accepted that gravitational radiation would interact with the substance of his antenna in the same way as electrostatic forces. He also put it beyond question, at least for the time being, that the insertion of a localized pulse into one end of a bar antenna would have a similar effect as the insertion of energy into the bar as a whole from a source at a great distance.

These assumptions might seem so slight as to be hard to dispute; yet as has been shown in this chapter, there were times when at least informal discussion took place as to whether gravitational force might be coupling more effectively than expected with the matter of the bar through the release of latent energy via mysterious mechanisms such as "proliferation" or the release of "metastable states." These mechanisms might not be affected by a local force in the same way as they would by a distant force even failing the existence of something special about gravitational radiation per se.

ANOTHER ALTERNATIVE SURROGATE

Lest these considerations seem too bizarre, it should be pointed out that direct gravitational force as a surrogate signal has been favored, first in the Sinsky experiment and subsequently using simpler methods. Peter Aplin, whom I interviewed in 1980, was planning to use gravitational calibration once more. He intended to use the fluctuating gravitational attraction induced by a small spinning bar of material located close to the antenna. This would give a bigger and more easily seen effect than Sinsky's vibrating bar but would still use the medium of gravity.

[1980] *Collins:* What is the advantage of the spinning bar calibration over electrostatic calibration?

Aplin: Well, since it couples gravitationally to the antenna, it does give you a somewhat more basic measurement—if you like—it's still not really what you want. It still doesn't duplicate the effect of gravitational radiation, because it's a near-field effect, and the spinning bar really only couples to one end of the thing instead of coupling uniformly to the entire antenna. So this is the limitation of this sort of approach. The spinning bar is more appropriate with something like a Weber resonant antenna, where you can more nearly couple to the antenna...

Collins: How certain can you be that electrostatic calibration pulses are acting as an exact analogue of gravity?

Aplin: Oh, they're not. They're certainly not.... From simple measurement [using electrostatic calibration]... I know precisely the force that I'm applying... and I can calculate the size of the signal that I ought to get out of the transducers and that's all. But it doesn't mimic the effect of a gravitational wave on the antenna. And that's true whether it's this sort of antenna or whether it's a resonant bar. The fact is that the gravitational wave interacts with all parts of the antenna, with all of the mass of the thing, and there's just no way of reproducing that—at least there's no way I'm able to think up of producing that effect...

What you are trying to do with electrostatic calibration is to check your theoretical calculations.... What you can't test in this way is the theoretical calculation that tells you precisely what happens when a gravitational wave of a certain amplitude hits the antenna.

For Aplin, with his more complex antenna and his idea for a different method of calibration, the assumptions underlying electrostatic calibration were worth analyzing and circumventing if possible. He had thought of a way of avoiding the need for electrostatic impulses, using instead changes in the gravitational attraction of a local mass. He was, however, still unhappy with the need to use a localized source rather than a powerful distant source that would more nearly mimic effects of gravitational radiation on his antenna.

Actually, Aplin's bar and his calibration experiment were never completed, but the same gravitational method of calibration, as I was to discover, was used by the Japanese experimenter Hiromasa Hirakawa, who built a low-frequency antenna to look for the Crab pulsar. And the same method would be used in conjunction with the EXPLORER cryogenic bar in Geneva. In Geneva a rotating mass was mounted on a movable frame, and it was shown that this bar could detect the fluctuating gravitational field with signal strength consistent with the distance of the calibrator from the bar. The actions of these research groups demonstrate that had he been determined enough, Weber could have held out longer against the conclusions that were drawn from the electrostatic calibration. In accepting the scientific legitimacy of electrostatic calibration for his gravitational antennae, he accepted constraints on his freedom to interpret results. After he agreed to calibrate his bar electrostatically, freedom of interpretation, at least in the near term, was limited to pulse profiles rather than the quality or nature of the signals.

Making Weber calibrate his apparatus with electrostatic pulses was one way in which his critics ensured that gravitational radiation remained a force that belonged well within the ambit of physics as it was understood at the time, and that one set of deeply set stones were not removed from the dam. It ensured the continuity of physics—the maintenance of the links between past and future. Calibration is not simply a technical procedure for closing debate by providing an external criterion of competence. Insofar as it does work in this way, it does so by controlling interpretative freedom. Insofar as this control on interpretation is maintained through calibration, it is the control, not the "test of a test" itself, that breaks the circle of the experimenter's regress.

CHAPTER 11

FORGOTTEN WAVES

The scientific debate over gravitational waves in the larger part of the community was now over. If this was a standard history of science, Joe Weber would now become invisible. But Weber did not give up; he was to continue to champion his findings for another 25 years. His efforts would not meet with success, but they show what it is possible to do within the scientific community, and they reveal what that community does to control its mavericks. The smaller splashes and extinctions to come are as interesting sociologically as the noisier events of earlier years. To begin, I will describe at some length Weber's next paper, which repeats many of his previous arguments. The point of the description is, once again, not to talk about the science but to better understand the meaning of the science's eventual fate.

In 1976, Weber and his collaborators published a fourteen-page review paper in *Physical Review D*. As is standardly revealed beneath the author's byline, the first draft of this paper was received by the journal on December 16, 1974; the revised manuscript arrived on April 26, 1976, a sixteen-month delay. One can see that much agony and argument had gone into the paper's preparation.

COINCIDENCES ARGONNE-MARYLAND DEC 15-25, 1973

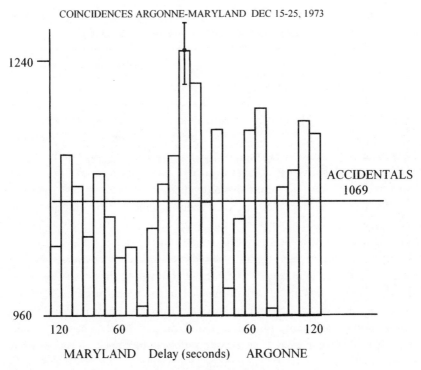

Fig. 11.1 Delay histogram from Weber's 1976 paper (fig. 3).

In this paper, Weber reviews and analyzes his recent findings. In spite of their almost unanimous rejection by the scientific community, the National Science Foundation had continued to give him funds so that he could reanalyze his results as thoroughly as possible. No doubt it was hoped that Weber would discover errors deep in his data and that the controversy would end. Had this happened, Joe Weber would have been much more of a central figure in the field of gravitational wave detection as it has developed since the 1970s. But Weber's 1976 paper was uncompromising. In it we find twenty delay histograms, of which most look like figure 11.1, a rough rendering of figure 3 in the original paper—a figure that I selected arbitrarily.

As can be seen, this figure covers coincident bursts of energy in the bars at Argonne and at Maryland between December 15 and 25, 1973. There are eight such figures (figures 2–8 in the original), each representing the same data distributed into the same number of boxes on the same horizontal scale but with a different threshold each time. As the article explains quite correctly,

> An important aspect of the search window is the set of thresholds employed for a coincidence experiment. Suppose there are small numbers of moderately strong signals. Setting thresholds too low will give a very large accidentals rate with large fluctuations which mask the signal. Setting thresholds too high will result in no coincidences at all.[1]

The eight histograms cover what Weber refers to as "signal space in the threshold dimension." Each histogram has a clear peak at zero delay; sometimes it is higher and sometimes lower, but in every case it is obvious to the naked eye. The calculated statistical significance of these peaks varies from 2.1 4.8 standard deviations, an impressive outcome.

The paper also includes one of Weber's attempts—proliferation, previously discussed in chapter 9—to address the accusation that he was seeing too many pulses to make sense. He points out that what his computer algorithm counts as a signal is a threshold crossing. If a bar's average energy state is raised by the impact of a gravitational wave and the bar's subsequent ringing, he explains, even a small increase in the background noise of the detector could cause a threshold crossing. Thus the small fluctuations in noise that are normally unnoticed could cause a whole series of apparent signals, so that what is really one pulse could appear as many—what Weber calls proliferation. Proliferation, Weber claims, increases the number of apparent signals. He says that his rivals' computer programs measure only sudden changes in the signal in any direction and miss out on the proliferation bonus; though these programs might be thought to be more efficient because they look for all energy changes, they are actually less efficient.[2] We will see other explanations of the high number of apparent gravitational waves as the story unfolds.

---

1. Lee et al. 1976, p. 895. I am afraid, however, that I cannot square this with Garwin's claim, made in chapter 9, that as Weber's threshold falls the number of his zero-delay coincidences *and the confidence level of his statistics* get better! At this point my attempt to maintain symmetry falls. But I am not a scientist; maybe I am just not ingenious enough to see the way around this problem. Unfortunately, as with some other points, Weber is no longer alive to ask.

2. In interviews, respondents told me this was Weber "violating the second law of thermodynamics." I find it hard to disagree. Here is another point at which I find it hard to turn off my residual scientific instincts and retain "symmetry." I cannot "get around" a remark made by James Levine in a letter he wrote to Weber in 1974, pointing out that the proliferation effect should have exactly the same result on the noise and signal, affecting the statistics in no way. (The mean level of the nonzero bins in the delay histogram would go up as much as the zero bin, leaving things as they were.) Furthermore, Garwin points out elsewhere that proliferation of this kind would degrade the time resolution of the detectors and therefore make it harder to find "zero-delay" signals amidst the noise. So, if Joe Weber were still alive, I would need to ask him how he would respond to these seemingly fatal criticisms.

Weber also develops his explanation of why his detector can see certain events to which his rivals' devices are blind. He says the problem is that "[t]he spectral character of the signals which cause the coincidences is unknown. A number of physicists have assumed that the signals consist of short bursts, less that 50 msec [milliseconds] long." He explains that his rivals' reading of calibration tests on different data processing algorithms, even if valid for short bursts, might not reveal the true picture for signals with different signatures. Proliferation means that a few very large signals would confound the comparison. Then there are signals that "sweep through the detector bandwidth" (p. 899). His procedures would be more sensitive to these, too.

Weber repeats the argument that electrostatic calibrations are very difficult because the electromagnetic pulse can affect the amplifiers directly without being mediated by the bar's vibrations, giving a spuriously high response to the calibration and a spuriously high measured sensitivity of the bar. He states, "Extreme care is required to avoid introducing some of the [calibrating] pulse energy directly into the electronics" (p. 897). Another possible problem with his rivals' algorithm is its extreme sensitivity to temperature variations in the antenna. "Gustaf Rydbeck has performed computer simulations which suggest that very small variations of temperature may be sufficient to modify the computed relative sensitivity of algorithms A and B [his own and his rivals'] and lead to the observations reported here" (p. 899). A series of delay histograms then show the responsiveness of Weber's bar to calibration pulses.

Weber admits to the problem of the computer error, albeit somewhat gracelessly.

> Copies of our magnetic tape 217 (June 1–5, 1973) were sent to other laboratories in August 1973 to check programs and resolve differences. A program error was discovered by D. H. Douglass. Unfortunately, Douglass employed insufficiently long time delays and his calculation of the accidentals rate was, in our opinion, in error. As a result he and other physicists concluded incorrectly, we believe, that there was no significant zero-delay excess on tape 217. (p. 900)

This, presumably, was the "much worse computer error" that he earlier reported Douglass to have made. Weber tells us that the first fully correct analysis of tape 217 was carried out by a Paul C. Joss of MIT in April 1974. He says that "several additional months of correspondence were required to obtain reasonable agreement among four groups concerning the data on tape 217"

(p. 900). We are presented with delay histograms representing tape 217 and other data, now corrected for various errors that others have pointed out. In each case the zero-delay bin stands proud, sometimes signifying a result with a statistical significance as high as 5.7 standard deviations.

It had been said that the telephone link between Argonne and Maryland could have been a cause of spurious coincidences—perhaps electrical noise at one site was being transmitted down the telephone link and causing coincident pulses at the other site. Weber explains that for one period, June 5–13, 1973, the telephone link was disconnected and synchronization was achieved through clocks alone. He provides a delay histogram for the six days of data collected when the clocks were disconnected. Again we are shown a delay histogram for the data (figure 24). The zero-delay excess stands out clearly, but no statistical significance is provided. We are told, however, that a threshold was chosen so as to maximize the zero-delay excess.[3]

More results for periods running through to October 1974, with their accompanying delay histograms, are then reported. There were some periods of quiescence, but most periods show a clear zero-delay excess.

Weber then uses his idea of proliferation in a more explicit way, defining an event as threshold crossing accompanied by at least three more threshold crossings within 30 seconds. This analysis yields a very fine delay histogram for data collected between May 21 and June 25, 1974. Next, Weber discusses exchanges of data with other groups. "A number of data exchanges are in progress. At the request of other research groups some of the results may not be published" (p. 904). A little later he says, "We believe that another data exchange does give statistically significant results with a 4-standard-deviation zero-delay excess. The other research group has not agreed to publish at this time" (p. 904). These references are to the exchange of data with Tony Tyson at Bell Labs mentioned in chapter 6.

Weber also covers the embarrassing four-hour episode but with little evidence of contrition.

> Preliminary results of an exchange of magnetic-tape data involving antennas at 710 Hz [frequency response] and 1661 Hz have been discussed at meeting without claims for a positive result. A 2.6-standard-deviation effect was

3. Read at this distance in time, this analysis is not very convincing and seems to have allowed the analyst the advantage of post-hoc statistical manipulation; one would have liked to see the results for other thresholds.

observed at a delay originally thought to be zero. Subsequently it was shown
to have occurred at a time delay of 4 hours. A slight larger effect was
observed at correct zero delay. Since neither result exceeds 3 standard
deviations, both may be purely statistical fluctuations. (p. 904)

In the Discussion section of the article, Weber tells us that "[a]n event
rate cannot be considered a universal, invariant law of nature since almost
nothing is known about the source" (p. 904). As he explains, since the source
might be highly variable, the variation in the observed event rates, from 5.7
standard deviations over four days, to zero, and the variation with change of
threshold, are to be expected. Weber explains that missing days in the data
record for particular periods are entirely due to power failures or excessive
tape-writing errors. He says that the apparatuses of other groups, which
found no signals, are either smaller or use a data-processing algorithm
designed for short signals. He believes his results suggest that the signals
are long, thereby invalidating many aspects of the comparisons between
his experiments and those of others.

FORGETTING

In 1977, Weber published his last paper in the professional journals pertaining to the "phase" of his career in which he tried to defend his findings
without straying too far from orthodoxy. It was a short letter to *Nature,*
received on October 5, 1976, and accepted for publication on January, 19,
1977, in which he simply restates the precautions he had taken to eliminate
all human bias and computer errors from his analysis. He concludes that
the zero-delay excess remained at a level of statistical significance of 5.6
standard deviations.

Self-citations aside, according to the *Science Citation Index,* the 1976 paper was never cited by anyone (though Tyson cited it in 1982, I could not
find the entry in the *SCI*), and the 1977 paper was cited once, in 1979,
in a geophysics journal.[4] Thus, in 1976 and 1977, Weber published major
defenses of his work in the world's most important journals, answering all
his critics, as he saw it; but the papers were close to invisible. In terms of
their impact, they might just as well not have been written.

4. Jensen 1979.

THE LAST PIGEONS RETURN TO THE ROOST

In the meantime, the field of experimental gravitational radiation detection had not been standing still. The exciting topic at the conferences and in the journals was the attempt to build resonant bars that would suppress background noise by running at liquid helium temperatures or below. These experimental efforts were directed at looking for the much less easily seen fluxes of gravitational radiation that were predicted by the cosmologists and astrophysicists. I will use a neologism for these gravitational waves: LowVGR, which stands for Low-Visibility Gravitational Radiation. HighVGR—High-Visibility Gravitational Radiation—is what all the experiments I have described so far were looking for.

Though the focus of the science had shifted, there was still unfinished business in HighVGR. There were groups who had been working and reporting their results on the conference circuit without formal publication. The last two or three of these results now trickled into print, though these reports were essentially ritualistic; they were irrelevant to consensus formation in the core set, which, as we have seen, had already gelled. In April 1975, *Journal of Physics A* received a submission from W. D. Allen and C. Christodoulides, who had built two bars on the "split-bar" principle in the UK.[5] The bars both weighed 625 kilograms and had a main frequency of 1180 Hertz.[6] One bar was built near the University of Reading and the other at the Rutherford Laboratory, giving them a separation of 30 kilometers (a separation, note, that is fine for a negative experiment but insufficient to justify a positive result, since at that distance there is a considerable chance of spurious effects impacting both bars). They found nothing of interest in their delay histograms and concluded, "unless Weber is detecting pulses of a narrow spectrum and near his frequency of 1661 Hertz, all attempts to verify his claims of gravitational radiation have failed" (p. 1733).

On May 12, 1975, *Physical Review Letters* received a submission from the groups headed by Douglass and Tyson at Rochester and Bell Labs, Holmdel.[7] Their two 710-Hertz bars each weighed 3700 kilograms. They display their delay histograms, which show no zero-delay peaks, and describe themselves as having set an upper limit to the possible flux of gravitational waves. "As indicated, this 710-Hz upper limit is several orders of magnitude below

5. Allen and Christodoulides 1975.

6. But claimed to be able to see signals up to 300 Hertz on either side away from the main resonance.

7. Douglass et al. 1975.

the claimed event rate of the University of Maryland groups at 1660 Hz"
(p. 482).

The last of the papers which set the experimental results of room-temperature bars against the results of Weber was published as late as 1982.[8] In this paper, Tyson and his collaborators report the results of a total of 440 days of observations with a single detector, setting a new upper limit to the flux of gravitational waves that can be seen. As they argue, one can set an upper limit to the size of the flux with a single detector even though one cannot prove the existence of gravitational waves in this way (we will return to this point in chapter 40). The team conclude that they have developed new ways of "vetoing" local disturbances, and that during their 440 days of observations there is only one "event" which they cannot explain as being caused by a nongravitational source. As for this one unexplained event, they mention that another group—running a low-temperature bar at Stanford University—failed to see it, and so it, too, cannot have been caused by a gravitational wave. In other words, they use the coincidence technique to veto their one gravitational wave candidate. Tyson's earlier claims, that Weber's results can be explained away as being of geomagnetic origin, are not mentioned in this paper.

More interestingly from our point of view, the paper reviews the earlier results of experiments conducted by various groups. The parameters of fourteen resonant bar experiments—including Weber's of 1970 and of 1976—are listed, with estimates of the sensitivities of all the apparatuses or reports of the outcome calibration tests. The following table 11.1 is a simplified version of their summary table presented on page 1210.

We can use the table to reveal just how fragile the estimates of competing sensitivities were and to provide a handy reference for many of the points made throughout this section. Notice that the Silvano Bonazzola experiments carried out in Paris are not mentioned. Notice, too, that in the Noise Temperature column there are many tilde ($\sim$) symbols. Noise temperature is a measure of sensitivity, because it shows the extent to which background noise has been calmed: the smaller this number, the more sensitive the detector; a tilde indicates that the noise temperature has been estimated rather than directly measured by inserting calibration pulses. Estimates of sensitivity were continually disputed by Weber, and he also disputed the significance of sensitivity measurements for the waveforms he believed he might be detecting. Even in 1982, we find Tyson needing to report that the first number in the column, 630, has been calculated by

8. Brown, Mills, and Tyson 1982.

Table 11.1 Summary of bar characteristics

	Location of detector	Date of last report	Frequency	Mass	Noise temperature °K	F(v)	Alg	Duration days
1	Maryland/Argonne	1970	1660	1.7	>630	970	1	180
2	Maryland/Argonne	1976	1660	1.7	~>22	~>40	1a	300
3	Maryland/Argonne	1976	1660	1.7	~22	~40	2	270
4	Glasgow/Glasgow	1975	900–1100	0.3	4.5	35	1b	500
5	Moscow/ISR	1973	1640	1.3	~280	500	1b	10
6	IBM	1974	1660	0.5	18.5	87	2	27
7	Reading/Rutherford	1975	1070–1300	0.63	~12	~44	1b	210
8	Tokyo/Tokyo	1975	145	.078	10	300	2	80
9	Munich/Frascati	1975	1645	1.7	7.5	12	2	150
10	Stanford	1980	842	4.8	0.02	0.01	2	210
11	Bell Labs	1973	710	3.6	~50	33	1b	89
12	Bell Labs/Rochester	1975	710	3.6	18	12	2	87
13	Bell Labs/Rochester	1975	710	3.6	9	6	2	33
14	Bell Labs	1972	710	3.6	6	3.8	2	440

Adapted from Brown, Mills, and Tyson 1982, p. 1210.

another scientist (Ho Jung Paik) to be 240. We also find in footnote 18 that Tyson admits making a mistake in the calculation of the sensitivity of one of his early experiments; this was pointed out to him by a Russian team.[9]

In the next column [F(v)] is another number representing sensitivity. It is the "minimum detection level," a number worked out by Rainer Weiss in 1978;[10] again, the lower the number the better. It is worth noting that the relationship between the sensitivity of the experiments looks rather different under Weiss's measure as compared with the previous column. For example, there is less difference between Weber's apparatus and that of the Glasgow group; the Braginsky experiment looks even less of an effective competitor; the IBM detector looks worse than Weber's rather than better. So we can see that there is still plenty of room for argument about sensitivity even though it has to be admitted that there are significant experiments with much better performance than Weber's.

The column labeled "Alg" is, in the original, labeled "Detection Algorithm"; I have simplified the notation in the column. In my notation, "2" represents the phase-change, or "linear," algorithm increasingly favored by most of Weber's competitors, while 1, 1a, and 1b are variations on the algorithm favored by Weber. Tyson explains that algorithm 2 is the more

9. Braginsky et al. 1973, p. 271.
10. Weiss 1979.

sensitive to calibration pulses. Of course, if Weber were right about the strangeness of the waveforms he was seeing, then few of the other experiments would bear directly on his claims. Tyson, quite properly, refers to the 1974 Tel Aviv conference at which the efficacy of different algorithms was disputed.[11]

All this, of course, is irrelevant to the fate of Weber's findings; it merely indicates the high degree of interpretative flexibility still available. As though to demonstrate this very point, the paper described above was published almost immediately after the first of a series of papers by Weber and his collaborators that would mark an attempt to revive HighVGR.[12] The Weber et al. paper was published too late for any discussion of it by Tyson et al., or for it to be included in Tyson's listing of previous experimental work. But this paper was part of that phase of scientific work called theory and experiment—and in principle it could have turned the whole of Tyson's analysis into an anachronism. As may be anticipated, it did not.

11. Shaviv and Rosen 1975.
12. Ferrari et al., 1982. The associated events are described in chapters 19–21.

CHAPTER 12

HOW WAVES SPREAD

Our project is the study of the propagation of ripples in social space-time and how they spread through the concentric rings of the target diagram. In the different rings, ripples spread via different media. In the core, the important media are personal interactions and the conference circuit. Slightly further out from the core, the technically interested scientific community gets its information from the journals. Nowadays, the journals are part of the "literary technology" of virtual witnessing, rapidly being superseded by the electronic preprint.[1] A little further out are the journals read by a wider section of the scientifically literate public.

Here we can briefly examine two of these media, the scientific journals and the popular journals. I will examine the spread of ripples in the scientific journals by using the *Science Citation Index*, and I will take *New Scientist* as my example of a more-popular outlet, because it has enough mentions of gravitational wave work to reveal the way opinion changed.

1. Shapin and Schaffer 1987.

THE *SCIENCE CITATION INDEX*

When scientists write papers, they mention previous work that is relevant to their own. In other words, they "cite" the papers of other authors. The *Science Citation Index* collects and lists all these citations, noting where the citing was done and which papers were cited. One measure of a scientist's degree of recognition, notoriety, or whatever, is the number of citations that other scientists make to his or her work.

The *Science Citation Index (SCI)* covering the years 1965–69 shows Weber being cited a total of about 170 times—a respectable but not outstanding number of citations (see table 12.1). During this time, as we have seen, Weber was working out the principles of gravitational waves and wave detectors. He was also making his contributions to the idea of the maser and doing other less important work. It should be borne in mind that the *SCI* lags a bit behind a scientist's degree of recognition, because it takes time for other scientists to write and publish the papers that cite an exciting or important piece of work, and time for the *Index* to collect and publish the results. Unsurprisingly, the 1967 and 1968 papers in *Physical Review Letters* received only fourteen citations between them in the first five-year period, and the 1969 paper had not yet made an appearance in the *SCI's* returns by this date.

Table 12.1 shows citations to a selection of Weber's papers in *Physical Review Letters*. As for citations recorded to all Weber's works, the *SCI* for the years 1970–74 shows a sudden jump, which reflects the first real impact of Weber's gravitational wave findings. For these years there are about 700 citations, with about 600 of them referring to the gravitational wave detection period. As we can see, about 330 of these are to the *Physical Review Letters* papers, with the 1969 paper becoming a "star" reference, with 140

Table 12.1 Citations to Weber's papers: *Phys Rev Lett*, 5-year cumulations

Year	Vol	65–69	70–74	75–79	80–84	85–89
1967	18	9	12	7	5	3
1968	20	5	16	6	2	0
1969	22	–	140	49	16	15
1970	24	–	67	22	6	0
1970	25	–	97	26	5	0
1973	31	–	–	21	3	0
Total		14	332	131	37	18
All citations (approx)		170	700	330	140	100

citations all to itself. We can see from this that during the early 1970s, Weber was one of the most famous physicists in the world.

Somewhere around 1974 or 1975, Weber's claims, as we have seen, ceased to be believed by the majority of physicists. Not surprisingly, however, the decline in credibility is not reflected in the *Science Citation Index* until 1978; the *SCI* for 1975–79 still shows a total of about 330 citations to Weber's works. The *Physical Review Letters* papers up to and including those published in 1970 received 110 citations in this period, with the 1969 paper still being the standard reference, with 49 citations.

For the 1980–84 period, Weber's total citations had more than halved again to about 140, with only 50 or so citations to work produced in the heyday of the detection claims. Of these 50, 34 were to the *Physical Review Letters* papers up to 1970 (the 1969 paper gaining 16). Thus by this period, Weber's citation pattern had fallen back to that of a good but not outstandingly famous scientist. He was still recognized as the pioneer of what it was hoped would one day become the new science of gravitational radiation astronomy, and his early papers and his textbook were still admired; it was only the empirical claims (and later theoretical claims) that were suspect. The pattern continues in 1985–89, with most of the 100 or so references being to work from the early 1960s or earlier.

There has been a huge growth of interest in gravitational waves during the more recent decades, and therefore a much greater source literature of potential citers. Even taking that into account, however, the maintenance of Weber's aggregate citations continues to reflect his pioneering status. The movement to citation of his less contentious works is indicative. Indeed, one might argue that his recognition *is* the huge growth in the body of potential citers—he founded the field. From 1993 until the middle of 1998, the *SCI* records more than 200 such citations.

NEW SCIENTIST'S HEADLINES

The story of the first phase of gravitational wave detection and its impact on the outer rings of the target diagram can be seen in *New Scientist's* headlines and reports; here I provide the complete sequence of headlines. In the list the significance, as I indicate, sometimes lies in the fact that Weber's name is not mentioned at all in the report.

> 1967 The first signs of gravitational waves
> 1968 Further indications of gravitational waves

1969 Gravitational waves from space

1970 Gravitational waves may come from our Galaxy

1971a Concordes of the cosmos may create gravity booms

1971b Is the universe nearly dead?

1971c When an irresistible force meets an immovable object

1972a British astronomers tune in to cosmic vibrations

1972b Cutting the Galaxy's losses

1972c Gravitational waves come down to earth

1973a "Gravity waves" may simply be geomagnetic effects

1973b Another nail in the coffin of gravity waves

1973c Weber's "gravitational waves" need not be galactic

1975a Frozen scintillation of distant sources [Weber not mentioned]

1975b Gravity waves may power X-ray binary [Weber not mentioned]

1975c No gravity waves to be found in Berkshire

1976 More on gravitational wave front [Weber not mentioned]

1977 Gravity waves should be detected soon [Weber not mentioned]

Very roughly, from 1967 to 1970, *New Scientist* reports Weber's exciting new findings and their growing credibility. From 1971 to the second report in 1972 come a series of discussions of the cosmological implications of the high fluxes, with theoreticians arguing for and against the possibility of the findings; some are ready to discover mechanisms that will make Weber's results more rather than less feasible, while others think the results are not credible. Article 1972c has a slightly upbeat conclusion, looking forward to the results of the new British detectors. From 1973, however, it is downhill all the way: "Gravity waves" start to appear in scare quotes; alternative mechanisms for the production of Weber's coincidences are put forward; the sidereal correlation is explained away; no gravity waves are found in one place after another; and Weber starts to become invisible. Attention turns to teams trying to build much more sensitive detectors, and Weber was no longer thought to be in the race.

The material gathered here supports, if only weakly, the argument put forward in my introductory chapters—that knowledge waves gain strength as they move from the center. In the core set and its supporting conference circuit, news of controversial new findings such as Weber's will slowly gather credibility. If the consensus switches, as it did in Weber's case, the findings will rapidly lose credibility in the core. But neither the certainty that the findings are true nor the certainty that the findings are false will ever be as marked in the core set as they are in the popularizing journals. This is not to say that individuals will not hold polarized views, but in the core set there will be equally strong counterviews, so that the whole set of scientists always exhibits a residue of ambivalence missing in the more popular accounts. The specialist scientific journals will give an impression somewhere between that of the core set and that of the popular journals—they will seem to indicate greater positive certainty when the results are on the way up, and greater negative certainty when they are on the way down. New technologies for detecting gravitational waves were now grabbing the headlines; to these technologies we now turn our attention.

PART II

TWO NEW TECHNOLOGIES

CHAPTER 13

THE START OF CRYOGENICS

If it's green, it's biology; if it's yellow and smells, it's chemistry; if it works, its engineering; if it doesn't work, it's physics.

—Italian physicist

What scientists want is to get on with their scientific lives. There is only so much one can do in a day, so like the rest of us, scientists have to choose how to distribute their time and attention. By the early 1970s, Weber's ingenuity had succeeded in changing the order of scientific thinking enough to make others invest time, energy, and financial resources into his ideas. As we have seen, before that period hardly any of the other scientists put more than a proportion of their energies into this work. This was in contrast with Weber, for whom the search for gravitational radiation was a full-time obsession.

For some scientists, such as Silvano Bonazzola, the work was of so little importance that they did not even write it up properly, so they have disappeared from the published history of the time.

Others, such as Richard Garwin and James Levine, came into the field with a specific purpose in mind, invested the minimum effort to achieve that purpose efficiently, and got out again. Others, such as Tony Tyson, continued experiments with a negative goal in mind right through to the bitter end. But many of those inspired by Weber to begin experimental gravity work now wanted to find the waves, either by building more sensitive bars—usually by lowering their working temperature—or by making bigger interferometers.[1] In this part of the book I deal with these two new technologies, the cryogenic bars and the interferometers.

Resonant bars cooled to liquid helium temperatures and below—cryogenic bars—promised to be far more sensitive than room-temperature bars, and it appeared that, at worst, the Weber claims would be well and truly verified or laid to rest when these were working. But the cryogenic bars promised more than refutation or confirmation of Weber; they promised to be just about sensitive enough to find levels of radiation compatible with the astrophysicists' predictions of much less easily seen fluxes of gravitational radiation—so they promised to do positive work even if Weber was wrong.

EARLY DAYS IN CRYOGENICS

Stanford University is the key player in the shift to cryogenics. No room-temperature bar was ever built at Stanford; they set out to work at low temperatures from the very beginning. The name most closely associated with this program is William "Bill" Fairbank, a Stanford professor.

Fairbank, who died in 1989, had at least one thing in common with Weber. He built his reputation on doing impossible or nearly impossible experiments, not every one of which worked out. He was already very well known for low-temperature work and attempts to measure tiny effects predicted by general relativity, but became notorious in the late 1980s for claiming that he had detected free quarks. Quarks are supposedly the elements of which the elements of atoms are constituted. Thus, both protons

---

1. Of the groups or individuals so far mentioned, Bonnazzola abandoned the field; the Maryland group began to build cryogenic bars as well as continuing its room-temperature work; the Glasgow group would soon turn to interferometry; the Moscow group would try bars made of sapphire and then become interferometer enthusiasts; the IBM team left the field as planned; the Reading-Rutherford team left the field; the Tokyo groups turned to interferometry; the Munich group took up interferometry; the Frascati group moved on to advanced cryogenics; the Stanford group developed cryogenics, turning to interferometry much later; Bell Labs eventually abandoned the field; and Rochester kept an interest in cryogenics but did not build anything.

and neutrons are supposed to be made up of three quarks, the particular mixture of quark types giving the particles their character. Theory says that quarks can never be found on their own, which is why electrical charges are never found in smaller sizes than the charge on the electron or proton. While the charges on quarks are measured in units one-third the size of the charge on an electron, theory says that they can be arranged only in ways that add up to whole units—the negative charge on the electron or the positive charge on the proton—or zero.

The consequence of the existence of free quarks would be that one-third and two-third units of charge could be found "in the wild." In Robert Millikan's notebooks, there are intriguing indications that he had found and ignored such "fractional charges" during the course of his famous experiment at Caltech in the early years of the century to prove that electrical charge could not be subdivided below a certain size; in the 1970s Fairbank, too, seemed to be finding them, but his claims were widely contested. In 1982, a group of physicists met to hold a conference in honor of Fairbank's sixty-fifth birthday; the flavor of the dispute can be gauged from a comment in the introduction of the celebratory book: "I close these remarks with a word of caution to the physicists of the world. Simply because William Fairbank's laboratory is the only one—so far—to obtain definitive evidence for fractional electronic charges does not mean they do not exist. Quite to the contrary." It is now concluded by the overwhelming majority that Fairbank was wrong.[2]

Though Fairbank is thought of as the pioneer of cryogenic bars, another William—William "Bill" Hamilton—also deserves a good share of the credit. Hamilton received his Ph.D. from Stanford about 1960 and stayed on for ten years as a National Science Foundation Fellow and assistant professor. In the 1960s, he and Fairbank would spend their early hours in the lab, discussing science. Hamilton had the idea of building a 150-foot-long resonant bar that would be able to detect the low-frequency gravitational wave emissions from the Crab pulsar by integrating the signal over about a day's observations. It was an ambitious project but not outrageous, given that Fairbank was planning to build a 500-foot superconducting accelerator. A 150-foot bar could not be suspended by a wire around its middle, but cryogenics provided the possibility of suspension with superconducting

---

2. An account of Millikan's experiment discussing the anomalies in his notebooks that could indicate the presence of fractional charges can be found in Holton 1978. Pickering 1981b provides a sociological account of the controversy. The book containing the quotation, which comes from the Introduction, is Gordy 1988, p. 18.

magnets. Fairbank and Hamilton, however, decided that this project, though feasible, was just too complicated.

By 1970 Hamilton had left Stanford to take up a permanent post at Louisiana State University (LSU) in Baton Rouge. At that time he and Fairbank again started thinking about low-temperature resonant bars with the aim of building a device having similar dimensions to Weber's bar. (The members of Hamilton's LSU group actually did much of the design work on the bars built at both Stanford and Louisiana.) In November 1972, Fairbank explained his rationale to me. He said that low temperature was the only way to improve the sensitivity of the bars, because longer bars would vibrate at too low a frequency; hence the need to lower the bar temperature. Low temperature also meant that one could use a superconducting shield, which eliminates stray electromagnetic radiation better than a room-temperature shield. Moreover, one could levitate the bar with superconducting magnets, which would be the best form of seismic isolation. Furthermore, the bar would be supported along its entire length rather than from a wire around its middle, thereby eliminating creep and noise from the vibrating wire. Fairbank thought he could gain a sensitivity increase of 10^{-5} with these methods.[3]

But low-temperature work is difficult, and this project turned out to be much harder than anyone expected. In 1988, Hamilton, with the advantage of hindsight, jokingly set out two principles of low-temperature work. There is Fairbank's rule, "Any experiment is better if it is done at low temperature," and Hamilton's corollary, "Any experiment will be harder if it is done at low temperature"[4] As Hamilton remarked, the subsequent experiment was to prove both principles. This is why the Stanford results make no appearance in our earlier story until they were mentioned in Tyson's 1982 paper.

The Stanford initiative was the indirect ancestor of another group whose entire effort for many years would be cryogenic bars. This was the team founded by David Blair at the University of Western Australia (UWA) in Perth. Blair worked with Hamilton as a postdoctoral fellow at LSU in the middle 1970s and visited Stanford frequently during this time.

The plan to cool resonant bars to liquid helium temperatures was either conceived independently in Italy or through the well-known physicist Edoardo Amaldi's connection to Fairbank (though as was pointed out to

---

3. This rationale in Fairbank's own words can be found at WEBQUOTE under "Fairbank on the Start of His Program."

4. Hamilton 1988.

me, the idea is an obvious one). In the early 1970s, an Italian group visited Stanford and LSU, where Hamilton had just arrived. The group included Guido Pizzella, who will figure large in our story, and Massimo Cerdonio, who will reappear strongly in the late 1990s; the figurehead was Amaldi. This group, whose leadership was to pass from Amaldi to Pizzella, began work immediately with an optimism not dissimilar to that of Fairbank and Hamilton.

Liquid helium boils at 4° on the Kelvin scale of temperature. The Kelvin scale has the same intervals as the centigrade, or Celsius, scale, but its zero point is 273 degrees below zero on those scales. Minus 273 degrees in centigrade/Celsius is as cold as anything can get and is known as absolute zero. Four Kelvin is four degrees above absolute zero. If a bar is simply cooled with liquid helium, 4° will be its working temperature. But one can go lower, to about 2°, by partially evacuating the helium tank, thereby lowering the boiling point of the helium. To work at four Kelvin or two Kelvin is to work with "cryogenics." But the cryogenic pioneers had calculated that to be sure of seeing supernovas in other galaxies, they would have to go to lower temperatures still. They needed to get down to three millidegrees to gain the maximum reasonable sensitivity; this enters the realm of "ultracryogenics." And this was the kind of impossible goal favored by Fairbank. It could be achieved by using a device known as a dilution refrigerator to take over once the liquid helium had done its job. Fairbank's Stanford program was always intended to reach millidegree temperatures rather than liquid helium temperatures, and this is one reason that his program went so slowly. Hamilton opted for the easier liquid helium and accepted 2° as his target. Millidegree temperatures would not be achieved in such devices until the 1990s, and then by Pizzella at Frascati, near Rome. It was also achieved later by Massimo Cerdonio and his group in Legnaro, near Padua, with a detector whose major components were copied from Pizzella's device.

Fairbank explained to me that he had not set out to test Weber's ideas but to see the theoretically predicted fluxes. To do this, his team would need to improve their sensitivity over Weber's detector by five or six orders of magnitude (a factor of 100,000 to 1,000,000).[5] As he said in 1972,

> [1972] We think it's important to go to this 10^5 or $[10]^6$, so that's why we
> designed from the beginning to go to a few millidegrees. Lots of people asked

---

5. This account is based heavily on Fairbank's 1972 interview, because I did not meet up with Bill Hamilton until my 1975 round of interviews.

us from the beginning why didn't we just go to liquid helium temperatures, and then at some later date go on to a few millidegrees. And the answer is that it's very expensive to build a detector—to build the Dewar system and everything—and so we had a very strong incentive from the beginning to build into it the possibility of going all this way. That's why we made the bar that big—it's difficult to make the bar bigger, but we just pushed everything to get this 10^6 in principle.... [W]e aren't in quite as much of a hurry to check Weber as some of the people working at room temperature. In other words, it really doesn't affect our objective what happens to Weber's data right now.

I told Fairbank that other scientists had criticized his approach for being too complex and expensive, but he replied—wrongly as it turned out—that it would not be as hard as other people thought because of his long experience.[6]

In 1975, when I interviewed Fairbank a second time, his program was under way. He had built a prototype bar about half the size of Joe Weber's which, apparently, he had succeeded in supporting by magnetic levitation at a temperature of 2°.

> [1975] [W]e've completely concentrated on a smaller system which could go in an accelerator Dewar [that is, a "vacuum flask" of the kind readily available, because many had been made for the superconducting accelerator]...and we have cooled that down about three or four times and gotten over the problems of floating it and...and each time we've made quite major improvements. And the last time we saw signals which were the theoretically predicted noise level...
>
> Now, there were still things that weren't right about the system. The resonant detector [the new design of transducer—see below] was coupled in only 3 or 4%, whereas we can make it couple 50%, and the Q of the bar with glued material on it, has now dropped to about 50,000. And this detector should have a sensitivity of about a millidegree, at two degrees, and that would be another factor of 100....
>
> We're starting to assemble the big system, and we don't really have enough manpower and money to work on two at the same time....Extrapolating this, we should have a sensitivity of about 10^7 over room-temperature detectors, and that's enough to see almost any collapse in our galaxy.

6. See WEBQUOTE under "Fairbank on Experience."

The plan was to make a much bigger bar, three feet in diameter and ten feet long. Fairbank hoped that if Hamilton at LSU, Blair at UWA, and the Italians also had instruments, their joint "observatory" would have a long-enough baseline to be able to pinpoint the origin of the signal in the sky. The idea was—and this is a principle that applies just as much to today's efforts—that the difference between the time of arrival of a signal at distant sites would indicate the direction of the source. Fairbank talked as though he, too, were planning to use a niobium bar rather than the "traditional" aluminum alloy or, at least, that he had been very much involved in the idea for using that material—an idea subsequently implemented at UWA. Additionally, in the middle 1970s, Blair was working at LSU and cooperating closely.[7] This multiteam partnership is recognized in a conference paper presented in 1973 and jointly authored by a group from Stanford (including Ho Jung Paik, who will be discussed below) and one from LSU (including David Blair).[8]

Though Fairbank's three-millidegree strategy might have been reasonable in principle, as indicated, the subsequent story of the cooled bars illustrates the difference between a design idea and its successful implementation. It took about twenty years to make cooled bars work steadily and reliably at something close to their expected sensitivity, and the experimenters have never been able to put into practice all of Fairbank's plans.

"Q," or quality factor, is related to the length of time for which an object will ring. A bell that rings for a long time after being struck has a high "Q," and vice versa. "Q" is a vitally important feature of all designs of gravitational wave detectors. In the case of resonant bars, the longer they resonate the better. Much of the delay in implementing the cryogenic bar idea was a consequence of Fairbank's ambition, particularly his choice of magnetic levitation. As the bars got bigger and heavier, schemes to make them levitate either did not work or ruined the "Q." It would eventually turn out that superconducting magnetic levitation could not be made to work satisfactorily for bars of the size that Fairbank and the others needed to use, but this was far from obvious at the time.

7. At this time, other groups, such as Douglass's in Rochester and Braginsky's in Moscow, were planning to abandon very big bars and use sapphire to make smaller bars with very high quality factors. Fairbank thought that they would gain nothing unless they could find a way of coupling this "Q" to an amplifier and transducer, and he felt that on this front sapphire may not be the way forward. No sapphire bars were completed to the point of producing interesting results so far as I know; Braginsky was to switch to interferometry, and Douglass's group experimental effort became very much reduced in size.

8. Boughn et al. 1974. Weber was reporting his spurious 2.6-standard-deviation coincidence run with Rochester at the same conference.

Superconducting magnetic levitation works because magnetic fields are generated on the surface of the suspended superconductor which are a mirror image of an applied field; the applied field and the mirror-image field repel each other. The system is stable because as the relationship between supporting magnet and supported object changes, the mirror-image field automatically readjusts so as to match the applied field exactly. A early problem was that the aluminum alloy used for the bars was not superconducting above one Kelvin, and bars would not reach a temperature as low as this for decades. Furthermore, at one Kelvin and below, aluminum is not good at maintaining the strong mirror-image magnetic fields needed to support large masses. Therefore, the bars had to be coated with a better superconductor. Initial efforts used either coils on the surface of the bar, superconducting metal deposited on the surface, or metal sheets glued to the surface. The first two worked for small masses but not for large.

The third method was tried by Bill Hamilton at LSU, and he showed experimentally that with sheets of niobium-titanium superconductor glued to the bar, the necessary mass could be lifted. This method (as suggested by David Blair) might well have destroyed the "Q" of the bar, preventing its working as a sensitive gravitational wave detector, but Hamilton's research never progressed enough to for him make the necessary measurements.[9] He reported that he worked on this method of magnetic levitation for six years before discontinuing it.

> [2000] I still have the cradle and the superconducting coils and things like that here, and we built the whole thing and then we attempted to raise the bar. And one of the magnets broke, and I said, "How in the hell am I going to figure out where it broke: they're all wired in series," and I said, "There's just no way we can do this sensibly and continue this levitation business."

But Hamilton pointed out that magnetic levitation was not the only problem, and eventually he had to give up on many complex features of his design. For example, he persisted in trying new methods to support the bar. The traditional method, invented by Weber, is a wire around the middle, but a wire support seems like a crude mechanism and has the great

---

9. For this passage I rely on David Blair's 1997 popular book (written with McNamara), and also on a telephone conversation with Bill Hamilton. There is no argument about the fate of magnetic levitation; I simply describe the consensus without trying to open up any potential doubts about its failure. A scientist respondent tells me, however, that magnetic levitation was a red herring—it would not have added significantly to the sensitivity of the bars even it could be made to work.

disadvantage of vibrating like a violin string, causing noise. Hamilton used what became known as a "dead bug": a small metal device like an upended little table with short legs.[10] The bar is meant to rest on the four legs—which look like the legs of a dead bug. Hamilton could not get the dead-bug support to work, however.

> [2000] There's a tremendous amount of pride in being a young man. And you have to do everything yourself. And it has to be original. And so we wasted not just a bunch of time on magnetic levitation, but we also wasted a bunch of time on what we called "dead bug" supports.... We diddled with a lot of harebrained schemes, probably for another two or three years. And actually I think it was first the success of the Italians by hanging their bar around the middle that we said, "OK, we're going to do that." And as a matter of fact, I made a very conscious decision when I said, "Anything that works!—To hell with this pride business—anything that works I'm gonna copy it—and if we can make it better, we will." And so the current ALLEGRO support is really adapted completely from the Italians—except we said we could make it better.

Hamilton refers to his apparatus as ALLEGRO. It became the fashion among those who built cryogenic bars to give them names, sometimes in the form of acronyms that were strained and stretched to make a classic-sounding word. Thus, in addition to ALLEGRO, we will encounter EXPLORER (an Italian-built bar located in Geneva), NIOBE (David Blair's bar at UWA, featuring a bar made from niobium), NAUTILUS (the millidegree bar at Frascati), AURIGA (the millidegree bar at Legnaro). and the stillborn TIGA (a huge spherical detector which was to built at LSU).

The only person who was successful at supporting a resonant bar with anything other than Weber's original circumferential wire, or something closely related, was David Blair,[11] who called his support a "catherine wheel." Blair's design was like Hamilton's dead bug, but the "tabletop"—the back of the bug, as it were—was replaced with a central disc with four arms spiraling outward—like the pattern made by a catherine wheel. In turn, the "legs" were attached to the ends of the spiral arms. The catherine

10. I am told that J. P. Richard actually invented the "dead bug" idea and that David Douglass gave it its name.
11. Braginsky's Moscow group suspended their bar from ribbons of aluminum locked into the upper surface of the bar with dovetail joints.

wheel design allows the legs to move sideways in two directions so that both the longitudinal vibrations and the complementary "breathing" in-and-out vibrations of the bar could be accommodated by the legs without stiffness or friction. To this day, Blair's bar rests on a catherine wheel support. Nevertheless, though he managed to make this innovation work, we will see that Blair also went through a process of progressive simplification of his design.

At Stanford, the initial goal of building the best possible design was maintained, but things went badly. One key player in the Stanford team, Robin Giffard, left, and Fairbank died in 1989.[12] Peter Michelson was now given charge of the experiment and pressed ahead with a perfectionistic approach in the style of Fairbank. (Albert Michelson, of the Michelson-Morley experiment, is Peter Michelson's great-uncle.) The Stanford group did go on to construct a relatively straightforward and successful device running at 3°, and this did take some data, which was compared with the data from the Louisiana bar and the Italian bar; but they then tried cooling to below liquid helium temperatures while attempting to implement a very complicated and cleverly analyzed suspension system based on cantilever springs.[13] This project dragged on and on, absorbing more and more money and more and more time. In 1995, Peter Michelson, then the leader of the Stanford team, put the matter in rather dry terms: "[W]e developed some new vibration insulators that worked at ultralow temperature—we'd done a lot of engineering effort, but, as I say, it took longer than we thought." I have heard much more colorful descriptions of this "extra time" from other members of the gravitational wave community.

The Stanford 3° bar was damaged in the 1989 San Francisco earthquake. Again in Peter Michelson's words:

[1995] It was very severely damaged. It was at low temperature during the earthquake. The day of the earthquake, I remember running into the lab and I thought, "Gee, it looks OK" and then I thought, "Whoops—it wasn't"—it was just barely hanging together, and a lot of the instrumentation inside had been severely damaged.... So at that point we had decided to put most of our effort into the ultralow-temperature system, and then that went along and some of the development effort took longer than we thought.

12. I understand Giffard was denied tenure.
13. Amaldi et al., 1989.

The whole bar effort at Stanford was closed down in 1994, and the team began to concentrate on interferometers instead. It is likely that NSF had by this time become unwilling to continue to provide the extra resources that the Stanford bar was absorbing in the absence of signs that the project was coming any nearer to a successful end.

David Blair compared the eventual successful outcome of his simple device with Stanford's pursuit of complexity and subsequent failure.

[1998] Stanford embarked on something highly elaborate, very complex, with similar basic performance to what we achieved, but done with very complicated bonded elements—they were perhaps trying to solve a slightly more complex problem than us, but not really any different. But they abandoned the whole thing, and the reason for the abandonment was that they were going in for these complex solutions that were too complicated and too difficult. Years ago I guess we learned the hard way—but fortunately without having to abandon the project; just redirect the project—to go for most simplicity.

And what happened there was that when I came here I had this scheme which was actually suggested by Braginsky, to have a niobium bar—he hadn't suggested a niobium bar, but he had suggested that, at the end of the bar, to have a magnetically levitated transducer, hovering at the end of the bar, and read out the vibrations with the magnetically levitated transducer. And we built a small one of those—a very beautiful piece of technology—it didn't look beautiful, but it was.... It was a superconducting magnetically levitated bar next to a magnetically levitated transducer, all controlled in all degrees of freedom—you know, one hovering there and the other one hovering next to it, and control systems to bring the one up to the other. And we measured the vibration, and in this small model we measured the lowest noise temperature that had ever been observed in a gravity wave detector at the time, and that was in about 1979 I think.

And then we tried to scale it up to the big system, and that was fantastically difficult. Partly it was difficult because all the systems we needed to be monitoring these magnetically levitated things just meant the quantity of electronics involved was enormous, and to keep everything working well—and lots of it was superconducting electronics, too—and you only had to have one failure in a superconducting joint and you were in trouble.

So that, I guess, was when we finally said, "Look, we've got to get simple and simplify everything." And I guess there were two things we learned, and one was to simplify and the other was to make the vibration isolation performance way better than you would calculate you needed to allow for things to work less well than you expected.

David Blair had started this work in Perth in the late 1970s after working for three years as a postdoctoral fellow with Bill Hamilton in LSU. He chose to make his whole bar out of the very expensive metal niobium, which is superconducting at liquid helium temperature. This property makes it ideal for magnetic levitation in that no extra coils or sheets need to be glued to the surface. Also, niobium has a very high "Q"; it will ring for days if undisturbed. Blair built up to a prototype one meter long and ten centimeters in diameter, which, as he explains above, was magnetically levitated and very sensitive for its size. But he found that the limit for the diameter of superconducting magnetically supported niobium was twenty centimeters, whereas the typical big aluminum bars of the day were one meter in diameter. The problem was that as the mass increased, the mirror-image field began to penetrate into the bar, destroying the superconductivity.

As a result, Blair's group, like the other teams, had to abandon magnetic levitation, though they persevered with niobium because of its high quality factor. They now built a much bigger liquid-helium-temperature niobium bar employing a number of original features; the metal alone cost a quarter of a million dollars but was probably a good investment for the university. Partly because they had been thinking in terms of superconducting levitation, however, the acoustic isolation as implemented was poor, and it took them ten years to reach the sensitivity of the ten-centimeter magnetically levitated prototype that they had built. Thus NIOBE, as it was called, was not made to work at reasonable sensitivity until 1993, though the program had been started in the late 1970s.

The Italian groups were the most professional and most well funded. Edoardo Amaldi, the instigator of the work, could command considerable resources. He had contacts with Fairbank, although as I understand it, an Italian low-temperature physicist named Correli was the very first to attempt to build a low-temperature bar in Italy. The housing for the cold part of the Correli experiment—effectively a large thermos flask or Dewar, also known as a cryostat—collapsed when the experiment was first operated, however. Amaldi also recruited to the effort Guido Pizzella, who for many years would be the dominant figure in the Italian experimental program. Pizzella explained that, inspired first by Weber and then by Fairbank, they started to construct a cryogenic antenna called EXPLORER (later to be installed in Geneva). They soon realized, however, that "they had to learn many things," so they built two prototypes: a small cryogenic device, ALTAIR, and a room-temperature antenna, GEOGRAV.[14]

The Italian group was the only team of researchers who could afford to build a series of prototypes. One influential American scientist told me that when he visited the Frascati group, he realized that the American projects, such as that at Stanford, would have been far more efficient if, like the Italians, they had spent money on industrial design and careful prototyping right from the beginning rather than trying to do everything themselves.

> [1995] The Italians have, maybe, five times the funding and twenty times the manpower to make a similar system work.... I wish to hell that we had done something like that instead of what we did at Stanford. Because they [the Italians] contracted out all the plumbing to industry.... It's beautiful and clean, whereas Michelson and two postdocs were trying to do it all themselves.... It's to Pizzella's credit that he did a much more professional job.

In 1995, when these comments were made, the Italian group was completing NAUTILUS, the first ultralow-temperature device, which will be described in the next chapters.

Another group that began the long march to cryogenic detectors was Joe Weber's team at the University of Maryland. The project was never to be completed, however, and I will postpone discussion of it to part III.

TRANSDUCERS

We find ourselves continually discussing the differences between one resonant bar and another. Nearly every experimenter wants to make their bar the best by incorporating in it design features that take it beyond what already exists. Alas, many of these turn out to cause trouble and delay. We have seen bars made of many materials, from aluminum alloy (some of it, as in Douglass's effort, rumored to have given trouble with creep) to niobium to sapphire. We have seen disputes about the efficacy of different kinds of amplifiers. We have seen the failure of all but Blair to break away from

---

14. For the developments described in Pizzella's own words, see WEBQUOTE under "Pizzella on the Start of His Program."

wire suspension or its close equivalents, with the attempts at magnetic levitation consuming huge amounts of time and energy. We have seen arguments over computer programs and signal extraction algorithms, and if we went into detail, we would find more arguments over the way to handle data. One thing that is worth a little more detail, however, because it will play an important part in what follows, and because it reveals humankind's ingenuity, is that part of a resonant design known as the transducer. In a resonant bar, a transducer is what transforms the mechanical movements of the bar into electrical signals that can be amplified and processed.

Joe Weber, it will be recalled, used rings of piezoelectric crystals glued around the center of the bar for his transducer. Peter Aplin invented the "split-bar" detector in which the bar was cut in half across its long axis and a sheet of piezoelectric material was glued between the halves, the design used by Drever and the Glasgow group. Guido Pizzella cut a slot in his room-temperature bar and fitted the piezocrystals into it. Garwin hung a small weight against the end surface of his bar with a piezocrystal squeezed between the end and the weight, a design saving the need for unpredictable glue. Braginsky attached "horns" to the upper edges of the ends of his bar, the ends of the horns nearly touching at the center, treating the gap between the ends of the horns as a variable capacitor. Blair mounted a (relatively) fixed metal plate near the end of the bar and used the gap between it and the vibrating end of the bar as a variable microwave cavity.

The other cryogenic bar groups, however, all adopted a brilliant idea thought up by Ho Jung Paik, a onetime graduate student of Fairbank's at Stanford. Paik's transducer gives a sense of the ingenuity of physicists and will figure significantly again in part IV. The principle of the invention is the "double pendulum." David Blair in his popular book suggests that the principle can easily be demonstrated at home by hanging a heavy weight from a piece of string and then a lighter weight from the heavy one. If one starts the heavy weight swinging by giving it a little push, the light weight will initially remain still, but then it, too, will start to swing. The light weight will swing more and more vigorously until eventually it has absorbed all the energy of the heavy weight's swings, the heavy weight becoming still. At this point the light weight will be swinging much more vigorously than the heavy weight ever was, because all the energy has transferred itself into the lighter object.[15]

---

15. Left alone, the energy will slowly return again to the heavy pendulum (Blair and McNamara 1997).

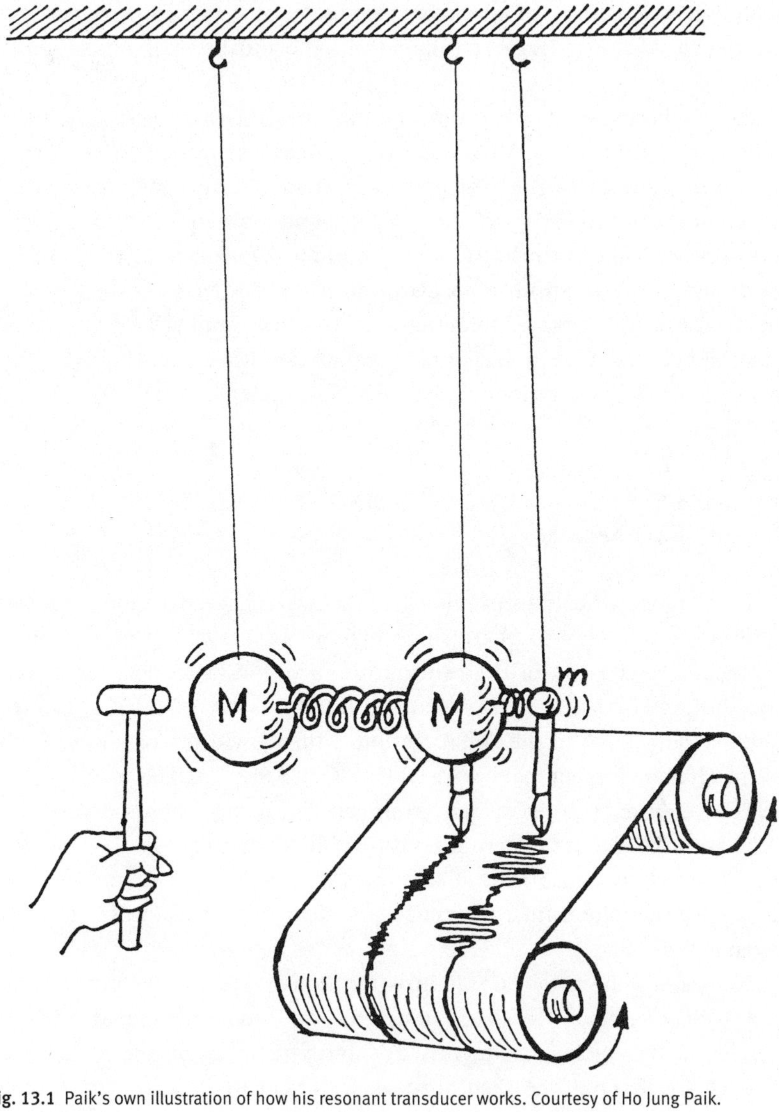

Fig. 13.1 Paik's own illustration of how his resonant transducer works. Courtesy of Ho Jung Paik.

Paik's idea was to use this principle to transform the tiny vibrations at the end of the vibrating bar into much larger movements of a lighter object, the point being that big movements are easier to measure than small ones. Paik's own hand-drawn diagram (fig. 13.1) represents the principle in a pair of linked chart recorder pens. In modern practice, the design used resembles a metal mushroom "growing" out of the end of the resonant bar. The small oscillations in the end of the bar travel down the mushroom's stalk and turn into much larger resonances in the mushroom's much lighter disc.

Paik is best known for inventing this transducer, but he also wrote what many treat as a definitive survey of the sensitivities of the competing resonating systems of the Weber era. We have already seen this analysis quoted in Tyson's 1982 paper.[16] Later, Paik would move to Weber's laboratory and work with him in his attempts to build a cryogenic system as well as continuing to try to improve his own transducer design. For example, with Jean-Paul Richard, he tried to develop it further by adding an extra stage— a still lighter mushroom growing on top of the other one. Unfortunately, Paik's career in gravitational waves had less salience after he moved to Maryland.

CURRENT CRYOGENICS

The first cryogenic bars would not approach their intended sensitivity and reliability until the mid-1990s, some fifteen years after the project began in earnest. By the end of the century, there would be five cryogenic bars contributing to an international data exchange known as IGEC, International Gravitational Event Collaboration, which I will discuss in more detail in chapter 25. The bars were ALLEGRO at LSU; NAUTILUS at Frascati; EXPLORER, run by the Frascati group but located in Geneva; NIOBE at UWA; and AURIGA at Legnaro near Padua. By 2001, all these detectors had been working on and off for at least a year and some for many years. Let me put some color into the account by describing the locales of these instruments.

Louisiana State University is in Baton Rouge, about a 90-minute drive west of New Orleans. The Baton Rouge airport is not frequented much by the airline on which I "frequently fly," and after a lot of long delays and layovers, I have finally worked out that it's easier to fly into New Orleans, rent a car there, and drive to Baton Rouge. Nowadays I try to stay at the university's Faculty Club, which is inexpensive and always courteous. The guest rooms are large but showing signs of age, and the facility lacks the amenities of a modern hotel, but I have found "Louie's," an old-fashioned all-night diner, close by. If I am on a short trip from Europe, I try to stay on European time, so I get up at 2 or 3 a.m. and drive to Louie's to eat breakfast with the

---

16. This is Paik 1974; Paik 1976 describes the transducer.

rest of the night people. I sit there feeling smug, enjoying the quintessentially American experience, as in that famous painting by Edward Hopper. Louisiana is the only place I have been to in the USA where the regular filter coffee is worth drinking; the first cups of Louie's coffee in the morning are a treat. I have not been to Baton Rouge in the height of summer, but I'm told that it's grim. When I visit, the weather is either pleasant or raining.

LSU's wooded campus is in the modern style, neat but not striking, with big concrete arches and the like. The concrete is covered with brown gravel, so all the buildings are light brown. The LSU resonant bar ALLEGRO sits in a large ground-floor room. The laboratory moved in 2001, and the new facility is a bit less gloomy and feels a little less "tucked away" than the old one. Once inside, it looks like most other bar laboratories: the overall aesthetic is "grunge." Dominating the room is the vacuum chamber enclosing the bar, surrounded by vacuum pumps; cryogenic pipes with frost on them; shiny electrical shielding tape, some silver, some gold, some lead-colored; and with suspension components and all kinds of other wires and instruments emerging from it. The rest of the lab contains scattered bits and pieces of apparatus, some looking dusty, derelict, and abandoned, some looking as though they remain the object of someone's attention. Tucked in here and there are scientists' desks, covered in papers and maybe scraps of electrical components, with computers humming away. The walls are covered in diagrams.

Frascati can be approached by car or train. It lies in the picturesque hills to the east of Rome where the pope has his summer retreat. "Frascati" is also the physics community's colloquial term for the network of Italian government research laboratories clustered below the historic little town. If you fly in and then rent a car, a nerve-racking encounter with the crowded highway skirting the city, the Gran Raccordo Anulare, takes you from one of the Rome airports to the Frascati exit. The junction is a roundabout below the road, which was to inspire a sociological insight (see chapter 22).

Frascati is a pleasant town dominated by a huge, semiderelict villa. It is not a gem, like the towns and villages of Tuscany, but if it were anywhere else but Italy, you would think of it, with its steep cobbled streets, restaurants, and cafes, as a place to which you would want to return. Frascati is famous, of course, for its eponymous light white wine. Istituto Nazionale di Fisica Nucleare, the government-sponsored high-energy physics research laboratory, is located a few kilometers outside the town. All entrances are guarded by security personnel, and visitors need to deposit their passports and pick up badges. INFN consists of scattered buildings with lumps of old but once pioneering apparatus set out at the sides of the roads here and there.

The laboratory that houses NAUTILUS is like a large warehouse, far bigger than the lab in Baton Rouge or the equivalent in Perth, Australia. The warehouse is, nowadays, divided by a partition, with the other half being used by another experiment. The gravitational wave side of the space is dominated by NAUTILUS's vacuum chamber in an unfortunate shade of green. Suspended above NAUTILUS, like a kind of huge sunshade, is a flat array of cosmic ray detectors. The lab's very high ceiling features girders for cranes. I remember being one of the last in the lab one evening and watching the junior member of the team, a young, slight female member of the team, Viviana Fafone, use a crane to lift a huge container of liquid helium into place to refill the cryogenic gas reservoir of the bar.[17] Lying on the ground against the outside wall of the Frascati "warehouse" are a couple of spare aluminum bars waiting for emergencies. After lying there for years, one of these is to be shortened slightly and used as the resonator to look for the gravitational waves from the remnant of supernova 1987A.

I have seen EXPLORER only once. It is located in Geneva, in the same overall building as the European Organisation for Nuclear Research (CERN).

AURIGA, which is near Padua, takes its basic design from NAUTILUS; from the outside, the two devices seem identical except for their color. Whereas NAUTILUS is an unpleasant green, AURIGA is painted a cheerful burnt yellow. I tried to tease the Frascati group by telling them that if only they repainted the vacuum chamber of NAUTILUS, all their troubles would be over. AURIGA is located in its own laboratory at the site of Padua University's physics park in Legnaro, fifteen kilometers outside Padua. As with INFN, there is a security gate. No passport is demanded, but it is necessary to have someone bring you in. The laboratory is much less gloomy and cluttered than the usual accommodations and seems to have plenty of purpose-built offices as well as a roomy, high-ceilinged space for the detector itself. For a number of years, AURIGA was the most forthrightly ambitious of the bar groups. In due course I will relate some of the upbeat comments of one of the AURIGA scientists who typified the group. But AURIGA's performance, though sound and solid, does not seem outstandingly better across the range than the bars of the other groups. My one visit to AURIGA was the occasion of fieldwork disaster, as I was stricken with a

---

17. Italian physicists are, on the face of it, unaware of gender divisions; females seem not so much "women physicists" as women who happen to be physicists. I understand that the "glass ceiling" is nonetheless there, however.

serious traveler's ailment. The occasion fills me with embarrassment to this day.

NIOBE is at the University of Western Australia; its geographical location, and what follows from such isolation, is described in appendix III.1. Perth is a city small enough to walk around in. Its North Bridge area is full of ethnic restaurants and at night is full of life and energy. Nearby are splendid beaches on the edge of the Indian Ocean, and not far away is Freemantle, Perth's port and hippy hangout, a bit like a small California town. In fact, Western Australia overall is very like a clean, quiet, and civilized version of California that has avoided total domination by the motor car.

UWA is approached by driving a couple of kilometers from central Perth along the Swan River estuary, which is wide enough to look like a small inland sea flanked by a narrow beach and green meadows. UWA's campus is like a refined version of LSU—large concrete buildings, some of which have a sense of architecture about them, surrounding green swards. I like the UWA campus: the birdcalls are exotic, the trees are huge yet have a European shape to them, few cars are in evidence because they are banished to the periphery, and the concrete is a pleasant pinkish color. One morning, when it had rained overnight but the sun had greeted the day, the whole place had a wonderfully clean yet exotic feel and smell to it. The physics department, which houses NIOBE, is like university physics departments everywhere: a rabbit warren cluttered with new and old equipment and mementos to past achievements and mistakes, and scientists scraping along, trying to do great things with diminishing resources.

We now return to NAUTILUS, describing its trials and tribulations as a case study in cryogenics.

CHAPTER 14

NAUTILUS

Vittorio Pallottino of the Frascati team gave me a diagram (fig. 14.1) he had drawn in 1973, at the beginning of the Italian experimental program. In it, he likened the building of gravitational wave detectors to a horse race. As well as being a joke, the diagram, he said, pointed out "the difficulties of the task and a nice (as well as typical) way to overcome them, i.e. to always devise something more complex, requiring more time, and uncontrollable."[1]

We need to understand why it took so long to make the cryogenic bars work. Why did it take twenty years to go from idea to a device that approached the design specification and that could be kept in a state sensitive to signals for long, unbroken periods? We have already gained some sense of the difficulty of gravity

---

1. This is my rendering of a photocopy of a photocopy, Vittorio having long lost the original. The comments and the note are from his letter to me of April 3, 1996, and an email of January 2003. He explained that "attivita equestri" means activities related to horses and races and implies, among other things, abandoning a futile task; "concorso ippico" means race meeting in general (as well as a specific fashionable Rome meeting); and "gloria, fama, eterna e prebende," the final destination, means "glory, eternal fame and money," with a somewhat antique or baroque connotation.

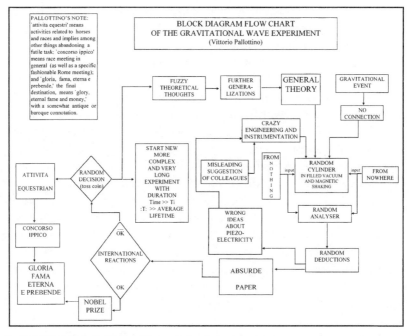

Fig. 14.1 Vittorio Pallottino's view of the gravitational-wave horse race.

experiments from Sinsky's work; in the case of cryogenic bars, bigger teams took still longer to accomplish their objectives.

The aim here is twofold: the first object is to continue the work of describing science as a practical, hands-on activity that we began with Sinsky; we want to "de-sterilize" science. Second, in the confrontation between the bars and the interferometers that was to follow, it was claimed by the bar proponents that the idea of an interferometer was being set against the reality of a cryogenic bar, without taking into account the unforeseen difficulties that would affect interferometers just as surely as it had affected their own project. At the time of this writing, the big interferometers are just beginning their "shakedown" period, and it will soon be known whether there will be unforeseen difficulties or, more to the point, how long it will take to solve the inevitable unforeseen difficulties. If the interferometer teams can overcome the problems with their apparatus more quickly than the bar groups, it would be nice to know how. To explore the path from idea to accomplishment in cryogenic bars, we will look at a case study of one bar, NAUTILUS, admittedly the most complicated one. NAUTILUS's troubles, as they were experienced over a short period of its development, can stand for the troubles of all the bars.

In 1995, the internationally held view seemed to be that the best established gravity wave detector program in the world, though one which would soon be overtaken by interferometer developments, was that of the Italian team based in Rome. In particular, the EXPLORER detector, built by them but located at CERN in Geneva, was thought to be the best understood and engineered device. The same team, however, was in the process of completing NAUTILUS, which ought to be a far more sensitive antenna. One may be sure that the acronym for the device came before its name, "Nuova Antenna a Ultrabassa Temperatura per Investigare nel Lontano Universo le Supernovae," rather than after. NAUTILUS's home was a large, warehouse-like laboratory in the Istituto Nazionale di Fisica Nucleare.

NAUTILUS was intended to be the most sensitive resonant-bar gravity wave detector in the world, but, of course, in 1995 it was not known if it was a gravity wave detector at all. At the time these sentences were written— about 2 p.m. on October 12, 1995—no one knew if NAUTILUS was the most sensitive *anything* in the world; all we knew was that it was *potentially* the most sensitive something or other. On that day, the bar which sits at the heart of NAUTILUS was at a temperature of 38.41 Kelvin, that is, $38.41°$ above absolute zero.

Absolute zero is the coldest anything can ever be, because at that temperature everything is absolutely still. (Water freezes at about 273 Kelvin.) NAUTILUS had been colder than this between February and May of 1995. In fact, between these dates the bar reached temperatures lower than 0.1 Kelvin (100 mK = milliKelvin), and within a week or two of October 12, 1995, it would be as cold as this again. What we did not know was whether NAUTILUS would be as sensitive a detector as it ought to be. The last time it was this cold it was not much good at all.

NAUTILUS AND COLD

To be exact, 4.2 Kelvin is the temperature at which liquid helium boils, so to cool something to 4.2 Kelvin, it is enough to bathe it in liquid helium. But the temperature can be forced down further by evacuating the space above the helium and turning the liquid gas into a "superfluid"; this allows operations down to 2 Kelvin. Helium has the lowest boiling point of any gas, and these are the lowest practical temperatures for most large-scale experiments; yet liquid helium, it should be borne in mind, is very expensive, and it is not used carelessly. Consequently, the workhorse cooling liquid found in the physics laboratory is liquid nitrogen, which is much cheaper because the atmosphere of full of this gas, and it has to be cooled only to 77.4 Kelvin to condense into a liquid. This temperature is still extremely cold—a rubber tube dipped into liquid nitrogen becomes so brittle it will shatter like glass.

One says it is easy to get to liquid helium temperatures, but that is only the case if you have some liquid helium; helium was not liquefied until 1911, and industrial quantities were unavailable until it was discovered in natural gas in the 1950s. Large-scale liquid helium–based experiments could not be done in the Frascati area until a liquefying apparatus was built on site in 1989.[2] In Italy, liquid helium in large quantities costs about $8 per liter, and NAUTILUS requires 4000 liters to cool it to 4.2 Kelvin; thereafter it uses about 60 liters per day.[3] The NAUTILUS team recirculate and reliquefy the helium they use, losing only about 10% of the throughput.

The techniques used to cool objects below two Kelvin are normally reserved for small laboratory experiments—such as investigations of the properties of matter at very low temperatures, for which purpose only small volumes need to be cooled down. Decreasing temperature below 2 Kelvin requires an altogether different technique known as a dilution refrigerator, a complex device that depends on the cunning interplay of the liquid and gas phases of ordinary liquid helium and its lighter cousin, helium 3.[4] (Helium 3 and helium 4 are related in the same way as light and heavy water or the dangerous and stable types of uranium; it is a matter of heavier and lighter isotopes of the same element.) A dilution refrigerator has three receptacles at different temperatures. The warmest "pot" is at about 1.2 Kelvin and contains liquid helium 4 that has been cooled below 2 Kelvin by feedback from the refrigeration stages which follow it. The second-warmest pot is at about 0.7 Kelvin and contains liquid helium 3, with helium 3 gas above it. The coldest pot, which is where the minor miracle occurs, contains a mixture of liquid helium 4 and liquid helium 3, with helium 3, which has been made to behave as though it were a gas, *below* it! Pumping at this helium 3 "quasi gas" produces the ultralow temperatures. The dilution refrigerator made for NAUTILUS is the largest in the world and was specially built by Oxford Instruments in collaboration with the Italian team.

When it is cooled down, NAUTILUS's bar is the coldest large object in the world. Actually, failing the existence of intelligent extraterrestrial

2. By Professor Ivo Modena, a pioneer of cryogenic techniques.

3. In America there is a much greater abundance of helium in the locally found natural gas, and this makes liquid helium much cheaper. The Americans, however, consider liquid helium a strategic material, and therefore will not export it.

4. I am especially grateful to Viviana Fafone for explaining this to me so lucidly.

civilizations doing the same kind of work, it is the coldest large object in the universe. The background temperature of the universe—the temperature of anything left to cool down in the farthest reaches of deep space—is three Kelvin, so there is good reason to think that there is nothing colder than three Kelvin unless it has been deliberately made so. NAUTILUS's bar weighs 2350 kilograms (which is about 2.6 old-fashioned English tons). The idea is to cool the bar to around 0.05 Kelvin. Temperature is a measure of the random motion of atoms and molecules. At 0.05 Kelvin almost nothing moves almost never, so at this temperature the bar should be undisturbed by even the tiniest movement of its fabric.

THE STRUCTURE OF NAUTILUS

The foregoing description gives us some idea of what is needed to build a device that can detect a movement of less than the diameter of an atomic nucleus. Nevertheless, even a technical description obscures the *work* that goes into realizing the design; that this design can be sketched out in a few paragraphs is dangerously misleading. That the technicalities can be described even in a whole chapter is also dangerously misleading, but it is a bit less misleading than describing it in a few paragraphs. I'll start my richer description of gravitational wave detectors with a fairy tale.

> In the morning they asked her how she slept. "Dreadfully," she replied. "There was something hard in the bed that kept me awake all night and I'm covered in bruises."

Thus spoke the princess in Hans Christian Andersen's story "The Princess and the Pea."

In Andersen's tale, the royal court faces a dilemma: a true princess must be discovered from among the many pretenders who vie for the prince's hand. Every aspirant maiden is put to sleep in a bed with twenty mattresses covered by twenty eiderdowns; beneath the bottom mattress is a pea. Every maiden sleeps soundly except for one, who answers with a tale of sleepless torment. Only a real princess, with a real princess's heritage, would be so sensitive; she is the one contestant of truly royal blood.

The job of the builder of a gravitational wave detector is to construct, first, a "princess"—a device of such exquisite sensitivity that it could detect the merest grain of dust under the royal court's mattresses. If the "princess" is that sensitive, then there is a chance that she will waken to the merest caress that is a gravitational wave. Second, the detector builder's task is to improve the mattresses the court provides, making them so soft and fine that even a princess who would waken to the ethereal embrace of a gravitational wave would sleep through the night with a coconut in place of the pea. In the case of resonant detectors, *making the princess sensitive* involves making the resonating bars large and heavy, from a material with the highest possible "Q" (quality factor), and attaching to them the best possible transducers. In turn, the transducers need to be linked to the best possible amplifiers and the best possible data analysis algorithms. *Making the princess sleep soundly* until the gravitational wave prince arrives involves, the construction of the mattresses, and it is on this that we concentrate here.

The first gravity wave detectors ran at room temperature, and their bars slept upon only one "mattress" and two "pillows," as one might say. Each aluminum bar was housed in a vacuum chamber to quiet sound and to shut out disturbances caused by changes in temperature. This hollow, cylindrical vacuum chamber was the prototype for all subsequent designs; modern resonant detectors have more layers of insulating cylinders nested within one another. The vacuum chamber itself rested on a pile of lead and rubber layers to minimize the effect of vibration in the ground, while inside the cylinder a second pile of lead and rubber sheets supported the wire which in turn supported the bar.

NAUTILUS has seven concentric cylinders and a suspension system based on the same principle as those used for Weber's bars. Look at NAUTILUS and what you see is the outer cylinder—green painted and lying on its side. The cylinder is twice the height of a person and about three times the length. Its center section is fatter than the two end sections and houses the refrigeration equipment. Above the cylinder is a platform reached by a staircase; from this platform, many pipes enter and leave the fat section of the cylinder. Some of the pipes, the ones that take away the "warm" helium for recycling, are coated with a thick layer of frost. Mounted higher still, near the ceiling of the laboratory, is a set of "streamer tubes" designed to respond to cosmic rays. Should the "princess" respond to cosmic rays—which cannot be kept out with "mattresses"—this will be recognized.

Other devices look out for other forces that cannot be kept out completely, however good the insulation. Major seismic disturbances shake everything, and large electromagnetic fields, such as are caused by lightning, may also produce a false "awakening."

NAULTILUS's outer cylinder is made of stainless steel about two centimeters thick. It in turn sits on pneumatic supports alongside some rigid legs. It was found that the pneumatic supports when used alone allowed the whole massive instrument to rock, so the rigid legs are used in combination to minimize this movement. Resting on top of the outer cylinder are what look like four cylindrical wastepaper bins upside down. These bins contain stacks of alternating steel and rubber layers. From the top layer of each bin, though all this is invisible when NAUTILUS is assembled, a wire runs down through the center of the stack and through the wall of the outer cylinder. The next three concentric cylinders are hung on these four wires, so each is insulated from vibrations in the outer cylinder by the stacks of steel and rubber. This simple design, pioneered by Joe Weber's team, has been used subsequently by everyone in the bar business. With its accessories and streamer tubes mounted, NAUTILUS is the size of a substantial two-story cabin.

The outer cylinder of NAUTILUS is at room temperature; it is essentially a vacuum container like that used by Joe Weber, but much bigger. The air is pumped from the interior of the outer cylinder so that there can be little conduction of heat from the outside and so that sounds or other disturbances which might be transmitted through the air are eliminated as much as possible.

The next concentric cylinder—suspended on the wires coming from the stacks—is made of aluminum. When NAUTILUS is cool, this cylinder is kept at a temperature of 100 Kelvin (recall that water freezes at about 273 Kelvin).

Inside this is the third cylinder, suspended in the same way. It, too, is made of aluminum, but it is kept at twenty Kelvin. The two aluminum cylinders just described are cooled by exhausted helium gas as it leaves the colder parts of the apparatus. The tube carrying the spent gas winds around the cylinders in a serpentine fashion, encouraging them to yield their heat to the helium as it passes back to the reliquefying plant. The purpose of the two aluminum cylinders is to help keep the heart of the apparatus cool. All of the first three cylinders are, in a sense, in the same "space," because although they are at different equilibrium temperatures, there are places for gases to pass between their walls. Thus, when the outer cylinder is evacuated, the inner two cylinders are evacuated also.

The fourth concentric cylinder is the start of the true inner heart of the apparatus. Granted, it too is suspended by the wires coming down from the stacks of steel and rubber, but there is no connection between the space inside it and the space outside. This cylinder is more complicated than all the rest. It has a central fat section—which accounts for the central widening of NAUTILUS that can be seen from the outside—and two end sections, which are bolted onto it. The central section is made of stainless steel in the form of a cylindrical—or more correctly, "annular"—tank; the tank itself consists of one cylinder inside another with the ends of the space between them blocked off. The tank is perhaps five feet in diameter on the inside, seven feet in diameter on the outside, and eight feet long. It acts as the main source of bulk cooling for the apparatus. Capable of holding 2000 liters of liquid, it is normally partly filled with liquid helium. Thus, this tank is kept at 4.2 Kelvin, or 2 Kelvin when pumped and the helium is superfluid. The two end caps bolted to the tank are made of copper.

NAUTILUS's fifth cylinder is made from specially constructed and welded copper. This is suspended not from the stacks but, like a pendulum, from titanium rods fixed into the roof of the fourth cylinder. When the apparatus is cool, the fifth cylinder is kept at one Kelvin.

From this point forward, the sixth and seventh cylinders are almost like Russian nesting dolls in relation to the fifth. Each is made from copper, specially constructed and welded so as to be highly conductive to heat; and each is hung like a pendulum inside the other. The sixth cylinder is meant to reach 0.2 Kelvin; the seventh, 0.05 Kelvin or lower. The idea of this cascade of pendulums is to isolate each inner space from any vibrations still felt by the cylinder outside it. A pendulum may swing slowly, but it does not vibrate rapidly, which is key. Gravitational waves of the kind that resonant bars are meant to detect have a high frequency in comparison with the frequency of the swinging of a heavy pendulum. Therefore, the designers of these detectors do not have to worry too much if low-frequency noise enters the system—it does not make the bar respond. The art is to eliminate high-frequency noise, which might disguise itself as a gravitational wave.

Very important in this regard is the isolation of the inner cylinders from the liquid helium tank. The very evaporation of the helium makes some noise that falls within the dangerous frequency range, and the cascade of pendulums is meant to silence it. There is no other way of damping noise inside the liquid helium tank, because any ordinary cushioning material, such as air or rubber, would be frozen solid at this temperature, rendering it no more useful as a damper than a piece of glass.

The resonant bar itself is suspended on a copper wire hung from the roof of the seventh cylinder. Effectively, the bar is the fourth pendulum in the series. It can be cooled to about the same temperature as the last cylinder, though if we continue to think of it as a princess, her head and feet will always be a little warmer than her middle.

To understand why the bar will never be evenly cool all over, one must understand something about what heat is like when there is not much of it. Heat is the random movement of atoms and molecules. The atoms and molecules in a hot object are bouncing about much faster than they are in a cold one. When you put a hot thing up against a cold thing, the atoms and molecules in the hot one crash into the atoms and molecules in the cold one and start them moving faster; that is why the cold thing starts to get hotter. Each time an atom or molecule in the hot thing crashes against an atom or molecule in the cold thing, it itself will slow down a bit. That's why the hot thing gets cooler as the cold thing gets hotter. This mechanism of heat transfer is called conduction.

As every beginning scientist learns, heat travels in three forms: conduction, convection, and radiation. The hotter things are, the more important is radiation; a red-hot piece of iron radiates quite a lot of its heat away in the form of red light and infrared radiation. An atomic bomb emits nearly all its heat as radiation—the blinding "liquid" light (as a Hiroshima survivor referred to it) that sets fire to everything within a large radius. Radiation plays very little part in the transmission of heat when things are very cold, and we can forget about it as a useful means of cooling at the sort of temperatures found in the middle of NAUTILUS (though it is important to guard against radiation warming the outer layers).

Convection is the physical transport of a hot liquid or gas from one place to another. If the vacuum in the chambers of NAUTILUS is a good one, very little convection will take place, because there is so little gas that could move from one place to another. Therefore, we can forget about that form of heat transmission, too.

Conduction is thus the only useful way to move heat about in the coldest parts of NAUTILUS. To cool the apparatus, heat must be *conducted* from the warm parts to the cold parts, such as the liquid helium container. There has to be something to conduct the heat away, and in a 4.2-Kelvin system this would be the last few atoms of helium gas left bouncing around in the vacuum chambers—the exchange gas, as it is called. But once you get below the liquefying point of helium—and the operating temperature of the core of NAUTILUS is well below this point—there is no gas left to conduct the heat away. The only thing that can conduct the last few wisps of heat away from the body of the princess is the wire that suspends her. That wire is made of copper to make heat conduction a little speedier. But the wire is wrapped around the middle of the bar, which means that the middle will always be the coolest place: the only place that will ever quite catch up to the coldness generated by the last pot of the dilution refrigerator.

The temperatures of all these last three cylinders are, as I have said, too low to allow for conduction cooling via residual gas, and their suspension rods are made of a poorly conducting material to maintain the three copper cylinders' different temperatures—the temperatures of the three "pots" of the dilution refrigerator. The bar and the last cylinder are meant to be at the same temperature as the coldest pot. For these reasons, a special means of heat conduction has to be introduced to allow the last three cylinders to be cooled. The three cylinders are linked to the three pots of the dilution refrigerator with copper braids, which are meant to conduct heat well while not conducting any vibrations—braids are soft. It remains to be seen if this is a completely effective way of doing things or whether it will turn out to be a channel for noise that cannot be eliminated.

This problem of the braids indicates one of the difficulties not yet mentioned. The princess is useless if there is no way for her to communicate with the outside world, should the "prince" of gravitational radiation approach. There must be something attached to the bar to get the signals out, but whatever is attached to get things out can also act as a channel to get unwanted things in. Thus, the transducer attached to the end of the bar—in this case, the kind of resonant transducer designed by Ho Jung Paik—reduces the detector's efficiency and could represent an additional source of noise. Part of the art of the detector design is to reduce these problems to a minimum.[5]

The transducer has to feed into an amplifier. The kind of amplifier used by all groups who have built liquid-helium-temperature detectors is know as a SQUID, which stands for Superconducting Quantum Interference Device. Again, an amplifier is not just something that can sit connected to the bar without disturbing it. While the amplifier "hears" the sounds made by the bar, inevitably, it also whispers back; this is known as "back action." A SQUID, which will work only at liquid helium temperatures and below, is the best amplifier known for reducing the effect of back action. Back action can never be eliminated entirely, however, because this would result in us knowing more about the exact shape of the world than quantum theory will allow. Quantum theory tells us that we are only allowed to know the position of very, very small things in terms of probabilities. The aim of good amplifier design, then, is to reach the "quantum limit." It may be that, with much work, the low temperatures at the heart of NAUTILUS will allow this limit to be approached. But first other, cruder, sources of noise must be eliminated.

5. For a useful technical discussion, see Bassan 1994, sec. 2.2; see also sec. 3 of the same work.

AN ENGINEERING HISTORY OF NAUTILUS

In 1983, the design for NAUTILUS was begun. Fortunately at that time, CERN, the Frascati team's outpost, was also the scientific base of Tapio Niinikoski, a Finn who was one of the world's foremost experts on dilution refrigerators. Niinikoski was available to help design the unprecedentedly large dilution refrigerator to be used by NAUTILUS. By 1984, Guido Pizzella had divided up the design work among his team, and by 1985 the design had advanced sufficiently for him to ask for financial support for the work. By 1986, funding provided by the Istituto Nazionale di Fisica Nucleare was approved, and construction went ahead in 1987 and 1988. In 1989, at CERN, NAUTILUS was assembled for the first time, and it was experimentally cooled to below 4 Kelvin though without the detection apparatus being present.

In 1991, NAUTILUS was cooled a second time, now to below 0.1 Kelvin, with the bar and the detector present in the inner tank. On this cooling it was discovered that though the device performed excellently from a cryogenic point of view—that is, the refrigeration worked as hoped—there was far too much mechanical noise to contemplate using NAUTILUS as a useful gravity wave detector.

It was clear that NAUTILUS would have to be taken apart and rebuilt. The decision was made to move it from CERN to the laboratories at Frascati, near Rome. The CERN location involved a great deal of expensive travel for the Rome-based scientists, and a degree of national pride made it seem more appropriate that an Italian detector should be located in Italy. By this time, Frascati had built a helium liquefying plant, so for the first time it was possible to contemplate siting a large and very cold experiment near Rome.

In reassembling NAUTILUS at Frascati, a number of changes were made in an effort to eliminate noise of mechanical origin. Bellows sections were inserted in some of the stainless steel pipes used to transport helium to and from the helium tank, and the suspension systems for the three inner copper cylinders were changed. Previously, the last three cylinders were like nesting Russian dolls—each was suspended from the last by four titanium wires. This design transmitted too much mechanical vibration from the outside. Therefore, the outer one's wires were cut, and each had a heavy lead-block damper placed in the middle of it. The next shield's titanium cables were replaced with stainless steel ropes, each about one centimeter in diameter. Ropes are flexible, so, once more, this design should have represented an improvement in vibration damping.

Finally, the suspension between the penultimate and the innermost cylinder was changed from four titanium wires to two, each of which was wrapped around the cylinder rather than being bolted into it, to save the trouble of adjusting the tension in four separate wires. Equalizing the tension in four wires is like shortening the legs of a four-legged table so that they are even. The way it is done in the case of NAUTILUS is to flick the cables with a finger and listen to the resulting sound. The human ear, it turns out, is the most delicate instrument when it comes to detecting differences in the frequency of a plucked string that correspond to tiny differences in tension. To equalize the tension in four strings in this way takes between 30 minutes and an hour; substituting the last stage of four strings with two wire "cradles" saves some trouble in assembly.

It might be asked why a different adjustment was used at each stage of the Russian doll: why lead dampers at the first stage, stainless steel ropes at the second, and switching from four strings to two cradles at the third? The answer is pragmatic. There was no room to use lead dampers for the penultimate stage of suspension, so stainless steel ropes were used instead. It turned out that the lead weights were effective in increasing noise attenuation, but the stainless steel ropes were not. Nevertheless, there was no point in reversing the change. And why not change from four strings to two cradles at the earlier stages? Well, this modification was a matter of convenience rather than performance, and because of the way NAUTILUS was constructed, it would have been harder to do the same thing in the other places. Thus, even at the heart of the apparatus, design proceeded in a way that would be familiar to any handyman.

In 1992, NAUTILUS was reassembled in Frascati with the innovations incorporated. In early 1993, it was cooled for the third time in its life, but it was soon discovered that the SQUID amplifier was not working; the detector had to be warmed and taken apart once more.

The SQUID repaired, NAUTILUS was cooled again, for the fourth time altogether, in late 1993 and kept cool for several months into 1994. Once more, problems were found; there was still too much mechanical noise. NAUTILUS is based on an ambitious mechanical design as well as an ambitious cryogenic design. It is the only gravitational wave detector in the world that sits on a platform that can be rotated.[6] But as a result, this platform is also less solid than the sort of foundation on which other gravity

6. This is no longer true; the LSU bar, ALLEGRO, when it was shifted to a new laboratory, was mounted on air pads that could be used to rotate it. I assisted in one such rotation in 2002 as well as mopping out the lab, which had been flooded.

wave detectors sit. So confident were the designers of NAUTILUS that they could eliminate mechanical noise coming from the outside that the noisy, vibrating vacuum pumps were mounted on this same platform. It was found that this feature of the design was overambitious; the mechanical attenuation was insufficient to damp out the vibrations of pumps placed in such close contact with the antenna. Thus, an immediate innovation was to remove the pumps to a place beyond the rotating platform. Once this was done, NAUTILUS became, for the first time, a more sensitive device than its predecessor, EXPLORER. Unfortunately, it was also discovered that the refrigerator had a leak. Once more NAUTILUS had to be dismantled.

The fifth cooling took place between October 1994 and June 1995. Now NAUTILUS was clearly a more sensitive device than EXPLORER or any other putative gravitational wave detector in the world, but there were still problems to be overcome.

Gravitational waves that can be seen on detectors like NAUTILUS have their frequency measured in hundreds or thousands of cycles per second. This means that low frequency noise is not fatal and can be allowed to remain. For example, it is recognized that there is no way of eliminating low frequency seismic disturbances, which will rock the whole apparatus. But the big problem that slowly made itself known was that NAUTILUS seemed to contain mechanisms for turning harmless low frequencies into harmful high frequencies. What seemed to be happening was that slow movements made components within the device squeak or creak. To use the jargon, low frequencies were being "upconverted" to high frequencies, and this was fatal to the experiment.

Two sources of upconversion were considered causes of most of the trouble. The bar was rocking slightly on its suspension wire; as it rocked, it scratched against the wire, producing enough high-frequency noise to drown out the effect of a potential gravitational wave. Second, it was considered that there might be some upconversion through friction in the copper braids that were used to conduct heat from the bar and the inner cylinders to the three pots of the dilution refrigerator.

In June of 1995, the team decided that the apparatus must be warmed up and disassembled in order to adjust the wire. Instead of having the bar sit on a wire that went straight upward from its sides, the wire would turn in at a three-degree angle on each side. In other words, it would wrap just a little further around the bar and, with luck, grip just tightly enough on the princess's waist to prevent the rocking motion and the scratching noise. At the same time, the number of copper braids would be reduced from eighteen to ten. These tiny, pragmatic changes, like every previous warming and cooling episode, were to cost the NAUTILUS team four months. By the time NAUTILUS was cool again, something like two years of its life would have been spent doing nothing but warming it up and cooling it down. Why four months to warm and cool an apparatus like this? Let us look still more closely at what it meant to make this minor modification to the suspension wire.

NAUTILUS Adjusted

Before NAUTILUS could be opened, all its parts had to warm to room temperature. If it was opened while too cold, water vapor from the air could condense on its electronics and other sensitive parts and do great damage. The sheer bulk of NAUTILUS meant that it took a full month before the last inner copper end pieces could be removed even though the caps on each outer cylinder in turn were unbolted as soon as that section had warmed; this allowed the heat from the air to penetrate to the core of the device more quickly, but it still took a month to warm the middle.[7] Once disassembled, it took only a couple of weeks to make the planned modifications and just one more week to bolt everything back together again, for a total of seven weeks thus far.

The next job was to pump out the air. Vacuum pumping a volume as large as NAUTILUS is no trivial matter. Three weeks were taken pumping out the air to get the inside back to an acceptable state of vacuum. There is no point in pouring in cold liquid gases while heat can enter via air between the cylinders. So far, ten weeks have elapsed.

Once the air was pumped out to a satisfactory level, the apparatus was precooled with liquid nitrogen. Cooling a system the size of NAUTILUS is a slow business for the same reason as vacuum pumping something of this bulk is a slow business. It was expected that precooling would take three weeks, but it actually lasted four, because the vessel containing liquid nitrogen that was delivered to the laboratory was almost empty. A week was taken up in rectifying the mistake. We have reached fourteen weeks.

An additional week was taken in cooling from liquid nitrogen temperatures to liquid helium temperatures (fifteen weeks so far). At this stage, a little exchange gas is left in the apparatus to aid conduction cooling from the central parts to the helium tank. It was partway through this process that I paid my second visit to Frascati, watching the experiment cooling

---

7. I paid my first visit to Frascati just before the last copper end cap was removed during this warming up process.

from about 50 Kelvin to 5 Kelvin over two days. Every night the process was monitored; the slightest mistake, and it would take another four months to warm and cool again. Too many such incidents, and funding might be jeopardized.

Failing the detection of gravity waves, funding depends on the demonstrated technical virtuosity of the "instrumenteers." And what does this technical virtuosity comprise? In addition to writing learned theoretical papers, it also requires one to be there half the night swinging liquid helium tanks around with a crane, connecting up a new one every few hours, and supervising the drip of freezing liquid into the heart of the apparatus so that cooling takes place at just the right speed.

By Friday, October 13, 1995, NAUTILUS reached liquid helium temperatures, and the team began to pump out the residual exchange gas from the experimental spaces in the deep heart of the apparatus. There was no purpose left for the residual helium gas to serve; in the temperature regime to which the bar was headed, the remains of the gas could cause only trouble.

By Saturday, the bar had warmed a little to six Kelvin. This was because the exchange gas was no longer doing the job of cooling, but this slight rise in temperature was of no consequence. It was as though the muscles of the animal were being gathered for the next leap forward. Indeed, it is a little easier to pump out the residue of a slightly warmer gas, because the molecules are less sluggish. Pumping continued through Sunday.

On Monday, October 16, pumping began on the main liquid helium vessel to turn the helium into a superfluid. By the end of the day its temperature had fallen to about three Kelvin. Also on the sixteenth, the first pot of the dilution refrigerator was filled with liquid helium. All day long, the bar remained at a temperature of about six to seven Kelvin.

By Tuesday, the bar was beginning to cool below the normal temperature of liquid helium; by the end of the day it was about three Kelvin. By now the vacuum inside the experimental heart of NAUTILUS was like the vacuum that exists in outer space.

By the end of Wednesday, October 18, a stable temperature regime had been reached—the last step before the final cooling. The main liquid helium vessel had reached its superfluid temperature of two Kelvin while the bar, aided by the dilution refrigerator, sat at 1.4 Kelvin. These temperatures would be held while the working of the supercooled amplifier—the SQUID and the transducer—were checked and the whole apparatus examined for sources of noise.

On Thursday, the transducer was energized to ten volts, and the team began to take data from the system. There was too much noise. On Friday, the voltage across the transducer was increased to 100, and a first program of electrical shielding began to try to reduce any noise that was caused by stray electromagnetic fields. This program continued over the weekend as more data was taken, all of it disappointing, though not unexpectedly so, given previous experience.

By Monday, October 23, the liquid helium tank held about 1000 liters of liquid gas, the SQUID amplifier was acting as it should, and the transducer on the end of the bar was turning its movements into electrical signals in the way expected. The team had a working something or other rather than a failing something or other. But whatever the something or other was, it was still noisy; the princess would not sleep.

Everything that makes NAUTILUS especially advanced also provides an opportunity for trouble. In Joe Weber's early designs, the antenna looked from the outside like a simple thing. What one could see was a fairly plain cylinder. In the case of NAUTILUS, the top of the cylinder is like a forest of pipes and wires; it is as though NAUTILUS is surmounted by a kind of oil field wellhead. Whereas all Joe Weber needed was a vacuum pipe and some wires to record the signal, NAUTILUS needs great tubes and valves for the liquid gases, more tubes and valves for the dilution refrigerator, and "Christmas trees" of wires for the many monitors which register the temperature in all the many separate cavities in the apparatus. Each of these inward and outward passageways provides its own opportunity for the entry of unwanted energy, and the team were trying to close these doors.

Now a second, more intensive effort was begun to shield everything against stray electric fields. Vacuum and cold call for neoprene sealing washers, while electrical wires require plastic or rubber insulation, and every pipe or wire that entered NAUTILUS's body through a piece of plastic offered an entryway for unwanted electrical noise. The team were trying to cover every nonmetallic orifice, washer, or seal with metalized tape. A layer of copper tape might come first, followed by a layer of sticky aluminum tape, with sheets of aluminum kitchen foil wrapped around the whole lot for good measure. NAUTILUS was being decorated in shiny paper, looking more and more like Christmas as the day wore on. The tape was stuck on as you or I might do it: cut a strip, press it down, bend it over the curves, and try to flatten out the creases with the back of a pair of scissors. When has one done enough? Who knows, so stick on some more; it can do no harm. "Why are you putting on so many layers when the top layer means you cannot see if the bottom layer is still secure?" I asked Eugenio Coccia, one of NAUTILUS's chief scientists. He replied, "We found out last time that this worked." Here was the coldest large thing in the universe being tended by techniques, to use the old cliché, no different from cooking. All physics is like this: at some point people just have to know how to do things with their hands, because there is too much in the world to reduce to figures, calculate, and plan ahead for.

Attempts to shield the apparatus continued into Tuesday, October 24, but by the end of the day it seemed that these attempts to close the last doors were having no effect at all. NAUTILUS was just as noisy as before, and the team looked glum. As Coccia put it in his dry way, "The results are not very satisfactory."

There is something important to understand about experimental science that even the most sophisticated observers often overlook. When we think of a machine, we normally think of something reliable, like our car or our computer. But these are very special kinds of machines, designed to repeat simple tasks over and over without any idiosyncrasy and civilized by years of use and development. Furthermore, their users have been trained so deeply in their handling that any residual quirks are known and circumvented as though they were part of the users' body or mind. We tend to think of all kinds of machines, including experimental apparatus, as like cars and computers. Indeed, the foundation of the commonsense view of experimental science—that anyone can repeat the same observations with the same apparatus—turns on thinking of this sort. But most machines, especially the new kinds of machines that are used on the frontiers of experimental science, are like cats—they can't be held still against their will.

In the first study of experimental science that I ever conducted—a study of the building of a new kind of laser—I watched as a laser builder tried to reproduce on one side of the laboratory a carbon copy of a laser that he had built himself, and that sat happily working on the other side the laboratory. It took him days of frustrating adjustments to make work a copy of his own hand-built machine. Here in Frascati we were seeing a still more extraordinary thing: the very same piece of apparatus—NAUTILUS—had been taken apart and reassembled with only the smallest of adjustments, and now it refused to act like the same NAUTILUS. The old NAUTILUS, though it had a residue of unacceptable noise, was at least as quiet as EXPLORER; but, reassembled, the old NAUTILUS had become a squirming monster.

But this is not physics talk. Physicists would say there must be something different about the reassembled NAUTILUS; something of significance had been changed that had been overlooked and must be found. On Tuesday evening, October 24, Guido Pizzella, the overall boss of the project, told me that it was crucial in experimental physics to have patience. The following Wednesday morning, Coccia remarked to me that the most important quality of an experimental physicist is perseverance.

On October 25, the team was wondering whether their efforts at completing the shielding for stray electromagnetic fields might not have been misdirected. Perhaps the source of noise was not electromagnetic interference but seismic disturbances. One discovers in moments such as this that the radius of the circle within which one casts one's net to catch the difference between this time and last time gets wider and wider. One possibility I heard being discussed was that the crucial difference had nothing to do with the inner heart of NAUTILUS but turned instead on the presence or absence of the upper layer of streamer tubes intended to indicate the existence of cosmic ray showers. This upper layer of tubes rests upon steel legs, which are supported by the rotating platform. The tubes were not in place on October 25 for the very good reason that they reduce the headroom above NAUTILUS and make it very difficult to do the shielding work on the "wellhead" that had been carried out for the last several days. But the upper bank of streamer tubes weighs four tons, and it may be that this bank was providing a necessary degree of stability to the rotating platform—enough stability to damp out vibrations. It may be that these vibrations, undamped by the bank of streamer tubes, are causing the trouble. The next step was to reduce the voltage across the transducer in an attempt to ascertain whether it was seismic or electromagnetic disturbances that were causing the noise.

By October 26, it was still not clear whether the excessive tossing and turning of the princess was caused by electromagnetic or mechanical input. That morning, the pressure inside the pneumatic supports was increased so as to raise NAUTILUS slightly off its rigid legs. More layers of sticky copper and aluminum strip were added later in the day.

Guido Pizzella told me that day that he personally did not think that the trouble was being caused by mechanical vibration—in expressing this opinion he was disagreeing with many of his team. His own view was that NAUTILUS was not really any more noisy now than it had been last time it was assembled. The trouble was that that this time the experienced team had managed to get the transducer and the SQUID working very quickly. What they were now seeing was the relaxation of strains within the apparatus as a whole. For example, the liquid helium tank would be straining as the metal slowly settled down into the shape it would finally adopt at two Kelvin. All the metal parts had to relax into the smaller sizes that went with lower temperatures. Pizzella believed that the last time NAUTILUS had been cooled, they had not noticed all this relaxation noise, because, by accident rather than design, the transducer and SQUID were not well adjusted soon enough to detect this noisy period; consequently, he thought the noisy period would pass by itself. Nevertheless, the team hoped to mount the four tons of streamer tubes on top of NAUTILUS just to make sure that as much as possible was in the same state as last time, working on the principle "only change one thing at a time."

One might describe this phase of operations as pragmatism tempered by experience; it was interesting to see Pizzella's cheerful confidence in contrast with the gloomy aspects of some other members of the team.

By the morning of Friday, October 27, NAUTILUS seemed to be slowly settling. Maybe the streamer tubes would not be mounted until the following week, since NAUTILUS ought to be allowed to settle as far as it would before making a further change. Coccia told me he believed the pumping up of the pneumatic suspension was the crucial change, and this meant the team now had some control over NAUTILUS; they could increase or reduce pressure at will and see the effect of it. Of course, it might also be that the Pizzella thesis of thermal relaxation was the right one and the pneumatic pressure was irrelevant. Whether either or both were correct, the point was that NAUTILUS was beginning to calm down and turn back into the NAUTILUS they knew before.

On that morning, my last day at Frascati for a while, I began to notice smiles on the faces of the team. If I had brought my smile meter with me, I would have been able to judge NAUTILUS's noise levels during the week by the angles of the corners of Eugenio Coccia's and Viviana Fafone's mouths. Now I began to feel a little more free to make my joke about NAUTILUS's problem possibly being the nauseous shade of the green paint on the outer cylinder. Through all this, Pizzella's mouth was not a good indicator of progress; it was always smiling as befits the mouth of a person responsible for the morale of a team—and I must say I benefited from the smile, too.

I left Frascati on this occasion with the assurance from the team that they would keep a daily diary of NAUTILUS's exploits, an agreement I reinforced with a quick visit in the last week of March of 1996.[8] At that time I discovered that NAUTILUS was still having terrible trouble with unexplained noise. The team believed the problem was still "upconversion," probably mainly due to the copper braids that linked the bar to the

8. The diary was never much use. As Eugenio Coccia was to tell me in 1996, "I have to apologize because of the diary. You can consider this the result of an experiment. So the statistics are poor, but a physicist with many things to do cannot . . . Because you touched me when you said, 'if there is a discovery then this diary will be an invaluable document.' You touched me at that moment. But then, day by day, the things to do are—maybe this is wrong."

innermost cylinder so that the last vestiges of heat could flow away. The copper braid was soft; its individual strands might still be rubbing against one another as the braid flexed, thereby producing a high frequency where everything had been designed to produce low frequencies only.

In July I visited the AURIGA site in Legnaro, near Padua, where I did discuss the matter of upconversion. Knowing of Frascati's problems, instead of braids the AURIGA team were using solid copper ribbons at the final stage, with their ends split into several separate strands to provide flexibility without rubbing.

CHAPTER 15

NAUTILUS, NOVEMBER 1996 TO JUNE 1998

In November 1996, I returned to Frascati. I asked the team if they believed that their colleagues at AURIGA in Legnaro had solved the upconversion problem with the switch from braids to ribbon. I was told that was is too early to tell, since the Legnaro team had their problems, too. They had not yet managed to get their bar and their SQUID working together, so they still had a noise level that was too high to reveal any beneficial effect of copper ribbon. Frascati would wait until the Legnaro team had solved the problem before committing themselves to a change to ribbon.

In any case, by November NAUTILUS seemed to be calming down further. One reason was that the team was accepting that their design was not as good as they had thought. The earlier problems, we might say, were a matter of design bravado: if you believe your own propaganda about the effectiveness of your design, then you believe that nothing smaller than a football will disturb the princess; the whole idea of the careful design of all the cylinders nested one within another was to keep out disturbances, so it follows that one can be careless about what happens beneath and around the apparatus. NAUTILUS as a whole was, then, set down with a flamboyant disregard for the immediate environment—which is understandable, for only if the cylinders, suspensions, and so forth do not meet their specification should one need to worry about the immediate environment.

Let's remind ourselves of the structure of NAUTILUS. It sits upon a platform that can be rotated so that the whole thing can be aligned in different directions if desired. This platform has a metal structure that can make a solid connection with the ground. On the platform, sitting on pneumatic dampers, rests the outer casing of NAUTILUS. The dampers can be pumped up or deflated, which raises or lowers the whole device. When deflated, NAUTILUS rest on solid supports. Thus NAUTILUS can be supported in four different ways: it can be lowered onto the rigid supports or raised on the dampers, and the platform can be rotated freely on bearings or fixed rigidly to the ground. In the early days, so confident were the designers about the built-in noise insulation that they placed the noisy vacuum pumps on the platform, but they had since removed them to the solid floor beyond. The first attempts to quiet NAUTILUS had been made with the dampers lowered and the platform rigidly supported.

Above NAUTILUS and welded directly to the outer green cylinder was a working platform supporting the wellhead of wires and pipes. Every few days, someone had to climb onto the platform and hook up one of the pipes to an insulated tank full of liquid helium so that the helium tank could be refilled. During the refilling operation, it was accepted that NAUTILUS would be too noisy to see data. It seemed to me that fixing this upper platform directly to the outer cylinder demonstrated substantial confidence in the noise suppression inside the nest of cylinders, since the platform was at best a kind of acoustic antenna, able to pick and transmit all manner of sounds and vibrations from the laboratory directly into the outer tank. It was also a problem during liquid helium refills, because someone had to clamber over the device itself.

In October 1995, I asked Guido Pizzella whether he was concerned that NAUTILUS would be too closely coupled to the various sources of external noise. He replied that he was not worried.

> [1995] *Pizzella:* No! I don't think it is sensitive to external noise. I think if anything it is sensitive to internal noise.
>
> *Collins:* Such as the cryostat.
>
> *Pizzella:* Yes . . .

Collins: So you still have confidence in the original design?—That there is enough
attenuation inside?

Pizzella: Yes! Until I convince myself that it is wrong, we have to... there is still
no indication that it is wrong.

I asked then if it would be a good thing to raise NAUTILUS on its pneumatic
dampers, but Pizzella said that it would make no difference.

By November 1996, however, Eugenio Coccia said that he thought that
the most important thing the team had done to quiet NAUTILUS was to
raise the pneumatic dampers. They also decided to try a run with the
metal supporting framework for the platform removed. Coccia explained,
"We thought that after having taken away all the sources of vibration from
the platform—pumps and so on—then it was worth it to test the situation
again with the platform completely free."

Now they found that to refill the cryostat with liquid helium, it was
necessary to lower NAUTILUS by deflating the dampers, because anyone
clambering about on the platform would rock the machine and disturb
everything within it. Unfortunately, even lowering and raising the device
on the dampers appeared to disturb it undesirably.

[1996] *Coccia:* It is very sensitive to any change you make. Even we think that if
you lower and then go up, the system is slightly different.

Collins: New stresses?

Coccia: I don't know—maybe—it is difficult to say. Even if it were of glass and you
could see inside, I think it would be difficult to say what happens. One of the
things is that if you move the system, then the detector has different stable
positions, I think, and all these stable position settings are not equivalent. In
the sense that one position is causing more noise than another position...

Since we are in an unexplored region of measurement, we cannot find a
solution in the books or in the literature—I mean—we have to try—most of
the work is based on a trial-and-error improvement work. So, we found that
with the NAUTILUS on the dampers, and free from any support, and without
disturbing it—without touching—without raising and lowering it—then the
system has more probability of working quietly.

Therefore, the team had decided to remove the upper platform and rebuild
it without any direct contact with the cylinder.

They were also building a wall, which included a lead lining, right
across the large warehouselike building that housed NAUTILUS. The building was shared, and the occupant of the far end made a lot of noise and

could not be expected to baby the gravitational wave detector in the same way as its builders. So strangers and, with luck, strange noises, would be kept out.

Another odd thing is that the noise in NAUTILUS seems to be correlated with temperature. Strangely, NAUTILUS is most noisy during the warmest part of the day—about 5 or 6 p.m. This is not a matter of the absolute temperature at that time of the day but of the change from colder to hotter. So the team decided to try to control the temperature of the huge space in which NAUTILUS is housed.

The temperature effect seems enigmatic in a device meant to hold its temperature to a few milliKelvin. But there are pipes going in and out of the apparatus, and these will undoubtedly feel a change of environment. Viviana Fafone guessed the trouble was with one pipe that connected the Dewar (vacuum flask) to the one-Kelvin pot. Now the team always pump out the pipe sheath before loading the liquid helium. She said that before, the warmth was penetrating the poor vacuum in the sheath and was either causing the helium in the pipe to boil, or causing nonlaminar flow in the helium; either way, it was introducing noise into the heart of the apparatus. Since the team began to pump out this pipe sheath more carefully before filling, she thought that the big noise problem had been solved.

I sense that there is some unspoken difference of opinion among the team. It was Fafone who found the tiny change in pressure in the one-Kelvin pot that correlates with temperature—"I was lucky—I found it" [laughter]— and this is her favorite hypothesis about the source of the noise. Coccia, however, did not mention this as the source of noise; his favorite hypothesis still seems to be strains emanating from slight changes of alignment within the parts of the apparatus associated with movements of the whole thing. He is also a strong supporter of temperature control in the whole laboratory, perhaps because he is less certain than Fafone that the pipe sheath was the source of the problem. They cannot be sure of picking the correct cause of the noise, because the ramifying skein of causality is too complex to allow them to isolate one variable at a time.

A very important parameter of a detector is its "duty cycle"—the proportion of time it is usable as a detector rather than undergoing maintenance or in too noisy a state to detect gravitational waves. I asked Coccia about NAUTILUS's duty cycle.

> [1996] *Coccia:* Now let me make one thing clear so that the situation is understood. What is the sensitivity of the detector, and what is its duty cycle? We tried to give some number that can transmit a feeling about the detector.

So, we consider the detector is working properly if it can see with signal-to-noise ratio of ten—with a good signal-to-noise ratio—an event considered a standard event—which means a gravitational collapse—supernova explosion—in the center of the galaxy, in which 10^{-2} solar masses are converted to gravitational waves. . . . This 1% is considered by many people to be a good number. . . . This event is not so frequent from our present knowledge.

Collins: What is the rate?

Coccia: The rate would be one per five years—some people think one per 10 years.

Collins: Even this is quite a lot, isn't it? Because visible supernovae are infrequent.

Coccia: Visible supernovae, in our galaxy, are one per 100 years, or 200 years, but not all the galaxy is visible, and then there are supernovae that are silent—do not emit a lot of photons, but could emit a lot of gravitational waves. So at the end, these estimates are very uncertain, because we have not yet detected gravitational waves; however, there are detectors of neutrinos, for instance, for which sensitivity is limited to our galaxy . . . all these experiments can detect solar collapses, and they say their sensitivity is limited to our galaxy.

So if we consider the detector is working properly—if it's sensitivity is better than that—then we consider the antenna is working as a gravitational wave detector—and we can calculate the duty cycle—how much time the detector stays in this condition. So if we do that, then we have a duty cycle today of 60%.

That gives some dignity also to our efforts, because hearing that we are trying this and doing that and we have extra noise, one can have the impression that we have so many problems; but actually, if we leave the detector working, we have, of course, hours in which the working is not good because we are refilling with helium, because we have extra noises, and so on. But if we consider the detector is working when it can see clearly a supernova in our galaxy, then this is 60% of time. This is a strain of better than 10^{-18}.[1]

Coccia also told me, wistfully, that his profession is a difficult one.

[1996] It is very hard, this business. And I think that if one starts [with an attitude of arrogance]: "We are people of high-energy physics. We are good managers and very good physicists. And now we will build and do beautiful things." It is not so easy. Everyone who enters this field at the beginning is, of course, very optimistic. But life is hard, in this field especially, and one can

---

1. That is, it can detect a change in the bar's length of one million, million, millionth of that length.

get the impression that we are not professional enough; but if anyone really starts working in this field, after a few years they will become more modest and they will realise that this is very, very difficult research.... We are in an unexplored region—also an unexplored region of bad luck. This is all new.

NAUTILUS, as it would turn out, would not fulfill its promise for a long time. It would eventually run for extended periods at a few millidegrees with a reasonable duty cycle, but only in the twenty-first century would its sensitivity creep to a better level than that of the old EXPLORER antenna. Whatever the technology, all the cryogenic bars turned out to be about the same sensitivity. In 1998, Guido Pizzella was ready to admit to me that he now thought ultracryogenics was a mistake—technically. It was far more trouble than it was worth. Politically it was a sensible move, because no one would have funded another ordinary cryogenic bar. For similar reasons, he wouldn't now remove the dilution refrigerator and rebuild the bar—it would be politically unacceptable to the future of the group. So this whole program had been a disappointment to him even though NAUTILUS has been working very well since June 1998.[2] He believed there were two changes that had finally brought success: the switch to AURIGA-style copper ribbons instead of the braids, and the replacement of a "very silly" wire feeding the 300 volts to the transducer. Pizzella told me it may well be the second change that made all the difference.

2. To abandon ultracryogenics would make sense only if the dilution refrigerator were removed so that it did not introduce noise from the outside; but this would mean they would have to learn to understand a new machine. Cryogenics has been such a disappointment that Pizzella encouraged a southern Italian group to build a very large room-temperature device. If a room-temperature device were large enough, it could be as sensitive as a cryogenic device but much less trouble (so long as the transducer was appropriate—an interferometer, however, still would be the best). The group did not get the funding.

CHAPTER 16

THE SPHERES

What was the next big step for resonant masses after ultracryogenics, given that Bill Fairbank's dream—ever lower temperatures—seemed to have run its course? The answer was resonant spheres. Spheres, three meters in diameter and weighing 100 tons or so instead of the two tons that bars weigh, would have many advantages.

When a bar is hit by a gravitational wave, it rings; most noticeably, it expands and contracts along its length. Because the biggest change in the bar's mass occurs along its length, modern transducers are placed on the end of the bar, at the point of largest movement. But the fact that the largest movement occurs along the length of the bar is simply because that is the longest part of the bar. By contrast, a spherical detector, can resonate with equal magnitude along many axes and in many ways. For example, the sphere can grow widen at its equator while it shortens between the poles, and vice versa; it can "breathe" in and out as a whole; it can grow longer between "Europe" and "Japan" as it grows shorter between "India" and "America"; and so forth. Moreover, it can exhibit five different kinds of resonance, each equally measurable.

So where should the transducers be placed? It turns out that every vibration can be measured and extracted by looking at resonances on five axes, which requires six transducers distributed over the surface of the sphere. And it turns out that the output of these six transducers, when appropriately analyzed, can be read backward to pinpoint the exact way the force that caused the vibrations had interacted with the sphere; hence the direction of the source and the character of the wave can be known. Directional sensitivity gives spheres a great advantage over every other kind of gravitational wave detector.

This point is sometimes misunderstood. Regarding direction, two related qualities are associated with a gravitational wave (or any other) detector. There is directional sensitivity—the ability to discern the source of a disturbance—and there is "omnidirectionality"—which means good sensitivity whichever way the apparatus is oriented in respect of the source. Spheres have both omnidirectionality and directional sensitivity—they are equally sensitive in all directions, and they can locate the source of any signal. Interferometers have pretty good omnidirectionality but almost no directional sensitivity, which is why triangulation between four interferometers a long distance apart is needed to determine the direction of a source; bars have a degree of directional sensitivity that is given to them precisely because they are *not omnidirectional.*

Lack of omindirectionality can be used to give directional sensitivity. Thus, Joe Weber could claim to have pinpointed the source of his gravitational waves (the sidereal anisotropy), because his bars were more sensitive at some times of the day, when they were in certain orientations, to waves coming from the center of the galaxy. Omnidirectionality is a good thing if you want to build a detector that is going to be sensitive to gravitational waves all the time, wherever they are coming from, but not such a good thing if it is not combined with directional sensitivity. To reiterate, the spheres, which would be cooled to two or four Kelvin, would be much more sensitive than the cryogenic bars because of their greatly increased mass; more efficient because of their omnidirectionality; and better than any other kind of detector in terms of directional sensitivity. That is why they were the obvious next step.

Where did the idea of using a sphere come from? I do not know who first had the idea of building a spherical detector in the laboratory; but the first analysis of the impact of a gravitational wave on a sphere dates at least to Robert Forward, in a paper he published in 1971. The first attempted use of a sphere to detect gravitational waves goes back to the 1970s as well; it is just that the sphere in this case was Earth. In Forward's 1971 paper, yet another advantage of spheres is mentioned that their protagonists were to discuss from time to time without making much progress. This is a spherical detector's ability to distinguish between certain competing theories of gravitation. The theories can be distinguished by looking at the way the sphere's "breathing mode," its overall expansion and contraction, behaves.

Four groups became seriously interested in building spheres. They were the two existing groups at LSU and Frascati, and two new groups in Holland and Brazil, respectively. Of these, the LSU group were the first to make practical progress and the clear winners in terms of incisive acronyms. They named their sphere TIGA, which stands for Truncated Icosohedral Gravitational Wave Antenna. To appreciate how far ahead they were in the acronym game, one has to understand that Louisiana State University's mascot is a tiger, and that, to rub things in, a live tiger prowls in a small cage on campus. As to the rest of the words, the "sphere" to be built by LSU was not a complete sphere but a twenty-sided (icosohedral) polyhedron. That is to say, twenty flat faces (truncations) were machined onto a sphere for mounting transducers, all the information that is there can be extracted, and the acronym unfolds with irresistible logic.

In my judgment, coming second in the acronym competition were a brand-new group located at the National Institute for Nuclear Physics and High Energy Physics (NIKHEF), a high-energy physics research establishment in a suburb of Amsterdam. Their acronym was GRAIL, which connoted not only a holy yet intensely difficult quest, but also Gravitational Radiation Antenna In the Lowlands. The GRAIL groups planned to build a three-meter sphere of copper-aluminum alloy. This places the Frascati Italians, led by Eugenio Coccia, third in terms of acronyms, though a clear second in terms of effort and advancement. Their antenna was known as Sorgente Ferroeletricca di Elettroni Robust A-2 (SFERA) and was to weigh 100 tons, to be cooled to twenty milliKelvin, and to have a strain sensitivity of 3×10^{-24}.

The sphere groups were somewhat less separate than this description makes them appear. The LSU team were instrumental in providing a prototype sphere for Frascati, and Giorgio Frossatti, a dilution refrigerator expert from the University of Leiden—and a Netherlands-based physicist in spite of his Italian-sounding name—led an individual effort, promoting the sphere projects and joining up with anyone who wanted to use his know-how and expertise. Frossatti was determined to go ahead whatever happened to the other groups, and he had investigated and tested sphere materials and fabrication. He believed copper-aluminum alloy spheres could be fabricated in one piece by firms who made ships' propellers, and he was determined to build a small one of these before advancing, possibly in collaboration with the GRAIL group, to a three-meter sphere which he expected to cost $30 million.

Making a three-meter metal sphere is not easy. Quite simply, there are no facilities in the world for casting blocks of aluminum alloy this large, so new methods and/or materials had to be found. Frossatti thought that whole pieces could be cast by the ship companies, but others favored explosive bonding of separate sheets of metal, building up to the sphere size slab by slab. In explosive bonding, one slab is placed on top of another, with a sheet of explosive on top of that. Next, the explosive is detonated at one end and the explosive wave traverses the two slabs, forcing the top one down onto the bottom one with sufficient force to weld the two together. As the explosive wave travels from one side to the other, impurities, such as oxides, are expelled ahead of the traveling bond, so in theory the result is as good as if the whole two slabs had been cast as one. Unfortunately, in practice, the experimenters found that some imperfect areas of bonding remained, but that they would not necessarily damage the "Q" of the bar fatally.

The LSU group were pressing ahead with the development of prototype TIGAs, led in their effort by Warren Johnson. About 80 centimeters seemed to be the accepted diameter for a prototype sphere, and at LSU one could find an 84-centimeter truncated icosohedron, while a sister model, manufactured by the same firm, was sent to Frascati. The LSU prototype was used to show that hammer taps on any face could be located by analyzing the readouts from the transducers.

The hard-won experience of the LSU and Frascati groups made them sure that they knew how to build spheres and how to cool them without excess sources of noise. The Dutch GRAIL group, coming from a background of high-energy physics, were used to executing large projects, and possessed the (misplaced?) confidence of this physics culture. The spheres, it was calculated, would be at least as sensitive as the planned interferometers, the other rival technology, if in a narrower waveband.

Because resonators vibrate preferentially around their natural frequency, they tend to be narrow-band instruments; small spheres will resonate at higher frequencies than large ones. But the groups planned to build "xylophone" arrays of spheres of different sizes, to cover the whole waveband of interest, once the concept had been proved with the first large sphere. Furthermore, the spheres promised to be much cheaper to build than interferometers. Estimates varied, but one-tenth of the cost of interferometers seemed to be at the high end of these estimates. Thus, the resonant

detector program could look forward with optimism to a burgeoning research and development program that should be relatively easy to fund,
followed by the implementation of arrays of spheres in gravitational observatories with excellent directional discrimination, and sensitivity at least
equal to that of any other technology. Before describing the fate of the
spheres, I will introduce the other new technology—interferometers.

CHAPTER 17

THE START OF INTERFEROMETRY

At the turn of the millennium, one would have had to bet that interferometry would be the technique that would make the first direct detection of gravitational waves; certainly the worldwide interferometer development effort dwarfs all previous attempts to see the waves, the Americans alone spending in the region of $300 million. The first people to think of using interferometers as a means of detecting gravitational waves seem to have been a pair of Russians, M. Gerstenshtein and V. I. Pustovoit, who published a paper on the idea in Russian in 1962. Independently, Weber and his students considered the idea in 1964.[1] The first person actually to build such a device was Robert Forward, Weber's copioneer, but it was too small to see the waves. Forward, as we can now see, was perhaps the most important member of Weber's team when it came to the constructing the first resonant bar; was the first to analyze a sphere as a detector; and was the first to build an interferometric detector.

---

1. Gerstenshtein and Pusotovoit, 1963 [in English]; Thorne 1989, unpublished.

It is Rainer Weiss of MIT who is now almost universally credited with the conceptualization of an interferometer capable of seeing the kind of flux of gravitational waves the theorists believed might be there to be seen. There are, perhaps, three reasons why so much of the credit is given to Weiss despite Forward's having long preceded him in actually building a device, and Weiss's invention having been anticipated at least twice and possibly four times.[2] First, Weiss was the person who first analyzed the mode of best operation, the sensitivity, and the noise sources that allowed him to estimate the appropriate scale—kilometer arm lengths—for an interferometer that could detect predicted sources. Second, many of Forward's ideas reached him from Weiss via a mutual acquaintance—Phil Chapman of Arco-Everett—who seems to have shuttled between the two in the early 1970s. The third reason is more sociological: Weiss is still a leading figure in the interferometric search for gravitational waves and has inducted many of the rest of the team into the project. This means he is more salient when this team, now the dominant force in international interferometric gravitational wave detection, think about the sources of their enterprise. This is not to say that Weiss is not justly credited with the principal input to large-scale interferometry, just that it would be harder for anyone else to get the recognition. This sociological reason does, almost inevitably, lead to the contributions of others being less salient in scientists' minds; the same forces work against, say, Ron Drever's contributions to interferometery being fully appreciated and, for that matter, the survival of Joe Weber's reputation as a pioneer of gravitational wave detection as a whole. None of this is Weiss's fault. Indeed, one might make the contrary case: because Weiss has always been bad about publishing and publicizing his ideas, it took a longer time for his work to bring him acclaim than might have been the case.

WHAT IS AN INTERFEROMETER?

An interferometer is a device that looks at changes in the way two light rays *interfere* with each other. A beam of light is shone onto a "beam splitter"

2. Weiss was born in Berlin in 1932. He obtained bachelor's and doctoral degrees from MIT in 1955 and 1962, respectively. After short periods at Tufts University and Princeton University, he joined the faculty of MIT in 1964 and remained there henceforward. According to Thorne (1989, unpublished), Weiss came up with the idea of the interferometric detector in 1969 without knowing that Weber had already had the idea. Weiss himself acknowledges Felix Pirani and Phillip Chapman (see below), and we also have the Russians as precursors of the idea.

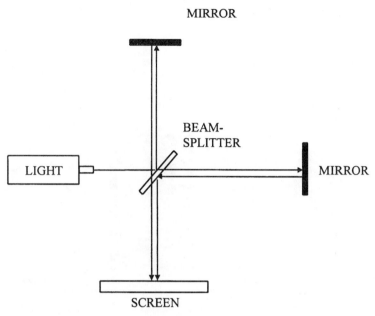

Fig. 17.1 Schematic diagram of a simple interferometer.

such as a half-silvered mirror, which converts it to two rays, usually at right angles to each other. The rays are then reflected from mirrors and recombined (see fig. 17.1). The way the rays recombine indicates whether they have had different experiences on their journeys. If the apparatus is arranged so that the paths are differently affected by some changing force, then one can see those changes and measure that force. When one uses an interferometer to look for gravitational waves, one arranges that the two rays have different experiences as a wave passes through the interferometer.

Interferometry was invented in the 1880s to measure something quite different. The most famous interference experiment was carried out in 1887 by Albert Michelson and Edward Morley. Michelson and Morley were trying to measure the speed of Earth as it traveled through space. It was then thought that space was filled with something called the "aether," the medium through which light rays were thought to be transmitted, in the same way that air is a medium through which sound waves are transmitted. Michelson and Morley split a beam of light into two, sent the separated rays out at right angles to each other, bounced them back with mirrors, and checked for "interference" in the recombined beam. If one beam had been affected one way by Earth's speeding through the aether—perhaps

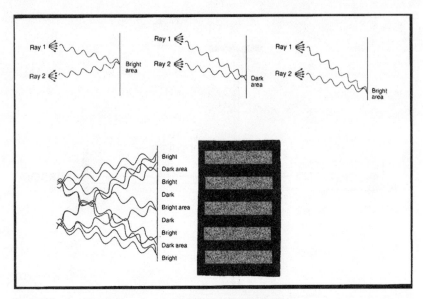

Fig. 17.2 Two rays hit a screen and produce interference fringes.

that beam had gone back and forth in the direction of travel, while the other had traveled across Earth's direction of travel—the pattern of interference should reveal the different experiences of the beams. This pattern, if properly analyzed, would reveal the speed and direction of Earth's motion. Disappointingly, Michelson and Morley could not find any effect. This result was later taken to be one of the most important experimental confirmations of the special theory of relativity, and it is one of the most famous experiments in the history of science.[3]

Michelson and Morley recombined the rays by shining them onto a screen. Such a screen reveals a series of "interference fringes"—a series of light and dark bands (see fig. 17.2). The effect is due to the light *waves* in each half of the beam alternately reinforcing each other—the bright bands— and canceling each other out—the dark bands. This is a simple geometrical consequence of the superimposition of two wave motions: as one moves across the field upon which the rays converge, the path length of each ray changes slightly. For example, the left-hand ray (Ray 1 in fig. 17.2) has to travel a certain distance to reach the left-hand side of the illuminated area. To reach a point a little to the right, it will have to travel slightly less far;

3. For a more complete account of the meaning and controversy surrounding the Michelson-Morley experiment, See Collins and Pinch 1993/1998.

to reach a point on the far right-hand side of the field, it will have to travel a little further.

Light rays can be thought of as electromagnetic "waves." They undulate, and like any other wave they have regular series of peaks and troughs. Let us imagine we freeze the action of the light streaming onto the screen at an instant of time. At that instant, let us say that ray 1, where it strikes the left-hand edge of the screen, is at a peak. Then, when we move a little distance from the left-hand edge, ray 1 will be at a trough when it hits the screen. We say that ray 1 is in a different "phase" at the two points. Remember, everything is frozen, so this is just a consequence of geometry. If we roll the film forward for a period that would allow half a wavelength more light to pass, where we saw a peak we would see a trough, and where we saw a trough we would see a peak—the phase of the light at each point would be reversed. Of course, we can't see the peaks and troughs in a light wave; if all we had was ray 1, we would see an even field of light.

Now consider ray 2 alone. The same geometry applies, so that at any frozen moment in time, ray 2 strikes the screen at some points at a peak and some points at a trough. But because ray 2 is following a slightly different path than ray 1, on some parts of the screen a peak of ray 1 will coincide with a peak of ray 2, and on some parts a peak will coincide with a trough. Where the phases of the two rays are the same, the screen will look twice as bright—the energies of both beams will add together—but where they are "out of phase"—peaks combining with troughs—they will cancel each other out and the result will be darkness. If we unfreeze the film and let it roll forward at normal speed, the bright and dark bands remain stationary. At any one point, as ray 1's trough becomes a peak, ray 2's peak becomes a trough, so that points of reinforcement and cancellation stay in the same place.

The slightest change of one path length in relation to the other, however, will affect the pattern. Suppose the path length of *one* beam changes enough to make the impact point of a peak become the impact point of a trough, so what was once reinforcement will become cancellation; dark bands will become bright bands and vice versa. It will look as though the fringe pattern has moved sideways across the screen by the width of one band (half a fringe).[4] Smaller changes in arm length will result in smaller shifts of the pattern. The interferometer is an immensely sensitive device, because changes in geometry of the order of a wavelength of light are easily seen. Even with a small and crude interferometer, changes in relative

---

4. A "fringe" is thought of as a light plus a dark band; with a once bounce interferometer, a half-fringe displacement will be caused by a change of arm length of a quarter of a wavelength.

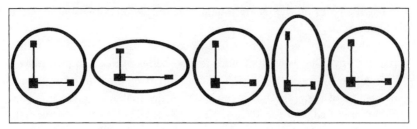

Fig. 17.3 Stretching and squeezings of an interferometer as a gravitational wave passes.

arm length of about one-thousandth of the diameter of a human hair can be seen with the naked eye.

A gravitational wave ought, momentarily, to shorten one arm of an interferometer and lengthen the other; then, as the gravitational wave phase is reversed, the opposite should happen (see fig. 17.3). So, if a gravitational wave comes by, the interference fringes on such an apparatus ought to move backward and forward across the screen at the same frequency as the wave. That said, because gravitational waves are so weak, a modern interferometer has to be about a billion times more sensitive than the one constructed by Michelson and Morley. How, then, does the modern design differ from the 1887 version?

The sensitivity of the instrument will depend on the length of the paths along which the two rays travel. If the paths are very short, then a percentage change to the lengths—a strain—will not affect the rays as much as if the arms are long; the longer the paths, the more marked the effect of any relative disturbance will be. The 1887 interferometer was mounted on a five-square-foot sandstone block floating in a trough of mercury so that it could be rotated. Michelson and Morley effectively increased their path length to eleven feet or so by reflecting the light in each arm backward and forward a number of times before recombination. The modern devices have arms up to four kilometers long and use many reflections to produce path lengths in the region of 100 kilometers.

A second feature of an interferometer is that, like all the instruments described in this book, it will be easily affected by outside disturbances precisely because of its high sensitivity. In fact, Michelson conducted his first interferometer experiments in Germany in 1881. He first tried the experiment in Berlin but found he had to move it to the quieter location of Potsdam, because the fringes were destroyed by vibrations. Even in Potsdam the fringe pattern could be broken up simply by stamping on the ground 100 meters from the laboratory. It was some years before Michelson and

Morley found a way to make the experiment sufficiently insensitive to outside disturbances, even to observe fringes. The modern interferometers have to be exquisitely isolated.

Again, the brighter the beams the better. The brighter the fringes are, the more the contrast between the dark and bright, the more readily will it be possible to see slight changes. We will come to a set of more subtle reasons in due course, but "the brighter the better" is an intuitive principle that carries through into the most sophisticated interferometer designs.

In the same way, it seems clear that the light used should be as pure and stable as possible so that the fringes are sharp and steady, affected only by changes in the path length, not by changes in the source. In the modern devices, coherent light from powerful lasers is cleaned and stabilized further by one or more "mode cleaners" before being projected into the apparatus.

Bob Forward

Bob Forward built his interferometer in Malibu, California, where he worked for his long-term employer, Hughes Aircraft. Forward's antenna had a folded path length totaling about eight meters. Speaking in 1972, he recalled how it was that he came to build an interferometer, referring along the way to the three small resonant bar detectors he had already installed in California.

> [1972] About that time a friend of mine named Phillip Chapman, who's a scientist-astronaut—he was—he's now working for Avco. He came to me with . . . an idea that I'd heard at least a half-dozen times before, and that was instead of connecting two masses by strain—why not measure their relative motion with a laser.
>
> I started to say to him the same thing I'd said to many people before—the gravity antennas that are in Maryland, made in Maryland, can measure displacements of the order of 10^{-15} to 10^{-16} meters. A wavelength of light is 10^{-6} meters—and "you're trying to tell me that you can measure to a 10^{-9}th of a fringe! It's easier to use piezoelectrics."
>
> And then, while I was telling him that, I realized that Mr. Moss and Frank Goodman [colleagues at Hughes] had been working on a laser ranging system—a Doppler ranging system—in which they measured vibrational motion to the order of an angstrom, which is sort of like [10^{-10}] meters—and

so they'd gone a factor of 10^4 without very much trouble. And then once you begin to look at it, you begin to realize that just because you're using a wavelength of light doesn't meant that you're limited to measuring a fraction of it—really what you're doing is you're measuring the Doppler shift [a slight change in the wavelength of the light caused by movement of the reflecting mirror].

So all of a sudden, er, I realized that this was not only possible but had real potential. Because I had been, while I had these three [resonant bar] systems running, carrying one guy after another, coming to work for me during the summer, and I gave them an assignment of a design: "Come up with the idea for a wideband gravity antenna." 'Cause Weber seems to have something we can't use as an astrophysical tool; we can't use it as a telescope, because it only gives you information about one narrow piece of the spectrum.

Forward here refers to one of the great advantages of an interferometer over a resonant detector, other things being equal: precisely that it is not resonant. A resonant detector tends to be sensitive only around the frequency at which it resonates, whereas a nonresonant detector can be "broadband"— that is, sensitive at many frequencies and able to resolve the exact form of the wave that disturbs it. This is immensely important for astrophysics, because it makes it possible to recognize the particular "signature" of different cosmic events, such as two neutron stars spiraling into each other. To oversimplify, the resonant detectors know little more than that they have been hit by an energetic signal; a broadband detector will be able to record what that signal looks like.

Forward went on to explain to me that Phil Chapman "walks in off the street and tells me how to do it"—that is, build a sensitive interferometer— and how Chapman had funded the project with a "pot of money." He also gave credit to Gaylord Moss at Hughes for his collaboration. At that time, Forward was gathering data, and he described their rigorous shift pattern intended to gather about 100 hours of data; the "chirps, clicks, and tones" that they were hearing; the difficulty of separating them from stray noises; and his intention to compare notes with Weber.[5]

Forward published the results of his initial technical developments in 1971 in a paper that also acknowledges the contributions of Chapman, Weiss, and Weber. In 1978, Forward was to publish a paper describing his initial results.[6]

5. A evocative transcript of Forward's actual words can be found at WEBQUOTE under "Forward on Early Days in Interferometry."

6. Moss, Miller, and Forward. 1971, p. 2496; Forward 1978.

WEISS SETS OUT THE CASE FOR BIG INTERFEROMETERS

Weiss started to think about the interferometric detection of gravitational waves in about 1969. When I interviewed him in 1972, he discussed his plans to build a gravitational-wave-detecting interferometer.

> [1972] We've been stewing about this thing for, probably, about three years, but never doing it. And we're starting now—mainly because we were on another experiment that was highly interesting to us, and we didn't want to get bogged down in something that looked like such a marginal thing in the first place. And so, now all of a sudden that experiment's over, I haven't got a better idea for doing anything else, so we're gonna do that. But mostly because it looks like a slick way of doing it.

RLE Report

What has come to be seen as the classic source document for large-scale gravitational interferometry is a section of a quarterly progress report issued by MIT's Lincoln Research Laboratory of Electronics (RLE) in 1972. In the relevant section of the report, whose principal author was Weiss, the origins of interferometry are attributed to a thought experiment by Felix Pirani, to Phillip Chapman, and to an undergraduate seminar run by Weiss.[7] In addition, it sets out and systematically analyzes for the first time the potential sources of noise that would limit the performance of an interferometric gravitational antenna. One respondent told me that in this report, Weiss effectively "invented LIGO"—the $300-million Laser Interferometer Gravitational-Wave Observatory that is nowadays the center of gravity of gravitational wave research. But Weiss never published this report or any part of it—an oversight many respondents thought led to the difficulties he experienced in gaining early recognition for his ideas and support for his proposed research program.

Weiss is quite explicit about his self-denying attitude toward publication.

> [1975] My tendency is never to publish an idea. I like to publish an experiment that's finished. And—you know—there's so many ideas you have that other people come up with later, that it's maybe not a good policy. But I think that in the end you can live with yourself better.

---

7. Weiss 1972, p. 58.

I'll tell you what's wrong with that—with publishing an idea. There's some poor bastard, who also had the idea, but later. And he sweats it out and actually makes it work. He makes a commitment, spends three years of his life—or longer—and then publishes it. And he has to give homage to that son-of-a-bitch who had the idea and didn't do anything about it. I think that's disreputable. That just happens to be the way I look at it.

The guy who should get the credit, really, is the guy who made it go. So that's the reason I don't like to publish every harebrained idea I have.[8]

Later, in part IV, we will explore Weiss's attitude toward publication at greater length under the heading of "evidential individualism." The point to notice here is that it contrasts with Forward's attitude, which was to build something and publish the results.[9] Weiss explained the provenance of his project as follows:

[1972] We're starting on our own thing, and the thing is that there, there's a little saga on its own. If you are going to visit Forward, his thing and the thing we want to do are based on the same idea [the laser interferometer]. . . .

Let me put it now in a positive sense, instead of checking on Weber. My intent was never to check on Weber. The thing that excited me the most was the pulsars. . . . And so we're breaking our heads finding a way of looking for the pulsar in the Crab Nebula—the radiation from that—and that's really the major intent of this antenna that we want to build here. And it's a big project, inasmuch as it has to be big, and it might turn into a space project eventually because of what problems we might run into because of the earth noise.

However, since then I've learned a lot more, too. It turns out that, prompted by the Weber business, people who had normally done, sort of, bullshit relativity—the kind of stuff where you calculate what is useless—you know, the mathematics journals are full of it—a lot of people started putting effort into finding out if there could be a source like Weber was seeing, and they uncovered all sorts of possible mechanisms which would make gravitational radiation from collapsing stars, which all have at best a total energy that comes out of them of probably one part in a thousand of what Weber's seeing. And that's an impetus alone for making a much better system

---

8. Social scientists might like to give their own discipline some thought in the light of this sentiment.

9. Something of the flavor of Weiss's and Forward's attitudes toward collaboration can be seen in quotations from my 1972 interviews at WEBQUOTE under "Weiss, Forward, and Collaboration."

than Weber has even if Weber's proved wrong. You see, to open up the field of what I would call gravitational astronomy...

And so this has nothing to do with Weber's work that the people want to look for this—at least certainly that's not my viewpoint...

Collins: How long do you think it will take to get it built?

Weiss: It probably won't take so long to build, but it will probably take forever [to make it work properly]—it might come to a juncture in a year or so when we will decide it ain't worth it. That's what may happen, because we will find out all sorts of problems that we didn't think of. That's what might happen.

However, I'll tell you one thing, and this is a crucial question that you can ask of all these people, and that is, will they quit when Weber has been proved wrong? And in our case I'm not going to. And that's because I do think there is something to the field of gravitational astronomy. I don't see there being too much of a field in trying to prove Weber right or wrong. That's different.

Collins: But at a thousandth of the intensity?

Weiss: There might be something once every few years at that level....

Collins: What about collaboration in respect of looking for coincidences?

Weiss: When it happens and we have something, then of course I will be very interested in getting data links between everybody around. But we're so far from that—it really hasn't happened yet.

Collins: Are you going to set up two of these?

Weiss: No—first one and then if it—no. I hope it's just one and somebody else. I don't know who it will be.—For example, the superconducting guys, these Hamilton-Fairbank guys might be a good bunch to collaborate with.

Forward was never approached for help in the Weiss project and was soon to disappear from the interferometer business. When I interviewed him 1997, he was still bitter about the fact that he and Weiss had not managed to collaborate.

Weiss's Research Proposals, 1972 and 1974

In 1972, Weiss submitted a proposal to the National Science Foundation that failed to attract funding. The sole evidence that I have been able to gather about this proposal is a statement in a letter written by Weiss in 1976 (see below). The relevant remark in the letter is as follows:

By the end of 1972...I applied to the N.S.F. for research support. The proposal to the N.S.F. was unfavorably reviewed at the time most likely because it was too big a step from acoustic gravitational wave detectors...

We can assume that the 1972 proposal was based heavily on the 1972 RLE report, but I have not been able to locate a copy of the proposal.

In 1974, Weiss submitted another project proposal to the NSF for continuing the construction of a prototype interferometer with arms nine meters in length; it was to be funded to the tune of the $53,000 requested.[10] This proposal, which I have read, certainly was based on the 1972 RLE report, and such elements as the discussion of the noise sources are nearly identical to those in the report. In this proposal Weiss says,

> Preliminary work has been going on at M.I.T. for several years with the support of the Joint Services Electronics Program and the M.I.T. Sloan Fund [JSEP]. The JSEP has terminated support for the project as of June 1974 as it cannot justify the relevance of gravitational research to its own program. (p. 2)

Later he states,

> Initial work on such antennas began at M.I.T. in 1970 as part of several senior thesis projects following the development of an experiment in 1967 that demonstrated shot noise limited interrogation of fringes in a laser illuminated Michelson interferometer. [Footnote 14 here refers to G. Blum and R. Weiss, Phys. Rev. 155, 1412 (1967). (p 6). The proposal refers specifically to the doctoral thesis work of a student, Kingston Owens.] (p. 6)

The proposal goes on to distance itself from Forward's already completed device, referring to it as giving rise to a "premature publication."

The proposal was submitted in August and requested a January 1975 startup. It was sent to Peter Kafka, a member of the German group at the Max Planck Institute in Munich, to be refereed. Kafka showed the document to other members of the group. To Kafka's embarrassment, these others became very enthusiastic and decided they would like to proceed with a similar design. The German group believe that Weiss was angered by this outcome, and may even have thought initially that they had acted unprofessionally. They also believe he may have thought that Kafka had given a negative assessment of the proposal and then encouraged the German group to go ahead with the same idea.[11] Kafka explained to me in

---

10. "Interferometric Broad Band Gravitational Antenna." Grant letter received at MIT on May 21, 1975, assigning grant identification number MPS75-04033.

11. I have been unable to check the accuracy of these matters with Weiss.

1976 that, being a theorist, he had passed the proposal on to his experimental colleagues for advice and had been shocked when they took it up.[12]

Though Weiss's 1974 proposal asked for a January startup, the acceptance letter was not received until May 1975. The events described above are certainly consistent with the notion that the NSF's more favorable response to this, Weiss's second proposal, had to do with the enthusiasm of the Germans, but it would be hard to be certain.

The best-known names among the early German interferometer enthusiasts are Maischberger, Billing, Schilling, and Winkler of the Max Planck Institute, and they did indeed press ahead. The Munich prototype interferometer, which had a three-meter path length and was based closely on Weiss's ideas, was the best in the world for many years.[13] It seems that nothing deeply unprofessional had been done, but it was ironic that Weiss initially found it harder to obtain support for a design that he had invented than did the Germans.

About this time, Ron Drever, then at Glasgow University, also began to turn his attention to interferometric techniques. Drever's first use of interferometry was in measuring the separation between two massive bars, a compromise borne out of not having enough money to buy vacuum tubes and being given the bars and their vacuum tanks by the group at the University of Reading. The bars were the two halves of the Reading group's split bar antenna.[14] His own group did not start to build an interferometer proper until 1976, its design seemingly benefiting from Weiss's inspiration at least indirectly: according to Drever, the more direct influence on the project came from Forward, but Drever is not sure when he first saw Weiss's early analysis.[15]

NOISE SOURCES

As I have said, the 1972 RLE report and the 1974 application contained an identical, or nearly identical, description of the noise sources, and hence

---

12. Kafka died in December 2000; my 1976 interview was the last conversation I had with him. His actual words, describing his embarrassment, can be found at WEBQUOTE under "Kafka Referees Weiss's Proposal."

13. A 30-meter prototype was constructed in Garching, near Munich, in 1983.

14. Private conversation with Ron Drever, January 22, 2003.

15. Private conversation with Ron Drever, January 22, 2003. A number of others (e.g., Thorne; see chapter 18), remember Drever being much enthused by interferometry as a result of a meeting in Erice, Italy, in 1976.

limitations and technical requirements, of a maximally sensitive interferometer. This analysis was to form the basis of all subsequent discussions of interferometry, so I will make a small technical diversion and lay out this framework here.

I'll work here from the interferometry section beginning on page 54 of the 1972 RLE report, which is the document that has historical priority. Section 5, "Noise Sources in the Antenna," begins on page 59. It is broken up into subheadings enumerated by lowercase letters *a* through *i*. The technical introduction provided by this analysis will stand us in good stead throughout the discussion of interferometry, so I will briefly describe the contents of each section (which is, of course, filled with technical detail in the original).

a. Amplitude Noise in the Laser Output Power

This source of noise is the fluctuation, or perceived fluctuation, in the power of the laser light source. It relates, according to Weiss, to "laser shot noise," the unavoidable statistical variation in the emission rate of photons from the laser, and to the variation in the likelihood that a photon finally arriving at the detector will be registered—again a matter of statistics. As with all these noise sources, they are noticeable only because they compare with the extraordinarily small disturbances that we are trying to see.

b. Laser Phase Noise or Frequency Instability

This has to do with fluctuations in the frequency of the laser.

c. Mechanical Thermal Noise in the Antenna

This source of noise is caused by unwanted movement of the mirrors due to the "Brownian motion" of the mirrors and of the materials that make up the mirrors. The solution is to make sure that the natural frequency of the elements of the interferometer lies in a much lower and/or much higher frequency range than the frequency of an expected gravitational wave, so that this source of noise will not be mistaken for a gravitational wave.

d. Radiation-Pressure Noise from the Laser Light

This is movement in the mirrors caused by fluctuations in the pressure of the radiation impinging on them when the laser power varies.

e. Seismic Noise

Noise coming through the ground is to be eliminated by suitable design of suspensions for high frequencies and by feedback mechanisms to compensate for low-frequency, large-scale movements.

f. Thermal-Gradient Noise

The largest source of this noise is caused by the light beams heating the mirror surface, which in turn heats any residual molecules of gas in the (inevitably) imperfectly evacuated beam tubes. Residual molecules hitting the heated mirror surface will rebound more energetically than those striking the colder back face, and this could move the mirrors in unwanted ways.[16]

g. Cosmic Ray Noise

This is noise produced by the impact of cosmic rays on the antenna. Weiss envisages the need to place the mirrors underground to avoid it!

h. Gravitational Gradient Noise

This noise is caused by changes in the gravitational field, which affect the mirrors. For example, if the air pressure changes around the interferometer, pockets of air will change their density and therefore their gravitational pull on the mirrors. The same will apply to density fluctuations in the ground or movements of anything massive (including people) in the vicinity of the detector. Again, these tiny forces are only of interest because of the envisaged sensitivity of the detector.

i. Electric Field and Magnetic Field Noise

Although the antenna would be shielded from electrical and magnetic sources, no shielding can be perfect.

WEISS'S 1975 CONTINUATION, 1976 LETTER, AND STUDENTS

Weiss's 1975 project did not run smoothly. The easiest way to understand the history is to read the remarkably frank summary Weiss gave in 1976

---

16. The exact mechanism for this force, which is the force responsible for the spinning of toy, windmill-like, "radiometers" in response to sunlight, seems not to be fully understood to this day.

to Richard Isaacson of the NSF, in a letter asking for an extension to the grant.[17] In this letter, he complains about the difficulty of attracting dedicated graduate students to work on such a difficult project.

> Gravitation research, although viewed as fascinating, is considered too hard and unfortunately profitless not only by the average student but also by much of the physics faculty. In short, the atmosphere if not outright hostile to such research is certainly skeptical.

Weiss explains that he was on the point of giving up but is excited by the idea of a space-based interferometer; has found a good graduate student; and is confident that interferometry is becoming more attractive, as indicated by the interest of Ron Drever, among others.

> [D]amn little actual progress on the prototype antenna has been made in the past year aside form some minor work on fast electronics for the fringe detection system. I know that on the basis of the experimental progress alone a renewal request at this time would not survive a review procedure. On the other hand some real progress in the theoretical developments has been made and that now with the active interest of another capable physicist, progress in the experimental work on the prototype antenna can be expected.

The troubles alluded to in this letter echo themes that Weiss returned to many times in discussions with me. The first was the hostility and shortsightedness of the MIT academic administration when it came to the search for gravitational radiation. Weiss believed that much of this may have arisen out of the controversy over Joe Weber's claims, which made it appear to the MIT establishment that the whole field was built on insecure foundations. Thus, in 1995, in the context of another controversy, Weiss referred to the early days.

> [1995] Anyway, so what we had to fight was that problem, the problem that this was a field for crazy people, that was number one, and none of the academic places liked it, MIT in particular. They looked at this and they said, "This was for the birds." Here I was pushing like hell to do this by another technique [interferometry as opposed to resonant bars], and I couldn't get any backing in this place, intellectual backing. And reason was, in part, because it had this very speckled history, and it looked, also, very difficult to do.

17. The complete letter can be found at WEBQUOTE under "Weiss's 1976 Letter to Isaacson."

As we will see in later parts of the book, Weiss's reading of the situation was to color his approach to subsequent debates. It also seems almost certain that the administration's unwillingness to back Weiss and gravitational science wholeheartedly would lead to MIT becoming the de facto junior partner in the collaboration with California Institute of Technology that was to come.

The second problem to which Weiss refers in his letter is the difficulty that his graduate students experienced with unsympathetic MIT faculty members on their examination panels. He felt strongly that some of his students were belittled by faculty who considered that the only worthwhile piece of physics was one that made a discovery, not developed an instrument that would one day make a discovery. Weiss felt humiliated and angry at the indignities suffered by his students. It was for these reasons that he found it hard to recruit students for his project, because it was almost certain that the prototype instruments would not see any new phenomena.

> [1975] [Y]ou're in a physics department. And this is—you can call it, if you want, a sociological problem . . . what's recognized as respectable physics is not engineering. And building something and showing that it behaves as you say it does, and not having made a measurement with it of something new, doesn't count really as any accomplishment.

INTERFEROMETERS, WEISS, AND WEBER

As we have seen, Weiss wanted to build an interferometer capable of detecting the gravitational radiation emitted by the Crab pulsar,[18] a rotating neutron star that will emit gravitational waves if it is asymmetric. He believed that Weber's work had led to the development of theoretical models that brought the so-called burst sources, or impulsive sources, to salience. Talking to me in 1975, he gave Weber still more credit.

> [1975] Weber turned out to have been right in one regard, namely, there seems to have been a reason to look for fast events. I would never have thought of that. Never in my furthest mind would I have thought that you should look for things—that in astrophysics there could be events that would

---

18. The same aim as the Japanese group who built a resonant antenna specially built for the same purpose (Hirakawa and Nariha 1975).

take milliseconds. And I would always have thought you'd look for something
that was slow—because everything is slow...and so it fell right in with this
business of the black holes, and the possible discovery now of black holes by
X-ray astronomy.

Weiss, then, accepted Weber's influence on the field, although he was less
flattering about the reasons behind it.

> [1975] So all in all, he wasn't dead wrong—he bumbled into it, I suppose. I
> know the reason why he did it at high frequency—he did it because that's
> what's possible. To make a long-period system is very hard.

In the 1974 proposal, however, Weiss was still talking of detecting what
are known nowadays as periodic sources, pulsars and binary stars in steady
orbit around each other. The nine-meter prototype described in the proposal
was meant to lead to a next stage featuring a one-kilometer arm length
device, which he says would be capable of seeing the Crab pulsar if the
periodic signal were integrated over three to six months.[19] In the document,
Weiss also looks forward to a much longer baseline interferometer in space
which could see the Crab pulsar in a matter of a few hours of integration.

The proposal also contains an appendix of possible sources, most of
which are still favorites today. But the infrequency of occurrence of the
potential burst sources suggests that the prime interest is still a search for
periodic sources such as pulsars, as is clear from his comments as late as
1975.

> [1975] I always had it in mind that you want to look for things where there
> ought to be gravitational radiation....The right thing to do is look for the
> gravitational radiation from binary stars. The trouble there is that the
> frequency at which you expect to see these waves is so low that it is very
> hard to do something on the ground. You can't make a nice suspension for
> something that is at a 40-minute period, or an 80-minute period, or a
> couple-of-hour period....You want to do it in space.

One can see that Weiss is not thinking about the burst source that is the
modern favorite, the last moments in the life of inspiraling binary stars,

---

19. The signal from continuous sources with a repeating pattern can be integrated over a
period as long as one likes by adding their tiny effects on the detector's response over and over
again. This will be explained at greater length in due course.

but binary stars slowly rotating around each other and emitting very little gravitational radiation such as can be seen best with a space-based system. This, in 1975, is where he sees gravitational radiation research going. He explained that no one was much interested in merely detecting gravitational waves, because everyone knew they were there; space-based astrophysics was more compelling.

> My view now is to demonstrate on the ground that the noise performance of an interferometric system is just what you'd calculate; then, if one's still interested, go on and in another few years convince someone in the space agency that this is something that they ought to put money into and see if they can see their way clear to putting up an interferometric antenna [in space].[20]

It is interesting that Weiss has since become the principal experimental champion of ground-based detection and, as we will see, one of the vociferous advocates of the search for unknown sources; but in 1975 he was certain that the best way forward was to look for specific sources with known signatures using space-based detectors.

20. See also WEBQUOTE, under "Weiss Wants to Look for Known Sources."

CHAPTER 18

CALTECH ENTERS THE GAME

KIP THORNE

Sometime in the mid-1970s, Weiss found himself together at the same meeting with Kip Thorne, the well-known Caltech theoretician. Thorne had been an enthusiast for gravitational radiation from the outset, giving much encouragement to Weber. In 1975, Weiss said to me,

> [1975] Kip Thorne, who is one of the pushers for black hole physics and things like that, keeps coming up with new estimates for how big a strain you can get for a *gigantic* black hole colliding with another *gigantic* black hole. He keeps impelling this business so that people won't lose interest in it. At least that's how I interpret what he's doing, though he may have better reasons.

Thorne had become very close to the Russian physicist Vladimir Braginsky, who was the first to challenge to Weber's experimental findings and who planned to push the sensitivity of gravitational wave detectors higher using sapphire crystals as resonators.

Certainly Braginsky affected Thorne's thinking in various ways, and it is likely that in the early days Thorne would have been no more convinced that interferometry was the way forward than Braginsky. Indeed, in a book cowritten by Thorne (with Charles Misner and John Wheeler and generally known as MTW), which has become a standard text on relativity, there is a passage that explains why interferometry cannot possibly work.[1]

One of Braginsky's most important contributions to the whole business of gravitational wave detectors was his thinking of them as quantum systems. Normally, the quantum world is thought of as quite different from the world of large experimental apparatuses; quantum effects take place at the scale of the subatomic particle, and manifest themselves in ways that can be used in the design of transistors and similar very small devices. (The SQUID amplifier used on the cryogenic bars is based on ideas belonging to quantum theory.) But Braginsky realized that as gravitational wave detectors became increasingly sensitive they would enter the quantum regime. He believed that to make the devices sensitive beyond a certain point would require a technique that sidestepped quantum uncertainties; the term *Quantum Non-Demolition* [QND] was invented to refer to this technique.

Present-day devices are well short of the kind of sensitivity that requires them to take advantage of QND, but Braginsky was to show that interferometers could avoid the problems of the quantum regime more easily than resonant bars. When choosing their research trajectory, physicists could be persuaded that it was better to go forward with a technique that had the potential for continued increases in sensitivity well beyond current designs even if the need would not be realized for a decade or two. Perhaps it was this kind of argument that led people like Thorne and Ron Drever to become enthusiastic about interferometers. Certainly Thorne told me that he believed that in 1976, at a conference in Erice, Italy, Drever had decided to switch his main efforts to interferometry as a result of discussions that compared the future of bars and interferometers and presented the quantum problem for the first time in public.

Certainly by 1980 or so, Thorne, too, had changed his mind and had become an interferometer enthusiast. Weiss believed it might well have been he who changed Thorne's mind as a result of their meeting in the late 1970s. He explained that after being invited to the same review panel in Washington, both he and Thorne found themselves wandering the streets of

---

1. Misner, Thorne, and Wheeler 1970, p. 1014: "[S]uch detectors have so low a sensitivity that they are of little experimental interest."

the city looking for a hotel room. Washington's hotels were heavily booked, so they were forced to share. Weiss said his first reaction to this forced intimacy was ambivalent, but it was to be the start of a long professional relationship and a close friendship.

> [2000] [W]e spent the whole night talking. And it was quite evident what Kip was up to.... He had in mind that here was Caltech—a very prestigious relativity group that he ran. But they felt that they were not contributing in any way, or did not have a vital experimental role in that. And he was shopping around and looking for directions to go in.
>
> He had also started talking to Vladimir [Braginsky], because they were close buddies, and they were coming to the decision—Kip had said they looked at every field of experimental relativity: cosmic background studies; Eotvos experiments; all these kind of experiments. And I convinced him that they should look at interferometric detection. And I suggested they should go and look at Drever as a person to get. [At that time Braginsky favored bars]...And so what happened is that I then said to Kip, "That [the bar method] doesn't have a future." And he hadn't thought about that. If you look in his book, for example, MTW, you will find that there's a problem, which says, "This [interferometers] is an impossible way of doing it"...it's one of the problems of his book. So: "Yes, it's feasible—it's an idea—but it will never have enough sensitivity to do this." It's right in MTW, for everyone to see.

By this time Thorne had convinced Caltech that they should start a major experimental program to detect gravitational waves, and Weiss says he suggested Drever as the principal researcher to head up the team. Thorne, as we will see, did recommend the appointment of Drever, and his account of his own conversion to interferometry is that Drever convinced him that this was the way forward.

Drever had starting using interferometric techniques with two half-bars as test masses, but in the mid-1970s he started to build interferometers at Glasgow University similar to the modern devices. By the end of the 1970s, he was leading a team at Glasgow that had half-completed a ten-meter interferometer. Then, in 1979, he was invited to head up the team at Caltech, where he accepted a part-time post. He continued to work on the Glasgow ten-meter device for several years while simultaneously building a 40-meter interferometer at Caltech. In 1983, he took up a permanent post at Caltech, and Jim Hough took over the leadership of the Glasgow group. We will learn much more of these and subsequent events.

WEISS'S 1978 PROPOSAL

Weiss's NSF funding continued with a continuation grant awarded November 1978 (for a December start), for a term of two years at about $130,000 per year.[2] It was extended in March 1980. The proposal includes this comment: "I have slowly come to the realization that this type of research is best done by secure (possibly foolish) faculty and young post-doctorates of a gambling bent." Weiss then goes on to talk further about arm lengths. He has become more ambitious, claiming that a serious instrument will need arms of ten kilometers!

On page 28 of the proposal, he considers the possibility of joint work with Caltech. In this proposal Robert Forward's work is not mentioned,[3] but Weiss does refer to the work of the Munich group and of Drever.

> At present, the group around Billing and Winkler at the Max Planck Institute in Munich (triggered by our earlier NSF proposal) have, I believe, made substantial progress with a system quite similar to ours. Ron Drever in Glasgow, Scotland has begun a small-scale project to use interferometry as a means of measuring oscillation in a split bar system.

A referee echoes the point about the comparative fast progress in Europe.

> Techniques first proposed by Rainer Weiss more than 6 years ago are now being used in the prototype interferometric gravity wave detectors which are being enthusiastically assembled in Europe. It is therefore interesting to learn why progress at MIT has been fitful, and a relief to know that no hidden snag has surfaced.

All the referees of the project think that isolation of a long-baseline interferometer on Earth is too difficult. One is forthright in saying that ground-based systems will not work, so the proposal should be modified to go toward space-based development work or be restricted to research on isolation systems, leaving the main project until that problem has been solved. Several of the referees say that the problem of building a short detector is hard enough, and that the project should not turn on the promise of being able to build gigantic ones in the future. As we have seen, however, the grant was awarded.

2. PHY 7824274: Interferometric Broad Band Gravitational Antenna.
3. A point upon which a referee reflects critically.

INTERIM SUMMARY

We have now reached a point in the early 1980s where we have all the initial interferometer teams in place, so it is sensible to take stock. The idea of using interferometry to detect gravitational radiation was invented by a Russian group or by Weber, and the first interferometer intended to be a gravitational detector was built by Forward at Hughes. Forward, however, probably learned at least some of the key ideas for implementing an interferometer in this way from Phil Chapman, who in turn had interacted strongly with Rainer Weiss at MIT. The Munich team gained their principal impetus through reading an NSF proposal submitted by Weiss in 1972 and, subsequently, reading Weiss's insightful Research Laboratory of Electronics report. The RLE report (and probably the 1972 proposal) foresaw most of the fundamental noise sources that would limit a practical interferometer and, consequently, the fundamentals of a practical design.

Thorne, a theoretician, had boosted enthusiasm for the detection of gravitational radiation and was determined to start an experimental program at Caltech; but until persuaded otherwise by Weiss and/or Drever, he imagined the future lay with bars. Thorne recruited Drever to Caltech, ensuring that interferometers would be Caltech's route.

By the early 1980s, Forward's work in interferometry had ceased. There was a small experimental interferometer at MIT developed by a team led by Weiss, and an interferometer at Munich's Max Planck Institute that was the most successful in the world at that time. The Glasgow group were building a ten-meter interferometer, and its leader, Ron Drever, was also leading a team building a 40-meter device at Caltech that would become the most sensitive in the world by the mid-1980s, and would convince Vladimir Braginsky of Moscow State University that the future lay with interferometry rather than resonant sapphire crystals. In 1981, Weiss proposed to the NSF that he should be funded to develop plans for a kilometer-scale interferometer—a size that he thought would have a chance of detecting the flux of gravitational radiation thought reasonable by the astrophysicists.

In 1970, Jim Hough had joined the Glasgow group as Drever's graduate student, working initially on the bars project. In 1976, he began working on interferometers with Drever, and in 1983 became the leader of the group when Drever took up his full-time appointment in Caltech. In 1986, Hough was to develop plans for building a three-kilometer interferometer. At the end of the 1970s, the Munich group were also thinking in terms of a large-scale device; and the teams, both discovering a shortage of funds in their respective countries, combined efforts to try to gain support for a joint British-German device.

Still later, French and Italian teams were to come together with the aim of building a three-kilometer detector. The first intimation of this in the American literature was a remark in the "Blue Book" produced by Weiss and his group in 1983 (see below):[4]

> We have learned recently that a group is being formed by A. Brillet at the Laboratoire de l'Horlorge Atomique at Paris-Sud to develop a prototype electromagnetically coupled antenna [i.e., interferometer], the state of this project is not known to us. (p. I-10)

All these large-scale proposals almost certainly followed Weiss's initial work on large-scale detectors, which we will now examine.

WEISS'S 1981 PROPOSAL

Weiss's successful 1981 proposal to the NSF was to continue work on his 1.5-meter prototype and to conduct a three-year study of what would be needed to build a ten-kilometer device.[5] Parts of the proposal, such as the following passage from pages 3–5, are written in a flowing style, not at all what one might expect from such a document. Weiss cleverly starts by imagining a device so huge that it would be bound to detect gravitational waves on almost any theoretical assumption, and insists that the only thing preventing the building of such an instrument immediately is finances. Overcoming objections in principle in this neat way, he then suggests that the key problems are obtaining sufficient sensitivity in a smaller and more affordable device.

> [T]he problems that must be solved to achieve interesting sensitivities [in interferometers] are primarily ones of scaling rather than of developing fundamentally new and difficult technologies. One can say this in a very bold way. It is now, and has been for several years, *entirely in our capability* to construct a gravitational wave antenna with an interesting sensitivity [strain] 10^{-21} at one kHz, if only one was willing to spend the large amount of

---

4. Linsay, Saulson, and Weiss 1983.

5. PHY 8109581: Interferometric Broad Band Gravitational Antenna. The cost was about $1.3 million.

money involved. There would be no subtlety in such a system, and precious little application of new technologies that we are working on in the laboratory prototypes at MIT and elsewhere.

We are not proposing the following, but imagine such a system: a square 100 km on a side, constructed of thick wall stainless steel tubing 42" in diameter, including the diagonals [that is, there would be a square of tubes, and the diagonals of the square would also be fitted with tubes]. The system is buried 200 meters at its midpoint to accommodate the curvature of the Earth. There are 400 miles of tubing in such a configuration. The tubing is supported every ten meters on servo mounted cradles. The system is evacuated to 10^{-7} mm using 10^5 ion pumps. The system includes 6 interferometers encompassing all pairs of adjacent differences....The point made by the above presentation is the following: Look at the requirements individually. They are all within our present technical capability, even in a commercial setting. However, the cost of such an installation ranges between 150 and 200 million dollars,[6] a major U.S. investment in science rivaling medium size space missions and probably too large, considering the limited applicability of such an installation to one branch of science, even though the rationale for such an installation could embrace earth science and other relativity experiments, in fact might even have military applications.

The realization that such a system is possible, but outside the range of economic reality, does color and affect in a deep way the strategy for pressing the current effort. If one follows the natural desire to see the field come to fruition in a timely manner, it would drive a commitment to the idea, and suggest a strategy to analyze the individual cost factors and optimize the antenna as a system. This may sound like administrative-ese, but it is in fact the core of this proposal. In plain English, what are the areas to work on to bring the cost of such an installation down by a factor of 10 to 20?

In an earlier part of the introduction (p. 2), Weiss has looked beyond the proposal, saying, "If the results of the study [of the ten-kilometer device] are favorable, we would like to begin construction of the large antenna."

Interestingly, Weiss proposed to subcontract much of the engineering study to industrial firms, Arthur D. Little and Stone and Webster. These firms had experience in the practice of building large technological

---

6. The figures need to be roughly doubled to match the changes in the Consumer Price Index for the year 2000.

installations, studying their thermal and other properties, and surveying sites for their suitability for installation of sensitive facilities.

The proposal also indicates two features that would loom large as the American interferometer project grew from prototypes to conceptual designs to large-scale real projects. On page 25, Weiss wrote, "We hope the study will be the basis for the formation of an NSF sponsored Science Steering Group for Gravitational Astronomy, composed of all interested parties, as, if such a large project becomes a reality, it will require the support and wisdom of the entire community." In other words, Weiss was already seeing the need for national cooperation.

Earlier, on pages 11 and 12, Weiss discusses what would become one of the causes of friction when that national cooperative structure was finally formed. This was the technical matter of the way the light was to be bounced backward and forward between the interferometer mirrors. The two possibilities, which will be explained in detail in a later section, were the "delay line," favored by Weiss, or the "Fabry-Perot cavity," favored by Drever.

> At the outset, one already knows some of the advantages and limitations of these designs. The multi-pass delay lines may experience multiple backscattering of the beam, first analyzed by Ron Drever. . . . Another more serious difficulty with multi-pass dealy [sic] lines . . . is that the tube diameter is not minimized. . . . The delay lines do have the significant attribute that their properties are independent of wavelength, which permits their use with frequency unstabilized high power laser sources . . .
>
> The Fabry-Perot cavity has the advantage of having the minimum beam diameter, and probably less difficulty with mirror scattering, important in small length, high finesse systems.[7] It imposes a constraint on the frequency width of the light source, and therefore on the maximum useable power.

The NSF archives reveal that at this time, members of the agency thought the cost of the ten-kilometer square array with one diagonal would be about $19 million and that the big approach was important because of competition from Scotland, Germany, and the USSR. One reviewer of the

---

7. That is, Weiss is suggesting, as he does elsewhere in the document, that the problems so far encountered, and the suggested solutions, will be different once the interferometers are scaled up. ["High finesse" implies that the beam bounces backward and forward many times.] A larger-scale system, it is implied, might have fewer bounces. Weiss has said on page 6, "A good example [of the difference between large and small scale] is the backscattering problems on the mirrors, which would be a decisive problem in a small interferometer, and of far less consequence in a larger one."

proposal writes that the nation will never spend more than $10 million on a gravitational wave detector!

THE "BLUE BOOK"

In October 1983, the MIT group presented to the NSF the report based on these studies of large-scale interferometry. This document has become known as "The Blue Book" because of the blue covers in which it was issued.[8] Three of its authors were from MIT; two of these contributors, Weiss and Peter Saulson, were to stay with the project until the present day.[9] The one "contributor" from Caltech—Stan Whitcomb—continues to figure prominently in interferometry. The report, which contains sections on engineering, budgets, potential sites for large interferometers, and the science itself, also acknowledges the industrial consultancy of Arthur D. Little Corporation of Cambridge, Massachusetts, and Stone and Webster Engineering Corporation of Boston.

The first section of the Blue Book is a superbly written, sixteen-page overview of the problems of large-scale gravitational interferometry. Physicists who write well write *really* well, and Weiss, who was the principal author of both this introduction (though not every section of the Blue Book) and the earlier research proposal that I have quoted from at length above, can be seen to be such a physicist.[10] In the introduction, MIT is described as the driving force of the project, and that

> [i]n the later phases of the study, the advice and criticism of the Cal Tech
> Gravitational Research Group were sought. Parts of this study have benefited
> from this interaction, however the tyranny of the schedule to complete this
> document have left substantial parts of it without their constructive view.
> (p. I-1)

The conclusion anticipated in the introduction is that it was now feasible to build two interferometers that would provide a million-fold increase in

---

8. The Blue Book is Linsay, Saulson, and Weiss 1983.

9. The third MIT author listed was Paul Linsay; he was listed first in alphabetical order.

10. Other physicists whose superb style I have encountered at first hand include Steven Weinberg and David Mermin. The same applies to talk: physicists who speak well speak *really* well. Weiss is perhaps the most colorful and quotable speaker of all the respondents in my study, but my prize for stunningly brilliant spoken presentation of ideas—and that includes the ideas that one is currently forming in one's own head—goes to Gary Sanders, the current LIGO project manager who we will encounter somewhat later in the book.

sensitivity to gravitational radiation over existing technology, with much scope for further improvement.

The introduction brings a new finding into the argument, the famous binary star system studied by Russel A. Hulse and Joseph H. Taylor known as PSR 1913+16.[11] Hulse and Taylor won the Nobel Prize in physics in 1993 for studying the slowly decaying orbit of a pulsar and a neutron star and finding, among other things, that the rate of the decay over a decade or so exactly matched Einsteinian predictions for the consequences of energy loss through gravitational radiation. This was the first evidence for the existence of gravitational waves. One might think that this result would weaken the case for further tests of the idea of gravitational waves, even though it was an indirect observation; but it is used by physicists to strengthen the justification for a direct detection, especially where much more extreme stellar and cosmic events are involved. As the Blue Book says,

> This is the first and only time that the existence of gravitational radiation has been demonstrated empirically, a fact which puts the entire field of research on a much sounder basis than it was only a few years ago. (p. II-3)

It is also interesting to see the way the Blue Book distances itself from Weber's contribution. Weiss believed, and continues to believe, that the Weber controversy left the field with a damaging legacy. After a passage describing relativity theory's prediction of the existence of gravitational waves the document goes on with the following:

> Much of this was on the mind of Joseph Weber when he initiated the search for gravitational radiation using acoustic (bar) detectors in the mid 1960s. Setting aside the controversy raised by the first experiments, it became clear, after a world wide effort, that nature was not as generous as might have been indicated in early experiments. In a broad sense this realization was the beginning of the search in earnest; the arguments for carrying out the search remained persuasive, the sources were just not that strong nor, in hindsight, was it reasonable to expect that they could have been. (p. I-6)

This is, perhaps, a trifle unfair to Weber, who realized from the beginning that it was "unreasonable" for his sources to be that strong; also, the notion

---

11. Weisberg and Taylor 1981; Taylor and Weisberg 1982; Hulse and Taylor 1975; Bartusiak 2000.

that the search "in earnest" only began after Weber's work was over is a little ungenerous, but it can be understood under the circumstances.

A large part of the scientific portion of the Blue Book is taken up with the evaluation of different optical systems. Stan Whitcomb is given a separate section, written over his own name, in which the characteristics of Drever's favored Fabry-Perot arrangement are set out.

Drever is credited in this text with two other new ideas (which I will discuss in greater detail in chapter 29). The first is an arrangement of the interferometer that makes it specially sensitive to sources of a particular narrow frequency of gravitational radiation, because the light in both arms is effectively made to resonate at that frequency. This embryonic idea was later to be developed by others and become known as signal recycling, and was first implemented in the British-German interferometer known as GEO600.

The other new idea credited to Drever, and now used in most large interferometers, is known nowadays as power recycling. In this setup, most of the light from the interferometer that does not carry a gravitational wave signal is fed back into the interferometer to boost the effective light power well beyond the power of the actual laser. As we have seen, and I will explain in more detail in chapter 29, the sensitivity of an interferometer depends on the light power.

Sources of Gravitational Waves

The Blue Book also contains an elaborated discussion of potential sources of gravitational waves, written in the same frank and amusing style (though here the principal author was Peter Saulson). The study of sources is the business of cosmology and astrophysics, both of which have a very shaky observational base—or sometimes, as in the case of black holes, or the origins of the universe, almost no observational base at all. In any case, observations of the way the galaxies hang together strongly suggest that what is visible in the skies is 10% or less of what is actually there. This estimate comes from the inferred gravitational attraction of the components of the heavens, one for another, which is insufficient to be accounted for by what we can detect. The very openness of the skies to varied interpretation by theorists is what made it possible for Joe Weber to be taken seriously at all and, as we will see, what made it possible for related claims to be made in the late 1990s and beyond. In the survey of wave sources, Weiss is disarmingly clear about what we might call the "Wild West" quality of cosmology and astrophysics, and in the end exploits it to the full in the Blue Book. The argument has survived, and to this day Weiss is the most vocal of those who claim that a search of the heavens with instruments of hugely increased sensitivity must be expected to discover something unexpected.

The Blue Book's section on sources divides them up three ways, in what has now become the conventional approach: we have what Weiss calls "impulsive sources," often called "burst sources," such as supernovas, which were the kind of brief events that Weber looked for, and of which a modern favorite is the last few seconds of the life of an inspiraling binary system; "continuous sources," which are the much less powerful emissions of, say, a pulsar, which is a rotating asymmetric star, or a close but not dying binary system, whose signals, though weak, can be integrated over a very long time period because they are steadily emitted; and stochastic sources, which have no well-defined pattern and include the aggregate of many distant burst sources going off like "popcorn," as the modern metaphor goes, and the gravitational equivalent of the electromagnetic radiation left over from the big bang—the "cosmic background radiation."

To know the likelihood of being able to detect a source with a detector of known sensitivity, one needs to know two things: the amount of radiation it will emit—the province of astrophysics—and, for burst and continuous sources, the population of such things of varying strength in the heavens at observable distances. Both numbers are open to huge uncertainties in the majority of cases, with only inspiraling binary neutron stars, boosted by the work of Taylor and Hulse, giving rise to estimates for which the term *sound* might be used without embarrassment. The population of supernovas can be estimated reasonably, but estimates of the likelihood of their emitting detectable fluxes of gravitational radiation continues to vary alarmingly. The Blue Book suggests that something similar applies to pulsars.

Black holes can give rise to large fluxes of gravitational radiation when they are born, when they absorb other objects, when they complete a lifetime of spiraling around another black hole or some other kind of star, and so forth. But the Blue Book says (this is 1983), "many skeptics remain unconvinced that black holes exist" (p. II-15). It goes on,

> The remaining facts needed to make a prediction of signal strengths are the densities of the populations of black holes of various masses. It is here that our knowledge of astrophysics fails us. It is certainly not possible to rule out zero density [i.e., complete absence] for all masses. The models which are considered most plausible at any given time are subject to the whims of astrophysical fashion . . . (p. II-16)

And, of another black-hole-related source, the Blue Book notes,

> Our ignorance on this subject is too vast to make confident predictions, yet opportunities for a far-reaching discovery are nowhere greater than in the case of black holes. (p. II-17)

On the matter of stochastic sources, the authors give this argument:

> The likelihood that the missing mass [the invisible part of the universe] is in supermassive black holes is, at the moment, a question to be answered only by personal inclination. Put more positively, the search for stochastic gravitational background provides a unique means to search for a possible component of the universe which would otherwise be invisible. (p. II-18)

The Discussion section of this part of the document (pp. II-19–21) sets the tone for years to come. It says that there are many uncertainties, but "reasonable argument" predicts that there should be detectable radiation. Furthermore,

> one might consider the uncertainty as a virtue, for it shows how much there is to learn from the study of gravitational radiation. (p. II-19)

It goes on to claim that only well-known astrophysical processes have been studied by those making predictions about numbers of sources, but there are many other kinds of sources. The information carried by gravitational waves is quite different from that carried by electromagnetic radiation and can reveal otherwise hidden process. The way to make discoveries is to use radically new ways of looking at the universe, and the new devices will be a huge improvement over existing detectors. Finally, we cannot claim to understand relativity until gravitational waves have been detected.[12]

There are three things to note about the passage. The first we have already mentioned: Weiss's view that one is bound to see something if one looks at the heavens with a much more sensitive instrument. I've argued with Weiss about this. It is certainly true that every time astronomers have looked at the heavens with a more powerful telescope, or a telescope

---

12. The complete discussion can be found at WEBQUOTE under "The Blue Book Discussion."

that explores some previously unexplored region of the electromagnetic spectrum, unexpected things have been discovered. But those cases involved a large but incremental improvement to an existing and already proven technology. In the case of gravitational wave detection, nothing has been seen at all, and so the argument by extension from the previous successes in the electromagnetic spectrum do not work: the increase in sensitivity may be a large number, but one is "multiplying by zero," as it were. Weiss has responded that it is only with the implementation of the large interferometers that we will have reached a level of sensitivity where it is reasonable to expect to see something, and that is why the argument works; in my terms, his nonzero multiplier comes from theory.

The other two things to be noted relate to research funding in science. One has already been mentioned: in embarking on a new, expensive, step-function enterprise such as the large-scale interferometry program, the justification for funding has to come from theory. I will argue later that the relationship between theory and empirical findings varies over time in physical science, and that during the justification phase of a brand-new technology such as this one, theory is bound to be dominant; it is theory alone that can say in advance that when the apparatus has been built, it will do something useful. We will see how at a certain point the dominance of theory brought about by the funding argument weakened the claim of certain embryonic empirical findings coming out of the old bar detector technology.

The third thing to be noted is the nice ambivalence between the use of theory as a warrant because of the certainty of what it says and the use of theory as a warrant because of its uncertainties. In the above passage we see theory being used to justify the building of interferometers of a certain sensitivity because we can estimate that such devices will see gravitational waves; but we also see theory being used to justify the building of interferometers because there are so many unknowns that there are bound to be interesting surprises to come.

This, indeed, would be exactly how the theory would come to be used in the funding debate. What would become LIGO would be justified because it should be able to see occasional bursts of energy from inspiraling binary neutron stars—which at the time of the big funding debate were the best understood source of gravitational radiation. But scientists such as Kip Thorne, who argued this case, also told me that they did not believe inspiraling binary neutron stars would be the first sources detected; Thorne favored inspiraling binary black holes, which would be a more powerful source but about which too little was known to produce any kind of even quasi guarantee. In the meantime, the third instrument in the justification orchestra was the still more general "we must expect the unexpected."

We will return to this issue at a later point, but in order to understand the relationship between LIGO and the resonant-bar community, which I will describe in the next parts of the book, it is important to remember that LIGO was born against a great deal of opposition. Interferometry was the only gravitational wave detection technology that needed funds of such an order as to make it, as we might say, "public property." The other technologies could do their business in the relative privacy of the scientific community and the funding agencies, whereas LIGO had to be seen to be a good thing by scientists in different specialisms and by the nonexpert communities in the outer rings of the target diagram; it was so big it had to rely on these others' support. And there were many of these others who did not want it; they believed it would take funds from their own enterprises, and they believed that it would not work. Thus, not only could LIGO afford few mistakes, it felt it had to try to keep the whole gravitational wave community "in line" to avoid guilt by association. The history of gravitational wave detection was not on their side and made it all too easy for others to brand the field an occupation for "crazies," to use Weiss's term. If any of the wider community of gravitational wave scientists were seen to fail again, then LIGO, they felt, would be seen to be failing, too. This, I will argue, had an impact on the relationship between LIGO and the bar community.

This relationship was by no means straightforward. Bill Fairbank, Guido Pizzella and Edoardo Amaldi, and, of course, Bill Hamilton—that is, representatives of all the successful cryogenic bar groups of the early 1980s—had decided not to oppose LIGO head on. The only resonant bar proponent who decided to make such a confrontation was Joe Weber. But Weber, one might have thought, was no longer credible after 1975. How could he assault the huge interferometry program? He could assault it for the reason given above: his attack was mounted not within the group of core scientists, where, as we will see, it would have been given little attention, but via the outer groups of politicians and others on whom LIGO had to rely because of its size and consequent need for funds. And his assault had to be taken seriously because of opposition to LIGO within the wider scientific community. Weber, in other words, though now an outcast from the core-group of gravitational wave scientists, could still exploit the outer rings shown in the target diagram.

LOOKING FORWARD AND LOOKING BACK

Writing in the early years of the twenty-first century, and trying to look a decade or so into the future, cryogenic bars are probably going to be said to have played a minor role in the history of gravitational radiation astronomy. Cryogenic bars fall between the beginning of the search—Weber and room-temperature detectors—and what is likely to be its end—the large interferometers. Unless they somehow manage to detect some large cosmic event in the next year or two, bars will be said to have detected nothing and to have been a technological backwater; they will be said, to borrow Winston Churchill's phrase, to have been "the end of the beginning." Indeed, writing this quasi history, it is hard work to keep a sense of their near twenty-year dominance and the optimism which drove, and in some cases still drives, their builders and operators. For a long time, bars were the only operating technology.

When at Stanford in 1972, I spoke with Bill Fairbank, the pioneer of cryogenic bars. I had already interviewed interferometry pioneers Rainer Weiss at MIT and Bob Forward at Hughes Aircraft. Consequently, though it was a minor theme of my work at the time, I talked with Fairbank about interferometers. It was quite clear from our conversation that they played no role in his thinking about the future of gravitational wave detection. Forward's design was of laboratory size and Fairbank explained, quite rightly, that the path length was too small to allow Forward to see anything. The design envisaged by Weiss was unknown to Fairbank. I could not enlighten him, because I did not understand it either. I was concentrating on Weber and his troubles.

In any case, I didn't grasp what large-scale interferometry was about until about the mid-1990s. The crucial conceptual breakthrough for me, a sociologist, came with the notion that the mirrors of an interferometer were free-floating "test masses" rather than integral parts of a single apparatus. Thus, the resonant devices have to be insulated from the ground so that their vibrations are not confounded with seismic noise. When one looks at the design of Forward's interferometer, it does not immediately strike one that it is very different in principle from the bars; to the untrained eye it seems to be, as it were, an elaborate resonant detector that does not resonate. Forward himself described it as a "broad band" version of a gravitational wave detector, so the fundamental difference does not immediately strike one. Indeed, thinking of interferometers as continuous with previous detectors, the whole idea seems crazy: "How," I asked myself, "are they going to insulate a multikilometer device from seismic noise when it is so hard to insulate a bench-top or laboratory-sized device?" I only ceased to ask this question when I came to understand that the end mirrors of the interferometers are meant to be independent test masses floating in space rather than component parts of an integrated apparatus.

Needless to say, Fairbank was a far more sophisticated thinker than I in these matters; but listening to him as he spoke to me in 1972, it appeared that he still thought that experiments using light as a medium were of two types: there were the experiments that used interferometry to look at Earth-Moon displacements and the like, and there was Ron Drever's Glasgow-based device, which, as far as I can make out, he (like me) thought used interferometry as a transducer in what was effectively a split-bar design.[13]

Interferometry is so dominant nowadays that it is important to realize that these observations were far from obvious in 1972. It is important to remember that Weiss could not get funding to embark on his initial prototype until 1975. Weiss guessed that the reason for this was that the cryogenic bars appeared to everyone to be the obvious way forward. As Weiss said in 1976, referring to his unsuccessful 1972 application, "The proposal to the NSF was unfavorably reviewed at the time most likely because it was too big a step from acoustic gravitational wave detectors."

What is sure is that when the interferometry program began in the mid-1970s, it was the newcomer knocking on the door of a bar program that had been running for a decade and which represented the frontiers of the technology. So what was happening in the old technology?

13. Only years later did Ron Drever explain to me that the Glasgow device was an interferometer with free test masses which just happened to use the ends of a scrap split bar as test masses.

PART III

BAR WARS

CHAPTER 19

THE SCIENCE OF THE LIFE AFTER DEATH
OF ROOM-TEMPERATURE BARS

CORE SETS AND CORE GROUPS

The most surprising thing going on while cryogenic bars were reaching maturity and interferometry was picking up momentum was that Joe Weber was trying to revitalize his room-temperature bar program. Weber was founding what we can call a rejected science.

The *core set* of a science is made up of those scientists deeply involved in experimentation or theorization that is directly relevant to a scientific controversy.[1] It is often quite small—perhaps a dozen scientists or a half-dozen groups. They comprise a "set" rather than a "group," because the members may disagree so violently that there may be few social ties between them.

But when a scientific controversy ends—achieves "closure," in the jargon—there are winners and losers. If the new claims are rejected outright by the majority, the winners write their "told-you-so" books and papers and go back to their previous scientific

---

1. Collins 1981a, 1992.

305

lives; for them, there is a resumption of business as usual. As for the losers, they may disappear, too, or they may form a rejected science—a determined rearguard action that the mainstream cannot accept or will not understand. Another possibility is that a new postclosure science develops that is a modification of the science at the center of the core-set controversy. When this happens, the mavericks are expelled and a *core group* takes over; a core group can be expected to be much more socially solidaristic than a core set, for they have a more unified aim. A core set, we might say, is like an explosive chemical reaction: after the explosion has subsided, everything may have evaporated and there may be nothing left behind; there may be a core group doing a new kind of orthodox science; or there may be the hard, dense cinder of a rejected science—a group of scientists obstinately refusing to give up their ideas in spite of the crushing consensus that surrounds them. How is "life after death" possible after a resounding defeat such as was suffered by Weber's room-temperature claims?[2]

In Western democracies, we think of the natural sciences as providing the prime example of a social institution designed to discover empirically verifiable truth. In part I, we looked at the way one claim to truth was extinguished by scientific institutions. Now we will find that the scientific institutions of the late twentieth century had not extinguished the old truth entirely; they could be said to have allowed the simultaneous existence of more than one "truth." Even an esoteric branch of physics such as gravitational wave detection seems to exhibit surprising tolerance and plurality in what it publishes and what it funds—but, as we shall see, only so long as no serious damage is being done!

THE LIFE AFTER DEATH OF HIGHVGR

Weber continued to be funded by the National Science Foundation after 1975, but on a much-reduced basis. He was given enough money, however, to

---

2. Those who refuse to accept the consensus are described here as constituting a rejected *science* rather than a set of misguided individuals, because they use the tools and procedures of science within the institutions and cognitive world of science. To conclude, on the basis of their observable activities, that this group were not pursuing a particular strand of science would be a cognitive judgment, and this is rarely the business of the sociologist. It almost goes without saying that many of the scientists in the mainstream of gravitational radiation research thought this group had stepped beyond the bounds of science, but we are concerned with the way the group continued to function as a science in spite of these judgments. An alternative approach is to study the mechanisms by which groups of scientists become defined as insiders or outsiders (Gieryn 1983, 1999), or to stress the haziness of such boundaries (Simon 1999). The approach adopted here is different.

enable him to try to become completely clear in his data analysis. From 1975 to 1978, Weber was funded by NSF at about $50,000 per annum, growing to $160,000–$320,000 yearly by 1983, by which time he had a group working on the new cryogenic technology. Richard Isaacson, program director for gravitational physics at the NSF, told me that they made every effort to help Weber resolve the conflict between his results and those of the rest of the gravitational wave research community.

> [1999] Marcel Bardon, then director of the Physics Division, saw to it that Joe never lost all support. He insisted that in the face of a controversy the NSF not choose sides. Rather, he believed that the NSF's role was to support Joe to help resolve the conflict. Consequently, whenever Joe sent in a request with a broad new scope, Marcel got together with Joe, the program director handling the proposal that year, and Howard Laster, chairman of the Maryland Department of Physics and Astronomy. Together they negotiated with Joe to write a revised, focused, and fundable proposal. For several years this resulted in a lower funding level to support his redoing and extending data analysis to allow him to publish the most complete documentation possible. Later [new collaborators] got involved, the budget expanded, and the group pressed on to try to build more-sensitive cryogenic apparatus in the hope of settling things that way.[3]

As we saw in chapter 11, Weber's first post-1975 publication was his summary and defense of his previous experimental observations in a 1976 issue of *Physical Review D*. As remarked earlier, this paper represented the end of the first Weber era rather than anything new, and it received little or no notice. But then Weber started the move to cryogenics. When I talked to him in 1995, he saw it this way:

> [1995] [I was told that] we wouldn't get any money unless we pursued that [new technology] approach, and at the time there was no experimental data and I saw no reason why that approach wasn't good . . .

Weber, then, reluctantly accepted the need to press forward with a cryogenic bar program. Ho Jung Paik joined him from Stanford about 1977 or 1978, bringing his experience of cryogenics and his expertise at designing bar-end resonant transducers. Weber's new team also included Jean-Paul Richard, another expert on the new transducers, whose plan was to add

---

3. E-mailed May 20, 1999.

an extra, intermediate stage to the mushroom resonators—a mushroom on top of a mushroom, as it were.

Plans were made to build two cryogenic bars. The vacuum tanks were ordered and put in place in the golf-course laboratory. (The tanks were still there when I visited the laboratory in the late 1990s.) But as time went on, strains developed in the Maryland team set up to build the LowVGR (Low-Visibility Gravitational Radiation) detectors. Part of the problem seems to have been Weber's continued energetic support of his old findings, while his new, younger colleagues wanted to press ahead with their own ideas, based on the much more widely accepted theories of what the flux of gravitational radiation should be. In our terms, the new members of the team wanted to investigate LowVGR, the current topic in the community in which they had grown up scientifically, while Weber's heart was still in HighVGR, not cryogenics.

In 1997, I asked Ho Jung Paik if he had gone to Maryland with the expectation of starting a brand-new program of cryogenic detection that had nothing to do with the old Weber results and subsequently had been disappointed. He replied,

> [1997] Yes—but he [Weber] had his room-temperature detector going all the time. And they were doing coincidence experiments with Rome....And in 1982 they published some new results.
>
> Whenever I had the chance, I cautioned Joe against saying too strong things about his earlier work. I said, "Be patient—let's get the cryogenic antenna working with ten times or 100 times better signal-to-noise ratio. If you really saw this, this will stand out as a strong signal. That will take care of the problem. No matter what you say, if we don't find it later, then it's all useless—time will tell."
>
> But he was still going around saying everyone else was stupid, and he was still the only one who did the job right, and so on.

In 1982, Weber coauthored a publication reporting new coincidences between an Italian bar, ALTAIR, and his room-temperature device in Maryland. The Italian authors, who were listed first, were V. Ferrari and Guido Pizzella from the Frascati group. ALTAIR was a cryogenic bar but was unusual in that it did not take advantage of the end-of-bar mushroom transducer, which was discussed in chapter 13. Instead it used a variant of Weber's piezoelectric crystals at the center. ALTAIR had a hole machined into it with a rectangular cross section in which the piezocrystal was a tight fit. This gave very good coupling between the bar and the crystal and avoided the problems of glued joints with their stresses and strains. This paper found a level of significance in the coincidences between the two bars of 3.6 standard deviations.[4] Since one of the bars was running at room temperature, this was a claim supporting HighVGR. But as it happens, no one took much notice of this paper.

Those attending the Third Marcel Grossmann Meeting (Shanghai, 1982) were to realize that Weber was reaching deeper into the dam of assumptions on which his critics depended. He was now embarking on a quite different kind of defense of HighVGR—a quantum theory of gravitational radiation antenna[e]. In his new theory, Weber claimed that the interaction between gravitational waves and resonant-bar detectors was far stronger than had been believed. In physicists' language, he was saying that a resonant bar presented a far larger "cross section" to incoming gravitational waves than current theories implied, and this meant that it was so sensitive that HighVGR—High-*Visibility* Gravitational Radiation—no longer implied high *fluxes* of gravitational radiation. As we have seen, the old ideas about cross sections meant that Weber's findings implied that far too much of the mass of the universe (many orders of magnitude too much) was being converted to energy in the form of gravitational waves. He had tried to solve the problem with the ideas of triggered metastable states and proliferation, but neither of these had convinced anyone. Now he had a more fundamental theory: the small fluxes thought to be compatible with astrophysics and cosmology were now compatible with his findings, too. Weber was to use the phrase "resolution of past controversy" in talks and papers on his new theory.[5] Indeed, he would adopt my terminology to explain to me that he considered his design of room-temperature bar to be a "low-visibility gravitational wave detector."

Before developing the new theory, Weber had made no irreparable break with the rest of the gravitational wave community; he thought he was the better experimenter—the others thought that claim was incorrect—but nothing shockingly radical had been declared. Up to that time, as we have seen, he continued to accept the majority view on many issues such as the broad theory of the detectors, the possibility of calibrating them with electromagnetic pulses, and the generally accepted theory of the transducers which favored the end-of-bar, resonant mushrooms being developed by his own team. All this changed with the new theory. The new theory turned on a quantum level analysis of the behavior of atoms within the solid material

---

4. Ferrari et al. 1982.

5. For example, a lecture given at Erice in 1991, and two publications: Weber 1992a and 1992b.

of the bars as they interacted with gravitational waves. As Weber put it to me in 1993,

> [1993] This is the twentieth century, and a bar isn't just two lumps of matter connected by a spring, but a bar is a large number of atoms coupled by chemical forces and described by the modern twentieth-century quantum theory. And so why not roll up one's sleeves and use the most advanced techniques for calculating the cross section. Well, I did this and got a quite different cross section.

He recalculated the interaction between the atoms in the crystalline parts of solid material and the weak forces that impinged upon them, and concluded that

> [t]he model of an antenna as an ensemble of interacting particles has a much larger cross-section than a classical continuous elastic solid of the same mass and dimensions.[6]

Weber was suggesting that the elements of a solid acted in a more coherent way than had previously been believed, so that much of the substance of a bar could act in concert when the bar was impacted by a weak force.[7] Coherence is more typical of the behavior of atoms in a maser or laser than with atoms in an ordinary solid. The bars, he claimed, were six to nine orders of magnitude (a million to a billion times) more sensitive than the orthodox theory—the theory which Weber himself had first elaborated—would suggest.

An additional consequence of the new theory applied to the detection of neutrinos, the almost massless particles that stream through space and solid matter with few detectable consequences. To see neutrinos emitted by the Sun, a great tank of perchlorethylene cleaning fluid was set up by Ray Davis of Brookhaven National Laboratory in the Homestake goldmine in South Dakota. Occasionally the passage of neutrino would cause one of the atoms of chlorine in this huge tank to mutate into an atom of argon, and the whole tank had to be searched for the few atoms of argon that

6. Page 4 in Weber 1984b.

7. An incidental consequence of this claim, if true, was that none of the methods of *calibrating* the bars that had been used to date could now be considered effective. This is interesting sociologically (for sociological analysis of calibration, see chapter 10). I asked Weber about this, but the answer was not clear. Our discussion can be found at WEBQUOTE under "Weber Discusses Calibration under the New Cross Section."

would indicate detections.[8] Now Weber claimed that a crystal that one could hold in one's hand could do the job of Davis's tank or the other large and complex detectors used to see neutrinos emitted by nuclear reactors. Weber's theory implied an improvement in cross section of 21 orders of magnitude (a thousand, million, million, million).

The new theory represented a radical break with both gravitational radiation research and other areas of orthodox physics; Weber was pulling out one of the large stones near the foundation of the dam. For example, the consequences of the removal of this stone for our understanding of the interaction between various weak forces and matter were enormous, and it appeared that the physical effects would be huge and unacceptable. Thus, if neutrinos were so easily absorbed by solid matter, astronomical bodies would no longer be transparent to them, and the energetics and dynamics of planets, asteroids, and so forth would, on the face of it, become incompatible with well-established observations.

To anticipate the scientific argument that was to come, Weber's claim, if it made any sense at all, applied only to solid matter in which the atoms were arranged in an orderly fashion, as in a crystal. It applied, then, only to coherent domains. A bar of metal is not a crystal, though it does have many crystals within it. One phase of the argument would turn, then, on the extent of coherent domains in metals and ordinary matter. Interestingly, the argument could work both ways. For example, one critique of the new theory as it was to apply to the detection of neutrinos was, as I have said, that the dynamics of heavenly bodies would be very different if they were absorbing neutrinos in the way that Weber's new theory suggested rather than being nearly transparent to them. Weber could counter that the material of planets and so forth contained few coherent domains, so this effect did not follow. On the other hand, critics could argue that bars of metal contain little in the way of coherent domains, so that even if the coherence idea was valid, there was no way it could make the room-temperature detectors especially sensitive. But Weber was to counter with the idea that when a metal bar is struck, its ringing puts the atoms into an orderly pattern—a pure quantum state that can be maintained.

> [1993] If you take a second every day to interact with it, you can restore it to
> a pure quantum state, by a second a day putting a signal into it. Well, if you
> put numbers in that, you discover that firstly you don't have the normal
> Brownian motion, because it's not the situation of maximum entropy, it's in

8. Pinch 1986.

a pure state—it doesn't wander off through phase space; it just sits in the part of phase space that's consistent with the quantum state that you've put it into.[9]

Weber, then, was distancing himself further from the mainstream physics community with his new theory. He believed he could do the job that the interferometers were proposing to do, but for a fraction of the cost. In the normal way of things, the new theory probably would have been ignored by the members of the gravitational wave core group. One model of scientific progress takes it that it is the job of every scientist to eliminate personally every doubt about every competing claim, but it is an unworkable model. To eliminate every loophole would simply take too long and divert too much attention from the rest of science. Human ingenuity is open-ended; as soon as one loophole in an argument or the interpretation of an experiment has been closed, the determined maverick can open another. To attempt to eliminate every one would be like the task of Tantalus: the goal would recede just as scientists thought they were about to reach it.[10] What is left is to reach agreement "for all practical purposes," and this usually means that after scientists with heterodox ideas or claims have been given a reasonable "run for their money," they are ignored; the attitude is, in the words of a scientist commenting on the reaction to Weber but from outside the field, "Ho, hum—more of this!" And from inside the field one scientist, himself a cryogenic bar pioneer, said to me,

> [1994] [M]y own feeling was that it [the new theory] was not correct, and everybody I talked to, simple plumbers, experimentalists like myself, no one believed it. And so I think we just sort of let it settle.

9. He went on:
   [Y]ou find that the sensitivity has just gone up astronomically. Now the thing that I like about this is that if I'm asked to isolate something that I can hold in my hand from the environment, it takes apparatus the size of a room, and I can put acoustic filters, electromagnetic shields, and isolate it enormously from the environment and operate under those conditions. Now, you can't do that with a $211 million interferometer—it's a several-mile hole that's coupled to the Earth, and I can't imagine isolating a vacuum chamber several miles long from the Earth [Weber misunderstands the interferometers; the vacuum tube does not need to be isolated.]...and again, the cost involved in putting this thing together and instrumenting it is, as I said, hundreds of thousands of dollars, not hundreds of millions. [419]...That was what I proposed to the National Science Foundation, but it was turned down by them.

   10. And hence the dangerous enthusiasm of scientists to get friendly outsiders, such as stage magicians in the case of parapsychology and water memory, to do their dirty work for them. See also Collins and Pinch 1979.

Of course, different groups of scientists have different views about when it is time to let things settle, and what amounts to "agreement for all practical purposes." The reasons Weber was not simply ignored were threefold: first, the neutrino element in the story was new and a new group of scientists were ready to take it seriously; second, Weber found a group of theoretical allies for his new theories, and friendly experimenters whose findings seemed to reinforce his ideas; third, Weber decided to attack LIGO on the basis of his new ideas, which meant that he had to be taken seriously by the mainstream if only for a short while. The details are tangled and follow no clear chronological sequence, but I will deal first with the theoretical and empirical work.

EMPIRICAL CONFIRMATIONS

The 1982 coincidences between ALTAIR and the Maryland room-temperature device would turn out to be interpretable as strong empirical support for the new theory, but it was not presented as such in the publications. It was advanced as just another empirical paper in the sequence of positive results that no member of the core group was now taking seriously. And it was almost entirely ignored, as we might expect.

In 1987, however, a supernova exploded in a nearby galaxy. As it happened, none of the cryogenic bars were running at the time, but the room-temperature bars in Maryland and Italy (GEOGRAV) were "on air." Weber and his Italian colleagues (the group led by Guido Pizzella) claimed, in a paper that was eventually published in 1989, to have seen gravitational waves emitted by the supernova in the form of coincident signals on these room-temperature bars. This empirical claim made an immediate impact, and I will discuss its fate in due course.

Note that there was no mention of the new theory in the 1989 paper; it had not been endorsed (or rejected) by the Pizzella group. The paper said only that, in the context of existing theory, the detected signals implied that 20,000 solar masses had been converted to gravitational wave energy in the course of the explosion—a ridiculous conclusion that meant that something unexplained was going on. The new theory could have made the observations less ridiculous, because it allowed these "insensitive" bars to be counted as sensitive enough to see the output of the supernova if it had been equivalent to the conversion of only a fraction of a solar mass. In 1993 (and again in 1995), Weber insisted that the pulse heights had been calculated before the supernova happened.

> [1993] And what's important is that the cross-section numbers given here [indicating a publication] in 1986 before the supernova, if you put numbers in those formulas, it predicts the pulse heights we observed during the supernova.... And I think that's of extreme importance. The theory published before the supernova predicted the pulse heights observed during the supernova. That's much more important than getting the pulse heights, publishing a theory after the supernova, and saying, "Theory and experiment agree."[11]

Unlike the 1982 claims, these findings did not go unchallenged, but I will return to the details later.

THE NEW THEORETICAL ALLIANCE

The supernova (SN) 1987A claims caused new attention to be paid to Weber's new theory, though it was not mentioned in the 1989 paper. In 1988, an Italian theorist, Giuliano Preparata of the University of Milan, was asked by senior Italian colleagues to look into the theory "just in case." Preparata rapidly concluded that it was incorrect and began to make this clear in conference presentations. But then, in another of the almost operatic events that characterize this field, he changed his mind. The story is so striking that I will tell it mostly in Preparata's words.

> [1995] So these people [discoverers of the apparent 1987A gravitational wave flux] were presenting in La Thuile [a conference center on the Italian lakes] in 1988, in the February of 1988—I have a vivid recollection of this—so in 1988 they were presenting this.... and they were presenting the first carefully analyzed coincidence events of the two antennas, Maryland and Rome—OK? And they were concluding—Pizzella was concluding, with a probability of 10^{-7}, that this was not an artifact. They have seen different events.

---

11. There are two effects: what Weber says is more or less correct for the pulse heights, but nothing before 1987 predicted the multiple pulses of neutrinos that the 1989 paper claimed were seen. (For the similar comment in 1995, see WEBQUOTE under "Weber on SN1987A Sequence of Events.")

Then Joe [Weber] stood up, and he said, "Yes—the fact that we've seen events doesn't mean that the mass of the supernova was twenty or thirty thousand solar masses like it is implied by the classical cross section. I have a new method of calculating the cross section that in fact leads to an enhancement of a factor of 10^9, and then this becomes 10^6, and so on and so forth. And this... implies that in fact only a third of a solar mass went into [gravitational waves]..." Which is perfectly OK.

People boo-hooed, and so on and so forth. I said, "I don't understand what's going on—his calculation seems to me reasonable."

Then, Amaldi—the old Amaldi—asked me, "What do you think?" I say, "Well—this looks perfectly OK—I mean at that point, you know, he has done the calculation and he has seen this factor."

And then he said, "Well, you see, we don't believe it"—because Cabibbo, who is a big man in Rome, has looked into the calculation, and he thinks that he doesn't understand it, blah, blah, blah.

So I said, "Well, I think, anyhow, a very clear answer can be given to this question." So he said, "We need this [answer] very soon." So, well, I think in a couple of weeks I can come to Rome and give a seminar on that.

OK, so. And then I came back from La Thuile, and I worked three or four days—that was '88—I was still keen on this—and I worked out the solid state anybody works with, and I said, "No way." Joe Weber cannot have this and I explained [why].

Preparata published his negative comments in *Nuovo Cimento* in the same year, writing,

Our conclusions share the simplicity of the calculations performed in this paper.

For the case of a realistic gravitational antenna, we have seen that its response to an incoming gravitational wave is accurately described by the classical analysis... the quantum calculation presenting no appreciable difference.

On the other hand, quantum mechanics does provide in general a dramatically different description of the interaction between wave and antenna, as Weber contends, but this deviation is only relevant for wave frequencies more than a billion times higher than those involved in a realistic gravitational process. Thus, Weber's point, though correct, is unfortunately seen to be physically irrelevant.[12]

12. Preparata 1988.

After his change of mind, he found that his old arguments were being used against him.

> [1995] Actually, these arguments [the ones I used in *Nuovo Cimento*] are the ones which are now used by the older people to criticize me without—because in this paper I showed that the resonant cross section cannot work with the single atoms like Joe Weber wants.... With the classical cross section there is no coherence to the classical cross section.... And all this ... the Russian with the Kip Thorne [Braginsky] and so on and so on, they have exactly used my arguments against me.

A little later in the interview, he continued,

> [1995] All right, then for me it's over. Then I was invited to a meeting in Perth.... I was invited just because ... and of course I was the hero of all this community, because I really [showed that Weber was wrong].
>
> I didn't know Weber, you see, before, so [sounds ashamed] ... I mean, I have this attitude towards science: I always decouple the ideas from the man. So I take an idea, and I just put it there, and I think—but this is not true, you know, this is not true. This calculation should not be done this way, because this is not what happens. Because I was making the picture of a crystal, you know, just strung together, you know, by little strings, and so on and so forth—so that's it.
>
> Then, a year after, just about the fall of 1989 ... the cold fusion has already passed, blah, blah. I was deeply involved in cold fusion at that time—I understood coherence at that time—I mean, how it works in condensed matter. And so I was ready to redo the calculation even though I had already forgotten [about it].
>
> And then I got an email from Guido Pizzella ... and he said, "Giuliano, I am analyzing further data ... and this thing doesn't go away, the effect doesn't go away. Are you really sure about your first paper?" ... So that was the question.
> The answer [was], "You know what, Guido—I'm not sure anymore. I'm gonna redo the calculation in the way that I see."
>
> And I did the second calculation, which is published in *Modern Physics Letters* and which is reported in this chapter of the book [indicating a volume]. And [I reached] a totally different conclusion.
>
> And with this paper I send immediately ... to Joe with a letter apologizing. But, you know, I was making a calculation following everybody else's condensed-matter physics—crystal physics—but now that I have my own

[way of doing things], then I see that he is right. And of course he was delighted and everything like that.

[Preparata here explained that his way of calculating the cross section, while it reaches the same conclusion as Weber's method, is based on more-profound physics, and that Weber had to some extent reluctantly accepted this.]

> [1995] So—just to make this long story short, after a year I became convinced—and having gone through cold fusion—I was now convinced that Joe Weber and Guido Pizzella were right. And I published this paper. This paper didn't make it into *Physical Review Letters*—I remember that the comment was very stupid—I showed it to Rocky Kolb because I was visiting Fermilab at that time, and he said, "Alright, Giuliano, don't worry, I'm gonna pass it to this Singapore review"—because he said it was a very good thing—and he published it immediately.[13]

Preparata's approach to the new cross section was not exactly the same as Weber's, but it turned on the theory of *quantum coherence* or *superradiance*. Quantum coherence is a small, independent field on the edge of physics that deals with quantum effects on macroscopic objects. In 1998, one of its leading proponents told me, "There aren't any better field theorists in the world than Giuliano Preparata." Note, then, that while Preparata's reputation in the core gravitational radiation community was to sink as low as it was possible to go, it remained high for another group of well-placed physicists. In the second group he was considered the same brilliant, upstanding man as he had been in the eyes of the mainstream before he tripped over Weber. Again, for a time, there were two physics.

What is quantum coherence? Quantum effects are normally thought to apply only to microscopic objects on the scale of the atom, but in certain circumstances they can be made to show themselves macroscopically when the arrangements are just right. For example, a laser is arranged in such a way that quantum effects reveal themselves macroscopically. *Superradiance*, the other related term, also has a perfectly respectable scientific pedigree under appropriate circumstances.

Preparata worked with a small group of Italian colleagues who were publishing a regular stream of books and papers on quantum coherence

---

13. Preparata 1990.

and superradiance.[14] It was the range of conditions under which these special effects were supposed to take place that constituted the heresy. One can see from this interview extract that Preparata's conversion to Weber's view came not from his fascination with gravitational waves but from an interest in cold fusion, a still more heterodox scientific claim. In 1989, two chemists then working at the University of Utah, Stanley Pons and Martin Fleischman, announced that they had discovered that controlled fusion between hydrogen atoms could be made to happen in a test tube instead of enormous and hugely expensive machines. This claim was rapidly crushed as far as most scientists were concerned, but in 1995 cold fusion was still going strong in a Toyota-funded, purpose-built laboratory run by Pons and Fleischman in the South of France, and in many other locations, including the Milan base of Giuliano Preparata. When I interviewed Preparata in 1995, none other than Martin Fleischman was present and contributed throughout the interview! We will see that the connection with cold fusion and other heterodoxies is important in this story, but, interestingly, Preparata's loss of credibility in the gravitational wave core group had nothing to do with these other connections; my respondents did not know about them.

Preparata's version of Weber's theory also made sense of the biological transfer effect in water associated with the name of Jacques Benveniste, another appalling heresy according to the mainstream scientific community.[15] For a short period, there seemed a chance that these other rejected sciences would absorb Weber into their new social network. Meetings were held and funds were sought to support a group of sciences held together by Preparata's ideas and a common feeling of exclusion by a powerful and self-interested establishment. I'll call this group the Quantum Coherence Heresy Group (QCHG).

The driving political principle of QCHG was that the new ideas promised to accomplish old scientific ambitions much more cheaply than existing research programs. For instance, if cold fusion worked, it would replace the enormously expensive hot-fusion program with tabletop experiments; if the Weber/Preparata gravitational cross section was correct, it would cut the cost of gravitational wave astronomy by a factor of more than 500. The new network justified itself in part by the view that big science was wasting taxpayers' money when cheaper solutions were available, and they hoped

14. For example, Bressani, Del Guidice, and Preparata 1992; Bressani, Minettie, and Zenoni 1992; Preparata 1990, 1992.

15. Which also continues to live on after its official demise.

that influential people in the US Republican Party would fund their efforts. The tone can be judged from remarks in a letter from the Italian group written in November 1994. This was sent to a potential Republican funding source, Richard J. Fox, associated with various radical programs at Temple University in Philadelphia. For example, "[Our] view is not only orthogonal to the dominant science of the Academic Government establishment, but also to a large number of so-called 'heretical' sciences that have not foregone the basic atomistic view of their rival orthodox science."

Weber certainly attended this group's meetings, of which there were at least two. One was held in London in 1995, and the participants included Weber, Martin Fleischman, the founder of cold-fusion research, Preparata, and "Dick" Fox. In March 1995, Fox wrote to Weber, thanking him for his participation and explaining, "We have taken the first of many steps toward the recognition and broad acceptance of a new scientific paradigm based on 'Coherence in Matter and Life'"; moves toward the founding of a journal with that name were also described.[16] What unified the scientists and their potential benefactors, as far as I understand it, was that under the new science, the same discoveries could be made for much less money than the American taxpayer was currently paying. In the end, however, the financing of the new network did not materialize.

Some sense of the strange relationship of quantum coherence and the rest of physics, perhaps partly exacerbated by the heretical waters toward which Preparata and his group were leading it, can be gained from one of my own experiences as a interviewer. In a very friendly and helpful phone call, I arranged to travel some distance to interview a leading quantum coherence scientist. When I arrived, I discovered the whole atmosphere had changed; he refused to talk to me. Eventually I succeeded in having a somewhat stilted and initially very unfriendly conversation.

As mentioned, none of those in the gravitational wave field whom I asked about Preparata and his work knew of his cold-fusion and other heterodox connections, and they also did not know of the temporary closeness of Weber with what they might have thought of as the mainstream of pathological science. It is surprising how disconnected the individual "villages" of physics can be. But what the core group knew of Preparata's theories was quite enough to damn him in their eyes anyway. He was considered by the famous leader of experimental physics, Eduaordo Amaldi, to be the right man to pronounce on Weber's new theory on behalf of the

---

16. Letter dated March 1, 1995, on Institute for Bio-Information Research notepaper and transmitted by fax.

rest of the theoretical community. Moreover, according to his own account, he had been on a sharp upward trajectory when he was younger, having been a research associate at Princeton and a research fellow at Harvard, as well as having had a tenure track post at New York University by the age of 28 and a full professorship in Italy, to which he had decided to return by 1975 at the age of 31. As we have seen, Preparata was also considered by members of the quantum coherence groups to be a brilliant physicist. But when, after the events of the late 1980s, I asked one powerful member of the gravitational wave core group his opinion of Preparata's work, he replied with a single, ironic word: "Who?"

"Center-Crystal Instrumentation": Transducers in the New Theory

Weber's new theory could be seen to have empirical support in the SN1987A observations. Remember, the experimentalists did not publish any support for the new theory; they preferred their findings to stand on their own, to be interpreted by others as they saw fit. But in private they were prepared to speculate that Weber might be right. We also see that Weber's ideas were given theoretical support by Preparata and his team.

Unfortunately, things now become still more complicated, and I am unable to sort out the exact sequence of events. Weber's theory turned out, according to him, to have yet another consequence: only bars with transducers mounted in the center would be able to detect the full effect that the new theory had uncovered. Under this ramification of the new theory, resonant transducers—the mushrooms at the end of the bars that were being developed in his laboratory and that even his supporters in Italy were using on their latest cryogenic devices—prevented the bars from seeing what they should. The "center-crystal instrumentation" theory, as Weber would come to refer to it, said that the signal reverberates backward and forward along the bar as it resonates, energizing the nonresonant central piezotransducers at every pass, but that this effect would be damped by the long resonance period of the mushroom transducers. Here is how Weber explained it to me in 1995:

> [1995] Now once the group of pulses is formed, it doesn't suddenly die away when the supernova stops pulsing, but pulses bounce back and forth from end to end. Now with center-crystal instrumentation, the center crystal sees the wave each time it goes past, so that the center crystal sees first this

[drawing a diagram on the board] and then half a cycle later it sees the, sort of, mirror image of it, and then half a cycle later it sees that [drawing]. So what the bar is, is a kind of transducer that converts energy over a wide band of frequencies into a fundamental component within the bandwidth of the electronics. So that although the electronics is narrowly tuned to 1660, the bar is wide open over a much wider band of frequencies than any of the critics imagine....

Well, the thing is, if you put a resonant detector on the end and you do the analysis, you conclude that if you wait long enough, this energy will get into the resonant detector, but you have to wait a very long time and that effects the statistics, where with the center crystal you see every pass, but you don't see every pass with the end crystal detector.

And here is how this idea was presented in a 1996 publication:

Greater sensitivity has been claimed for bar antennas with small harmonic-oscillator systems at the end of the bar. Significant numbers of coincidences have not been observed with antennas having end-of-bar instrumentation.... [G]roups of phonons...may oscillate from end to end of the bar. Center-crystal instrumentation observes each pass. End-of-bar instrumentations require a long time for significant amounts of such energy to be transferred to the bar end harmonic oscillator. For these reasons, it appears unlikely (but not impossible) that the results reported here will be confirmed by gravitational-radiation antennas with instrumentation significantly different from center-crystal instrumentation.[17]

As far as I can make out, this element of the theory was not developed until about the time of SN1987A, but in interviews Weber spoke as though it had been part of the original 1984 work. I believe that in conversation with him I was able to get him to agree that the critique of end-of-bar transducers did come after the 1987A result even though the rest of the theory came earlier.

Center-crystal instrumentation was needed to explain a series of empirical findings. It explained why the cryogenic bars, fitted with resonant transducers and, on the face of it, many times more sensitive than Weber's bar, saw nothing while he saw many pulses. It fit with the fact that the 1982 coincidences and the 1987 supernova coincidences had been seen on bars with "center-crystal instrumentation"; ALTAIR, the Italian bar involved

17. Weber and Radak 1996, pp. 691–92.

in 1982, in spite of being cryogenically cooled, had a piezocrystal embedded in its middle rather than a resonant mushroom at the end, while both the 1987 bars were room-temperature, center-crystal-instrumented bars. As Weber said to me in 1995, "All of the ... input of the 1982 paper, of course, with the Rome and the 1980 Texas symposium, those first detectors were all center crystal instrumented. ... So at this point I don't know of any significant data that's been obtained with a detector that wasn't center crystal instrumented."

The Italian Reaction to the Center-Crystal Theory

Preparata initially had nothing to say about the role of transducers and seemed, if anything, to verge toward the skeptical. Thus, in 1995 he told me,

> [1995] Well, I think that Joe has a point, However, it's not very clear that in the end that due to the fact that they work at much lower temperatures, you know the disadvantage of that configuration, transducer configuration, cannot be sort of compensated by the much lower noise. ...
>
> I understand, however, I don't think that the situation is so bad. I think Joe Weber has a better configuration for the transducer, but I don't think the situation with the end transducer is that bad. ... If they don't have any signal, he needs to say, "Oh! this is very bad and this explains it," but actually I think now it's coming out that in fact, and you will hear this from Pizzella, that he has—something that he's very afraid of, and you should be very careful.

In other words, Preparata was telling me that though Weber's arguments seemed plausible, he thought any difference could be compensated for by the cryogenics. Furthermore, he had just heard that Pizzella believed he had seen some signals with his end-transducer cryogenic bar, and this was reinforcing his lukewarm feelings about Weber's critique of end transducers.

So far as I know, Preparata never wrote anything about the transducer claim, but at a conference in 1997 he told me that he had changed his mind and now believed Weber was right about the center crystals.[18] This had happened as a result of an "interrogation" by Weber to which he had been subjected a year or so previously, when Weber had paid a visit to University

---

18. The Eighth Marcel Grossman Conference held in Jerusalem in June 1997; I spoke with Preparata on this topic on June 26.

of Milan, Preparata's home institution. It is, however, relatively easy for theoreticians to change their minds; all they have to be prepared to do is admit they were wrong, and start inscribing different things on paper. For an experimentalist it can be much more difficult; an experimentalist may have to start building again from scratch. In addition, an experimentalist needs funds to build with, and funds are controlled by agencies who have to refer to a wider community. In the terminology of the target diagram, theorists with tenure can choose the centrality of their location on the target diagram at will. An experimentalist has to have at least some contact with the second ring, for the sake of the experimental funds that some of its members will be providing.

This situation is nicely explained by members of the Italian experimental team from Frascati, the only ones who were likely to consider making such a change, because they were the only experimentalists prepared to entertain Weber's theoretical ideas in the first place. In 1995, I asked Guido Pizzella why he did not modify his apparatus to take account of Weber's ideas by removing the end transducer and instead mounting center crystals. He replied, "I[I]n general, suppose we tried to do that [now]. It is not an easy decision. It requires a lot of work. We have to warm up [the apparatus], take the bar out, bring it to the shop, do a very careful machining."

I pressed him on the success of ALTAIR:

Collins: So you would not consider using the center crystals because you do not believe Weber's theory?

Pizzella: No—Weber has two theories. One is the cross section, which many people say is wrong. The other is using piezoelectric ceramics is better, because it takes all the frequencies. Yeah—But let me say it is not a matter that I don't use it because I don't believe in Weber's theory. If it cost to me, not much, I would do it, being an experimentalist, but to do that would take two or three years. So it has a high cost. So before doing that I must have given up all other possibilities. And we are still fighting to run the system as it is.

It is not like doing a calculation. If I were a theoretician, I could say, "Weber is right," or "Maybe he is right," so let me calculate and spend some time doing a calculation. But here it is tremendous—the entire group changes, everything changes, including the data acquisition.

In 1995 I asked similar questions of Eugenio Coccia, then a relatively junior member of the team.[19] He answered as follows:

19. By 2003, Coccia had become director of Italy's important Grand Sasso Laboratory.

[1995] I think that following the standard theory, you should put a resonant transducer at the end, because in this way you pick up more energy than with a crystal at the center.

Following alternative theories, like the one of Giuliano Preparata—I should say the one of Joseph Weber—then the two situations are qualitatively different, since the piezoelectric ceramic is a wide-band transducer and so can catch the energy of all the frequencies of the bar. Following alternative theories, there is a lot of energy released in the bar due to gravitational waves, and one can measure the energy with the piezoelectric crystal. And with the resonant transducer at the end, one cannot, since the transducer at the end is resonant itself, and so it takes only the energy of one frequency.

Nevertheless, I should say that our funding agency expects us to reach a high sensitivity to gravitational waves, as they are now following standard theory of the interaction of gravitational waves and the resonant mass. So I think we are doing the right thing, actually.

I think, however, we should not be blind, and also give some attention to these other theories. But we are not theorists. So it is difficult for us to take a holiday for one or two years and study these theories—you see. So I think what we do is . . . we are making a standard experiment, let us say.

I pressed Coccia further, and he continued,

Well—I think that here there is a lack of communication between theoreticians and experimentalists, and even between theoreticians themselves. We are not philosophers, we are physicists, so at the end we should agree which is the correct theory. So if two theories predict different things, one should really measure everything and say which measurements agree with which theories and so on. But today, these alternative theories have no credibility, so it seems quite . . . it is very hard to start an experiment for these alternative theories, and perhaps only a person who has already reached the top of his career and is convinced that this alternative theory is good, and does not mind what other people say—they will do that. But most of the people of this group have studied textbooks in which gravitational radiation is treated in the standard way, and the interaction between gravitational radiation and matter is treated in the standard way, and it seems natural to perform this experiment, and a little crazy to do something else—you see.

However, I have to remark that I don't know exactly why these alternative theories are not considered serious theories. They are simply ignored, and they are ignored with heavy words . . . "rubbish" . . . "garbage." Maybe there is an intuition behind all that, because Joseph Weber is a very intuitive man. He is an inventor. He has made, I think, of the same matter as Columbus and Marconi. So people that try something and it looks perhaps that behind the darkness there is some glimmer of light, and even then they don't know why exactly, they follow, but they feel they have to follow this . . .

Joseph Weber once said, "Only dead fish follow the stream." So surely he's not a dead fish, but it is hard for a young person to follow his road when everybody around thinks this is not serious.

The Italians, as we can see, could not contemplate changing their whole experimental program just to test Weber's ideas out, however much sympathy they had for them. They were locked into an experimental program and a set of institutions with their own financial and cognitive momentum. What Coccia has nicely described is the felt social force of a dominant paradigm.

Gamma Ray Bursts and Gravitational Waves

It is little known, and never published as far as I can discover, that about 1991 the Frascati group and the Louisiana group, using their cryogenic bars, conducted a joint search for coincidences between gravitational waves and gamma ray bursts. Gamma ray bursts are one of the most extraordinary astronomical discoveries of recent years. It appears that Earth is bombarded once or twice a day with an intense burst of these rays. The sources of these beams are distributed across every part of the sky independent of strength, which suggests that they do not come from our galaxy. Because of their strength and apparently "nonlocal" origin, the sources of these bursts are a puzzle. It is tempting to try to correlate the gamma ray bursts with gravitational waves, the only trouble being that hardly anyone thinks that gravitational waves from distant sources are strong enough to detect with existing antennae.

It was NASA's Goddard Space Flight Center, in the suburbs of Washington, DC, that funded the Frascati-Louisiana project. A report in its files, authored by Guido Pizzella and Bill Hamilton among others, reveals that it found no correlations. This is not surprising, considering that Pizzella and Hamilton could not agree that they had found any gravitational waves in the first place. There were no funds forthcoming when they applied to continue the search.

But this was not the end of the matter. The Goddard Space Flight Center then funded Joe Weber to conduct a search for gamma ray–gravitational wave coincidences using his data.[20] And Weber did find coincidences. Here is how he described matters to me in 1995:

> [1995] Well—um—as I said, I think the situation has changed, and I'd be happy to tell you the reason for the change in status, but I'd have to swear you to secrecy. I hate to swear people to secrecy [*Collins:* You got it.], but I'd like to swear you to secrecy for about two years and tell you what happened. [*Collins:* Yes, please.]
>
> Well, firstly, when I was properly fired by both my universities at age 70 and all my space was taken away, then I decided I wouldn't commit suicide, or violate the laws, but I'd enjoy the best years of my gorgeous wife [laughter] and try and see if I could change things....
>
> Well, I had the following thought two years ago. There are about two of these [gamma ray] events a day, and when I do gravity wave experiments I see about two "out-of-statistics" pulses per day.[21] Maybe they're correlated!
>
> So I went to NASA-Goddard here with my proposal, and they said, "That's a better idea than you think, because we thought of it first."
>
> I said, "Well, I'm honored."
>
> They said, "We went to your competitors because we know they have more sensitive detectors than you do, and they found nothing."
>
> So I said, "Well—here I am—would you like to give us a little bit of money to look into the matter." And they said, "Well, we'll give you $10,000 and that's all. And after the $10,000 is expended, then you're on your own."
>
> Well, so with the $10,000 in hand I went to my associate chairman and said, "I have been guarding tapes practically with my life since you started taking space away, and what I want is a competent computer programmer to read these tapes."
>
> And they said, "Well, we've got a brilliant and temporarily unemployed computer programmer, and we strongly recommend him."
>
> So I hired this highly recommended person, and, well, he was a brilliant guy. And I'll show you some of the data he produced, and out of the first 80 pulses—pulse times—he looked at, the gamma ray pulse was close to a large gravity wave antenna pulse—twenty of the first 80.
>
> Well, I know that sort of thing is controversial, and after we consumed the $10,000, I gathered the data on the twenty and went to NASA headquarters, and the first official I talked to—I won't tell you his

---

20. Grant NAG 5-2571.
21. An "out-of-statistics" pulse is something that cannot be explained by random fluctuations.

name—encouraged me to hold a press conference, insist I had made a major discovery, and publish in a rapid publication journal.

Somehow this didn't strike me as the appropriate thing to do, and I kept battering away at the system and finally, about a month ago, received some additional money. . . .

And NASA said, "[U]pgrade your bars, send us the data to Goddard in real time, so we can check in real time and see whether or not we see pulses when we get gamma ray bursts from our observatory."

Well, when they said this, I jumped for joy. It was the nicest thing that I've heard for twenty years. There are four scientists at NASA-Goddard, of high reputation, who have asked to be collaborators.

Now, when I've mentioned this to some critical people here [at the University of Maryland], I've got a rather hostile response from them: "There can't be anything in it." "It's all a fraud." But the reason I'm optimistic is the following: four distinguished scientists at NASA-Goddard are coming in as collaborators.

Now, the correlations will either be there or they won't. OK—suppose they're there—it's not a matter of asking for $360 million to build a LIGO interferometer to look for one or two events per century. These things are occurring every day. If we see correlations, I can go up to my apparatus, turn a screw, and see if that makes the correlations go away. I can add filters and see if the filters make the correlations better or worse. I can start doing science the way Galileo said it should be done, not the way Aristotle said it should be done.[22] And I can do this quietly—and I beg you—I can do this quietly with four people at Goddard—four critical distinguished people looking over my shoulder, checking everything I do. And I don't have money to hire an independent programmer, but I figure I don't need an independent programmer, because there are four independent guys getting large salaries from the federal government who are interested.

So for all those reasons I'm much more optimistic than I've been in twenty years that something sensible will come out of this 37-year effort. . . . It is $20,000 to buy equipment so that the data can be sent to Goddard in real time.

In 1996, Weber offered me a sense of how it felt to be back doing such science.

---

22. This is a motif which I suspect is taken from the discourse of Martin Fleischman, the founder of cold fusion. He pressed it upon me several times during our conversations. I believe it is meant to stress that experiment is the crucial element in science rather than a priori reasoning.

[1996] When I get back to my office [after lunch], I'll look at the email to get the latest gamma ray burst data from NASA-Goddard. Then on Monday I'll sit in front of my PC and look at my data disks and search for correlations. Sometimes they are there, sometimes not.

In 1996, Weber published the data from this search in the journal *Il Nuovo Cimento*. There he claimed that "[t]he probability that the correlations are accidental is estimated as approximately 6.10^{-5}" (6 chances in 100,000).[23]

Weber, being aware that Hamilton and Pizzella had failed to find any gamma ray coincidences with what were, on the face of it, much more sensitive devices, was bound to feel that his center-crystal instrumentation theory had to be right. The results of the two searches could be made to seem compatible—barring questions about the basic competence of his ally Pizzella—only if the center-crystal theory was correct. Pizzella's bar had to be insensitive as a result of the installation of the end-of-bar resonant transducer. Thus, Weber was now locked into this view. The world was consistent: only bars with center-crystal instrumentation had ever seen any gravitational waves.

23. Weber and Radak 1996, p. 687.

Plate 1 Joe Weber at the start of his career. Photograph courtesy of the University of Maryland and Virginia Trimble.

Plate 2 Albert Einstein's likeness looks down on a Weber bar sans vacuum tank in the Einstein Centenary Exhibition at the Smithsonian Institution in 1979. Photograph courtesy of the Smithsonian Institution.

Uncredited photographs were taken by the author.

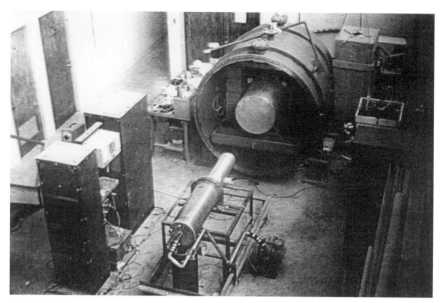

Plate 3 Joel Sinsky's calibration experiment with vacuum shields partly removed. Source unknown.

Plate 4 Joe Weber at the height of his fame. Photograph courtesy of the American Institute of Physics, Emilio Segrè Visual Archives.

Plate 5 Weber in 1993 at Irvine, holding a crystal for detecting neutrinos.

Plate 6 Bob Forward in the late 1990s.

Plate 7 Freeman Dyson in the late 1990s.

Plate 8 Richard Garwin in the mid-1990s.

Plate 9 Giuliano Preparata in the mid-1990s.

Plate 10 Tony Tyson in the mid-1990s.

Plate 11 Richard Isaacson at the inauguration of the Laser Interferometer Gravitational-Wave Observatory (LIGO) in November 1999.

Plate 12 Warren Johnson (left) and Bill Hamilton in front of a disassembled ALLEGRO. Photograph courtesy of Louisiana State University Physics and Astronomy.

Plate 13 Guido Pizzella in front of a diagram of NAUTILUS (Nuova Antenna a Ultrabassa Temperatura per Investigare nel Lontano Universo le Supernovae).

Plate 14 Pia Astone, leading statistician of the Frascati group.

Plate 15 Eugenio Coccia in front of NAUTILUS (Nuova Antenna a Ultrabassa Temperatura per Investigare nel Lontano Universo le Supernovae). Photograph courtesy of Eugenio Coccia.

Plate 16 Viviana Fafone, who now runs the NAUTILUS detector.

Plate 17 From left: Vittorio Palladino, David Blair, and Guido Pizzella in the corridor at the Pisa conference, discussing Blair's upcoming announcement of the Perth-Rome coincidences.

Plate 18 Heated meeting of the International Gravitational Event Collaboration (IGEC) in Perth, Australia, in July 2001; Massimo Cerdonio was the chair.

Plate 19 Peter Saulson (left) listens to Rai Weiss in the late 1990s.

Plate 20 Ron Drever. Photograph courtesy of the Palomar Observatory/California Institute of Technology.

Plate 21 Kip Thorne. Photograph courtesy of the Palomar Observatory/California Institute of Technology.

Plate 22 Ron Drever working in his new basement laboratory in the late 1990s.

Plate 23 From left: Kip Thorne, Ron Drever, and Robbie Vogt in the 40-meter lab. Photograph courtesy of the Palomar Observatory/California Institute of Technology.

Plate 24 Vladimir Braginsky in the late 1990s.

Plate 25 Riccardo de Salvo at the Laser Interferometer Gravitational-Wave Observatory (LIGO) in the early 2000s. The notice on VIRGO stationery beside his head is a progress report to the council which reads, "Different actions and declarations from Riccardo de Salvo (formerly a VIRGO engineer, now a LIGO physicist) have not contributed to improve the relationship between the two projects." De Salvo has awarded it a First Place ribbon.

Plate 26 Jim Hough, leader of the British side of the GEO project, greasing a thread behind his ear.

Plate 27 Karsten Danzmann, leader of German side of the GEO project.

Plate 28 Alain Brillet (left) and Adalberto Giazotto, the French and Italian leaders of the VIRGO project.

Plate 29 David Blair repairs a punctured tire in the bush on the way to the site of the Australian International Gravitational Observatory (AIGO)—soon to enter service in Gingin, an hour's drive north of Perth, as a test facility for high-powered lasers.

Plate 30 Peter Aufmuth outside an end station at GEO600; note the old radiator propping a door.

Plate 31 "Black Spaghetti": the patchboard at the TAMA interferometer in Tokyo. The skein of wires gives some indication of the number and complexity of the feedback circuits.

Plate 32 Control room at LIGO.

Plate 33 Control room at VIRGO.

Plate 34 Bob Spero, leader of the 40-meter group. Photograph courtesy of Bob Spero.

Plate 35 The new management at LIGO: Barry Barish (left) and Gary Sanders. Photograph by Felipe Dupouy. Photograph courtesy of the LIGO project, California Institute of Technology.

Plate 36 Clean-assembly procedure at LIGO. Photograph courtesy of the LIGO project, California Institute of Technology.

Plate 37 Welding the spiral beam tube for LIGO.

Plate 38 Nergis Malvalvalla (left) and David Shoemaker making LIGO work in September 2000.

Plate 39 Aerial view of the LIGO installation in Hanford, Washington. The arrow indicates a midstation for the 2-kilometer interferometer. Photograph courtesy of the LIGO project, California Institute of Technology.

Plate 40 Aerial view of the LIGO installation in Livingston, Louisiana. Photograph courtesy of the LIGO project, California Institute of Technology.

CHAPTER 20

SCIENTIFIC INSTITUTIONS AND LIFE AFTER DEATH

FOUR COMPONENTS IN THE SURVIVAL OF A REJECTED SCIENCE

We have seen how the science of HighVGR could be made to hold together so long as empirical findings were given authority over theoretical argument. But science needs resources in addition to ideas. What else does it take to keep a science going?[1] How did HighVGR keep going for twenty years after 1975?

A reasonable-scale, experimentally based science needs *refereed and other publication outlets and material resources including laboratories*. A *science* is more than an individual's eccentricity, so it must also have a *network of active colleagues*. Finally, to survive in the long term, a science needs a *body of graduate students* to carry the ideas through to the next generation. We will look at the survival of HighVGR in terms of these four headings.

1. For a complex classification, see Whitley (1984); see also Simon (1999), who uses a similar scheme to the one used here.

Publication "after Death"

In the next chapter, I will look at the responses to the HighVGR papers published by Weber and his associates, but let us first ask how any papers at all were published by this rejected science. I find that scientists are sometimes puzzled by all the fuss that is made over publication and responses to it. I have been slow to understand why, but it is quite easy to understand if one were to imagine the same fuss being made over a publication in a sociology journal. Given that I have a low opinion of much of what gets published in the social science literature, I, too, would be puzzled by someone who wanted to go into great detail about why and how this or that piece of work was published. But it is different in the natural sciences. The natural sciences have a special role in public life—they are supposed to be exact. Spokespersons for science who appear in public are always trading on the special nature of scientific publications, especially the much vaunted peer review. Therefore it is always interesting to know that they are not exact. Furthermore, this was a high-profile disputed field. These papers, one might have thought, were not part of the run-of-the-mill 95% of scientific publications that interest practically no one; yet in some cases, this was how they were treated.

Let me then try to explain the publication of five papers in high-prestige journals. There is the 1976 summary paper that was published in *Physical Review D*. There was a paper published in 1981, in *Physics Letters A*, claiming that the sensitivity of detectors could be enhanced through pulse proliferation. There was the report of new coincidences between the Italian ALTAIR detector and the University of Maryland detector, also published in *Physical Review D*. And there were two papers outlining the consequence of Weber's new theory for a different field entirely—the detection of neutrinos; both of these were also published in the *Physical Review*.[2] The first journal paper on the neutrino idea, which also included discussion on the consequences of the theory for gravitational waves, was published in *Foundations of Physics*.[3] Weber was also publishing papers in conference proceedings, edited volumes, and the like, which are less likely to be refereed by key members of mainstream core groups; but we are mainly concerned with the mainstream journals.

---

2. The papers are Weber et al. 1976; Weber 1981; Ferrari et al. 1982; and Weber 1985, 1988.

3. Weber 1984a. *Foundations of Physics*, which is the journal corresponding to this reference, is less prestigious than *Physical Review*, so I will not discuss that paper in any detail.

It seems that the first kind of explanation for the existence of these publications is simply that the physics journals can be very tolerant.[4] Weber was still widely respected as a scientist. Even before his controversial work on gravitational waves he had published on the idea of what became known as the "maser"—the microwave predecessor to the laser—and, as I have said, some believed he was a credible candidate for a share in the Nobel Prize for the laser. Furthermore, he was still respected as the pioneer of the field of direct gravitational wave observation, and his early publications, in which the principles for the search for gravitational waves were first set out, were still recognized as original and sound.

Kip Thorne put his interpretation of the policy of the physics journals as follows when I spoke with him in 1995:

> [1995] In physics it seems much easier to get things published that people are skeptical about, and get them published in reputable journals, than it is in softer areas of science. Perhaps because people view it as less dangerous, because you know there will be confirming experiments or confirming theory. It will be proved right or wrong, and if somebody who is relatively eminent, particularly, wants to stick his neck out, a large fraction of the referees will argue for a while and try to get the paper improved, cleaned up, made as solid as possible and then will say—OK, go ahead and publish it, it's your neck. In [a certain social scientific field with which I am personally acquainted], that just doesn't happen. People block each other from publishing unless it meets community consensus.
>
> So I'm not surprised that a paper of that sort would get published. It's almost guaranteed that it would get published after a few rounds of refereeing and arguing.

This is not to say that the physics journals will offer equal charity to anyone. Giuliano Preparata could not get his version of Weber's idea into the mainstream journals, but Preparata had no reputation to fall back on.

---

4. Zuckerman and Merton (1971) explain that, for the data gathered about the time they were writing, probability of acceptance of a paper, given an initial submission in physics, was between .75 and .9, whereas in sociology or economics it was more like .2 to .3. They stress that this figure relates to initial submissions rather eventual success—that is, it is not just a matter of the overall rejections being higher in the social sciences, just that one is likely to get rejected more often before acceptance.

Table 20.1 Weber's publication outlets

		57–59	60–64	65–69	70–74	75–79	80–84	85–89	90–94
1	Main journals	1	3	11	8	2	2	0	0
2	Early *Nuovo Cimento*	0	3	0	2	0	–	–	–
3	Late *Nuovo Cimento*	–	–	–	–	–	0	1	0
4	Proceedings	1	4	2	4	2	2	5	0
5	Edited collections	0	2	0	1	3	1	2	10
6	Popular magazines	0	1	1	2	0	0	0	0

This explains how Weber was able to publish HighVGR papers up to 1984.[5] After this, however, things changed. The overall pattern of Weber's publications is represented in table 20.1, where we can see from row 1 that he published nothing on gravitational radiation in the mainstream journals after 1984 (he published 27 "mainstream" articles on gravitational waves up to 1984). We can see from row 6 that he ceased to publish papers in outlets such as *Scientific American* and *Physics Today* after 1974.

Weber did continue to publish in the Italian-edited *Il Nuovo Cimento*. In fact, two very (sociologically) significant papers were published there: the 1989 report with the Italian team on the gravitational wave output of supernova 1987A, and Weber's 1996 claim to have seen correlations between gravitational waves and gamma ray bursts.[6] But *Il Nuovo Cimento* had changed its status during the course of the years examined. In the early period, the journal seemed to be just one of the places where reports and discussions of "run-of-the-mill" gravitational wave results appeared.[7] Nowadays, however, members of the gravitational wave core group do not publish in *Il Nuovo Cimento* and they seem no longer to read it, though it remains, unmistakably, a physics journal.[8] *Il Nuovo Cimento* was one of the places that did continue to publish Giuliano Preparata's papers; the second two of his key publications which I have described above were published in that journal. *Il Nuovo Cimento* aside, the bulk of Weber's publications on gravitational radiation

5. His 1984a publication is in a journal that is not in the very first rank, so it can be seen as a turning point.

6. The publication list on which this table is based was given to me by Professor Weber and ceases at 1994. For that reason, the second key *Il Nuovo Cimento* paper does not appear in the table. For some reason, the 1989 *Il Nuovo Cimento* publication does not appear on Weber's curriculum vitae; I have added it for the purposes of analysis.

7. For example, Bramanti and Maischberger 1972; Bramanti, Maischberger, and Parkinson 1973; Levine 1974; Bertotti and Cavaliere 1972; Billing et al. 1975.

8. It would be difficult to pin down the exact moment when *Il Nuovo Cimento* changed its policy and status, but since there are no Weber publications in that journal between 1972 and 1989, we can safely place the transition anywhere in that period without affecting our conclusions.

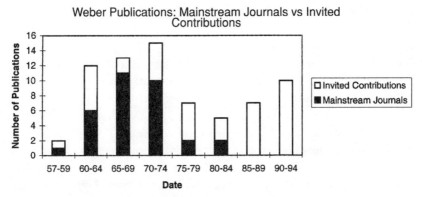

Fig. 20.1 Weber's publication outlets in graphic form.

after 1975 are found in conference proceedings and, increasingly, edited collections.[9] The ten edited collection publications found in the 1990–94 period indicate, to some extent, his movement into publication by editors belonging to the network actively supporting Weber's heterodox ideas—what I have called the Quantum Coherence Heresy Group.[10] This, of course, was a network that was marginalized from the gravitational wave core group.

A clear picture of the trend in the gravitational wave papers emerges if we chart the sum of rows 1 and 2 (main journals) together with the sum of rows 4 and 5 (which I will call invited contributions) (fig. 20.1).

Why did things change in this way? Unfortunately, we cannot get at the detail of the individual refereeing decisions for these papers, but it is probable that eventually the patience of mainstream journal editors was exhausted, and referees who would once have been charitable became more influenced by the continuing overwhelming opposition to the ideas being presented, even though they had been given "a good run for their money." We have better evidence for this kind of change of heart in the treatment of another set of Weber's papers—those addressing the consequences of the new theory for neutrino detection.

9. The dividing line between conference proceedings and edited collections is hard to draw clearly, because some edited collections emerge from conference proceedings and all conference proceedings are edited by someone. Nevertheless, our crude division works well enough for the purpose at hand.

10. Sonnert and Holton (1995) find that those in physics, mathematics, chemistry, and engineering typically publish about 85% of their papers in refereed journals. Up to 1974, Weber published about 67% of his papers in mainstream journals, while after 1974 the percentage drops to about 14%.

Neutrinos under the New Theory

As explained above, according to Weber, a crystal analyzed according to his quantum theory interpretation would have 21 orders of magnitude greater sensitivity to neutrinos than had previously been thought, and if he was right, it could be used in place of the massive and hugely expensive neutrino detectors then thought necessary. One can see that this idea fit very nicely with the motif of the Quantum Coherence Heresy Group—that big science was robbing the taxpayer. According to the group, with a bit of imagination, the same science could be done for a fraction of the cost. For this very reason, the group believed, the neutrino claim would run into heavyweight opposition.[11] Weber himself said to me in 1993 that the big-spending neutrino experimenters "regard me as a threat."

Nevertheless, the theory developed by Weber was not a trivial one and could not be dismissed out of hand. Apparently some very well-known physicists had looked at the early neutrino presentations and could not find any mistake. Even those who did find mistakes admitted that it took them a very long time to root them out. Weber was, then, allowed to stick his neck out in the mainstream journals.

Nine papers concerning the new neutrino cross section ideas are found in Weber's publication list. Journal articles were published in 1984, 1985, and 1988; proceedings contributions were published in 1986 and 1987; an edited-volume contribution was published in 1991, with three more in 1992. Thus the pattern of publication is similar to that of the gravitational wave papers, but the turning point from high-prestige to low-prestige outlets comes a few years later. In respect of the neutrino core group as opposed to the gravitational wave core group, Weber's ideas were new in 1985 and relatively new even 1988, so he had not yet exhausted the tolerance of the mainstream.

The neutrino paper published in 1985 attracted a great deal of criticism, and I conducted telephone interviews with the critics, many of whose comments fit the model presented here. One remark from a neutrino scientist makes the point nicely.

> [1998] The gravity waves—everyone's tired of hearing about it, and there were too many counterarguments. So if somebody publishes something about

---

11. Barry Barish, the Gravitational Wave International Committee chair who knew the events at first hand, assured me that Weber's claims had no chance to be taken seriously given the opposition it was up against.

gravity waves, it's gonna be small. In our case, it was the first we had seen of this; everybody saw it at the same time; they published it. Now—as you notice—I haven't even noticed [the subsequent 1988 neutrino paper].

...If you go back twenty years ago, you saw a lot of negative comments about the gravity wave stuff. And so I think what's happened is "Ho, hum—more of this!" Now the reason that a lot of interest in this neutrino thing came about is that it was quite new. And it was a reported experiment, and I think that's a difference right there.

This critic is explaining the difference in *reaction* to the published papers, a point I will deal with at length; but his comments can also be taken to apply to the willingness of editors to publish neutrino claims but not gravitational wave claims in the mid-1980s.

I also have a comment from one of the referees who recommended publication of the 1988 paper. At the time of the interview, the 1994 and 1995 theoretical papers had been heavily criticized in print—there are no less than seven published rebuttals. The 1988 paper, however, reported an experiment.

[1998] So the reason I recommended, eventually, the publication of that paper was because, although it seemed impossible, he was presenting data, he did have a calibration, and he did see a difference between the data and the calibration, and he describes his apparatus. So you can't not publish it just because you don't believe it can be possible. Because there are other examples like seeing strange particles from Cygnus X3 in the sky and so on, where people have published even though they have not been substantiated subsequently. So what we've got here is somewhat extraordinary....

There were referees who said, "Don't publish that paper," but I said that you'd better publish it, because with my changes he's obeyed all the rules.

But unfortunately, the thing I'd missed is that he subsequently used that—because when he gave a talk and people said they didn't believe it, he then said to them, "Well, it's been published in a refereed journal." So he was using that to advance his case. So I've got a whole file in front of me on this Weber stuff...

I was the culprit in getting that published, simply because I thought in all fairness he claims he's got an effect there, calibrated, and not there with a control bit of lead, and therefore it obeys all the rules of publishing, unless you just say, "Oh, it can't be true"—a very dangerous thing to say.[12]

12. The source of this quotation is a telephone interview I conducted on May 31, 1998.

That was the last neutrino paper to be published in mainstream journals. What is especially interesting is that papers on gravitational waves and those on neutrino detection were subsequently printed in the *same* conference proceedings, suggesting still more strongly that the later publications were made possible by the existence of a specialist network.

Funding "after Death"

Nearly a quarter of a century after the debate about HighVGR might have been thought to have been closed, and right up to the time of his death, Weber maintained two room-temperature gravity wave detectors at the laboratory near the golf course of the University of Maryland; though officially retired, he still had offices in two universities—Maryland and UC Irvine; and at least until 1997, he was still submitting research proposals to the National Science Foundation and other funding agencies. In 1995, I asked Weber how he was supporting this activity.

> *Collins:* Who pays the day-to-day running costs of the Maryland bars?
> *Weber:* They're in a building on campus which is maintained by the Physical Plant Department of the University of Maryland, and the Physical Plant Department pays someone to check the thermal controls—air conditioners, things like this do require maintenance, and the state has been doing this.
>
> It's University of Maryland money that maintains this. Now, it turns out that due to my advanced age, essentially almost my entire income comes from state retirement funds and federal Social Security, so that I don't need any money to operate. Now, because of the fact that I'm receiving state retirement funds and social security, even if I work 100 hours a week, it's illegal for me to receive more than $10,000 per year from the State of Maryland. And when I start teaching again for the honor program, they will probably pay me some sum like that.
>
> Now, from the standpoint of the experimental program that's quite important, because if I were in industry, and suppose I were 50 years old—if I were 50 years old in industry, they would pay a person of my experience about $100,000 pa, and the industrial overhead is about 300%. So it would cost about $400,000 pa for me to keep the experiment running.... Practically, it costs nothing.
>
> *Collins:* What about little things such as a new transducer?
> *Weber:* Well, I've been paying for those things out of my own pocket.

For a few months, Weber had also paid the salary for a postdoc to look for correlations between the gravitational waves and gamma ray bursts. I never went to Weber's house, but I noted that when he drove me from the Maryland campus to the golf-course laboratory, we traveled in a very old car. But this was when Weber was 76 years old; the 1975 "watershed" had happened twenty years earlier. What had happened in the interim to enable him to do the work which had generated all those publications?

As we have seen, in the early post-1975 days the bulk of Weber's gravitational-wave research funding came from the NSF. His NSF funding was cut to about $50,000 per year for a period after the 1975 watershed, but not to zero. The NSF's input increased to $150,000–$200,000 once Weber had agreed to set out a cryogenic bar program. While it lasted, funds for the new cryogenic program provided the infrastructure to maintain his old room-temperature bars.

We saw that Weber's continued announcement of new findings and new theories relating to HighVGR helped to create an uneasy atmosphere within the Maryland team.[13] But his interests in the exotic also made it more and more difficult for the NSF to fund him. Richard Isaacson, program director for gravitational physics at the NSF, told me the story of the final cancellation of Weber's funding.

[1995] *Isaacson:* The last grant that was made to Joe. Let me tell you what the situation was. It ended June 30th, 1989.

Collins: As recently as that!

Isaacson: Well, we funded Joe for a long time. And the last twelve months of that—from April 15th '88 that's basically when I said, "We're pulling the plug, but we're giving you a year to phase out the operation." We continued to fund Joe for a long time, together with his colleagues at Maryland who are still there—who he recruited—Richard and Paik.

After all the scandal and it became clear, years earlier, that there was a controversy and that Joe probably was—was in the wrong: I think we brought in groups who looked it over, and I think the foundation knew that it was highly unlikely he was seeing anything. But he was funded for quite a few years. First in a focused way to try and ensure that he got all of the resources he needed to present to the scientific community his case. You know, to get the data, to organize it, to analyze it, to publish it. So he was being funded,

---

13. There seem to be at least two other reasons for the friction within the Maryland team, however. Ho Jung Paik and Jean-Paul Richard, the two researchers, did not get on with each other, and Paik was spending a lot of time on a completely independent research program.

basically, to put as much information out as he could, for a few years. That's where we thought the problem was—completely in the data analysis—and we wanted him to make his analysis and present it and to have the normal course of science decide on its validity.

But Joe was an incredibly gifted experimentalist, and he was terrific at building instrumentation; he invented the field. So after that phase, we got proposals from him many, many times.

Only a subset of those proposals was funded, but as Isaacson said,

Now clearly if you are seeing weak signals, a signal-to-noise ratio of 1, one way to tell whether they are there or not is to improve the apparatus, so for a while we funded him for a bunch of instrumentation improvements....

It started out where he was working with Richard and Paik, and three of them were going to build a cryogenic system. Joe was going to handle the cryogenics, Paik was going to work on point contact squids as amplifiers, and Richard was going to be building the multimode transducer. So the overall system sounded interesting.

This grant was intended to run for four years from January 1986, but it was eventually terminated one year early, as it became clear that the team as a whole were not pulling together sufficiently well to achieve the milestones that as were appropriate to a project of this kind. As has been explained above, this was partly because of strains within the team, and partly because Weber's heart was still in his old claims.

Richard Isaacson, the man who took ultimate responsibility for the termination of Weber's NSF support, did his doctoral studies at the University of Maryland, with Weber as chairman of his Ph.D. oral examination committee; if life imitates art, gravitational wave research imitates Greek tragedy, with Isaacson as Oedipus!

Pascalianism

Among the many proposals that Weber sent to the NSF, there were some for work on experiments to test his unorthodox neutrino claims. NSF turned these down.[14] NSF may have given up on Weber, but he had other sources. The US scientific research funding system provides many opportunities for unconventional work if you know how to exploit it. For example, at

---

14. I have been made aware of this, and many other things involving the NSF, from sources outside the agency. NSF has always been exact and proper in releasing information to me for public exposure.

Marshall Space Flight Center in Huntsville, Alabama, considerable sums of discretionary money and related resources—in the millions of dollars—have been spent on very high-risk programs, such as antigravity research. The existence of this kind of research, lavishly funded as it is, angers scientists working within the university system. Again, Robert Forward, one of the pioneers in Weber's program whom we have encountered several times already, was for many years employed by Hughes Aircraft to explore completely wild ideas outside the mainstream of science; he, too, invented an (in principle) antigravity device. Industrial firms also have sponsored the continued search for cold fusion, and we will see the role of military funding and other non-NSF establishments below. None of this kind of activity could be contemplated within an NSF framework.

Expenditure on scientific research by a military agency, certain kinds of industrial firms, or a high-spending agency such as NASA (National Aeronautics and Space Administration) fits a quite different model from that of the National Science Foundation. First, the budgets for NSF-sponsored research are usually small compared with the budgets that these other agencies and firms handle; second, the cost of a missed opportunity or, in the case of the military, the danger of missing a development in weapons technology, far outweighs the risk of spending money on a fruitless venture.

Let's call this the "Pascalian" model of funding. The philosopher Blaise Pascal said everyone should believe in a deity because of the odds: very roughly, the cost of believing was small, while the consequences of not believing—a trip to hell—were potentially great. The irony is that if you want to do some very high-risk research—the kind for which success is a significantly less likely outcome than failure—it is better to approach the agencies responsible for applied science, such as the military or NASA, which can afford to have a Pascalian element in their less carefully scrutinized activities, rather than the agencies responsible for pure science, such as the NSF. It was Pascalian funding that now began to support Weber's work.

Weber's new theory not only applied to gravitational waves but neutrino detection, as I have explained. The work growing out of this aspect of the new theory was funded by the US Navy, so I was told, but the acknowledgments in Weber's papers are to other defense organizations.[15] Neutrino

---

15. The acknowledgment given in Weber 1984a and 1985 is to the Air Force, contract F-49620-81C-0024 and the National Science Foundation. Weber 1988 acknowledges the NSF; DARPA, the Defense Nuclear Agency and Strategic Defense Initiative; and the "Office of Innovative Science and Technology, managed by the Harry Diamond Laboratories," but contract numbers are not given. It may well be that Weber tended to acknowledge all agencies that were giving him any kind of background support.

detection is normally a big science involving very large detectors, whereas Weber believed his new theory showed that a crystal small enough to be held in the hand would be at least as sensitive as the big devices. An easy way to detect neutrinos would make it possible to spot the nuclear reactors of hidden submarines. Defense agencies, one might speculate, could not afford to take even a slight risk that Weber was right; it would be just as important to know whether the enemy had the potential to spot your submarines as to be able to spot theirs, and even if the experiments failed, this would be an important piece of military information.

Later the Navy funded a local group at Johns Hopkins University in the Baltimore area to check Weber's results.[16] They disagreed with Weber, and his funding soon ceased.

Weber also managed to obtain funds from two other non-NSF sources. The prospect of being able to detect neutrinos with ease appealed to the US Defense Nuclear Agency, part of whose job is to monitor the nuclear programs of foreign states. According to the DNA, Weber approached it in the first instance, explaining his ideas; but in line with the standard procedure, it advertised in the *Commerce Business Daily*, inviting bids for work on "Research and Development of New Methods for Detection of Antineutrinos from Nuclear Reactors."[17] In its call for bids, the agency promised to send out a supplemental package to applicants which described Weber's work: "Applicable documents to be included in the supplemental package are: 1) Final report from the Defense Nuclear Agency (DNA) contract DNA001-77-C-0223 entitled 'Experimental Report on Scattering,' by Dr. Joseph Weber, University of Maryland, 31 Dec 83; and 2) Technical paper entitled 'Chopper for Neutrinos and Antineutrinos' by Dr. Joseph Weber, University of Maryland, 31 Oct 89" (pp. 1–2). It was clear, then, that applicants would be expected to follow Weber's lead. So here was a government agency taking Weber very seriously indeed, even as he was being rejected by the gravitational wave community. Weber was delighted by this, and one may well imagine that it gave him a great boost. As it turned out, he was the only applicant for these funds. His application was reviewed by one outside referee,

16. The Baltimore group regularly did work for the US Navy; their decisive paper is Franson and Jacobs 1992. The work was supported by US Navy contract N00039-89-C-0001. I would like to know much more about the informal interchanges that were going on about this time; I have no evidence to this effect, but it could be that the published papers tell only a small part of the story.

17. The reference number given is SOL DNA001-95-R-0043 POC, and the contract negotiator is given as Sandy Bednoski with contracting officer as Edward Archer. The "solicitation was synopsized in the CBD on 22 May 1995." The deadline for submissions is given as July 7, 1995. Weber told me that this notice came to him "out of the blue," but I suspect he misremembered.

who considered the *Physical Review* papers, on the face of it, were hard to fault.

The grant from the DNA enabled Weber to buy some apparatus, such as improved signal recording devices, but it was cut short long before the initially agreed end point. According to DNA spokespersons, this was either because changes of personnel implied changes of priority, or because Weber was originally funded from some end-of-budget unspent funds which ran out.

In the case of both the Navy and the DNA, much larger sums were initially promised to Weber than were eventually delivered. It is possible that he was at first able to obtain funds from non-NSF sources relatively easily because the reviewers were not closely tied to gravitational wave core-group-related networks. After a time, however, it might have been that news of Weber's new sources of funding got around, and powerful scientists drew the history of his work to the attention of the relevant people. I cannot prove this; I can only show that the US physics community is sufficiently large and heterogeneous as to leave space for one physicist not to know what another is doing. I can show this in an interview I did with a scientist who had published a paper critical of Weber's neutrino claims.

> [1998] *Respondent:* We'd published our paper; we'd read Harry's [critical] paper, and OK, everything is fine: "forget it." Now all of a sudden I get a second proposal from the NSF—the second proposal comes directly from the NSF—and this is for solar neutrinos, using the words right out of our paper, and he's going to sell this now for solar neutrinos. [Y] was here... and we said, "How can he do this?" He refers to this paper, and he also refers to our long paper, and so we started looking through it, and we found out it was just a sophomoric error. After that I just lost any interest in it.
>
> *Collins:* Did you know he eventually got a grant from the US Navy and carried on doing research on this?
>
> *Respondent:* Jesus!
>
> *Collins:* He published a paper in 1988 giving an experimental result in *Phys Rev D.*
>
> *Respondent:* You're kidding! You got a reference on that? You see, that's a long time after I was up in it. I left ONR [Office of Naval Research]. It was about '87 or '88 when I had an opportunity to shoot this down. [Collins provides reference to paper]
>
> *Respondent:* Good God!

This interview gives a sense of the heterogeneity even within the single field of neutrino detection. It shows how it can be that the details of a reputation gained in one area may easily not be known in another area. And this also means that panels of referees can fail to overlap, and that where a discretionary element is involved in funding, it is easy for money to be made available for one who has set him- or herself "beyond the pale" in another areas of physics, or even in the same area of physics. We can see, perhaps, why from Richard Garwin's point of view the place to publish a refutation of Weber was *Physics Today*, rather than the more narrowly professional journals which would be read by specialists alone. Scientific specialisms, as I said at the outset, are deep and narrow, like crevasses. The outer rings of the target diagram are the bridges between these crevasses; to stretch the analogy, the target has one set of outer rings but many bull's-eyes. Or, to put it the other way around, every specialism (bull's-eye) is set with the same set of outer rings. This gives a role to the generalist, quasi-popular, journal—transferring reputations from one narrow crevasse to another. In this case it was only partly effective, as Weber did manage to get a hearing in the neutrino field, at least for a period. Of course, where Pascalian funding is concerned, even a reputation for outrageousness may not be fatal.

By 1987, Weber's claims had been damned by another very powerful group fronted by Freeman Dyson, Weber's old ally from the Princeton Institute for Advanced Studies days. Dyson led a summer study on the problem for JASON, a powerful group of freelance science advisors originally founded as a division of the Institute for Defense Analysis, and one of whose members was Dick Garwin.[18] In 1987, JASON produced a report criticizing Weber's neutrino work.[19] Both the heterogeneity and the Pascalian element in defense funding are nicely revealed in quotations from the first pages of the report.

> JASON became involved with the problem of neutrino detection as a result of a proposal submitted to DARPA by the Raytheon Corporation in 1984. We continued to be involved in 1985 as a result of a proposal to OPNAV-095 by Professor Joseph Weber of the University of Maryland. We reviewed both proposals for their respective sponsors and advised against their funding. As a response to these contentious proceedings, DARPA asked us to write a general assessment of the state of the art of neutrino detection, to explain in

18. JASON is not an acronym, though I've been told that wags sometimes say it stands for "July, August, September, October, November," since the group meets in July but often does not finish its reports until November. An alternative is "Junior Achievers Somewhat Older Now." Garwin appears not have been involved in JASON's neutrino study.

19. Callan, Dyson, and Treiman 1987.

general terms why the claims of Raytheon and Joseph Weber could not be correct. The present Primer is intended to provide such an assessment.

...Our reply to the Weber proposal is contained in document JSR-85-210, submitted to OPNAV-095 in July 1985. Our judgement was that both proposals were flawed by gross errors in theoretical analysis. (p. 3-1)...

The cross-section implied by his [Weber's] tritium results were about 10^{20} times larger than the cross-section predicted by orthodox physical theory. If such large cross-section were real, it would be an easy matter to detect neutrinos emitted by submarine reactors at distances of hundreds of kilometers.

Weber's claims naturally caused concern among responsible officials in the Navy. The officials, quite rightly, took these claims seriously. They saw one distinguished professor of physics with a published paper making these claims, and a number of other distinguished professors of physics saying in private that the claims were nonsense. How could the Navy tell who was right? If it should happen that Weber was right, it would be a matter of life and death for the submarines. So JASON was asked to study the question thoroughly and dispassionately. It was not enough to state our opinion that Weber's results were incredible. One Navy official said to us: "Didn't Lord Rutherford say that the idea of a practical use for nuclear energy was moonshine? And are you JASON professors smarter than Rutherford?" To justify our belief that Weber was wrong, we had to go back to fundamentals and work through the theory of the interaction of neutrinos with crystals from the beginning. We had to establish firm mathematical upper limits to the possible magnitude of neutrino cross-sections. (p. 3-3)

In the case of the neutrinos and the NASA-Goddard funding, the initial grants were made through more or less informal mechanisms. It is tempting to suggest that the agency's subsequent drawing back from commitment also resulted from relatively informal mechanisms as the news circulated and pressures on the agencies developed.

As we have seen, the other source of funds that Weber was able to use after his NSF grant ran out was NASA's Goddard Space Flight Center. But it was not NASA-Goddard's peer review committee that funded the work. Weber should have had little chance of finding the correlations, given that Hamilton and Pizzella's apparently much more sensitive devices, which Goddard had also contracted to do a search for gamma ray correlations, had failed. Consequently, NASA-Goddard's $10,000 grant to Weber came out of discretionary funds. NASA-Goddard is located in the same sector of the Washington, DC, suburbs as the University of Maryland, which is essentially its local university, and there seems to have been a strong personal network between it and the Maryland physics department.

It was this $10,000 that enabled Weber to produce the data that led to his 1996 publication on the coincidences between gravitational waves and gamma ray bursts. We see that there was further support for analysis to have been provided by NASA-Goddard scientists, but what seems to have happened is that Weber preferred to do that analysis himself; by 1996, contacts between him and NASA-Goddard seemed to have become thin. Thus, when I conducted interviews in 1998, my respondents at NASA-Goddard said that though they had heard about it, they had not seen the 1996 paper that Weber had published concerning the correlations, the search for which they had themselves funded. It was I who had to tell them that Weber was the sole author.

Networks of Colleagues "after Death"

The theme of tolerance up to a point applies equally to the continuing existence of networks of colleagues after the main scientific work of a scientist is widely perceived to have been discredited. Even before he began his work in gravitational radiation, Weber was a well-respected scientist, noted for his contributions to the invention of the maser and laser. As a result of his early gravitational wave work, for a period between about 1968 and 1972 he was one of the most famous scientists in the world, with many believing his results were about to cause a revolution in our understanding of astrophysics as well as provide spectacular confirmation of the general theory of relativity. In these early days he was, according to the accounts of many respondents, a brilliant original thinker and a witty and amusing expositor of ideas. As the Australian gravitational wave physicist David Blair says of a seminar given by Weber in Louisiana in 1974—"Never before and never since has mathematics on the blackboard brought tears of laughter to my eyes."[20] Others who visited him in person were to become longstanding allies. For example, the leader of the Italian experimental group that was to publish with Weber as late as 1989, Guido Pizzella, had a long visit at the Maryland laboratory at the outset of his program in the early 1970s and learned to respect Weber's abilities and judgment. In the early days, Weber won debates with staunch critics about the bandwidth of bars and the sensitivity of amplifiers, and recollection of these victories reinforces Weber's colleagues' faith in him. These early alliances seem to have lasted,

20. Blair 1991, p. 19.

buttressed by the common experience of marginalization. And almost everyone agreed that whatever Weber was or was not, he was a brilliant inventor of experimental schemes. Thus two respondents said respectively, [1995] "Joe was an incredibly gifted experimentalist, and he was terrific at building instrumentation; he invented the field" and [1999] "That was the joyful thing about him—he was somebody from outside—he didn't think like us."

This meant that for a long time, Weber's academic colleagues would stand by him even if they no longer believed his results. A poignant exchange of letters between Weber and Freeman Dyson gives a sense of these relationships. It was Dyson who had arranged for the electrical engineer with degrees from relatively obscure colleges to spend time at the Princeton Institute for Advanced Studies—perhaps the most prestigious research institute in the world—because of the joy of Weber's fresh insights. But as time went on, Dyson, who was initially on Weber's side in refusing to accept the sheer theoretical impossibility of his claims, began trying to persuade Weber to think again while not withdrawing his friendship.

> Dear Joe,
>
> I have been watching with fear and anguish the ruin of our hopes. I feel a considerable personal responsibility for having advised you in the past to "stick your neck out." Now I still consider you a great man unkindly treated by fate, and I am anxious to save whatever can be saved. So I offer you my advice again for what it is worth.
>
> A great man is not afraid to admit publicly that he has made a mistake and has changed his mind. I know you are a man of integrity. You are strong enough to admit that you are wrong. If you do this, your enemies will rejoice but your friends will rejoice even more. You will save yourself as a scientist, and you will find that those whose respect is worth having will respect you for it.
>
> I write now briefly because long explanations will not make the message clearer. Whatever you decide, I will not turn my back on you.
>
> With all good wishes,
>
> Yours ever
>
> Freeman[21]

Weber's response was to send more and more details of his empirical findings.

21. Freeman Dyson to Weber, June 5, 1974 (letter from Dyson's private archives).

Friendship is one kind of support, but to build a science one needs colleagues who will endorse and use one's findings so as to build a community of fellow believers and workers; this is what making a science means. In this, Weber was fortunate in having practical support (as well as personal support) from the Italian group based in Frascati. The Italians were the best resourced of all those working on cryogenic bars, and in interviews Weber was always quick to stress that the Frascati group confirmed all his early findings and that he had letters to prove it. This kind of endorsement, along with the coauthored paper of 1982 which claimed to see fluxes consistent with HighVGR, could be thought of as the start of an elemental new community.

Still more important than these early confirmations was Pizzella's collaboration in the high-profile SN1987A claims, which were based on data from room-temperature bars. Weber and the Rome group claimed to have jointly seen signals consistent with gravitational waves coming from the visible supernova of 1987. Pizzella weakened his own position substantially in respect of the core group in doing this work and pressing the claims forward, but it gave salience, or notoriety, to HighVGR and room-temperature technology more than a decade after closure. The eventual outcome was further marginalization, but to have such a staunch ally continued to help Weber avoid complete isolation from the rest of the scientific community.

The Pizzella group, though they published two positive empirical papers along with Weber, provided neither support nor opposition regarding Weber's theoretical claim that bars had a cross section many orders of magnitude greater than orthodox theory suggested. Indeed, as Weber developed the second strand of his theory, which stressed the importance of "center-crystal instrumentation," the Italian experiments could no longer support him, as he was effectively attacking the sensitivity of his friends' detectors. Certainly, they were not prepared to accept the marginalization that would be involved in even asking for the resources to redesign and rebuild their already recalcitrant instruments.

As regards the theory, the crucial actor was Giuliano Preparata; he connected Weber to the Italian theoretical physics community. As Weber told me in 1993, it was Preparata's influence that was "the origin of my being invited to the Italian Winter School [at Erice] every year for the last several years." Later, Preparata forged a link between Weber and his own group of physicist mavericks, who were to put together many of the collected volumes in which Weber would publish his last papers; and they came close to developing a flourishing new network for him—what I have called the Quantum Coherence Heresy.[22]

Graduate Students and the Next Generation

Planck's dictum states, very roughly, that ideas that lose credibility die only as their upholders die: "Science progresses funeral by funeral." For a rejected science to outlast Planck's gloomy prediction, graduate students must get jobs in universities so that they can carry the radical ideas on to the next generation. In the case of Weber himself, there are no loyal graduate students taking up the work; they all went on to do other things. Even in the case of the wider network, we can be fairly sure that there will be no graduate students. Even if others were to take up ideas associated with the new theories about cross sections, they will not be in a position to do empirical work on HighVGR. Unless something happens that turns the whole scientific community around, such as the discovery of huge fluxes of gravitational waves by the new generation of detectors (meaning that HighVGR is no longer rejected), HighVGR will not survive Weber. We can say this with such certainty because the whole professional employment market now favors the new "big science" of interferometric detection, and there will be no jobs for anyone supporting the search for HighVGR. Indeed, even the *noninterferometric* search for *Low*VGR is struggling to survive.

When it comes to survival into new generations, we see the institutions that support an initial plurality break down, because new entrants to a field, having the least power of all the actors within it, are unlikely to be able to establish themselves on the basis of rejected heterodoxy. For a rejected science to survive across the generations, special institutions are needed (as in the case of parapsychology for example). The graduate students have to "follow the money," and this does not lead in the direction of HighVGR.

This is not to say that the theoretical "spin-off" from HighVGR—the ideas behind the cross section as developed by the Preparata network—may not survive into the long term. Theoretical heterodoxies are far less demanding of resources, and far less obtrusive than experimental heterodoxies; they can be hidden behind mainstream work and have no need to reach beyond a narrow set of believers.[23]

22. I understand that they, too, would meet at Erice.

23. And indeed there is evidence that new theories of the cross section continue to arise. Two papers by Brautti and Picca published in 2002 (a and b) develop a corrected version of a

HighVGR as a Rejected Science

HighVGR, as we have seen, was slowly forced to reach ever further toward less prestigious institutions and networks as it struggled to survive. In chapter 12, using the *Science Citation Index*, we followed Weber's trajectory as a scientist by looking at the way his work was referred to by other scientists. We can use what I'll call "cell diagrams" (fig. 20.2) to gain a more impressionistic picture of what happened to Weber and HighVGR, as well as the whole field of gravitational wave detection. The cell diagrams are cross sections of the central core of the imaginary time cylinder (see the introduction).

In the early 1970s, we have a standard core set with opposition between Joe Weber and the other room-temperature bar builders. Weber was the dominant member of the field at this point. By the early 1980s, he was being squeezed out and had only the Frascati group supporting his experimental work. (In the diagram I use "Pizzella" to represent the whole Frascati contribution.) The big effort was now in cryogenics; for example, Weber aside, in America only Tyson was still discussing room-temperature results, and he was simply writing up the completion of his program somewhat ritualistically. Interferometers were just making an entry into the field as a whole, becoming visible in the work of Rainer Weiss and Ron Drever but having no significant results.

By the early 1990s, the Caltech 40-meter prototype built by Drever had revealed the potential of large interferometers, but these results were still well behind what was expected of the cryogenic bars, which were at last becoming reliable. Crucially, however, LIGO and the other big interferometers, which had obtained their congressional funding, were being supported to a far greater extent than any other technology. In the early 1990s, then, we

---

Weber/Preparata coherence-type theory referring to room-temperature bars. They argue that room-temperature bars are about five orders of magnitude more sensitive than conventional theories would suggest, and they use this to defend the supernova 1987A claims. (Brautti and Picca list their institutional affiliation as Istituto Nazionale di Fisica Nucleare and Physics Department, Bari, Italy.) In 2003, a paper with a similar conclusion for all aluminum bars claimed an enhanced cross section of about 7×10^3 (Srivastava, Widom, and Pizzella, 2003; the authors are from Northeastern University and University of Perugia and, of course, INFN for Pizzella). This paper cites the enhanced cross section to make sense of both the 1987A results and the 2002 *Classical and Quantum Gravity* results. Alan Widom's name appeared on various pieces of documentation associated with the Quantum Coherence Heresy. I believe yet another such paper has been submitted for publication. These papers are largely invisible in the core group. To date, no one has mentioned these papers in public. I stumbled across the Brautti and Picca papers during a search for citations to Preparata. I learned of the existence of the Pizzella et al. paper from Warren Johnson during a meeting of the LIGO Scientific Collaboration in March 2003; he mentioned it in passing with a metaphorical shrug. The third paper I learned of through a private communication.

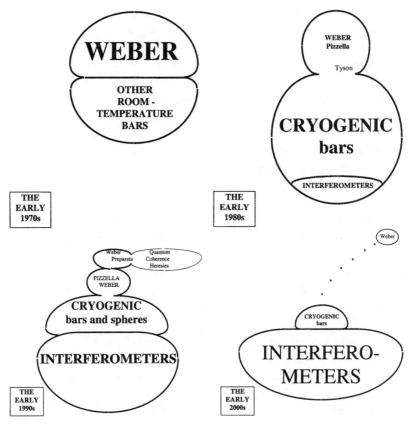

Fig. 20.2 Cell diagrams showing how alliances changed.

had the first intimation of the dominance of interferometry to come. Nevertheless, the resonant-mass program still seemed to have a bright future as a minor player (in funding terms) in the form of resonant spheres. The funding of LIGO could still be seen as promising a net increase in activity to the field as a whole, rather than as a threat to other approaches within it. Weber, with his theoretical ally Preparata, had marginalized themselves further with the new theory of sensitivity of bars and, as we will see, with an attack on LIGO which had been roundly and visibly defeated. No one in the gravitational wave community knew that Weber, through his collaboration with Preparata, risked more isolation by flirting with "Quantum Coherence Heretics" from maverick fields remote from gravitational radiation. The Frascati group linked Weber to the rest of the community but had, to some extent, marginalized themselves in the process—especially with the supernova 1987A claims. Frascati was, however, a major player still in the cryogenic bar effort with EXPLORER and NAUTILUS; if the diagram allowed, Pizzella's name would appear boldly in the cryogenic "cell."

By the early 2000s, interferometry was dominant. The last of Weber's claimed new findings—the correlations between gravitational waves and gamma ray bursts—had been published in 1996 and, as we will see, entirely ignored. Thus by the turn of the century, Weber was floating alone, no longer connected to the core. He had even isolated himself theoretically from his Italian supporters with his center-crystal instrumentation claims. As a measure of Weber's isolation, I found to my astonishment that in the late 1990s it was I who had to tell Pizzella and Coccia about Weber's 1996 paper; they did not know about it.[24]

By this time, Pizzella and certain of his colleagues had been sidelined as a result of supernova 1987A and some other positive findings they were to advance and which I will discuss below. The cryogenic sphere program had been more or less killed, and the remaining cryogenic bar program had organized itself into an international cooperative that had adopted interpretative conventions, for its data almost guaranteed to ensure that it would cause no trouble for the interferometers. To be unflattering, I might say that this group had become a client of interferometry, one of its main roles being to suppress maverick claims from Pizzella and the like.

And everyone was getting older. Weber had died in December 2000, setting the scene for his possible rehabilitation as a respected figurehead, the outcome of which we await; Preparata also died in 2000 at the young age of 57, with much less chance of a revival of his reputation. Pizzella had withdrawn from the front line and passed spokesmanship for his group to Eugenio Coccia, who was more likely to be cautious. David Blair, who had been one of Pizzella's allies in one of the attempts to get low-energy cryogenic bar coincidences taken seriously (see chapter 22), was now putting most of his efforts into trying to build a large interferometer in Australia. In sum, there was no one left within the field who had both the desire and the energy to challenge the dominance of interferometry and its interpretations of the heavens.

COMPARISON WITH OTHER FIELDS

Another way of understanding what is represented in the cell diagrams is to compare HighVGR with other rejected sciences and the way they adapt

24. As a respondent pointed out to me, this also implied that Weber had isolated himself by not sending out preprints or reprints.

to the difficulties they encounter. Using the same four categories as the organizing framework—publication, funding, networks of colleagues, and graduate students—we will look at another half-dozen rejected sciences.

Scientific parapsychology is a small-scale activity practiced in perhaps a dozen laboratories throughout the world, most of them in university departments. For example, Edinburgh University has a professor of parapsychology and a group of two or three researchers; Princeton University's engineering department includes a professor whose interests have turned in the direction of the paranormal; Cornell University's psychology department includes an active parapsychologist; and so forth. Parapsychologists in these kinds of institutions tend to run long-lasting, painstaking experiments looking, for example, at the ability of subjects to guess the identity of hidden symbols with a very slight above-chance success rate. Any positive outcome tends to emerge after long statistical analysis, and there is nothing spectacular about this science.

Linus Pauling, the Nobel Prize–winning chemist, became notorious in later life by recommending huge doses of vitamin C as a dietary supplement capable of fending off the common cold and cancer. His theoretical arguments and experimental program in favor of this idea effectively founded a small science that was violently attacked by mainstream medical scientists. James McConnell believed that memory was encoded in chemicals in the brain which could be transferred between organisms. He first claimed that worms trained to run mazes transferred their training to other worms when ground up and fed to them. Later, the same claim was made regarding rats' brains. Cold fusion, another heterodox claim, has been described in chapter 19. The AIDS causation heresy is a theoretical dispute centered on Peter Duesberg, who claimed, and still claims, that AIDS is not caused by the HIV virus. What we refer to as "star crushing" is a theoretical controversy from within the field of gravitational radiation research: two scientists claimed that their mathematical models of inspiraling binary neutron stars showed that before they merged, their mutually interacting gravitational fields would cause them to be crushed so much that they would become black holes, making it much more difficult to model the gravitational wave signature of such a system. This claim, too, gave rise to a controversy.[25]

---

25. For sociological analyses of these controversies, see the following. For scientific parapsychology, Collins and Pinch 1979, 1982; for the vitamin C controversy, Richards 1991; for memory transfer, Travis 1980, 1981 and Collins and Pinch 1993/1998; for cold fusion, Simon 1999 and Collins and Pinch 1993/1998; for the Duesberg heresy, Epstein 1996; and for star crushing, Kennefick 2000. Simon considers at length the definition of a rejected science; he argues that the borderline between pre- and postrejection sciences is not sharp in time or social space. Thus he prefers to describe

We are interested in the different ways the central scientists in these fields adapted to their rejection by the core group. Some of the differences between them stem from cost and technological factors, which can confound the sociological analysis. Thus of the seven cases in all, including HighVGR, only the latter needs sizable laboratories; the others can bootleg apparatus and laboratory space, or use it after regular working hours.[26] The theoretical debates need only papers, word processors, and, perhaps, the supercomputers that are already available in the working environment. Furthermore, some of these sciences can make use of the help of amateurs, while others cannot. This is a consequence of the complexity and size of the apparatus or the extent to which the theory is esoteric. That said, let us compare adaptations to rejection.

Publication

We have seen the way that HighVGR claims continued to be published after HighVGR was no longer acceptable in the core group. At first Weber's ideas were tolerated within the mainstream journals, but he later had to publish in journals less heavily committed to the mainstream consensus—or so the evidence suggests. Finally, he moved toward publication in conference proceedings and edited collections. The same sequence applies to his HighVGR-related claims about neutrinos except that everything happens a decade later because a new community is involved.

The spectrum of publication outlets, then, makes it possible for there to be more than one type of physics.[27] Nevertheless, in spite of the shift in style of outlets, all Weber's publications, right until the end, were recognizably within the realm of professional physics. The same applies in

---

the current state of cold-fusion research, his chosen case study, as "undead science." Simon's analysis is important for pointing out that rejected sciences are not just streams running parallel to the main body of scientific work; their postrejection life is indelibly marked by the terrible experience of exclusion from the mainstream of scientific society. But in pointing this out he shows that it is not hard to recognize a rejected science in practice in spite of theoretical difficulties of definition; his own work shows that rejected sciences are distinct and recognizable kinds of activity (at least insofar as any social phenomena are distinct and recognizable).

26. Simon 1999.

27. The plurality in the publications of a high-prestige "hard science" such as physics is not often noted. Baldi (1998) claims to have shown that the results of the microscopic studies are wrong in this respect. Perhaps studies that adopt a "representative-survey" approach reflect practices typical of orderly scientific progress rather than the small pockets of science where passions are high. I would argue that though they are statistically unrepresentative, the small-scale studies are more epistemically significant: they show the contribution that science can make to difficult decision making where the stakes are important.

the star-crushing debate: all the heretical papers were published in main-stream journals, though the delay between submission and publication grew longer and longer as answers to an increasingly powerful body of critics were demanded.

The majority of the work in scientific parapsychology, by contrast, is published in its own specialist journals. These follow the form of the mainstream scientific journals—highly statistical with little or no appeal to the nonspecialist—but they have very little overlap with mainstream specialisms in content, readership, or authorship. On the rare occasions when parapsychology papers are accepted by mainstream journals, they are generally given conspicuously special treatment to show that "something unusual is happening."[28] Duesberg's heretical account of the causation of AIDS was published in a mainstream journal, *Cancer Research*, and then in the *Proceedings of the National Academy of Sciences*, making use of the special publication rights granted academy members. His claim that the HIV virus did not cause AIDS would not have survived regular refereeing procedures in the mainstream journals beyond its initial announcements, and it was published in spite of the negative comments of the *Proceedings'* referees; the wide coverage the claim was receiving in the popular press made it more difficult for editors to reject the publication, thereby blurring the boundaries of core group and general public. Subsequent submissions to the *Proceedings* were nevertheless rejected, and Duesberg has been forced to publish in less well-recognized journals and to write for a mass audience.[29] There is also an amateurish newsletter associated with his views.

Researchers of high-dose vitamin C fought hard to gain a few early pub-lications in science journals but could not get published in the medical journals to which their field's founders aspired. After these initial publi-cations, Pauling and his colleagues found themselves forced to publish in holistic and fringe medicine journals rather than the mainstream.

The controversial "memory transfer" field set out by publishing some results in a specially created "jokey" journal known as the *Worm Runner's Digest*. Cold fusion, after initial publications in mainstream journals, is now happy to support an amateurish newsletter which includes contributions from nonprofessionals. Thus we see a range of possible publication policies for a rejected science, and we see that on this scale HighVGR was always conservative; Weber never published a newsletter or tried to publish outside recognizable scientific journals.

28. Collins and Pinch 1979.
29. Steve Epstein, private communication, 1999.

Material Resources

Turning from publication to funding, we can usefully describe HighVGR's post-1975 funding as passing through four stages. Stage 1 turned on the tolerance of the NSF for a pioneer and its desire to see his results reconciled with orthodoxy; this is like the first stage of tolerance under the Publication heading. Stage 2 turned on Weber's accepting the new technology and carrying on his HighVGR work on the back of funds for LowVGR work—the mechanism that has, in effect, funded Italian and Australian experimental findings which have given support to Weber. Stage 3 was the Pascalian stage combined with relatively unscrutinized discretionary funding. Stage 4, coming after even Pascalian funds ran out, turned on infrastructure support from Weber's university. He was lucky in that once room-temperature bars are "on air" and working satisfactorily, they are not expensive to maintain. The normal maintenance facilities of his own university department were sufficient to keep Weber's bars going since the mid-1990s. Thus, on the basis of this sample of one, it seems that the modern American university still has enough resources to provide space for scientists with reputation and determination, even if they are now widely seen as having departed from the correct path.

Again, these sources are all of a relatively conservative nature—the entire spectrum draws on public tax revenues from agencies whose main business is the support of orthodox science and technology. HighVGR flirted with more-"big-P," politically motivated funding sources associated with the Quantum Coherence Heresy, but in the end, by accident or design, it did not take them up. HighVGR also remained firmly located in university laboratories.

The "star crushers" needed little in the way of resources, because this was primarily a theoretical debate. They did need computing time, but they had the necessary access through their ordinary work. Duesberg had his grant applications to the National Institutes of Health turned down after he became a dissenter. I am unaware of subsequent sources of funds, though he has the resources of a major university to call on.

Postrejection cold fusion research also has been funded from Pascalian sources, but in this case, private firms such as Toyota, Canon, and Pirelli have been the donors. This funding took cold fusion researchers out of universities and into private laboratories. Pascalian funds supplied by the military also have been used to support parapsychology, notably the notorious "remote viewing" experiments at Stanford Research Institute. But parapsychology has stepped further from the center than either High-VGR or cold fusion in its wide use of private bequests (for example, the Edinburgh chair is funded by a bequest from the well-known novelist Arthur Koestler, author of *Darkness at Noon*). Nevertheless, though it has not always succeeded, parapsychology has tried to spend the money, from whatever source, in research institutes associated with universities.

So far as I know, memory transfer, not a terribly expensive science, obtained all its funding from regular funding institutions. High-dose vitamin C research, on the other hand, exhibits a mixed pattern. Pauling founded the Linus Pauling Institute (LPI), which obtained some money from the National Cancer Institute for relatively uncontroversial work, but the controversial work has been funded by fringe medical institutions such as would never fund orthodox medical science. Since Pauling's death, however, the LPI has relocated at Oregon State University (Pauling's original university) and now has an endowed chair and a staff of researchers and students supported by gifts from individuals, private corporations and foundations, and by Oregon State University.[30] Thus, in terms of funding and institutional location, high-dose vitamin C is similar to parapsychology.

Networks of Support

The pattern of publication and of funding for HighVGR could be said to be repeated where its networks of colleagues are concerned. There was no sharp change after the 1975 watershed, but support and opportunities slowly drop away, and Weber turned to less central institutions and groups as time marched on. Nevertheless, support came from within the physics community aside from a brief flirtation with more radical sciences, such as cold fusion. There have never been any amateurs or popular groups associated with HighVGR.

The same applies to the esoteric star-crushing controversy. Cold fusion, memory transfer, the Duesberg heresy, and parapsychology have all entrained an enthusiastic following of amateurs while being more or less successful in retaining networks within their parent sciences. High-dose vitamin C draws colleagues from within its parent sciences and from the fringe sciences.

30. Evelleen Richards, private communication, 1999.

The Next Generation

When we come to graduate students we find that, unlike the cases of publication, funding, and the support of colleagues, there was no gradual demise for HighVGR. Weber simply had no loyal body of academically successful graduate students. Thus, it looks as though experimental work on room-temperature bars has died with Weber.

Of those I have discussed, the two rejected sciences that are likely to hang on within the institutions of science beyond the initial generation are parapsychology, which has already proved its longevity, and high-dose vitamin C. Parapsychology has set up an alternative employment market along with its alternative journals. The parallel employment market is funded by private bequests of a size that it is attractive to universities, and thus parapsychology has managed to maintain a foothold in the mainstream through a smattering of graduate studentships and endowed faculty positions. High-dose vitamin C also has a university base, adequate sources of nonpublic funding, and students. Apparently, however, graduate students are avoiding the Duesberg heresy and the star-crushing debate.[31]

DISCUSSION

The strategies of the various rejected sciences have come to be the way they are by a variety of routes. Certain choices have been cut off because of technological constraints, as discussed above. Some choices, such as sources of funds and publication outlets, have been forced on the scientists. Others, such as McConnell's jokey newsletter and cold fusion's amateur publications and amateur networks, seem to have been deliberately chosen to "cock a snook" at orthodox science.

An important caveat in respect of this framework is that the analyst must recognize a science as rejected before drawing any conclusions about the adaptation style of the science. Amateur helpers, newsletters, new journals, industrial funding, private bequests, and so forth mean one thing in combination with a mainstream science and something else in a rejected science. Thus, many successful sciences have established themselves by founding their own journals; others run newsletters and less professional publications. Many sciences draw on private bequests and funding from the military and industry; for example, Bell Laboratories and IBM

31. Epstein and Kennefick, private communications, 1999.

funded early gravitational-wave detectors, which produced the *negative* results that supported the mainstream attack on Weber. Sciences such as paleontology, botany, and astronomy benefit enormously from the help of amateurs.

If a science is strong, it can embrace the help of nonscientists without risk to its status as a science. A rejected science affords such luxuries at its peril: the more it associates with nonscientific institutions, the more it defines itself as a nonscience. Hence, in these cases the rejected scientists walk a tightrope between staying near the center and losing the resources to do experimental work, and aligning themselves with nonscientific institutions and encouraging others to define them out of the category of science.[32] Perhaps the most interesting case of the rejected sciences in this regard is parapsychology, which has embraced many nonscientific elements in its quest for survival but has managed to retain tenuous university affiliations for many years; high-dose vitamin C appears to be following suit.

HighVGR, we can now see, was always conservative, even though the bulk of our analysis has shown it straying further and further from the mainstream; the furthest it strayed was to join alternative networks within physics. Joe Weber wanted only one thing: for his findings to be reaccepted by the mainstream community. Of course, in respect of this community, he had begun as an outsider; after the heady middle years, that is not how he would have wanted to end.

---

32. Crane (1976) compares the "reward systems" of cultural enterprises, and concludes that a danger is loss of control over the reward system. The danger, however, varies with the strength of the science. Perhaps the most bizarre case of this phenomenon is one I have mentioned earlier: the recruitment of stage magicians into the very heart of science's cognitive recognition system; stage magicians have been applauded as they rule on whether paranormal phenomena are possible, or even whether water has the kind of polymeric structure that can give rise to a "memory" useful in homeopathic medicine. Sectors of the mainstream seem to feel so secure that they are prepared to hand over elements of their defining professional responsibility to outsiders while most of the rest of the scientific community applauds.

CHAPTER 21

ROOM-TEMPERATURE BARS AND THE POLICY REGRESS

In chapter 19 we looked at the way HighVGR marginalized itself from the core group of scientists with its increasingly heterodox claims; it moved from finding fluxes of gravitational radiation too high to be compatible with astrophysical theory, to developing a new theory of bar cross sections that flew in the face of consensus physics, to the extraordinary claim that old-fashioned center-crystal instrumentation, now used by Weber alone, was more sensitive than the more modern kinds of transducers used by every other group. In chapter 20 we looked at the way HighVGR reached out to new institutions and networks as it became marginalized, and we compared its adaptations postrejection to those of other rejected sciences. But all this was happening within a context, the growing power of the big science of interferometry.

The growth of interferometry marginalized and weakened not only Weber's heresy but the whole bar program, affecting even those cryogenic bars which never associated themselves with Weber's work in any way. The cuckoo of interferometry was to thrust all the other fledglings out of the nest. Interferometry began as the new "client" technology, the main body of experience being in cryogenic bars. Only by the early 1980s was the growth potential of the cuckoo beginning to be understood. This can be seen nicely in the following story told to me by Bill Hamilton of the LSU team.

> [1996] Now, to put this into perspective, years ago, when LIGO [Laser Interferometer Gravitational-Wave Observatory] was first trying to get started, there was serious concern by several people: I can remember conversation with Bill Fairbank, I can remember conversations with Guido [Pizzella] about should we try to shoot down LIGO. Bill Fairbank said to me, "I know that I can get it stopped," but he [also] said, "I don't think it would be a good idea to do it," because . . . "It will bring more money into the field." And he said, "If there's more money there, then we ought to be able to get more money off of the side"—If there's more money in gravity . . . there ought to be automatically more money for bar detectors. Fairbank said that to me, and it made sense.
>
> Guido said he and Amaldi had had similar conversations, and they said "No"—they felt LIGO would be good for the field, so we shouldn't try to kill it. [Hesitation] The problem with something like LIGO is that when you begin to get the attention of people who are terribly naive of the problems in the field, then the first question they ask is why are you keeping these bar detectors around when they are not nearly as sensitive as LIGO, and you could better use the money developing LIGO.

Here Hamilton is astutely pointing to the way LIGO, because of its size, would draw the attention of the outer rings of the target diagram to the whole gravitational wave detection program, including the bars. Those in the outer rings, he realized, would be hard to persuade of the need for two technologies.

This was easy to see in 1996 but not in the 1980s. We often fail to understand the long-term significance of small changes; a tsunami is scarcely visible in the deep ocean. The bar builders had not yet realized that the barely visible swell of interferometry would not turn into a wave on which everyone could surf; it was a tidal wave that would swamp them when it reached the shore.[1] On December 8, 1996, in the foyer of a conference in Cambridge, Massachusetts, Hamilton told me about Fairbank's and the Italians' once optimistic thoughts; at that very moment, in Washington, DC, a committee of the National Science Foundation was writing a report that

1. I borrow this beautiful metaphor from Michaels 1997.

would effectively end significant funding for resonant mass technology in the United States. I knew this was happening, but I couldn't tell Hamilton. The upshot would be the end of large-scale cryogenic sphere programs not only in America but also in Italy and Holland.

WEBER'S ATTACK ON INTERFEROMETERS

Joe Weber decided to try to use the outer rings of the target to the opposite effect—to destroy interferometry. His new theory of bar detector cross section was an implicit attack on it. Why should the US government spend hundreds of millions of dollars on interferometers when a cheap bar detector would do the job as well or better? The new theory dealt with the way whole groups of atoms reacted to the impact of external forces, but there are no groups of atoms to react in an interferometer. An interferometer is almost entirely made of space—it is simply light bouncing off the surfaces of distant mirrors.[2]

The lack of progress on the cryogenic bars must have been the main reason for shutting off funds from the Weber group. Let me put this in a counterfactual way: if Weber and his team had been making good progress on their cryogenic bars, it would have been very hard to cut off his funds. As it was, were NSF to support his new ideas, effectively they would be telling the world there was a cheaper way to do the job of the interferometers. When SN1987A threatened to bring the new ideas to prominence, questions were asked.[3] NSF decided that even if Weber were to find more data supporting his new theories, his track record would mean it would have no credibility within the scientific community, and it could use this justification not to support him further given the failure of his cryogenic bar work. NSF, we might say, engaged in some sociological analysis of its own: by anticipating the trajectory in social space-time of any future claims by Weber, it could feel secure in shutting him down.

In any case, Weber was not content with mounting an implicit attack on LIGO; he also mounted an explicit attack. At the time when LIGO was fighting for funds in Congress, he wrote a number of letters to people in

2. The second strand of Weber's theory, which had to do with the way signals reverberated back and forth in the resonant mass, was equally inapplicable to interferometers (as it was to cryogenic bars with end-of-bar transducers).

3. A copy of a *New Scientist* report of the Rome-Maryland observations of gravitational waves associated with NSF1987A circulated in the NSF with a covering note expressing alarm. An anonymous view of the problems caused for NSF by the SN1987A claims can be found at WEBQUOTE under "Another View of the 1987A Results."

positions of responsibility, arguing that the US government was wasting its money. He wrote to powerful members of the scientific community and opponents of LIGO, asking them to lobby NSF personnel on his behalf. He was determined, as one might say, to realize the immense political potential of the new ideas as well as the scientific potential; and this involved his carrying the argument into a whole new set of institutions in the outer rings, far removed from the research core.

For example, in July 1992 Weber wrote to a congressional staffer who worked for the committee chaired by Senator Barbara Mikulski. Mikulski's committee was responsible for the NSF; moreover, she represented Maryland. This ensured that the letter received some attention even though, as the staffer explained, he would receive a dozen or so complaining letters every day. In his letter, Weber explained that he had recast the theory of bar sensitivity, published the findings, and made observations that proved that LIGO was a waste of money.

> The 1984–1986 analyses were published and constituted a prediction that we would be able to observe pulses from Supernova 1987A....
>
> Our antenna and the University of Rome antenna with similar instrumentation were operating, and observed 12 large coincident pulses during the Supernova 1987A observational period....
>
> These data are confirming evidence that properly instrumented bars are much more sensitive relative to interferometers than scientists had thought. A small (100 kilogram) available silicon crystal can be instrumented to have sensitivity 10^{10} better than LIGO at a total cost of less than $500,000.

Inside and Outside the Core: The "Policy Regress"

Here is one way of looking at the problem of science policy: "Ignorant patrons worry about getting their money's worth for their delegation of funds to the researchers." Here is its complement: "Expert researchers face the ... unenviable task of performing for patrons who might not appreciate it." As the author says, "The asymmetry of information between those who would conduct research and those who would govern it presents the central problem of science policy."[4] As I have argued, when science becomes big, the locus of decision making moves outwards, away from the expert community. Joe Weber, a well-qualified and famous scientist, was

4. Guston 2000, pp. 14, 17.

claiming that funds delegated to researchers were being unwisely spent, yet the Congress-based personnel he wrote to would have no direct way of judging his claims. Still, a decision has to be made. What happens in these circumstances?

In this case, the problem was referred to experts in the NSF, but that solves no problems if it is the NSF's actions that are in question. I asked an NSF spokesperson, "Why should the person in Congress trust a reply from NSF when the person who is writing the letter is saying that NSF is 'screwed up'?" He replied,

> [1996] The answer is that you build up a long-term relationship. You interact
> every year with the same congressional committees and staffers, and you not
> only tell them the good things, but you tell them about serious problems,
> very early on, as in LIGO, when we knew we had management problems [see
> below], and we knew we had to replace the director, and the project was
> having trouble committing funds as fast they should have, and so on. The
> first thing we did, after going up the internal chain of NSF, we went and
> explained it all to Congress before somebody else did. And we told them
> what we were going to do to fix the situation—how long it was going to take
> and what the steps were—and we would report back as we went through this
> whole process. And then we proceeded to march through all of that. And so
> they knew what was going on, and they knew how remedies were
> progressing all the way through, and so they had some confidence . . . and so
> long as we stayed on track, then they maintained their confidence in what
> we were doing. So they could stick with projects that were in trouble if they
> knew there was a plan to stay on track.

I was also told that the letter sent back by NSF would have to make sense to the congressional personnel.

> [1996] If it sounded like it was plausible and we knew what we were talking
> about, that would be part of their understanding. And they would be
> encouraged to trust this reply because of the traditions of trust that had
> been built up over the years.
>
> And they meet people all the time making extravagant claims for and
> against things, and they get some sense of what sort of thing is within the
> normal range of criticism and when people are over the edge. They may not
> be able to tell the technical details of the article, but I think they have a
> sense of character in people. That's another clue to them about who to pay
> attention to.

> Someone like Joe Weber, who is an established professor at a major university in Maryland, and writing to, perhaps, somebody on the staff of the Maryland delegation—they're gonna pay attention to him, they're gonna treat it seriously, and that would be part of the reason they go through the process of asking the NSF for a response, and looking and evaluating the response and seeing whether that makes sense; but in the end they're not technical experts generally, and they could call on a few other people as well as us. I think if they really think there's an issue, they can pursue it.

One can interpret what is going on here in terms of the experimenter's regress. Policy makers cannot judge the papers or experiments on purely scientific criteria, and so they have to reach for some "nonscientific" criteria in order to make the judgments. In this case, the nonscientific criteria are first, trust—built up over a long relationship—and second, what we might call discrimination. Discrimination is the ability to judge the quality and integrity of a person's work from the way the person presents him- or herself.[5]

But if the policy maker's dilemma—let us call it the policy regress—is the same as the experimenter's regress, what differentiates the policy maker from the expert? The answer is not very much in terms of logic, but a great deal in terms of experience. The scientist cannot be guaranteed to make correct judgments, not even correct scientific judgments; but until the world goes mad with populist fervor, we would surely prefer scientific judgments to be made by those with scientific experience rather than those without. Thus, the experimenter's regress is like the policy regress but less of a cause for anxiety.[6]

What might the NSF have said to Senator Mikulski and her staff? We can make a fairly reliable guess. They would have said that Weber had made early major contributions to the field; they would have said that over the past decade, Weber's work had not continued at this same high level of distinction, and a recent grant had been prematurely terminated; they would have said that over more recent years, Weber had devoted much time to unorthodox theoretical calculations, producing a body of work that had been extensively reviewed by experts and found to be inadequate; and they would have criticized Weber's collaboration on the supernova, which had

---

5. This reaching out for other criteria in policy decisions as a substitute for expert judgment has been alluded to by Stephen Turner. For discrimination as a category of expertise, see Collins and Evans 2002.

6. See Collins and Evans 2002 for more related to this.

not been accepted as correct by knowledgeable workers in the field, since orthodox theory would suggest that the source could not have emitted energy equivalent to 2400 solar masses.

Of course, Weber would have tried to argue against this had he been able to get a hearing, but trust relations with the NSF would have made the response fruitless. Decisions have to be made and this is not a bad way of making them, even though it can never guarantee that the decision was correct.

FORMAL AND INFORMAL CONTROLS

So in spite of the fact that Weber was publishing in the mainstream and then the less mainstream physics journals, and in spite of the fact that he was running experiments and finding phenomena, the mechanisms of informal control operated by the core group prevented him from being taken seriously. The normal mechanism of informal control in this kind of case is absence of any comment at all; the heretic is simply ignored whether there are publications in the journals or not. The scientific community can afford to be tolerant within its formal institutions because its informal control is so powerful. Indeed, in the case of Weber it was only because he had made his claims visible outside the core group of scientists that they needed to do more than ignore him. The point is that the fuss was being made by Weber and his colleagues just about the time that LIGO was looking for funds—that is why what he said to those outside the core became important. In fact, it became so important that formal controls began to be exercised as well as informal ones. If we study the core group's reaction to the heterodox papers, we can see the switch taking place.

RESPONSE TO THE HETERODOX PAPERS

We know that about half the papers in the scientific journals are never cited by anyone—they just disappear without a trace, because they deal with small and uninteresting matters—but Weber's papers are of a different kind. If they were taken at face value, they would have caused a revolution on physics, so the lack of attention paid to some of them cannot be explained by their triviality. In chapter 12, we noted that Weber's last empirical paper before the 1975 watershed was published in 1973 in *Physical Review* and was cited 21 times during the 1975–79 period and 3 times during the 1980–84 period. We also noted that in 1976 he published another major empirically based paper in *Physical Review*, which, up to 1990, was never cited by anybody. This is what we should expect in this kind of case: scientists are given "a run for their money," but after this, even if they are able to continue to publish heterodox claims in mainstream journals, informal controls come into play; their work is simply ignored. There are anomalies, however, in the response to Weber's work.

One apparent anomaly that turns out not to be easy to understand is the response to the neutrino papers by the neutrino community. They were still paying attention to Weber in the middle 1980s, but, this period over, they ignored his 1988 empirical publication. The neutrino "anomaly" is really another instance of the "good run for your money" rule.

But there is another curiosity in the response to Weber's post-1975 work that requires explanation in terms of the context of the big science discussed above. What I will now show is that Weber's work was ignored just as we would expect it to be, except near the time of the LIGO funding battle, when it suddenly revived in terms of the negative attention paid to it in the constitutive forum and the imposition of formal controls as well as informal. To use a bodily metaphor, we have to separate the physical form of the scientific paper—its publication in the journal with all the results it reports—from its "soul." For the core group it is as though certain papers, which look completely healthy from the outside, are in a vegetative state; the scientific soul has departed even though the life support machine of the journal continues to function. Most of Weber's post-1975 papers were treated by the core group as if in a vegetative state, but a few of them seemed suddenly to come back to life.

For convenience, to illustrate these themes, we will examine four landmarks represented by publications: the 1982 paper; the 1984 and subsequent papers on cross section and neutrinos; the 1989 paper that reported events coincident with the visible supernova that was seen in 1987; and the 1996 paper, which claimed to find coincidences between gamma ray bursts and gravitational waves.[7] The 1982 and 1996 publications were ignored—they were treated as though they were devoid of a scientific soul. On the other hand, the 1989 supernova paper was heavily criticized from the outset, while the 1984, 1986, and 1991 papers were initially ignored but then treated as though they had suddenly sat up and started talking.

---

7. The papers are Ferrari et al. 1982; Weber 1984a, 1986, 1991 (among others); Aglietta et al. 1989; and Weber and Radak 1996.

I'll treat the papers in the order that brings out this contrast most clearly—
1996, 1982, 1989, and 1984/86/91.

1996: Coincidences with Gamma Ray Bursts

Some 21 years after the search for HighVGR had been abandoned by the scientific community as a whole, Weber published his paper claiming that his gravitational wave detector coincidences were themselves coincident with "gamma ray bursts." The paper claimed that there were only six chances in 100,000 that the results were due to chance.

The reception of this paper is the cleanest case of all in terms of my argument. Here was a paper promising to link and partially solve two of the great astronomical enigmas of our time—gamma ray bursts and gravitational waves. What impact did the paper have? Quite simply, up to the end of 1998, apart from the editor of *Nuovo Cimento,* the journal which published the paper, I could not find a single person in the core group who had read it, and I managed to find only seven persons who had heard of its existence in a vague kind of way.[8] In many ways, these seven represented the most remarkable phenomenon, for they knew of the existence of this potentially stunning finding in their own field, but did not think the paper worth reading. It goes without saying that I never came across anyone discussing the paper in corridors, conference presentations, or cafes. Effectively, within the core group of scientists the paper was invisible. Unsurprisingly, the *SCI* reveals no citations to the paper up to the middle of 1998 and only a couple of ritualistic mentions to date.

Let me describe some specific people whom I asked about the paper. I asked one person who was not in the gravitational wave core set or core group. He was, however, one of the most important scientists in the world working on gamma ray bursts. He had never heard of the paper.

I spoke to the two principals at NASA involved in funding Weber's gamma ray coincidence research and who had supplied the gamma ray burst data that provided one side of the correlation. One of these knew that the paper was out but had not read it. I supplied him with the reference; the other said he knew nothing of it. They both asked me to tell them whether their names were on the paper (this was in May 1998).

I asked the program director for gravitational physics at the National Science Foundation, I asked the author of the principal textbook in the

---

8. These numbers are increasing all the time as I talk to more people at conferences and so forth, but the picture does not change.

field of gravitational wave research, and I asked the director of LIGO; none of them had heard of the paper or the work. None of them were either surprised or particularly interested when I told them about it.[9]

While giving a talk in a university physics department that is one of the most important in the world in terms of the theory of gravitational waves, a member of the audience, himself a very significant theorist, told me that he had heard of the paper but had not read it. I was able to remark to the audience that this was "sociology in action."

In September 1998, I gave a talk on the topic at the British Association for the Advancement of Science. An eminent scientist in the audience suggested that though he would not have read the paper in *Nuovo Cimento,* he might well have read it if it had been published in *Physical Review.* Journal of publication is, of course, an important variable, as quotations within this paper reveal, but it does not seem to be the dominant cause in this case. We may note, however, that Weber's 1982 paper (see below) published in *Physical Review* was ignored, while the 1989 paper (see below) published in *Nuovo Cimento* was not.

When I gave a talk in Glasgow, a European member of the audience said he knew the paper well, but it turned out to be a confusion of dates; the reference was to a paper published by Weber in 1969, not 1996.

In August 1998, sitting with a group of physicists at dinner during a conference in Boulder, Colorado, I was describing the invisibility of the 1996 paper—which none of them had heard of. One physicist said confidently that he bet that "scientist X" had read it, because that scientist had earlier tried to test the same idea. I bet a dollar that he had not. Not long after, scientist X appeared, and we were able to call him over and question him. I won the bet, much to everyone's surprise.

Even more remarkable, though it does not bear upon the major thesis, Weber's principal Italian experimental collaborators, supporters, and coauthors, Guido Pizzella and Eugenio Coccia, had not read the paper. One had heard of the paper but not read it; the other had not even heard of it. Of course, Weber was not anxious to draw the paper to their attention, as it was effectively a critique of their design of transducer and the sensitivity of their detectors.

At first this total lack of awareness of such an important claim by the founder of the gravitational wave field is an extraordinary state of affairs; and yet, looked at from another point of view, it is easy to understand.

9. The author of the principal textbook did read the paper (and work out a detailed criticism of it) in April 2002, as a result of encountering an embarrassing question after a talk at Maryland.

When I said to scientists that I was astonished that such a paper should appear without anyone knowing of its existence, they merely shrugged. It is simply that over the years, the work of Weber has become so marginalized in the core group that no one cares about the paper or its potential impact any longer.[10]

The 1982 Coincidences

Fourteen years earlier, in 1982, *Physical Review* published the paper claiming that a statistically significant number of coincident excitations had been found between the detectors run by Pizzella's team and Weber. The last paragraph of the paper concludes that the excess of coincidences over noise was associated with a significance level of 3.6 standard deviations.

Like the 1996 paper, the 1982 publication fits the interpretative death model well if not so strikingly. In spite of the fact that the paper was published in a prestigious mainstream journal, it was never challenged. According to the *SCI*, up to mid-1998, the paper has been cited only three times. The only citation that appeared near its date of publication was in the paper by Tony Tyson of Bell Labs, which was essentially a ritualistic "tidying up" and which just happened to get published as late as 1982. The other two citations are recent: one was in a 1995 paper published by members of the Italian group, some of whose members shared the 1982 authorship with Weber; one was in Weber's own 1996 paper.[11]

Another way to gauge the lack of impact of the 1982 paper is in scientists' recollections. From 1995 I started asking scientists about that paper. The following quotations are taken from six interviews with gravitational wave researchers who began their research in the 1960s or '70s and are still active.

> [1995 (1)] *Collins:* Can you remember the 1982 paper?
>
> *Respondent 1:* ... Have you had views from other people on this?
>
> *Collins:* Yeah! ...
>
> *Respondent 1:* Because this is a whole subject, this one ... I don't know how far you've got or if you've been getting the general picture. That this is basically

---

10. Lest the nonreading of papers be overinterpreted, it should be pointed out that published papers are not the main source of physicists' information these days. The main source is talk at conferences and other networks. Physicists read when they need details, it was suggested to me. Nevertheless, the case is still extraordinary, perhaps precisely because in this case the normal networks, including electronic preprint exchange (which Weber did not use), were not delivering much in the way of information.

11. The Tyson and Italian papers are Brown, Mills, and Tyson 1982 and Frasca and Papa 1995.

probably wrong, and also more of the general picture have you seen—I don't know if it's ever been published—some of the—there was some careful analysis done of this, which I think is very important, because they show that it's not significant. Are you aware of that?

Collins: There's been analysis of the 1987 one, but I was not aware of any reanalysis of the 1982 one.

Respondent 1: What was the 1982 one that you're talking about?

Collins: It was Pizzella running his cryogenic bar in Geneva, looking for coincidences with Weber in Maryland.

Respondent 1: Oh! I don't know much about that. I haven't paid much attention to that. Look, at the time it looked to me—most of this struck me as unconvincing, and so after a bit, because there was so much showing that it wasn't convincing statistically. I was thinking it was the more recent things.

[1995 (2)] *Collins:* The first event that I'm interested in is a paper published in 1982, a paper by Weber, Lee, Pizzella, and Ferrari in *Phys Rev D,* claiming that they've seen coincidences between a Maryland bar and a Rome bar. Do you remember that paper?

Respondent 2: Yeah, I vaguely remember that paper. I haven't looked at it in a long time. . . .

Collins: You suspect that you read it at the time?

Respondent 2: Oh, yeah! I did read it.

Collins: Can you remember what it said it found?

Respondent 2: [Long pause] I believe—there was a period of time when, you know, they looked at the data, they do this threshold of cross-correlation, where they saw some evidence of a zero time-delay excess in the correlation.

Collins: Do you remember the level of significance?

Respondent 2: No, I don't remember the statistical significance. I remember—I must have concluded that it wasn't very significant. I mean, there were subsequent claims that I'm actually more familiar with to do with [supernova] 1987A, which in fact I don't believe were significant at all . . .

Collins: I mean, can you remember what you thought about it at the time?

Respondent 2: [long pause] Um, I may have that paper confused with—there was a period of time when the Italian group was publishing some papers claiming some coupling to the normal mode oscillations of the Earth—I don't think it was necessarily that paper—that was a different thing.

[1995 (3)] *Collins:* In 1982 there was a paper published by Weber and Pizzella claiming to see some gravitational radiation. Do you remember that paper?

Respondent 3: The '82 one—they had an excess of coincidences—I don't remember it much; much more attention was paid to the stuff on the supernova 1987A, which came later, but I do remember, remember, that—not well.

Collins: So, I mean, from the fact that you don't remember it very well, it sounds as though it didn't have a big impact on the community.

Respondent 3: It didn't have much impact on me. I'm afraid that I am much influenced by theorists' prejudices about what is plausible out there, and also by the degree to which this field has shown an ability to show excess coincidences when, in fact, much deeper examinations later showed that they were not there, and I was just skeptical of this at the time. It didn't really smell likely to be real gravitational waves, and so while I recall it being presented, probably I first heard it presented at Texas Symposium on Relativistic Astrophysics in Chicago, which would have been around '82 or so, and I discussed it with people at the time, and I just wasn't ready to get excited by it.

Collins: Right. Now, do you happen to remember what level of significance they reported?

Respondent 3: No, I don't, I don't remember. I have much better memory of the 87A...

I think most people's attitude was, "This is curious, but it will work itself out in the end"—were very skeptical, and that was that.

[1995 (4)] *Collins:* Do you remember the paper published by Lee, Weber, Pizzella, and Ferrari in '82?

Respondent 4: Vaguely.

Collins: Do you remember what it said?

Respondent 4: I think they were claiming coincidences—this was between Italy and one of their detectors—and they . . . were claiming . . . to have some excess events. And a lot of us looked at that data, and it also looked pathological. It had the smell of pathology about it, in the sense that they didn't come up and tell us when the events were. They didn't come up and do a full statistical analysis, it was simply a preliminary report, as I remember it—is that about right? Is that about right?

Collins: It was in *Phys Rev D.* . . . Do you remember its level of statistical significance?

Respondent 4: It didn't impress us. I don't remember the details, but we can go and look at it together. I don't remember the details, but it wasn't impressive.

[1995 (5)] *Collins:* In 1982, a paper was published by Lee, Weber, Ferrari, and Pizzella saying they'd found some coincidences. Do you remember this paper?

Respondent 5: Yeah!

Collins: Do you remember reading it?

Respondent 5: No—I just know about it.

Collins: Do you remember what level of significance they claimed?

Respondent 5: No.

Collins: What did you make of it at the time?

Respondent 5: I thought it was bullshit. I'm sorry to say, I by that time had become so polarized that I didn't trust anything.... And I said to myself if it isn't going to be a joint experiment of... people of the caliber of [named experimenters], I was not going to pay any attention to it. I'm sorry, I had a real prejudice.

Later in the interview:

I've been weaseling around, but I have not read that paper, because I knew from conferences that there was going to be nothing like what I was looking for.

[1995 (6)] *Collins:* I'm just trying to see if anybody remembers the '82 paper. For example, do you remember what the significance level was?

Respondent 6: No. I guess I didn't pay much attention to it. [Laughter]

We see that none of these respondents could remember the level of statistical significance reported in the 1982 paper even though they were deeply involved in the field at the time. We also see that some of them tended to confuse my queries with the much later and much higher salience 1989 paper concerning supernova 1987A.[12] Although the 1982 paper was a report of a new measurement, it was seemingly treated as devoid of information.

The View from Another Planet

These remarks contrast nicely with the comments of another scientist, whom I met by chance. The Marcel Grossman relativity conference is a large meeting that attracts physicists interested in all aspects of general relativity. Over breakfast at the meeting in Jerusalem, I found myself chatting with a physicist from outside the field of gravitational wave research. I'll call him the Marcel Grossman Scientist. He volunteered his view of the 1982 paper. He had read it "cold" in the *Physical Review,* and found it entirely

---

12. Memories can be "actively constructed" (see Lynch and Bogen 1996).

convincing. It had, as far as he was concerned, settled the matter; gravity waves had been detected by resonant bars.

The view of this scientist is important, because it again illustrates our theme about the difference between insiders and outsiders in terms of their understanding of the difference between body and soul. Suppose you were from a neighboring planet, and you wanted to learn about the current state of terrestrial science. You decide to read the peer-reviewed literature, bringing with you knowledge equivalent to a science degree or two from our best universities. In this case, you would read Weber's papers in the core journals in just the same way as the Marcel Grossman Scientist. You would have no clue that the soul of the papers had departed (according to the core group), since the body would look so healthy. To know the predominant interpretation of a paper, you need to be able to do more than read it; you need to be a member of the core group or have access to their views. Policy makers in the outer rings of the target diagram are in the same position as an alien or the Marcel Grossman Scientist; hence the policy regress and the need for trust in the judgments of the core group.

When things get tricky, however, trust can be replaced by more direct action. The core group controls the formal institutions, and if necessary it can use them. The core group can, as it were, kill the body, preempting the need for a subtle understanding of the soul. This is what happened in the case of the papers published in the middle 1980s. Later, by 1996, LIGO was already under construction, the danger from the "policy aliens" had passed, and implicit rejection was adequate once more. Now I shall document this analysis in detail.

1989: Supernova 1987A

The reception of a paper published in 1989 is a very "clean" case at the opposite end of the scale from the reception of the 1996 paper. It remains one of the great scandals of the gravity wave field that none of the cryogenic bars were running at the time of the 1987 supernova (SN1987A), and so what were purported to be the most sensitive antennae then available missed what could be the best opportunity to see gravitational waves for many years.[13] The only bars "on the air" at the time were Weber's room

---

13. Supernovas of this magnitude happen on average about once every 30 years. Actually, calculations suggest that even the cryogenic bars would not have seen the waves, so the disaster may not be quite so great in terms of physics. Nevertheless, for reasons of politics, no one would want to see such a debacle repeated, and all the bar groups now try to run their devices on an agreed schedule for downtime so that at least two of them are always "on air."

temperature detector and an old detector belonging to Guido Pizzella's group in Italy, also running at room temperature.

Examining and analyzing their data, Weber, Pizzella, and their coauthors concluded that they had seen coincidences between their bars, which were themselves coincident with the flux of neutrinos emanating from the supernova.[14] A paper recording the events was submitted to *Physical Review D*. In terms of the interpretative-death-with-lingering-life-for-the-body-of-publications metaphor, we see the first unusual event: *the paper was rejected*. There was a long dispute and the paper was resubmitted, but the authors could not get it published. In the end, it was submitted to *Il Nuovo Cimento*, which published it.[15]

On the face of it, the 1989 paper, concerning Supernova 1987A, is not much different from the 1982 coincidence paper. It claimed that coincidences had been seen on two room-temperature (i.e., somewhat insensitive, according to the orthodox view) detectors which, after a few adjustments, correlated with the flux of neutrinos emitted by supernova 1987A.[16] The neutrino fluxes had been seen by neutrino detectors at a level that was roughly consistent with conventional theoretical expectations. Nevertheless, if the bars were responding to gravitational waves according to the standard theory, the finding made no theoretical sense, because, again, far too much energy must have been emitted by the supernova. Furthermore, the same two groups were involved as in the 1982 publication, with the same two principal authors. Both the 1982 and 1989 papers claimed only that coincidences had been seen, not gravitational waves.

> It should be stressed that there is not now sufficient information about either backgrounds or sources to confront observations with theoretical predictions from gravitational radiation theory. This is because the observation of a small background of coincidental excitations tells us nothing concerning the origin. Detection is statistical. There is no way of separating the coincidences which are due to chance and those due to external excitations. And we cannot be sure what fraction of the external excitations are of terrestrial or nongravitational origin.... Only when the nature of the backgrounds is understood will it be possible to employ the antennas for gravitational radiation astronomy. (p. 421)

14. This mention of neutrinos has nothing to do with the matter of neutrino cross sections and Weber's experiments to detect neutrinos that are discussed elsewhere.

15. Aglietta et al. 1989.

16. A respondent remarked that the "few adjustments" were somewhat substantial and contrived.

The 1989 paper included a calculation based on the standard theory, showing that the energy required if the signals recorded were gravity waves corresponded to the "abnormal figure" of 2400 solar masses. (A reasonable figure in terms of standard theory would be a fraction of a solar mass.) The paper went on to say that

> [t]he objection mentioned above about the amplitude of the signals recorded by the g.w. [gravitational wave] antennas still remains....But the existence of correlations appears to us so well founded from both the observational and the statistical points of view that we publish them, with the proviso that when we talk of g.w., or even of neutrinos [the direct influence of which were a putative but equally unlikely cause of the coincidences], we refer to the events recorded by the corresponding detectors, without neither presuming nor excluding that a part or all these events are actually due to physical g.w. or physical neutrinos. (p. 77)

In spite of this broad similarity of content and level of claim, continuing reaction to the 1989 paper was very different from that of the 1982 paper.

The first difference, as I have mentioned, is that *Physical Review,* which published the 1982 paper, rejected the 1989 paper, and that this rejection followed a second submission which attempted to address the extensive criticisms of the first set of referees. Perhaps not too much should be read into rejection taken on its own, because editorial and refereeing teams change and there is a lot of "noise" in the system. Among my respondents were some who would never agree to the publication of a paper they considered incorrect, while others said their duty stopped at advising authors of what they perceive to be their mistakes. I asked a respondent to comment on the publication of this paper in contrast with the 1982 paper.

> [1995] *Collins:* I'm rather surprised that the 1982 paper got published...
> *Respondent A:* Why do you not publish a paper?
> *Collins:* Because the results are incredible—that's one reason.
> *Respondent A:* But what if it's true?—What if it's true, what if it's false?—If it's true and you suppress it, it's terrible. If it's false, and these guys are not saying they're seeing anything, they're saying they have a problem, then, in the end, if the problem is that they're stupid, then that comes down on their head. And when something becomes controversial, I'd say that's the most responsible way to deal with it.

Collins: OK, but *Phys Rev D*—

Respondent A: It probably went through lots and lots of cycles with reviewers to get the appropriate language for parts—I think—go ahead.

Collins: I mean, but *Phys Rev D* didn't publish the supernova paper. One of the things I'm trying to compare is the reception of the 1982 paper and the supernova paper a few years later.

Respondent A: Er—the supernova paper said that they saw gravity waves? Didn't it? Is that right?

Collins: Er—I think it said they saw excess coincidences.

Respondent A: Excess coincidences—Aha—[slightly embarrassed laughter]. Well—[long pause]—yeah, I suspect I know who the reviewer was at *Phys Rev*. And he was—it was so unbelievable. There, you know, for a supernova it's very hard to get radiation out. And, there were a variety of flaws with it—you know, the timing, the relative speeds of gravity waves and neutrinos and light. Er—yeah. We-ell, OK, I don't—I think there may have been an earlier version of the paper that said they were seeing gravity waves, and it may have gotten toned down by the time it was actually published . . .

But, of course, it was published, in *Physical Review*.

In spite of some scientists' view that those who stick their necks out ought to be published, one might still argue that the 1982 paper was lucky in its referees and the 1989 paper was not. I have seen the three unattributed referees' reports on the resubmitted 1989 paper (apparently the paper went back to the original referees). Two reports were forthright rejections, while the third was a thirteen-line response that confesses to "the troubling nature of the manuscript" and "agonizing over what to say about it." This referee also wrote as follows:

> The problem is that I really do not believe the results. Yes, the authors have largely attempted honestly to answer the criticisms. I have a few quibbles still . . . [but] . . . The larger problem is that I do believe that the results presented by the authors are in some way fundamentally flawed.

The outcome of the refereeing process was consistent with subsequent events. Having been rejected by *Physical Review*, the paper was published in the Italian journal *Il Nuovo Cimento*, which has a more open publication policy that many American scientists believe has reduced its prestige. One eminent respondent told me,

[1995] *Phys Rev D* is the journal of preference for this field by a very large margin, and that's where people look for things. That's where the original 87A paper should have been published....

...Nobody pays attention to *Nuovo Cimento* these days. These things are sociological, I guess [laughter]. There is not much of interest published there any more.

But now events take on a still more distinctive complexion compared to 1982. Even though the paper had been pushed into what was perceived as the "marginal" *Il Nuovo Cimento*, the 1989 paper *was* attacked in print (by C. A. Dickson and Bernard F. Schutz) and, surprisingly, the attack was published by *Physical Review*.[17] So *Physical Review* had refused the positive paper after two submissions and seen the paper "sidelined," yet were still willing to publish a refutation. To publish a refutation of a paper published in a different journal is odd—why did the editors not tell the authors to submit their rebuttal to *Il Nuovo Cimento*?

Most physicists, whether inside or outside the field of gravitational radiation research, consider this sequence of events unusual and unfortunate irrespective of whether they think the original paper was right or wrong. One important physicist who was strongly in favor of the publication of the Dickson and Schutz rebuttal told me that it would have been better if both it and the original 1989 paper had been published in *Physical Review*.

[1995] I think there was great interest in and great importance in Schutz's paper to the community and in the community that I think there is strong feeling that that kind of paper—as well as the original paper—ought to be in *Phys Rev D*.

Finally, the original authors wrote a response to the published critique and submitted it to *Physical Review;* the response was also rejected.[18] One

---

17. Dickson and Schutz 1995. For the record, the challenges to the 1989 paper were based on the a priori implausibility of the outcome of the energy calculation, and the relative plausibility of an explanation in terms of inadvertent statistical manipulation of the data; it was said that too many thresholds or other conditions were tried before a statistically significant result emerged from the analysis. It could not have been a trivial matter, however; as a well-informed respondent put it, [1995] "[T]here had been a long controversy over that paper—great puzzlement is a better way to say it... real puzzlement, because people who had looked at it—good people who had looked at it—had not been able to really take the statistical analysis apart and, er, seriously come in and question."

18. It was eventually to be promulgated as *Internal Report No. 1088, 20 May 1997* (Rome: University of Rome, La Sapienza).

might have thought that the claim to have seen gravitational waves associated with SN1987A would have been treated as scientifically brain-dead, but here it was eliciting reactions appropriate to a fully fit heavyweight boxer. The one anonymous referee whose comment was used to justify the *Physical Review*'s rejection of the response to the rebuttal wrote,

> Even if there was something interesting in this analysis, the atmosphere on this topic has been so poisoned and the tone of this latest contribution is so polemical that no one will believe it. The authors must begin to realize this; further writing on this topic will only erode their reputations, without advancing in the slightest our knowledge of the issues.

Again we see a scientist acting as a sociologist. But why was there such a poisonous atmosphere? Why was the polemic ever allowed to get going, since it takes two to make an argument?

The strong reaction to the 1989 paper, as compared with the total absence of reaction to the 1982 claims and to Weber's 1996 paper, can be explained once more in terms of background developments in interferometry. The field of gravitational wave detection was no longer one in which a few deeply concerned scientists could be expected to draw the "right" conclusions as a result of their socialization into the core-group; unsocialized policy makers and hostile astronomers now had an interest.[19] As one scientist put it,

> [1995] I thought it was outrageous. I was very pissed off. I thought they'd learnt their lesson.... I was scared to death that this was gonna kill LIGO.
>
> *Collins:* How would it kill LIGO?
>
> *Respondent:* Because it was more of all this fanaticism and craziness. You see, I lived in this hostile environment here, at [Tech], and they were just gonna say, "See, you guys are all crazy, and here's more of this bullshit that you guys are presenting to us. How could you do that."... I was hoping it would disappear.
>
> *Collins:* So because you think... wow, this is amazing... so because, let's say, [named resonant bar researchers] made another one of these crazy claims, let's call it, this endangered LIGO?
>
> *Respondent:* Yes.
>
> *Collins:* Just explain it to me again, because I find it very hard to comprehend.

19. Bear in mind, however, that SN1987A was a very big and exciting event in astronomy and astrophysics, which helped to give the heterodox claims additional salience.

Respondent: Well, we were fighting with the astronomers still at the time, very much so, about the value of doing LIGO in the beginning, and we were continually being tainted by "You guys are all nuts!" And this just was further evidence of it. So, I mean, they lumped us together with that.

Collins: Yeah, but people surely would have known enough to know that you were separate groups.

Respondent: Doesn't matter. This is a field. "Look—it's pathological science again." That's where it came up again. You see. Now, maybe I overreacted, but I felt that . . . if [named resonant bar researcher group] does something cuckoo . . . we're all gonna get it.

Bernard Schutz, principal author of the rebuttal and a major theorist associated with interferometry, talked with me in 1996. He accounted for the reaction to the 1989 paper in terms of the reflection back from circles of influence having to do with funding LIGO. He said that it seemed to him that the paper was another "Weber affair" in the making and that "we had to police ourselves in a way, that there had to be a critique of this coming from within the gravitational wave community. And I spoke to a few other people in the field, and they all agreed that that had to be done."

Schutz explained that "the field was just becoming reputable again as we were pushing the interferometers as respectable, as having a respectable chance of getting to the right sensitivity, and so there was a lot more money at stake now. Funding agencies were beginning to say we could imagine spending a lot of money on this. To have a bar detector group come up and say that they had seen gravitational waves, then to have that shot down in flames would probably have been very damaging to the prospects of extracting money from funding agencies for the interferometers."

As Schutz made clear, this was not because he thought anyone would believe Weber's claim that he could find the waves more cheaply: "It seemed to me it would affect the funding, but not because people felt in a sense that interferometers were doing something that could be done more easily by somebody else, but in the opposite way, that interferometers were participating in a field that was full of crazy people." When I suggested that anyone ought to be able to see that the Weber claims and interferometry were two completely different enterprises, Schutz responded that

[1995] this bears on the sociology of the approval process for the interferometers, so there's a much bigger question here of competition between different fields for funds and the need to convince people who are not in the gravitational radiation field that it's worth spending money on this project.... I met many of my astronomy colleagues who were by nature opposed to the idea anyway, and who would have been very happy to have been able to point to my bar colleagues as examples of bad scientists who—you know—if—if there are these guys in gravitational waves doing that, how do we know your other colleagues in the interferometry field are not going to be just as crazy. And we put all this money into something and maybe you see something, maybe you don't, but how do we know when you come and claim to have seen something that you're not just as crazy as everyone else?

That was an argument I was afraid of having to deal with, so that was the reason, the main motivation, in my case for writing this paper.

Schutz felt, then, that an incredible claim that was published and *not seen to be refuted* by the gravitational wave community itself would give ammunition to the enemies of the whole field of gravitational research who populate the second ring of the target diagram—the scientists who feared they would be starved of funds by the interferometer effort. Schutz told me that his sensitivity was shared by Rich Isaacson of NSF: "Rich was very happy when [he learned the critical paper was to be written].... So were [some of the names principal scientists at the American and British interferometer groups]."

Earlier I posed a question to you, the reader: why did the editors of *Physical Review* not tell the authors of the rebuttal of the *Nuovo Cimento* paper to submit it to *Nuovo Cimento*? The answer is, they did! In a letter dated June 23, 1992, the editors wrote, "It would be appropriate for your work to appear in these journals which contained the papers being criticized rather than in *Physical Review D* which has published none of this work." The editors, however, were persuaded to change their minds. A letter was sent to them by Schutz, the principal author of the rebuttal, pointing out that many American-based gravitational wave scientists had told him that they were anxious to see the rebuttal in print. He also pointed out the fundamental nature of the disagreements, and that the 1989 paper was being taken by Weber to bolster the idea that "new physics" was needed to understand gravitational wave detectors. A second letter supporting publication was sent to the editors by one of the leaders of the LIGO effort on November 15, 1992. This was described to me by Schutz: "[An important scientist] wrote a letter to the editor supporting [the] assertion that this was a global issue, not one to do with bad selection of articles for journals, but one that clearly

affected the United States because of LIGO, and that it was appropriate for
Phys Rev. So *Phys Rev* did publish it."[20]

Finally, I tested my whole hypothesis on Schutz:

> [1995] *Collins:* [L]et me give you my theory so you can see where I'm going—I
>   believe that within the inner networks of scientists, the people who really
>   know what's going on, often a very small number ... know what's right
>   and what's wrong, and there often isn't a need to write a paper to rebut
>   something.
>
> *Schutz:* Yeah, that's right, that's the point of view that the editor of *Phys Rev* took,
>   I think.
>
> *Collins:* So, in this particular case, the fact that [all those people wanted it
>   published]—I want to explain that that bears a relationship to what was
>   going on with the funding of LIGO at the time. That's the hypothesis I'm
>   trying to pin down.
>
> *Schutz:* Oh, yeah! We all saw it as related, and we all saw the publication of this
>   as helpful in the case for LIGO.[21]

These last extracts can act as the conclusion to this section. The 1989 paper
was met with a strong published reaction because there was a nascent
big science fighting for funds. "Crazy" publications without rebuttal in the
constitutive forum could be a problem when "alien" policy makers were
looking on.

The New Cross Section: 1984 Onward

Analysis of the treatment of the cross-section papers is complicated by the
fact that they were directed at two different communities: the gravitational
wave community and the neutrino detection community. The responses of
these two communities have to be pulled apart.

Weber's 1984 paper refers to both neutrinos and gravitational waves;
the subsequent publications deal with each separately. Thus the conference
papers of 1986 and 1991 deal only with cross sections for gravitational
waves, while two papers published in *Physical Review* deal with cross sections
for neutrinos.

The 1984 paper has attracted twelve citations in its lifetime, but this
result is misleading unless we notice that the first three of these citations in

---

20. I have seen all the relevant correspondence.

21. The full and very evocative extract from the interview can be found at WEBQUOTE under
"Schutz on Why He Attacked the SN1987A Claims."

chronological order are from papers that discuss only the neutrino aspect of the 1984 paper. Thus, though these papers are critical, granting a scientific life and soul to the 1984 paper, nothing is implied about the life and soul of the gravitational wave claims therein; the gravitational wave community did not respond until later. Further, by 1987 there were no less than seven published rebuttals of the cross-section-for-neutrino claim represented in the two papers published in 1984 and 1985, but these had all come from the neutrino community; the gravitational wave community was still treating the papers as if in a vegetative state.

The first criticism of the 1984 paper that came from the gravitational wave community was not published until 1988. It seems, then, that the response to the papers on enhanced cross section for gravitational waves comes in two phases: the first few years of the cross-section claim exhibit interpretative death, just like the 1976 and the 1982 papers; this, as I have established, is the expected reaction, and it is the subsequent invigoration that represents the anomaly.

Once more we can use physicists' recollections to support this idea. Thus, some of my respondents, when first questioned, thought that Weber's new cross section was invented only *after* 1987, in order to account for the supernova claim. In the following extract, two respondents are being interviewed together and start off talking to each other about the SN1987A claim.

[1995] *Respondent i:* Was the n-squared [new cross section] involved?

Respondent ii: Well that came afterwards. I think that—this is only my impression, but I don't know how long he'd been thinking about that; but my impression was that was invented, sort of, after the facts to help explain that he'd actually seen something [in 1987].

Collins: The supernova?

Respondent ii: Yeah!

Collins: That's wrong, actually.

Respondent ii: He was thinking of it before?

Collins: He talked about his new cross section in 1984.

Respondents: OK! Alright!

These respondents, as with the ones that follow, had hardly registered the existence of the cross-section claim before it became important for a much larger debate. Something similar is seen in the following two interview extracts.

[1995] *Respondent iii:* [The 1987A claim is] compounded by the fact that at the time, Weber had come up with a new theory of the cross section of a gravitational wave antenna, which I think was really absolutely wrong—that his calculation was flawed. And, that would have been needed to explain if you believed the signals were real, if you believed they were gravity waves from supernova 1987A, [it] would have likely required this cross section, but I think it just made no physical sense at all.

[1995] *Respondent iv:* [first few minutes of interview] He's broken with virtually every physicist I know on what's called the cross section. And that all comes from this [business] of the supernova 1987A.
[approx. 40 min. into interview] He wants to get credibility for the supernova detection—based on the new calculation of the cross section.
Collins: [approx 50 min. into interview] Weber's new cross section is the second event [I want to talk about].
Respondent iv: Yeah! OK! I spent a lot of time with Joe on that.
Collins: So 1984—
Respondent iv: Is that when he did that?
Collins: Yeah! So it's actually before the supernova.
Respondent iv: Oh—he did that before? I didn't realize that. OK.
Collins: Because you said earlier on that he'd done this in response to the supernova to justify it, but actually his cross section was '84.
Respondent iv: I see. OK.

What I am trying to show here is that respondents have no memory of the fact that the cross-section claim was initially made three to four years before the SN1987A claims began to be made. In other words, within the gravitational wave community, this claim and the associated paper were so invisible—so scientifically soulless—as not to be remembered until later high-profile events made it salient.

I am arguing that within the core group, the transformation of the theory from an interpretative corpse to a dangerous enemy had to do, once more, with the transformation of gravitational wave detection from a small science to a big science. The first sign of this was the request by Amaldi to Giuliano Preparata to look into the matter. He initially concluded, it will be recalled, that the Weber theory was incorrect, and he proceeded to make this clear at conference presentations and in a speedy publication in 1998. Thereafter, as we saw, Preparata changed sides and became Weber's strongest supporter and a principal proponent of the idea that quantum coherence has macroscopic effects of the sort that Weber was claiming. As a result, Preparata became rapidly and thoroughly marginalized in respect of the gravitational radiation core group.

Preparata's change of heart left Weber's cross-section claims unchallenged, but by this time the stakes had become so high in the light of the funding battle over LIGO that they could not be left this way; simply ignoring them was no longer good enough now that those outside the core group were taking an interest. As a result of pressure from concerned colleagues, Caltech theorist Kip Thorne also analyzed the work, reached the conclusion that it was incorrect, and began to attack the cross-section claim in conferences.

Thorne's critique was never promulgated in a journal, however. As an admirer of Weber's path-breaking work, he would have preferred not to confront him at all on the matter of the cross section but felt impelled to do so because of the potentially damaging effect it would have on LIGO. In this case, the groups of outsiders who are the cause for concern were not the hostile astronomers who might have lobbied Congress, but the members of Congress themselves, who among other things were being directly lobbied by Weber, as in his letter to Senator Mikulski's aide. Thorne told me in 1995, "We were in the early stages of trying to get LIGO developed and approved, and so our funding officer in Washington [Richard Isaacson] was rather concerned about this and the impact on LIGO. If Joe [Weber] is right, then there is no need to develop such a big expensive instrument."

But Thorne did not want to attack Weber in public.

> I did not want to do to him what [another scientist] had done [a forthright public attack][22].... [T]his became a sort of minuet that occurred conference after conference between Joe and me. Joe would get up and talk about this, and I would reply in that way—a brief reply. I felt my credibility in this was such that people would listen, and that would be enough. Which Richard Isaacson felt was not enough, because it was not just the community that was involved. And he felt that something had to be written.

He explained that Isaacson's anxiety was caused by the wider group of people who were now looking on.

> Because LIGO is such a big project, and the most expensive project that NSF has ever done, it was inadequate to just have me as a member of the community stand up and say that Joe was wrong and people within the

22. I'd guess it is Garwin who is being referred to.

community being pretty clear on it. They [the outsiders] could see that as well. He felt it had to be in writing so it could be shown to people in writing.[23]

Thorne explained that at the time we were discussing this, he felt he could now relax because LIGO's funding had been secured, but "if this were a significant issue in these questions, then I might think differently; but I don't see any need to fight those battles in public when the community has reached a consensus on it."[24]

Though Thorne did not publish his argument in a journal, in 1992 another important gravitational radiation theorist did. I asked Leonid Grischuk when he decided to publish a response to Weber's and Preparata's (later) cross-section ideas. He replied,

> [1996] When it became clear that it creates difficulties. It creates difficulties in the sense that, though I always suspected that he was wrong, it did not create too much harm until he really started to lobby American congressmen—I don't know the details, but I found from the literature and from the reaction of other people that it became not simply a scientific matter, but it was an argument—the implication was why should we build some larger systems while it is enough to work with these solid bar antennas and actually to claim that it was already detected. So since I know that scientifically it cannot be true, and since it became politically important, I decided to write something....
>
> Political motivation did play a role. It did play a role because I notice that people are getting interested, people are getting confused, and if I do not clarify this, they may wrongly assume that this is the correct formula and probably would decide not to build LIGO for instance—you know—and so on.[25]

Finally, and somewhat more tenuously, we can show less directly that at least some of the scientists involved in disputing the Weber cross-section claim were more concerned with the impact of their work on the outside

---

23. Thorne's remarks were eventually published in a conference proceedings (Thorne 1992b), but this is less of an attack than it would have been if it had been published in a journal.

24. A longer section of this interview extract can be found at WEBQUOTE under "Thorne on his Reaction to Weber's New Cross-Section."

25. Grischuk stressed that he had tried to do an objective analysis in spite of the quasi-political motivation, and that he considered his analysis the only one to have explored the issues thoroughly and decisively.

world than with the science itself. Thus, in the 1995 Dickson-Schutz paper criticizing the Pizzella-Weber supernova claim, the authors included a footnote as follows:

> The claim that gravitational radiation detectors actually have a much larger cross-section than we have taken here, e.g. J. Weber, *Found. Phys.* 14, 12 (1984), is wrong, as has been shown by Thorne, in *Recent Advances in General Relativity, proceedings of a conference in honour of E. T. Newman,* edited by A. Janis and J. Porter (Birkhauser Boston, in press), and L. P. Grischuk, *Phys Rev D* 50, 7154 (1994); and so does not offer a way out of these problems.[26]

Note that Dickson and Schutz do not refer to either the old or the new Preparata findings in this footnote, but only to Weber. Note also that they get the reference to the Grischuk critical paper wrong. The paper they refer to is something else entirely, published some two years later. This suggests that they were not engaged in a close reading of the Grischuk paper when they wrote their footnote and that for them, these papers were seen as what we might call "political markers." A political marker we can define as a paper that may not be read carefully—except by those it criticizes—but that can serve a political purpose by being spoken *about*. Policy makers and other outsiders who may have been told that such-and-such a technique can detect gravity waves for 1% of the cost, or that such-and-such a bizarre claim shows that the gravity wave community is still not to be trusted, can reply that the technique or the claim has been definitively shown to be wanting in the *Physical Review,* and "Here is the reference." Not every *consumer* of a scientific paper has to read it!

Furthermore, we can note the complete absence of acclamatory references to Preparata's 1988 critical paper. Ironically, Preparata was the first to publish a refutation of Weber's gravity wave cross-section claims in the international journals, but his analysis was rendered useless for political purposes because he changed sides by 1989. We need to think hard about what this means. Is it the content of Preparata's 1988 paper which counts, or is it Preparata the author which counts when we assess the 1988 paper?

It might also be worth noting that neither Thorne nor Dickson and Schutz take on Preparata's later support for Weber (though Grischuk did). Thorne's analysis of the cross section, and Dickson and Schutz's mention of it in a footnote, refer to Weber alone, even though Preparata's analysis came

---

26. Dickson and Schutz 1995, p. 2668.

later and was, arguably, more technically advanced. Again, it might be that writing from Italy and using his new analysis to support cold fusion as well as enhanced cross sections for gravitational wave detectors, Preparata was considered too marginal a person to need explicit refutation for the sake of the outsiders; the important outsiders were not going to take him seriously even if they knew of his existence.

We can get some indication of the relative degree of recognition of Preparata and Weber from replies to an email questionnaire that I sent to those who attended a conference on gravity waves in Pisa in the spring of 1996. Of 29 replies, twelve were from Italians and seventeen from non-Italians. Respondents were asked questions about Weber's and Preparata's ideas about cross sections. They were given the option of answering that they did not know what I was referring to. In respect of Weber's claims, two out of twelve Italians did not know what I was referring to, compared with three out of seventeen non-Italians—roughly the same proportion. Regarding Preparata's theory, however, while four out of twelve Italians took the "don't know" option, eleven out of seventeen non-Italians confessed to it.[27]

FOUR MODELS OF DECISION MAKING IN SCIENCE

In chapter 19, I discussed the demise of HighVGR in terms of the growing cleavage between it and the mainstream consensus. Weber began with heterodox claims, moved to a heterodox theory, and then progressed to a still more heterodox implication of that theory—the need for center-crystal instrumentation. These moves cut him off from more and more potential allies and left him with almost no connection at all to the gravitational wave community.

In chapter 20, I looked at this again from the standpoint of the institutions and networks that supported his work. In the immediate post-1975 period, journals and funding institutions gave him similar support to that which he had received before the watershed; but the publication outlets and sources of funds slowly faded away, forcing him to seek support further and further from the center.

In this chapter, we have looked at the way the opposition to HighVGR took an active rather than a passive form as the big science of interferometry

27. The response to this email questionnaire was poor, and I do not recommend it as a method in the sociology of scientific knowledge.

became the dominant context against which all gravitational-wave policy decisions were made.

The last three chapters have examined different aspects of the stilling of the ripples in social space-time. There could also have been another chapter, immediately preceding chapter 19, in which the scientific case would have been argued. In that chapter it would have been shown that Weber's science went more and more wrong. I could not have written that chapter, because I am not a good enough scientist; and I would not have wanted to write that chapter, because that is not the business of this book. Here, however, is how a scientist might have summed up the conclusion of that protochapter; indeed, this is a quotation from a respondent:

> [1995] [Weber's recent claims are] nonsense from the point of view of physics, and part of the reason that Joe has never been able to rescue his reputation among physicists is he lives his life in this highly embarrassing way. That is, he will say anything that allows him to say at the end of the paragraph, "And this is how come I think I've detected gravity waves." And the arguments he gives in favor of so-called center crystals—and each of those words has weird meanings to them—they're nonsense, and they don't convince anybody. And that's true also for his argument that a perfect crystal has a much larger cross section for gravitational waves. You can't literally say no, and I think there are one or two other papers from marginal people who say they agree, but in effect—essentially—they convince no one. I mean, they do not correspond to anyone else's understanding of—I shouldn't say freshman physics—but undergraduate physics.

These are the words of a central player in the gravitational wave physics game, yet even here one sees why the protochapter, the pure science account, would not be enough. What we know of science, as I suggested at the very outset of the book, is nearly all hearsay. In this physicist's words, one see it clearly. Weber's claims are nonsense from the point of view of physics; that is said unequivocally. And yet Weber is no fool. And we see that the flaws in the cross-section argument are wrong, because they don't "correspond to anyone else's understanding of . . . undergraduate physics," and are such that "you can't literally say no." Well, of course not. Only three physicists ever did the hard work of really analyzing Weber's cross section and finding the flaw, and one of those changed his mind later. In one way or another, every other physicist is going on his or her trust in the reputation of those who have done the analysis. And the chances are that if they did go back to the original source, even if they were very, very good, they would have a hard time finding a flaw that would stand up against the determined arguments of a Weber or a Preparata. Even those who did do the analysis, and whose business this is, did not find it a trivial task.[28]

And so the protochapter would need some backup from chapter 19, which talked of the marginalization of Weber from his potential allies. This chapter showed why those who, when faced with the choice of trusting the analysis of one group rather than another, would have had that choice simplified by the increasingly obvious fact that Weber had few others trusting him. And this again would be reinforced by the analysis of chapter 20, which showed that major institutions were trusting Weber less and less. And this again would be reinforced by the analysis of this chapter, which shows how it was easier to disregard Weber because he had been "officially" refuted. And one did not need to read the refutations to know this; one had only to know that the refutations had taken place "in the peer-reviewed journals," while the initial claims had eventually been relegated to less central journals. That is a powerful example of scientific argument.

Thus it seems to me that whether or not you need the protochapter to understand the change in Weber's fortunes, you certainly need the subsequent chapters. But to reiterate an earlier point, I want to establish less than this. I do not really want to argue about the relative status of these different accounts, only that there is a job to be done to produce accounts of the kind found in the subsequent chapters, and that this job is best done by concentrating on the mechanisms displayed in these chapters and not on the rightness or wrongness of the science.

And to reiterate yet another early point, this is not a criticism of science; it can only be seen as a criticism if science is thought of in terms of the abstracted model that we might call scientism, or scientific fundamentalism. In that model, experiments are conducted according to recipes—what I have elsewhere called the algorithmic model—and each individual scientist has the capacity to encompass the whole history of scientific theorizing and data gathering right up to the present instant. In our model, experiments are matters of the transfer of skills among the members of a community—the enculturational model—and knowing science is to belong to a

---

28. A respondent complains that many of them did go back to the original paper and found the flaws. But could they really have done a *thorough* job—the kind of job that could be presented in public, with one's reputation staked upon the conclusion—given the difficulties encountered and the time needed by even the best theoretical physicists?

community, share its "cherished assumptions," and be part of a ramifying nexus of trust relations.

Ironically, *it was Weber* who cleaved most desperately to scientism. From Weber's point of view, he assiduously followed the rules of good scientific practice under a model in which the individual alone can correspond directly with nature, requiring exposure to the criticism of colleagues only to eliminate error. Starting from a position of widely acclaimed scientific virtuosity, he continued to develop theories, carry out experiments, and expose himself to criticism by publishing the results in peer-reviewed journals. Weber believed he had found gravitational wave events that correlate with neutrinos emitted from a supernova and which correlate with gamma ray bursts; at the same time, his theory makes his observations compatible with the general theory of relativity and standard astrophysical models. As he put it to me,

> [1993] It seems to me that there are two hypotheses. [First,] I've done an experiment and observed something with center-crystal instrumentation, and the Bell Labs builds a copy with center-crystal instrumentation and a coincidence experiment produces something. Rome builds something with center-crystal instrumentation that produces something. A hundred million dollars has been spent by the end accelerometer instrumentation—it's produced nothing. Therefore you start by assuming I'm completely wrong—all my work is worthless, because the more sensitive approach has produced nothing.
>
> There's another possible interpretation. The other possible interpretation is . . . that the end accelerometer approach is worthless and they don't know what they were doing, and my data are valid. And although no one seems to adopt that, it seems to be that just as a basis of objective looking at evidence, that's a reasonable conclusion.

As Weber might have seen it, "What more could a scientist do?"[29]

If science did make progress according to this model, it would indeed be hard to understand why Weber's most recent work, given an imprimatur by being funded by NASA, published in a physics journal, justified by a theory, and reporting what seem to be remarkably interesting results, has been ignored. The point is, of course, that science cannot make progress

---

29. For an evocative repetition of the comment made to me in 1995, see WEBQUOTE under "Weber on His Work as an Empirical Test of Resonant Transducers."

in the way this model suggests, because every finding, every theory, and every paper has to be interpreted.[30]

Not only does the necessity of interpretation provide the space for the continued existence of rejected sciences; it also allows those rejected sciences to be seen as irredeemably flawed by the core group. Ironically, only under this kind of interpretative model of science can Weber's actions be classified as *deviant*. Ironically, *Weber was a deviant because he cleaved rigidly to the right of a scientist to continue to speak directly with nature,* rejecting the long and painfully established consensual, collective interpretation.

To turn to the mainstream, the need for interpretation of results that, perforce, will include rejection of certain heterodoxies is not a matter of bias but of the underdetermination of interpretations by theories and experiments. There is no need to posit a conspiracy to explain why the core group or the mainstream of the scientific community ignores maverick claims after an initial examination; there is need only for there to be more scientific work to do than time to do it in. And there will always be more scientific work to do than time to do it in, because the time needed to check, definitively, every conceivable alternative interpretation of events is indefinitely long; human ingenuity at inventing new interpretative possibilities ensures that this is so.[31] If science is to make progress, then experiments and calculations are not enough—someone, or some community, has to make judgments and choices. The sociology of scientific knowledge shows how and why different groups sometimes make different choices.[32]

Formal and Informal Controls Again

In the case of HighVGR and, one might speculate, in physics as a whole, tolerance within the mainstream journals and funders, and the wider

30. Compare Merton's (1957) norm of organized skepticism. Under the Mertonian scheme, organized skepticism unmasks deviance through replication and the like; under the framework offered here, organized skepticism can deliver only an interpretation; it cannot disprove.

31. See earlier remarks on the role of stage magicians. Karl Popper's (1959) criterion of falsifiability is another approach to the problem: scientists who are too ingenious at inventing ways of preserving their findings are said to have rendered them unfalsifiable.

32. None of this, of course, lessens the scope for conspiracies if they are there; it is just that the existence of interpretations justified by networks of judgment and trust is not a necessary condition for the existence of a conspiracy. Incidentally, there is little here that has not already been said about other kinds of formal organizations. To go back to the Wittgensteinian (1953) point, rules do not contain the rules for their own application, and this means that nearly all apparently smooth-running bureaucracies are lubricated by informal practices, which allow for interpretation of the rules. Among humans, only very rarely, as in the case of, say, military drill on the parade ground, do we encounter what might be called uninterpreted rule-following (Collins and Kusch 1998). Different "Patterns of Industrial Bureaucracy" (Gouldner 1954) emerge from different arrangements of the informal and the formal, and in science, too, we find different patterns.

plurality within the publication and funding systems, can be understood once we also understand the role of core sets and core groups. I would argue that it was because the gravitational wave core group was so strong that it could allow Weber's heterodox publications to appear in the mainstream journals and allow him to take resources from other institutions. The core group could control the interpretation of the papers and research findings through the mutual understandings of its members; they could make clear what was to be ignored; they all knew that if scientist X had studied the matter and was happy that it was flawed, then they should be happy, too, even if scientist Y thought that scientist X was wrong. The community was based on tight social networks continually reinforced by face-to-face meetings at conferences. The right interpretation was also enforced by the central position of certain funding agencies; we have seen how they have to control funds to match the mainstream view once the consensus has formed.

If, as the philosopher Wittgenstein advises, when we want to understand the meaning of an idea, we look to its use, we find that hardly anyone was reading the heterodox papers; and even if they were, it was almost impossible to act upon them. The heterodox work was, then, "meaningless" within the core group. Weber may have claimed that he was seeing waves in physical space-time, but there were no waves to be found in social space-time.

We have seen, however, that when significant assessors began to enter from the outer rings of the target diagram, so that Weber could "make waves" in a new medium, then formal controls were put in place, and journals refused and/or rebutted heterodox publications. Weber's waves could have propagated through the social space-time of the outer rings even though the medium of the inner spaces would not transmit them. For this reason, inner space had to allow itself to acknowledge the waves temporarily, if only to reflect them back whence they came.[33]

33. Where core groups are weaker, formal institutions are likely to be a more important means of control. This suggestion seems to fit with theories of "social capital" (Coleman 1988; Bourdieu 1997), in which core groups are constituted of stronger or weaker ties in different sciences. To prevent a science going out of control, or to avoid scientists having to spend all their time trying to eliminate every loophole that supports any unorthodox approach, journals and funding agencies would be more anxious to suppress heresy at the outset, where core groups are weak. Perhaps we can now see why the peer-review system for physics journals is less combative than that of the social sciences—intolerance in the formal institutions goes with lack of control over interpretation in the informal institutions. Of course, to establish this comparison properly, we would need to compare the treatment afforded to heterodoxies in different sciences in the same detail as we have examined HighVGR.

CHAPTER 22

SCIENTIFIC CULTURES

Type I Errors, Type II Errors, and Joe Weber

There is yet one more way in which we can analyze the relation-ship between the core group, now increasingly dominated by big interferometry, and the maverick bar scientists. Evidential cultures which have to do with scientists' approach toward the use of ev-idence in general, rather than their views about specific pieces of theory or data.[1] I will introduce the notion with a final visit to the world of Joe Weber, and then examine the interactions be-tween Italian, Australian, and American cryogenic bar groups and the interferometer community.

The easiest way into the notion of evidential cultures is through a simple idea drawn from statistics—the difference be-tween "type I" and "type II" errors. Scientists use statistics to try to separate "signal" from "noise" when the signal-to-noise ratio is low, asking themselves, Could this piece of data, which stands out above

---

1. Knorr-Cetina (1999) talks of "epistemic cultures" within whole sciences. My evi-dential cultures are differences within one science.

the general level of noise, be merely a chance configuration of extra-loud noise? They ask how often chance alone would have produced such an apparent signal and decide that if this would have happened only rarely due to chance, then the apparent event really was a signal; whereas if it could have happened frequently due to chance, then it probably was not a signal.

There are no arithmetical procedures for choosing numbers to represent "rarely" and "frequently," however. Which is to say there are no such procedures for assigning the number implying "signal" and the number implying "noise." Indeed, different sciences choose different numbers. For example, in the social sciences, if you can conclude that a certain "signal" would only have arisen by chance five times in 100 (a two-standard-deviation level), then you publish your finding as a result. In physics, however, you would want to be able to say that the result would have arisen by chance only one time in 1000 before you were ready to publish (a three-standard-deviation result), and you might demand even more statistical significance before starting to take the data seriously (as in fields such as high-energy physics). In the earlier argument, when Weber seemed to find a 2.6-standard-deviation result (roughly one chance in 100 that it could have been due to chance) in what turned out to be noncoincident runs between himself and the University of Rochester, he was later able to claim that 2.6 standard deviations was not statistically significant (in physics), so nothing positive had really been claimed at all. 2.6 would be a great number in sociology or psychology!

Each time one makes the choice about the statistics—the choice about whether or not the signal really was a signal—one risks making two complementary kinds of mistakes. One might make a type I error or a type II error. Here are the definitions:

- *Type I error:* you say there is a significant effect when there really isn't; a false positive.
- *Type II error:* you say there isn't a significant effect when there really is; a false negative.

Individual physicists, and groups of physicists, differ in their preference for the kind of error they would rather commit; the choice is between being very conservative and making more type II errors and being less conservative and making more type I errors. Weber, as we have seen, was a scientist of the second kind: he was relatively adventurous in his interpretation of his data, so if he were to err, he would rather make a type I error than a type II error.

Let me risk some pop psychology and suggest reasons this might be. First, Weber's apprenticeship in physics was not an orthodox one. He was an electrical engineer who was not trained within the physics establishment and would not, therefore, have absorbed its conservative values. For example, at one point Weber told me that he did not know what "double blind" experiments were until long into his gravitational wave detection career; it also seems that he did not appreciate the generally perceived need to avoid all possible accusations of human bias in data analysis until this was explained to him by high-energy physicist Luis Alvarez.

> [1995] When this work started, I was approached by the late Luis Alvarez, who said that it was very important to do the experiments in such a way that you cannot be accused of biasing the data. And he said that you should record everything you can on tape and have an independent person do the data analysis. . . . And also he suggested double-blind experiments: for example, having people put signals in at times only known to them, and seeing if the program finds a signal.

Second, as an outsider without the immaculate credentials which only elite institutions can provide, his best hope of gaining notice, as he was soon to find out, was to do adventurous physics. Third, and perhaps most important, during World War II Weber served as an officer in charge of the technical equipment on a submarine chaser. In wartime, it can be fatal to make type II errors, but it is never fatal to make type I errors. We have here something similar to what I have described as "Pascalianism" in the funding system. If there is a signal, one had better find it; and if it turns out to be false, little has been lost. The way to do the job is to search the dials and "tune in" to any vestigial signal that indicates its presence. What Weber was accused of by many of his antagonists was "tuning in" to gravitational waves: that is, adjusting the parameters of his apparatus and data processing until a signal stood out clearly. This is just what you should do aboard a submarine chaser, but it looks like statistical massage to a high-energy physicist. A physicist explained to me how he saw things.

> [1995] Joe would come into the laboratory—he'd twist all the knobs until he finally got the signal. And then he'd take data. And then he would analyze the data: he would define what he would call a threshold. And he'd try different values for the thresholds. He would have algorithms for a signal—maybe you square the amplitude, maybe you multiply the

things ... he would have twelve different ways of creating something. And then thresholding it twenty different ways. And then go over the same data set. And in the end, out of these thousands of combinations there would be a peak that would appear, and he would say, "Aha—we've found something." And [someone] knowing statistics from nuclear physics would say, "Joe—this is not a Gaussian process—this is not normal—when you say there's a three-standard-deviation effect, that's not right, because you've gone through the data so many times." And Joe would say, "But—What do you mean? When I was working, trying to find a radar signal in the Second World War, anything was legal, we could try any trick so long as we could grab a signal." And [someone] said, "Yeah Joe, but somebody was sending a signal." And Joe never understood that.

Scientists told me in outraged tones that when Weber was asked why he could see signals when no one else could, he would say that it was because he tried harder; that is just the right action to take on a warship, but the scientists saw it as finding signals that were not there.

Here ends our last substantive visit to the work and life of Joe Weber; we now turn to cryogenic bars.

THREE DIMENSIONS OF EVIDENTIAL CULTURES

It is December 1996. At the Legal Seafoods Restaurant on the MIT campus, two groups of experimenters are arguing about how gravitational radiation data should be interpreted. The data they are talking about are coincident bursts of energy effecting on their detectors. The team from Frascati have suggested that it is reasonable to publish reports that say that coincidences have been seen even if they do not claim that they are gravitational waves; the group from Louisiana State University (LSU) disagree and would prefer to publish nothing.

> *LSU2:* That's a difficult thing—it's not the—. . . it's a reasonable thing to do, but it's not what most physicists want to see. Normally, you get to the end of an experiment and you give a result: "I didn't find the X-particle, and my results are consistent with 'X-particles will not be produced this way because their cross section must be less than something.'" That's a negative result, and a positive result is "I saw the thing." . . .
>
> If you say, "I saw something very unusual that doesn't seem as though it could be due to chance," then what do you do—that's the gray area—you're

not making a negative claim, you're not making a positive claim, you're somehow in the middle.

Frascati1: You know, one consequence of all this is the following—[it is] very strange. That, you do a coincidence experiment; if you find nothing—in other words, if you find a number of coincidences equal to the number of accidentals—then you publish, and you give the upper limit [i.e. you say what the maximum flux of gravity waves can be]. If you find an excess [a positive result], you don't publish—that's the conclusion [laughter]. Don't you think so?

LSU1: There is a danger of that—there is.

Frascati1: If you find nothing, you publish; if you find an excess, you don't publish—I mean an excess with minimum probability of 1% or 1 per 100. You see, the only excuse not to publish is if one is afraid that by publishing there might be bad consequences, like not getting money from the funding agency, or things of that sort. Then you can be excused for not publishing, but it is a bad thing. But—of course—

LSU1: Well, we have been in that situation, too—in years past—I don't think we are anymore. But—there's—

LSU2: The problem is particularly difficult for us. It's not the normal problem in science, I don't think—frankly. Or at least a fair fraction of science. If we really did see a gravitational wave at our current level of sensitivity, it would be in conflict with what many people believe are well-founded, established facts. It might be in conflict with the general theory of relativity, which is not—

Frascati1: No—!

LSU2: Well, if we saw gravity waves, what possible conclusions can you draw? What is the theory?—

Frascati1: Well, first point is that a coincidence excess might be due to another phenomenon, not gravitational waves—a very important phenomenon, even, at this point, even more important, that nobody knows about.

LSU2: That's a second possibility, that there's unanticipated physics operating. Well, that would be—that's a good reason to publish if you've got evidence that's pretty clear. But in my opinion you have to be pretty certain that the chance of this being ordinary accidental behavior has to be pretty small to make it worth it to even start to believe such a thing.

Frascati1: But the definition of small is not easy....I think each one follows his own philosophy. My personal philosophy is the following: if I have a situation where I don't know whether I should do something or not to do it, I do it. Because the outcome is better than no outcome. Now this does not apply—

LSU1: I was gonna say that that could lead to overpopulation of the world. [laughter]

Frascati1: Because, if you don't do it, it's nothing—if you do it, it's something.

Let us call the two positions represented in this conversation open and closed evidential cultures. (It is partly the peculiar nature of gravitational wave detection science that makes the clash of evidential cultures expose itself so readily.) When science is done by individuals or by small teams, experiments are worked up, data are taken, analyzed, and interpreted, and then "results" are published, all under the control of one laboratory. The boundary between the closed world of the laboratory and the open world of journals and conferences can be policed by scientists who either work on their own or are responsible for leading a physically and morally integrated team to a consensus. Even while the science remains within the core set or core group, the difference between developing an experiment, evaluating the data, and announcing the results for further evaluation beyond the laboratory is clear to the scientists. The idea of the laboratory as a place where the consequences of Nature's agency are observed and reported with a minimum of human intervention is maintained by keeping the laboratory closed, which is why "distance lends enchantment." But in the case of gravitational radiation data from burst sources, because findings are sets of coincident readings, responsibility for initial interpretation is likely to be the prerogative of more than one team.[2] A result has to come from more than one bar because of the need for coincidences. There cannot be any such thing as a finding until the data streams have been combined. All science is, in the last resort, communal; but in the search for bursts of gravitational waves using resonant bars, collaboration across laboratories is nearly always a *condition* for the *initial* announcement of a credible result.

Such an intrinsically intimate relationship between diverse groups of scientists is very unusual. There are, of course, many examples of replication by other groups, or collaborations in which the findings of different laboratories are aggregated, but in this case there are no positive findings at all until the raw outputs of the groups are combined. Consequently, clashing research styles will cause the laboratories to pull against each other like convicts from a chain gang trying to run in opposite directions. Once the social analyst knows what to look for, the predicament is striking.

Evidential Collectivism and Evidential Individualism

Evidential cultures can be usefully separated into three dimensions: evidential collectivism versus evidential individualism; high versus low evidential significance; and high versus low evidential thresholds.

2. Only the Italian bar group at Frascati and LIGO have more than one detector of their own.

At some point the results of an experiment have to be made public. "Evidential individualists" believe that it is the job of the individual or group of authors to take full responsibility for the validity and meaning of a scientific result before it leaves the laboratory. Under this philosophy, as much responsibility as possible is gathered into the individual or named group of researchers. In contrast, "evidential collectivists" believe that it is the job of the scientific collective—the whole network of other individuals and laboratories—to join the process of assessment of results from an early stage. Under this philosophy, the responsibility for establishing the truth is seen as widely shared rather than gathered into identifiable persons or groups. In the case of both individualists and collectivists, the ultimate arbiter is, of course, the scientific community; but the evidential individualist will consider it a matter of professional failure if the community rejects results sent forth from the laboratory, while the collectivist thinks discussion by the core set or core group—whether the outcome is acceptance, rejection, or a demand for further clarification—to be a normal part of the scientific process.[3]

As far as I know, the individualism-collectivism dimension has not been thought about before, either in science or in the rest of social life, but it seems to be a choice that is found in many of our activities. Looking at these other areas helps us to see what is going on in science. For example, the two philosophies can be compared to styles of car driving. The first duty of a British driver, or an American driver, is to "follow the rules" so as not to upset the equanimity of others who are driving acceptably—this is an individualist ethos. Rome, on the other hand, represents a case of driver collectivism. On the crowded highway skirting the city, I once stopped my car on a busy roundabout, blocking a lane completely while I conspicuously consulted a map. In Britain, such behavior would be almost unthinkable; every passing driver would hoot and gesticulate, exhibiting what has become known as road rage. In Rome, the approaching traffic simply drove around the obstacle I presented without a second glance.[4]

3. Evidential collectivism differs from the norm of "organized skepticism" (Merton 1942). Although a collective activity, organized skepticism is meant to *check* the validity of *individuals'* findings rather than interpret them. In any case, the idea of organized skepticism has its roots in the supposed ready reproducibility of scientific findings rather than the intentionality of scientists' actions. For a discussion of the complexity of replication, understood through the notion of the "experimenter's regress," see Collins 1985/1992. The Mertonian norm of "communism" refers to the collective *ownership* of results, not the collective *establishment* of results.

4. Another such driving story was told to me by an American physicist. He explained that once on Rome he found his car had been parked by someone else, parallel to the curb, nose to tail with the cars at either end, on a road with a dense and slow-moving stream of traffic squeezing

In sum, in this part of Italy, responsibility for avoiding accidents and ensuring a smooth flow of traffic is passed to the community of drivers; this works well without engendering fury. In Britain and America, the responsibility remains with the individual, and departures from the norms are met with sanctions.

That this is a matter of driving norms rather than national character is evident, because the relationship between motorists and pedestrians is reversed in Italy and America. In most of the United States, drivers take it as their responsibility to avoid damaging pedestrians, who seem to think it their right to cause any amount of inconvenience to the driver. In California, the traffic will immediately stop for a pedestrian who steps off the curb, or even approaches the curb in a meaningful way, with drivers halting many yards back and politely waving the walker across the road.[5] In Britain, it is the pedestrian who always defers to the car on pain of injury or, at least, indignation, and the same applies in Italy.

It has been put to me that the terms *individualism* and *collectivism* are here being used the wrong way around, and it is the Roman drivers who are the irresponsible individualists, while the American and British drivers are far less antisocial. But there is no such thing as a society made up entirely of individualists; everyone is following one set of rules or another, whether explicit or implicit. The deep point is that in Rome it is the community that is given much more responsibility for maintaining order, so the individual can act what with what looks like carelessness; the collectivity "repairs" any potential disruption before it becomes serious. In the British-American approach, it is individuals who are meant to retain much more responsibility for the smooth organization of traffic-society.

As I have suggested, these alternative ways of organizing society in science and driving reflect similar choices made throughout social and political life; after all, what we are talking about is different ways of dividing labor, in this case cognitive labor. Consider, then, that there are two ways to organize an undergraduate course in a highly politicized subject. One can insist that every teacher present an unbiased course; in this case if,

---

by right next to the parked cars. But as his initial maneuvers indicated his desire to leave the parking space, the traffic compressed itself and moved across the road in such a way as to leave him room to extricate himself and join the traffic stream. The initial act of parking was, one might say, antisocial, illustrating, as he initially thought, the careless individualism of Roman drivers, but the cooperation of the community of drivers made it into a reasonable act.

5. This point cannot be generalized to whole nations without risk—pedestrians and drivers have a quite different relationship in Boston.

for example, teachers favor a Marxist approach, they must also put the counterarguments within their course. Alternatively, one can allow each teacher to teach according to his or her convictions, producing lively courses and good classroom interactions, but make sure that the students are exposed to other teachers with different views so that the faculty as a whole is balanced. The first of these solutions is what we might call pedagogically individualist; the second is pedagogically collectivist in the sense I am using the terms here.[6]

To return to gravitational radiation, I am arguing that the Frascatian approach (shared with the laboratory in Perth, Australia, to be introduced below), is set within the kind of collective ethos that is found among drivers in and around Rome. The Louisianan approach to science is better compared to the driving style found on the majority of British and American roads.

Sociology would be a lot easier if social life arranged itself neatly. Unfortunately, evidential collectivism is not the uniform style of Italian science and/or Australian science, or even any single laboratory within those countries. For example, the Frascati laboratory includes many members who are as much evidential individualists as most Americans, and the discussion about whether to publish the contentious findings was carried forward as much within the Frascati laboratory as between it and the rest of the world. For this reason, I will henceforward refer to the subset of Frascatians who shared the evidential culture under discussion as the "Frascati Team," while retaining the usage "Frascati group" for the whole laboratory. To complicate the picture further, there may well be groups of evidential collectivists scattered within the dominant culture of evidential individualism of Britain and America. For example, it was suggested to me that Bill Fairbank was an evidential collectivist, ready to issue bold claims from his laboratory, such as the announcement that he had found free quarks (chapter 13). On the other hand, Joe Weber does not seem to fit the pattern of evidential collectivism, though he is at the extreme end of another dimension of evidential cultures to be discussed below. He was never ready to accept that he might be shown to be wrong by the community, so when he issued his results it

6. Consider again that those who insist that reflexivity is a vital part of the sociological analysis of scientific knowledge are individualists in the sense used here, because they believe that it is the duty of the *individual* to produce a complete analysis: the analysis must include not only a discussion of the social influences on the science under examination but also an analysis of the social influences on the analyst. The collectivist (such as myself, in this instance) believes it is satisfactory to complete the analysis of the social influences on the science, leaving other members of the community to analyze, if they are interested, the social influences on the individual (as in Ashmore 1989).

was not in the spirit of passing them to the community to assess; it was in the spirit of the evidential individualist who felt he had done enough to be sure he was right.

That said, it is interesting that the principal Italian physics journal, *Il Nuovo Cimento*, seems to express a collectivist philosophy. I asked its editor about his policy regarding the gamma ray correlation paper published by Weber and Radak in 1996. The editor told me,

> [1997] Well, this is a complex issue, and it is very much a matter of the judgment of the editor in this case.... My opinion as an editor is that not all the papers have to be necessarily correct, but we must make every effort possible to publish papers which are not wrong–a priori wrong; if a paper is [obviously] wrong and we publish, this is unacceptable. But if a paper is not obviously wrong, and it is on a topic which is important, I think it can stimulate discussion. And the important thing is finally to have an answer–to promote the discussion and to reach a definite answer.
>
> Now, in this sense, even if something is not absolutely proved to be final, it could generate other people to look, in order to disprove it, for example–to stimulate a search...
>
> This is a general approach that I have been following. There are, for example, other cases of theories which are not settled, which I decided to publish anyway in order to provoke–to generate discussion....
>
> So long as things are not done randomly, but are done by an editor in order to converge to an answer, that is OK.

It is tempting to speculate about the historical roots of these phenomena. Steven Shapin examined the origins of scientists' actions in the norms of English gentlemanly behavior in the seventeenth century. The norms described by Shapin can, with a little stretching, be mapped onto evidential individualism. Questioning another gentleman's assertions could be construed as branding him a liar–a charge that could lead to a dueling challenge. The prohibition on "giving the lie" imposed a reciprocal obligation, making the validity of observations very much the responsibility of the individual even while obeisance was given to the idea that the community was the ultimate arbiter. Thus, it may be that in exploring the relationship between English gentlemanly norms and science, Shapin has discovered the origins of evidential individualism. It follows that other national traditions or religious cultures may have given rise to different expectations of the way scientists should act, the traces of which are also still visible in modern science. For example, Leopold Infeld in his autobiography remarks that the attitude of pre-war English journals was "better no paper than a wrong paper," while German journals felt "better a wrong paper than no paper at all."[7]

Evidential Significance

Now let us turn to the second dimension of evidential cultures. In his description of the search for solar neutrinos, Trevor Pinch was the first to discuss what I am treating as one dimension of a larger phenomenon.[8] As Pinch explained, experimenters try to detect neutrinos through their ability to transmute the chlorine atoms in a tank of cleaning fluid into radioactive argon. Putatively, radioactive argon atoms betray themselves as marks on a chart recorder. The experimenters can report this as "marks on a chart" (which is the only thing they see "directly"). More interestingly, they can do more in the way of imputing meaning to these marks. The amount of meaning they impute—the length of the inferential chain—is a matter of judgment. Thus the marks can be said to be evidence of the existence of radioactive argon, solar neutrinos, events taking place in the sun, or whatever. This is an increasing scale of *evidential significance*. The higher the evidential significance, that is, the longer the chain of inference, the more important the findings. But the more important the claim, the greater the risk of engendering opposition; and the longer the chain of inference, the more ways there are of being wrong. Thus, a choice of high evidential significance entails what we might call high interpretative risk, while claiming low evidential significance involves only low interpretative risk.

In a similar manner, in gravitational radiation detection the same coincidences can be reported simply as coincidences or as gravitational waves. A claim to have seen gravitational waves is far harder to support and far less likely to be credible than a claim to have seen some coincidences. In the Legal Seafoods conversation, we can see that the Frascati Team is happy to announce unadorned coincidences, whereas the Louisiana group wants to announce nothing unless they are sure they have gravitational waves. The Louisiana group are demanding a higher level of evidential significance before a result is counted as worthy of announcement.

---

7. Shapin 1994; Infeld 1941, p. 190. Dear (1990) discusses the Protestant and Catholic influence on the different approaches to experimental natural philosophy in Britain and the European continent.

8. Pinch 1981, 1986.

Evidential Threshold

A third independent dimension of evidential culture is the choice of evidential threshold, which we have already discussed in the context of Weber's work. It is a question of how much statistical risk the scientist wants to take. Irrespective of whether the collectivity or the individual is thought to be the proper locus of scientific findings, and irrespective of the degree of interpretative risk accepted, a scientific culture might be more risk averse or less in terms of the level of statistical certainty. As explained above, the social sciences tend to publish results with lower statistical significance than is counted as acceptable in physics. Thus, in our language, the social sciences have a low evidential threshold, while high-energy physics has a high evidential threshold.

One might argue that the whole field of gravitational radiation research would never have started were it not for Joe Weber's high-risk—in both interpretative and statistical senses—reporting of putative phenomena. Most scientists having a long acquaintance with gravitational wave physics would agree that he founded a field as a result of pressing forward obstinately with his high-risk claims.

In the Legal Seafoods conversation, and in the section on signal thresholds to be discussed below, one sees clearly that the Frascati Team endorses a strategy involving high statistical risk—that is, low evidential thresholds—together with low interpretative risk (low evidential significance). The Louisiana group reveals the opposite position.

OPEN AND CLOSED EVIDENTIAL CULTURES

Putting together these three dimensions enables us to imagine an "evidential culture space." In figure 22.1, the Louisianans are at one corner of the imaginary space—low collectivism, high significance, and high threshold—with the Frascati Team at the opposite corner—high collectivism, low significance, and low threshold.

What all three dimensions have in common is that the position adopted by the Frascatians on all of them tends to early release of relatively unprocessed data, whereas the position adopted by the Louisianans causes them to restrict access to their results until they has been much more highly processed: The Frascatians have an open evidential culture; the Louisianans, a closed evidential culture. These two corners of evidential culture space also represent the change in approach from the early days of gravitational

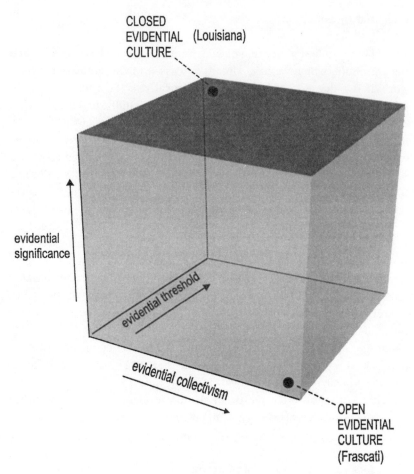

Fig. 22.1 Evidential-culture space.

radiation detection to the present; once it was reasonable to "point a de-tector at the sky" and report what was seen; nowadays what it is legitimate to report is very much more constrained by theoretical considerations.

The tension between the corners of the evidential culture space was experienced within the Frascati laboratory itself, and the polar positions there were described to me as the symptoms of, respectively, an *experimental animus* and a *mathematical animus*. These are useful labels in this case, because of the current role of theory in holding back speculation.[9]

---

9. In a field such as biology, it might well be experiment that holds back "irresponsible" theorizing, so the terms *mathematical animus* and *experimental animus* would label different points in the same space.

The experimental animus is very well illustrated by SN1987A paper. In it, the authors state that the energy in the waves they claimed to see was outside the realms of theoretical possibility. Furthermore, the levels of statistical significance were not outstanding. Nevertheless, the Frascati Team believed that this data should be published and looked at by the rest of the community; as far as they were concerned, something had been found, and it could not be wished away. In contrast, the mathematical animus, taking experiment to be the servant of theory, would have suppressed the data.

Currently, the most widely accepted theory paints a picture in which, in terms of the emission of "visible" gravitational waves, the sky is almost completely black. Under this theory it is virtually impossible for the cryogenic bars to see anything except extraordinarily rare events such as might occur a few times a century. But in harmony with their evidential culture and institutional position, the Frascati Team felt that theorists' claims about the structure of the heavens should not dominate experimental work.

> [1995] [I]f we find something, it is because there is something new. Because Nature is kind to us.

> [1996] I think that—we are aiming to make a discovery. And no discovery can be made if one has a mind which is within a frame that does not give the freedom to think something different, something new. It seems to me so obvious. Actually, we started this experiment with this idea in mind, because I said that already, and I repeat it now—if the gravitational waves that are expected are exactly those which are predicted by the physics, nobody will see gravitational waves from now for the next 50 years. I hope we will both live long enough to find out. So, even if the laser [interferometry] people don't say that, I think everybody hopes that the situation is different.[10]

Again, the Frascati Team's evidential culture is well set out in its attitude toward publication, as expressed in the following quotations, recorded on separate occasions in 1996. Both reflect a leaning toward low evidential significance, the first combining it with high evidential collectivism, the second with readiness to adopt a low evidential threshold.

10. Sam Finn pointed out to me that it is possible to have an adventurous view of the cosmos and still be conservative in terms of evidential practice; it certainly is true that one does not follow strictly from the other, though without an adventurous view of the cosmos, one is less likely to want to release results early.

> *Frascati1:* You should say, "Look"—no, it's a difference of how you publish. You should say, "Look, we found this result—we inform you—so if it's the case that you find the same results, you will report it to us." You should not say that you found gravitational waves.... Other people should say that.

> *Frascati2:* There is a coincidence excess over what is expected. Now, is this coincidence excess enough to make an extraordinary claim or not? What is the probability that the excess is by chance? Calculating this, if it is a few percent, I understand. Is a few percent enough or not? Well, maybe it is not enough to conjecture that these are gravitational waves, because to do that you need extraordinary evidence. But it is enough to publish the results, saying, honestly, "We have these results."

In contrast with the Frascati Team, a member of the LSU group provided the following by way of explanation of their unwillingness to be associated with anything but the most secure results.

> *LSU1:* [O]n thinking about it, I'm pretty sure that foremost in my mind was, there is no way that I want to do anything that is going to have me labeled as a crackpot.... I suspect that underlying all of this is the desire to really say, "Well, OK, if my work is going to be significant, I've got to make sure that no one thinks that it's crazy."

Irving Langmuir Again

In chapter 9, we first encountered the notorious paper by Irving Langmuir entitled "Pathological Science." We described how Richard Garwin sent a copy to Joe Weber, urging him to read it so that he could see his own scientific practices fit best with Langmuir's pathologies. In this debate Langmuir, unsurprisingly, surfaced once more. The desire not be cast as a pathological scientist—not to be "labeled as a crackpot"—is strong among physicists, partly because of the influence of Langmuir-type thinking. In the Legal Seafoods discussion, Langmuir was invoked in that vague way that signifies its mythological role. In this conversation no one can remember the name of either the author or the paper, but they all know what is being referred to. ("Pathological Science" was reprinted in *Physics Today*.)

> [1996] *LSU1:* There have been a number of famous instances in physics where people thought they had something that was right at a threshold, and they always turned out to be threshold phenomena—that's what scares me.
> *Frascati1:* But in many cases they made mistakes. For instance...

LSU1: Well, did you see that article—it appeared in *Physics Today*—um.

Frascati1: Oh, yes, many years ago, you mentioned it to me.

LSU1: Well, a couple of years ago.

Frascati1: No, four or five years ago—you mentioned it to me.

LSU1: Well, maybe. What was his name? The vacuum man at GE. He went back in history and looked at several events from the 1890s on to the present day, the latest one he had was in the 1930s with an apparatus that could supposedly affect the atomic number of things by putting them, if I remember correctly, in a weak magnetic field or something like that. And, um, then there were a lot of papers published on it that later went away, but it depended on the behavior of something just when it was going above threshold, in all of the things that he spoke about. What was his name?

Unidentified speaker: I know who you mean, the guy who did all the—

LSU1: —he did all the surface physics.... um, I will think of it, it will come to me eventually. Anyway, you read the article, and I would say if there's anything that's influenced me on this, its been that I am worried about things that are just a little bit above the threshold.

Frascati1: But, you see the problem with that article is that he makes a selection of the data. [greeted with laughter] In other words, he's showing a special case. Now, I remember Amaldi, 40 times, told me the story, when they started in Rome, with Fermi . . . when they did experiments with radioactive materials in 1933 or '34. They missed—they thought they had fission, but they were not courageous enough to—not enough fantasy—in spite of having Enrico Fermi with them—that they indeed were observing such an important phenomenon, like the fission of an atom, and a couple of years later this was discovered by other people in Germany. But they had that possibility in their hands. Another example is the discovery of the new particles by Carlo Rubbia at CERN. You know that there were two experiments doing the same thing—one was headed by Rubbia, and the other was headed by a very brilliant French physicist—I have forgotten his name—he is a CERN physicist, a very good one. And the French group hesitated because they wanted to do all possible checks. But the Rubbia group, pushed to publish in spite of the fact that the evidence was not so big, so they published—

LSU1: So they published and they were right.

Frascati1: And the other people didn't do it. So one has to be careful.[11]

Citing Langmuir supports a closed evidential culture; citing Rubbia (and in this case, Amaldi) supports an open evidential culture.

11. For an account of the Carlo Rubbia incident, see Krige 2001.

Intention and Interpretation

The low evidential significance strategy would be fine if the development phase of an experiment could be kept clearly separate from the process of making a claim about the existence of new phenomena. As a member of another sympathetic group put it,

> [1997] [I]t is only by poking around at the data that you sometimes manage to make a discovery. Because of reproducibility, results must be the same or similar in the next experiment—and because of the ability to repeat the experiments . . . data massage is not as easy as [some] like to claim.

This respondent is saying that if a discovery is to be made when the "signal" is well hidden in the "noise," then it is necessary to have an exploratory phase in which data are pushed and poked in the attempt to make it yield a secret—just as with Joe Weber tuning his radar set to find a submarine, or tuning his gravitational wave detector to find a gravitational wave. This respondent is also saying that the exploratory and the later confirmatory phases can be kept separate. But keeping report and interpretation distinct is not easy, because the meaning of a paper is in the eye of the beholder. It is not always the case that papers are read or interpreted in the same way as the author intends.

> [1996] *LSU:* You were telling me that the paper you wrote with Weber, you did not claim gravitational waves. Everyone thought you did, but—
>
> *Frascati1:* Well, at CERN they said, "Oh—you publish a paper with Weber—Oh! Oh!" [I said,] "Well—have you read the paper?" [They said,] "No—just the title." I forget what it was, but it gave the impression that we found something. Look, I spent six months looking at the data—I did a data analysis. Why shouldn't I publish what I found?

The Frascati respondent is saying, "All we did was publish something suggestive, and others immediately assumed it was meant as far more than this even though they did not read it." Another passage from the Legal Seafoods conversation makes the same point.

> [1996] *Collins:* Let me intervene in this again. What [Frascati1] was saying earlier was that he doesn't want to publish this as a claim for gravity waves, but just publish as—you know—"We found this data." But what you're worried about—tell me if I'm right—is that in fact people won't read the paper this

way. People will read the paper as a claim for gravity waves whatever it
says. . . .

Unidentified speaker: Sure—people want to reduce it to "yes" or "no."

Collins: People will reduce it to "yes" or "no."

LSU2: That would be my guess.

Collins: OK, but hang on. Let me ask this question: Why are you worried about
that? . . . why are you worried if everybody does read it in this way? You know,
if it's not written but they read it, what's the problem?

LSU2: I'm thinking about it. . . . One is, you could certainly expect a lot of ridicule
from your colleagues at some point—maybe not overt but covert—if they find
the result somehow ridiculous. I mean, you know, everyone values the good
opinion of their colleagues—

Frascati1: —This is true.

There are no Americans among the gravitational wave research community
still alive who are sympathetic to an open evidential culture, and some
of them, perhaps in keeping with a strong individualistic ethos, despise it.
The most forceful expression of this point was put to me as follows:

> [1995] It was mostly at these general relativity meetings. . . . They would give
> their presentations in such a way that they would lead you, they would show
> you this data, and they would show you the events, and they would show you
> some statistics they'd done, but never enough of it so that you could really
> get your arms around it. And they left you with this tantalizing notion that
> they could go either way. They either could make a claim that they had
> detected something if they had wanted to, if they had gone the next step in
> their presentation, or they could back off and say, "Well, yeah, maybe the
> statistics isn't good enough." And they left you at that critical
> juncture. . . . What would happen is that you had to draw the inference. Now
> that gave them the freedom at any point later on, or maybe even . . . to say,
> "Well, if we choose to say this, we have detected it, or if we choose to
> interpret that way, we haven't detected it, because the statistics isn't good
> enough." It was this . . . ambiguity—OK? That got to me—OK?

In another context the same scientist said, "When one publishes data and a
result, it has a pedigree associated with it, and your reputation as a scientist
rests with how well you have analyzed and interpreted it." The worry here is
that an "open" research claim could too easily be used to the researcher's
advantage—allowing a claim on the Nobel Prize to be made without the
risk of being wrong.

Less forceful Americans, such as the members of the LSU group, also disagreed strongly with the Frascati Team's approach.

> [1996] [W]e just don't agree on the theory of data analysis...when he says, "I'm not claiming gravity waves, I'm just putting out what I have." And that's perfectly acceptable, except for the fact that, if you look at Weber's early papers..."I wasn't claiming gravity waves, I just said what I had," and it is that legacy, I guess, in American gravitational wave physics, that has me very, very cautious of doing that kind of publishing....And that's never gonna change, because it's just a different way of looking at the physics we do and we just have to understand that we, as individuals, are different. And that's where your problem comes, actually, as a sociologist, and you're trying to say, "OK, what is it that enters into these things?" And it is much more than just the science, as you know.

EVIDENTIAL CULTURES IN THE CONTEXT OF INTERFEROMETRY

Chapters 19, 20, and 21 moved us from discussion of the increasingly heterodox nature of Joe Weber's science to the institutions and networks that supported the work, and thence to the context in which the work and the institutions were set. We now make a similar move. In the above sections, we have looked at what I have called different evidential cultures; now I want to show how they fit in a wider context. We move from the core to the political and policy ring of the target diagram.

I am going to discuss the context of three sets of scientists using cryogenic bars to detect gravitational waves: the Frascati Team, in this case using their EXPLORER bar; the Louisiana (LSU) group, with ALLEGRO; and the group at the University of Western Australia in Perth, led by David Blair, running the niobium cryogenic bar, NIOBE.

The long-standing LSU group[12] has always received its funding from the National Science Foundation. Its members are typical research-active university faculty. The Frascati Team is located within the complex of government-supported research facilities at Frascati, near Rome, and draws its members from various universities. Its funding, based heavily on civil-service principles, seems lavish compared with the American group; and once the funds are granted, there appears to be much more local and long-term control over how money is spent. The Perth group is the only bar group in Australia

---

12. The LSU group is the only state-of-the-art resonant bar group left in the United States.

and typically scrapes along on what funding it can get from the relatively poorly resourced Australian Research Council.

Let us ask first what we might expect the relationship between the bar community and the interferometry community to be in terms of their preferences for type I versus type II errors—false positives and false negatives: is the major risk not finding a signal that might be there, or is it claiming to find a signal that is not there? So long as the bar groups are on air, while the interferometers are still not up and running, the two technological approaches occupy the two poles of the dichotomy. The potential damage caused to the interferometers by false claims from the bars is their principal worry; they have little to gain should the bars really detect something. The bars, on the other hand, are in a race against time. They are losing credibility to the big interferometry program and will be overtaken in sensitivity once the interferometers are on air. Their very survival as a credible technology, at least in the short and medium term, could depend on their finding signals fast. The bar groups ought to prefer type I errors, while the interferometer groups would prefer the bar groups to lean toward type II errors.

And yet this is not quite what we see. Up to early 1997, the time at which this account is set, the LSU group had never been party to any claim to have seen signals consistent with gravitational waves; over two decades they had offered only upper limits: "Gravitational waves can be no stronger than such and such because our bar is of such and such a sensitivity and has seen nothing." The Frascati laboratory, on the other hand, did what we would expect. They had issued the two positive reports jointly with Weber in Maryland that I have discussed at length: the one published in 1982 giving the 3.6-standard-deviation result and the one published in 1989 concerning SN1987A.

As 1996 turned into 1997, the Perth group's actions also began to match what we would expect; they joined with the Frascati Team to report another positive finding. By February 1997, the Perth group had made at least two conference announcements to the effect that they had recently found strongly suggestive coincidences between their bar and that of Frascati. These claims ran into strong opposition.

We have, then, two questions to answer: Why was the Louisiana group so conservative when it would seem that its interests would be better served by being less risk averse? And why did anyone take any notice of Frascati-Perth claims as late as 1997? The second question arises because, as we have seen, Weber's 1996 paper was ignored by the establishment, and I have suggested that this was because LIGO was already securely funded.

To begin to answer the second question first, though LIGO was being built by 1996, the funding was always felt to be at risk. Under the US system, additional expenditure has to be approved every year, and the demise of the Superconducting Super Collider, after substantial outlay, left researchers feeling vulnerable; a number of physicists migrated to LIGO from the dead Superconducting Super Collider project with this memory burned deep. In any case, LIGO itself is just the first step in the gravitational wave detection program for interferometers. Among the scientists it is known as LIGO I, with the plans for LIGO II already being well advanced. Thus LIGO scientists felt that the future of LIGO could still be jeopardized by what they saw as the potential for further reckless reporting of incorrect results by members of the bar community. Weber, however, was no longer a threat, as his credibility with the policy makers and others had been destroyed; we might say he was a coconspirator in the demise of his threat to LIGO because of the way he isolated himself from everyone. This is not so true of the other bar groups.

The answer to the second question and the answer to the first have something in common, and this is that the bar groups did not isolate themselves from LIGO. Guido Pizzella may have been ready to sacrifice himself in the same way as Weber, but this was not true even of all of those in Frascati. Blair's position on interferometers was somewhat ambivalent—while announcing the new coincidences with Rome, he was developing prototypes and trying to raise the funds to build a major interferometer in Australia. As I have already argued, gravitational wave science is unusual in that all the teams need one another to produce the coincidences that will give their findings credibility, but Bill Hamilton and Warren Johnson, the key players in Louisiana, saw LIGO as a major part of their future—and this, I would suggest, may have played a part in their choice of a conservative strategy. The relationship between Hamilton and the other bar groups continued to link maverick bar claims to the interests of LIGO, making them harder to ignore.

Methodological Aside: The "Antiforensic Principle"

In the sections that follow, I am going to try to explain why Bill Hamilton and Warren Johnson made the choices they made, with more explanations of this kind to be found in the rest of the book. But it would be wrong to think of what I am trying to do as the kind of attribution of motives for individuals' actions that are the stock in trade of courts of law and the like. What I am talking about is institutional and cultural pressures. I simply do not have the resources to do the work of the legal system in uncovering the internal states of individuals, and in any case courts of law often get it wrong. This is one element in what I call the "Antiforensic Principle." One should not, then, fall into the trap of thinking that sociological fieldwork has much to do with detective work; one should remember that sociology is about collectivities and cultures, not individuals, and that the actions of individuals should, at best, illuminate and illustrate those wider forces rather become the object of investigation. Here, then, the strongest claim I am making is that there is a homology between the actions of the members of the Louisiana and Frascati groups and their respective social settings.

The history of scientific heroism is the history of scientists resisting institutional pressures, and there is no reason to suppose that the principal actors who speak from these pages are not just as capable of resisting external pressure as their famous forebears. Of course, given that all scientists discussed here would readily concede that in the last resort, blinding is the only sure way to eliminate unconscious bias in data analysis, they would also have to accept the possibility that structural forces *might* affect their larger judgments in subtle and invisible ways. It would be very difficult to turn this "might" into anything definite, as the complex workings of law courts reveal. Fortunately, we do not have to prove the matter one way or another for any individual scientist; we have only to describe the pressures and the plausible patterns of action within the scientific community. That this is the way we *should* proceed is key to the Antiforensic Principle.

The Louisiana Group and LIGO

The link between the Louisiana group and the interferometers had several strands. As we have seen, Hamilton had once believed his group would benefit from spin-off research from the funding of LIGO. Warren Johnson in particular had made his commitment concrete by finding one of the sites for the two LIGO interferometers in Livingston Parish, near Baton Rouge, the hometown of LSU. LSU could visualize its physics department rising to a more prominent place through its association with the prestigious and high-powered LIGO activity on its doorstep. As Hamilton told me,

> [1994] [I]t is our hope that since we're already active in the gravity business, that if they come in and build a LIGO detector here, that this is going to enhance the program here, and it will. We've said that we would make some faculty appointments in gravitational physics and so on.... We see our position being with new faculty hires...who would be actively involved in using the detector and also quite naturally with people running it—it's about

> 30 miles from here...there's nothing over there [in Livingston
> Parish]—absolutely nothing. So you're not going to find your Ph.D.-level people
> living in the country town of Satsuma—they'll base themselves around here.
> We've got good computing facilities—I think it's just going to be natural.

Graduate students aside, Warren Johnson especially, and Bill Hamilton to a degree, did seek and to some extent find positions working with LIGO while continuing to work on their own bar. The last twist was that ALLEGRO was redesigned so that it could be reoriented to be parallel or antiparallel with LIGO's most sensitive direction, thereby acting as an extra detector in the LIGO network.

This, then, was a special reason for the Louisiana group not to want to upset the new mainstream, but the same considerations now apply to all gravitational wave groups in America and elsewhere. LIGO, because of its sheer size, dominates the employment market for gravitational wave scientists. One youngish physicist explained to me his attraction to LIGO in the following way:

> [1997] I'm certainly just following my gut. I'm certainly following the
> interferometers. That's where the money is, that's where the people are,
> that's where the scientific opportunity is, so that's where I concentrate most
> of my attention. It's just that—it's one of those unconscious choices that—it's
> a value judgment. It's the kind of value judgment that people make when,
> you know, you decide whether to have children or not, or things like that.
> They often don't make those judgments very explicitly, but they know they're
> based on some kind of understanding of things.

The tie-in to the mainstream interpretation of things also worked in a more direct way. Hamilton was funded by the same agency as LIGO, so whatever the reality, he felt he had to err on the side of caution. Under the American funding system, decisions are made as much by powerful program directors as by the committees they appoint. There are good and bad aspects to the procedure. On the whole, scientists prefer NSF program directors, who are well-informed members of the relevant scientific community, to committees. Furthermore, individual program directors are more likely to make braver funding decisions—for example, concerning interdisciplinary work or adventurous lines of research—than committees ruled by majorities. The program director in this case was, of course, Rich Isaacson. Isaacson, like other top NSF personnel, did his best not to lose touch with the core group. Thus he occasionally took research sabbaticals to perform theoretical work at the research front. In the early days, Isaacson was a respected theorist himself, developing a calculation that, as Hamilton pointed out, "was one of the really fundamental ones to show that gravity waves should exist."

The downside is that program directors' preferences are understood by the scientific community, so when their decisions go against a favorite project, disappointment may be personalized, for accusations of bias are more easily directed at an individual than at a committee. As American scientists put it to me,

> [1994] [I]f there is any one single individual in my opinion, in the United States, that has influenced the direction and the attitudes that people have on gravity waves, it would probably be Rich Isaacson, because he has been the man in charge of funding research.

> [1996] Isaacson does have very definite views, and by being in the position of power that he is, and by the way that the American funding system works, why, he has a good bit to say on what work gets done—no way to avoid it.

Members of the LSU group did report to me that they once felt that their project would have been in jeopardy should they have reported anything too dubious.

> [1996] *Collins:* You said earlier that there was a stage when you felt that if you had released findings, you felt your grant might be in jeopardy.
> *LSU1:* I know, absolutely, it was, earlier. There wasn't any temptation, because very early on our experiment didn't work well enough [for me] to be happy with it, but had I tried to say that there was anything, we never would have—it was one of those cases where, as [another member of the group] said, it would be so improbable that everyone would say, "This man cannot be trusted."

At the time of the events described here, the Louisianans were continuing to seek new funding in an atmosphere of increasing difficulty for resonant bar research. Thus, insofar as US science funding does discourage speculative reporting in this area, the pressures have not gone away.

Tying oneself in with LIGO also meant taking on what the Frascatians called a theoretical animus. This is because theory is much more important to LIGO than to the bar groups. As we have seen, a big science initiative such as LIGO must promise success to the government and its representatives. I have argued that there is a "policy regress" in this case because policy makers are still less able to judge the validity of claims made by scientists than scientists themselves. Policy makers will, therefore, reach for other grounds on which to make their judgments. But for them to make any judgments at all, there first has to be some science for them to believe or disbelieve! LIGO represents a step-function increase in funds for gravitational radiation research and promises a step-function increase in sensitivity; the new machine is meant to be good enough to guarantee (possibly in a later version) not only the detection of gravitational radiation if it is there but the founding of a new field of gravitational astronomy by using a brand-new technology. Because this is not a matter of gradual development based on existing technology, as would be the case, for example, with the building of a bigger telescope, there is only one source for such a guarantee—theory. It has to be theory that scientists refer to when they assure policy makers that their ideas will lead to success.

In the course of the search for gravitational radiation, from the late 1960s onward the importance of theory has grown. In the early days, expenditure was small, and it was quite reasonable to build something on a purely speculative basis. But with LIGO needing such large sums of money, the funding has to be done on a much responsible foundation, so theory has come to dominate experiment, at least in America. It is theory that has justified the spending of hundreds of millions of dollars, and it is only theory that can reassure the policy makers that major expenditure gives us the only chance of observing and understanding any glimmers of gravitational waves that disturb the gloom.

Probably the most important piece of theory is the model of the gravitational emissions coming from inspiraling binary neutron stars pioneered by Kip Thorne and Bernard Schutz. This moved the principal source away from supernovas, which are unpredictable because the gravitational energy they generate depends on how asymmetrical they are—an unknown. It was shown, on the other hand, that the output of inspiraling binary neutron stars could be closely modeled. In 1996 at a conference in Pisa, I managed to record a snatch of conversation between knowledgeable physicists talking about this feature of their field.

> [1996] *Ron Drever:* [E]arly on, everybody was going for supernovae, then it really was there were more reliable calculations [i.e., more reliable calculation came along later]. I think that it was felt that you wouldn't argue for something like LIGO without having much more reliable calculations. I don't think that they were anything like reliable ones [before]...it was a political reason that forced that [the generation of more reliable calculation].

Bernard Schutz: That's very true.

Drever: That's a kind of unusual one [unusual case]. It's a political reason that switched that [way of doing the theoretical estimates].

Bob Spero: I don't know. I think the emphasis would have switched anyway, but certainly politics brought it out.

Schutz: I think politics brought it out as a key argument. I think without coalescing binaries we wouldn't have the detector. Without the coalescing binaries we wouldn't have had confidence.

These sentiments were greeted with general agreement around the table.

Theory is integral to the interferometry project in other ways. For example, at the levels of sensitivity predicted for the first detections, using precalculated templates (e.g., for inspiraling binary neutron stars) will be the best way to have confidence that a real signal has been extracted from the noise. Again, instead of just looking to see what is there, data will not count unless they match a template—a template representing a theoretical consensus of how the universe works. If the templates are to have credibility, maverick theories must be suppressed—and this includes any theory that would allow the cryogenic bars to detect waves.[13]

Finally, theory predicts that the ultimate limit of sensitivity, set by quantum theory, begins to impinge on the bars at a point just where the signals should begin to increase significantly in number. The quantum limit for interferometers, on the other hand, is further off. Thus has quantum theory helped to change the balance between the bars and the interferometers even though both devices are so far from reaching the ultimate limit that this is not currently a matter of any practical importance. Theory, in sum, has aggregated influence in many ways; in this field, theory, once the servant of experiment, has become like a massive star with a gravitational field that affects the shape of everything near it.

Another element of what has been called epistemic culture that goes along with LIGO is the ways of thinking of high-energy physics.[14] Even from the early days, Weber's data analysis practices were unfavorably compared with those of high-energy physics. Nowadays, high-energy physics' influence on gravitational wave research has become marked, as LIGO's management and much of its personnel have been drawn from its community (see below).

13. See Kennefick 2000 for an example of maverick theory suppression in the case of inspiraling binary neutron stars.

14. The term *epistemic culture* is borrowed from Knorr-Cetina 1999; she means the different sciences go about claiming truth, reacting to failure, and so forth.

Sharon Traweek's anthropological study of high-energy physicists reveals some of the features of that community that encourage evidential individualism.[15] For example, she repeats a story of a physicist who prematurely reported the discovery of an "exotic meson" before it had been confirmed with sufficient statistical rigor, consequently angering the community and losing a promotion (p. 118). In addition, she reports that high-energy physicists say that even if one has absolute trust in an experimentalist, one should multiply his claimed amount of possible error by at least 3 (p. 117). Traweek also reports that American high-energy physicists from different groups do not talk to each other very much. Even within physics there is a marked hierarchy and a noticeable level of distrust among groups. One respondent, himself a fully qualified lifelong physicist, had worked in a high-energy physics team, but as someone who made the accelerators work rather than someone who analyzed results. His "real physicist" colleagues would not refer to him as a physicist:

> [1999] High-energy physicists and also nuclear physicists have a tendency to
> call themselves just "physicists," because all other branches of physics do not
> involve physics. The real frontiers of physics [for these people] are in
> subatomic physics.

I was told by this respondent that when he failed to gain approval for a project, he had heard it said that "the trouble with this project is that it is not led by a physicist."

The bar groups experienced this arrogance as the brash new wave of interferometer scientists first entered the field. Many bar researchers felt bullied by the interferometer teams. It was said that the newcomers treated themselves as an elite, showing no respect for the networks or expertise that had been built up painfully over the years. The high-energy physicists seemed to treat the bar scientists as an unsuccessful and outdated set of craft workers overdue for replacement by real physicists; because these were newcomers, there were no personal alliances to moderate these attitudes. As one director of LIGO once said to me, "No one knows who Joe Weber is these days."

Even long-established interferometer scientists were not immune to this treatment. In the early days, a GEO scientist told me that the VIRGO group were very inhospitable when he visited them. At one conference, Alain Brillet, the leader of the French side of the project who came from a

---

15. Traweek 1988.

different background, had felt bound to apologize to him for the unpleasant treatment to which he had been subjected by the more arrogant elements of the VIRGO team.

It was also suggested from more than once source that high-energy physicists are used to people "buying in" to a project. You may partake in the fame and prestige of the outcome of an experiment only if you pay your way—the relationships were essentially contractual. Gravity wave research had been much more of a family affair, but now it was becoming a business.

And the newcomers had not yet experienced the problems of experimental work in gravitational wave detection. As one resonant bar scientist told me, high-energy physicists live in a world where everything is calculated in advance. They know that they need so much money to build magnets and detectors of such-and-such a size to detect such-and-such particles, and they go ahead and do it. The newcomers had not yet tried to build experimental apparatus in their new field, so they knew nothing of the difficulties to come. They could juxtapose their clear ideas with the manifestly unclear practices of the resonant bars, and the comparison was bound to be unfavorable. The paper version of some new device has well-ordered timetables and clearly calculated levels of sensitivity. The paper version can do everything that it is calculated it will do. Real devices, however, manifestly do not do what they are supposed to do according to the timetable set for them, and the bars had demonstrated this "in spades," taking ten years to complete programs which it had been anticipated would take only one year. The bar people felt that their very experience gave them an enormous *disadvantage* in the competition. I discussed the matter with a team who had become involved in a program to build cryogenic spherical detectors.

[1995] *Collins:* What do you think of the LIGO project?

R1: Do you really want to know? [laughter] I don't think they have an idea what they're doing.... [T]he current problem with the resonant bar gravity wave effort has been overpromising—right from the start. And the task that was undertaken was unrealistic for the amount of resources that were available, and so people labored in small groups for a decade or better and weren't able to meet the originally promised milestones, and it just led to all sorts of discomfort for everyone involved, including the people at NSF who put the money up because they had to justify to their bosses, and so everyone was in a tight spot.

And I think that LIGO is just overpromising on a grand scale. And I believe that there are parts of the technology that they're just guessing at, basically...

> R2: . . . It's a political problem to get it started, then you solve the scientific
>   problems as they come up. They may not ever be able to solve the problems.
> R1: The sensitivity of phase 1 is not terribly interesting. It's just barely getting to
>   the point where they might be able to see something. But the sources that
>   they're really excited about, like these coalescing binary stars, they're not
>   gonna be able to see that until they reach their phase-2 sensitivity, and to
>   reach phase 2 there are advances required that they don't even know the
>   physics yet.

There was, then, a difference in approach between high-energy physicists and certain resonant bar groups having to do with the "normalcy" and success of high-energy physics as a science. Over many years, as a result of experimental bravado combined with engineering excellence, new particle after new particle has been discovered. A certain pattern of expectations for the nature and duration of the experimental cycle has been built up and with it a well-defined image of what constitutes a mistake arising out of an insufficiently rigorous approach. The relationship between theory and experiment has been fruitful, with the experimentalists repeatedly showing that, given the resources, they can build the machines that will detect anything that theorists say it is possible to see.

The high-energy physics experience also encourages scientists to believe that they have an almost complete picture of the world. Not only does theory predict accurately, but theoretical models of physics machines seem to be complete or potentially complete. In other words, there are not thought to be many significant unknowns left in the world of experimental practice. In the tension between the bar groups and interferometry, high-energy physics would again stress that experiments were being carried out in a world that we know rather than a world full of potential surprises. This is a more generalized version of the idea that known theory is reliable.

In contrast, the whole history of gravitational wave research up to the late 1990s has been a matter of looking for something that ought not to be there or, more recently, probably won't be there, and hoping for a pleasant surprise. All the ingenuity of the scientists has been directed at struggling to keep going while trying to do the impossible rather than regularly doing the possible. Any signals have to be extracted from noise using any and every means; the problem has been not to drive away false signals but to maintain some grounds for hope. One can see how all the pressures I have already described as accompanying the funding of LIGO would favor the high-energy physics approach.

These, then, are the pressures that would be felt by any bar group, such as Hamilton's, who wanted to maintain working relations with the LIGO project. The Frascati group had no such ambitions; they saw their future in resonant mass technology.

The Frascati Team and LIGO

At the time of the Perth-Rome coincidences, the members of the Frascati Team were making optimistic plans for building resonant spheres. The Frascati laboratory was relatively rich for a resonant mass setup, and its funding appeared to be secure so long as it was seen as *productive*—that is, producing results. The duty to publish that the Frascati laboratory experiences is understood both within and without their laboratory. Thus Hamilton told me, [1996] "[Guido Pizzella] argues that he's had people funding his research for all of these years, and he really owes them papers."

Frascati member Eugenio Coccia said,

> [1996] [Y]ou cannot decide not to publish the results of these experiments. There are many people, working hard, taking data, making the detector work—then you cannot expect that people don't want to publish their results. They should be published, otherwise why are we doing these experiments?

Furthermore, though there is less money overall in Italy, the two kinds of research, resonant masses and interferometers, do not seem to be in such direct competition with each other. Thus, at the time of the events to be described, a second ultralow-temperature resonant bar detector had just been completed in Legnaro, near Padua, even while the Americans were shutting down the options for the future of their one remaining resonant bar group.

As far as job prospects for younger people are concerned, then, there is less pressure in Italy for a rush into interferometry. It must be said that some members of the Frascati group share the mathematical animus. Many of these are junior and may be more influenced by long-term career forecasts and professional networks. The interferometer groups may eventually be a source of employment for them just because of the interferometry community's size. Furthermore, partly because all searches for gravitational bursts are coincidence experiments, the whole community is becoming increasingly globalized, and the cultural influence of the US community is bound to be felt by physicists who frequently hold temporary fellowships away from their own laboratories. Furthermore, they interact with all the members of the gravitational research community several times a year at conferences—interacting, for purely demographic reasons, with far more interferometer scientists than resonant bar scientists. Globalization is, of course, well advanced among high-energy physicists, who now form a sizable proportion of gravitational wave researchers on both sides of the Atlantic. Still, any pressures on the Frascati group were long term, whereas those on the Louisiana group were more immediate.

If we turn to the role played by theory, we again see that the Frascati Team did not feel themselves bound by the American consensus; on the contrary, they adhered to an *experimental animus.* In terms of astrophysics, the sky seems gravitationally brighter over Frascati than over the United States. For example, one member of the Frascati Team speculated that relatively high energies—enough to provide a half-dozen or so pulses a year—might be emitted by coalescing MACHOs (MAssive Condensed Halo Objects). Such objects that would be invisible because they do not emit infrared, light, or shorter wavelengths might make up part of the "dark matter" component of the universe. These massive but invisible stars could fill the area around our galaxy. This speculation contrasted with the LSU group, who seemed happy to accept the calculations of the major theory groups that stand behind LIGO, and concentrated on defending their program within the set of assumptions on which LIGO was founded. Louisiana argued that resonant masses were valuable because they could be made to be so sensitive within narrow bands that they could compete in observing certain kinds of special sources; the group also looked to the still further improved sensitivity of the generations of resonant spheres that they believed were still to come, and the much greater directional sensitivity of resonant technology. Arguments like this could be conducted almost entirely within the framework of the levels of wave energy and the cosmic scenarios informing the development of LIGO.

Finally, the Frascati group felt themselves far less constrained by what seemed normal to high-energy physicists. This is what Guido Pizzella said on the matter of the delay histogram technique, which he called "experimental probability."

> [1995] No. The concept of standard deviation is used by most physicists, particularly the high-energy physicists. It is not the concept which I think is appropriate for our experiment.... But if I convert this to standard deviations, and say four standard deviations, then any high-energy physicist will say "not enough," because they think in a different way.

INVOLUNTARY BLINDING AND THE PERTH-ROME COINCIDENCES

As I have said, the search for coincidences compels disparate groups to work together if they are to produce credible gravitational wave data. In this case, the groups forced together have very different propensities as far as risk taking is concerned, and very different evidential cultures.[16] In spite of the contrasting institutional, financial, and cultural backgrounds of the Frascati Team and the Louisiana group, they had to cooperate to produce results. How could Louisiana maintain its preferred strategy when the Frascati Team could take Louisiana's data stream, compare it with their own under their preferred analytic protocols, and in a spirit of evidential collectivism issue high-risk reports with low evidential significance? Even worse, how could the Louisiana group be certain that the findings generated by the partners were not generated in part by statistical massage such as was now generally believed to have been the source of Weber's early findings and the SN1987A results?

In chapter 9, I explained that statistical massage is usually not deliberate. I also explained that to understand the meaning of a figure for statistical significance, one needs to know the whole hidden history of a piece of data analysis. Furthermore, in Joe Weber's history as a naval commander earlier in this chapter, we saw that one person's statistical massage is another's assiduous exploration of the data. Thus, statistical massage is very hard to recognize. The contributions of an open evidential culture and unconscious statistical massage are very hard to separate, and they become more tangled as data analysis becomes more sophisticated.

Before turning to the resolution of these problems enforced by LSU, let us explore the room for statistical maneuver a little more deeply. The data coming from a pair of gravitational wave detectors consist of simultaneous signals defined one way or another that can be displayed on a delay histogram. The trouble with gravitational waves is that they cannot be turned off; since they lose only an infinitesimal proportion of their energy to matter, they penetrate all imaginable shields. Conducting a detection experiment on a phenomenon that cannot be turned off is difficult, because a control situation, when no signal should be present, cannot be compared with an experimental situation, when any signals might show themselves; yet this is standard scientific practice. One might look at the notorious sidereal

---

16. Note that the principal members of the LSU group and the Frascati group have been friends for many years, not only professionally but at the level of personal visits between families; the drama to be described was played out with a minimum of personal animosity.

correlation as a partial solution to this problem: it compared periods when the gravitational waves from the center of the galaxy were "turned off" due to the orientation of the bar, with periods when they were "turned on." And that is why the sidereal correlation was so powerful.

The invention of the delay histogram can be seen as another partial solution to the problem of turning gravitational waves off. Remember, the height of the zero-delay excess in the delay histogram—the extra height of the zero-delay bin over the other bins—shows how many gravitational wave events (or other events mimicking gravitational waves) have been experienced by a pair of bars. Once we think of a gravitational wave experiment as comprising two detectors rather than one, and we think of the effect of gravitational waves as being coincident excitations of two detectors rather than excitations of a single detector, we have a way of turning off the effects of gravitational radiation on the experimental apparatus. The effects are turned off in all the nonzero delay bins and turned on in the zero-delay bin.[17] Unfortunately, this is not a complete substitute for true shielding, because it is not really gravitational radiation signals that are being turned on and off but the influence of anything that can affect both bars simultaneously. Thus it has been suggested that two bars separated by the width of the USA might be influenced by such things as the simultaneous switching on of many electric pumps across the country in response to coordinated toilet flushes during the commercial breaks in *Monday Night Football* telecasts. Nevertheless, it is the invention of the delay histogram, with its ability to turn the gravitational waves off, that makes the experiment possible.

The delay histogram method of data analysis is very seductive. Even a small zero-delay excess, when set among a large array of delay bins, looks convincing. The experimenter is bound to ask, "If there is nothing there, why did the data-analysis algorithm choose this particular central bin to be the highest one?" A quotation from a 1975 interview with Joe Weber is revealing.

> [1975] More recently, two Japanese scientists have repeated the experiment, again using the amplitude algorithm. They did see the largest number of coincidences in the zero-delay bin, with a factor of between one and one-and-one-half standard deviations. By the standards of modern physics, that's not a significant effect. Nonetheless, the fact that their computer discovered the largest number of counts at zero-delay may be significant.

17. The interpretation of the sidereal correlation and the delay histogram protocol as a substitute for shielding is my own.

Of course, the separation of the detectors and the finite speed of gravitational waves make the meaning of *coincidence* less than exact.[18] There is room to debate the width of the coincidence bins, and such "negotiations" create the opportunity for statistical massage. Furthermore, a delicate apparatus such as a gravitational wave detector will have noisy, insensitive periods and quiet, sensitive periods: the outside world of cosmic ray showers, tides, and earthquakes may vary in its intrusiveness; the immediate environment, both human and physical, may affect the apparatus in many ways; and the internal components such as seals, electrical connections, liquid gases, and the metal itself may harden, crack, leak, cease to insulate, and stretch and strain. And yet the tiniest change in any of these things affects the functioning of the immensely delicate detector; consequently, signal filters have to be continually monitored and adjusted. As Hamilton explained,

> [1996] If our apparatus is not working correctly, as we can tell—the noise temperature going up—we'll do a new calibration and build a new filter, and then on the basis of that filter we would begin recording events again...each filter would be different for a different apparatus. And this is something that [named group] have not understood. [Named group] have felt that they could build a filter on a theoretical model and that would be good, and in theory you could do it; but in actual fact you don't understand the actual operation of the apparatus well enough. So it is more complex than one would initially think.

Thus, even this most conservative of experimenters recognized the need for continual monitoring and readjustment of all the apparatuses in the light of their performance, a practice hard to separate from post-hoc manipulation.

Then there is the threshold dilemma (chapter 4). The Lousianans wanted the threshold for what could count as a potential signal set high, whereas the Frascatians wanted it set low to maximize the chance of detecting faint

18. Apart from the physics of the meaning of coincidence that I have just described, simple, accurate timekeeping seemed to be a source of trouble over and over again in this field. One would like to be able to compare the time at which events occurred to within an accuracy of a hundredth of a second or so, but at some sites, clocks were used that could drift by a second or two per day before they were readjusted. One group, so it was reported to me, used their computer clocks, correcting them a few times a day by telephoning a local radio station. When I expressed surprise at the problems encountered with timekeeping, a respondent explained, "[E]veryone was working hard just to make their experiment work, and there was little thought given, in the early years, of just how coincidences would be done."

signals. The Louisianans argued that the threshold should be set so high that even a single excess event in the zero-delay box would stand out. They maintained, accepting the constraints of theory, that expecting to see even one event a year on antennae with their sensitivity was extremely optimistic, and that the threshold should be set accordingly. Furthermore, they wanted to minimize the opportunity to "dig down" and pull something out of the data that was not really there.

The Frascati Team argued that finding one event would be meaningless anyway, because a single event could always result from chance, in which case it could have no statistical credibility. Such an observation would, therefore, require replication; and if gravitational events are as rare as they are expected to be for instruments of that sensitivity, decades might pass before such a thing would appear again. The high-threshold, low-event-frequency search program was, as they saw it, a recipe for finding nothing that could be believed until long after the interferometry program was completed and the bars had become irrelevant. In any case, if the threshold was set high, in theory the amount of energy required to produce the single event would be so great that astrophysicists would refuse to accept a finding as valid for that very reason. In sum, the Frascati Team argued that if the resonant bar program were to have any chance of success, it must be based on the assumption that there was something unexpected in the heavens—something not predicted by the theory—and that they should set the threshold with that in mind. The threshold should be set such that any zero-delay excess implied a frequency of events sufficient to allow repeated observations within a reasonable fraction of an experimenter's lifetime. This meant that most of the noise should be treated as potential signal and exposed to scrutiny; it should not be assumed to be noise at the outset and hidden behind a high, meaningless threshold.[19]

The "Louisiana Protocol"

These varying data analysis preferences arising from disparate evidential cultures and institutional settings were irresolvable. The prisoners were chained together at the ankle but trying to run in opposite directions. What could Louisiana do to prevent their data being used, as they saw it, recklessly?

19. Bear in mind the reservations expressed in chapter 4, where the threshold dilemma was first discussed. It may be that setting a low threshold brings only a small improvement in the likelihood of seeing a real signal.

In the early 1990s, the LSU group introduced an intriguing new data exchange protocol to try to get around the problem. I will refer to it as the "Louisiana Protocol." Bill Hamilton was one of its architects, but Warren Johnson, Hamilton's second in command and the other principal architect, would become its most obstinate supporter.

The trick was that LSU decided to release data from their own experiment only in a disguised form. Initially, in 1991, they began to release two data streams, one genuine and one false, without saying which was which. Later, they began to release their data streams in the form of a continuous loop with 1000 "starting points" marked on it. Of these 1000, only 1 was genuine, while the other 999 were false. They would not say which was the true starting point. Any other groups receiving their data had to find out for themselves which point represented the start of the data stream.

To solve this puzzle, other groups would have to construct a delay histogram with 1000 bins corresponding to the 1000 potential starting points on the Louisiana data loop. Of these bins, 999 would be the result of comparing data trains with a time offset, and only 1 would exhibit zero delay. If a group discovered that one bin out of 1000 was higher than the rest, they might choose to claim that they had discovered a zero-delay excess—that is, a genuine signal. But they would not know if this bin was really the zero-delay bin or if it was a bin with an excess due purely to coincidences in the noise. To know whether they were right, they would have to depend on the Louisiana group. If the second group were to announce a "result" without consulting Louisiana to find out if they had found the right starting point, they would risk making a public fool of themselves. In effect, the Louisiana data exchange protocol forces any group wishing to cooperate with them in a coincidence search to run their experiment "blind," whether they like it or not. Because they are blind, they cannot do post-hoc data massage either consciously or unconsciously.

Initially, the implementation of this protocol, with only two data sets in the first instance, was mutually agreed by LSU and Frascati. It also appears that the Frascati group was happy at first with the extension to 1000 starting points, since any result based on this protocol would look sound. Later, however, the Frascati Team became much less happy. They considered that to continue with the protocol served the interests of the strongly risk-averse interferometer groups. A member of the Frascati group put it this way, overlooking, perhaps, the group's initial ready agreement to the protocol:

[1995] If we write a paper on coincidences, we might damage the LIGO project. So [the NSF gravitational physics program director] asked the Louisiana people to be very careful in writing these papers. So we have a lot of difficulties in exchanging the data with Louisiana. And finally we reach this agreement:... So when searching coincidences—if we find coincidences with these wrong files, it certainly is not good. This is a very clever point.

In fact, NSF Program Director Rich Isaacson did give strong support to the new protocol. As he put it,

[1995] I had talks with the people at Louisiana before they got involved with data exchanges. I said, "You've gotta protect yourself, you don't want to risk errors." And they knew that, too. They were asking, "How much garbage shall we put in with the good stuff? Should we put in a factor of 5 or 10 or 100?" And I was ultraconservative, and I suggested massive amounts of nonsense. And I think they got a practical compromise.

In a later interview, he went on to amplify as follows:

[1996] *Collins:* Who invented that LSU protocol?
Isaacson: Well, [a member of the LSU group] was talking about this, and I remember having a discussion with him where I asked him what would the chance be if you were exchanging your records blind.... "What's the chance that you can get a false alarm?" And he came up with a number which is a few percent. And I said that he should be very careful, and that if I were in his shoes I would certainly not want it to be more than 1 in 10,000 or 100,000. And he said, "Well, you have this problem, you have to do data analysis—massive amounts of data analysis, and it gets to be a headache." And I said, "Who cares [laughter], we're playing for big stakes, and we don't want false alarms." But I think he was getting similar advice from lots of people.... He worked out some procedure...

The Advantages of Involuntary Blinding

As far as physicists inclined toward a closed evidential culture are concerned, the major benefit of the compulsory blinding is that no chaos can emerge from irresponsible interpretation of the Louisiana data: anyone can use it (in principle), but it would be a foolish person who broadcast a result of a coincidence analysis with the LSU data without checking first that what he or she believed to be the zero-delay box was indeed the zero-delay box.

Thus LSU was effectively enforcing its own evidential culture—compelling the struggling chain-gang convicts to run in their chosen direction.

Of course, if a blinded analyst does find the right box, this is powerful confirmation that the effect is real and not a result of post-hoc statistical massage. Any post-hoc tampering to increase the height of a favored bin may be enhancing the wrong one! The procedure seems, then, an ideal way of eliminating both "crazies" and unconscious biases.[20] A third party, a partner in the laser interferometry effort, put it this way:

> [1995] If he's doing his statistics right and claiming a coincidence at a believable level, he'll have no problem picking out which of the thousand is the right beginning time. But if he's fudging things—even unconsciously—then [he'll] fudge on the wrong one, or something, or won't even be motivated to start fudging because it will get lost in a morass. I think it's very clever. You know, not all social or psychological problems have technological solutions, but I think in this case someone was clever enough to have come up with a technical solution for a psychological problem. So I think it's beautiful.
>
> ... [S]cientists from other fields would say, "Oh, look, those gravity wave people can never do anything right, here's another piece of crap from them.".... Claims that are believable—whether they're true or not, of course, is very subtle, as you guys [sociologists] know—but they have to be believable. So this is a way of getting around the biggest hurdle of believability, namely the incredible temptation [to fall] into ... massaging the statistics. This will make it virtually impossible; and furthermore, it makes it certifiably virtually impossible for anybody from outside who would look at it, that's why I think it's so clever. So that either there won't be a false find, or there might be a believable claim—either way it's better than being almost certain of an unbelievable claim.

It is true that under this protocol the Frascati group has made a few wrong guesses about the starting point. They have, however, checked these out privately with the Louisiana group before publishing. Thus from the point of view of anyone not working in gravitational waves, it looks as though the protocol is excellent; it looks especially good to anyone working on interferometers. As one scientist put it to me—stressing the view of the interferometer community,

20. "Crazies" include complete outsiders who might take the data from any pair of laboratories, perhaps via the Internet, and analyze them in their own preferred way without knowing anything about the experimental apparatus.

[1996] *Respondent:* And they've had several times when [a member of the Frascati group] has said, "Number 62"—that's the one, and they [Louisiana] said, "Nope!"

Collins: That's happened a few times, has it?

Respondent: Yep!...."Oh, well, it wasn't 62, it was 54."...."Now wait a minute...how many guesses are you going to get before we invalidate the procedure?"

The Disadvantages of Involuntary Blinding

The Frascati Team confirmed that they had indeed made some wrong guesses. Is involuntary blinding, then, simply a brilliant solution to a "psychological" (that is, sociological) problem? Not according to the Frascati Team; for them, cutting risk to near zero is not the best policy.

First, the 1000-starting-point procedure imposes a lot of work on the recipient of the data, because he or she must construct a 1000-box histogram—in the normal way, a delay histogram would contain many fewer boxes than this. More serious, a member of the Frascati Team pointed out that the only effect that an analyst can discover under this protocol is one that is strong enough to have occurred by chance less than 1 time in 1000 (i.e., more than three standard deviations above background). If the effect is less strong than this, one or more of the non-zero-delay bins is likely to show a higher peak than the zero-delay bin purely as a result of chance. For example, if there is a genuine effect such that there is a 1% chance of its being due to chance (2.5 standard deviations—a result that would be considered more than adequate in the social sciences), the 1000-starting-point protocol is likely to disguise it among approximately ten results of equal or greater apparent significance. Thus work on a 1% signal is impossible. In physics, 2.5 standard deviations is not much and would not normally be thought to amount to a significant result, but this is not a normal field of physics; we know we are going to stay near or within the noise for a long time to come, and therefore we know it will be a long time before we escape from the developmental phase of the experiment.

A member of the Frascati Team argued that it is necessary to continue to understand the phenomenon and develop the apparatus in appropriate directions, and for this purpose it is necessary to work with any signals that might be there, however poor their statistical significance. [1996] "I still think it's important that we find 1% excess, that you cannot find it in that way because you require better than one per 1000. That's the problem." To enhance the signal in a situation where the signal and the apparatus are so ill understood that it is impossible to design the best filters and set other experimental parameters to their optimum level by prior theoretical considerations, one should adjust the parameters in such a way as to enhance the height of the zero-delay bin, it was argued. In other words, one has to "tune" the apparatus, in the same manner as Joe Weber had—and to do this one has to know which is the zero-delay bin!

In any case, according to the Frascati Team, there are yet more disadvantages to blind analysis. Once one has a signal to work with, one might be able to enhance its value as evidence in ways other than simply boosting the signal-to-noise ratio. One might find, for example, that not only does one have coincidences—albeit at a low level of statistical significance—but the ratios of the levels of energy of the coincident signals are consistent between the two detectors; this would be further evidence of a common source. Or one might find that the signal peak, low though it is in absolute terms, correlates with the times when the antenna is pointed in some particular direction—putatively a strong source of gravitational waves (as with the sidereal correlation). These are ways of boosting the credibility of a signal, but only after one has a vestigial signal to work with; and it can be done well only if the vestigial signal is in the public domain. From the point of view of evidential collectivism, compulsory blinding is a mechanism for suppressing the development of instruments and data analysis techniques.

The Perth-Rome Coincidences

The tensions and differences of view I have described were to find a new focus as a result of the discovery of the apparent coincidences between the output of the Frascati EXPLORER antenna and Perth's NIOBE. The new signals, like all those that had been announced, imply an energy level that requires what the scientists refer to as "new physics" or "new cosmology" to make sense; in short, they are as difficult to accept as all the previous claims. Nevertheless, at a conference in Pisa held in March 1996, the leader of the Perth group suggested that they be announced; the leader of the Frascati Team, who was organizing the session, was less enthusiastic. I recorded their presession conversation as they talked agitatedly in the corridor.

> [1996] *Blair:* Because we're not [scheduled] to be giving any talks . . . in that bar session, and . . . I don't want to give the impression that we're not working on the bars. . . . And so I thought it might be an opportunity just to say that we're seeing some interesting coincidences—we're trying to understand it—

Pizzella:—If they ask me, I will make "no comment."...Because, you see, we are in this difficult situation, since we are publishing papers on the supernova....

Blair: I think it's not right...because of the fuss that there's been in the past, to deny the data—

Pizzella felt too bruised by the reception of the supernova claims to risk controversy again, but Blair went ahead and offered the results to a skeptical audience. Blair explained to me much later,

[1996] [W]e were not going to be bullied by people who have their own agenda. We believed that what we had seen was reasonable and interesting, and that you should tell the story as the story goes—as it unfolds.

Blair was not going to be accused of "colonial cringe."[21]

Even though the evidence for coincidences between Perth and Frascati grew stronger over subsequent months, the results were never going to be sympathetically received. Nevertheless, the Frascati Team once more felt it was time to publish something positive. They believed that credibility would be increased if the results of the LSU group, positive or not, were included in any report. That is the context of the Legal Seafoods conversation.

We have now understood enough of the technical, institutional, national, and cultural environments of these groups to understand the state of the argument in December 1996, when the Legal Seafoods conversation took place. The immediately pressing point of the conversation, and of various other discussions that took place at the same data analysis meeting, was whether the LSU group would agree to their names and their data being included in a joint publication that would include a report of the coincident signals between Frascati and Perth.[22] The Frascati group also wanted the Louisiana protocol relaxed for the sake of future collaboration and development, and they wanted the threshold for potential signals to be set low.

21. For "colonial cringe," see appendix III.1. Blair would be attacked at this meeting by Massimo Cerdonio, who said, effectively, that the coincidences were due to noise caused by strains in the metal. For a brief account, see the next chapter.

22. The event was the first Gravitational Wave Data Analysis Conference, held in Boston December 6–8, 1996. The LSU group, it should be stressed, were not suppressing positive data that would support the Rome-Perth coincidences. But it was generally agreed that even the presentation of neutral data by the LSU group, in the same paper as the positive reports by the Rome-Perth collaboration, would boost the credibility of the whole exercise. As it was, Frascati was becoming marginalized.

Bill Hamilton was trying to make up his mind what to do. His position was becoming less clear-cut in spite of the pressures toward the retention of uninterpreted data. The LSU group had been running detectors for two decades with very little to show for it. For some time they had been seeking funds from NSF for the new generation of more-sensitive spherical resonant detectors. They were unsure whether this development would go forward, and if there were nowhere for resonant bars to go, Hamilton would be approaching retirement with little to show for a lifetime's dedicated work.[23] As he put it in discussion with me,

> [1996] *Hamilton:* And it may be that [we have] been too cautious, er, in saying we want this to be a real field—and we want to develop the science—the science is important stuff to do, because no one really understands gravity yet; they don't understand why it's the way it is, so since we can begin to do these experiments, we have to do them, we have to do them responsibly. And I can't say that I looked at it this way when I started, but on the other hand, on thinking about it, I'm pretty sure that foremost in my mind was, there is no way that I want to do anything that is going to have me labeled as a crackpot.
>
> ... [W]hen someone then begins to find experimental things that may or may not be there, then again, you've got to be very careful, and I suspect that underlying all of this is the desire to really say, "Well, OK, if my work is going to be significant, I've got to make sure that no one thinks that it's crazy."
>
> *Collins:* Yeah, you've got a gamble to make, though, haven't you. Because you can always make absolutely sure that nobody thinks you're a crackpot by never finding anything.
>
> *Hamilton:* Oh—that's true—absolutely—and Weber accuses me of that. In years past, Weber has said, "There are two mistakes you can make—one is to find things—anything—you know, to find coincidences or something like that given any data"; and—I've said this at a number of talks I have given—you can make the other error, and that is to try too hard, and that is what Weber accused me of. He said, "You can also try too hard to not see events." And I've been very aware of that.

Later, discussing whether LSU would agree to publish with Frascati, the conversation went on.

---

23. At the time of the interviews reported here, the LSU group did not know that the US resonant mass program would not be allowed to develop.

[1996] *Collins:* So [to add your name] will add more weight to the paper, but you must also have a worry that if the paper is dismissed, it's gonna take credibility away from you?

Hamilton: Yeah, but we can stand the shot now; that we can stand. But on the other hand, maybe there is something there—I am very aware of that, and so do you throw away 25 years of working on this. And this is not thinking of a Nobel Prize in this...it's just saying, OK, well, we've not published much—do you want your 25 years of work just to quietly disappear—while other people go ahead and get on with advancing the field.

Now we know what was going on in Legal Seafoods. The snatch of conversation was the tip of an iceberg whose hidden base was the theory and politics of gravitational wave detection in the latter part of the 1990s. The iceberg would eventually sink the bars and leave the interferometers as the dominant technology.

CHAPTER 23

RESONANT TECHNOLOGY AND THE NATIONAL
SCIENCE FOUNDATION REVIEW

THE CHANGING RELATIONSHIP BETWEEN
BARS AND INTERFEROMETERS

Over the course of a few years, the relationship between resonant technology and interferometers would evolve. Eventually, interferometers would come to dominate gravitational wave detection, with only a small role left for the diminishing number of bars. But as we saw, in the early days the bar groups believed they could benefit from the new big money going into interferometry. As Guido Pizzella of the Frascati group told me,

> [1995] . . . I can show you the letters I wrote to the senator in favor of LIGO [the Laser Interferometer Gravitational-Wave Observatory project], after all, and the letters I got from Kip Thorne and other people, thanking me for these letters. So, they are good friends.

Thorne, of Caltech, agreed:

> [1995] Through that whole rocky period, we've had review after review after review [of the interferometers], and the bar people have

always been significant players. And the bar people were always very
supportive, [with no one] except Joe Weber trying to stop it.

Weber and some of his colleagues thought that the "interferometeers"
(my term) were scared of his claims because he could do the same science
for very much less money. Weber had told me in 1993, "Of course, if the
1984 cross section is right, it will discourage the United States Congress
from giving Caltech their $211,000,000"; and in 1995, "Any data which
comes from a much less expensive approach to the problem, with Congress
in a budget-cutting balance of view . . . will probably result in Congress simply terminating LIGO. That's my own feeling." And at least one supporter
of Weber believed, [1995] "Now that Joe has shown that the aluminum bar
antenna can get much better results for a hundredth of the price, then it
[LIGO] is not a very good idea."[1] As we have seen, the inteferometeers did
react to this possibility by confronting the cross-section claims head on,
but this period was over once interferometry had been funded.

This is not to say that even after this the "resonateers" were convinced
that the relative sums of money going into the two technologies were appropriate. We can see the view from both sides.

[1995] *Collins:* If you could stop money going into LIGO and have it going into
resonant bars instead, you'd do it?

R2: Yeah! In a New York minute. [everything happens faster in New York]

Collins: So the story I hear from the LIGO people—that there's no real tension
between the groups—. . . you don't think that's true?

R1: Well, they're right, because we people don't hate the LIGO people. We wish
them luck. We just think their plans are unrealistic. They're starting to
realize that uh—They don't believe that actually deep down that our people
have a chance, but they haven't studied it. They started up working with
LIGO and never with the bars—and their guru Kip Thorne has written some
rather [widely] read review articles attempting to analyze both points of view
and hasn't done the bar analysis well.

[1995] . . . [S]ome players in the bar area, particularly, say, [named scientists from
more than one group], are jealous of the amounts of money that are going
into the interferometers. They feel if they could get amounts of money that
were comparable, they could do comparably well; and they had fears,
particularly in the United States, that their efforts would simply be turned

1. See also WEBQUOTE under "Another View of the 1987A Results."

off in the era of the interferometers, despite the fact that they feel that they have a good shot at being the first at discovering the gravitational waves and doing gravitational wave science.

The period when that jealousy was displayed in public was now over, however, because LIGO had become a fact of life. And the interferometeers could afford to be gracious in victory.

> [1996] *Respondent:* [T]here were heated exchanges at conferences up until, about, I would have said, 1991/1992. Until it was clear that the interferometers were going to be funded; then there wasn't any point. Then the bar groups were rather looking for support, to make sure that we would support them to keep their funding. And that we certainly all have done ... when vetting applications—whenever possible to point out the value of the bars and that the funding shouldn't be stopped unless they wanted to change.[2]

Once LIGO was funded, the bar groups had to concentrate on the future—particularly the next step, which was to be the resonant sphere program. Now, however, cooperation rather than confrontation seemed to be the way forward, at least in America.[3] Bill Hamilton of the LSU group put it to me as follows:

> [1994] *Hamilton:* Well, they were saying why bother, and then we definitely showed them that if you consider looking at these [spheres] as a—not as competition but as a complement to LIGO, then you can see that the whole LIGO concept becomes a lot more viable. Not only that, but if you can make a great big one of these [spheres] for 500 thousand dollars, you've got a detector to verify whatever you see for 5 million dollars a shot. That means ... you might want to make two or three of these at each frequency. And then you look at coincidences between those, and you look for the signal, then, that LIGO gets around that time, so you might use these as a

---

2. The interferometeers also tried to mend fences between themselves and the resonant bar groups. Thus, Bernard Schutz's group at Cardiff, who had ruthlessly attacked the SN1987A results, invited Pia Astone, Frascati's data analyst, to Cardiff for some months. The incident is described by Schutz at WEBQUOTE under "Schutz on Pia Astone's Visit to Cardiff."

3. Both communities of physicists accepted that, now that things were reasonably settled, a united front against the rest of the world was better than division and argument. At the Pisa conference in 1996, Bernard Schutz told me, "[Is it time] for the two groups [to be] making a joint case that there should be a lot of money put into those people? I think the feeling is that a joint case is far more likely to yield benefits for both sides. Better than a fight would be, and I think that's really why people, er, try to look like friends and so on."

> trigger for the LIGO to let the LIGO guys then try to look for the shape of the wave form and things like that.
>
> *Collins:* Because of their bigger bandwidth and bandwidth sensitivity.
>
> *Hamilton:* Yes . . .
>
> *Collins:* But, I mean, let me be brutal: it's a damn good move to say this isn't in competition with you, it's going to help you, but you must really think that people would be a lot better off with these [spheres] than LIGO at $200 million.
>
> *Hamilton:* Oh, yes. But that's a juggernaut that we can't stop, and I'm not sure that we should. Because LIGO's going to be.

In sum, the relationship between bars and interferometers was a complex one.[4] Weber and colleagues had tried to confront the new technology and had been defeated. The cryogenic bar groups had accepted interferometers, hoping there would be some advantage for them, but jealous and ready to argue against the imbalance of resources. Once interferometers had become a fait accompli, both sides put on a display of cooperation. In Italy, however, the Frascati group were still trying to find ways of making their data more convincing, and their claims would continue to irritate the interferometeers into the new millennium.[5] The Louisiana group endorsed cooperation with more enthusiasm, hoping to move forward with their sphere program. How did they fare?

THE NSF REVIEW

In December 1996, in Washington, DC, the National Science Foundation held a meeting of its chosen advisors to report on the progress of the gravitational wave detection program. I was present for the whole of that

---

4. There is also a cross-cutting *scientific* argument that says that it would be good thing if bars detected gravitational waves, because this would confirm the prospects for LIGO's gravitational astronomy. For quotes to this effect, see WEBQUOTE under "What if Bars Could See Gravitational Waves?" This argument works well if the bars detected waves compatible with LIGO's conservative view of the heavens, but this is a near impossibility. (Note, however, the countervailing view of Thorne in the same section of WEBQUOTE.)

5. This, remember, is the Frascati group alone. Massimo Cerdonio of the Legnaro group took a different line, perhaps because (as I have been told) he wanted to become a major player in European interferometry. Certainly, Cerdonio led the attack on the Perth-Rome coincidences at the Pisa conference mentioned at the end of chapter 22. See WEBQUOTE under "Cerdonio Attacks the Perth-Rome Coincidences." Cerdonio would become the chair of the International Gravitational Event Collaboration (IGEC), a committee that coordinated the results from all bar groups, and would run it very conservatively.

meeting except for the last day, when the report was to be written; but by that time it was entirely clear what the committee's view would be: resonant masses were not the future of gravitational wave detection in the USA.

The review committee's brief was to advise on general research and education priorities in the gravitational physics program for the next five years. Nevertheless, I am going to concentrate on one particular aspect of its work: the discussion of the future of resonant masses. Even then, the committee was not reviewing any particular proposal. The American sphere program, TIGA, was intended to be built by a consortium centered on LSU. The TIGA project would eventually be presented to NSF in the form of an unsolicited proposal, which was to be reviewed in the normal way by anonymous referees; it would be rejected. Thus, in terms of *formal* procedures, the committee's role was minimal, and even as an advisory committee it provided just one piece of advice among many. On the other hand, I have stressed throughout that one of the NSF's strengths is that it is more than a responsive organization; it has taken a significant proactive role in the development of gravitational wave research. One aspect of its proactivity is its pruning of the potentially ever-ramifying tree of scientific and technological possibilities; where there is scarcity, and there always is scarcity, to allow all branches to grow is to allow them all to die. At the very least, then, this committee was contributing significantly to the pool of advice about the way the proactive brief should be handled.

Sociologists notice the informal aspects of decision making that are visible in meetings such as this. Peer review and the like, as we have seen over and over again, are a matter of judgment as much as calculation. Here, the background of values and expectations on which judgments rest was being cemented. Scientifically trained decision makers, on the other hand, are more aware of the formal procedures that, in part, constitute their professional identity.[6]

I made careful and extensive notes at the meeting, as it was considered inappropriate for me to use a sound recorder. I feel, however, that I should confess my biases. The date was the end of 1996. I was yet to become enamored with the interferometers. Up to this point, I had spent nearly all

6. An NSF spokesperson suggests that I have exaggerated the role of this committee in particular. He points out that even if this committee played as large a role as I impute to it, there would still have been opportunities for a low-level sphere program to be funded if it had been aimed at more modest research and development rather than the immediate construction of giant devices. He also suggests that the committee did nothing to prevent the continuation of other resonant mass groups such as Joe Weber's University of Maryland group. To understand why these other less politically visible possibilities were cut off, one would have to examine the contingencies of grant proposal and review in a more detailed way. Here I deal with large processes.

my research time with bar researchers. The people I had come to know well and who had been kind to me were in the cryogenic bar business, people like Bill Hamilton and Guido Pizzella; these were the people with whom I had held heartfelt conversations about their predicaments. The interferometry crowd seemed comparatively brash—fat-cat newcomers with overweening confidence in their abilities, often born from experience in high-energy physics that would, so everyone told me, prove entirely misplaced once the real experiments began. I am repeating sentiments that I found among the bar community, though let me hasten to add that these should not be attributed specifically to Hamilton or Pizzella—I am reporting what was in the air. And to explain my interests further, I had spent a lot of time learning to understand the resonant mass technology, and I felt I could grasp it. Interferometer science and technology, on the other hand, seemed alien and complex. It was obvious what you had to do with a bar—cool it and isolate it—and nearly every failure was a failure of isolation or uniformity of cooling. The interferometers had power recycling, Fabry-Perot cavities, mode cleaners, Pockels Cells, and so on and so on—these were all still strange to me. So as far as my investment in my own research was concerned, both technical and personal, it was much better if resonant masses continued to have an important place in the field. And again, it would make for better sociology if there were real and continuing competition between two technologies. And then again, as a British sociologist I tend to lean toward the underdog.

In sum, as I watched the beginning of the end of the resonant mass program in the USA I felt I, too, was under attack; I did not like the particular piece of pruning I was watching. As a social scientist I should have been neutral, but my sympathies were with the masses. Looking back at it now that I have as much understanding and appreciation of the interferometers as I once had for bars, I still think my account of the proceedings is accurate, but I think I should give the reader a chance to discount my analysis.

YOUTH AND CHOICE

What was going on here? In his autobiography *Timebends: A Life,* Arthur Miller reflects on the meaning of youth and describes it as "a refusal to surrender the infinitude of options that we at least imagined we had"[7] Miller says youth is characterized by a sense of possibility that disappears

---

7. Miller 1987, p. 70.

once the choices we actually make narrow the future. If this is how we think of aging, sciences, too, have phases of youthfulness and maturity, phases which, however, are not necessarily correlated with the age of the scientist. Joe Weber, once young in years and youthful in his scientific work, toward the end of his life still exhibited that refusal to surrender options so characteristic of the determinedly young. Weber would not give up skittish interpretations of his experiment in spite of the massive scientific mortgage that was by then the almost invariable condition of a future in gravitational wave science. Likewise, in Giuliano Preparata the fires of theoretical youth burned undiminished till his death.

The persuasive rhetoric of these passages reads only one way. Who can resist the allure of the youthful adventurer? What is more, this romantic way of seeing science has deep roots in the foundation of the experimental method. When Robert Boyle was insisting that his newly invented air pump could reveal matters of fact, he was arguing that the individual scientist could speak truly of the world of nature irrespective of what was believed by the king and all his court. Disrespect for authority is built right into the idea of science. So how can we not sympathize with a rebel?

And yet the agony remains, because if all we have is sympathy for rebels, then we have no science. Without pruning the options, the tree of science will die. The work of cutting off options, as much as the work of inventing them, is, then, the work of knowledge making, and this makes institutions such as the National Science Foundation as integral to science as the Joe Webers. The decisions made by these institutions will never be completely justifiable, and sometimes they will not be exactly right. Someone will always resent them, sometimes with justification. Yet without those decisions there will be no more scientific history.

Let me say it one more way: a science without youthful indiscretion and skittish adventurism is a worthy march toward a too-predictable goal; a science with nothing but youthful indiscretion and skittish adventurism is a bright but useless plaything. The whole story of gravitational wave detection is about how these two ways of being in the world, sometimes represented by two kinds of scientists, work, often in disharmony, to make science. This is not good or bad—it just is.

THE MEETING

Concentrating on the resonant masses, I could once more give an account of the purely scientific arguments that were presented at the NSF review meeting: the bars and their potential successors, the spheres, were found to be technically inferior to the interferometers.[8] There were no sources of gravitational waves that the spheres would see that the interferometers would not see as well or better. And the interferometers had a long-term future, reaching toward levels of sensitivity where the spheres could not go. Furthermore, the estimate of the spheres' sensitivity was based on technical advances that had not yet been made, whereas experience of the cryogenic bars suggested that each of these advances would take much longer than imagined. The spheres might cost a tenth as much as LIGO or less, but there was no clear advantage to be gained from building them.

But let me take a liberty with my position and pretend to be a quasi scientist. I'll try to use such expertise as I gained from being around the bar people for a long time and hearing every argument.[9] That seemed enough expertise to enable me to follow the NSF committee's discussions and sometimes feel less than satisfied. The committee, of course, was not made up of core experimentalists lifted out of the deep crevasse that is gravitational wave detection, but scientists with long experience in related experimental fields or theoretical analysis—people in the second ring out if the target diagram is centered on the gravitational-wave core group.

The way the discussion worked was that Rich Isaacson, himself a theoretician though deeply knowledgeable in the ways of experiment, turned out to be the chief source of information about the spheres. Isaacson found himself setting right the deeper misunderstandings of members of the panel and answering their elementary questions; quite often, I thought I could have spoken for him. But Isaacson knew all along how this would come out: he had, effectively, chosen the members of the panel, and he did not defend the spheres but presented a distanced view: "this is what they say" rather than "this is how it is." So here we go! I will switch to present tense to give a sense of the flow of debate.

I do not find the technical arguments against the spheres entirely convincing. I keep wanting to say things such as "But what the bar people would say to that is 'X, Y, Z'" when I think Isaacson has not fully delivered the goods. Furthermore, the future of an unknown technology, kilometer-scale interferometry, is being staked against the future of a very much better-known technology in which many previously unimagined technical problems have been given the chance to reveal themselves; the problems of the interferometers are yet to be discovered in practice. Imagined problems

8. For a paper arguing the scientific case in some detail, see Harry, Houser, and Strain 2002.
9. I am using my "interactional expertise"—See chapter 42 below and Collins and Evans 2002.

rarely match up to actual problems, so to some extent an idealized future is being unfairly set against an all-too-real past. Oddly, though, the argument is presented as though things are just the other way around: that the sensitivity of an ideal sphere is being set against the much more cautious estimates of LIGO I, with it somehow being forgotten that it could be a long time before we know how long it will take LIGO I to achieve its sensitivity.[10] LIGO I has a full year's "shake-down" in its time budget, but we do not know if that one year will turn into ten—as it had in the case of the bars.

The bar people see it the other way; all of them believe that the interferometers will run into a host of unanticipated problems just as they did with their cryogenic bars. For example, one sphere proponent had said to me,

> [1996] This technology of the resonant detectors is more reliable in a sense. It has been developed, and the only detectors that are able to measure gravity waves at the sensitivity of 10^{-19} even are resonant bars.
>
> All the prototype interferometers were less sensitive, so the [bar] technology is more developed and more reliable, and so the future developments like the sphere are based on solid technological guidelines, so it would be wrong in my opinion not to optimize the technology based on resonant mass detectors. Optimize means having the largest possible spherical mass—which gives many possibilities for measuring bursts, stochastic backgrounds, chirps, direction of the source—I think it would be wrong not to build these detectors and to rely only on the large interferometer detectors, which still have to prove that their technology can used for really competitive detectors.

But in the committee it is assumed that the interferometer program has an advantage over the bars in this regard. When I ask why the interferometers won't have these troubles, I am told it is because of the difference in numbers of people. The bars, I am told, were put together by small teams in an amateurish way; the interferometers are being put together by large, professional, high-energy-physics-style teams. And all this is true; it just has not yet been proved that it will make a difference. I think to myself, "It may have been proved that big teams can build an atomic bomb, but that is really not so hard; and it may have been proved that big teams can

---

10. There was much longer to go in 1996 than there is now, and many problems have been solved; even if it turns out that LIGO I reaches its design targets in good time, this will not affect the knowledge of those who were making judgments in 1996; hindsight has to be avoided in this kind of analysis.

build particle accelerators, but that is an established science; and neither example proves that big teams can build interferometers of the sensitivity required."[11]

Let me not exaggerate. Isaacson presents the overall problem fair and square: "LIGO is an 800-pound gorilla which is driving everything because of its cost and its promise for the future," and that this is a high-risk strategy in case it goes wrong. Therefore it is necessary to at least consider other approaches. What my notebook shows is missing from the discussion is the concept of the "xylophone." One of the criticisms of the sphere concept is its narrow bandwidth compared with LIGO's. Thus, even if a sphere can be made as sensitive as an interferometer at about its resonant frequency, it still could detect sources only in those narrow bands. But the sphere people have an answer to this: the "xylophone" an array of spheres of different sizes with different resonant frequencies. They were not there to press that case even though it could be found in the literature, where graphs showed xylophones dipping below the sensitivity limits of LIGO at a series of points.[12]

What my notebook does not show is whether the sphere's advantage in terms of directional sensitivity is really pushed home, and I do not recall any extended discussion of the its advantage in distinguishing between different theories of gravity, though this is a small thing anyway.

More important, the discussion is predicated upon the idea that LIGO will be building new and more sensitive interferometers every two to three years at a cost of $10 million to $20 million each, thus rapidly outstripping the spheres' sensitivity, which was comparable with LIGO I at best.[13]

The review committee agrees that it would be a good thing to have an entirely different kind of detector, such as a resonant mass, to confirm what LIGO might see; but the great advantage of directional sensitivity that the spheres could confer in the long period before a worldwide array of four interferometers was built is never satisfactorily discussed, as I see it.

One of the panel members wants to close down all bar work except for that at LSU, but here Isaacson steps in to defend a continuation of

---

11. We will deal with this issue at length in part IV.

12. An important counterargument is that by the time you have built a xylophone, you are approaching the cost of an interferometer.

13. To take advantage of hindsight, this has proved entirely wrong: a new interferometer in the old installation will cost more than $100 million, and it looks as though the 2- to 3-year interval is hopelessly optimistic, as it will take that time to install and commission a new interferometer once it is fully designed. So that is one way in which the real future has already bitten back at the quasi future presented at the meeting.

transducer developments in order to maintain a reservoir of talent and experience in these matters and to make it easier to maintain cooperative links with Europe. It is thought that resonant sphere projects are pretty certain to be run in Europe, and this is an important consideration to the committee.[14] The same reason is given for not "mothballing" the bars: apparatus can be mothballed, but brains cannot.

I note that the expert committee is not sure whether the proposed spheres are solid or hollow (they are solid), nor even why there is a half-length interferometer at the LIGO installation at Hanford, Washington—both of which I understand. My notebook contains the remark that listening to the discussion would be an incredibly frustrating experience for a sphere proponent, were one to be here. I note the irony that Rich Isaacson is the one who is forced into the role of the defender of the spheres, whereas I know he is merely playing devil's advocate—and plays the role as well as he does only because he knows the devil will not win (as I see it).

I also observe, with some fascination, the way the theorists here—the astrophysicists—talk about their work. I note that they continually borrow the language of experiment. They talk of "discovering" and of getting "physical insight" from working with paper and pencil (that is, computer models). They are working on the problem of the collision of two black holes, and the outcome of that modeling process will be, as represented in their language, as *real* as data generated by an experiment. It is as though what experimenters do and what theorists do are roughly the same, but the experimenters need a lot more money. I feel there is a lack of insight into the trials and tribulations of real experiment. Though, on the other hand, since the black hole modeling problem has still not been solved, it might be that theory is not so different to experiment after all—it, too, takes much longer than expected.[15]

As the discussion draws to a close we begin to see how it will come out. It is agreed that the LSU bar should be kept going, because there is just a chance of seeing something given that the Italians are on the air with bars and coming on air with more. But to ask for the tens of millions needed for the spheres will be hard in the context of LIGO; it will be a "hard sell," in Isaacson's words. He says that there is a lot of science funded by the NSF that is guaranteed to produce first-class results, and if he asks for

14. Another case in which the future was to bite back almost immediately, as the European large-sphere projects were also cancelled.

15. And this has turned out to be the case: the black-hole-black-hole inspiral problem has proved intractable so far. For illuminating parallels between computer modeling and experiment, see Kennefick 2000 and MacKenzie 2001.

$20 million for science not guaranteed to produce anything, they'll ask, "Why?"

A panel member says, "I think your community would be seen as completely irresponsible to go forward with this . . . " and there is general agreement around the table. Remarks include "It would look nuts" and "The rest of the physics community would land on us like a ton of bricks."

Everyone agrees that things will be different if LIGO actually sees anything. And in any case, sphere technology is developing in Europe. Isaacson wonders if it will be possible to fund American groups to be users of European resonant sphere facilities.

The committee decides that the existing bar technology should be kept going and that Americans interested in new developments should hook up with the Europeans. TIGA, the LSU sphere project, is now dead as far as this committee is concerned.

Next, the committee discusses the thousand-starting-point "compulsory-blinding" protocol and agrees that it is a good thing. I write in my notebook, "Will this protocol be used for LIGO?" The committee resolves any doubts by agreeing that if LIGO does see something, the TIGA project can be rapidly revived by a new generation of physicists, because its technological awareness would have been maintained via European links.

We see now that the discussion of TIGA has been held in terms of its opportunity cost for the whole of gravitational physics, which is what this panel has been set up to discuss, and each member of the panel is now well placed to say how he or she would spend any available money on their own projects.

My notebook records, "Very sad business seeing a field killed. So many good ideas and hopes, and careers, going down the tubes."[16]

The NSF report that emerged from the panel was a reasoned scientific document, but in this, its form and meaning were related as in any other scientific paper: there is a social unfolding of events in the laboratory/committee, and there is a retrospectively reconstructed rational ordering of those events. I cannot avoid the conclusion that whatever one's prejudice regarding ways of explaining scientific change, one would have to say there was more than science going on here. The reason the bars had no chance in this committee was simple—it was politically impossible to continue to fund an old technology for detecting gravitational waves when

---

16. Let me stress again that in *formal* terms, the committee only advised; it did not kill anything.

such a powerful case had only just been made for a new and extremely expensive technology to supersede it if gravitational waves were to be found. To continue to develop the old bars in this context would be politically insane. The case for the interferometers superseding the bars had been made in public, and for this committee to decide any other way would be to say, effectively, that "the case was not so powerful as we made it out to be." And given that the rest of the scientific community was already resentful of the size of the slice of funding cake being taken by LIGO, and that many other cake eaters sat on the panel, only one result was possible.

The committee report was eventually broadcast on the Internet. Its crucial paragraph reads as follows:

> Over the past decade, the improvements in low temperature acoustic gravitational wave detectors have been substantial. Currently the cryogenic, acoustic detector at LSU is the most sensitive in the world. It is important that it be funded at an appropriate level to ensure its continued operation, with emphasis on maximizing coincident "on-air" time with detectors in Europe and Australia. This support should continue until the sensitivity of the LSU detector is surpassed. However, in light of the heavy commitment of resources to the LIGO program, the Panel recommends that initiatives for major improvements of existing detectors or development of new generation acoustic detectors should not be undertaken.[17]

When LIGO was initially fighting for funds, the complaint of NSF scientists was that it would take resources from their projects and programs. The LIGOists resisted this argument, pointing out that special allocations from Congress were going to support LIGO, and that the net effect would be an increase in science funding as a whole. The irony is that the one piece of science that was probably robbed of funding by the LIGO project was gravitational wave detection via very large resonant masses. Were there no interferometry program I would bet, and give good odds, that the spheres would have been funded!

But I might be wrong: it may be, as has pressed upon me, that the sphere proponents never organized themselves well enough to justify even the one-tenth of the LIGO money they were asking for. The LSU-based proposal

---

17. Report of an NSF Special Emphasis Panel on Gravitational Physics, Washington, December 6–7, 1996. NSF staff present for some or all of the meeting. David Berley, Robert Eisenstein, Richard Isaacson. NSF consultants—Harry Collins, Dr. Arthur Komar, Syracuse University. Panel members: Eric G Adelberger; Abhay Ashtekar (Chair); Beverly T. Berger; Stephen P. Boughn; Eanna E. Flanagan; Ken Nordtvedt; David T. Wilkinson; Jeffry H. Winicour.

was for a multicenter cooperative rather than the much tighter style of organization represented by LIGO. Nevertheless, if spheres were the only option, then the NSF might have put the same kind of pressure on the bar program to reorganize its effort as NSF put on the MIT and Caltech groups before they became LIGO, and on the LIGO program as it developed. The history of LIGO, which we will see unfolded later in the book, shows that all kinds of organizational problems can be overcome if the scientific hunger and the administrative will are there. I am betting that the sphere program would have been knocked into shape if it had been the only option.

We have seen in this committee meeting one way in which the tree of scientific possibilities is pruned. Another way is with talk. We now go on to visit a series of conferences, both preceding and following the NSF review meeting, where still more pruning was effected.

CHAPTER 24

RIPPLES AND CONFERENCES

Physicists spend a lot of time at conferences. The International Conference on Gravitational Waves: Sources and Detectors was held near Pisa between March 19 and 23, 1996. Like most of the attendees, I flew in the day before, and spent a very pleasant few hours in the first sunshine I had seen in several months, wandering around the city and inspecting the Leaning Tower.

Sociologists have their conferences, too. What I usually do when I arrive at a sociologists' conference is look up the list of delegates, telephone some people I know, and ask if they want to join me for dinner. That way one spends the first evening having a nice meal and catching up on gossip and other peoples' doings over the course of a few beers. At the physics conference, of course, I was an observer, largely on my own.

Pisa was chosen as the site for this conference because VIRGO, the European laser interferometer, was to be built just a few kilometers away. The organizers had arranged for buses to take the delegates from their Pisa hotels to Cascina, the small town where the meeting would be held and where VIRGO would be built. From about 8:20 to 8:35 the following morning, small groups of men and a few women could be seen standing around in front of Pisa railway station, wearing that vacant look that comes from being in an awkward social situation. We were probably but not absolutely certainly waiting for the same bus to take us to the conference. Were we supposed to know each other or not? To avoid embarrassment, we had to act like any other set of strangers, no one catching anyone else's eye.

The buses pulled up, and as each person embarked on a manifestly individual trek to one or the other he or she passed through a literal as well as metaphorical door, becoming one of a group of colleagues going to a conference. Still feeling pretty lonely, I boarded, too, and was delighted when my pals from Frascati stepped onto the same bus and, recognizing me, shook my hand and exchanged some pleasantries as they passed down the aisle. I had now gained a little status that I could use in my work; I was somebody, not nobody—somebody that physicists spoke to and someone whose hand was worth shaking.

But while there was little at stake for me, there were a number of *physicists* who were visibly not "part of the crowd." As at every conference, there are stars who are never seen alone, always surrounded by admiring "clients," and there are those sad people who seem never to be part of any group. I felt sorry for the lonely physicists whose manifest isolation spoke volumes.

Scientific meetings are often held in pleasant places—Aspen, Florence, San Francisco, Elba. A lot has been written about their purpose. On the face of things, what happens is that person after person stands up at a lectern and presents the results of his or her latest research; this way, the scientific community can keep up with recent advances in the field. This function of conferences is not very important, however. Most of the manifest claims the lecturers make are provisional, and if they were important, they would have circulated through the "network" already, through phone calls or the Internet. The formal aspect of conferences—the transmission of information through lecture presentations—is useful only to newcomers or others having marginal roles in the proceedings, such as, in this case, myself. At Pisa I learned a lot about what was going on from the actual contents of the presentations, but I suspected that the only others who gained much from this were the graduate students and other outsiders who were not yet thoroughly "plugged in" to the network.

As the Internet expands, more and more people are saying that it is time to put an end to these expensive little holidays for scientists in pleasant places. But conferences are vital. The chat in the bars and corridors is what matters. Little groups talk animatedly about their current work and potential collaborations. Face-to-face communication is extraordinarily efficient—so much can be transmitted with the proper eye contact, body movement, hand contact, and so forth. This is where tokens of trust are exchanged, the trust that holds the whole scientific community together.[1] That is why, though it does not look like it, the delegates start work the moment they share those beers on the first evening! So important is the personal stuff that Gary Sanders, LIGO's project manager, once flew nineteen hours to a conference in Perth, stayed one day, and flew nineteen hours back again without staying the night in Australia! That couldn't have been fun, but truth is made with trust, and trust with personal contact.

Conferences, then, are one of the crucial locations where truth is made. On the one hand, there is a literal, unsubtle side to this idea. It was at conferences where Kip Thorne stood up time after time and explained to everyone that Joe Weber's cross-section calculation was wrong. Thorne did not publish his analysis, he just stood up and talked about it; and the audience, whether they followed the details of the calculations or not, could see that someone whose career had been established by his contributions to theory was saying that someone else, whose career had not been made by contributions to theory, was wrong. Being regular conference-goers, everyone in the physics community would have seen the presentation at least once, and those formerly lacking the confidence to be completely sure about Weber's contested idea would know that they could safely forget about the new cross section. If anyone was to blame for its being dismissed prematurely, it would be Thorne, not them. Thorne's demeanor—calm, considered, avuncular—was more important than his mathematics, which most would not have followed in detail.

The less literal side to truth making is still more interesting. Conferences are the places where the community learns the *etiquette* of today's truth; it learns what words and usages are properly uttered in polite company. Thus, in conference after conference, Joe Weber would stand up and present his papers, explaining that he had found gravity waves long ago, and the delegates learned that the right response was to quietly move on to the next paper. And later, conferences would happen without the physical presence of Joe Weber or even his virtual presence in the vibrations of the airwaves that constitute words. In my first day at the Pisa conference, during which I listened to every paper, Weber's name was mentioned just

---

1. At this very conference I had a conversation that defused the distrust that David Blair had for me. Blair had been reading the usual science-wars, antisociologist diatribes, and when we first sat down to talk it was clear that he was uncomfortable and suspicious; in fact he told me so. By the end of the conversation, however, we were nearly mates.

once, in passing. One important paper began with a history of the field, and this was how Joe Weber's contribution was projected into the aural space of the auditorium:

> [1996] On the experimental front 30 years ago, [at the start of] observations, I remember how astronomers were scratching their heads—How could it possibly be true with this kind of event rate and this kind of energy yield? You would have seen this [about three words inaudible]. It was only twenty years ago that... people started building interferometers.

In this spoken paragraph, not only is the work of Weber written off but also the significance of the whole resonant bar effort, including the cryogenic program. So now we knew that Weber's contribution was no longer to be mentioned except, perhaps, by some people as an honor to *the man* who was the pioneer but never to the supposed *scientific findings*. The young people especially could learn that Joe Weber was not even history. He was an amusing footnote at best—a diversion. He had diverted the search for gravity waves from its true path, and he was "diverting"—a story or an anecdote about him was always good for taking the mind off serious business. A question I try to ask myself whenever I go to a gravitational-wave conference these days is, What would happen if someone stood up and said they thought Joe Weber was right? Imagining the scene, one can get a sense of the power of the group. It would be like someone standing on a chair and showing their backside.

And what else was being established? No black hole has ever been seen, and some scientists refuse to believe in them at all. I have sat in a lecture where a scientist from the floor pointed out that according to the theory of relativity, the conditions of the formation of black holes were such that time would stand still for any observer watching a black hole form, and that, therefore, they could not be seen. Yet at the Pisa conference, black holes were as comfortable and familiar as cups and saucers. The modalities surrounding the term *black hole* were those having to do with certainty. The theory of black holes was a matter of fact; this or that feature of black holes has not been postulated but "discovered." Theorists have hijacked the discourse of discovery, and nowhere was it more apparent than here. Theorizing and computer modeling, when it is conducted by consulting adults in public, can transmute calculations into real stuff, a whole new slant on the notion of "social construction of reality." Philosophers have tried to define the real: Ian Hacking says that he thinks electrons are "real" because experimentalists talk of spraying electrons, and "if you can spray them they're real."[2] He should have been in Pisa to see what could be done with black holes—everyone was spraying them all over the place. They were using spit.

In Pisa, black holes were far more real than Giuliano Preparata, whose name was never mentioned once. Preparata had been swallowed up in a social black hole with no escape. The discourse throughout this conference allowed no syllable to survive that might imply that there was such a thing as strange detector cross sections.

But other kinds of reality were being manipulated. In particular, the quantities of gravitational radiation in the sky fluctuated wildly around the sensitivity of the next generation of gravity wave detectors. At this conference, no one said the interferometers might not detect anything; the question was how much each design might see, and which design would see which phenomenon best. Upper limits were the usual way of presenting data, and the upper limits somehow became transposed into the actuality.[3] In the discourse, the predicted performance of any detector that had not yet been built was its actual performance.

This mode of speech extended even to LIGO II, which would need scientific breakthroughs as yet unimagined if it was to work. On every graph and every calculation of sensitivity, there was a solid line representing the performance of LIGO II. LIGO I, six years from completion, seemed almost obsolete in this environment. As for the detectors currently on air, for nearly all the delegates except those who actually had a stake in them, they were an irrelevance. At one point Guido Pizzella remarked despairingly that all the comparisons of sensitivity between bars and other devices were matters of comparing things that were already running—after two decades of development—with things that were going to be running just as soon as the hardware could be put together—as though building a gravity wave detector was like building a garden shed. But this made little difference. What counted was the future—the brave new world of huge machines. When NIOBE group leader David Blair announced the Perth-Rome coincidences to this meeting, he had no chance of being heard.

2. See Galison 1997, chap. 8 for a description of the first use of a discourse of "discovery" applied to computer modeling of atomic weapons. Hacking 1983 refers to electrons being sprayed.

3. See chapter 40 below.

CHAPTER 25

THREE MORE CONFERENCES AND A FUNERAL

The turn of the year 1996–97 was a fraught time for the Perth-Rome coincidences. At a conference in Brazil at the end of May 1996, NIOBE group leader David Blair talked about his results, and the LSU group found themselves resisting them. Warren Johnson, a key member of the Louisiana team, explained from the floor that he did not trust the data, because Blair had made too many statistical "cuts" on them after he had both data sets in hand. At the winter conference in Aspen in January-February 1997, from which Blair was absent, Bill Hamilton showed a "viewgraph" that stated, "The coincidence between Rome and Australia has disappeared—it was an artifact of the filtering," with this gloss: "The good news is that the coincidences have gone away." It seems that Blair had been persuaded to use a different filter on the data, causing the statistical significance to diminish from .1% level to only 8%. That is, instead of there being one chance in 1000 that the results were due to a chance coincidence of peaks of noise, under the new filter there were eight chances in 100, which does not look good in terms of physics.

When Hamilton's "good news" remarks were reported back to Blair, he was angry; this is not what he had said. In the game of "telephone" ("Chinese whispers") that comprises reports of what went on at conferences, a mildly negative comment can be blown entirely out of proportion. It is certainly the case that others who had either been to Aspen or heard reports from the conference told me that the outcome was that Blair had withdrawn his claims, while in fact he had not done so, at least as late as June 1997.

One third party had heard a variety of interesting rumors concerning the coincidences. He had been told that Blair had been strongly advised not to publish anything based on the coincidences but had seemed hard to convince. He also had heard that Blair had submitted a paper concerning the coincidences to *Nature,* but it had been rejected. Moreover, he had heard that the data had indeed been passed through a new filter and had apparently disappeared—the "good news" reported by Hamilton at Aspen—but that Blair did not agree that it was such good news.

MARCEL GROSSMAN 8, JERUSALEM

By the time of the Eighth Marcel Grossman Conference on General Relativity, held in Jerusalem June 22–26, 1997, the members of the Louisiana, Rome, and Perth groups seemed to be back on good terms, even though the Rome-Perth coincidences were to be given another airing at a plenary session on the twenty-fifth.

ALLEGRO and the Continuous Waves

Before that, Hamilton presented his paper to a small group in one of the stifling parallel-session rooms. Parallel sessions enjoy a different status from plenary sessions. Those who attend a parallel session tend to be members of a "club" that has a particularly strong interest in the narrow topic being presented. Here a presenter can take a few more risks, first, because he or she is likely to be among friends, and second, because members of the audience are likely to have enough expertise to make the right judgments concerning extravagant claims—they will understand just how much weight the presenter is intending to give to a claim, and they will know that he or she might be throwing out something in order to gauge the reactions of the aficionados. Plenary sessions, on the other hand, are attended by the wider physics community, and there a subspeciality holds the spotlight. What Bill Hamilton had to say in the parallel session came as quite a surprise, at least to me.

The 30-year history of the resonant bar has been a history of struggling to eliminate outside disturbances and maintain the delicate equilibria that keep a bar in a good enough state to detect a signal. As we have seen, once cryogenics entered the scene, everything became harder: the bars became more sensitive, so that allowable thresholds of noise became much lower; any adjustment to the hardware required a warm-up and cool-down that would take months; and the cryogenic technology, with its pumps, bubbling gases, and stress-relieving creaks of contracting metal, was itself a major additional source of noise. Thus it was not until the 1990s that the cryogenic bars could stay on air for significant lengths of time. Before then, the "duty cycle" of all these devices was poor. By 1997, however, LSU's ALLEGRO had run for some long, uninterrupted periods, and we now saw what could be done under these circumstances.

Some years previous, one of Hamilton's new graduate students, Evan Mauceli, had needed a project, and Hamilton had assigned him to work out how to search a small part of the sky for continuous wave (CW) signals. Nearly all previous resonant bar results had been the real or spurious signatures of "bursts" of radiation, such as would be associated with supernovas or inspirals. To identify these bursts, coincidences between well-separated bars were vital if the results were even to begin to have credibility. By contrast, CW sources can, in principle, be seen with a single detector.

A CW source might be a pulsar—a rapidly rotating neutron star. The star has to be gravitationally asymmetrical to produce gravitational waves. Pulsars are asymmetrical in some ways; otherwise, they would not look like lighthouses, with a single beam of light sweeping across the sky as they spin, so it is a good bet that they satisfy the conditions for producing gravitational radiation. Now, the gravitational radiation emitted by a pulsar is going to be very weak indeed compared with that emitted by one of the "standard" cosmic catastrophes. A normal cosmic catastrophe converts sensible proportions of solar masses to energy to power the gravitational beams it emits, whereas a spinning neutron star maintains an almost constant rate, so it cannot be losing very much mass or emitting much energy. Nevertheless, as far as the bars are concerned, the smallness of the energy output is counterbalanced by its concentration into a narrow frequency band. The frequency of the gravitational waves emitted by a spinning source in cycles per second is double the number of revolutions per second of the source. If the rate of revolution of the source is equal to the resonant frequency of the bar or one of its close harmonics (that is, if it is half the frequency of the bar, a quarter the frequency, twice the frequency, or some such), a high proportion of the energy of the source that hits the bar can be gathered in. This is the usual principle of using resonance as a way of integrating energy. Thus, although a continuous source does not give out much energy, the right CW source—one that is spinning at the right speed—concentrates all its energy into the right frequency band for the right bar.

In this role, resonant bars lose one of their disadvantages in respect of interferometers. The relatively broad bandwidth of the interferometers and the relatively low frequency range at which they are sensitive are of no special advantage for CW source searches, because the detection apparatus has to be tune to the exact frequency (or one of its harmonics) at which the source emits.[1] For CW sources, it all comes down to absolute sensitivity and the amount of time you can spend on air. The detector can increase the salience and statistical significance of a CW signal by looking at the same narrowly defined patch of sky day after day.

Hamilton and Mauceli had chosen to look at a patch of sky below the plane of the galaxy in which there were believed to be an abundance of relatively close-by rotating sources. Hamilton explained to me that this patch was about the size of a little fingernail at arm's length.

One does not point a gravitational radiation detector like a telescope; it remains fixed in its position on the ground, scanning the sky with its most sensitive orientation as Earth rotates. Gravity wave detectors are not like telescopes, where you see what you are looking at only when you are looking straight at it, the sensitivity waxes and wanes. To "look" at a certain point of the sky with a gravitational wave detector, you look for signals that wax and wane in a way that is characteristic of that point. Any signal that does not get stronger or weaker at the predicted rate as the most sensitive orientation of the detector passes across that section of the sky either must be coming from somewhere else or must be noise. Whatever signal comes out of the detector with the wrong sort of wax and wane can be thrown away.

The other thing that has to change as Earth rotates is the apparent frequency of the signal. The frequency must seem higher as that part of Earth

1. In due course, it will be possible to tune interferometers to a narrow frequency using "signal recycling" (see chapter 28).

with the bar on it moves toward the source, driven either by the rotation of Earth on its axis or the movement of Earth in its orbit around the Sun. Likewise, when the bar is moving away from the source as a result of either or both causes, the frequency should go down. These are very small effects and they can be quite complicated, but they are exactly predictable. We may say, then, that any signal coming from the part of the sky being searched must have the correct, complex pattern of "Doppler shifting," and anything that appears to be a signal but does not have the right characteristics either must be coming from somewhere else in the sky or must be noise. Anything having the wrong Doppler shift profile can also be thrown away.

Thus, to look at a portion of the sky as big as a little fingernail at arm's length, one does not point the gravitational wave detector; one does lots of complicated filtering of existing detector output using a computer. One can take the same output and reprocess it into a search for wave sources in any other patch of sky that one would have preferred to have been looking at. The limitation is computer power and processing time.

Given that one is looking over and over again at the same patch of sky just within the narrow sensitive frequency band of the detector and adding up any signal that fits the same well-defined Doppler shifting pattern day after day, weak signals can be enhanced. Given that one is throwing away any signal that does not reliably fit that well-defined pattern, nearly all the noise can be extracted. All in all, the detector becomes, effectively, a much more sensitive instrument than it is when looking for burst sources so long as it can look for long enough.

In Jerusalem, Hamilton showed a graph that revealed the outcome of Mauceli's data analysis. It contained three month's worth of data, representing a total of 30 days of usable complete sweeps of the relevant area. It demonstrated that working in this way, the detector could see strains between 10^{-23} and 10^{-24}. Suddenly, the bars had become very sensitive devices. This strain sensitivity was only available for the very weak CW sources, so it was still insufficient to detect the expected flux of gravitational radiation. A new record upper limit to this flux could be set, however. The graph that Hamilton placed on the overhead projector showed strain sensitivity on the vertical axis and frequency on the horizontal. The curve of data was a thick fuzzy line in the shape of a very shallow U, dipping to about 5×10^{-24} strain sensitivity right on the detector's most sensitive frequency of 920.3 cycles per second and rising to about 10^{-23} at the low end—919.8 cycles per second—and at the high end—920.7 cycles per second. An arrow drawn with a blue felt-tip marker indicated a spike standing up a little higher from the fuzzy line than the rest of the "noise." Hamilton said,

> [1997] I put on arrow on there, and you might say, "Gee, there might be one
> suspicious guy there." And we can check that easily enough—we can run
> another three months' data and see if this reappears. If it's still there in
> another three months, then we're concerned.

When I spoke to Hamilton later, he said that by the time of next week's
Amaldi conference, he hoped that Mauceli would have completed a similar
search directed at the center of the galaxy to use as a comparison.

Hamilton and the Politics of TIGA

That part of his talk over, Hamilton decided to make a more overtly political
plea for the survival of the resonant mass program, concentrating on the
potential of LSU's TIGA project, which everyone now knew had been killed
by the NSF. He began his remarks with an apology.

> [1997] Now, here's the beginning of the politics, so you can leave if you want
> to. Let me point out a couple of statements and the advantage of what we
> call the TIGA detector—a spherical detector. And most of these, I've made the
> points before, so I'll just put it up there for you to look at.

Hamilton then showed various pictures and diagrams which showed results from LSU's small, prototype, room-temperature sphere. These revealed
that the potential TIGA's software could tell where the prototype had been
struck with a little hammer. The claim for the directionality of the spheres
had been demonstrated. Then Hamilton put up an estimate of the sensitivity for TIGA compared with LIGO. He went on to explain that the whole
sphere program was in jeopardy even though the LSU group believed that
spheres were an important adjunct to interferometers. The spheres had better directional discrimination and were an entirely different technology;
they were cheap; and the more detectors the better.

This was the most highly charged conference session I had attended.
Hamilton, normally a very quietly spoken, cautious man, had spoken out
in a way that must have been completely foreign to him and very painful.
One could see the emotional strain on his face and in his body language.[2]
Was Hamilton leaning more toward the evidential culture favored by the
Frascati group in the face of his abandonment by the NSF? He must have
been on the cusp of making such a decision. He had seen in the starkest

2. For Hamilton's emotional talk and the beginning of the question session, see WEBQUOTE
under "Hamilton Defends Resonant Technology at the Jerusalem Marcel Grossman Meeting in 1997."

terms the consequences for his program because of its failure to see anything. And now he had the suggestive data from the three-month search
for a continuous source. If he was ready to stand behind this claim in a
Pizzella-Blair-like way, it would imply that the bars were in serious business
six years or more before the interferometers would be on air.

In a week's time, at the Amaldi conference in Geneva, we would know
whether Hamilton was ready to risk making some type I errors. At least,
that's the way I saw it.

When I asked Hamilton about this, he put a slightly different gloss on it.
He said that all he was trying to show was how sensitive the bars were when
they were used to search for continuous sources at a well-defined frequency.
He was still not in the business of making ill-supported claims, and I should
not read too much into the blue arrows pointing to the little unexplained
spikes on the curve. The crucial thing was the 10^{-24} strain sensitivity that
had been achieved even if there were no signals. This showed that the bars
were potentially very useful for checking results from the interferometers.

Another feature of this session was that I seemed to be far more excited
about Hamilton's potential CW sources than were his fellow bar builders,
such as Pizzella or Blair. I never did manage to pin this down and can only
make a stab at a hypothesis. Hamilton had spent his life acting as what must
have seemed sometimes a "fifth column" in respect of the bar community;
his and Warren Johnson's interpretations were always on the conservative
side, and the Louisiana Protocol put the Louisiana group in charge of what
could be said in terms of coincident observations. Yet here was Hamilton,
potentially on the point of making the first ever confirmable gravitational
wave observation claim, and he was doing it all on his own. That's the
beauty of CW observations—you do not need coincidences. Granted, you do
need eventual confirmation, but it is clear who is making the discovery
and who is doing the confirmation. With burst sources, remember, there is
no observation until there are coincidences, but with CW an observation
can be made with one detector. If Hamilton should turn out to be right,
it would be a bitter pill for those other groups who had run with an open
evidential culture in the hope of making an early observation.

Blair Makes the Case for Resonant Masses

As a performance, the Blair plenary session at the Marcel Grossman Conference was brilliant. First pleasing the audience with a series of whistles
mimicking the frequency patterns of different types of gravitational wave
events, Blair settled into a convincing description of the potential of bar

technology. As regards NIOBE, the cryogenic detector run by the University of Western Australia in Perth, Blair expressed his particular pride in the wireless linkage between its transducer and amplifier. Again, the audience was delighted by his description of the problems of wires as set out on an overhead projector slide.

> [1997] *Theorist:* "A wire is a one dimensional conducting fibre of zero stiffness and zero mass"
>
> *Experimentalist:* "A wire is a multimode acoustic resonator of ill-defined shape, ill-defined boundary conditions, with unpredictable mechanical and acoustic properties, but which invariably has an acoustic resonance at a critically catastrophic location"
>
> *Problem:* A few experimentalists believe the theoretical definition![3]

Blair then went on to express in concise form the way the bars should now be conceived—not only as relatively insensitive burst detectors, but as competitive detectors of continuous wave sources and the stochastic background.

The Stochastic Background

With their increasing reliability, the bars could do more than look for the coincidences that represent unpredictable bursts of gravitational energy. I have already described the use of bars for CW sources, but astrophysicists predict there ought to be another continuous source—the stochastic background.

The notion of a stochastic background is familiar in the microwave region of the electromagnetic spectrum. The whole universe is bathed in microwaves left over from the big bang. When these microwaves were first seen, they were taken to be a noisy background hiss disturbing certain radio reception devices. Almost immediately, however, the three-degree microwave background became a discovery, yielding information about the condition of the universe shortly after it was formed. It is because of these waves that NAUTILUS, the millidegree bar at Frascati, can be said to be the coldest large object in the universe (see chapter 14).

It turns out that there ought to be an equivalent gravitational radiation background from two sources. First there is a big bang remnant similar to

---

3. This description nicely captures one of the features of my first-ever case study in the sociology of science described in chapter 6, where what looked simply like a "wire" joining two components on a circuit diagram turned out to need a certain inductance. The problem of the color of wires is mentioned in the same chapter.

the microwave remnant. Second, because the universe is filled with exploding supernovas at various distances, their gravitational energy taken together will provide another source of gravitational hiss. (It will be at a higher frequency than the big bang remnant, which, as we will see later, is important.)

To search for this background hiss, one does not use the widely separated detectors that are most suitable for burst sources, and one cannot use the single detectors with Doppler shifts that can detect CW sources—one uses two detectors that are located quite close together. The background hiss is very slight, and as far as one detector is concerned, it is indistinguishable from noise. But if it is genuine background hiss rather than locally produced random noise, it will be correlated on two detectors if they are close to each other. If the detectors are far apart, there will be a poor correlation, because the hiss will not be in phase at the two sites and becomes harder to detect. Thus, when it is said that it is better if two detectors are close together to detect the background radiation, this means close in terms of the typical wavelengths of the assorted gravitational waves that make up the background hiss. About 100 kilometers was said to be a good distance; this would be sufficient separation to eliminate some common sources of noise—such as traffic movement at the two sites—but near enough to ensure that phase differences would not be great, because the wavelengths of interest are large compared with the separation.

Naturally, if the bars are close, they will be subject to some of the same noise sources—for example, seismic or electromagnetic disturbances—but what is being looked for in this case is a similarity in the fine structure of continuous noise rather than the sudden increases in energy that characterize both burst sources and major seismic and electrical upsets. For the reliable detection of rare burst sources, which could be confused with large terrestrial events, separation is important; the resulting phase differences can be investigated once the correlated increases in energy have been measured. That said, continuous background gravitational radiation will be hard to separate from continuous background seismic or electrical noise, but the correlation of close detectors remains the preferred method.

Blair showed an overhead projector transparency that set the three capabilities of bar detectors side by side—bursts in the left column, continuous waves in the second, and stochastic background in the third. The bars were turning from single-purpose detection devices into astronomical instruments; their output was becoming potentially much richer and more competitive with interferometers.

Rumors of My Death Have Been Greatly Exaggerated

Blair now began to talk about the technicalities of extracting signals from burst sources in a coincidence experiment. Taking us step by step through the decisions that had to be made about choice of threshold level and choice of filter, he showed that each of these decisions affected the number of signals that would be seen. Blair demonstrated this by showing the impact of the choices on the number of artificial calibration pulses that were extracted under the different signal-processing regimes. One could see that if signal-to-noise ratio was low, different processing regimes highlighted more or fewer calibration pulses with apparently different timings, so that a coincidence search would yield bigger or smaller subsets of the calibration pulses depending on the filters used by the different detectors.

What became clear was that so long as signal-to-noise ratio was low, the choice of filter used for processing the signal was not a neutral decision. Different filters would select different subsets of noise as candidate events. Indeed, the Rome group had proved that the same detector output, run through two different filters, seemed not to correlate with itself.

Blair moved toward the potentially embarrassing part of the talk with some panache. He explained that early searches, in which he had used something called the ZOP filter, had given a marked coincidence excess with the Rome group's EXPLORER detector based at CERN in Geneva (which used a Weiner filter). These had found a zero-delay excess with a probability of only one in 1000 that it could be due to chance alignments of noise. He showed the result on an overhead projector transparency listing time delays out to $+/-200$ seconds, on which the zero-delay bin stood clear of all the rest.

He explained that the Perth and the Rome groups had tried everything to "try to make the zero-delay excess go away," but it remained in spite of repeated manipulations.

When, however, he had changed to what should have been the more effective Weiner filter, the zero-delay excess fell back nearly into the noise category. Though the zero-delay bin was still high, the relevant overhead projector transparency showed that it was just one among many bins of similar heights. Blair explained that the probability of its being a false positive was 8%. It had already been established that different filters should produce different results on weak signals, so the conclusion must be that the initial anomaly still had to be explained, and the jury was still out regarding its significance. In the light of further data gathering and analysis, it may turn out to have been a statistical fluke, or it may turn out to have been the harbinger of the first discovery of gravitational radiation or some other ill-understood effect.

It is important to be clear about what was going on when Blair changed from the ZOP to the Weiner filter. The number of above-threshold events that are detected on a single bar is, as we have explained, always in the hands of the analyst. With any filter, setting the threshold "just so" will produce whatever number of "events" are desired—one big one per day, 100 smaller ones per day, or 1000 still smaller ones per day. Let us say that both Perth and Rome always set their thresholds so that each of them "saw" around 100 events per day. There is no way of telling whether these events are caused by some external agency such as gravity waves, or are just chance peaks in the noise until the delay histogram reveals a zero-delay excess. It should not be thought that the Perth and Rome filters being different on the early runs made those runs less reliable. Each device needs its own filtering algorithm matched to its own characteristics, and any mismatches in the characteristics of the two machines should, if anything, have made the finding of a zero-delay excess less rather than more likely. Yet the early runs with the ZOP and Weiner filters showed a significant zero-delay excess, and just because that excess had disappeared when a different pair of filters was used could not, of itself, show that it was spurious—a zero-delay excess, even with only one pair of filters, was still an unlikely occurrence. In a hostile climate, of course, it tended to reduce the credibility of the initial claim far more drastically than, in strict terms, it reduced the statistical improbability.

If we try to imagine that we have nothing before us but the two delay histograms showing the 1/1000 chance and the 8/100 chance respectively, we might say that the true odds of the combined result being due to chance was about 1/500. But from the critics' point of view, the Weiner filter result showed that something funny had been going on when the first result was produced. For example, Warren Johnson of the LSU group complained that Blair had done too many statistical cuts on the data after he had the Rome coincidences in hand, and Ron Drever of Caltech had been reported as saying something similar. The view of the critics was, then, that the apparently unlikely first delay histogram was the outcome of unconscious statistical massage. Given the demonstrated lack of correlation between the outputs of different sets of filters, there is no strictly logical way of getting from the second, poor, result to the conclusion that the first result was the outcome of statistical bias, but the rhetorical force was strong. In the face of this, Blair was doing a good job in retaining his dignity.

Impressions of the Blair Talk

As we have seen, one cannot analyze science without having in mind the crucial boundary between the core set and outsiders represented on the target diagram. Here we see the phenomenon again. Members of the core set would not have had their mind changed in any respect by Blair's talk, though they might have been pleased that he did such a good job of selling the multiple capabilities of the bars and intrigued at the way he talked of the disappearance of the coincidences under the Weiner filter regime without having to admit too much fault. A graceful exit could now be anticipated. Outside the core set, however, things looked different. I fell into casual conversation with an American physicist from a different field of research who, when he learned of my interests, was full of praise for the Blair plenary. He complained that there seemed to be only one gravitational wave show in town as far as the Americans were concerned—LIGO. In America, "LIGO, LIGO, LIGO," that was all you ever heard about, but Blair had done a good job in showing that the bars are good, too. This was something this physicist had not realized until now. He said that the trouble was that Americans were too good at selling their product. He asked me to notice how good Americans were at selling Coca-Cola—they were the best people in the world at selling things. In America, LIGO was like Coca-Cola; it had been very well sold—perhaps too well sold. As a result of Blair's talk, he could see that there was another product on the market.

To go back to the core, a member of the LIGO community felt deeply disappointed that Blair had shown the slide with the zero-delay peak surrounded by others of equal height. He said that the slide was not something that should have been shown at a plenary session, where there might have been people who did not know how to read it, or people who did not know that these were not the kinds of statistics normally presented in the field.

AMALDI AT CERN

Less than one week later, most of the gravitational wave community, together with some newcomers, reassembled for the Second Edoardo Amaldi Conference at the European Organization for Nuclear Research (CERN) in Geneva. General relativity theorist Clifford Will—"Cliff Will," as he is universally known—gave a plenary session that advertised another use for the resonant spheres. Will pointed out an imaginary line running three-quarters of the way around the very large lecture theater and stated that a shelf of that length would be needed to hold all the monographs that argued for an alternative to Einstein's version of general relativity. He explained that a spherical detector could test for a certain polarization of gravitational radiation that would decide between certain of these theories (this was anticipated, as I have explained, in Bob Forward's 1971 paper). Spheres, with Will's help, could become a useful tool for doing more than affirming the existence of gravitational waves; they could be set in the evidential context of still more of the science of general relativity.[4] Will appears to be one of the few theoretician allies of the spheres.

CERN was a much smaller conference than MG8, with no parallel sessions; every talk was given in a large lecture theater. Here everyone was an expert, and every word uttered would be weighed and understood by other experts in both the physics and the politics of the field. The smallness and unity of the conference meant that it was easy and natural for small discussion groups to form in the corridors. I spent a lot of time on the fringes of informal gatherings of "resonateers," listening as they talked through their problems; the conversation was not always to the benefit of the "interferometeers."

In these conversations, Warren Johnson stuck resolutely to his idea that the easy way out of the complexities would be to look only at events that were robust enough to show up under any combination of reasonable filters. This meant looking for only a few large signals. Blair's practice was to choose an event threshold to be 10 times the mean value of the noise, since this gave 100 above-threshold events per day. The LSU group had to choose a ratio of 11.5 to get the same event rate but, according to Johnson, would have preferred a chosen ratio of 15, which could have given them only about one event per day. Blair's view was that they should accept many events and give their search credibility by looking for multiple coincidences between many detectors. But Johnson had reproduced Blair's post-hoc statistical analysis procedure and had managed to find a pattern in what should have been random data; it convinced him that the Blair procedure could not be trusted.

Next, Blair and Johnson swapped war stories. Johnson started to tell a story about a dispute over people looking into the sky and seeing clusters of stars and arguing whether the clusters were really there or whether it was just the tendency of the eye to organize random elements into patterns—like the constellations. Blair trumped him by pointing out that the outcome of that controversy was that astronomers had proved that the clusters really

---

4. For "evidential context," see chapter 22.

were there. The Rome group argued that a single event would never be believed anyway.

The Flight from Bars

Physicists have a saying: "follow the money." By the middle of 1997, there was no question where the money was—interferometers, and the crowd was growing. Warren Johnson had found the Livingston, Louisiana, site for the second LIGO interferometer; now he was moving into mirror suspension design. He hoped soon to be spending 50% of his time, officially, on LIGO work. Johnson spent some time at the Amaldi Conference talking to Kip Thorne about suspensions. Even Bill Hamilton had spent his sabbatical at Caltech, working on the forty-meter interferometer. He said that the only reason he did not move into interferometers was that it would spell the end of bars. "They're waiting for him to retire," his wife said, "but he's not going to retire." Even David Blair was making plans and handing out shiny brochures for AIGO—the Australian International Gravitational Observatory. He was trying to enlist help to persuade the Australian government to invest in it as a millennium project.

At the conference, the bar community were deeply hurt by a paper given by Bruce Allen, a theorist, on the detection of the stochastic background. In his paper, Allen calculated the chances of pairs of detectors seeing the background. He worked out the combined sensitivity of the various combinations of detectors: LIGO at Hanford, Washington; LIGO at Livingston, Louisiana; the French-Italian VIRGO; the British-German GEO600; and the Japanese 300-meter interferometer, TAMA. He showed that even though these detectors were separated by large distances, by making assumptions about the wavelength of the background radiation one could account for the phase difference that different interferometers would encounter and still do a cross-correlation, even though the optimum close spacing had not been realized. Allen's paper had the general title "The Stochastic Background: Sources and Detection," but he did not mention bars! As far as Allen's talk was concerned, the bars no longer existed.

There were some mutedly angry questions from the floor following the talk, the last two being from Blair and Hamilton. Blair asked about the common sources of noise in two interferometers reducing any confidence one might have in a common signal; the technology was too similar, and what was needed was correlations between different types of device. Hamilton simply asked him if he had considered interferometer-bar coincidences; Allen prevaricated.

Later I asked Allen why he had presented his talk without mentioning bars. He first said that he was too short of time to discuss the bars. Then he pointed out that since no two bars ran at exactly the same frequency, it was not possible for them to do cross-correlations. I pointed out that Vittorio Pallottino of the Frascati team, in a paper given that very morning, had explained that NAUTILUS and EXPLORER had been tuned to exactly the same frequency and used to carry out a cross-correlation search which had set an upper limit on background of 10^{-22}. Allen then agreed that he had probably excluded the bars because he had decided that interferometers were where the future of physics was heading.

The next day, David Blair started off his presentation with a very clear statement that there was a stochastic supernova background of a frequency such that widely separated interferometers could not detect it, and that, as Bernard Schutz of Cardiff had argued much earlier, the best way to see this was with a combination of bars/spheres and interferometers side by side. Further, he claimed, this approach avoided the problem of common technology between the interferometers. Blair pointed out that by accident or design, there were bars located at just about the right distance from the major interferometers to optimize the search. He did not mention Allen's talk, but to the initiated the context was clear.

In the question session, Allen asked first whether coincidences between bars could arise from noncosmic sources (Blair responded that lightning had been eliminated and invited Allen to offer suggestions), and second whether Blair could be sure that a supernova background really existed (Blair explained that it depended on one's model of the universe). Again, to the initiated it was clear that this was a battle in a war between interferometers and bars.

Allen's approach typified what the corridor groups complained of—the theorists were just not thinking about bars any longer. They were no longer interested in giving the resonateers the help they needed with questions of data analysis or, *inter alia*, recognition and legitimation. And this was acutely frustrating, because the resonateers were the only people with any data—the interferometeers would not have data for another five years at best.

Bill Hamilton's Talk at Amaldi

As at Jerusalem, Hamilton's CERN talk consisted of an introduction and two parts. The first section covered the continuous wave (CW) observations, the second, was an advertisement for the bars and spheres.

Also as at Jerusalem, the conference atmosphere was charged, even more so if anything. Hamilton was defending the technology on which he had spent his entire scientific career. He remarked for the benefit of interferometer people, "I hope that you will see from this talk today that there are some detectors working that are already pretty damn good, and that are trying to detect gravity waves now."

He explained that he was becoming more optimistic, because the resonant bar groups had agreed to unify their efforts by setting up a cooperative for the exchange of data. He hoped that when the interferometers started to gather data, they would join in.

Hamilton said his optimism was further bolstered by the plans to build large resonant spheres in Italy and the Netherlands, and possibly Brazil. He would explain a little later that problems of fabrication of large spheres would be solved by explosive bonding of sheets of aluminum alloy.

He then expressed his belief that even the existing bars would be able to be markedly improved by concentrating more on the Superconducting Interference Device (SQUID) detectors, which should give them a strain sensitivity 10^{-21} over a wider bandwidth. He felt he had to show how sensitive the bars were to protect their future.

> Now one thing that we have been seeing in the United States is a very large feeling among a substantial number of people that resonant detectors have had their day, and they're not really very sensitive, and we can't push them very much further.... If we're seeing this kind of thing in the United States, I would not be at all surprised that it begins to strike some of the rest of you. In fact, I've already heard that the politicians are beginning to make noises about "We're putting too much work into gravity wave detection and we should"—You've heard the arguments—"There's only a finite amount of money. If you're going to do this, then you have to cut off that." And this is the kind of thing that we're beginning to see, I believe, in our [resonant mass] end of things.

He complained that none of the theorists (except Sam Finn) show any interest in analysis bar data even though they are the only data that exist and even though the bars are more sensitive than people think (thereby showing that the data analysts truly are "following the money").

Then Hamilton returned to the continuous waves that his group had been searching for and that he discussed at the Jerusalem conference. As at Jerusalem, he put up a slide that showed the detector touching a strain sensitivity a little bit worse than 10^{-24}. He pointed to the spikes on the curve that we were shown in Jerusalem, but he said,

> [That is] what we would expect to see if there was a CW source—we might expect to see something like that [indicates spikes on the curve]. Is that a gravity wave? It is not!

This time, Hamilton did not pick out the spike with a blue arrow. He went on to dismiss the spikes on statistical grounds. His only claim was for the sensitivity of the bars.

> I hope that these numbers—I didn't see anyone faint out there from seeing numbers this small—but these, if you've been looking at this kind of thing, these are damned impressive. These detectors are pretty good.

Next, Hamilton returned to the future of the spheres, putting up a projected sensitivity curve for the financially moribund TIGA and showing that it was superior to LIGO I (in its narrow frequency band).

> [W]ithin the US, an advisory committee recommended against additional funding. And, as a matter of fact, they went further than that. They implied that we really ought to be trying to cut resonant detectors out altogether. And I think—my own opinion is that that advisory committee's advice was not very good [especially as]... the TIGA proposal was quite well reviewed scientifically.

He put up an overhead transparency setting out "The advantages of TIGA," including

> Omnidirectional and direction sensitive; Highest spectral sensitivity; Complementary technology to interferometers; Independent confirmation; Relatively low cost.

The final slide went up. It was labeled "Should we give up?" The answers were

> No!—The advantages are compelling and the cost is not as high as stated
> A true collaboration will be taken seriously—The design work developed by LSU can save time and money
> Strong and persuasive arguments can change minds

There was prolonged applause, followed by a couple of desultory questions and then a final round of applause.[5] For a time, Hamilton has made the future of resonant spheres and the prospect of European collaboration seem real. But in this case, the ripples would soon cease to propagate; Hamilton's optimism is misplaced.

The Bar Alliance

The bar community closed ranks. On the morning of July 4, 1997, in a committee room at CERN, four groups signed up as members of the International Gravitational Event Collaboration (IGEC). IGEC draws together data from four groups running five cryogenic bars. There is ALLEGRO at LSU; the Perth bar, NIOBE; Frascati's NAUTILUS, near Rome, and EXPLORER, at CERN; and AURIGA, soon to come on air in Legnaro, near Padua. When I jokingly remarked that the Fourth of July, being Independence Day, might not be the most propitious moment for the signing of a cooperative agreement, I was equally good humoredly told to shut up.

Much of the discussion of the nature of the upcoming agreement and the wording of the potential document took place in the corridors before the meeting itself; I overheard some of what was going on and talked about it with the scientists. The key question in my mind was what would happen to the Louisiana Protocol and compulsory blinding. It will be remembered that Warren Johnson was the protocol's most determined advocate. At one point during one of these corridor discussions I asked Hamilton, standing with a group of others, what was going to happen (Johnson was not there).

> *Collins:* Right, gentlemen, I'm now going to use my shit-stirring license. What about the 1000-hidden-starting-point protocol for a five-detector coincidence experiment?
> *Hamilton:* Oh, I think we have to abandon that.
> *Blair:* Does Warren agree?
> *Hamilton:* Er—I think he will; I think he will.
> *Blair:* [Laughter]

Next morning at breakfast, I asked the same question of Warren Johnson. He said that he would try to get them to accept the blinded protocol,

---

5. For the complete talk, see WEBQUOTE under "Hamilton Defends Resonant Technology at the CERN Amaldi Meeting in 1997."

but that he doubted he'd be able to persuade people to go along. He did not seem too upset about this.

As always, the story is not simple. It would be nice to report that now that the bars were under pressure, even Johnson was ready to relax his ideas about protocol; but sociology does not take place in the laboratory, and two things were happening at once. On the one hand, the bars were under pressure; but on the other, a five-way coincidence is much less open to statistical manipulation than a two way coincidence, so there was also a nice scientific rationale for dropping the blinding. A five-way conspiracy was, perhaps, even more improbable than a two-way statistical massage.

Nevertheless, the protagonists themselves were conscious that both sorts of argument might play a part in the decision to drop the blinded protocol. A member of the Australian group, Mike Tobar, discussed it with me freely. I asked him whether he thought the LSU team was becoming more relaxed in its attitude toward publishing.

> [1997] *Tobar:* [T]here's a lot of politics involved, you know; the US people say, "Oh, you can't publish it—this will make the LIGO people think this will defame all of gravity waves"—they've got all these political answers. It basically becomes a political decision as far as I can see, and I don't care much about the political side.... And if you look at it as science, it's science. And it's not a positive or a negative affirmation of gravity waves—it's some work we did. And the reason it might be rejected is because of politics—all the people from the US get really concerned. You know, "This could really damage us if there's another incident like Joe Weber." Which seems to me a bit sort of funny....
>
> ... [M]y impression is that they're hesitant towards it because LIGO have been. I think now that they've probably got to pull their head out of the sand because their whole funding's in jeopardy anyway. They reckoned they were not in favor of it because their funding was in difficulty, but it's in jeopardy anyway. So who can trust the people working on LIGO? They might want to see the resonant bars get faded out, so I reckon that now there has to be a real big international effort to do the coincidence analysis for their sake....
>
> I think we have to make this effort to get the resonant bar detectors on line to do a five-way coincidence. The Louisiana people have a different attitude. They want to have all these safeguards and waste a lot of time. I don't think we should muck around with that.

The signatories to the agreement were Hamilton and Johnson for Louisiana, Pizzella and Pallottino for Rome, Blair and Tobar for Perth, and

Massimo Cerdonio and Stefano Vitale representing the Legnaro team. The groups met to discuss the protocol between 8:00 and 9:00 a.m. on July 3 and between about 8:30 and 9:00 on the fourth. Early on the third, Warren Johnson, the principal proponent of the Louisiana Protocol, was asked to make his move. If he maintained his position, it could have led to a crisis. I recorded what was said.

> [1997] *Pizzella:* I want to hear Warren...
>
> *Johnson:* As some of you know, I've been an advocate for some time of doing a so-called blind exchange. Because this is the way that I propose to evaluate accidentals. If one does the analysis a multiple number of times—any analysis that one wishes, making any cuts on the data, but with multiple data sets, then, at the end, if you can pick the correct data out of the many incorrect data sets, then it gives a sort of a confidence that I don't see how to get any other way.

After some technical discussion, Hamilton said,

> [I]f we exchange data between five sites, and we have a semi-infinite number of time shifts, it is just impossible—absolutely an impossible situation. I can see the blind exchange between two sites. When it involves more than two, it just does not make sense.
>
> *Johnson:* You mean because it's cumbersome? Because no one will know how to use it.
>
> *Hamilton:* I don't have any idea [exasperated tone of voice] how to do anything with something like this. [widespread laughter]
>
> *Johnson:* OK, I propose—A proposal—we did a thousand shifts before. The shifts were—there was a list—it was like a set of keys to apply to the data— decryption keys, if you like—and there were a thousand keys, and only one of them was the right one and the rest were wrong. So what we do, you alter the number of keys that each groups has. Instead of a thousand, that's too many. A thousand to the fifth is way too many combinations. So the principle would be to have the number of keys that each groups holds be much smaller so that—call it "n"—so that n to the fifth is still a thousand. Then there's still only a thousand possibilities.
>
> *Vitale:* I am afraid we are still mixing up two [issues].
>
> *Johnson:* No—you can't [reconstruct] the sweater once you've [unravelled it]...
>
> *Pizzella:* I want to consider the practical case. Suppose you do that, and I look at my computer. I know my own file and I have four other files on the base, and I don't know which is the time for them. So what is it I do?

Johnson: You have a list with a thousand possibilities and you try each of them.

Pizzella: So, suppose each one is ten possibilities—so ten to the fourth is ten thousand. So I try all possible cases. OK? And then I find one is a greater success. And then I ask the other people if that was [it]—well, from a theoretical point of view it seems—but I don't think—

Vitale: Again, don't you think that having five different sites providing data and five groups independently elaborating the data, somehow takes the place of a blind exchange against a false conclusion? Because a blind exchange is not a protocol to do with exchange, it's a protocol against a false conclusion. So we have five groups that, at the first stage, are listing their conclusions independently from the data of all other groups. I think this is, maybe, a safer procedure than blind exchange.

Hamilton: I agree.

Blair: I agree.

Johnson: I've lost [laughter]—I've lost all support for this proposal [loud laughter]. But let me predict the following problem, which is that at some point down the road we will have a disagreement. Some people, maybe only one or two, will look at the data and say—and do something, and say, "Ah—we've found it." And the rest of us will say—[many cross-cutting interjections].

Cerdonio: . . . [this point belongs in the next part of our discussion] We have to agree what happens in that case, of course—that's a big point. . . . At some point we must discuss what we do with the data . . . in this statement we agree first that we exchange real data. Now the second point is, what is the data?

Pizzella: So the first point is solved.

I think it is easy to see what Johnson was asking for. There are four groups in IGEC. We will treat EXPLORER and NAUTILUS as one instrument for the sake of simplicity and also because it follows the logic of what needed to be done. If Johnson's idea had been accepted, each group would have to combine its own data with the data from three other groups with unknown starting points. Let us say that each group issued its data in a loop with ten possible starting points. Then any one group assembling that data, already knowing their own starting point, would have $10 \times 10 \times 10 = 1000$ ways of putting the data together. From here on, the procedure is very like that of the Louisiana Protocol. If a group found that one of those 1000 ways of combining the other groups' data with their own produced a statistically significant excess which appeared to be at zero-delay, they would not be able to feel confident about issuing a statement without consulting with the other groups and finding out if indeed that one combination represented the true starting point in each case. As far as I can see, that would have been a technically workable solution, but it would have been logistically clumsy; as with the Louisiana Protocol, it would still be much more work to look through 1000 sets of data than one, and then, if something positive was found, it would be necessary for each of the three other groups to respond quickly and responsibly.

In any case, the whole idea would have been predicated on lack of trust just at the point where the groups were trying to set up a trusting cooperative arrangement. There are a possible six two-way relationships among four groups, and to feel the need for such an arrangement would imply that there were twelve potential nexuses of distrust. When the Louisiana Protocol was set up, there were only two groups, and the only nexus of distrust ran from Louisiana to Rome. Here, we might say, it was much easier and more reasonable to substitute the social improbability of the breakdown of so many trust relationships for the statistical improbability of a revised version of the Louisiana Protocol. That, I believe, is the best way to understand why Johnson was shot down.

The possibility of binary disagreements was still anticipated, but the group decided they would deal with them in due course. Part of that discussion had already been held; it was a matter of how "raw" the data should be when they were exchanged. Cerdonio was anxious to develop a data-processing algorithm that was shared among all the groups and therefore wanted to exchange data that were as raw as possible, minimizing the amount of data processing (and statistical manipulation) that would take place in any one laboratory. In the end, however, it was agreed that at this stage anyway, each group would supply time-stamped "events," using its own judgment about what constituted an "event." (The choice between exchanging raw versus processed data will be discussed at greater length in chapter 38.) The point had been made nicely in an earlier discussion in the corridor.

[1997] *Blair:* Each group has been independently deciding what is a good event list.

Hamilton: Our argument on that has been—I stand ready to be corrected if I'm wrong—that we understand our detector better than any of you, and I don't know anything about your detector—how your data is taken, so I can't make any kind of intelligent decision about what is a good event on your detector and what is not, and so that way we have thought that every group knows its own detector best and gives its best candidates for events—however many we decide on—100–1000—I don't care, and then we exchange those lists.

The choice of this approach speaks to the notion of tacit knowledge—one group cannot possibly design the right signal-processing procedure for a device with which they have no experience. This was the logic of the agreement, with the safeguard remaining that with five separate groups, any statistical massage in one group would be very unlikely to coincide with that in all the other groups. The document that was signed included the following clauses:

> For each operating antenna the individual groups will select, with a threshold matched to the noise of the detector, of order of 100 events per day. The technique to produce the events will be chosen by the individual groups on the basis of their experience and the type of noise of their detector.
>
> The groups are committed to set up procedures to reduce spurious candidate events.

This style of collaboration still contains the potential for disagreement over sharing credit for a discovery. Suppose a subset of the bars found coincidences, but a few did not see anything. The group tried to talk through a solution but failed to find a clear conclusion.

There were easy cases that could be dealt with. For instance, it was agreed that the Legnaro group, who were still not on air, would not be joint authors of any paper that resulted from retrospective analysis of data generated before they had a working detector.

A bit less clear was what would happen if a coincidence was found among a subset of detectors at a time when one of more the others was off-air—for maintenance or because it was encountering high noise levels. There seemed to be a vague agreement that missing groups would not be co-authors of the positive publication, but this was not completely firmed up.

As might be anticipated, more worrying was the situation in which some groups disagreed with the positive interpretation of the majority—perhaps because their own detectors had seen nothing and they would not agree that they were experiencing an insensitive period, or perhaps because they did not agree with the way the data analysis had been done. In the case of this eventuality, it seemed to be agreed that no one can stop anyone else publishing but that the consequences can be dire for anyone who went ahead in the face of strong disagreement. Good sense would prevail, it was suggested, because public disagreement would be bound to reduce the credibility of the claim of a subset. What everyone agreed was that no group should make public the output of the device of *any other group* without that group's agreement.

It's a Wrap

It fell to Bernard Schutz, perhaps second only to Thorne as a theorist of the field and supporter of interferometry, to sum up the proceedings of the whole Amaldi meeting, and hence the state of play in 1997. Schutz, it should be remembered, was someone for whom peace among the competing groups was important. He had even made a case for the effectiveness of sphere-interferometer combinations for detecting the stochastic background.[6]

The resonateers saw to it that Schutz had a copy of the IGEC document and had been apprised of its significance before he spoke. Schutz, true to form, praised the bars' ability to see CW sources. The only reason they had not looked earlier was the lack of continuous time on air. But as he noted, a bar and an interferometer are very similar in the case of CW, because the bandwidth of the source is small. Schutz said, "So we have a lot to learn in the interferometer world from what's happened in bars on this subject now, and I think that's very encouraging."

But when he discussed the Alliance, the bar groups were to be disappointed. Schutz's skepticism about the bars' ability to detect burst sources was clear, and IGEC was about burst sources. Schutz could not ignore what had been going on among the resonateers, but he presented it to the audience as a very low-key item.

> I've just been told about a very positive development here—the formation of an international gravitational event collaboration among the different bar groups to exchange data, and look for a serious event. That's very commendable.

And that was all he had to say.

The final few comments of Schutz's talk, which apart from routine thank-yous were the last official comments in the meeting, instruct us on how to smooth out the world. I present them in full.

> I'm coming to the end of what I want to say, and I guess it's appropriate to focus on the ultimate goal.
>
> Bill Hamilton reminded us of this—that we're not just here building detectors, or constructing data analysis systems, or having fun . . . we're here

---

6. Which was the outcome of the discussions reported at WEBQUOTE under "Schutz on Pia Astone's Visit to Cardiff."

to detect gravitational waves. However, I think it's fair to say that although we all share that goal, we would define it in different ways, and there isn't complete agreement on what detecting gravitational waves means.

Sam Finn in his talk tried to emphasize the importance of our assumptions, and making them explicit. I think, even beyond that, it is important to make the goals explicit. For instance, I would guess that—I'm speaking for my own personal view of what the early goals for the detection of gravitational radiation are—is that we're at the stage where we're still trying to make our first confirmed detection of gravitational waves. The highest priority must be confidence in that detection, and once we're confident we've recognized a gravitational wave—once we understand some of the characteristics of the first gravitational wave sources that we see—then we can start with population studies, and we change our attitude to gravitational waves—it becomes more of a gathering system where maybe you accept a few false detections in order to get a larger number of real detections. But in the beginning we have to accept that we are going to be missing almost all the gravitational waves.

Sam Finn said—here we are—"We want to detect a gravitational wave, what's the criterion?" How do we know a gravitational wave has passed through our apparatus. Well, I'm very sure a gravitational wave has passed through me several times while I've been standing here and giving this talk. It's not a question of detecting every possible gravitational wave. We have to set ourselves a confidence limit and agree that we will miss gravitational waves that are below that. And we set that limit according to the performance of our detectors and according to our ability to convince ourselves that whatever's above them really is a gravitational wave.

At the moment, things that are less easy to quantify, but, I think, are very important, are the alternative explanations for whatever coincidences we might see. And I had a conversation with David Blair after he showed the analysis—er, the preliminary analysis of coincident data that had an excess of a few events at zero time delay [the Perth-Rome coincidences]. I think in this conversation it appeared that really one's confidence in whether that's a gravitational wave detection is nothing to do with one's confidence about whether statistically that's a significant coincidence rate. It's then a question of whether you can exclude all other plausible explanations and be left only with the possibility that that's a gravitational wave. And I think the community out there that we're trying to convince—the physics and astronomy community—will come up with every possible explanation—erm, alternative explanation—particularly if what we are seeing is something just near the threshold, that's weak, and we see a few of them and we're arguing that we have a population instead of one really important event. So I think it's a very difficult thing, and my own feeling is that we have to accept that we're missing populations that are just at noise level or just below the noise level, and we're not going to be able to detect a real detection of gravitational waves until we have one big single event that's just—that you can't argue away—that's just there. And we won't get to that point by having a population of coincidences that can't be explained but each one of which is insignificant in itself.

It could not be made much clearer that the official view, as represented by Schutz, was a conservative view. No one was going to believe that gravitational waves have been seen, or encouraged to think that gravitational waves have been seen, unless they were well out of the noise and made sense according to cosmological theory. A high population of weak events is more easily explained as something other than gravitational waves, and that is what the rest of the physics community would believe. Schutz had set out the problem for the future, and it is a sociological problem—not how to detect gravitational waves, but how to convince the wider community that they have been detected. In Schutz's view, IGEC was not going to be able to manage this, however good the statistics. So much for the cooperative.

POSTSCRIPT

Jump forward to November 1999. In Louisiana, LIGO is holding a grand ceremonial inauguration. Blair has been chosen to present the work of the bar groups. He puts up an overhead transparency showing the progress of IGEC. The operational periods of each of the cryogenic bars is represented by a block of color. We see that two or three bars are always "live," checking their output against each other, waiting for coincidences. The final column in the illustration shows the number of coincidences between bars. Triumphantly, Blair points to the contents of that column and announces, "Zero, zero, zero, zero, zero." The bars have succeeded, he seems to say, in mastering their technology and in mastering their ambition. Their future lies in setting upper limits and understanding noise sources—spuria—for the interferometers. The future of the bars, even for an erstwhile champion like Blair, is to find zero.

People come up to Blair afterwards and tell him what a great talk he'd given. Guido Pizzella must be somewhere in the wings.

CHAPTER 26

THE DOWNTRODDEN MASSES

At the Amaldi Meeting at CERN, Hamilton had said he was much more optimistic about the future of the resonant-mass program for two reasons. One was the formation of IGEC, the new bar co-operative, and the other was what appeared to be a rosy future for big spheres in Europe and Brazil; he looked forward to cooperating with the other sphere programs so long as there was little or no funding for spheres in the USA.

On the IGEC front, his reasons for optimism were, in one sense, well founded. IGEC is still functioning and publishing papers that accumulate their results. But theirs is a Pyrrhic victory; the standard for a sound detection has been set so high by the leaders of the group that, barring something extraordinary, they have defined themselves to be a footnote in the history of gravitational radiation. At the end of the year 2000, an American interferometer scientist, who understood my analysis of these matters, told me that he thought Bill Hamilton had "gone crazy" in his caution about detecting signals. He himself believed that IGEC had set the threshold so high that it was impossible for them to see anything. He said he had asked Hamilton why he did this, and Hamilton replied that it was because he could not trust his colleagues in the IGEC group. Hamilton had told him that he thought LIGO would be able to work with a lower signal threshold, because each member of that team could trust the other. Once more, statistics and trust were interchangeable. Later I asked Bill Hamilton about what I had heard.

> [2000] *Collins:* This concerns the relationship between setting threshold and seeing a signal. The higher the threshold, the less chance of seeing a spurious signal but the less chance of seeing a real signal, too. [X] says that you've now set the threshold so high that your chances of seeing a real signal are now negligible.
>
> *Hamilton:* Possibly.
>
> *Collins:* Am I reporting this correctly to you?
>
> *Hamilton:* He said it's a thing you need to be concerned about. Where we chose to set our threshold was based on the best information that we have of how often events were likely to occur that we would detect. In other words we said, "If we expect to have one supernova every ten years, then we want to set our threshold such that we have less"—and I would say "much less—chance of getting an accidental coincidence once every ten years." And we know the statistics of our individual detectors, so we know how often they give events, and so we set our threshold high enough so that when you consider all the detectors together, the probability is low that you get an accidental more often than once every ten years.

We see clearly in this statement that IGEC has chosen to accept the "dark sky" model of the universe favored by astrophysicists which was the foundation of both the building of LIGO and the switch of funding resources away from bars.

> *Collins:* It means you can only see a strong—that source once every ten years has to be strong. And that's the price you pay.
>
> *Hamilton:* That's correct.
>
> *Collins:* So if I understand it, then, you're probably going to set a more demanding threshold than LIGO is [going to set].
>
> *Hamilton:* I don't know. LIGO is just beginning to face up to this problem. The problems they're talking about in here are problems we faced ten years ago. And it's one of these things that no one ever listens if it doesn't apply to them directly. And so now, unfortunately, they're having to reinvent the wheel. Hopefully we can help them along a little....

Collins: Let me see if I have this right from [X]. He said that you may have to set a more demanding threshold than LIGO set, because your community is less coherent than LIGO's community.

Hamilton: I think that's right, though the incoherence is rapidly leaking out of our community. People are not so eager to see gravity waves in absolutely every [little?] noise as they were in the past.[1]

Interestingly, at this very same year-2000 meeting, a few months before LIGO was to make its first data runs, I began to hear tremendous enthusiasm from the interferometeers for the unknown—for what might be out there that no one knows about! When I put it to X that this confidence was unfounded, given what we already knew of the heavens, he told me, "You've been listening too hard to the theorists," and that I should not believe them. I was told by others that the interferometeers were now thinking about dropping their thresholds, and that is how it might quite easily come about that they would finish up with lower thresholds than the resonateers—with thresholds such as the resonateers were once roundly condemned for using.

THE END OF THE LARGE SPHERES

So Hamilton's first reason for optimism—IGEC—was hollow. What about the second reason, the spheres? (I last discussed resonant spherical detectors in chapter 16.) LSU had plans for building TIGA; in the Netherlands, the high-energy physics group at the National Institute for Nuclear Physics and High Energy Physics (NIKHEF) had joined with Georgio Frossatti and were planning GRAIL; the Frascati group were preparing to build SFERA; and the Brazilian group at the University of Sao Paulo had plans for a multiton sphere. In existence were two prototype spheres of about 80 centimeters diameter, one at LSU and one at Frascati, and there were many schematic drawings, calculations, and digitally produced overhead projector slides of what a completed sphere installation would look like, giving the spheres (forgive the oxymoron) an unmistakably concrete virtual reality. Spherical detectors, then, seemed to have some momentum abroad even though the American program had come to a halt.

Nevertheless, there were causes for concern on the international front, too. In April 1998, at a meeting of the Gravitational-Wave International

---

1. Hamilton was wrong about this; the Frascati group would continue to publish independent positive findings.

Committee (GWIC) in Livingston, Louisiana, Bill Hamilton had remarked that

> [1998] because of the size of the US science effort, anytime something happens here, people interpret it as the way the whole world should go.

Hamilton went on to ask for a political statement from GWIC explaining that the science of spheres was not being questioned by the NSF in its withdrawal of financial support. LIGO Director Barry Barish, as chair of GWIC, agreed that this might be possible if the GWIC community agreed. A spokesperson for the Netherlands GRAIL group who was at the meeting said,

> [1998] We have been discussing the GRAIL project, particularly with the science administrators. We have already been negatively influenced by the way these people read the lines of the Ashtekar report [Abhay Ashtekar chaired the NSF committee discussed above]. Actually, this very often happens because it's not a purely scientific group of people reading those reports. Actually, they ask as many questions [such as] "We read it and see that the resonant mass detector option for gravitational wave detection is actually outdated, or will be outdated soon by interferometers, so explain to us why you want to have this other option?" So there is lots of misunderstanding raised by the Ashtekar report. So it would be exceedingly important, particularly for the GRAIL project, that this group here discusses the scientific statement, and the political statement comes following on the scientific statement. And the scientific statement explains that it is important to continue this research [with spherical detectors]...I understand that NSF actually didn't intend to raise such ideas in the minds of administrators outside the USA—.

Richard Isaacson, NSF program director for gravitational physics, agreed: "If it might be helpful, the NSF can certainly write a letter clarifying what was going on..."

It does seem that the NSF report may have had much more far-reaching effects than its authors intended. My field notes show that most members of the bar community at this meeting wore the demeanor of beaten dogs, the one exception being the GRAIL group, who still believed they were in business. I noted that the papers given by Hamilton and Pizzella were received in an embarrassed silence, as though something salacious were being suggested. I noted that both NSF representatives were "looking a bit sheepish" throughout the proceedings. At one point, Hamilton referred to the NSF report, saying,

> An irresponsible report was circulated. This report was written out of a meeting held at the same time as the Boston Analysis Conference. It said that bars in America were not going to be supported.

Richard Isaacson butted in to put right any false impression. He said that NSF had assembled a special emphasis panel, which wrote a report containing a study of opportunities and funding possibilities, and that NSF was advised on what was possible. One of the conclusions, he explained, was that there was not enough money to spend tens of millions of dollars on another technology and that LIGO users had to take priority. He put the matter clearly and starkly.

> It was not conceivable that the US community would allow us to do both.

The NSF was therefore advised not to embark on a major new initiative. Nevertheless, it would continue to allow upgrades to the US bars until their sensitivity was far surpassed by other instruments. People who read the report as though it meant that NSF was getting out of supporting bars entirely were mistaken.

As the meeting unfolded, a small working party was set up to draft a possible endorsement from GWIC for the future of the resonant-mass program, and so far as I know, Isaacson wrote a supporting letter also. The minutes of that GWIC meeting, as posted on the World Wide Web, include the following:

> [1998] The NSF decision to not fund the spherical detector proposal made by the ALLEGRO group, together with the recommendation quoted above from the Special Emphasis Panel, has had a significant international impact. GWIC appointed a sub-committee (D. Blair, A. Brillet, and W. O. Hamilton) to draft a statement in support of the acoustic cryogenic detector effort. The committee reported back the following statement, which was adopted by the GWIC project members:
>
> *GWIC heard presentations from the resonant detector groups in the U.S., Italy, Australia, and the Netherlands. The bar detectors are operating continuously and reliably with good sensitivity. The operating detectors are exchanging data with the*

*intention of searching for burst sources by detecting coincidences. The proposed
spherical resonant detectors represent a major opportunity.*

*The proposed spherical detectors are complementary to the interferometers, and
offer the possibility of being more sensitive at higher frequencies. The spherical
detectors are sensitive to any polarization and are omni-directional. They offer the
possibility of locating the direction of a source.*

*A detection, at a moderate signal-to-noise ratio, will be greatly enhanced by
coincidence in several detectors, including interferometers.*

*A detection will be much more definitive if it is observed on detectors based on
completely different technologies. Improvement of the existing bar detectors both
increases the chance of detecting gravity waves from nearby sources, and also enables
the development of improved technologies for the spherical detectors.*

*We conclude that research on spherical detectors can be of large value to the
future of the field.*

But this, as it turns out, was not to affect the outcome of the GRAIL decision
or prevent the termination, in due course, of the Frascati group's large
sphere project.

As always, one can give a variety of explanations for the demise of the
sphere program. The technical argument has already been rehearsed when
I described the NSF committee meeting.[2] The spheres did not promise to
live up to the performance of LIGO, certainly not in the long term, and
probably not in the short term. They were narrow-band instruments, and
even though a "xylophone" of spheres could stretch the sensitivity out in
terms of frequency response, they would still not be able to do real astronomy, which required looking at waveforms. In any case, the xylophone
concept began to look expensive—it would start to approach the cost of
an interferometer. Another problem with spheres was their sensitivity to
cosmic rays. Cosmic rays impact upon the solid material, and if the lumps
of material are large enough, there will be so many significant cosmic ray
impacts that it will not be enough to "veto" them by detecting them with
other instruments; the impacts will be so frequent that one would be vetoing out too much of the potential signal. This means that the spheres
would probably have to be installed underground at increased expense,
which would also limit their usefulness for stochastic background searches
in coincidence, since it might not be possible to find underground sites
near interferometers.

2. See also Harry, Houser, and Strain 2002.

For what it is worth (not much), with one exception I find the technical arguments unconvincing as reasons for shutting down the sphere program in the light of the complementary nature of what the spheres could do. The one counterargument that seems convincing to me did not figure large in the debate, however.[3] The argument is that the interferometers can be tuned to detect at a narrow frequency with high sensitivity by "signal recycling," but more crucially, they are "frequency agile"–the specially sensitive frequency of an interferometer can be easily shifted to another place on the spectrum by altering the spacing of the mirrors, whereas the spheres' frequency is pretty well fixed by the size of the lumps of metal. Thus, if any specially interesting continuous wave source appears in the sky, radiating at a known frequency, the interferometers can be quickly tuned to listen to that frequency in particular, whereas the bars cannot.

Our analysis of the Ashtekar committee led to the conclusion that technical arguments were not, in the end, the crucial deciding factor in the funding debate. It would have been interesting to have sat in on the committees that made the decisions on the other sphere projects, too, but we have only some second-hand material, and that for GRAIL alone.

The GRAIL scientists, devastated by the eventual rejection of their proposal by the Dutch agencies, tried to explain it to me in terms of a number of specific problems. They felt they had not lobbied well enough in the higher echelons of the decision-making committees. They thought they had not made a convincing case that their large high-energy physics organization could properly manage the small universities with whom they would have to work if the various components, such as the dilution refrigerator and the sphere itself, were to be combined efficiently. They thought that their own backgrounds worked against them: they were high-energy accelerator physicists rather than detector physicists–that is, they were not physicists' physicists as far as the high-energy community was concerned. And they complained about the makeup of the final decision-making committee, which, in an echo of the NSF meeting, was constructed in such a way that a vote for their project meant a diminution of resources for the pet projects of the committee members themselves. They pointed out that lower-level committees had made the GRAIL project the top priority in the list that went forward to the final committee, and it was that committee of nonexperts in gravitation research that had reversed the prioritization. Once more, the relationships represented by the target diagram are important; as the locus of accountability moves outward, expertise is diluted

---

3. Though it was to be mentioned in Harry, Houser, and Strain 2002.

and choices about resource allocation are made in the face of competing pressures. The ability to "calculate" the answer to a technical problem decreases and, faced with the policymaker's regress, anything helps where a choice has to be made. What was known was that the Americans had decided against funding spheres.

It is tempting to think of events as unfolding like the collapse of a house of cards. The delicate balance of the argument for the spheres was destroyed by the removal of American support. Then the Dutch card fell, and there was nothing left for the Italian funding agency to rest on; they withdrew their funding, too. Later that year, Eugenio Coccia of the Frascati group was to tell me that Istituto Nazionale di Fisica Nucleare (INFN), the relevant Italian funding agency, had much less money than it once had, and now that the VIRGO interferometer was using up so much money it was becoming impossible to fund another project on spheres. Also, he said, the Italian funders were not willing to "go it alone."[4]

THE SMALL SPHERES

Even though the GRAIL project was now dead along with the large Italian sphere, Frossatti in the Netherlands and Coccia in Frascati, were still ready to push ahead with building small spheres, a little more than half a meter in diameter, which they could accomplish with minimal resources. The Brazilians had similar plans. Coccia, who has a dry wit, told me that the reason the spheres were getting smaller was that they were receding!

Frossatti believed that such a small sphere, even though it would run at a high frequency—about 4000 Hertz—could do real discovery work. Neutron stars in our own galaxy suffer the equivalent of earthquakes, which emit with a strength and frequency that should be detectable on a small sphere.

We must bear in mind that this was by no means the end of the bars. Hamilton and the others continued to develop their devices by improving the amplifiers and the transducers. In 1998, at a conference in Pisa, there was a talk by Giovanni Prodi from the Legnaro-Padua group that was afire with enthusiasm for the sensitivity that could be reached. At that time the Legnaro group believed they could make improvements to the tune of

---

4. This is a counterfactual argument; it may have been that the Dutch and Italian spheres would have been shut down even if LSU's TIGA had been funded.

three orders of magnitude on bars alone—an improvement still greater than that offered by scaling up to the size of the spheres.

At the time of this writing, LIGO and the other interferometers are experiencing a few difficulties, and it may be that the time window for bars to find something interesting is a little larger than had been expected. Perhaps there will be time for them to make these improvements before they are swamped by the sensitivity of the interferometers. The five-bar cooperative continues to run and publish papers, but these presentations of data remain upper limits. Cerdonio and Hamilton are making sure the evidential culture is a closed one and nothing untoward escapes from the collective; even Blair seems to have accepted that the future lies in inter-ferometry.[5] In 2000, the IGEC published its "First Search for Gravitational Wave Bursts with a Network of Detectors" in *Physical Review Letters,* report-ing no events detected but setting a new upper limit for the strength of the gravitational wave flux. In 2001 the Legnaro group published a search for "Correlation[s] between Gamma-Ray Bursts and Gravitational Waves,"[6] which also found nothing. The resonant masses had been forced to find a place somehow below the high ground of gravitational wave detection now occupied by the interferometers.

5. By 2003, Blair seemed to have accepted that NIOBE would have to be dismantled if he wanted to work on interferometers. On the other hand, a positive claim, issued unilaterally by the Frascati group, was published in 2002.

6. Allen et al. 2000; Tricarico et al. 2001.

CHAPTER 27

THE FUNDING OF LIGO AND ITS CONSEQUENCES

FUNDING

To a sociologist, there is something less than completely satisfying about the last chapter's account of the way resonant bars were superceded by interferometers. What has been described is the transformation of ways of speaking and acting, the transformation of one taken-for-granted reality into another, the transformation of a culture or form of life. But the sociologist always looks for more—some overarching explanation of why the form of life should have changed. Perhaps everything we have dealt with so far is what Karl Marx would have called "epiphenoma." What else might we look for to provide the sociological "meat and potatoes"? Perhaps we can follow Marx and look at material interests. Perhaps funding is the real phenomenon. Is there some "interest theory" of funding that might provide the materialist explanation for how things turned out? There are two obvious kinds of "materialist" explanations.[1]

---

1. Scientists, of course, look for different kinds of explanations. Kip Thorne told me in 2003 that it was all a matter of bandwidth.

1. *The pork barrel:* Influential senators wanted money coming into their states, and that made them decide to support the Laser Interferometer Gravitational-Wave Observatory (LIGO) project. There would be no equivalent support for the bars, because no large sums were involved.

2. *The technological frame:* LIGO and the interferometers fit better with the growing technologies that supported Star Wars and other parts of the military-industrial complex.

Let us try each of these out. I am going to argue that one cannot make a convincing story out of either. But this is not to say that funding issues did not influence what happened subsequently.

THE PORK BARREL

The first thing to note about LIGO is that, in the way politicians think about financial incentives, it was a small business. Insofar as senators had an interest in bringing a LIGO installation to their state, it was not for its direct input into the local economy. What they were interested in, if they were interested in LIGO at all, was improving the technological image of their constituency. This would make it easier to attract other cutting-edge technology projects, which might have more financial impact.[2]

But we have to keep reminding ourselves that that though LIGO was small fry in absolute terms (compare its $200 million with the Superconducting Super Collider's $8 billion), it was still the largest project ever funded by the National Science Foundation, and the NSF alone could not afford to provide the funds without wrecking the budget for the rest of the physical sciences which it was its responsibility to support. Therefore, LIGO had to go to Congress for a special allocation of funds. Nevertheless, the help of a powerful senator could be a double-edged sword. At least one respondent told me that NSF was wary about siting the LIGO interferometers in states represented by senators whose voices were especially powerful in Congress lest the special allocation be voted down, while the senators used their political muscle to insist that the device be built nonetheless.

2. Robbie Vogt, director of the LIGO project at the time, tells me this was explained to him by Senator Mitchell of Maine. The same motivation was made clear in the speech of Representative Richard H. Baker at the inauguration of LIGO in Livingston, Louisiana, in November 1999. To illustrate the need for more high technology in Louisiana, Baker told a joke against his own constituents (Thibodeaux and Boudreaux are typical Louisiana names): Thibodeaux and Boudreaux are out fishing, when Thibodeaux explains that he has a wonderful new piece of technology with him. He reveals a thermos flask. "And what does that do?" asks Boudreaux. "It keeps hot things hot and cold things cold," replies Thibodeaux. "So what have you got in it?" asks Boudreaux. "I've got a cup of hot coffee and an ice cream," explains Thibodeaux.

This would bring the promised benefits to their states while extracting the funds from NSF's own budget, with disastrous consequences for the rest of NSF's science.

Gathering as many stories together, including those of some central actors, it appears that the backing of powerful senators played an indeterminate role. The Hanford, Washington, interferometer site had been chosen early and had overwhelming scientific advantages along with a supportive senator, but the other choice on both scientific and political grounds was Maine. Choosing Maine would bring the powerful Senator Mitchell on board with the project. Nevertheless, Maine and Mitchell were dumped at the last minute—a potentially disastrous decision in terms of politics—and the Livingston, Louisiana, site was chosen instead. This brought in another powerful senator, Johnson, but only at the expense of upsetting Mitchell. The last-minute change, according to sources that "triangulate," was a matter of high-level congressional politics which had nothing to do with LIGO itself; someone powerful had a political grudge to settle with Mitchell.[3] In sum, people did not act as though an overwhelming imperative for LIGO's success was the carefully nurtured goodwill of senators, and the sums of money involved were not such as to make the kind of story out of this sort of materialist cause that would overwhelm other explanations for the change.

This does not mean that Mitchell and Johnson would not have been motivated to help LIGO through the Congress should it have been confirmed that there would be a benefit to the states they represented. We have to assume, however, that there were lots of other senators with lots of other interests vying for lots of other funds; LIGO would have had congressional enemies as well as friends, and we have failed to argue convincingly that the sums of money involved would ensure that there would be a preponderance of power among the friends. This does not mean that such arguments cannot and have not been convincingly mounted in the case of other larger technologies, only that each case must be argued on its own materialist merits, and there is a shortage of merit here.

BARS AND INTERFEROMETERS IN THE RIVER OF TECHNOLOGY

Now let us consider the positions of bars and interferometers in the broad rivers of technology in the wider world. Perhaps the real explanation for the shift from bars to interferometers is to be found in the way the two

---

3. My sources include Robbie Vogt and Walter Massey, who at the time was head of NSF.

technologies fit into other peoples' technological interests. This possibility will take a little longer to work through.

Interferometers need high-powered lasers, supermirrors, and the most exact means of pointing a laser beam in the desired direction. All these are technologies associated with projects such as Star Wars—the military development designed to shoot down ballistic missiles—and controlled laser-initiated fusion power. The bars had no such connection to the imperatives of the military-industrial complex.

I have heard it said that the spin-off opportunities for the optics industry were discussed at higher levels of the NSF when LIGO was first being considered for funding, and this may have been crucial. Consider this quotation from a scientist whose own program was shifting from bars to interferometers:

> [1995] One other thing that I think is important about LIGO that I've noticed. There's a much broader experimental and engineering community out there that can support the kind of things they're doing. And I think that's important because, for example, the laser source they're going to use for LIGO, you know they've been using argon-ion lasers, in fact they've recently decided to switch that technology ... of the laser that drives the whole thing, to a solid-state, diode-pumped technology. That technology development to build these much better, more efficient, more powerful lasers was not driven by LIGO, it was driven for completely different reasons. That just creates this synergism that there's a lot of things going on out in the optics industry that really feed LIGO in a very positive way. And on the cryogenic side I don't think that's true. You can build large-scale cryogenic systems, but it's not clear how many people are interested in the Q's of spheres that are three meters in diameter, and so forth, and the materials research to make that work. But there's been a lot of military work done on large-scale optics, you know, figured to a thousandth of a wavelength or better and so forth, and one can tap into a lot of that, and that's an important consideration, because ultimately it comes down to engineering and making these things work.
>
> ... We're right in the middle, in Stanford, of a whole Bay Area industry, and you go over to Livermore and they've worked on huge lasers and the optics and fabrication, and there's just all kinds of things going on.
>
> I think if LIGO had to invent all the technology they need based on the kind of funding [they're getting from] the NSF—if that were the only reason to develop these things—then it would never succeed. The reason it can succeed is because it can take advantage—there's a lot of opportunity to utilize what's going on out in the rest of the world.

One can think about the embedding of a program into the rivers of technology as having two components. One flows down the river from upstream and benefits the program, and the other flows out of the program and benefits others downstream, motivating them to support a program. As the quotation above makes clear, there was certainly a lot flowing downstream into LIGO.

Some of this flow we can talk about more specifically. Ron Drever, leader of the Caltech group, assures me that a conference paper on the subject of highly efficient "supermirrors" was published by a branch of the Los-Angeles-based firm Litton, makers of laser gyroscopes used in guidance systems for missiles. Litton coated the first mirrors for Caltech's 40-meter interferometer, with Caltech supplying performance monitoring techniques in exchange, and the teams developed a good relationship.

What also seems sure is that the pioneering Russian work on the qualities of sapphire, the mirror material canvassed for the next generation of interferometers (e.g., Advanced LIGO) emerges from Russian military developments, also in inertial-guidance laser gyroscopes for cruise missiles.

Yet one should not be too carried away by militaristic conspiracy thinking, since the first contract to be let for polishing supermirrors actually to be installed in LIGO went to a civilian agency outside the USA. This was the Commonwealth Scientific and Industrial lasers, Research Organisation (CSIRO) located in Sydney, Australia. It had gained its reputation by polishing mirrors for astronomical instruments such as solar observatories—though the possibilities may have become clear to the agency through its knowledge of prior military work.

Scientist Jesper Munch, working in the University of Adelaide in Australia, talked about his work to develop high-power stable lasers for the next generation of interferometers.

> [2000] I came out of a high-power laser program. I worked in the US in an aerospace company called TRW, and one of the things we worked on was really high-power lasers, mostly for military applications. For Star Wars . . . . that program built high-power lasers—real high-power.
>
> And that's a technology that was well developed. And you can use a lot of that. So it's really know-how and engineering that you have learned, and now the challenge is can you adapt such an approach to a relatively modest power of a few hundred watts.

He then went on to discuss laser "radars" that were used to sense the position and velocity of incoming missiles by using the missile as a mirror and bouncing light off it back to the receiver.

[2000] If you look at it from a technology point of view, we worked in the '70s—TRW did—on laser radars that had to have extremely long range—we're talking lots of thousands of kilometers and very fine velocity discrimination, and in some ways it was pushing the state of the art for the same kind of measurement. The mirror we were looking at was moving a lot faster, but we were looking for small deflections in that mirror, just like you are doing here. So the limit you were pushing laser technology at, in terms of the laser receivers, for that kind of pulse length was just as hard and just as difficult as gravitational interferometry. It's not that much different except that you could argue that gravitational waves is easier, because the mirror doesn't move—the one you're looking it. It's just sitting there—you know where to find it. So if you take a Michelson interferometer—whether or not it's looking at a suspended mirror in a vacuum or whether it's looking at something that's coming at you very fast, it's still a Michelson interferometer, and you're trying to make precision measurements on that mirror. And TRW was doing that in the midseventies.[4]

This respondent was careful to make it clear, however, that there was no direct application of one technology to the other. Thus, LIGO-type mirrors are much bigger and heavier than laser gyroscope mirrors and probably quite different from those likely to be used in any Star Wars program. And the lasers are likely to be very different, too. Star Wars lasers would be much more powerful and much less refined. He did concede that

[2000] . . . what the "old technology" did was to help develop the technology which later turned out to be useful and necessary for gravitational wave interferometry, including frequency-stable lasers, servo-control systems, power scaling, injection locking, coherent power amplifiers, et cetera. The laser physics problems and the techniques used were early versions of that faced by grav[itational] interferometry, and the know-how developed continues to be very useful, even if the application and the actual generic lasers were very different.

4. This is TRW's Space and Defense Group, in Redondo Beach, California. In the case of missile tracking, there is no "second arm," as in an interferometer. Changes in the signature of the beam reflected from the missile are compared with a fixed standard. These changes carry information about the velocity and deceleration of the missile. For missile tracking, a scaled-down version of the chemical "kill laser" was to be used. My respondent put it this way: "It was a low power, very stable and delicate laser. It was a low power version of the separate 'kill' laser, in that these were both chemical lasers, but there the comparison stopped. There were many orders of magnitude difference in power and frequency stability. The same technology was used because it was not limited in power scaling for later, and never realized, developments."

Munch's comments make it clear that we need to think more broadly than just in terms of direct applications.

> It's not obvious [how military technology and gravitational wave interferometers are related] except that it builds know-how and culture and knowing what to do and what not to do in this particular technological field . . . . You're building up your confidence basis. And the Star Wars program contributed to that in terms of your confidence in building lasers—not for this particular application—not directly applicable—but it's a technological confidence I think we're talking about . . . . Whether you can take the thing and use it directly or not, I think that would be taking too narrow a point of view.[5]

We turn, now, to what might flow out from the interferometer program and benefit those *downstream*. The 1986 NSF review committee included in its considerations the following comment written by one of the delegates during the period the meeting was being set up:

> It is clear that this technology [large-scale interferometry] has many other applications, and so low-cost elements of it might go ahead in any case, even in collaboration with Department of Defense funding sources, so long as there is no restriction on the flow of the information and technology to the scientific world. It would be counterproductive, in my opinion, to have these ingenious people working on technology for gravity waves which then went into a black hole of SDI [strategic defense initiative] without benefiting, in reality, either field.

This is not much in the way of evidence, but at least some downstream benefit was anticipated.

When we turn to cryogenic masses, we can find no major technologies that tie in with military technology. There are, however, some minor technologies associated with both interferometry and resonant mass technology that touch on wider interests. Let us step through these, summarizing the upstream and downstream relationships.

Both major detector technologies need first-class seismic isolation. So far as I know, neither benefited from significant contributions from upstream. That is not to say that they did not rely heavily on existing science

---

5. Munch conceded, however, that something similar flowed out of superconductivity research that was of broad benefit to the cryogenic bars.

and technology, only that there was no contemporary big industrial or military activity from which they could learn about such demanding seismic isolation in an interactive way. Downstream, it is possible that certain industries would gain from developments in gravitational wave science. I was told that microprocessor lithography—the engraving of very fine circuits—requires absolute stillness if accuracy is not to be compromised by vibration, and perhaps that industry could benefit from new developments. Music buffs also want their record decks to be vibration free, and there was certainly talk of some of the isolation systems developed for VIRGO and LIGO being transferred to this industry. But neither of these potential spin-offs would be likely to encourage politicians to fund either project, so neither technology gained in advantage from their vibration isolation.

Both technologies have to concern themselves with the quality factor of the materials of some of their components. These are masses in the case of resonant-mass technology and the mirrors and suspensions in the case of the interferometers. If the large spheres had been built, they might have been able to use and improve the upstream technology of explosive bonding, but this was never canvassed as anything but a problem to be solved for them alone, rather than a problem that needed solving for other applications, so the technology was unable to gain from much in the way of interaction with industry or the military in this regard. Smaller spheres, which are being built, have benefited from the technology associated with casting of ships' propellers, but this just seems like a fortunate convergence rather than anything that would determine a technological direction. As we have seen, however, the understanding of high "Q" in mirror materials gained from upstream developments in laser gyroscopes might benefit the same technology as developments flowed downstream.

Another requirement common to both technologies is data processing, but most of the problems are common to both technologies so do not discriminate between them.

The early cryogenic bars hoped to use superconducting magnetic levitation, but it did not work. Then tried to learn from upstream but to no avail, and have nothing to offer downstream. NAUTILUS, the millidegree bar at Frascati and the coldest large object in the universe, has now been joined by AURIGA at Legnaro; both use dilution refrigerator technology, which came to them from upstream. When Georgio Frossatti, who already owned a company making dilution refrigerators, joined the activity, it gave a big boost to the spheres. But this is a small activity in terms of the military-industrial complex. Downstream there is no one else who seems to need such large dilution refrigerators.

Another technology specific to resonant masses was the Supercon-ducting Quantum Interference Device (SQUID) amplifier. The resonateers learned about SQUIDs from developments in science, and it could well be that the improvement they have made will feed back into other applications that need superb amplifiers working at superconducting temperatures—of which there must be many. Once more, however, I never heard anyone trying to justify the resonant-mass technology on the basis of its SQUID technology spin-offs. Resonant masses also use the very clever multistage vibration transducers initially invented by Ho Jung Paik. As far as I can see, however, this is a technology destined to remain unique to resonant gravitational-wave detectors.

NIOBE, the superconducting bar located in Perth, was unusual in using a microwave cavity-based transducer, and this does seem to have found a number of scientific spin-offs that helped to keep the program funded. But, again, these were not such as would justify the building of the bar (and as of early 2003, it appears that NIOBE will be the first of the successful cryogenic bars to cease operation).

Turning to the interferometers, they need very large-scale, high-vacuum systems. This, however, seems to be a problem unique to the science. Expe-rience indicated extreme difficulty, and what came down to the interfer-ometers in this regard was a disadvantage if anything. As for downstream, no one else seems to need the technology.

A noteworthy technology coming out the interferometer program is the method of laser stabilization known as the Pound-Drever-Hall method, thought up specifically to allow large Fabry-Perot cavities to be controlled (see chapter 28). But even if this technique finds downstream application, it was not anticipated, so this development could not have played a causal role in the funding of LIGO.

Finally, it was suggested to me that an understanding of the science of "squeezed states of light," needed to circumvent the quantum limit on the interferometers, might find applications in due course, but this could not be a decisive argument for funding.

Reading down this somewhat arbitrary list, we can see once more that if there were any advantage to be gained from being embedded in a tech-nological frame, it fell to the interferometers, but there is too little here to enable us to claim that this was decisive in a way that did more than tip the balance of very similar forces. We cannot, then, find an overwhelming materialist explanation of events in the technological frame. As with the attempt to find an explanation based in the politics of science, we cannot conclude that this kind of explanation is never the right one, nor can we conclude anything about what was in the heads of individual decision makers; we can only conclude that this kind of argument needs to be made on a case-by-case basis, and in this case it is hard to make.

THE ARGUMENT ABOUT THE FUNDING OF LIGO

The answer to our question about how interferometers overtook bars and spheres has been anticipated in chapter 23, where I discussed the meeting of the NSF panel. The answer I have given is that the very funding of LIGO made it politically impossible to continue to fund resonant technology in the USA on anything but a small scale, and this had its effects worldwide. But we still have not answered the obstinate question of why LIGO was funded at all. We have tried out two materialist explanations and found them unconvincing. Perhaps there is no answer to the question of why LIGO was funded![6]

There are so many elements of LIGO close to or beyond the cutting edge of technology that many scientists believed it to be an irresponsible way to spend money dedicated to science. In the mid-1980s, LIGO was encountering opposition from vocal and powerful scientists. Among them was Richard Garwin, the outspoken opponent of Weber's claims. MIT's Rai Weiss (it seems) put into motion the idea that the LIGO question should be brought to a head and that there should be a wide-ranging review of its practicality under the auspices of the NSF. The outcome was a meeting which took place from November 10–14, 1986, and produced a "Report to the National Science Foundation by the Panel on Interferometric Observatories for Gravitational Waves" (January 1987). In general, and to some people's surprise, the sense of this report was favorable. The technical arguments presented at close quarters by the LIGO team were more convincing than had been expected; more problems had been anticipated and dealt with than had been realized. Garwin, however, who was present at the meeting for only one day, went away still believing that two interferometers were unnecessary and that the project should be developed more slowly.[7]

This meeting, though an immediate triumph for LIGO, did not still the opposition. On November 9, 1990, *Science* published a letter from Curtis

---

6. I am concentrating on LIGO rather than on the European interferometers because (a) the German-British GEO600 was funded at such a low level that there is no question to answer; (b) it is hard to find out anything about the funding of VIRGO; and (c) it is the fact of the funding of LIGO in particular from which the demise of the resonant technology followed.

7. Personal communication.

Michel of the Department of Space Physics and Astronomy at Rice University. Michel argued that the LIGO project was understaffed; that there was no reason to believe that it could achieve the promised increases in sensitivity, since no convincing reasons had been given for how this was to be managed; and that it was going ahead without sufficient public discussion.[8]

More of the flavor of the opposition can be seen in letters written to NSF officers. For example:

> May 6 1991: The LIGO project, which is being billed as a gravity wave *observatory*, is quite expensive and is not at all likely to be successful . . . . At this stage I do not think that LIGO would even be useful as an interesting experiment, although it would be an expensive one . . . .

> May 21 1991: The project seems to me to embody some of the worst excesses of big science. The knowledge that we can hope to gain is disproportionately small compared to the expense we are certain to incur. The most likely outcome will be an increasingly hostile environment for smaller science projects, whose contribution to our expanding knowledge of the universe is, by any standard, much larger.

Crucially, these arguments reached the professional and the national press. For example, in 1991 an article in *The Scientist: The Newspaper for the Science Professional* was entitled "Funding of Two Science Labs Revives Pork Barrel vs. Peer-Review Debate."

At the heart of the antagonism was the question of who would lose if LIGO were funded. The bulk of NSF's constituency were scientists doing solid, predictable, and valuable work with relatively small allocations of NSF funds. They believed that LIGO would cost so much that it would kill many of the tender but vigorous scientific saplings for which NSF was responsible. NSF argued that the special allocation of funds from Congress, without which it would not proceed, made sure that the saplings would be protected. It argued that the net effect would be an increase in the science budget in the short term that might well ramp up the whole budget to a higher level in the long term, by changing political expectations. This was countered by the claim that though a special allocation would cover the initial costs of LIGO, such a large enterprise would be bound to suck funds from its competitors as it needed extra money for unforeseen operational and developmental needs. Furthermore, if there were an emergency

8. *Science*, 250:739.

shortage of funds, politicians would not countenance the shutting down of such a large and visible project while it could still support itself by sucking sustenance from the still developing shoots around it.

The most powerful lobby among NSF's other "customers" was the astronomers and astrophysicists, and a group of them set out to kill LIGO's prospects. The astronomers' organized opposition to LIGO came to a head in evidence given to the hearing of the Subcommittee on Science of the Committee on Science, Space, and Technology of the House of Representatives on March 13, 1991. Tony Tyson of Bell Laboratories, part of whose speech to Congress I quoted at the beginning of the book, and who we have already encountered through his work on resonant bars, now found himself representing the astronomical community in its opposition to LIGO. Tyson caused rage among some LIGO supporters by "breaking ranks" in this way. Scientists generally support one another in their quests for funds; here was a scientist attacking his fellow scientists. It was, in the words of Robbie Vogt, then directing the LIGO project and quoted in *The Scientist,* "an unconscionable attack" on LIGO.[9]

One angry letter to Tyson from a prominent scientist said that killing LIGO would gain astronomy nothing; that it was not a zero-sum game; that there was lots of "pork-barreling" going on, and enormous sums being spent on things like the international space station, which made no scientific sense at all; and that the mere $200 million for LIGO could come from any of these sources even if it *were* a zero-sum game. Therefore, Tyson would do better to support all science projects, as that would be good for the whole of science. Nevertheless, he and other astronomers were undeterred.

Tyson had, among other things, reported to the House of Representatives on a rough survey of astronomers that he had conducted.[10] He is quoted in the *New York Times* of Tuesday, April 30, 1991, as follows:

> "I perused a list of about 2,000 astronomers and picked 70 who seemed to me to likely to have thought about LIGO," Dr. Tyson said in an interview. "I got 60 replies, and they ran 4 to 1 against LIGO. Most of the astrophysical

9. Mervis 1991, p. 11.
10. In 1995, Tyson was also to tell me,
   There's a famous astronomer in California who's made a public statement to the effect that if somebody came to her committee on the Keck Telescope, that would—if the observations were made on the Keck telescope—discover all of the astrophysics that LIGO could—that LIGO's advertised to discover—it wasn't clear to her whether they would actually get time, because it's so competitive. That was another way of saying that from her perspective as an astrophysicist, that the astrophysics that LIGO is going after is so-so. It's nice to know, maybe, but it's worth less than a million bucks.

community seems to feel it would be very difficult to get any important information from a gravity-wave signal even if one should be detected."

"... [A]t its present stage of development and sensitivity, I just don't think LIGO would have much chance of achieving its goals in the next few years..."[11]

The astronomers had already done severe damage to LIGO's prospects by failing to list the project in their regular Decadal Survey of astronomical priorities. *ScienceScope* of May 3, 1991, remarks on page 635, under the title "Ligo in Limbo,"

> The immediate future looks dim for the Laser Interferometry Gravitational Observatory... LIGO's cause wasn't helped by a National Academy of Sciences committee that last month failed to mention LIGO when it ranked astronomical funding priorities for the 1990s. In the eye of House Science Subcommittee chairman Rick Boucher (D-VA), this omission undercut the NSF's previously advanced argument that LIGO's expected astronomical benefits—in addition to its obvious value to physicists—helped justify the big expense. When Boucher asked the NAS committee for clarification, the reply was anything but reassuring: "[T]he secure scientific goals of LIGO for the 1990s are not astronomical."

The quotation at the end shows the astronomers disowning LIGO and claiming it is only of interest to physicists.

The most vocal representatives of the anti-LIGO movement seem to have been John Bahcall of the Institute for Advanced Studies in Princeton and Jerry Ostriker of Princeton University, both well-known theoretical astrophysicists.[12] An important pro-LIGO physicist complained to me, [1995] "The people who have been attacking LIGO are astronomers, who don't know what it's about and think that they own the sky."

Most LIGO physicists believe, correctly, that the animosity was exacerbated by their choice of title for the project, in particular the O in LIGO, which stands for "Observatory." To an astronomer, an observatory is

11. Brown 1991, p. C1. I looked at the results of this survey and allotted a pro- or anti-score to each comment, and found that my subjective assessment of the balance was 2:1 against LIGO rather than the 4:1 reported by Tyson. I was hoping to reproduce these anonymous remarks on WEBQUOTE so that readers could make their own judgment, but the survey, I found to my surprise, was considered a private matter. This is a pity and is not an isolated instance of unwillingness among the "anti-LIGO group" to release material evidence to this analyst.

12. I was told there was long-standing personal rivalry between Kip Thorne and one or the other of these men going back to their early days at Princeton.

something that observes, and when they build new observatories they build them on the basis of carefully worked-out plans for extending the regime of observation. A new and bigger telescope will look further out into the universe; look with improved definition and discrimination; explore a yet-unexplored, but quite predictable regime of the electromagnetic spectrum; or a new type of particle (such as the solar neutrinos with which Bahcall's name is most closely associated).[13] LIGO, by calling itself an observatory, seemed to be vying for a piece of this territory even though the best and soundest predictions of what it would see, at least in its first instantiation, were zero or close to zero. From the point of view of an astronomer, LIGO was just not good value for the money. It was going to spend sums comparable to those needed to build a new telescope and see next to nothing. In reality, they believed, LIGO was a high-risk exploratory physics experiment passing itself off as an astronomical instrument without accepting the corresponding responsibility to see things.

Astrophysicists such as Kip Thorne of Caltech teased the astronomers by referring to their enterprise as "electromagnetic astronomy," as though to say that traditional astronomy dealt with a small portion of the field's proper subject, which ought to include gravitational astronomy. And LIGO was sold, in part, as a superior instrument to the bar detectors on the long-term promise of its ability to do real astronomy. The promise was that it would be able to investigate the gravitational spectra of astronomical events, not just the bursts of energy they emitted. Analyzing gravitational spectra would reveal the detailed workings of astrophysical processes—the stuff of astronomy and astrophysics.

The Decadal Review of 1990 was chaired by Bahcall. As its name suggests, it is a big survey designed to prioritize future projects, and it is conducted by the astronomical community at the turn of every decade. The 1970 Decadal Review grapples with Joe Weber's claims and says that it is vital that they be tested. The 1980 review reveals optimism about the progress of the cryogenic bars. The 1990 review was the first to consider (or not to consider!) large-scale interferometry. Bahcall invited Thorne to serve on a subpanel reporting to that review to ensure that gravitational radiation was properly represented. But Thorne explained to me,

---

13. Bahcall is famous for predicting the solar neutrino signal of physical processes within the Sun and sticking to his model over decades when neutrino detectors failed to see the predicted flux (for an account, see Pinch 1986). In 2001, the problem was resolved in Bahcall's favor, as a result of a new consensus that neutrinos could change to a less detectable form during their journey from the Sun to Earth. A Nobel Prize was awarded for the work, but, surprisingly, Bahcall was not among the recipients.

[1999] Bahcall, at the beginning of that process, called me up and asked me to serve on a subcommittee that would be looking at LIGO. I told Bahcall that it would not be appropriate for that committee to be looking at LIGO. LIGO was not in the astronomers' purview; it was being done through NSF physics; it was being done by physicists; and it had [already] been blessed in a general way. That is, building big interferometers had been blessed by a previous *physics* survey committee.

Bahcall got quite angry with me and told me in no uncertain terms that I had made astronomical arguments for LIGO in trying to sell it, and therefore it was in some deep sense in competition with astronomical projects, and therefore they had the right to, and were going to, consider it in their prioritization procedure.

So I did serve on their subcommittee, as did Rai and Ron [Weiss and Drever], that sent a report forward to the parent committee. Within the parent committee, as I understand it from somebody who was on the parent committee, a straw ballot was made at the beginning of the process in which LIGO did extremely badly, and the motion was made by someone on the parent committee that they should not consider LIGO after all, for various reasons. And that passed and they dropped it.

This led to an extraordinary argument: the Decadal Review was cited in the arguments in Congress about LIGO's funding, and yet Thorne and others continued to insist that LIGO had not been considered in the review process. In other words, the argument turned not so much on the astronomers' assessment of the value of LIGO to astronomy, which was (or would have been) low, but on whether they had assessed it at all. The LIGO supporters said that the astronomers could not have ranked LIGO in the Decadal Review, because the Decadal Review had not considered it, whereas the astronomers said that LIGO had ranked "nowhere" in the Decadal Review. So this very important issue took on an Alice in Wonderland quality, with words meaning what those who uttered them wanted them to mean. A LIGO supporter told me,

[1999] After the parent committee issued its report, it—claims have been made by knowledgeable people, that Bahcall, in the process of doing briefings for people in Washington on the parent committee's report, told them that the parent committee had looked at LIGO (which was not included in the report), and that the parent committee had ranked LIGO very low.

There is no doubt, as Thorne says, that the Decadal Review did consider LIGO in the first instance. They considered it because the astronomers believed that LIGO was competing for their resources. They believed this for two specific reasons: first, LIGO had called itself an "observatory," not just a "facility," and an observatory is an astronomical term. Second, astronomers believed that the gravitational wave community had continually made claims for the likelihood of seeing sources that changed its status from a physical and engineering project to an astronomical project. They believed these claims were unfounded. This difference in perception is easily understood, because what LIGO supporters such as Thorne were ready to express informally was different from what they said in formal circumstances. For example, Rai Weiss is well known for expressing his confidence that the first instantiation of LIGO was bound to see some unexpected gravitational wave phenomenon from an unknown source. And though inspiraling binary neutron stars are the "banker" mentioned in formal settings, interferometer enthusiasts will discuss all kinds of other possible scenarios as optimistic guesses, especially when they are not being called to account. To illustrate again, in 1987 Kip Thorne allowed it to be put on record that he had made a bet with Jerry Ostriker that gravitational waves would be detected before the year 2000.[14] A bet is not the same as a formal presentation to a funding committee, however. Likewise, reading the reports of subcommittees reporting to successive Decadal Reviews, the likelihood of various generations of gravitational wave detectors succeeding is put quite optimistically; and even where no technical misstatement is made, what is being said is not always clear.[15]

14. Thorne 1987.

15. Consider the following from the 1980s Decadal Review of Astronomy (p. 11): "The detection of gravitational waves from cosmic events may be within reach during the 1980's [sic] if instruments currently in development or under consideration are placed in operation." In the same report (p. 93), we read that the sensitivity reached by detectors now in operation "is sufficient to detect gravity-wave bursts of the maximum strength consistent with conventional theories *at a rate of one per month*. In principle it could detect a wave burst from a very nonspherical supernova anywhere in our Galaxy" (My stress). To the casual reader, this suggests that we should soon be seeing gravitational waves, but what it is actually describing is not an estimate of probability but a possibility that is not ruled out by the laws of physics (that is, what is compatible with what Kip Thorne calls "our cherished beliefs"—the foundation stones in our dam described in chapter 5). The equivalent volume for the 1990s claims, "We expect the LIGO program to go ahead, with a substantial probability of direct detection of gravitational waves before or soon after the year 2000" (p. V-4). It also suggests (p. V-14),

> Of all sources in the LIGO's high-frequency band, the final inspiral of a binary neutron star is best understood [the waveform is known]: and binary pulsar observations have provided enough information about birth rates of binary neutron stars to pin down the distance to which one must look in order to see several coalescences per year. This distance is 100 Mpc, give or take a factor of a few. Advanced detectors in the LIGO should be able to detect the inspiral waves out to a distance of about 1000 Mpc, where the event rate is presumably many per year.

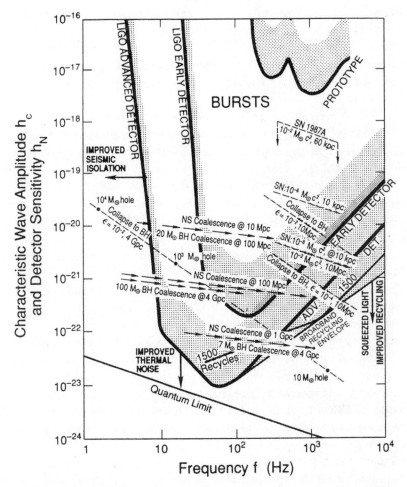

Fig. 27.1 LIGO sensitivity from 1989 grant proposal. Source unknown. Courtesy of Gary Sanders, LIGO project, California Institute of Technology.

Again, in the early 2000s I was present when a way of expressing source strengths and sensitivities that was open to misinterpretation came close to causing a real problem for LIGO. A senior reviewer from the wider scientific community showed a diagram taken from the 1989 LIGO proposal (fig. 27.1) to an influential group, insisting that it was culpably misleading and should discourage them from taking any such claim seriously. The plot in question

---

Reading these two statements taken together, it is easy to gain the impression that LIGO expects to see binary neutron star inspirals out to 100 Mpc in the early years of the twenty-first century. An extra phrase, explaining that early LIGO cannot see inspirals out to 100 Mpc but only out to 10 or 20 Mpc, would clarify the situation for the inexpert reader.

showed a variety of potential sources well above LIGO I's most optimistic sensitivity band (the solid line). What was not shown on the graph was the incidence rate of these sources. So the diagram was not incorrect (and the correct way to read it was explained in the text), but in itself the diagram made it look as though there were many sources that could be easily seen by LIGO I.[16]

Thorne's presentations to the National Science Board are also the subject of fevered dispute. He gave me his notes on the presentation of November 17, 1994, and it is easy to see how it could lead to varied interpretations. He displayed a graph of source strength estimates showing not only the well-understood inspiraling binary neutron stars, which fall below LIGO I's estimated sensitivity, but also his guesses about possible other sources—shown as a broad band well inside LIGO I's range. Furthermore, arrows point to the anticipated increases in LIGO's sensitivity with "No Major Upgrade Required." These arrows may represent the confidence of the time, but they have turned out to be entirely wrong. Thus, Thorne made no false promises about the well-understood sources, or the future of LIGO, but he also presented plenty of optimistic guesswork. That this is guesswork, and that it was made clear that it was guesswork, can be seen from his other documentation, but it is less clear to the reader looking from the outside.

Thorne found a way of assuring me that his presentation had not been misunderstood. He explained that he remembered that the National Science Board had spent a long time at that meeting discussing what would happen if LIGO I failed to see any events. It was clear, then, that the board members had not been led to a falsely optimistic view by his guesses. It is interesting—and this is a point I will return to in chapter 42—that the documents that Thorne was able to show me do not deliver historical truth. To discover what the documents meant to their recipients, we have to rely on a memory of an unrecorded discussion. Clearly there is ample room for outsiders to misunderstand what was going on.[17]

As I will discuss at greater length below, what is going on here is a constructive (perhaps innocently so) exploitation of the exact forms of words and the symbolic values of the forums in which sensitivity estimates are presented. In this world the exact meaning of statements changes radically

---

16. We do not need to argue about what the diagram really showed—I am accurately reporting how it was read by a senior scientist.

17. What Thorne described to me as the "official public presentation" of LIGO's sensitivity, the sort of thing that went into the official grant proposals, can be found in *Science* of April 17, 1992 (Abramovici et al. 1992). Here the estimates are very conservative, and there are no guesses on the graphs.

depending on whether they are thought of as presented in a metaphorical courtroom or a metaphorical corridor. The intensity of the argument over whether the Decadal Review *really* assessed LIGO seems bizarre, since it makes little difference so long as everyone knows that the panelists thought LIGO was no good anyway. Likewise, the insistence that the only things that matter in the way of gravitational radiation flux estimates are what would pass as evidence in a court of law seems strange when one knows that decisions that turn on the estimates are not being made in such a venue but in committees that deal with tacit expertises in the context of trust relations.[18]

Or am I wrong? Perhaps we can reach, once more, for the target diagram to explain what is going on. Perhaps the anger is about what is being presented to the outer rings, where there is little expertise and little connection to the trust relationships of scientists. If the Decadal Review did not look at LIGO in a formal sense, then to report the disenchantment of the panelists to Washington powerbrokers could be said to be divulging a confidence. Again, one can argue that the big committees are like courts of law, which, having no way of estimating the value of guesses, ought to make their decisions only on what has been formally presented (or ought not to be presented with guesses at all). Perhaps, then, the real anger is not about what different insiders to the argument believe but about how much of this core-set dispute should be exposed to the outer world of decision makers. There must be an interesting sociology of the rules of this game.[19]

Some strange claims were made by those opposing LIGO. In 1991, the *New York Times* quoted Jerry Ostriker as follows:

> Several prominent astronomers who strongly oppose the project refused to comment publicly saying [line of newspaper missing] ". . . should wait for someone to come up with a cheaper, more reliable approach to gravity waves," one said, "and we may already have a hint of the possibilities in the work of Demetrious Christodoulou."[20]

18. And the sense of self-justification is still so strong on both sides that, ten or more years after the events, parties in both camps refused to allow me to present certain pieces of evidence in this book because they believed I was misrepresenting the meaning of their actions.

19. There is also the problem illustrated by the game of "telephone," ("Chinese whispers"), where what was said to be said by whom can sound quite different to what was said when the information is passed down a chain of those with strong interests in what was said but who were not present when what was said was said.

20. Brown 1991, p. C5.

Christodoulou, a colleague of Ostriker's at Princeton, had written a paper suggesting that the passage of a gravitational wave would leave behind a permanent deformation in space-time.[21] The thrust of Ostriker's comment seems to be that such a one-time deformation, because it would be there permanently, would be more easy to detect than a momentary oscillation. This seems to be simply mistaken physics, as in the region of the barely detectable, signals that oscillate are much easier to detect that one-off transformations. As one of the LIGO physicists said to me,

> [1999] He [Ostriker] was quoted in the press as saying that the Christodoulu memory effect would mean that it would become much easier to detect gravitational waves because you have a permanent signal. Colleagues tried to educate him that DC signal is much harder to detect but he continued to push the idea. Of course, he didn't have any horse-sense about this, not being an experimenter. But Ostriker clearly had strong emotional views about this project, but he carried enormous weight.[22]

Continuing the story, in 1992, just as LIGO was on the point of being funded, the *New York Times* carried a piece by science journalist and scientific fraud specialist William Broad which included the following:

> The nation's top Federal agency for the support of basic scientific research is battling to reconcile the growing financial appetite of its biggest project with an overall budget crunch. The conflict is threatening to pull the plug on more than a hundred small-scale science projects across the country.
>
> "People here are in great pain," said Dr. Stuart A. Rice, dean of physical sciences at the University of Chicago. "Really senior people are having their grants cut."
>
> If the big project, a device intended to observe gravity waves, goes through and there are further reductions in smaller projects as a result, "it will be a calamity," he said.
>
> The squeeze is threatening a host of smaller scientific research projects with extinction.[23]

---

21. Christodoulou 1991; Thorne 1992a.

22. This is one of the very rare occasions in the book where I will take sides on a matter of physics which was a subject of dispute among physicists. Ostriker, if he was implying that the Christodoulou effect made the detection of gravitational waves cheaper or more reliable, was making a mistake about the nature of experiment.

23. Broad 1992, p. C1.

As we have seen, the LIGO supporters' counterargument was that they would be supported by a special fund from Congress, so there was no direct competition after all; but as late as 1999, Ostriker would tell me that the funds come from the same budget that supports graduate students and doctoral fellows. That money was being taken from physicists and put into the pockets of bulldozer drivers and the like, so he said.

John Bahcall's opposition had been ameliorated by the time of LIGO's official inauguration in November 1999. His specialism is neutrino detection, and that is also the field in which Barry Barish, LIGO's director, spends half of his time. This relationship may have played a part in Bahcall's presence at the inauguration. One important scientist, illustrating the deep bitterness that still existed, told me that he would not attend the first day of the meeting, because he did not want to be present when Bahcall spoke.

Bahcall gave an amusing address, joking about his past opposition, but eventually reached the point where he contrasted, unfavorably, his own search for neutrinos and the LIGO project. He noted that in the case of the solar neutrino search, the absence of neutrinos had been an interesting scientific puzzle, whereas an absence of gravitational waves will be merely a trouble. He noted the $300-million cost of LIGO as opposed to the $3 million spent on the neutrino search, and the 300 versus three persons needed to staff both endeavors. And, he remarked, the neutrino search had always been referred to as "an experiment," not an observatory, until it reached the point when large numbers of neutrinos were being seen. He said that LIGO could have done exactly the same science by calling itself an experiment instead of an observatory, and he noted that in the case of the neutrinos, whereas astronomers were not interested in it, they were not opposed to it. For many of these reasons, he thought that LIGO was going to have a much harder time than solar neutrino physics.

CONCLUSION ON CAUSALITY

So why was LIGO funded? I think that there is no sociological meat and potatoes in this story. The sociological interest is that the forces of all kinds were roughly evenly balanced, so the funding decision could have gone either way.

We could explore other kinds of potential causes, such as the skill at lobbying of the various parties. Vogt hired a professional lobbyist who arranged many meetings for him with congressional representatives, whom he charmed. But others, who ought to know, say that Vogt's meddling in

Washington affairs was dangerous and counterproductive, because no one knew what interests he represented. I have also been told that the crucial factor was that Washington needed big science because of its prestige; yet another report has it that it was all a matter of swapping favor for favor at the level of congressional staffers. And I have also been told, by a *very* big player in the game at the time the project was being funded, that (I paraphrase): "a congressional staffer explained that the cost of LIGO was only an accounting error in the size of budget they were dealing with."

As a sociologist who has looked hard for the big sociological story, I finally fall back on something like the last sentiment. The funding of LIGO was an immensely important issue to the scientific community, which works with the NSF, but it was not an immensely important issue to those who were actually providing the money. None of the energy and hard work that both the pro-LIGO side and the anti-LIGO side put into presenting their cases was wasted, but the net result was an even balance. Had either side faltered, then the others would have won; but since neither side did falter, it was a standoff, the outcome being a matter of comparatively mundane tradeoffs, probably at congressional staffer level. This should not be depressing to either party, for it was their work that made it a possibility that LIGO might be funded/not funded. This is the logic of the calculation of such outcomes: How can I be sure that this will be a great book? I don't know. How can I be sure that this will not be a great book? Don't write it! If the scientists didn't try, they were bound to fail; but the harder they tried, the more surely did they make themselves the playthings of history.[24]

So, there is no big story about the funding of LIGO except the story that put the funding on a knife edge. What is of sociological significance here is that sometimes things, even big things, just happen one way rather than another for no profound reason. But as soon as LIGO is funded, the story becomes big again. To use an old analogy, the flap of a butterfly's wing may have caused the hurricane, but this does not mean that the hurricane is no more potent than the flap of a butterfly's wing.

The very publicity surrounding the funding decision and the opposition to it had potency. The salience of the debate explains the relationship of interferometry, especially American interferometry, to the resonant-mass

---

24. Thorne offers an interesting counterhypothesis. He suggests that though my account may be correct for the funding in any one year, the scientific case for funding was so good that LIGO would have been funded eventually—even it had failed rather than succeeded on the penultimate occasion. On the other hand, I would argue that if the funding had been long delayed, the project would have been different (there would have been more prototyping), and the spheres might have survived.

projects. The actual and latent opposition to gravitational wave astronomy has made American interferometry ultrasensitive to any claim that might make the field look less worthy. Thus it can explain the treatment of Weber's mid-1980s papers and the treatment of the SN1987A results (chapter 21), the treatment of the Perth-Rome coincidences, the shrinking of the domain of open evidential culture, and the dominance of the theoretical animus over the experimental animus over the course of the funding period (chapter 22). It may even come to limit, pathologically, the degree of risk that LIGO will embrace in announcing its own results (though the experimental animus may be due for a revival). It explains, to some extent, the management upheavals that would beset the project at the end of the Vogt regime (see chapter 33), because LIGO had to be regarded as ultraclean in its handling of its resources and its managerial efficiency. I believe, as I have argued, that it was also the case that the very process of the funding of LIGO ended the future of the sphere program. Richard Isaacson described LIGO as an "800-lb. gorilla" (chapter 23), and we can see how it could be reasonably argued that the gorilla chewed off the future of large resonant masses.

In sum, the funding of LIGO illustrates how the outcome of a battle between two massively opposed forces in science can turn on something small, yet can nevertheless have widespread consequences. In this case, the consequences include future funding decisions for whole programs, the founding of a new astronomy, and the credibility of quite specific pieces of scientific data.

PART IV

THE INTERFEROMETERS AND THE INTERFEROMETEERS—FROM SMALL SCIENCE TO BIG SCIENCE

CHAPTER 28

MOVING TECHNOLOGY: WHAT IS IN A
LARGE INTERFEROMETER?

The interferometers were going to be built, and they were going to become the dominant technology for the direct detection of gravitational waves; earth would be shifted, buildings built, and enormous feats of engineering accomplished. In the case of the bigger interferometers, it was not only earth that would need to be moved but people. The way scientists worked would need to be transformed, and this was most marked in LIGO, the very biggest of the interferometer projects. Change one part of the world, and every part changes; move enough earth and enough people, and you change what is possible to believe about the sky. To understand the transformations, we need to understand interferometers.

REFINING THE NEW TECHNOLOGY

In chapter 17, I sketched out some of the design principles of the large interferometers. Now I take this further. A gravitational wave interferometer has a light source, a beam splitter, and reflecting surfaces on "test masses"—that is, mirrors, whose changing separation is to be measured. It may also have other mirrors, such as those used to bounce the light in the arms backward and forward many times. As we saw, it is desirable that a sensitive interferometer have the following features:

- It should have long arms.
- The path length of the beams should be still longer than the arms, so there should be some means of bouncing the rays backward and forward in each arm.
- The brighter the light the better.
- The light should be monochromatic (of a single wavelength) and coherent (of one phase).
- The crucial parts of the interferometer should be isolated from anything, other than gravitational waves, that will affect the path length.

We now explore each of these features in turn.

Arm Length

LIGO, at four kilometers, has the longest arm length of all the current generation of interferometers. The choice of length turns on a number of subtleties. In the gravitational wave frequency band, which detectors such as LIGO want to explore, an ideal length, other things being equal, would be a few hundred kilometers, but on the surface of Earth four kilometers is probably close to the optimum. One obvious reason for this is that the arms have to be straight, while the surface of Earth is curved. As the arms get longer the cost of the tunnels or embankments needed to keep the arms straight also increases. Another reason expense increases nonlinearly with arm length is that the light beams spread as they travel, and a longer path length needs a wider and more expensive beam tube. One of the ways in which the cost of GEO600, the German-British device, has been kept down is by using stainless-steel tubing of comparatively narrow diameter that can, therefore, have very thin walls.[1]

Longer arms are also more expensive because they are more difficult to build and maintain at the very high level of vacuum and cleanliness needed. The vacuum must be high and the inside scrupulously clean if the light is not to be dispersed and otherwise affected by stray molecules. Still

---

1. Unfortunately, this means that even if GEO had more space, it could not extend its arms much without increasing the diameter of its beam tube. GEO's beam tube walls are still thinner than they might be, because they are strengthened with corrugations.

worse, if the vacuum were poor, organic molecules could be deposited on the surfaces of mirrors, ruining their reflectivity. A good vacuum is hard to establish and maintain, and dirt is hard to control under the semi-industrial conditions required to build long beam tubes of a large diameter. The evacuated spaces required for LIGO are the largest high-vacuum enclosures ever made. GEO's task is much simpler and cheaper because, while the same conditions of vacuum and cleanliness have to be met, the volume to be evacuated and kept clean is only 3–4% that of a single LIGO interferometer.

There is a more subtle reason for not extending the arms indefinitely. The idea is to try to make the mirrors and all the other crucial parts of the apparatus behave as though they were floating in space. The foundational technique is to hang them from cunningly designed suspensions so that each mirror acts as the "bob" of the pendulum. If the suspension point of a pendulum vibrates horizontally at a high frequency, the pendulum bob remains still. We might say that it takes time for the message about the suspension having moved to reach the bob, and by the time the bob-mirror "knows" that the suspension has moved it will be back where it started. So pendulums are very good at filtering out high-frequency horizontal vibrations.

Vertical vibrations are harder to deal with. But since what we are measuring is the horizontal displacements caused by the passage of a gravitational wave, at a first approximation the vertical movements in the mirrors do not matter so much. In fact, in an instrument of such exquisite sensitivity, vertical movements should be eliminated, too, as far as possible, and the suspension systems are designed this way. Interferometer constructors have become much better at eliminating vertical vibrations transmitted from the ground, but removing the vertical component of thermal noise (for example, the miniscule changes in suspension wire length caused by tiny temperature fluctuations) seems almost impossible.

It is this remaining problem of vertical stabilization that sets the subtle limit on the arm length. The mirrors hanging from their pendulums at either end of an interferometer arm will hang toward the center of Earth. As the arm lengths get longer, the less parallel the pendulum suspensions will be. This is not something that can be eliminated by driving tunnels or building embankments to eliminate the curvature of the surface of Earth; it is a feature of the fact that the meaning of *down* changes as you move around on the surface. Thus "down" at the North Pole is at right angles to "down" at the equator. Even at distances of a few kilometers, the mirrors will be hanging at significantly different angles; and it follows from simple geometry that if they move downward, they will get closer together, and if they move upward, their separation will increase. The further the pendulum-mirrors are apart, the more will vertical displacements translate into horizontal displacements like this. Hence it is not worth going much beyond four kilometers in arm length.

The space-based interferometer being planned will have arm lengths of millions of kilometers. In space, arm length is limited only by the ability of a laser to shine across the distance without its light being impossibly dispersed; there are no tunnels or embankments needed, there is no curvature of Earth to account for, and there is far less in the way of vibration to worry about anyway, so no need for pendulum suspensions. Indeed, as we have seen, the whole idea of the pendulum suspensions is to make the mirrors behave as if they were floating in space. The space-based interferometer, then, is capable of enormous sensitivity compared with ground-based devices, because it can have enormous arm lengths; but for reasons that I will explain, extremely long arm lengths are good for very low frequency observations only. (In fact, other aspects of the space-based interferometer's design would make it still less sensitive in LIGO's frequency range.)

The longer the arm length of an interferometer, the harder it is to make it work. This is because the angle at which everything has to point at everything else becomes less and less tolerant to mistakes. To point a laser beam at a one-centimeter spot over a distance of 40 meters is one thing; to point it at a one-centimeter spot at a distance of 4000 meters is another. The angles will wander enough to stop the beam reflecting backward and forward in the right way far more often, because the tolerance is so much less. And every time it wanders it is going to be harder to get it back, because there is that much less time to react before it has wandered off again.

Path Length

Given a certain arm length, the path length is given by the number of reflections. To get a sufficient number of reflections (50 to 150 off each mirror) without distorting and degrading the light beams requires mirrors of extraordinary perfection and reflectivity. Mirrors must be finished to a geometry and surface perfection where defects on an atomic scale are taken into account. Their surfaces must have almost no losses. Mirror technology, however, is turning out to be one of the least problematic features of LIGO-like instruments.

Given that mirrors are perfect enough to reflect light many hundreds of times, why not lengthen the path indefinitely by using more and more reflections? There are two types of reasons—what we might call a techno-logical reason and a more fundamental reason.

The technological reason is that a folded path length does not do quite the same job as a straight path length even though the length is the same. The mirrors suffer from thermal noise—the bouncing around of their atoms within the glass and in its suspension—and the light bouncing backward and forward encounters this disturbance more frequently if the increased path length involves more encounters with the mirrors. Thus, the fewer bounces the better, and a shorter interferometer, alas, cannot compensate for its lack of arm length by increasing its path length with an indefinite number of foldings.

The more fundamental limit is not so much a limit on the number of reflections, but a restriction on overall path length—whether it is made by longer arms or shorter arms with more reflections. A gravitational wave has a characteristic frequency. As the wave passes it first lengthens arm A while shortening arm B and then lengthens arm B while shortening arm A. The switch of effects from arm A to arm B will take a certain length of time depending on the wavelength. If we are looking at a wave that has a frequency of 200 cycles per second (Hertz), it will switch from lengthening arm A and shortening arm B, to shortening arm A and lengthening arm B, and back again, in 1/200 of a second. Considering arm A alone for a moment, if the path length is such that it takes light 1/200 of a second to get to one end of arm A and back again, before it interferes with its twin ray, the amount of arm lengthening in A will be offset by the same amount of arm shortening as the second half of the wave comes by, and it will appear as if nothing has happened. Thus a path length that takes 1/200 of a second for light to traverse will mean that the interferometer is blind to gravitational waves having a frequency of 200 Hertz. A reasonable length for a path turns out to be such that light takes half a gravitational wave cycle to go to the end of the path and return. For a 200-Hertz wave, with light traveling at 300,000 kilometers per second, this means a path length of a bit less than 1000 kilometers—but let us call it 1000 kilometers. Given four-kilometer arms, this means an upper limit of about 125 bounces off each of the two mirrors.

Two Ways to Make Light Bounce

Michelson and Morley, when they built the first interferometers, bounced their light backward and forward off a series of fixed flat mirrors. GEO uses the same technique with just one bounce off an extra mirror, thereby doubling the path length. The initial design for a large interferometer, proposed by Rai Weiss and his team in the Blue Book, was a version of this technique, known as a "delay line" because it was originally used by early computer engineers to delay the passage of light and store it for a short time as a way of delaying an electrical signal. The delay line uses single large curved mirrors in place of the series of plane mirrors used by Michelson and Morley. One of the most important design decisions made prior to the actual funding of LIGO was to move away from the delay line idea championed by MIT to the Fabry-Perot cavity championed by Ron Drever at Caltech. To understand what happens in a delay line is quite easy: light shines in through a hole in one of the mirrors; it bounces between points on the surface of two curved mirrors until it returns to where it started and exits through the same hole.

The major disadvantage of the delay line design is that the mirrors have to be big enough to allow sufficient bounces to take place without the rays interfering with each other as they pass backward and forward. The prototype interferometer built at the Max Planck Institute in Garching, near Munich, used a delay line, and one of the "icons" in the field is a photograph of its green beam bouncing backward and forward many times. But the German team also discovered that this design had problems of light scattering that they could not solve.

The Fabry-Perot cavity avoids the problem of scattered light and can use smaller diameter mirrors. To understand what happens in a Fabry-Perot cavity is a little less straightforward. A Fabry-Perot cavity consists of two very slightly curved mirrors facing each other, bouncing light backward and forward. On the face of it, the light can bounce backward and forward an infinite number of times, but mirrors are not perfect. Assuming they were close to perfect, however, one can see that if one could get light into a Fabry-Perot cavity, it could stay there, bouncing backward and forward, for quite a long time.

But light has to be got into the cavity. To do this, the mirror at one of its ends is made less than fully reflecting. (The mirrors are thin coats of material of an exact thickness related to the wavelength of light on the surface of glass disks. Light can be reflected or pass through from either side of the disks depending on exactly how they are made.) The first mirror is made so that a certain amount of light is reflected and a certain amount passes through. This means that if a beam of light is shone onto the back of the first mirror, some will be reflected, and some will pass through into the space between the mirrors and bounce back and forth. But at each bounce a little light will come out again. Imagine that a very short, bright flash of light was aimed at the back of the first mirror. To a first approximation, some of it would enter the cavity and start bouncing backward and forward with a little coming out at each "pass." Thus after the initial flash, the front end of the cavity ought to glow with diminishing brightness for a little while. If the initial input was maintained at a steady state, the intensity of the light would build up until some equilibrium state was reached, with as much coming out again as was going in, but with any photon likely to have spent some time bouncing backward and forward inside the cavity before it came out again.

Moving away from the first approximation, exactly what happens in a Fabry-Perot cavity illuminated with monochromatic light depends on the relationship between the separation of the mirrors (the length of the cavity) and the wavelength of the light. The various beams being reflected from the mirrors and passing through them will be superimposed on each other and will either cancel each other out or reinforce each other, depending on the phase—peak and trough will cancel, while peak-peak and trough-trough will reinforce. In this way a Fabry-Perot cavity can be tuned by making it just the right length for the wavelength of the light you are interested in, thereby brightening rays which are in one phase and making those with other phases cancel out. On the one hand, this creates a problem for controlling the mirrors when the cavity is long—a problem to which Ron Drever made the largest contribution toward solving. On the other hand, a Fabry-Perot cavity can also be used to clean the light of unwanted phases, and large interferometers take advantage of this. The light from the laser is fed through one or more "mode-cleaners"—smaller Fabry-Perot cavities than the main one—before entering the "business end" of the devices.

To return to the main story, because a Fabry-Perot cavity stores light for a while, it can be thought of as very like a delay line. How long it holds the light will depend on how reflective the mirrors are—their "finesse." The greater the finesse, the longer the light will bounce back and forth before it emerges again. Mathematical analysis shows that the finesse of a Fabry-Perot cavity makes it equivalent to a delay line with a certain number of bounces and a certain path length. But because all the rays are superimposed, the Fabry-Perot design does not need big mirrors, and effective path length can be increased without increasing the size of the mirrors.

Making the Rays Brighter

As we have seen, "the brighter the better" is a good rule for the rays passing through a sensitive interferometer, because brightness averages out fluctuations in the beam and it gives a bright signal to look at. The first and most obvious way of brightening the rays is to use a more powerful laser. But building powerful lasers for continuous operation is still a black art even after years of development. Laser experts tell of firms who make lasers as alike each other as possible but only decide how to label them in terms of their power after they have been put together and tested. It is not possible to tell exactly how a powerful laser will perform purely from the design because there are too many ill-understood variables.

Power Recycling

There is, however, another way to brighten the beam in an interferometer—"power recycling" moves the design a further step from Michelson and Morley's simple concept. Nowadays one would not consider that looking at the fringes of an interference pattern with the naked eye was a satisfactory way of reading the output of an interferometer. The human eye is far too poor at seeing small changes in brightness compared with what can be done with instruments, and in any case, the human eye cannot measure and record the minute, complex, and rapidly varying signals which would indicate the passage of a gravitational wave. In a modern large interferometer, you do not see fringes. The beams are made parallel so that the beams at the point you observe are nearly out of phase, and what you see is slight variations in near darkness. Since the residual brightness is a function of the relative phase of the beams, you are looking at any fraction of a phase change in one beam as compared with another. You look at the out-of-phase darkness because it is there that it is easiest to see any small variations in light intensity. Seeing a tiny variation against a background of darkness is easier than seeing a tiny variation against a bright background.[2]

Somewhere else the beams are in phase, and there is a lot of light associated with this bright output. This light is "going to waste." In power recycling, this secondhand light is captured, fed back into the interferometer, and added to the new light bouncing around inside it.

This is not the same as adding an additional bounce to the new light. It had better not be the same as extra bounces, because extra bounces carry cost in terms of mirror noise and the length of time that the light takes to traverse its path. As I pointed out above, the amount of time the light "dwells" in the Fabry-Perot cavity must not exceed a certain limit if it is not to blind the interferometer to certain frequencies of gravitational radiation.

---

2. Actually, the analysis is a little more complicated and the darkest point is not always the optimum, but we can safely ignore the finer points.

But old light that has completed its journey through the interferometer and emerged at the "bright port" is simply light in phase, the mutually reinforcing part of the recombined beam. It is a nice, clean beam of light ready to be fed back into the interferometer; this is quite different from lengthening the path of a single ray. Things simply have to be arranged so that the old light (and light doesn't age—secondhand light is just as good as new light) is in phase with the new light that has just come from the mode cleaner. This *power recycling* is a way of increasing the brightness of the beam without needing a more powerful laser. All that is needed is another mirror.

Signal Recycling

Another technique that can improve sensitivity by doing something similar to brightening the beam involves brightening the signal. In effect, this means tuning the whole interferometer so that signals at a preset frequency are enhanced by recycling the light that has had that frequency impressed upon it. It is done by adding yet another mirror at the place where the beams are out of phase (the "dark port," to use the jargon), at a distance that will form a resonant cavity (like a Fabry-Perot cavity) for light of the specific frequency of interest. This technique is known as *signal recycling*. The disadvantage of signal recycling is that it reduces the bandwidth of the interferometer—this means that the apparatus can search best for signals at a particular frequency (or frequency band, depending on exactly how it is tuned), but the recycling makes the interferometer more sensitive to a lasting signal of this frequency. Signal recycling means adding not only the extra mirror and carefully tuning it but also adding an extra vacuum chamber to house it. The positions of the various elements are indicated in figure 28.1.

Isolation of Crucial Parts

These technical advances make a modern, large-scale interferometer far more complicated than the original Michelson-Morley design. There are now suspended mirrors for the mode cleaner or mode cleaners, a beam splitter, mirrors at the inboard and outboard end of the cavity, a mirror for power recycling, and a mirror for signal recycling. The practice is to mount each suspended component in its own vacuum chamber, separated from the others with a gate valve, so that each can be worked on separately if necessary without upsetting the vacuum in the rest of the system. This means there will be at least six or seven vacuum chambers at the corner

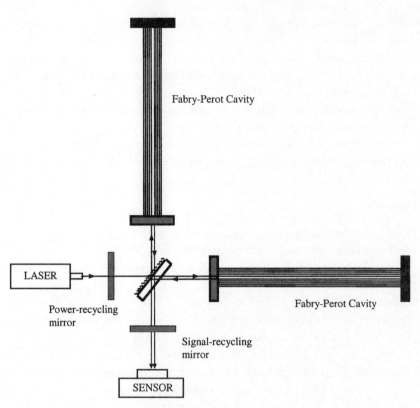

Fig. 28.1 Complex interferometer with Fabry-Perot cavities and both power and signal recycling.

station of each large interferometer (and double that number in the case of LIGO-Hanford, where two interferometers are housed in the same vacuum tubes). Housing these vacuum chambers generally means big buildings, and this is not to mention the vacuum tubes themselves and the "end stations" which hold the vacuum chambers for the outboard ends of the main Fabry-Perot cavities. Modern interferometers are big and complex.

CHAPTER 29

MOVING EARTH: THE SITES

TRANSFORMING THE LANDSCAPE

In these early years of the twenty-first century, five large-scale interferometers have been completed or are nearing completion. Because of its size, engineering a large interferometer installation
is a very different exercise from building a resonant bar detector.
A resonant bar detector is constructed inside a laboratory, but a
large interferometer is a self-contained structure that houses one
or more laboratories. It may also contain meeting rooms, storage
facilities, clean rooms, domestic services, and utilities. The larger
interferometers are built on virgin sites, usually some distance
from towns. Consequently, constructing an interferometer means
transforming a substantial part of the landscape, not putting an
instrument into a building.

In March 1996, I went to the gravitational wave conference
in Pisa that was described in chapter 24. Each morning, as I
explained, the delegates, who were housed in hotels in Pisa,
were bused the ten kilometers to the small town of Cascina,
near the future site of VIRGO—the Italian-French three-kilometer

interferometer. Up to this time, Cascina was known only for the manufacture of furniture and it had little to offer unless you wanted a cupboard made. Now it was the site of an international conference.

On the last evening of the conference—a Friday—the delegates took a detour on their way back to their hotels to visit the site of the nascent VIRGO. We traveled along ever more narrow roads lined with trees and deep ditches, the two big blue buses barely sliding past the trucks coming the other way. The conference had run late that day, so it was dusk when the buses took their final left turn onto a still narrower lane, drove another kilometer, and stopped at a place without any apparent distinguishing feature. We clambered down, crossed to the left side of the road, and stood along the edge of the ditch, staring across the fields while the buses drove away to find a place to turn. The flat fields slowly darkened to a blue-mauve. We enjoyed a moment of pleasing calm and silence, with no vehicles, no noise, and no overhead projector slides crammed with mathematics. But what were we supposed to be looking at?

Our host, the director of the VIRGO project, gathered us and, pointing out over the fields, explained that 500 meters from where we stood would be built the midstation of the interferometer. One three-kilometer arm would strike out parallel to the road, while the other would go directly away from us into the gloom.

A cynic might wonder about the point of taking the twisty detour to look out across this particular three kilometers of nearly invisible landscape. Any three kilometers of field would have looked the same. Indeed, if the buses had stopped two kilometers further down the road, or even if we had been looking at a field ten kilometers from the true site, it would have been all the same to us. And yet there was something moving about that moment. We could envision the vacuum tubes stretching away into the distance, and because we were looking at the ground rather than at a drawing or a plan, we gained a sense of the sheer size of the project. We could think about the fact that, if we started on foot at the end of one arm, it would take a whole hour of fast walking to get to the end of the other.

The very bareness of the field indicated the forces that were mustering. When someone renovates a house, they often take photographs at the start so that they can show what has been accomplished. Perhaps, in seeing this "before," we would better appreciate the "after." The sheer lack of anything to look at spoke to the political ability of project leaders—those who could harness the financial and political power to erect the modern equivalent of a pyramid.

During this roadside ritual, I stood next to Mark Coles, who had come to LIGO from the canceled Superconducting Super Collider project. He told me that he had built for the SSC the best magnetic diagnostic instruments in the world, and now they sat in cardboard boxes, never to be used. Huge amounts of money had been spent, but just a few things would be salvaged before the tunnels beneath the Texas plain were filled with sand. Coles told me that he had once before stood in just such a little group looking out across just such a field, only that time it was with the expectation that the site would become the SSC. In a still voice, he reminded those standing nearby that this ritual was only a beginning. Last time the ending was a sad one.

As of 2002, of the five "full-scale" interferometers nearing completion or "shaking down" toward optimum performance, two belong to the American LIGO project; each of these has arms four kilometers long (there is also a two-kilometer device installed within one of the American four-kilometer structures). The Japanese TAMA apparatus, in Tokyo, has 300-meter arms. The joint German-British GEO project has 600-meter arms and is located in a field near Hannover. VIRGO, the three-kilometer interferometer near Pisa whose site we had stared at in 1996, is a joint Italian-French project. There is also an embryonic interferometer—the Australian International Gravitational Observatory (AIGO)—soon to enter service in Gingin, an hour's drive north of Perth, as a test facility for high-powered lasers. And, of course, there are prototype interferometers; the longest, at Caltech, has 40-meter arms. Other than TAMA, I have visited all the sites, usually two or more times.

TAMA, though the smallest and theoretically the least sensitive of the "big" interferometers, was ahead of the others in its construction schedule. It was the first to show that such a long interferometer could be made to work at all, and that the noise could be driven down. The embryonic AIGO aside, which currently has no money to turn itself into a proper detector, VIRGO has been the slowest to advance and is not scheduled to come on air until 2003.

VIRGO'S TROUBLES

The first steps in building a large interferometer are to level the ground, build embankments to support the vacuum tubes if necessary, build roads, construct buildings, fabricate beam tubes, install vacuum chambers, evacuate the system and bake it out, hang mirrors, install and connect the electronics, and so forth. Only then can real commissioning start. In the case of VIRGO's schedule, the holdups began with the mundane. When we looked across the dusky fields near Pisa, what we did not know was that the title to all the land had not yet been acquired. The field comprised many small parcels of land with separate titleholders, and each of these had to be persuaded to cooperate before the beam tubes could be completed; gathering all the titles was to take a long time. And then VIRGO's main building turned out to be vulnerable to flooding; this had to be fixed. The installation also had trouble pressing ahead because of its management style. VIRGO was an international (French-Italian) project, but complicating matters further was its being a confederation of separate laboratories rather than a single project guided by strong leadership. It was hard to ensure, therefore, that every contributing laboratory was equally committed and that all were pulling in the same direction.

Given that all the interferometers are based on the same ideas, it is surprising how much variation there is from site to site. The two American installations are intended to be identical for the most part, and when one is inside one or the other building, it is easy to forget whether you are in the southeastern or northwestern part the USA. Even these installations differ, however, in subtle ways I will discuss below. Interferometers in other countries differ in more obvious ways.

AIGO—GINGIN

At each of the five sites the landscape is different, so you always remember where you are when you are outside any given installation. The most exotic surroundings are at what is now the test interferometer at Gingin, near Perth. That an interferometer is sited here at all is due to the drive and determination of David Blair, the pioneer of the NIOBE cryogenic detector at Perth's University of Western Australia (UWA) and sworn enemy of "colonial cringe." Blair is an extraordinarily energetic booster and publicist who has not endeared himself to everyone. Gingin, furthermore, is not a place that most Australian interferometer scientists would have chosen for their project, given that Western Australia is seen as remote and is, in fact, time-consuming and expensive to visit from Australia's main centers of population and scientific activity. That, of course, is one of the reasons Blair wanted the facility in that state—he wanted to help put UWA and all of Western Australia "on the map" and lessen the hegemony of the Eastern Australian scientific establishment.

The AIGO site was initially intended to house a full-scale, three-kilometer detector, but the Australian agencies have not provided the funds. The current compromise is the high-power test facility, which may be extended into a proper Southern Hemisphere detector once gravitational waves have been directly detected by others. The Australian establishment's view seems to be that there is no need to pioneer the technology independently in Australia when it is already being advanced in the Northern Hemisphere. Instead, Australia has filled a research niche—high-power lasers—in what is becoming a worldwide cooperative venture rather than a competition among independent national sites (a tendency which I will discuss further in chapter 39).

But Gingin is worth the visit. Once you have reached Perth, you have not reached Gingin, and once you have reached the area of Gingin, you have not reached the interferometer. The site is in the middle of the Australian bush. On March 7, 1998, a party consisting of David Blair, me, and Blair's two young sons was the first to visit the site in a wheeled vehicle, pushing down trees in UWA's Toyota Land Cruiser as we pioneered the route. In two weeks, Blair told me, there would be roads laid along the route we had opened out, and the corner station building would be finished by the end of the year.

The next time I saw the Gingin site was in 2001, when UWA hosted the annual Amaldi meeting on gravitational waves. This time dirt roads had been cleared along the path pioneered by the Land Cruiser, and the hundred or so delegates were taken to the site by bus. There we found buildings on a medium scale, a small museum, the shell of a visitor center that would house astronomical telescopes, and the center section of an interferometer, with arms leading into the bush but cut short after 40 meters. The bush still edged right up to the interferometer building, so whichever way you walked you were soon surrounded by plants—the weird Banksias dominating. Not long after this the roads were metaled, I was told. Here was a frontier being tamed, not by the railroads but by the demands of science and technology. (Though exactly what could and could not be done to the bush still turned on the word of the local native Australians.)

COMPARING VIRGO AND LIGO

AIGO is a large interferometer in embryo. The big contrast I want to draw is between LIGO and GEO600, the British-German device. But before we get to this, it is worth noting that though LIGO and VIRGO are roughly on the same scale—each LIGO interferometer has four-kilometer arms and VIRGO's are three kilometers in length—they are surprisingly different in concept, mode of construction, and appearance. Let us start with a sense of how deeply this philosophy of independence goes, because it bears on the question of international collaboration that I will discuss at length in chapters 38 and 39.

One of the most successful phases of the building of LIGO was the fabrication and preparation of the beam tubes. This task had been contracted out to a private firm—Chicago Bridge and Iron. CBI prepared a portable fabrication plant and assembled it first near the Hanford, Washington, site, producing all the tubes necessary for that interferometer; then they broke it down, transported it, and reassembled it near the Livingston, Louisiana, site to make the tubes there. The method of fabrication was to take a flat "ribbon" of carefully treated stainless steel and spiral it into a tube (look at the cardboard center of a toilet-tissue roll). In this case, the ribbon was about four feet wide, the diameter of the finished tube was also about four feet, and the length of the finished tubes was 65 feet. As the ribbon unrolled, its edge was welded to the adjoining piece in a continuous operation fed by a jig and powered rollers.

To visit one of these temporary beam tube fabrication plants was to see a remarkable operation. A half-dozen men welded steel ribbon to submillimeter accuracy. The plant had all the appearance of a metalworking industry, but amidst the chaos scrupulous cleanliness was observed. Any organic impurities on the metal would fill the evacuated tube with volatile molecules, ruining the vacuum and scattering the laser light. Thus, the welders wore white gloves to keep their sweat from the steel, and delicately swept the dust from the inside of the weld with little brushes made from dried grass. Dirty heavy industry was being executed with care reminiscent of the assembly of electronic components. During my tour, I touched the edge of one of the multiton coils of drab steel lying on the floor awaiting fabrication and immediately realized I had committed a faux pas. I had been more careless than the welders, who would never touch the steel with their bare hands. Luckily, each huge completed tube was carefully washed with hot cleaning solution before dispatch to the interferometer site.

To see a display of skills of this sort elevates the spirit. Strange though it may seem, there was a long and important period when Chicago Bridge and Iron was more important than Caltech theorist Kip Thorne, the chief booster of LIGO's scientific prospects, for building confidence in the future of gravitational astronomy. More than anything else, the successful fabrication of the beam tube seemed to make the detection of gravitational waves a reality.

The point I want to make is that the beam tube fabrication and cleaning, which was thought by some to be an insurmountable technological obstacle, was a huge success; the work proceeded on schedule, there were no unexpected sources of contamination, and no leaks marred the 50 miles of so of welds.[1] CBI then offered its proven services to the VIRGO project, but the offer was refused, and VIRGO's tubes were made locally by a different method: welds running the length of the tube joining the edges of rolled steel sheets. This, too, was a success, but it illustrates the desire of scientists to do things their own way and may reflect something deeper to do with national pride.

The housings of the beam tubes are quite different in LIGO and VIRGO. LIGO's beam tubes are mounted on concrete beds and covered with roughly made, unfinished, inverted U-section concrete covers. The covers are there to keep hunters' and ill-wishers' bullets out; already, sixteen bullet holes pock the Louisiana installation's buildings (though the animosity caused by this government project on traditional hunting land may now have been diffused). Inside the covers it is dark and cramped; the spaces are untidy with blown-in vegetation and are rapidly becoming populated with insect and animal life. Animals also nest in the insulation surrounding the tubes, left over from when they were heated to bake out gases in the metal.

In contrast, VIRGO's beam tube covers are wide, upright U-section concrete containers surmounted by a roof made of domed sections, which are a sandwich of expanded polystyrene between corrugated aluminum sheets. They are not bulletproof but do insulate against the heat of the sun, and the outside is painted a tasteful blue to blend, so the architect said, with the hills in the background. Inside the spaces are cool, clean, wide, and continuously lit along their length, like a three-kilometer-long underground railway station. You could run a very nice bicycle race alongside the beam tube; there would be just enough room for three cyclists to ride abreast. Furthermore, in VIRGO the bakeout insulation that surrounds the tubes is covered in stainless steel cladding, so animals and insects cannot nest inside it. The impression given by a VIRGO beam tube space is of modern engineering, whereas LIGO's feels like a covered wagon. (As we will see, this impression does not carry through the whole of LIGO.)

It is also the case that VIRGO's and LIGO's buildings look different in detail: the floor coverings in the end and center stations are different, as

---

1. Though there were some leaks in the joints between separate sections of tube.

are the architectural details. Where there are right angles at the corners of the LIGO buildings, there are curves at VIRGO; where there are smooth edges at LIGO there are rims at VIRGO; and so on. These smaller differences are, presumably, just differences in the local vernacular of industrial architecture, but with installations as delicate as these—where even the effect of the breeze on the buildings' walls has to be taken into account—the residual differences still seem surprising to an outsider.

Inside the buildings, LIGO's offices are smart, carpeted spaces, but VIRGO's are smarter, with marble on the staircases and ceramic tiles on the floors. Marble and ceramics are local products in Italy, and I was assured that because civil engineering is much cheaper in Italy than in the USA, the lavish spending on the beam tube covers and internal furnishings reflected little in terms of real difference in cost.

There is another kind of difference which, I believe, reflects badly on European science as compared with American science, and that is what the architecture says about relations with the public. LIGO's control room is a huge public space resembling the deck of the *Starship Enterprise,* including a front wall on which can be projected the image from the control monitors and a back wall made of glass. VIRGO's control room is about is as big as two average offices run together, with just enough room for the operators to sit at the banks of monitors. I asked both groups about these differences and was given mundane answers. VIRGO suffered a cut in budget just about the time that the first central facilities were being built, and that meant they had to build small. LIGO's control room was based on experience in high-energy physics, where it was found necessary to have many separate groups working simultaneously on different control functions without disturbing one another. But the underlying philosophy which I am trying to get at came out even as LIGO Project Manager Gary Sanders was explaining to me that there was no deep sociological significance in the size of the control room, merely practical considerations. He said that the back wall of the control room had been made of glass so that visitors could see what was going on from outside the control room without disturbing the controllers. Thus, from the outset, strangers were expected to visit, hence LIGO's control rooms were designed as public spaces. Indeed, with the giant monitor projections on the far wall and the glass partition at the back, it would be hard to imagine how such a control room could have been better designed to be publicly accessible. What is more, it is the first thing that the visitor entering the scientific part of the installation encounters. VIRGO's control room, on the other hand, is very much a private space. It is tucked away at the end of long corridors and up a narrow staircase with small landings leaving little room to stand. In Europe, science is seen as a very much more private activity than it is in the USA.[2] The difference between the public and private in these aspects of the design may not have been at the forefront of the designers' minds, but it is not unreasonable to think that it informed their decisions in a way so fundamental that it did not need to be made explicit.

A much deeper difference in design philosophy that comes out in these two installations is based on scientific considerations. I will deal with this at length in chapter 39 and merely mention it now. This is the difference in the principles behind the mirror suspensions. VIRGO decided that it would design mirror suspensions capable of cutting out unwanted vibrations even at very low frequencies—which are much harder to eliminate than high-frequency vibrations. The VIRGO design means the construction of especially tall towers to house the so-called super-attenuator suspensions, so a difference in the broad design of buildings was dictated. This is a choice that is integral to the scientific project and is, perhaps, more understandable than differences in choices about technology—such as the method of beam tube fabrication—where nothing scientific is at stake.

Of still more interest is VIRGO's choice of experimental strategy over prototyping. There has always been tension in LIGO over the role of prototypes such as the 40-meter device located at Caltech. As LIGO neared completion, the 40-meter prototype began to be less and less used; I will discuss this feature of LIGO's development later. VIRGO, however, decided to build no prototype at all—or, to be more accurate, they decided to use the interferometer as its own prototype. The argument was that, first, prototypes are different from full-scale machines in so many important respects that it is difficult to learn crucial lessons from them. Second, prototypes are time-consuming and expensive to build. They are nearly as difficult to build as full-size interferometers and, when they are finished, all you have is an expensive instrument that cannot do astrophysics; a prototype interferometer is not designed to see gravitational waves, making it hard to explain to outsiders why you need money for a more expensive version of a device that has just demonstrated that it cannot do the job.

Instead, it was argued by the VIRGO leadership that if the full-scale midstation and full-size input mirrors and suspensions were built, they could be used as a short interferometer housed entirely in the center station

---

2. I feel entitled to make this general claim, because it is backed up by my interactions with American and European funding agencies and respondents in general. It would be nice, of course, to see it confirmed in the design of other large scientific facilities in different countries.

building, and this could serve as the prototype. This is what has been done.[3] The input mirrors to what eventually will be the three-kilometer Fabry-Perot cavities have been replaced with small, fully reflecting mirrors to form an interferometer with arms only six meters in length. This was used to test many of the features of the full-scale device. For example, many of the potential difficulties with the feedback circuits and mechanisms—"the intelligence of the mirrors," as one of respondents put it—were solved with the short configuration. In the meantime, the end station supports and mirrors were assembled three kilometers away. When "the moment" came (in early 2003), the small, fully reflecting mirrors in the center station were replaced with full-size partially transmissive versions and a three-kilometer interferometer appeared—"simple."[4] Until then, this approach had suited VIRGO's construction schedule. The midstation could be built first and the building of arms and end stations delayed until all the land was acquired.

VIRGO, LIGO, and GEO

Somewhere on the large plain south of Pisa lies that part of the University of Pisa's physics department funded by the Istituto Nazionale di Fisica Nucleare (INFN)—the Italian state-funded nuclear physics agency. A dozen or so cheap, low office buildings and shedlike structures lie for no obvious reason in the middle of a field on a smallish road to nowhere in particular. One can walk across the site in a couple of minutes in any direction.[5]

Fieldwork with important physicists in Italy is enjoyable but more beset with scheduling problems than fieldwork in any other country I visit. In particular, I found trouble with the way time is structured. I visited Pisa in July 1996. My appointment was for 10:00 a.m., but various preoccupations prevented my respondent and me from sitting down to serious business

3. Alain Brillet, the leader of the French side of the team, seems most closely associated with this idea and the eventual decision.

4. There were actually a number of other differences between this prototype and the final version, so things may not be so simple. This is not yet clear at the time of writing. The input system for the light had not been completed, so this had not been tested with the rest of the interferometer. The laser power was much less than in the final device. The mirrors were much smaller. There were no Fabry-Perot cavities in the "prototype." There were no beam tubes with potential internal reflections. And, of course, there is the matter of maintaining the same degree of control over a much greater length.

5. That said, like other INFN sites that I visited, it does include a cheap, state-subsidized refectory—a small one in this case, but superb nonetheless. The excellence of the food produced by Italian canteen cooking is in some ways as deep a mystery to a British person as the collision of two black holes.

until 1:30 p.m. Fortunately, the break gave me an opportunity to drive the fifteen kilometers to the VIRGO construction site and back before the interview and to learn something of interest: Since I was taking the trip from INFN to the VIRGO site on my own, I had to ask directions, and I had terrible difficulty finding members of the VIRGO team located at the INFN building who (a) had ever been to the site, or (b) had been there recently enough to remember the route. Of the half-dozen I consulted, only one knew how to get there. I cannot think of a better way to convey the "hands-off" way of doing things in the case of VIRGO as compared with the other big project, LIGO; GEO, as we will see, is at the opposite end of the scale in this respect—in GEO, everything is "hands-on."

When I reached VIRGO, I did indeed find a construction site, pure and simple. There were three or four huge cranes, masses of reinforced concrete, a fleet of earthmoving trucks, and so forth. VIRGO, though it is a little smaller than LIGO, is very like it in scale and conception when compared with GEO. VIRGO costs about twenty times as much overall as GEO, and I would imagine the ratio of the cost of construction must be much higher. As Adalberto Giazotto, an ex-high-energy physicist and the leader of the Italian group said to me, high-energy physicists are not afraid to spend money. Here they were unashamedly constructing something equivalent to one of their big-physics machines, and unashamedly leaving it to the builders. I cannot imagine a German, British, or American experimental physicist not wanting to visit the site of their "baby" as often as they could, yet here fifteen kilometers was a long way away (of course, as installation of the crucial scientific components began there would have to be more visits). In the case of GEO, as we will see, there was no choice for the scientists but to spend time at their site, since it was they who were building the machine.

LIGO

Each LIGO interferometer is bigger than all other interferometers except its LIGO counterpart. The LIGO project is composed of two interferometers, each with two four-kilometer arms, and a third interferometer with two-kilometer arms housed within the same installation as the Hanford four-kilometer. The two four-kilometer detectors are separated by 2000 miles, and each in turn is distant from the home universities of the scientists by up to 3000 miles. To build LIGO, a total of sixteen kilometers of level and stable concrete foundations had to be laid, with sixteen kilometers of road alongside. On the foundation is mounted sixteen kilometers of vacuum pipe with special internal baffles to make it nonreflective, and sixteen kilometers of concrete protective covering. The covering is about fifteen feet high, with an inverted-U cross section. It is much too high to climb over, and a special bridge has had to be built at each site to enable vehicles to enter the inside of the enormous space enclosed by the arms. At the corner of the arms is a hangar-sized—cathedral-sized might be a better adjective—assembly hall on its own stable concrete slab. There are also smaller buildings, but each twice as high as a house, at the ends of the arms. Tens of very large stainless-steel vacuum chambers are needed to house the working parts of the apparatus, and most elements are separated by huge gate valves. This installation is by far the largest high-vacuum system that has ever been built. The internal area of stainless steel in the vacuum pipes and chambers is equivalent to a square about 270 yards on a side (fifteen football fields).

The two LIGO sites are located at opposite corners of the USA, one in the far northwest in Washington State and one in the hamlet of Livingston, east of Baton Rouge and about an hour's drive from New Orleans, Louisiana. The Washington State installation is on the Hanford Nuclear Reservation, a few miles from the reactor where the plutonium for the Nagasaki bomb was made. Approaching by car from the west, one travels through rich, irrigated agricultural land, some of it supporting picturesque hillside vineyards. As one gets nearer, however, the land changes to arid desert covered in scrub and tumbleweed. In my notes describing my first approach to the site, I mentioned that I noticed something odd about the desert: "overhead power lines began to crisscross the sky in remarkable numbers; the roads are too straight, too wide, too well-maintained; single-track railway lines appear at unexpected intervals; freshly painted signposts with obscure numbers point to nowhere. In among the nondescript, dusty, gritty khaki of the wasteland, the concrete domes of nuclear pressure vessels appear like weather-beaten skulls."

Later I was to spend a week or so at Hanford, and its strange beauty grew on me. The business side of the installation—the offices, laboratories, workshops, and so forth—are at the apex of the L formed by the interferometer's arms, and normal access to them is from the outside of the L. But if one crosses the bridge to the inside, the road stops suddenly and enters a private space which is little changed from the original landscape. I used to walk across this bridge just before sunset and sit down inside the apex of the L, where I would be completely on my own. The buildings and the concrete culverts cut off the continuous hum of machinery, so the only sounds came from nature. Every evening a glorious sunset filled the sky over Rattlesnake Mountain in the far distance, perhaps colored by the dust of the desert. The silence, the isolation, the muted colors of the landscape combined with the magnificent sunset were magical. Even the nuclear domes, mostly long disused, took on a weird beauty as familiarity made them less threatening.

A visitor approaching from the outside of the arms would, rather, experience both sites as smart, modern, "hi-tech" installations, nowadays served by smooth metaled roads. There are some postmodern architectural flourishes such as entry porches with curved concrete canopies, and the buildings are coordinated blue and gray. Each site also features a giant blue water tank with LIGO and the project logo—which is the first half of this book's logo—boldly painted in white. Separate warehouselike assembly areas for meetings have been built along with a clutch of maintenance buildings a short distance from the entrance structures. Considerable effort has been put into landscaping the sites immediately around the entrances. At Hanford the visitor will find attractive pebble and cactus gardens, while at Livingston the mud surrounding the buildings has been turned into a coarse lawn with decorative ponds, fountains, bushes, and flowers.

The construction of the Louisiana site, incidentally, was itself quite a bold feat of civil engineering. My notes on a visit in 1998 give an impression of the nature of the transformation.

> *Site visit to Livingston: 18 March 1997 c 9.30 am:* To get to the Livingston site, you have to drive east from Baton Rouge for 20 miles, then turn north onto country roads. Here we are in gun-rack country. The straight roads are narrow and carry very few vehicles. The drivers of the few pickup trucks that pass raise their hands to you in lazy acknowledgement. There are few enough vehicles to make a friendly wave a reasonably economic gesture. The roads are lined with thin pine woods, and as we enter and pass through Livingston they are bordered with cabinlike houses, some derelict or semiderelict, but each in its own acre.
>
> We come to the dirt road that leads to the LIGO site. I drive down it for a mile or so, noticing a cluster of young cattle sat down among the trees, like wildlife in an African savannah. Then I notice the first signs of machinery ahead, and suddenly the woods open out into a huge cleared space. As always with a site visit to an interferometer (the Hannover GEO site aside), the first impression is sheer size. The Livingston site looks especially big for two reasons. First, being cut from the woods, its edges and ends are defined by trees, so one can grasp its dimensions immediately. Second, the clearings in the woods simply are large. The Livingston site is very, very wet. As I was soon

to discover, I was quite lucky to have been able to drive down the dirt access road. Not long before it had been 2 feet deep in water in places. On either side of the road, the woods are covered in small pools, and streams crisscross the ground. Cutting through this tropical scene are the massive berms which have been constructed for LIGO. They are earth banks between 4 and 8 feet high (to take account of the slight slope of the site), and 50 feet wide at the top. The full width of the cut at its widest, however, is 300 feet. This is because the berm has to have shallow slopes, and on one side there is the "borrow ditch" from which earth has been taken to build the berm. The borrow ditch is filled with water, making quite a nice reflecting pool and adding to the tropical effect.

Livingston is not an ideal site. It is too wet and there is no bedrock to stabilize the buildings. The problem of drainage is a serious one, and the whole site had to be designed so as not to disturb the natural flow of water. Each berm is penetrated every so often by massive culverts allowing water to flow across the site in much the same way as it did before the berms were in place; without the culverts the berms would be 2.5-mile-wide dams. The mud and rain slowed down the making of the berms and the roads. Six months were allowed for that aspect of construction, but it took a year. In Hanford all they had to do to get a level roadbed was scrape the desert around a bit. Furthermore, in the summer the Livingston site will be plagued with insects; it is a superb breeding ground for mosquitoes, which will make it uncomfortable, and the insects will not make it any easier to construct clean rooms.

For a long time I tried to work out why the Livingston site, with its obvious disadvantages, was chosen for LIGO. It was hard to get a sensible story out of anyone still active in Washington, DC; I was always presented with the line that "many pairs of sites were looked at, and Hanford-Livingston was the most scientifically suitable." At the Livingston site, where the contradictions seem more immediately pressing, a more robust account was in the air: it was politics. Bennett Johnson, the senator from Louisiana, got in a competition with some other senator and decided that LIGO would be built in his state. He persuaded the Louisiana State Legislature to give LSU enough money to buy the land and lease it back to LIGO. Putting this account together with others, we can impute that the other senator was Mitchell from Maine, and that there were certainly some political favors being granted or grudges being repaid when Louisiana was chosen.

Gerry Stapfer, the site manager, kindly offered to drive me along one of the berms. Everywhere I looked, I saw effort and money. The top of the berm is wide enough to take a narrow road as well as the concrete bed for the beam tube itself. Soil for the roadbed had recently been "modified" to stabilize it. The top foot was scraped off and 6%–8% of cement was added and "rotavated" before it was rolled out again; this was an extra expense. It was not much in terms of the scale of the site, but when one realizes that five miles of this had to be done, any small modification is a major operation. The next steps were to lay a waterproof "cloth" over the modified soil and then lay gravel clay on top of that before finishing it with more gravel. After the construction was over and heavy trucks would no longer be using it, the road was to be finished with concrete for the use of the scientists.

The contrast between this modified earth and the still glutinous clay on either side of it was obvious. Every square foot of clay bore the imprints of the hooves of the cattle and wild horses that inhabit the woods on either side of the installation. Cow droppings seemed to symbolize nature's contempt for science; eventually, the site would have to be fenced to keep the cattle out. As we drove along we saw water everywhere. And as we returned, the four-wheel-drive vehicle got stuck, and Gerry had to reverse it out.

Each LIGO installation is of sufficient complexity to need staffing round the clock, and each has a permanent site manager. Now that the sites have been completed, you enter the front door of a LIGO installation and immediately find yourself in an auditorium big enough to seat about 150 people, and lined with amusing scientific displays and working models to please sightseers—both LIGO sites being strong on "outreach" to their respective communities. If you look carefully, you can see a ghost: the back of the auditorium chairs are marked with old stickers proclaiming that they are the property of the Superconducting Super Collider project in the state of Texas.

From the auditorium, doors lead off to the reception office, a conference room and library, a kitchen, a room full of workstations surrounded by offices, more offices, and thence into the labyrinthine heart of the science installation proper. The buildings have staircases that lead high onto the roof, from where you can see 30 miles in each direction. Or you can go into the clean room, where mirrors are prepared, or the control room, high ceilinged and banked with monitors, big enough to seat 30 for joint discussions and featuring giant screens for public display. Through another door is the computer room, housing banks of data processors. Opposite are the well-stocked workshops.

If you get the full tour, you will also enter the vast building that is the center station. Passing through a special room, you first put on overalls, overshoes, and goggles to guard against stray laser beams. The center station is the size of an aircraft hangar, its ceiling height determined by the need to enable cranes to lift the dozens of vacuum tanks—each as tall as a house and as wide as room—to twice their height for maintenance. Four-foot pipes leading outward through the walls of the hangar direct your curiosity to the beam tubes, stretching for four kilometers and, from outside, dominating the landscape with their concrete covers.

Twice now, in spite of myself, and taking me completely by surprise, I have seen LIGO from another perspective. Both times I was at Hanford and on my own. The first time was in October 1997, during a meeting when the central office space and auditorium were being used for the first time—the first time that the scientists could really see what they had wrought. I report my field notes verbatim:

> Later, walking by myself around the site in the silent drizzle, I had a moment of estrangement from the project, entering a kind of philosophical black hole—"a phenomenological epoche." Suddenly I saw that this was madness on the grandest scale! All this money, all this effort, all this steel, all this concrete—for what? To try and see movements smaller than the nucleus of an atom!
>
> After the initial delight in the achievement, the physicists too felt a little humbled and frightened. One or two of them remarked to me that "this had better work," or some such, and they said it without a chuckle.

The second time was when I found myself completely alone at the Hanford site in September 2000.

> No one had told me that it would be closed for Labor Day! So sat and drove around the site until 10.20 waiting to see if anyone would show up, which they didn't. It was a beautiful day—sun shining but not too hot—so it was not at all unpleasant. On the contrary, it was a strange and interesting sensation to be all alone in the middle of that great desert—the only person around the huge LIGO site. Once more the complete insanity of it struck me—all these millions of dollars to see nothing. From the outside it seems a strange kind of self-driving folly—one can see why so many other scientists are so annoyed about it.

The appropriate conclusion is, perhaps, the next paragraph from my 1997 field notes:

> Driving home from the site with one of my physicist confidantes I told him about my moment of estrangement. We agreed, however, that it was a mark of high culture to expend resources on such noble follies. If one could not do such things, then what was the point of generating wealth beyond the needs of survival? To stand and look at LIGO is to understand the old cliche: Big science builds the cathedrals of the modern age.

GEO

TAMA aside, GEO600 is the shortest of the interferometers that can reasonably be called gravitational wave detectors. I must have made about ten visits to one or other of the LIGO sites and spent about two months in or around them altogether, so I have become used to LIGO's scale. Sometimes I find it hard to recapture my sense of wonder at the extraordinariness of the installations. But I still get this sense, in "negative," as it were, when I visit the GEO600 site, at which I have spent much less time. After becoming attuned to LIGO, visiting GEO600 is like visiting the Grand Canyon in reverse. Each time you see the Grand Canyon, you realize that you have not been able to keep the concept of it in your head—its sheer size is a surprise every time. In the case of GEO600, its sheer smallness is impossible to grasp in the absence of actively seeing it—or at least impossible to grasp once you have accommodated to the scale of LIGO.

One of the reasons I have spent much less time at GEO than at LIGO is that its small size means one cannot visit without a guide. At LIGO one can simply "turn up" if one is known, as I am; go to kitchen and make a cup of coffee; head to a computer room and check email; and go to the library and look at documents; you can also have lunch with the guys and gals and shoot the breeze. A visit to a LIGO site has something in common with a visit to a small university department. Turn up "on spec" at GEO, however (if you can find it), and there is likely to be no one there. In any case, there isn't anywhere to sit down.

As you enter a LIGO site you have a growing sense of expectation—you are moving from one kind of ordinary space into a special kind of scientific space impressed upon the landscape. GEO600, on the other hand, is so invisible it is almost secret. Drive along a farm track, and a gate off to the side offers entrance to an orchard at the edge of a cornfield. GEO600 has no presence as an installation; it dominates no space and transforms no wilderness. It tucks itself into the edges of the farmers' fields. You enter the field and drive past a structure the size of a couple of garden sheds—the end station. You drive along a track to the center station and workshop, making sure not to collide with the fruit trees on your left. There is enough clear space to park half a dozen cars haphazardly. Three sheds stand together; one of them newly installed to satisfy visitors when GEO600 was made part of the nearby EXPO 2000 exhibition. One shed is the workshop housing a couple of staff members, who keep things going.

Originally, the British and German scientists each wanted to build their own three-kilometer interferometers, but they were forced to cooperate to build one scaled-down device because of insufficient funds to do anything more. GEO600 costs about $7 million in all, compared with LIGO's $360 million or so, though direct comparisons are difficult because of differences in accounting practices. The Americans feel they get a raw deal from comparisons of LIGO's high cost against GEO's few million, and they are probably right; but the difference in cost and scale is still enormous. Britain's contribution to the $7 million was that, when asked for a £2-million contribution, it offered £1 million, provided all the same science could be done for that price. The Scots-German—or should we say German-Scots—team had no choice but to agree. Thus, GEO600 is built in the tradition of "string and sealing wax." Karsten Danzmann, who led the German effort at that time, told me he was not sure whether GEO would work, but he was sure of one thing—it could not be built cheaper. Every Deutschmark had, as he put it, been spent twice. Referring to Jim Hough, his counterpart in the Glasgow team, he remarked, with no heed for political correctness, that it was "useful to have a Scot in the team" when one was working under the kind of financial constraints they had to meet.

GEO is, then, much more like a laboratory project than the bigger detectors. Far less of GEO's construction involved subcontractors, and the device was largely built by the scientists themselves. Various attempts at cooperation between the VIRGO project and the German-British group have collapsed, because GEO just isn't funded on a scale that makes it possible to combine it with the VIRGO style of work and organization.

Harald Lueck, my guide on one of trips, explained to me how the physicists personally designed the building, sought out and ordered the cheapest of materials and fittings, worked out ingenious ways of using standard builder's merchant accessories to serve as mobile suspension units for the vacuum tube, built the wooden hut which helped to keep dust and dirt out of the end building, and laid the 600 meters of aluminum rail from which the first tube would be suspended. "It only took us two and a half days," he said.

GEO600 is maturing as a site rather in the opposite way to LIGO. LIGO gets smarter and smarter, whereas GEO gets more and more decrepit and rundown; its location may be Hannover, but its looks like it would be more at home in the former East Germany. Dust and leaves blow into the buildings, and there is no air-conditioning except where the scientific apparatus makes it essential. The inner, inner, heart of the installation is separated from the rough and dirty outer sheds by a Plexiglas inner wall, built on timbers of the kind you get from your local do-it-yourself store. So there's a shed within a shed within a shed, and it is this that stops the outside getting all the way in.

The control room is a bit of space inside one of the sheds; it is big enough for four people, so long as no more than two of them want to sit down at once. There are the monitors and there the banks of electronics, all home-built by students. One looks through the half-timbered Plexiglas from the control room into the midstation workspace. The technicians are working on something: one is crouching under a desk to adjust a cable.

One of the end stations is near the farm-gate entrance. It has an outer and an inner door, with a rough concrete ramp, now overgrown with weeds, leading up to the outside. The outer door propped slightly ajar by a broom handle. This is because just inside is a portable air-conditioning unit blasting hot air from the inner sanctum, and the hot air has to have somewhere to go—out through the crack of the outer door. And that unit is fed by a pipe that snakes through the door of the inner room, so that door, too, has to be kept slightly ajar. So how is that door kept where it should be? It is propped by an old radiator leaning against it!

I should add that GEO600 is set among the most delightful experimental orchards. On the sunny evening I first visited it in 1996, I was able to sample many different and delicious varieties of cherries, gooseberries, blackcurrants, and redcurrants.

SCALE AND SCIENCE

The foregoing descriptions of the sites and work practices of the big interferometers are more than an indulgence. The "small and cheap" approach to these projects involves extra work and struggle for the scientists involved: they have to find ingenious solutions just to save money when they might be putting that creativity into advancing the technology, and they have to do building work when they might be doing experiments. No scientist, given the choice, would opt for this approach. Oddly enough, however, it is not the case that big scale and lavish funding are better than working on a shoestring in every respect. To explain this, we need to study the transformations that took place in the science of interferometry as it grew into a half-billion-dollar international project. I will concentrate largely on the history of the LIGO project, but the contrast with GEO should always be borne in mind.

Gravitational wave detection is a LIGO-centric business, and this book reflects it. As we saw in earlier chapters, the "gravitational pull" of LIGO on the resonant detectors was strong, and the same remains true of the other interferometry projects. Much of the time, the other groups' decisions are made in reaction to what LIGO is doing; they either collaborate with LIGO or compete with it. If LIGO were suddenly to disappear, the direction of the other projects would almost certainly change quite suddenly, just as the orbits of the planets would change if the Sun suddenly disappeared. For example, if LIGO were no longer there, the European projects would either expand in order to fulfill LIGO's current role, or they would get smaller as politicians concluded that the Americans' withdrawal showed that the work was unimportant. On the other hand, were the non-American projects to disappear suddenly, LIGO would probably continue pretty well unchanged.

But there are still better reasons for concentrating on LIGO in discussing the way sciences transform themselves. First, LIGO has undergone the biggest, and therefore most marked, transformation from small to big. Second, however, its very size and cost, as I will argue, have required its working life to be opened up to external scrutiny, exposing its transformations to the gaze of the public and to a commentator such as myself. Third, America is more open than other national cultures, which makes Americans easier to study.[6] None of this is to say that LIGO has always been

---

6. Each gravitational wave detection team has generously opened its doors to me, but American science has a special quality which reflects the very grain of American society—guilt about secrecy. Most American scientists' first instinct when you put a question to them is to ask themselves, "Is their any reason I should not answer this?" Most European scientists' first instinct is to ask themselves, "Is there any reason I should answer this?" I would not want my many delightful and open European respondents (nor my very few more circumspect American respondents) to mistake a generalization at the level of populations for a description of each individual (a mistake known in statistics as "the ecological fallacy"). But the fact remains that, for example, the National Science Foundation "leaned over backward" to give me access to the papers on the LIGO project, whereas anything I obtained from European funding agencies was a struggle; it was almost as though I were breaking the tacit rules even to utter a question, never mind get an answer. And it is much more than symbolic that in Britain, in spite of quite explicit promises by a succession of governments, we still have no freedom of information act. I have, then, been made to feel at home in American science (insofar as one can ever feel at home in a project which turns on turning the private

at the forefront of technological developments, and integral to the "plot," as it were, is an argument that its very size has forced it to make relatively conservative scientific decisions. This is one of the ways in which the "cheap and cheerful" approach of GEO has turned out to be an unexpected advantage.[7]

---

activities of powerful people into public property), and the American project has exposed itself most readily to my spotlight. Hence it is LIGO whose struggles and troubles have been most clearly described in these pages.

7. Were I to be attempting "historian's history" rather than "sociologist's history," the organizations of VIRGO, GEO, and TAMA would merit equal treatment, but here LIGO is the "case study within a case study" because it best reveals the sociological tensions.

CHAPTER 30

MOVING PEOPLE: FROM SMALL SCIENCE TO BIG SCIENCE

SMALL SCIENCE, BIG SCIENCE, AND LIGO

"Small science" is done by individuals or teams of about half a dozen or less. Small science can often allow a large degree of autonomy for the scientists. It does not cost much, so if it fails, it will not be seen to have damaged much else. This is most marked in the case of theorists, but even failed bench-top experiments are unlikely to upset anyone beyond a university department or a section within a funding agency or firm. Small groups or individual scientists can work on their own timetable and toward a goal of excellence defined by themselves. When science starts to spend big money and employ big teams, however, sections of society more and more distant from the researchers begin to take a share of the risk, and their interest will be alerted.[1]

---

1. For optimum size of teams in small science, see Martin, Skea, and Ling 1992; Johnston et al. 1993. For an argument for the need for autonomy within the "republic" of science, see Polanyi 1962. De-Solla-Price (1963) uses the term *big science* to refer to the exponential growth in numbers of *scientists*. Throughout this chapter, the term will be reserved for large projects. Capshew and Rader (1992) consider

When others are sharing the risk, departures from the preset timetable upon which a funding decision was based cause trouble; they mean increased spending and loss of confidence by the funding agencies and the politicians to whom they answer. It follows that in a big science, building to plan and schedule must sometimes take precedence over doing the very best science; technologies often have to be frozen before research scientists have reached consensus over the optimum design. Under these circumstances, it can be more efficient to give decision-making responsibility to team leaders who do not have the same emotional commitments to ideas as those who invented them. Furthermore, what was once done by scientific "craft work" in the laboratory may have to be handed over to outside industrial contractors accustomed to working on a large scale and to well-specified schedules and performance targets. As a scientific project becomes more organized it will also need to bring in new kinds of professionals, such as accountants and engineers, with career patterns and scientific values different from those of research scientists.[2]

In sum, small science is usually a private activity that can be rewarding to the scientists even when it does not bring immediate success. In contrast, big-spending science is usually a public activity for which orderly and timely success is the priority for the many parties involved and watching. A change from small science to big means a relative loss of autonomy and status for most of the scientists who live through the transition.[3] It is this transition as it took place in LIGO that we now examine.

---

the features of big science to be "money, manpower, machines, the media, and the military." As I will argue, certain sociologically interesting features of big science are absent in the case of the military. Other interesting discussions are to be found in Heilbron 1992; Kevles and Hood 1992; Hevly 1992; and the other articles in Galison and Hevly 1992 and Agar 1998.

2. University faculty often find that equally qualified people are paid better, but so long as they work in another sector of the economy—for example, industry or the financial markets—the difference is not experienced at first hand. When those others join the same organization as managers or designers, they become a "reference group," and this can give rise to a feeling of "relative deprivation" (Runciman 1966).

Agar (1998) discusses the tension between engineers and scientists in the Jodrell Bank radio telescope project, and Riordan (2001) describes the rise and fall of the Superconducting Super Collider project in terms of the tension between the culture of physicists and that of the military-industrial complex brought in to help build it. Another strong motif in Riordan's account is the need for external oversight of big projects. It is quite possible that some of the language and ways of thinking of the scientists described in these chapters were imported directly from the debacle of the SSC. (See also Galison 1997, pp. 614–18.)

3. For a wistful account of this kind of change in the wider society, see William H. Whyte Jr.'s 1957 book, *The Organization Man*. In part 5 of his book Whyte deals with the organization of scientists, entitling chapter 16 "The Fight against Genius." He explains that in the American high-technology firms of the era, genius was discouraged in favor of adherence to team values. He quotes a Monsanto promotional film: "No geniuses here; just a bunch of average Americans working

Big science comes in too many shapes and sizes to allow itself to be characterized as a whole.[4] Centralized big science often employs hundreds or even thousands of scientists, and on this scale LIGO is not very big. Most of LIGO's history turns on the activities of about 50 people, with hundreds of dispersed scientists becoming involved only in the most recent stages; by the time hundreds became involved, most of the growth traumas had been overcome.

But what counts as big money and large numbers of scientists is relative to the size of the agency that is supporting the project; organizations such as NASA, the US Department of Energy (DOE), or the military can absorb

---

together" (p. 235). Whyte's complaints were almost exactly echoed by those who found themselves uncomfortable under the new management.

Though in many ways the story of LIGO's growth mirrors the transformation of small businesses to large bureaucracies documented by Whyte, the disanalogy between scientific research projects and the commercial sector should also be borne in mind. Most members of a team attempting to do adventurous science—and this science has clung to the edge of scientific impossibility for more than forty years—are there out of choice rather than necessity and are likely to be deeply committed to the overall aim, however large the team grows and whatever the increase in its management overhead. In any case, even a big scientific project allows a much greater degree of autonomy for its members than almost any other kind of organized profession. If big science is to be compared with business, it is the small innovatory firm or the highly integrated modern enterprise that provides the more appropriate models, because in both these cases the "employees" are also part of the "management." (For examples of these types of organizations, see Goffee and Scase 1995.) The tensions between freedom and control that I describe below are situated, then, within a context far removed from those normally discussed under the headings of trust relations between "managers" and "employees."

Greiner (1998) suggests that organizations evolve through a series of revolutionary changes. But many characteristics of an adventurous scientific project are retained from the equivalent of Greiner's first stage even as the project changes. For example, under Greiner's scheme all stages after the first require special motivational schemes, because "new employees are not motivated by an intense dedication to the product or organization" (p. 60). Other features of Greiner's stage 1 that are retained are that "[c]ommunication among employees is frequent and informal," and that "[l]ong hours of work are rewarded by modest salaries and the promise of ownership benefits" (but in science the benefits will be a share of the discovery).

4. For example, Kevles and Hood (1992) distinguish between "centralized" big science, such as the Manhattan Project and the Apollo Program; "federal" big science, which collects and organizes data from dispersed sites; and "mixed" big science, which offers a big, centrally organized facility for the use of dispersed teams. All these cost big money but are likely to be quite different in their impact on the scientists who work in them.

If the big money is spent on a big device, such as a telescope, which is then used by smaller teams of scientists, the strongly coordinated period may be short. The history of the Jodrell Bank radio telescope (Agar 1998) seems to be an example, though; as Agar points out (private correspondence), the visibility and politics of a big science may continue to affect such a project even when a small group is left in charge.

High-energy physics is a big science, but many of those involved with it are dispersed in university organizations with only the activity near the large facility involving all aspects of big science (Knorr-Cetina 1999). On Kevles and Hood's schema, LIGO is centralized big science that is turning itself into mixed big science, though its actual expenditure, at $300 million to $400 million, is small compared with other centralized big science projects, and the number of scientists involved is also relatively small.

the same absolute expenditure with less internal impact or outside scrutiny than occurs in a smaller spending agency such as the National Science Foundation. A top-down, "managed" style of working may be so normal within a big agency that the shift of control over the fate of inventions from their creators to the team leaders is the expectation rather than a new and disturbing experience.[5] Thus, the setting up of a new big project within a big-spending organization is likely to have less of an impact and be less interesting as a social change than a case in which change in expenditure is smaller but the funding agency is proportionately smaller still. Again, if the project comes from within the defense sector, it may be further removed from public accountability and thus less affected by the need to be transparent to the outside world.

For all these reasons, not all shifts from the laboratory bench to big science are equally interesting. LIGO, however, is interesting for a number of reasons. LIGO's expenditure is large in respect of its funding institution—it is, as we have seen, the largest project ever funded by NSF. Gravitational wave science has grown from small to big over 30 or 40 years, with the main transformation happening in the 1990s, and many of those working in LIGO have lived through the shift. For many of these scientists, LIGO was their first experience of highly organized science; some flourished, some did not. Some left the project, or were parted from it less than voluntarily as a result of the transformation. Both LIGO and NSF are in the public domain, so the project attracts outside scrutiny from politicians, the media, and other scientists. LIGO's instrument is not clearly separable from LIGO's science, nor is its facility separable from its research, so the construction project cannot be so easily separated from the science as in some other expensive science projects.

Another important feature of gravitational wave science, as compared with other "big" sciences,—and this is a very important difference to the protagonists in the debates to be described—is that a project like LIGO is not growing incrementally from proven success. Big spending on big, publicly scrutinized projects is often justified because it represents a step up from smaller scale work which has demonstrated that the science will "work." Small accelerators have found new particles, so it is reasonable to predict that bigger accelerators will find more particles and so on; small telescopes have seen stars out to a certain distance, and it is almost certain that the next generation of bigger telescopes will see more stars out to greater distances and so forth; there are few big sciences that have not exhibited this

5. McCurdy 1993.

growth pattern.[6] But, as we have seen, the first small steps in gravitational wave detection are generally accounted to have been a failure, and the latest and most expensive gravitational wave detection technology is a step into territory that is less well mapped than is usually the case with big spending. We might say that LIGO is, effectively, taking small-science-type risks with big-science-type money. In the words of an NSF representative, the normal way for a project to grow is by a factor of 3 at each step, but LIGO is trying for two orders of magnitude (a factor of 100). The suddenness in the jump in size and technology has widened the scope for internal disagreement about how the project should be carried forward.

LIGO is, then, exposed to strong pressure from outsiders who would prefer to see the money spent on science with a more certain outcome; this has meant the project has had to be seen as especially orderly, which means it must be *transparent*. By "transparency," I mean having well-specified plans, schedules, and budgets that can be discussed by review committees who are answerable, via some chain of accountability, to the political sphere. In the long term, of course, what has to be visible is success; but the mistakes and departures from schedule and budget which are inevitable in a project of this complexity are easier to defend if the organization can give the impression of being largely under control.

I will describe the imperatives which outside scrutiny imposes on an expensive science, and the changes in work practices which go along with the growth of a science. More ambitiously, I will ask whether it is possible to distinguish, in some deep way, kinds of science that can only be done when the scientists are organized in some ways rather than others.

ORGANIZATIONAL AND COGNITIVE TENSIONS

From the early days of big science, it has been suggested that the demands of a big organization are at best different, and at worst antagonistic to the kind of scientific creativity that characterizes the best of small science. The switch in emphasis as science becomes more centrally organized is nicely documented by the historian of science, John Heilbron, who refers

---

6. The Manhattan Project was one: one could not be certain that a twenty-kiloton bomb would explode on the basis of 100-ton-equivalent atomic explosions and so forth, because there could be no such things; the whole massive industry for extracting the right amount of U235 or plutonium had to be set up to produce the first result. Even here, however, the existence of chain reactions had been demonstrated prior to the explosion—in direct gravitational wave detection, there is no equivalent of even a chain reaction.

to some of the founders of modern big science. For example, he quotes Ernest Lawrence writing a letter of recommendation in 1946 in which a new Ph.D.'s creative powers were not discussed, but he was described as "an energetic and effective member of a research team." A letter from Brookhaven National Laboratory in 1957 says to a potential employer, "His ability for independent research is about average, while his demonstrated ability to work congenially with others is outstanding." Heilbron also quotes Samuel Goudsmit, head of the Brookhaven Laboratory, who in 1957 wrote, in words that have great resonance for LIGO,

> In this new type of work experimental skill must be supplemented by personality traits which enhance and encourage much needed cooperative loyalty. . . . I feel that we must now deny its [the Cosmotron's] use to anyone whose emotional build-up might be detrimental to the cooperative spirit, no matter how good a physicist he is.

In 1967, a well-known discussion by Alvin Weinberg included the complaint that big science was "blunting science as an instrument for uncovering new knowledge."[7]

Could there be, as many scientists believe, a kind of science—let's call it "developing science"—that can be done well only by autonomous individuals or small teams using intuition and craft practices? And could it be that big-science practices can only be applied to "mature sciences"? Where large teams are needed because of the sheer size of a project, could it be that a high degree of coordination and routinization works only after a certain level of scientific maturity has been attained?

What can the history of disputes within LIGO tell us about this hypothesis? The question is complicated, because though LIGO as a whole has undergone a clear increase in coordination and routinization, there are still areas within it where autonomous work is allowed or encouraged. Like any bureaucracy, a big science must maintain a certain amount of flexibility and freedom for when plans go wrong. Studies of bureaucracies tell us that the appearance of carefully maintained order requires individuals to depart from behavior described by the rules so as to save the appearance of rules being followed; in the jargon, they "repair" the running of the organization continually in imaginative and creative ways. Mavericks are also needed in a scientific project to create the unforeseen, which gives the work somewhere to go once existing ambitions have been fulfilled. Thus, a project like

---

7. Heilbron 1992, p. 44; Weinberg 1967, pp. v–vi.

LIGO, however organized and planned it becomes, will not stamp out all autonomous activity. But of the creativity that remains, some will be directed toward the overall goal, and some will be localized and hedged around so that it does not damage the coordination needed in the rest of the project or tempt scientists back into the old, autonomous patterns of behavior.[8]

To understand the whole LIGO project, then, we need to break it down into different components. The big dispute that I will describe was not about the organizational trajectory of the whole thing but about whether there might be a still-developing science at the heart of LIGO, interferometry, which the managers treated as a mature science while it was still immature. Could it be that too much routinization too soon will turn out to have been harmful to the detection of gravitational waves, given that the science of interferometry is making such a large jump from an untested foundation? Some scientists think so.

To work this out is an ambitious goal. If there are in fact two kinds of science rather than one, they might be analyzed on a three-way cognitive schema as having different *psychological, temporal,* and *cultural dimensions.* A developing science might depend on creativity or intuitive insights that occur best when attention is not diverted by routine tasks; a mature science would do better in conditions of predictable, cumulative progress, taking an analytic approach to problems.[9] A developing science might need individuals to champion and push forward radical ideas and projects to the limit, which means leaving inventors to work on their own timetables; in a mature science, early freezing of designs, with inventors handing over control, would save wasted effort and allow an efficient division of labor, with the attention of specialists concentrated on pressing problems. The novelty needed by a developing science might arise more readily if scientists were relatively independent and could escape the social and cultural pressure associated with a dominant local "paradigm"—the local taken-for-granted ways of scientific thinking and doing; a mature science would gain from strong consensus within a solidaristic team.

8. LIGO insiders might recognize the role of, for example, Riccardo DeSalvo.

9. It is very hard to pin down the psychological advantages of small science—perhaps it is a romantic image left over from a golden age. Perhaps all kinds of science flourish well under routine management—creative people seem to be able to create under the most adverse of circumstances, and sometimes isolation within a routine nonscientific job can provide the freedom from cultural pressure that can come even with free and easy working conditions within a scientific team. After all, Albert Einstein is said to have made his principal breakthroughs in spite of the demands of the patent office and artists do well in their garrets whereas, for at least some people, the intense noiselessness of an isolated cell in an academic hermitage would stifle creativity. This is not to say that the temporal and cultural features of small science would not be significant.

Let us, then, assemble a vocabulary as a framework for our analysis of LIGO's history. We start with our hypothesis that *developing science* and *mature science* are not equally suitable for coordination and routinization. Developing science requires unpredictable leaps of imagination and the like and works best under an *autonomous* style of organization, while mature science is readily fitted to a *coordinated* style of organization when the size of the project makes this necessary. We expect a large, coordinated, mature science to be able to make predictable, cumulative progress based on routinization, teamwork, and freezing of designs to enable the efficient use of specialists, and this means that it can plan its work schedules and the plans can be externally scrutinized. In short, a mature science lends itself to *transparency*. By contrast, a developing science cannot predict in advance where it will go and therefore cannot be scrutinized to see if it is keeping to its goals and schedules; an autonomous science, then, tends toward *opacity*.

We will find that at least some LIGO scientists or ex-scientists believe, or believed, that these two kinds of science can be identified within the LIGO project as can the corresponding kinds of organization styles. Here, for example, are contrasting claims about the nature of interferometry; the contributors are referring to work on the LIGO 40-meter prototype.

> Every time we make a change in the interferometer—a change which we model and we think we understand—we spend six months trying to figure out why it doesn't work. To me that indicates that the whole art of interferometry is not yet well enough understood.

> Despite the image often projected . . . running these interferometers is not magic. It is amenable to analysis and systematic studies.

The cognitive claims were used to demand different kinds of organizational approaches.

> [T]here are times for doing things and times for seriously thinking about what you are doing. And they are both equally valuable. If you are standing just randomly twiddling knobs, it doesn't necessarily help you.

> I gave them models of how to put together a week, scheduling-in times when you turn things off and do some maintenance; scheduling blocks of time when you do quiet experiments; these are very simple examples.

A difference was summed up by a third party as the tension between

> being intuitive when confronted with a problem as opposed to being
> methodical...

These quotations show both the problem to be addressed and the difficulty of solving it: if scientists disagree about whether the same piece of research is mature enough to be routinized, it is hard for the social analyst to do any better. The problem is made worse because two of the themes of change are confounded. The imposition of schedules and teamwork demands both increased control over scientists' work patterns and a reduction in the status of scientists; these changes happen at the same time as the cognitive change is taking place. Few people enjoy the experience of their work being increasingly controlled, and no one likes to see their status diminish relative to others in the same organization. Therefore there is likely to be resistance to these changes for reasons that have nothing to do with the science *as science,* and it is hard to disentangle the resistance to organizational changes from a concern about scientific integrity. Unwittingly, scientists might present an argument about the undesirability of the cognitive change that is really rationalization for resisting the change in work practice.

Fortunately, as the story unfolds we will see the shift from autonomy to coordination that characterizes LIGO as a whole, repeating itself in LIGO's successor—Advanced LIGO. This reinforces the idea that autonomous science is needed at the beginning of a project. The reemergence of autonomous science at the first stages of Advanced LIGO suggests that the need for it is more than a rationalization.

Also, even in the case of the first stage of LIGO, the argument that I want to develop does not depend on pinning down the motives of individuals; we can still operate on the Antiforensic Principle (chapter 22). The fact is that though scientists argue about the nature and organization of specific pieces of science, they all accept the terms of the debate. Thus, arguments about the nature of the science can be taken to confirm the broad framing of the study—the conceptual world in which scientists find themselves.[10] The argument among the scientists is not about the terms

---

10. The "problem" of the argument among the scientists, which might confound the study, is thus seen as a "resource." This well-known sociological gambit is not unknown in physics. A famous case occurred when Arno Penzias and Robert Wilson found that a noise they could not eliminate from their radio antenna was the cosmic background radiation signal—a discovery for which they won the Nobel Prize (Mather and Boslough 1998/1996).

themselves but the size of the terrain of application of autonomous science in general and its relevance for interferometry in the late 1990s in particular. Thus, talking about the debate in general while leaving the particulars as open questions is, on the whole, good enough for the purposes of the analysis. In the next few years we will see whether LIGO can live up to its promises of reliability and sensitivity. This will represent a partial test of the protagonists' views, because if LIGO is successful, then, unless there has been some unexpected input to the science, it will show that the relatively routinized approach has worked. If LIGO fails to live up to its promises due to unforeseen complications in the science of interferometry, then the degree of coordination of work and the extent of routinization of the science will be open to question.

A failure will be only a partial test of the kind of science LIGO represents, because it is an asymmetrical test: LIGO's failure to meet its promises will not show that an increase in the autonomous science component would have done any better—that would be a counterfactual argument—only that the science was less mature than it was thought to be. The scientists seem ready for this test, since the current leadership promised to meet demanding and well-specified targets for sensitivity and duty cycle by 2002, while those who think that routinization has been premature predict that "in the next three years [that is, up to spring 2003], press releases from LIGO will be along the lines of 'first operation of LIGO interferometer,' 'uninterrupted simultaneous operation of two interferometers,' with no mention of working sensitivity."[11]

The Small Science of Gravitational Wave Interferometry

Part of the story of the founding and funding of LIGO has been told in earlier chapters. Here I will divide the history of LIGO into four management regimes. According to my analysis, the four regimes have favored, respectively, a small-science approach, followed by a kind of big-science approach that was unsuited to the public sector, then a fully implemented and transparent organization, then a more relaxed approach as the project absorbed a wider community. The final twist is the introduction of more

---

11. Email from "disaffected scientist" dated April 2000. There may be a methodological problem here. It could be that these statements have had an effect on the way my respondents present their work to the public; certainly there has been no shortage of mentions of working sensitivity, and this has been forcefully pointed out to me in the context of this prediction. But predictions from skeptics that are less subject to self-defeat will be presented below (for example, in chapter 41).

close control on the wider group as Advanced LIGO has begun to look for an additional $100 million to $150 million.

For reasons that will become clear, nearly all the events to be described took place at the Caltech end of the Caltech-MIT collaboration that formed the basis of LIGO. Thus, the MIT group receive relatively little attention here even though, as we saw in chapters 17 and 18, MIT is crucial to the science of gravitational wave detection. It is because LIGO management is located within what is now the bigger group at Caltech that most of the sociologically interesting events took place there.[12]

As we have seen, preliminary plans for very large and very expensive interferometers had been worked out by Rainer Weiss of MIT in the early 1970s. But since planning for large-scale work is not the same as performing it, the science of interferometric gravitational wave detection remained a small science until the late 1980s. The small science of laser interferometry was outstandingly successful in the late 1970s and early 1980s. For example, at Caltech, Ron Drever led a small team that built a prototype having 40-meter arms. It rapidly achieved substantial sensitivity, approaching or equaling that of the best resonant bars even though those had been under development for much longer. This success persuaded a number of experienced physicists to turn their attention to interferometry.

Dividing up the history of a scientific or technological development into discrete "inventions" is a dangerous business. Nevertheless, it is possible to make a statement to which most scientists in the field would agree. If we concern ourselves solely with large-scale interferometric gravitational wave detection rather than the pioneering efforts of Bob Forward, then Weiss set out the overall conception; and by the mid-1980s, Weiss and Drever between them had invented perhaps seven or eight of the ten or eleven crucial features found in the full-scale devices currently being built.[13] An

12. Members of the MIT group claim proudly that their lack of salience in the public domain results from their skill at managing the transition to coordinated science. (But they are a much smaller group.)

13. Sociologists of scientific knowledge or of technology should understand that this is a retrospectively constructed list—it is a scientist's accounting rather than a sociologist's analysis. Thus, I count the Fabry-Perot idea as a success for Drever; Weiss wanted to use a "delay line." Drever won the argument, and most very large interferometers now use his idea. Perhaps the delay line will make a comeback one day, but for the time being the proper accounting of contributions for the purposes of this chapter is best done without moving far from actors' categories. Risking the wrath of angels, it seems to me that Weiss's main contributions were an inspired and inspiring working out of the consequences and method of implementation of the principles of experimental design outlined by his mentor, Robert Dicke, whereas Drever's inventions emerged out of what is sometimes called "lateral thinking."

For scientist-readers, my rough list of inventions includes concentration near the dark fringe and the locking of the signal to that fringe, with the control signal needed for the locking as

immense amount of effort has gone into developing the big interferometers since the end of the 1980s, but scientists do not seem able to name many more discrete, big inventions. A feature of at least some of these ideas is that they were pushed through against opponents who said they were not sensible. For example, Drever's idea of using very long Fabry-Perot cavities seemed impossible until Drever coinvented what is now the very important and widely used method for controlling such devices.

The period from the late 1970s to the mid-1980s was, then, a period in which the small science of interferometric gravitational radiation detection displayed its glories; responsibility for small projects was left to creative individuals, and this paid off handsomely, reinforcing the value of autonomous small science in the eyes of many of the researchers.[14] We now leave that period behind as I begin to describe the way the project grew.

---

the output signal. These two interrelated ideas as well as the whole conception of how to achieve the kind of sensitivity needed in such a device belong to Weiss. Drever invented the use of the Fabry-Perot cavity and invented and codeveloped the Pound-Drever-Hall method for stabilizing it; he invented the idea of multiple interferometer beams in the same tube and the half-length interferometer realized with a midstation; he deserves at least half the credit for inventing power recycling and turning it into a practical proposition (Thorne helped him with the calculations), and he deserves considerable credit for inventing an early notion for "narrow-banding" an interferometer—"resonant recycling"—which became the very important and practical "signal recycling" in the hands of others, notably Brian Meers, who worked in Drever's Glasgow laboratory. Drever may also have first put forward the idea of what is now known as "wavefront sensing." The list of inventions might well be longer, but about these there seems little doubt.

These are the big discrete ideas. Scots, Germans, Italians, and Australians have invented radically new kinds of mirror suspension, and a still more advanced form of recycling called resonant side band extraction was invented by a Japanese graduate student, Jun Mizuno, while working in Germany. Innovations have also come from US LIGO Scientific Collaboration groups, notably a way of attaching mirror supports, which was invented at Stanford University. The Russians have led the way in new materials and new theoretical understandings of the devices. We also should not underestimate the value of more recent contributions to the integration and execution of LIGO, without which these ideas would have been stillborn. As one respondent put it, it is vital "that the engineering challenges are met, and that the design and implementation evolve during the design and building to optimize the product, as well as meet cost-schedule constraints." What I have said in this footnote almost certainly contains mistakes.

14. This is not to say that the members of the Caltech team were entirely happy with the micromanagement of their laboratory, but that is another story.

CHAPTER 31

THE BEGINNING OF COORDINATED SCIENCE

REGIME 1: THE TROIKA

We have visited the funding of LIGO as a big science several times already. I introduced the book with Tony Tyson's attempt to stop the funding. We looked at the funding in order to explain the head-on attacks on Weber's cross-section claims and the 1987 supernova events. I invoked the funding to explain the demise of the sphere program and the curtailing of cryogenic resonant bars. I described the way the big LIGO facilities look on the ground. Now I tell the story again as we explore the project's transition from small to big science. This time the perspective comes from inside the LIGO project.

To realize the conception of a kilometer-scale laser interferometer that could detect the fluxes of gravitational waves that astrophysicists predicted, and thus to realize the goal of establishing a new gravitational astronomy, big money (from the National Science Foundation's standpoint) would have to be spent. There was also the transition from small to big organization, the full meaning of which would be only slowly understood by many of the participants. As Richard Isaacson, NSF program director for gravitational physics, told me,

> [1995] This community is making a transition from individual entrepreneurial science to big science, with all of the pain and anxiety that communities go through. They adopt central facilities; they're giving up their individual control and identity and merging together and collaborating with people who a few years ago they could barely speak to. But it's what's happened in nuclear physics, high-energy physics, even atomic physics. And even people who are now doing biology have to go to big synchrotron facilities. It's just that the high cost of equipment forces communities to move up this learning curve. In fact, even theorists who are doing black-hole collision simulations on supercomputers in the US have gotten together....
>
> So, there's going to be this sociological shift in the community, and that's going to be wrenching.

The first move in the direction of big science was made by Rainer Weiss of MIT, whose experience in working with a big team looking at the cosmic background radiation made it clear to him that a US national effort would be needed.[1] Weiss made overtures to the Caltech group with the idea of combining their efforts and including a much wider group of universities, but he was resisted at first. Collaboration between the two major institutions was soon brought about by the NSF, however. It was funding Caltech, with its 40-meter device, and MIT, which was running a much smaller interferometer intended to investigate basic features of the design; and the foundation made it clear that there was a future for only a single joint project. Thus, what was to become LIGO was born in 1984. Between 1979 and 1984, NSF spending on research and development and the testing of prototypes at the two institutions had steadily increased, from roughly $300,000 to about $1 million per year. In 1985 the financial aid jumped to $1.65 million, and in 1986 and 1987 it was at about $2.6 million.

To realize the project, attempts were made to integrate the work at Caltech and MIT under a three-way leadership—Kip Thorne, Ron Drever, and Rai Weiss—that became known as "The Troika."[2] For a full understanding of events as they unfolded into the middle 1990s, one would need to know

1. Mather and Boslough 1998/1996.

2. Actually, the Troika seems to have started earlier in a loose way as a result of Weiss's discussions with Thorne before the NSF's enforcing collaboration. Rich Isaacson told me that at NSF they understood that the Troika was not likely to be successful, but they felt it was a necessary first step in allowing the participants to learn to work together. [2000] "We knew from the beginning that

something of the personalities of the principals involved. I believe the evolution of the project would have followed a broadly similar path whoever had been running things, but the way the changes were made and the strains within the different phases could have been different. Because of the personalities involved, the history of this case represents something close to the far tail of a distribution of styles of management transitions that runs from quiet evolution to a string of crises; in the academic world, nothing much more dramatic happens than happened in this case: there were rows, resentments, feuds, sackings, samizdats, sendings to Coventry, shouting matches in corridors, cabals, press reports, misreports, hearings, heartbreaks, buyings off, breakdowns of long-term friendships, multiple resignations, and destructions of careers. In sum, the transition suffered more than its fair share of tragedy and sadness.

The Troika was to be run through consensual agreement. When interviewed in 2000, Weiss said he had great misgivings about the organizational structure from the start, because it was too weak and because it was dominated by Caltech at the expense of MIT.[3] Rochus "Robbie" Vogt, who was to become the director of LIGO after the demise of the Troika, described the arrangement to me as follows:

> [1996] [I]t basically nullified any authority of anybody. All decisions had to be unanimous and were not decisions: they could be revoked at any time by one single person. And so it was an absolutely ridiculous memorandum of understanding between the three of them, and it couldn't work.

Weiss's explanation for this structure is that it was arranged, under Thorne's leadership, to mollify Drever, who would have preferred that Caltech work independently. It soon turned out that for Weiss and Drever to reach agreement was nearly impossible. All Drever's experience was in small projects, and all his many successes had emerged from relatively lonely effort, often in the face of opposition. Weiss's preferred model was also small science, and as he told a journalist in 1990, "most of us in this field hate big science,"[4] but he saw that the transformation was essential.

---

in order for this to be a large-scale project, it was going to require a more experienced management." He said the next step, when the Troika brought in Frank Schutz, an experienced project manager from the Jet Propulsion Laboratory, was more like the real thing.

3. An additional problem for Weiss was that MIT would give him no support because its senior administration always doubted the wisdom of the gravitational wave detection enterprise.

4. Waldrop 1990, p. 1106.

The problems were exacerbated because the scientific styles of Weiss and Drever were at opposite poles. As one scientist with deep experience of the period explained,

> [2000] The styles of the two, Ron and Rai, are so very different that it was just impossible for them to collaborate—they couldn't communicate, not even about technical things. If you look at them, Ron is extremely creative, very clever, he's a tremendously inventive person, but it tends to be almost all pictorial in his mind—he's very unmathematical. . . . Rai tends to be encyclopedic in his knowledge of physics, extremely rigorous, and not terribly intuitive. And so the two of them could never communicate. Ron would have a picture and Rai would say, "Well, show me how this works out, show me—you know—write down the equations," and Ron couldn't do that. And Rai would go away and do three pages of calculations overnight, with Bessel functions, and the whole business . . . you never approximate . . . you always carry the full mathematics. And he'd hand that to Ron and then say, "See, I've proven my assertion [for example, that the idea won't work]," and Ron would not know what to do with it. So the two of them couldn't communicate about anything, so the two of them disagreed about almost everything.[5]

This respondent reports that the Troika would often hold meetings lasting two days behind closed doors and emerge with next to no decisions made at the end of it.

Initially, the clash in styles worked in Drever's favor. A respondent with no reason to think warmly of Drever told me,

> [2000] [Ron's approach is] highly effective in some areas. I will say this: in most of the technical disagreements between Ron and Rai, Ron was right more often than not. There were more occasions when Ron's pictorial intuition stood up against Rai's mathematics than the other way round in spite of the fact that Rai's very good with that kind of stuff.[6]

But given this, and the fact that Weiss was arguing for a stronger organization and less uncontrolled speculation, the contributions of the protagonists

---

5. To present Weiss as a scientist without physical intuition is to caricature, as his virtually complete initial understanding of the concepts and noise sources in an interferometer demonstrated in his pioneering RLE report. The difference of temperament is still there, however. For a discussion of visual and mathematical approaches, see Galison 1997.

6. Lest this give an unbalanced picture, another respondent added on seeing the draft that "there were times when Rai's unique talents cut through confusion . . . "

became stereotyped. As Weiss himself agrees, he gained the reputation of a "naysayer": [2000] "I fundamentally became the *bete noire;* the guy who was killing people; the guy who was suppressing creativity. I wasn't doing that, but it appeared that way. Because I was saying, 'Look, we've got to make a decision on certain things.'"

Thus, using our three-way description of cognitive approach (see p. 552 in chapter 30), we could say that the difference between Weiss and Drever was *psychological* and *temporal:* Drever tended toward an intuitive approach as opposed to Weiss's analytic approach to a problem, and Weiss realized the need to make decisions, whereas Drever wanted to leave everything as open-ended as possible for as long as possible.

For at least the early period of the Troika, the decisions that were made tended to go against Weiss, especially as he was in the weakest position, having little support from his home institution and being the sole representative of that institution compared with the Troika's two-person Caltech contingent. To press the developing-science/mature-science theme, we might say that Drever's intuitive, inventive style was suited to the early days of the project. As we will see, however, in the long term, as LIGO became a matter of building to a design rather than freely inventing new ideas, Weiss's scientific style would come to dominate, with Drever's approach being more useful for thinking about generations of devices that would not be built for many years if at all. Furthermore, the coordinated approach that Weiss considered an inevitability whether he liked it or not would also come to pass.

Thorne, who tried to hold the Troika together, was a theorist, so the big/small dichotomy does not apply so readily. He explained, [1999] "I became the chairman fundamentally because with the two of them having such radically different views, there was no way that either one of them could be the chair. So I became the chair, but the chair with one and only one reason for being there, and that was to foster consensus." Nevertheless, as Thorne admits, [1999] "It was painful, it was slow. Decisions did get made, but they got made at a rate that was probably 5% of the rate at which they should have got made at."

I have not found anyone other than Drever who still believes the Troika could have worked. The best that anyone will say for it is that the same technical developments would have taken place even if the teams had been working independently in their own institutions. Such decisions as were being made were concerned with inventions made by Drever and Weiss prior to the establishment of the Troika. At least one opinion has it that setting up the Troika stopped the creativity that had been demonstrated before the merger. Thus it was said by one respondent that Drever's creative energies became diverted.

> [1999] His [Drever's] energies went into trying to protect his control of the project, since he felt very strongly that the project would fail if it weren't done the way he felt it had to be done.
>
> ...The great contributions he could be making could not be made inside this kind of organization.

Late in 1984, a project manager, Frank Schutz, was brought in to try to stiffen up the workings of the Troika but could not do anything in the face of the existing work practices. He was given a brief—to answer to the three-man committee rather than have it answer to him—which would have made it difficult to accomplish the transformation even if he had been otherwise suited to the job. His successor, Robbie Vogt, took over the project in 1987 (see below), and Schutz left shortly after. Schutz went on to become a highly ranked senior manager on a series of space projects. Vogt said of him, [1996] "[T]hat poor project manager—they just tore him to shreds; he wasn't allowed to do anything; it was terrible." Schutz, however, told me when I spoke with him in September 2000 that this assessment of his tenure at LIGO was too severe—although he had been frustrated. He didn't feel that he had been damaged by his experiences in the project even though he had not been as effective as he would have liked to have been. To him, my characterization of the change required as LIGO evolved (expressed in a draft paper) was accurate, and he had known that the organization needed a shock to transform itself; he felt that Vogt was better able to provide such a shock.

> [2000] My ability to convince the Troika that it needed to awake to the fact that this was a different sort of activity than simply a continuation of past research activities was inadequate to the job. The required transition was something for which they were intellectually unprepared and to which they were emotionally resistant, simply because it represented too much change. When Vogt was announced as the new "Czar," I hoped that he had the stature and drive to induce change. This happened, but with other untoward consequences for those involved.

Clearly, the Troika was not working well to produce either kind of science: on the one hand, it was failing to deliver the *organization* needed by a big, mature science; on the other, it was destroying the conditions for the kind of intuitive work which, arguably, is best for a developing science.

LIGO's Science

It is just about possible to imagine a rationale for the Troika. One respondent claimed that the consensus pattern that the Troika had adopted, allowing, as it did, individual members to retain decision-making power, kept the features of autonomous science within the big-science project. The attempt to retain some autonomy, though doomed to fail at this stage of the LIGO project, has to be understood in the context of LIGO's science. In terms of funding alone, LIGO is big in the context of the NSF. As I will explain in detail in the next chapter, the building of LIGO was, among other things, a major construction project built to standards of stability, cleanliness, and accuracy comparable only with high-energy physics projects, and vastly exceeding even that technology in terms of vacuum capacity. This part of the project, costing about $200 million, can be thought of as the facility.

Mounted inside the facility are the interferometers themselves. In the early designs there were to be up to six completely independent interferometers with their own signal mirrors, beam splitters, and so forth; the light beams would run side by side within the vacuum pipes. The original design philosophy, then, allowed for the possibility that small teams of scientists would eventually be able to run their "own" interferometers within the facility provided by the LIGO project. Thus, some hoped that a kind of small science would return once the facility-building project was completed. In the light of this thinking, we can make a little more sense of the Troika—it could be said to be have been trying to hold on to small science during what was seen as a big-science interlude. The facility, once it was completed, would be run by managers, while the old-style autonomous teams would reemerge to assemble and run their own interferometers.[7]

As we have seen, one of the great advantages of the technique that was eventually chosen to bounce the light backward and forward in the arms—Drever's Fabry-Perot cavity, which would be chosen over Weiss's preferred "delay line"—is that the diameter of the light path is small, and this

---

7. As suggested in chapter 30, the model here might have been something like work on a large telescope, where the machine itself is big, but there is nothing necessarily big about the teams of observers who use it for short periods to make their own specialist observations (Galison 1992). See also Smith (1992), who explains that a big instrument does not necessarily make for big science. Even high-energy physics has separate teams working on different kinds of detection devices for the same particle stream, but there even the detector teams tend to be large.

means more independent interferometers can be mounted alongside each other, holding out the promise of more locally controlled science. It was Drever who suggested multiple beams (though he said this was just a way of getting many interferometers for the price of one).

Quite a lot has happened to change this conception of how LIGO would work. First, cost constraints have meant that only one full-size interferometer is being installed at each site. Second, it has been discovered that even a single interferometer is not exactly small in terms of cost and installation time. Thus, the first upgrade, Advanced LIGO, which is being proposed now for installation in the late 2000s, will cost at least $100 million to $150 million for the three interferometers and require a downtime in the facility of between 16 and 27 months to enable installation. The notion of individuals or small teams coming in at their own pace and creatively fiddling about with bits of their own interferometer seems to be dead, at least for the next decade.[8]

The "big-science facility versus small-science interferometer" model needs to be qualified in one more way, because aspects of even the facility, the most conventional elements, represent great departures from the past owing to their scale or highly demanding specification. That said, the building of an interferometer with a very high sensitivity is a rather different matter from the building of a huge, clean vacuum facility. High vacuum has been seen on Earth before, even if only in smaller quantities. The kind of almost inconceivably tiny measurements that will have to be made to see the predicted flux of gravitational waves have never been made. The distinction between the facility and the interferometers may explain the desire of members of the team who, believing there were still deep and unsolved problems, preferred to hold on to a less-coordinated style of working.

Assuming that the astrophysicists are right about what is in the universe, and that there are no big surprises to come, even building the facility and installing an interferometer of the sensitivity promised is not going to be enough to realize LIGO's overall aim, gravitational astronomy.

8. When I first wrote this passage, the projected date was 2005 and the projected cost $100 million. An interferometer is not like a telescope in that it cannot be pointed at different regions of the sky. To be more exact, it cannot be physically pointed, though it can be "virtually" pointed by doing the data analysis in different ways. So we can imagine a future in which different teams will take the output and analyze it with different goals in mind, but they do not need to have to touch the instrument to do this. Looking forward still further, different interferometers in the same vacuum pipe could be tuned, using signal recycling, to different frequencies with the detection of different kinds of continuous signals in mind. One can imagine small teams putting forward proposals for runs at different frequencies.

To do gravitational astronomy, an interferometer must be capable of seeing many bursts of gravitational waves per year, not just an occasional burst. This means still more sensitivity so that distant sources come within range. From the outset of the LIGO project, it has been assumed that a second, or even third, generation of detectors would be needed to realize the new astronomy. A detector that can see ten times as far scans an Earth-centered sphere with a thousand times the volume, which, other things being equal, should contain a thousand times as many sources. Thus if the first interferometer manages to detect one signal in a two-year run, assuming this represents an average rate of observation, an interferometer ten times as sensitive should see 500 signals a year. Even if the first interferometer could see a signal of a kind that occurred only once in 30 years, a factor-of-10 increase in sensitivity would still allow observation of an astronomically satisfactory 30 or so events per year.

Hence there are three kinds of science involved in LIGO:

1. the relatively mature but still path-breaking science needed to build the facility;
2. the science needed for building the first full-scale interferometers for LIGO I; and
3. the still more adventurous science needed for Advanced LIGO and subsequent generations.

The Troika was going nowhere, but much of the trouble that was to follow was perceived by at least some of the protagonists as turning on disagreements over the priority to be given to the first kind of science and the organization of the last two.

The End of the Troika

The Troika, even if a rationale could be found for its managerial style, would not have survived under the kind of scrutiny that the potential for big spending was bringing to bear upon it. It worked too slowly and inefficiently to withstand exposure to the new groups who were sharing responsibility for LIGO and who had to defend it in front of a growing audience of skeptics. The demise of the Troika was effected by a panel convened under NSF auspices which met November 10 to 14, 1986. It decided, against at least some expectations, that the LIGO project as a whole was worthwhile. Its "Report to the National Science Foundation by the Panel on Interferometric Observatories for Gravitational Waves" (January 1987) recommended, however, that "an oversight committee be set up charged with overview of the scientific program, management of the facility and facility availability to outside groups" (p. 11) and that

> the project be headed by a director of scientific and engineering stature comparable to oversight committee members and the investigators. This director must be the final and single authority for decisions during construction and evolution into an operating observatory. Management by a steering group may have been adequate until now, but would not be appropriate for the construction and operation of a project of this size. (pp. 11–12)

In its summary of conclusions on page 13, it added, "Efforts should immediately be directed to providing such leadership."[9]

The recommendation that a strong leader be appointed may look like a response to managerial exigencies, but at the same time it is, perhaps, the most crucial change in the nature of the project; it meant that control over the choice and direction of scientific and technical ideas was taken away from those who were thinking them up. The major transformation in LIGO took place when the first manager with real power was appointed, and everything that follows can be described as variations in the way this change and its consequences were handled.

REGIME 2: THE SKUNK WORKS UNTIL THE DREVER AFFAIR

In 1987, Robbie Vogt became the new leader of the LIGO project. Vogt had followed what a respondent called "the familiar career path from physics to management." He had headed the cosmic ray laboratory at Caltech and been chair of the Department of Physics, Math, and Astronomy, but came to LIGO as immediate ex-provost. He had effectively been sacked as provost as a result of deep disagreements with Caltech's president. Vogt came with a reputation for obstinacy and had never done research on interferometry, but he had a good track record of extracting large sums from sponsors and was an experienced and successful supervisor of large physics projects. He was also the kind of strong team leader that the project now needed.[10]

9. Weiss told me that he persuaded the committee chair to make this its most central recommendation. An NSF spokesperson told me that the foundation set the panel up in part with this in mind, as all panels are convened with a good idea of what the outcome will be. "I'm sure it was no secret from the panel what the NSF thought."

10. Vogt did have a working physicist's understanding of interferometry, had taught courses in the subject, and would soon pick up more of the essentials.

The years immediately following Vogt's arrival saw NSF's funding increasing to about $4 million per year. The aim, of course, was to justify and install the management structure for the responsible and accountable spending of the hundreds of millions of dollars needed for the construction of the first full-scale observatory.

Under Vogt the project leapt forward. He sliced through the indecision that had paralyzed the Troika and began making choices, cutting off options that would otherwise have survived as the pet projects of the scientists at Caltech or MIT. This, of course, caused pain. Vogt found that the plans for LIGO he had inherited from the scientists would not be supported by NSF, so he recruited engineers to plan the construction professionally and provide a realistic cost estimate. As an NSF spokesperson said,

> [2000] That's something that I really think is his enormous contribution: conceptualizing the project and getting it in shape to pass peer review and scrutiny, to be sold to Congress, and so on, which all followed from putting engineers in place and going over everything and converting this from a piece of paper to what at the time I thought was a real construction blueprint.[11]

But Vogt also had to reduce the size of the project. Under NSF pressure, he cut the number of interferometers from six at each site to one full-length (four-kilometer) at each site plus an extra half-length (two-kilometer) interferometer within the Hanford, Washington, facility. Now the NSF had a cost estimate that it could represent with confidence. Vogt's plan was approved by the National Science Board, and it eventually went forward for congressional approval. Under Vogt's leadership, the project was funded—something that would almost certainly have been impossible under the Troika. But LIGO had now received a huge amount of political and media scrutiny from which spotlight it was never going to escape.

Insiders and Outsiders

Although Vogt had no research reputation in laser interferometry, he was now making the decisions. In general, scientists try to reach agreements on the basis of experimentation and calculation: they prefer to leave open all options until they are certain which is the right way to go. But the kind of certainty that many scientists believe is attainable—a certainty that will cause all, or nearly all, parties to reach agreement—can be very slow

---

11. Though the same respondents said that as his experience grew, he realized that what he had seen was nowhere near detailed enough.

in coming. The sociology of scientific knowledge has shown how long it can take to reach consensus within the scientific community over difficult issues, and the same applies on a smaller scale to the technological choices within a project.[12] Vogt closed off certain options that the scientists would have liked to leave open for longer; he had to do this if the project was to have any chance of running on a reasonable timetable.

Vogt was settling, for practical purposes, the issues that divided Weiss and Drever. Nearly all Vogt's initial technical decisions favored Drever. He pushed through one Drever idea after another against MIT opposition. It was Vogt who decided that LIGO should use Drever's favored narrow-beam, Fabry-Perot design, and this had a number of direct effects. Even after the multiple interferometer concept had been scotched on grounds of cost, Vogt was to agree with Drever's plan to fit a two-kilometer interferometer inside one of the four-kilometer installations.

With Caltech scientist Drever's ideas being accepted at the expense of Weiss's, it was bound to be hard for some scientists to separate Vogt's technical choices from his loyalty to Caltech. This first period of Vogt's leadership was, then, a bad time for the MIT team. It looks very much as though something like this was inevitable, because Vogt's strong leadership style meant that new classes of insiders and outsiders were defined. To be "inside," it was no longer enough to be within the institutional boundaries of the LIGO project; it was also necessary to be loyal to Vogt and Vogt's vision. Thus the new regime was always going to be excellent for the new class of insiders but bad for the new class of outsiders. In the first Vogt period, the new outsiders came from MIT, and Weiss's stereotype as a "naysayer" was reinforced further.

The description that was applied retrospectively by both Vogt and others to his regime was a skunk works. The Skunk Works is the nickname given to Lockheed's secret specialist military aircraft development installation in Burbank, California. The dictionary definition reads, "Often a secret experimental division, laboratory or project for producing innovative design or products in the computer or aerospace field." A book describing the Lockheed operation has helped to reinforce the mythological features of such an organization.[13]

12. Collins and Pinch 1993/1998, 1998/2002.

13. Rich and Janos 1994. The "Skunk Works" name derives, rather tortuously, from the *Li'l Abner* comic strip. The dictionary definition given above is that found in the *Random House Dictionary*, as quoted in that book. How close the style described in the book is to what actually happened in the Lockheed factory is, of course, open to question. A number of people have told me that Vogt took the already exaggerated account in the book and exaggerated it further in order to find a label for what he was doing.

A skunk works is said to have a number of key features. Its management is characterized by integrity: Lockheed, it is reported, sometimes voluntarily returned excess profits to the government or "pulled the plug" on potentially lucrative projects as soon as it became clear that they would not live up to the initial promise. It must be free from outside interference: There is enormous pride in taking shortcuts and avoiding cumbersome regulations within the overall "can-do" philosophy. Along with this goes determination to fulfill all promises and complete projects on time in spite of the absence of outside monitoring. An innovative atmosphere flourishes within a group that sees itself as a secret elite driven by intense loyalty to a strong leader, and high-pressure teamwork needed to accomplish goals, takes precedence over family life. According to Ben Rich, the second leader of the Lockheed installation, "A successful Skunk Works will always demand a strong leader and a work environment dominated by highly motivated employees. Given those two key ingredients, the Skunk Works will endure and remain unrivalled for advancing future technology" (p. 367).

In 1998, long after Vogt had left the LIGO project, he described his philosophy as follows: "You basically build a project and you build a wall around it and say, 'Throw the money over the wall.' And in n years, I break the wall down and deliver a beautiful thing—an airplane or a LIGO."[14]

Along with the skunk-works-like organizational style goes a promotions policy untypical of industry. To quote the last of the fourteen rules of Skunk Works founder "Kelly" Johnson, "Because only a few people will be used in engineering and most other areas, ways must be provided to reward good performance by pay not based on the number of personnel supervised" (p. 55). Johnson is given a paraphrased speech in the book, in which he criticizes another aerospace firm: "Hell, in the main plant they give raises on the basis of the more people being supervised; I give raises to the guy who supervises least. That means he's doing more and taking more responsibility" (p. 312).

In some respects, Robbie Vogt's management style was similar to that of Ben Rich, the second director of the Lockheed Skunk Works; he lobbied hard for resources, bolstered a very strong team identity, and paid his central team members well according to their innovatory talent rather than their supervisory responsibility.[15] For the LIGO scientists located at Caltech, these

---

14. Quote taken from Caltech's "house magazine" (Dietrich 1998), p. 14.

15. The skunk works *label* was applied only gradually as Vogt's management style began to conflict with NSF's demands. Vogt said he learned the management style from his mentors running large-scale physics projects in the years following World War II. We may imagine that the style had then been set by the Manhattan Project, and strong oversight would not be the norm. Vogt also

were halcyon days, and they became dedicated to Vogt. He obtained pay rises for them and made them feel valued even though they were still earning less than the engineers who now worked alongside them. One respondent who had enjoyed his work during these days told me that scientists can stand low pay and status so long as they are having fun. Robbie Vogt had told his scientists that the engineers get paid more, but the scientists have more fun—and this kind of idea was acceptable under his leadership. As in the Skunk Works ideology, people worked long if irregular hours, driven by the need to solve problems rather than by adherence to a schedule. Under these circumstances, the "insiders," most notably the research and development scientists in the Caltech laboratory, were happy.

I am going to argue that the skunk works management philosophy was bound to fail. As has been indicated, a source of trouble seems to have been that Vogt was unable to avoid the pathologies that this approach brings with it when it is uncompromisingly applied. Thus, intentionally or inadvertently, he alienated or lost contributors to the project who refused to declare themselves loyal to him and the team above all other loyalties. About half a dozen people left MIT about this time because of managerial malaise caused by the stresses and strains—at least one, and possibly more, as a direct result of Vogt's demand for loyalty.

But any project can survive the loss of some of its members and may even become stronger if the loss results in clearer goals. Indeed, the management regime that was to follow Vogt's was to lose a different group of personnel—including some of those who had been specially loyal to Vogt—in similarly strained circumstances, and they would argue that the result was an improvement in work practices. (This is a case that I will discuss at length in chapters 34 and 36.) Vogt's regime, however, was to suffer other problems, all related to the shift from small to big, publicly funded science.

The first and biggest of these problems was that Vogt and Drever fell out in the most calamitous way. In consequence, Drever was ejected from the project, causing a schism in the LIGO team, Caltech as a whole, and the gravitational wave community. Worse from LIGO's point of view, lurid accounts of the row, which gave rise to highly charged academic freedom hearings on the Caltech campus, found their way into the newspapers and science newsmagazines.

---

had experience with private funders, who, once more, would not expect the kind of transparency that NSF needed. Others say that the Skunk Works, even under Rich, was characterized by more delegation, structure, and schedules than Vogt's regime.

CHAPTER 32

THE DREVER AFFAIR

On July 6, 1992, LIGO Director Robbie Vogt sent an email to all members of the project.

> As of today, July 6 1992, Dr. Drever is no longer associated with the LIGO project.
>
> I have assured Dr. Drever that he may remove any personal possessions or files from the LIGO premises to his present office (355 W. Bridge), provided such acts are witnessed by a member of the LIGO staff.
>
> Dr. Drever is expected not to disturb LIGO staff at work and to refrain from entering LIGO premises.
>
> Robbie

Drever sent an immediate email response and a letter to "Colleagues at Caltech" dated July 12, 1992, which began,

> Dear Colleague,
> I am writing to inform you that I have been removed against my will from the LIGO project; I have been told that my keys to offices

and all laboratories and facilities associated with the LIGO project—including the 40-meter gravitational physics laboratory I created—will be taken from me; and the password to the LIGO computer I have been using has been changed so that I have no further access to the LIGO computer system. I am unaware of any valid reason for this summary expulsion, which was quite unexpected to me.

I attach, for your interest, a copy of an electronic mail message sent to "ALL@ligo," a list of more than 40 persons at Caltech and MIT (including undergraduate and graduate students—some of whom I may be asked to teach in class—postdocs, scientists, Faculty members, administrators, engineers, secretaries, and others). This message, which was sent out within about two hours of the verbal notification I received of my expulsion, is the only written evidence of it that I have seen.

The row that followed was of such depth and ferocity that the scar is still there—as much outside the LIGO project as within it. The turmoil engulfed not just the main protagonists but the whole Caltech campus. A scientist who had been involved in Caltech administration during the affair and whom I tried to interview many years later told me that he was torn between helping me to round out my account and hoping never to have to discuss the corrosive matter again; the latter emotion, he told me quietly, was far the stronger, and that was how our telephone conversation ended. Six years after the events, scientists still appeared scared to let me see the documents I requested; others told me that some of the records should never be seen again. Some documents I was allowed to look at for a brief time without taking notes or making any other kind of record.

Well-respected scientists with spotless reputations were said to have "lost their balance" during that period, and behaved toward their colleagues in ways that were beyond comprehension. The campus was thrown into a state of tension for months—even years. Graduate students would come across middle-aged, mild-mannered faculty members of great distinction, both on and off the campus, shouting at each other in the corridors. Life-long friendships fell apart, and the hurt remains. I was given to understand that this was the worst kind of thing that can happen in a modern university. More than half a decade after the row, the tone of my respondents was stilled, their body language that of those who had seen their friends commit atrocities in a combat zone.

Given the depth of the disagreement between the parties, it would be a foolish analyst who tried to reach any conclusion about who was right. At the time, a series of committees read every paper and listened to every argument, but their verdict did not convince anyone who was not already convinced; I could not possibly do better. What I can do, perhaps, is show that the argument was just an extreme symptom of something that was out of the control of either side: the shift from small to big science.

I now recount some of the incidents and causes for resentment that I have heard about. I have not done much in the way of "detective work" on these; I report merely what I have been told. For example, listed below are certain incidents to do with changes of locks or building design. I have not checked the dates, the building maintenance plans, or the intentions of the parties in making such changes as did take place. I supply true accounts of what has been told to me by the different parties. Where there are conflicting accounts or interpretations, I list these. This list is provided to indicate the depth of antagonism which held between the parties. What I want is to give a sense of how deep passions can go as a science transforms itself. That there are conflicting accounts should come as no surprise.[1]

How the 1992 Row Is Experienced

- Caltech promised Drever autonomy when he finally agreed to join Caltech full-time, but this promise was broken and Drever found himself merely a part of the much larger LIGO project working under someone else's direction.
- a) Drever gave firm and valuable guidance to his junior colleagues. b) Some younger members of the team experienced Drever's style of supervision as too stifling and intrusive; Drever treated junior colleagues as his personal assistants; Drever did not develop the kind of social sensibilities, especially American social sensibilities, that would allow him to spot the frustrations growing in the younger scientists.
- a) Drever felt that it was too soon to spend so much money on a large project when better scientific results could be obtained from smaller scale work. b) Drever did not act as though he understood that the time for new developments was over and the time for project building had come; he wanted his many alternative ideas to be fully evaluated before the design was finalized.

---

1. Originally, I wrote a much longer quasi-fictionalized account inspired by the famous Japanese film *Rashomon* and William Faulkner's *The Sound and the Fury*. Both *Rashomon* and the Faulkner novel deal with the marked difference in viewpoint of different parties or witnesses to an atrocity. Eventually, I was persuaded that this longer version would cause unnecessary pain to the parties and have replaced it with this telegraphic version.

- a) Drever was fond of championing a stream of alternative positions so as to push argument and analysis to the limits. b) Drever was indecisive; he might be completely certain of a view one day and equally certain of its opposite the next.
- a) Drever's initial refusal to join Caltech fulltime was a sign of his indecisiveness, damaging the work of both the Glasgow group and the Caltech group. b) It is sensible for someone of Drever's talents to try to spread his talents across two institutions, as many scientists do.
- a) Drever could not make small concessions to win larger points and did not know how to work as a member of a large team. b) Others did not appreciate Drever's willingness to adapt to large-team work and offer loyalty to the new management even though he did not like it very much.
- Political exigencies dictated that the initial bid for funds be for a full-scale project, and Drever's more cautious approach, irrespective of its scientific rationale and intention, seemed disloyal.
- a) Drever's publicly expressed views to the effect that more development work was needed before proceeding to the full-scale project were damaging; Vogt found he had an enemy within the project in addition to the multitude of enemies without. b) Others thought that Drever's belief that the leap to large funding was premature was causing him to be disloyal in public, but this was not true.
- At the height of the row, Vogt refused to allow Drever to be in the same room as himself, or, if he had to be there, he was supposed to sit outside of Vogt's line of sight.
- Vogt blocked the door between Drever's office and his secretary and photocopier.
- a) Vogt forbade Drever to discuss LIGO at a conference in Argentina; Drever's department head said he could go; Drever spoke at the conference (but did not talk about LIGO), and when he returned he found he was locked out of his office, the lock having been changed. b) The door lock on Drever's office was changed as a matter of routine; Drever was inconvenienced for a short time until a secretary returned from lunch with the new key.
- On his return from Argentina, Vogt, on behalf of the administration, summarily dismissed Drever from the project, sending a humiliating email to the whole team that warned everyone that Drever was to be allowed to collect his possessions only under supervision.
- a) It was Thorne who took the fateful step of first going to the Caltech administration and insisting that Drever be divorced from the LIGO project. b) It was Vogt who made certain that Drever would be divorced from the project. c) It was Jerry Negebauer, the head of the Physics Department, who

ordered Vogt to sack Drever on his return from Argentina; Vogt was taken by surprise, expecting and demanding only disciplinary action.

- A significant group of faculty felt that Drever had been disgracefully treated and helped him make a case for reinstatement. Caltech's Academic Freedom and Tenure Committee found that Drever's academic freedom had been violated on two counts.

- a) The Caltech administration would not put the LIGO project at risk by reinstating Drever; those who wished to see Drever sacked were placing the value of a scientific project above the rights and welfare of one of its most creative scientists. b) Drever's creativity had come to an end, and his behavior was threatening to put an end to a beautiful scientific enterprise. c) Vogt was determined that if Drever did not go, LIGO would be destroyed, and most people asked to comment were in fear of their future.

- Drever and colleagues produced the "Blue Book," which set out his grievances and was made widely available across the campus.[2]

- Drever's reinstatement was widely demanded at a series of public meetings that tore the small campus in two.

- Thorne organized the opposition to Drever, comprised of younger members of the LIGO team. Under his guidance, they produced the "Black Book," which countered the claims in the Blue Book. a) Thorne, in taking this course of action, had "lost his way." b) Thorne's opponents failed to understand the situation within LIGO that made Thorne's actions necessary. c) Old debts were being settled.

- The resignation of the Caltech president was demanded when Drever was not reinstated. Instead, a commission of enquiry was set up. a) The commission was packed with Vogt supporters and did not insist on Drever's reinstatement. b) The commission was fair, and the resignation of the president would have been an injustice that Thorne worked hard to prevent.

- Drever was never reinstated, but he was given a "sweetener" in the form of funds and space to support independent research, which he used toward building another interferometer; now there are two 40-meter interferometers on the tiny Caltech campus, about 200 yards from each other. For a long time, Drever would park his car by the door of the interferometer he had built, but was not allowed into it or any other LIGO meeting. Vogt continued as LIGO director. About two years after these incidents, Vogt was sacked from the directorship but continued in a more minor role for a year or two until

---

2. This "Blue Book" has, of course, nothing to do with the MIT team's Blue Book discussed earlier.

he resigned and moved on to other things.[3] Drever was offered reinstatement to the LIGO project by the new management but in a role he could not and was not meant to accept. Drever now attends LIGO meetings, but Vogt is no longer to be seen at them, spending most of his time as a consultant at Los Alamos.

- The events inflicted severe and lasting damage to all parties.

THE ANTIFORENSIC PRINCIPLE AT WORK

Opposing accounts can be treated as signifying very different things. The "postmodernist" view holds that there is no truth to the matter beyond the accounts themselves. There is the epistemological relativism approach, which maintains that it is impossible to know the truth even if there is one, since Freud we know that the individuals involved in an argument may be unaware of even their own intentions, never mind the intentions of others. The methodological relativism approach, which I adopt in the case of scientific disagreements, holds that even if there is an ascertainable truth in principle, it is not the business of the analyst to try to find it. But here I have the much more modest Antiforensic Principle in mind.

As explained, the Antiforensic Principle notes that even costly and elaborate institutions such as courts of law or, in this case, academic freedom committees, that are set up with the express purpose of reaching the truth are fallible. Indeed, courts of justice recognize that if something that counts as truth is to be extracted from the adversarial system, it will require the device of the jury or some such to bring a potentially endless argument to a forced conclusion. It is the jury that makes an artificial "truth of the matter" from the potentially eternal contestation of the adversaries. The sociologist, however, cannot approach even the flawed efficacy of a court of law. Instead, the Antiforensic Principle holds that sociological analysis should not turn on accurate identification of the internal states of specific individuals. Rather, the sociologist should identify the *range* of internal states that are available to the actors and that bear upon the social and cultural contexts and processes under examination. Biography and psychology are not sociology, yet it is biography and psychology that give rise to individuals' motives and choices.

3. In 2003, Vogt copied me on an email that he sent to the LIGO team, refusing to have his name on papers associated with the project, although the protocol allowed all contributors to its output to be listed. (Drever's name will be included.)

On the other hand, it is the proper subject of sociology—society—that provides the enabling and restricting range of ways of being and thinking in the world. Once we know the envelope of plausible thoughts and actions available to members of a society, we understand the society that envelope constitutes even if we do not know what is true of any particular individual.[4]

One can put this into a more technical framework. It may be that individual intentions can never be known, but this does not mean that nothing can be said about intentions. The distinction between a "type" and a "token" is useful here. A *type* of action is "opening the door." A *token* of that type of action is "John opening the door of his house at 3:00 p.m. on Sunday, January 27." Actors living through a drama such as that described here are concerned with *tokens:* Who did what to whom when? The business of the sociologist, on the other hand, as opposed to the historian or the jury, is *types:* What types of things were actors in group A doing to actors in group B? For the sociologist, the point is that not anyone can do any type of action at any time. For example, it is simply impossible for a member of a primitive tribe living in the Amazon to take out a mortgage on a hut; the kinds of actions that we can do depend on our social setting and even on our more immediate social circumstances.

The aim of this analysis is to elucidate the circumstances that made it possible to be accused of carrying out actions of a general type—"putting the project in political jeopardy," "imposing inappropriate work schedules," and so forth, rather than whether one particular act really was intentionally or even unintentionally jeopardizing, or whether the work schedules really were inappropriately imposed rather than appropriately. It is scientists, jurors, lovers, historians, and so forth who have to resolve the nature of the tokens to their satisfaction; sociologists need be concerned only with the set of possible types available to the actors and the way these change as

4. The principle, though without the label, has been introduced before. For example, Wright-Mills (1940) suggested that sociologists resolve the problem of identifying individual motives by talking only of the "vocabularies of motive" that provide the available repertoire of motive discourse that, in turn, indicates the nature of the social group in which the talk is embedded. Collins (1983a), in a paper called "The Meaning of Lies," argues that the analyst need not identify the true and false in respondents' accounts, only the plausible and implausible. A potentially effective lie contains as much information about what counts as a credible action in a society as the truth. The crucial dividing line for the social analyst is between truths and falsehoods on the one hand, and jokes and nonsense on the other. Lynch (1995) cites Wright-Mills, but argues further that since ascriptions of motive are part of the fabric of argument, no broad account of the motivating "ethos" of history, such as Weber's Protestant ethic thesis, can stand unqualified. Lynch's argument seems to present no obstacle here, as we are not concerned with broad historical change but the available and contested ways of looking at the world within a historical period. The argument is also the burden of the final paragraph of Collins 1998.

social structure changes.[5] What all this means is that though we may not be able to resolve with certainty the responsibility or culpability for particular incidents, we can say that such incidents *could* take place because of the changes that were taking place in the project. Clearly, assignation of praise or blame is not part of the sociological project conceived in this way.

Scientists often think the opposite way. Scientists tend to think that science is a business of individuals (or teams of individuals, which is the same thing, as far as this argument is concerned) up against nature. If a scientist admires a piece of scientific work, he or she will often attribute its success to the greatness of some individual. Some scientists fall into the trap of believing that great scientists are great men (or women), and that the ability to do great science reveals great strength, depth, and purity of character. In eulogizing dead colleagues, they will often refer to their goodness, their consideration for others, and, nowadays especially, their willingness to sacrifice some of their work for the sake of their family. Bad science, on the other hand, is caused by character flaws—dishonesty, for example, or weakness. Scientists talking of bad science often intimate that colleagues who hold minority views are crazy in one way or another, implying that it is the craziness that explains the deviant view.[6] We should not be surprised, therefore, were we to overhear something like this pastiche, which I have assembled from the kinds of remarks that have been made to me over the years:

Sociologist of Science: So, you see, my view is that Jim was under particular pressure to secure his respectability. He had to be seen as squeaky clean within the group of scientists who would be deciding whether he would get grants from the Einstein Foundation. This would be bound to make him less adventurous in announcing early results.

Scientist: Yes—ever since I've know him, Jim has been an awkward so-and-so when it comes to letting his team publish. You're right—he's paranoid.

---

5. One might equally say that social structure changes as the set of possible types of actions changes. That is why we call these formative actions (Collins and Kusch 1998).

6. For a striking example of the equation of personal integrity with scientific greatness, see the bad-tempered remarks of Max Perutz (1995) directed at the biographer of Pasteur, the late Gerry Geison (1995). For an exception which causes great embarrassment to the scientific community, see the autobiography of the "bad boy" Nobel laureate, Kary Mullis (2000).

On the issue of craziness, the sociologist is likely to take yet another view. The sociologist reasons that if you are rejected by your fellows hard enough and long enough, you are likely to become a little crazy, so the causal direction is the other way around—the structure gives rise to the personality.

Historian of Science: Though Jim is a brilliant man, one must understand that he has been given credit for work that he could not have done outside of a certain context. The foundation for the work for which he has been given credit was developed by many, many others over many, many years, and Jim simply found a neat way of expressing it and was in the right place at the right time.

Scientist: I couldn't agree more—not only is he a genius, and a great man, but he has been a great leader and an example to others.

Sociologist of Science: Walter's ideas, though we no longer accept them, were not so crazy at the time. The firm consensus now is that he was completely misguided, but he was never finally shown to be wrong with complete certainty—people just got tired of chasing down the novel arguments he offered. Don't forget that Walter had a group of supporters who went along with him and never accepted the apparent falsifications which the majority put forward.

Scientist: Yes, I agree, Walter became obsessed with his ideas to such an extent that he became separated from scientific life—he always was a little crazy. He had an unstable family life—did you know that? It's news to me that he had supporters—but there are always eccentrics waiting to come out of the woodwork; you should see some of the letters, written in green ink, that I get from people who think that relativity is wrong. By the way, have you seen that paper on pathological science by—whatshisname—the guy from General Electric or somewhere like that?

What we can say about the personalities of Vogt and Drever is that the events that were to unfold turned into melodrama because these two determined people were so deeply different. The transformation at LIGO would have happened anyway, but if the people had not been who they were it might have passed more quietly. Indeed, it is the boast of the MIT team that they accomplished a similar transformation from small- to big-science organization within their own group with no noticeable disturbance.

Let me just say, then, that Drever is a quintessentially "small" scientist (as well as being small and round in stature) in that he believes in being certain that the right scientific decision has been made before moving forward. He is the quintessential scientific inventor rather than developer of other peoples' ideas. As one of my respondents suggested,

> [1998] Ron has always prized cleverness above all else. Something like LIGO will use all of the cleverness he has to offer. But Ron wanted even more ... his imbalance in only prizing cleverness, and not reliability, working to schedule, or any sort of discipline, meant that Vogt's reliance on him could not be sustained.

Drever has no family of his own and appears to live for science. On one of my first visits to Caltech, by which time Ron must have been there for about fifteen years, I invited him to join me at the university's Red Door coffee shop. Caltech must be one of the smallest universities in the world; you can walk right across the campus in about five minutes. But Ron told me he had never been to the coffee shop. When I asked what kind of coffee he liked—"cafe-latte, espresso, cappuccino?"—he told me he didn't know the difference and that food was just chemical energy as far as he was concerned. In May 2002 I sat with Ron on the way to a conference and told him that I was going to put this story into the book if it was OK with him. "Why?" he said with genuine puzzlement. "What's interesting about that? I wouldn't go to there [the Red Door], because the coffee's more expensive than out of a machine." I laughed and laughed, but Ron still didn't get it. Drever retains a commendable resilience and cheerfulness in spite of what he has lived through.[7]

Vogt, by contrast, is a quintessentially "big" scientist (as well as being large and dominating in stature). Drever told me at one point (this is a paraphrase from notes taken in the late 1990s rather than a transcript): "Robbie brought me into his office and he started to shout at me. And he's a big man, you know. And at one point he came round from behind the desk and I thought he was going to hit me, so I ran out of the room." I asked, "So why didn't you just sit there and let him hit you? You could then have sued him and Caltech." Drever replied in his Scots accent, as though the answer were obvious, "Because I didna' wanna be hit!"

On the plane of social accomplishment, Vogt comes close to being the complement to Drever's social ingenue. For example, he knows how to charm important people. Here Vogt describes a successful interaction with an influential senator:

> [1996] [H]is staff would not let me near him, because they thought I was not powerful enough . . . [but we] found another lobbyist who had very good

---

7. March 4, 2003, and another endearing "Ronism" has me doubled up with laughter: Ron has been having serious lung trouble, though you would never know it from his bouncy demeanor and his willingness to argue relentlessly over dinner. One of his lungs has stopped working, and he's been driving back and forth to UCLA's medical specialists. Determined to get to the bottom of things, he's built his own lung-capacity measuring device from old buckets and bottles. Now he wants to be able to look at each lung separately. "I want to make an ultrasound imaging device for less than $20," he tells me. "Under $20?" I say. "Why not under $200?" Ron just looks puzzled. (June 13, 2003: Ron tells me that if American social sensibilities are as important as I say they are, then Caltech should have given him a book, when he first arrived, with them all set out. He assures me he would have read it assiduously!)

contacts with [the senator] and got me an appointment. And it was supposed to be a one-time affair for 20 minutes.

And I walked into [the senator's] office—an overpowering place—and there was [the senator] sitting behind a coffee table, and on his right was his chief staffer who was running the energy appropriations committee, and on the other side was another chief staffer. And they were sitting there like an audience watching me. And they all said, "You have 20 minutes—talk to the senator."

And I talked to the senator, and after 20 minutes the chief staffer stood up and said, "20 minutes are up, Senator, we have to go to our next appointment."

And the senator said, "No—cancel that appointment. I want to continue the discussion."

And for two hours we sat at the coffee table, and I was drawing pictures, and I had to explain to the senator why [when] we look back 15 billion years we don't see ourselves at the big bang, because 15 billion years ago everything was in a singularity...

Between July 30 and August 2, 1990, Vogt paid such about ten such visits to members of Congress and another handful of staffers.

Another of Vogt's skills was understanding the difference between doing science and building a project, again the opposite of Drever. Of Drever he said,

[1996] What frightened him about me was that I made decisions. He said to me, "How do you know that is the correct decision—we must keep all these options open." And I said, "That means an exponential [explosion] as we go forward." He did not understand the game we were in. We were funded to build an observatory, not to do independent research. And that is a discipline he could not accept.... He definitely felt I was a threat to the realization of his dream.

Whatever the rights and wrongs of the issue, it did lead to the removal from the project of the man who had been in it longer than almost anyone else, who had invented more of its discrete major components than anyone else, and whose whole life circled around it in a more intense and one-dimensional way than anyone else's. Unsurprisingly, the events were reported in the scientific press and beyond. For example, a account was carried by *Science,* which reported that Drever had been "thrown off the project, forced to turn in his keys, kicked out of the lab, and told he was persona non grata," says one Caltech faculty member familiar with the events.... Within hours of the dismissal, Vogt sent out an e-mail letter to the LIGO community saying that Drever was no longer associated with the project, would be allowed to remove his personal belongings from the LIGO offices only under staff supervision, and had been instructed not to enter LIGO premises or disturb project scientists. (p. 612)

The article went on to explain that in September 1992, Drever had filed a complaint with Caltech's Academic Freedom and Tenure Committee chaired by Steve Koonin and that its report, delivered in October 1992, sided with Drever.

Drever's separation had been without due process . . . Drever's academic freedom had been infringed when Vogt, in the words of a committee member, "strongly discouraged under threat of separation" Drever from attending two scientific meetings in which he was scheduled to talk about gravitational radiation research. Drever ignored Vogt's second warning and spoke at a meeting in Argentina. On his return Drever was fired.

Despite the committee's report in October, Caltech did not immediately reinstate Drever to the LIGO team, and several faculty members launched an effort to force the action. "I am really disappointed with the way the administration handled the situation" says Goldreich [Peter Goldreich is a Caltech astrophysicist who helped Drever prepare the case]. The committee report pointedly did not call for Drever's reinstatement, however—a deliberate omission, say committee members who talked to *Science*. The reason: Although the committee agreed that Drever's firing had been handled inappropriately, it could not decide whether LIGO had cause to remove him from the project. (p. 613)[8]

Reports like this also provided the opportunity for those who had expressed doubts about LIGO's initial funding to resurface. Thus, the same article quoted Jeremiah Ostriker, one of LIGO's most relentless critics, as saying, "I think it is a very large expenditure for a project that, according to its current specifications, has a small likelihood of detecting astronomical sources" (p. 614). The Drever affair could, then, have spelled the end of both Vogt's directorship and the LIGO project.

8. Travis 1993. I believe the story was also picked up by *Newsday*, but I cannot locate the reference.

CHAPTER 33

THE END OF THE SKUNK WORKS

In the public eye, Vogt, and LIGO as a whole, came out of very badly from the Drever affair, but in the end the Caltech management stood behind its chosen director. Vogt might have survived a lot longer had he not also been arguing with the NSF over the same kind of issue that had caused Drever to run afoul of him. This issue was the lack of transparency of Vogt's version of the skunk works. As Vogt told me,

> [1996] Here came a big philosophical difference between NSF and me. The way I had costed LIGO and the way I perceived LIGO—what it would take to make it work—it could not be done in the conventional style of a big project like CERN [European Organisation for Nuclear Research], for example. And I approved of the way CERN was built. But LIGO is a totally different animal. It is totally immature science; everything is improvization, and you cannot run it like a mature science project.... [The skunk works people] were able to do it because the bureaucracy had been completely eliminated. They gave them the money, they locked the door, and they said, "Do it," and it gets done.

Thus Vogt, at least initially, would not or could not submit to the level of outside monitoring that the National Science Foundation thought necessary to satisfy itself and the wider community. The skunk works ideology, at least as Vogt understood it, stresses lack of supervision, and Vogt seemed determined to keep outside interference to a minimum. The closed nature of the project to outside supervision was one of the chief factors leading to the next management upheaval. As we see in the quotation, Vogt justified the absence of planning transparency wanted by NSF on the grounds that the science itself was still developing and was not yet the kind of mature enterprise that could be run under a strongly coordinated organization. The relationship between Drever, Vogt, and the NSF was, then, one of transitivity: Drever drove Vogt to distraction by insisting that the science was too immature for designs to be frozen so soon, while Vogt made life impossible for the NSF by insisting that they did not understand how immature the science was and how unsuited it was to the kind of oversight the foundation said it needed!

Vogt seems not to have taken seriously the forces to which NSF was subject, even though he experienced them at first hand when he fought to get the project funded. The widely publicized debate over funding had made LIGO highly visible to outsiders. As a result, LIGO had felt it necessary to tame the resonant bars, and now the time had come for LIGO itself to be tamed. LIGO's need not to make mistakes, and to be seen not to be making mistakes, made it necessary to have every step of the program scrutinized, especially now that the Drever affair had thrown it into the limelight yet again.[1]

There have been only short periods when the LIGO project has felt financially secure. From the outset, plenty of scientists believed LIGO could not do what it promised. Still others thought that, even if it could be done, there were better ways to spend the money. Indeed, the LIGO management's first confrontation with a congressional committee resulted in a

1. Vogt said that his initial budget had to be cut to the bone to give LIGO any chance in the atmosphere of hostility in which it was first funded. This limited his freedom to employ outsiders such as accountants. Sapolsky (1972, chap. 4), taking the development of the Polaris submarine as his case study, describes the cost of providing management systems and their ineffectiveness even when they are believed to be all important. He, argues, however, that such systems are vital to protect a technological program from too much outside interference; they allow outside agencies to feel secure whether or not they are truly effective. This is what Vogt missed. But it is true, as he points out, that Barry Barish, his successor, was able to expand the LIGO budget considerably, having no responsibility for the Vogt budget put together in collaboration with the NSF. This allowed Barish both to pay for management systems and for more science. This accounts for some of the subsequent expansion but certainly not the growth of the worldwide collaboration which was accomplished under the much more open Barish regime.

resounding defeat. These dissenting scientists had not gone away, and they were not reticent about voicing their criticisms in public. As Mark Coles's remarks remind us (chapter 29), at the forefront of the mind of every physicist of the era was the cancellation of the Superconducting Super Collider project after hundreds of millions had been spent on tunnels, magnets, and other hardware. If that much money could be thrown away, there was nothing to stop LIGO being cut off at the knees even now that it was funded. Worse, written into the project from the outset was the idea of subsequent generations of advanced detectors, which would need still more millions of dollars. And LIGO, because it was so big by NSF standards, seemed certain to take funds from other NSF-supported scientists who would not be concerned with much bigger projects supported by the US Department of Energy or NASA.

How different the concerns of Congress were from those of the core scientific community can be gauged by some of the recorded comments made in the congressional hearing pertaining to LIGO funding. Thus, one representative congratulated Vogt on the earlier work Caltech had done on neutrino detection. He spoke of the LIGO experiment being conducted "to do about the same kind of thing, to determine the decay of neutrino stars."[2] Another member seemed to think that LIGO might reinforce the Hubble telescope. Another, referring to the black holes that, it had been claimed, LIGO would help us understand, asked, "When you get near one of these black holes, you made reference to—and anyone can answer this one—warped space-time—Do we have a sense of what this is, and can we commercialize that in any way?" It is this very "inexpertise" that channels the rest of the world directly into the core group of scientists when science becomes expensive and public. Ask for $100,000 and your application will be examined by your scientific peers with a fine-tooth comb; ask for $200 million and congresspeople without scientific qualifications will be the ultimate judges. Given that LIGO was still vulnerable, and that it was always going to be in a position where it had to ask for further allocations for advanced models, one can understand why the NSF felt that the highest standards of oversight were imperative. Even if a case could be mounted that Vogt's skunk works was an efficient way to build LIGO, NSF still had to be *seen* to be doing everything to make sure that not a penny was wasted, and that ruled out the skunk works approach.[3]

2. There is no such thing as a neutrino star, and neutrons and neutrinos are very different things.

3. See also Riordan 2001 for the same sentiments regarding the Superconducting Super Collider. That the pressures on LIGO have not gone away can be seen in recent press reports even as

THE END OF VOGT'S REGIME

Lest the emphasis on public accountability seem too cynical, it was also the case that the NSF felt it had reason to worry about whether the necessary degree of planning was actually being done. And there was another aspect to the problem. As we have seen, building a loyal team means defining an inside and an outside. Vogt's version of LIGO was not very good at attracting outside talent, and a project of this size had to grow if it was to succeed. Vogt simply did not have enough people in place to finish the job and did not seem able to expand the group fast enough. We will see that after Vogt departed, LIGO grew at a huge rate, and its current and future success has a lot to do with the amount of outside talent it has drawn into its orbit. But of the skunk works, an NSF spokesperson said to me,

> [2000] [The trouble was] his absolute refusal to open up the project to the scientific community. He had nothing like enough people to carry out the R&D [research and development] for all of the aspects of the science—to make it succeed.[4]

As matters came to a head, NSF appointed review committees made up of independent scientists which met in 1992 and 1993. They were forthright in their criticism of the project's lack of a clear management plan with milestones and its failure to recruit scientists from a wide institutional base, even feeling the need to insist that MIT be given more of a say in the management. What Vogt felt was that

> [1996] NSF started to harass me. They wanted a management plan. I wrote a
> management plan. But they said the management plan wasn't
> satisfactory.... Anyway, they had decided to get rid of me.
> *Collins:* Why was it that NSF wanted you out?
> *Vogt:* They wanted to cover their backs. And you do that by producing the
> paperwork. And I said, "I am not budgeted for that and I am not interested

---

LIGO is asking for funds for an advanced detector. In early 2002, a long article in *Scientific American* (Gibbs 2002) and another in *New Scientist* (Battersby 2002) described the project in less than entirely flattering terms. A number of the LIGO physicists were convinced that these had been inspired by one or more of those who had been most vociferous in attacking LIGO's initial funding. (And I can confirm that such sources were at least consulted during the preparation of the articles.)

4. Much of my account turns on the way that advanced LIGO science is now being done by the LIGO Scientific Collaboration—a far wider group of scientists than is employed directly by LIGO. An NSF spokesperson insists that NSF pressed for this kind of arrangement from the start and that from their point of view the evolution was less unplanned than my account implies.

in doing it. And for that I take full responsibility." I was not willing to give them what they asked for.

Vogt insisted to the NSF that all the plans were in his head. NSF felt that the success of LIGO was now being compromised by Vogt's approach; it could not know for sure whether the lack of paper plans was to be explained by Vogt's refusal to "do the paperwork," which was the counterpart of the plans in his head, or by a real absence of plans. If there were no plans at all, that could mean a still greater disaster than if it were simply that there were no plans for outsiders to scrutinize. The energy-sapping row with Drever had two consequences for the NSF's disagreement with Vogt. First, it drew Vogt's attention away from the project, so that he was even less able to provide a satisfactory account of LIGO to NSF's committees, and it certainly came to appear to the foundation that no satisfactory plans were in existence either inside or outside Vogt's head. Second, it weakened the Caltech management's determination to stay loyal to Vogt.

Eventually, NSF started to freeze the flow of LIGO money into the Caltech project, and the Caltech management—which had supported Vogt through the Drever row—folded. Vogt, as he told me, became the first principal investigator of a Caltech project ever to be fired from his job. But it should be noted that in spite of Vogt's defeat, Drever was never offered reinstatement on acceptable terms. There were no clear winners—everyone was, at best, maimed—but Vogt did not lose the battle with Drever so much as he lost the battle with the NSF. The Drever affair was not what brought Vogt down; though it may appear this way from the outside, it merely weakened his and Caltech's position in the argument with NSF.[5]

Vogt said that NSF "wanted to cover their backs," but to reiterate, this negative way of putting it fails to take account of NSF's position as the funder of its largest project ever. NSF does not generate its own income, so it is not completely autonomous. It may look autonomous to the scientist working on a small project, because each small funding proposal represents only

---

5. Richard Isaacson, the NSF director in charge of LIGO funding, terminated the NSF grant of Joe Weber, his thesis committee chair. When I described the Drever incident, I mentioned Greek tragedy and Oedipus in particular. Picking up the theme of Greek tragedy, a respondent suggested that Drever could be likened to Philoctetes and Vogt to Agamemnon. Philoctetes was a master archer who was betrayed and abandoned on a desert island when a snakebite festered and no one could stand his cries of pain; he was rescued when it was prophesied that only in his presence could Troy be made to fall. Agamemnon, to cut a long story short, had a tendency to overreach himself, was forced to sacrifice his daughter (Drever?) to win a favorable wind for his invasion of Troy, and was murdered by his queen (the NSF?). The queen would in turn be murdered by his son, so, presumably, the NSF needs to watch its back.

a small part of the foundation's budget, and it will never be scrutinized by anyone outside the scientific community (Senator Proxmire's "Golden Fleece" awards and the like aside). In the case of LIGO, however, because it was such a big project, NSF was itself a client of Congress and itself a potential victim of its own constituency. Not to be seen to be guarding its financial responsibility carefully would be to invite disaster, not only for the foundation but for the project itself. NSF custom and practice made it even more necessary for everything that happened in LIGO to be *visibly* under control and on a path to success. The Lockheed Skunk Works was not comparable in this regard: defense budgets are customarily vast, defense work is much further removed from the public domain, and defense projects do not compete for funds with a mass of university scientists each needing a hundred thousand dollars or so. As for other big sciences, high-energy physics is mainly funded not by NSF but by the US Department of Energy and works within a context where the typical project is large; NASA, too, is big science through and through. In these respects, as I explained at the outset of this section, LIGO was different.[6]

Even this way of looking at things casts NSF in too passive a role; it is not just a matter of reacting to outside pressures, because parts of the foundation, such as the physics directorate, take an active role in nurturing the science in their domain. Thus, from the outset, Richard Isaacson and others in the NSF management were committed to LIGO as physicists as well as administrators. As one central scientist put it,

> [2000] The thing that has been absolutely constant through this is the NSF's commitment. You know people always write about—you know—the scientists have these ideas—but it turns out in this case the "bureaucrats" had a tremendous influence on this thing. They were the marionette drivers—here, we were the puppets and they were pulling the strings on us. "Here—do this. Why don't you get a proposal. And why don't you do that." And they had their idea of how they would manage to get this thing through. It's a little scary—here you think you are an independent person, and here's Isaacson pulling this string and you move that foot, and "oh, wow!"

And so it would be wrong to think of NSF as merely battling Vogt for power as in a boardroom struggle. As the foundation saw it, it was trying

---

6. Of course, even high-energy physics and NASA are vulnerable to public scrutiny, as the cancellation of the Superconducting Super Collider and the reaction to the failure of the Mars probes reveal.

to shape LIGO in a way that would ensure the project's success both scientifically and politically. Vogt, perhaps encouraged by his experiences in lobbying congresspeople, believed that NSF was being unnecessarily intrusive and that he understood the political terrain better. And just as he wanted a loyal team within the project, he wanted to be trusted by NSF as a loyal team player. He felt that its demands for transparency revealed a lack of trust in his integrity, capability, and leadership skills and that they would lead to inefficiency. It might or might not have been possible to sustain this position in terms of the science—scientists have told me they believed that Vogt's regime would have succeeded and that good science would have been done under his leadership—but this, to reiterate, is to miss the point about good work *needing to be seen to be done* in an NSF project of this size. Without transparency, the whole project may never have lasted long enough to reveal whether Vogt's scientific management strategy was viable in the long term.

I have argued that a crucial step in the transition from autonomous to coordinated science is the passing of control over the freezing of designs from the scientists who invent them to the managers of projects; this was the step that the Vogt regime instituted. But a second necessary change, as a scientific project grows large, is the passing of a measure of control still further up, to agencies which have more contact with the wider scientific community and the political sphere; this is the consequence of the wider sharing of risk as a project grows. This was the step that Vogt would not take, and that is why his position and Drever's were like Russian dolls: in the middle was an unhappy Drever, trapped by the embrace of Vogt; but equally unhappy is Vogt, suffocated by the NSF. And the NSF? It is trapped inside the Russian doll of politics and its own constituency.[7]

Some later events show still more forcefully how important it is to make and meet clear plans in a project of this kind. In 1999, LIGO I was on schedule as far as facility building was concerned; but the interferometer was still incomplete, so it was not obvious that it would meet its design targets. And yet the team was already putting forward plans for spending $100 million on the more sensitive interferometers to be installed for Advanced LIGO. The demand for financial support was coming at this time because of the long lead time needed if enough research and development, and the fabrication of enough major parts, was to be completed for installation to begin in 2005 (the date still "on the table" in 1999). What could the LIGO

---

7. For a recent discussion of science funding in America which bears on some of these issues, see Guston 2000.

I team offer in the way of justification to the potential funders of Advanced LIGO? They did not have any scientific results, nor had they fulfilled any of the really demanding design goals that turn on the interferometer—they would not be in a position to reach their sensitivity and duty-cycle targets until 2002 at the earliest. Nevertheless, to wait for demonstrable results before pressing ahead with the next generation of detectors would have slowed the project to the point of risking the breakup of the large team of experts, and might have wasted years running what could be a technically sound device but one which would be too insensitive to detect gravitational waves. What the LIGO I project could offer, however, was a display of immaculate planning and management, with each preset technical milestone having been met (even though these were facility-building milestones only), and with each aspect of the project having been brought in on time and on budget. It was, in part, this display of management responsibility, virtuosity, and transparency in facility building that provided the justification for *considering* what might otherwise seem an outrageous request: still more big money for an unproved science.[8] Now let us see how LIGO regained so much credibility.

8. For a discussion of the way the basis of science funding decisions at the highest levels is shifted from esoteric to more mundane grounds, see Turner 1990.

CHAPTER 34

REGIME 3: THE COORDINATORS

FROM EXPLORATION TO ROUTINIZATION

In 1994, management of LIGO passed to a new team. Robbie Vogt was kept on as leader of the detector development group, though he was to sever his connection with the project entirely after another couple of years. Kip Thorne continued as the unofficial chief theorist of the whole enterprise and Rai Weiss, too, kept a principal role.

Ron Drever was offered the choice between reinstatement in a nonsupervisory role or a sum of money that would enable him to set up a separate 40-meter interferometer on which he could conduct independent research in another part of the campus. He chose the latter and continued to work on any idea he believed might bear fruit in the long term (for example, various designs of magnetic suspensions for the mirrors, an approach no one else found interesting in the face of the pressing work on the detectors that were to be built in the next few years). Drever now represented the quintessence of autonomous science—psychologically, temporally, and culturally—all three of our cognitive dimensions:

he could work as intuitively as he liked; he could press ahead with his own project for as long as he liked with no more external scrutiny than any other small-project scientist; and he was isolated from cultural pressure to conform to the LIGO team's taken-for-granted ways of doing things.

The overall leadership of LIGO now passed to Barry Barish, a Caltech particle physicist. Barish signed up for a half-time post on LIGO, wanting to keep up with his other work in neutrino detection. He brought with him another former high-energy physicist, Gary Sanders, to be his second in command and project manager. High-energy physics is the prime nursery for coordinated science skills in the modern scientific world, and the Barish-Sanders team had both the experience and the will to give NSF everything it wanted in terms of planning and transparency; [1] to Barish and Sanders, this pattern of working came naturally. The walls of the Caltech corridors came to be dominated by Gantt charts and the like, showing the critical paths that the project was to follow; the targets, milestones, and responsibilities of all the project members; and the times allotted for the completion of each phase of the project from beginning to end.

The new team inherited some goals from the existing regime and were clear about some new ones; there was now no chance of ambiguity. LIGO was to be assembled by 2001, it was to undergo a year's shakedown, and two years—2003 and 2004—would be dedicated to gathering scientific data with the interferometers. The sensitivity of the instrument and the expected duty cycle were taken over from the previous team and set out as promises. The duty cycle—the proportion of the time that the complete two-site detector was expected to be fully functional during the course of the two-year "science run"—was to be 75%, though only one year's worth of data was promised.[2] Of course, no promise could be made about the detection of gravitational waves, because the project managers could have no control over the sources.[3]

1. Barish had already served on a Caltech LIGO oversight committee, so he knew what was wanted. For a discussion of aspects of the culture of high-energy physics, see Traweek 1988; Knorr-Cetina 1999.

2. The difference can be squared, because there are scheduled downtimes for routine maintenance. The 75%, I believe, includes these downtimes, leaving 50% of time "on air."

3. Conservative calculations suggest that only with a great deal of luck will LIGO I detect any events. Rai Weiss, nevertheless, expressed optimism; he said that since, in electromagnetic astronomy, every time a new instrument went "on air" with increased sensitivity unexpected sources were discovered, the same would apply to LIGO I, it being orders of magnitude more sensitive than previous gravitational wave detectors. Whether Weiss turns out to be right or not, the argument seems fallacious: electromagnetic instruments start from a baseline of success, whereas gravitational wave instruments have never seen anything—we are multiplying by zero, as it were. Were the argument not fallacious, one could argue that Joe Weber's original instrument was likewise bound to have detected something because it was so much more sensitive than anything that had

LIGO's goals were now laid out with sufficient clarity for some scientists to suggest that the new leadership had effectively redefined the project's aims; it was no longer designed to initiate gravitational astronomy but to achieve reliability at a level of sensitivity that was too low to detect a significant flux. Since, from the beginning, LIGO's claim for funds had been premised on the need for an upgrade to do real gravitational astronomy, this apparent change of definition, if it took place, was subtle; it was not so much a change in the technical specifications for LIGO but a change in emphasis in the priorities for research and development.[4] This goes back to the balance of choice between freezing designs and building them versus continuing with exploratory science. One commentator expressed the problem well.

> [2000] I think there's an intellectual separation of function. If you can know
> that you're building something—an instrument—then you can get on with it.
> If you're trying to do the science, then you get into the [named person]
> confusion of always wanting to do the science and not knowing exactly how
> to build the best instrument. It's that tension. And so you don't have clear
> plans and goals that [can be implemented] at a given time, and you keep
> having more ideas. Both are necessary, but once you start on a large
> construction project, you need blueprints.

Of course, the long-term aim, to establish full-scale gravitational astronomy, was unchanged. Throughout the building of LIGO I, the potential for installing more-sensitive generations of interferometers had to be borne in mind.[5] It was just a matter of balance. For example, a decision was made to use single-stage pendulums in LIGO I even though multistage pendulums were under development. If the primary goal had been to get to a flourishing gravitational astronomy as quickly as possible, irrespective of financial and political context, then LIGO I *might* have been held up while more complex pendulums were researched. I do not say this would have been the right technical choice, just that choosing not to do it was the

---

gone before. Weiss's response when I made this point to him was that only now were we reaching levels of sensitivity that made the argument reasonable.

4. For discussions of the way success or failure of a technological trajectory can be redefined according to the way the meaning of the technology is interpreted, see work in the "new sociology of technology." One could see any redefinition of the early stage of the LIGO project as a sensitive interferometer-building exercise in the terms of the "Social Construction of Technology" idea (see Bijker, Hughes, and Pinch 1987; Bijker 1995).

5. At the time spoken of as LIGO II, III, etc.

kind of decision that allowed critical scientists to think of what was going on as redefinition of the project. The new leadership's choices tended to increase the chances that LIGO I would meet its promised sensitivity and duty-cycle goals on schedule, though the promises were thought by some to be too unambitious. As Barish explained to me,

[1999] When we began, we were pretty much under the gun. NSF was at the end of its rope and the project had to get on its feet quickly, and one thing we had to do was make sure we could do it in a way that would meet its goals on time—technically do things as promised and so forth.[6]

[1996] We listed a whole bunch of things that can be improvements. Why didn't we put them in the original LIGO—why? Simple! If we had put them in the original LIGO, they probably the wouldn't work; we'd spend all our time failing in another way, and that is, you turn this thing on and you don't know how to make it work, because you've screwed up on all these things. Because to go from an idea that you know is going to make a more sensitive detector to something that will really work well enough to work in this environment takes a lot of testing and finding the flaws with it, and so forth—in a friendly environment, not an unfriendly environment.—By unfriendly, I mean in the real LIGO, making this huge monster work.

So it has been our feeling that we have to be as absolutely conservative in every way in putting this thing together with what the NSF gave us. The worst thing would be to put together something that doesn't work. That's a much different kind of failure than putting together something that works, but the signals are weaker than that, so that you don't see signals.[7]

This approach made it possible for the LIGO I construction process to be moved toward a coordinated style, moment-to-moment judgments based on unfolding experience replaced with preset plans and rules of procedure. Barish and Sanders laid out plans, or set about trying to develop plans, for

---

6. This is certainly true. The project had got itself into serious trouble with the collapse of the Vogt regime and the adverse publicity over the sacking of Drever, its most creative scientist. It would have been easy for the project to have been shut down at this point. NSF spokespeople have told me that they were scrupulously honest with congressional representatives throughout the episode and felt that they were lucky that Congress had decided to do the right thing—partly in response to the display of integrity.

7. This interview continued, however, to stress that about 10% of the budget should continue to be spent on Advanced LIGO so that it could be put together as soon as LIGO I's science run was ended.

both facility and interferometers. The organizational approach associated with this transformation was described by Barish as follows:

> [1999] So we tried not to be particularly clever and to set up an organization that didn't take into account the peculiarities of the way physicists like to live or researchers like to think about problems, but was much more what I would call an organization which was put together to build a bridge.[8] I've said that from the beginning, it's completely traditional—it's exactly the way a construction company would go out and put together a complicated construction project—somebody on top—some people at the next level that report to them—people at the next level below that report to them. That at each of those levels the people have specific tasks, responsibilities, budgets, schedules, and so all we did was act like we were bridge builders. That's the scheme that works best even for building physics experiments, believe it or not, because I've seen a lot of them built, but it usually isn't done quite so rigidly, because you do like to take into account the fact you want some flexibility, and new ideas, and this and that. We pretty well shut that out, although we did change some things along the way. . . .[9] But mostly we just put our blinders on and moved ahead.

Barish and Sanders, then, changed the managerial flavor of LIGO when they took over. We can say with confidence that on the scale of autonomous to coordinated organization, Barish and Sanders were at the opposite end to the Troika, with Vogt somewhere in the middle.[10] Barish and Sanders, it must be said, were aware that their "bridge-building" approach would not see the whole project through. As Barish explained to me, [1999] "We knew from the beginning that that was not at all the kind of organization that you'd want when you start doing research, where the research in this case means turning on an interferometer." And by 1997, Barish had in place

8. This conversation was conducted in the context of a mutual recognition between interviewer and interviewee that there had been many complaints from disaffected physicists about the Barish-Sanders style (see below).

9. The wavelength of the laser was changed a year or two into the Barish regime.

10. The need for "repairs" remained, of course: there were problems such as surface treatment of the steel, leaks in the joints, failing gate valves in the vacuum system, poorly designed light baffles, and rubber parts that exuded water. These either could not be foreseen, or had to be solved by methods that could not be worked out in advance. (The problem of outgassing and the bake-out system for the vacuum pipes had already been solved while Vogt was in charge, and the contract for the manufacture of the beam tubes had been awarded to the extremely impressive and proficient Chicago Bridge and Iron.) But these are the kinds of breakdowns in routine that high-energy physicists are used to dealing with. For a very detailed discussion of the routine and the nonroutine in human actions and organizations, see Collins and Kusch 1998.

another kind of organization, referred to as the LIGO Laboratory as opposed to LIGO Project. This was to have more of a traditional R&D (research and development) flavor. The Laboratory began as what he called a "shadow organization," running in parallel to the Project with initially very few personnel. By the end of 1999, the facility had been completed, and all the staff had been absorbed into the Laboratory.[11]

THE SANDSTORM

Barish and Sanders had in mind, then, that the upheaval involved in moving to a bridge-building style of work would be a temporary one. Sanders said that he talked to the team and told them that

> [2000] the LIGO project is a sandstorm passing through a group of scientists. You have to hunker down and shelter yourselves against the pressure of the sandstorm and ride it out, but the sandstorm passes, . . . the sun comes out, and we act like scientists.

The move from the Vogt to the Barish-Sanders regime is most easily examined through the conflict that arose between the new management and the scientists at Caltech who were working on the 40-m interferometer, the detector built by Drever in the 1980s and rebuilt during Vogt's tenure.[12] This interferometer was the biggest prototype in the world, and it was this device that was initially taken to be the principal research and development

11. This involved no physical movements of personnel—merely a change of nomenclature and approach. Barish told me that the Project could not have been smoothly evolved into the Laboratory, because the former was such an "extreme" organization. Therefore, the Project ran by itself for three years with the Laboratory beginning afresh only in 1997.

12. We are talking of about half a dozen scientists here, but this is still a large number in the context of this kind of research. One mundane element in this case is that most of the scientists who left the 40-meter went to work for the neighboring Jet Propulsion Laboratory (JPL), a move in employment that was easy in that it demanded no upheaval in domestic life. JPL is in Pasadena, a few miles from Caltech, and comes under its overall administration. Because of the way it is funded, however, it can often pay higher salaries to research scientists than Caltech itself. Caltech staff and JPL staff often meet, and they can easily discuss their respective employment prospects. For these reasons, there is some bitterness on the part of Caltech about the ability of JPL to poach staff. If JPL had not been so close at hand, fewer people might have left LIGO, but the discontents would have been the same. One or two of the other scientists who left the project went to established faculty jobs elsewhere, and this is the kind of thing that is bound to happen—it had nothing to do with the push from within Caltech. MIT, situated 3000 miles from Caltech, was more or less insulated from these organizational upheavals and, in any case, rightly saw the change of regime as a positive move as far as their role in the project was concerned.

tool for LIGO. It was also intended to be the training ground for the growing team of LIGO scientists.

The members of the 40-meter team were the most experienced sizable collection of interferometer scientists in America, yet under the Barish-Sanders regime many of them were soon to resign from the project. The prospect of the promised calm after the sandstorm was not appealing to them. They believed, probably correctly, that what Barish and Sanders thought of as calm weather would still be stormy by the standards of the autonomous science they were used to.

Changing Work Practices

To some extent, the flight of the 40-meter personnel can be explained by mundane causes. When new management takes over an organization, existing staff will often feel that their positions are threatened. Barish and Sanders fired some of the junior technical staff who they thought were not doing the right kind of job, upsetting the 40-meter team scientists to whom they had previously answered.[13] And Barish and Sanders brought in their own people and placed them in authority over the 40-meter scientists.

This new way of organizing things was painful for at least some of the scientists. In 1996, one who is still with the project told me,

> [1996] Robbie met with the scientists every week—at least the Caltech scientists—met with them every week for a couple of hours—talked with them about the project, how it was going, and so on. The engineers primarily reported to [others], who then reported to Robbie. If you now look at the organization, you find . . . the facilities . . . engineers reporting directly to the project management.
>
> The scientists [now] report to [others]. And so, if you look at it, the scientists have been pushed from this rather privileged role, where Robbie viewed them as, in a sense, stockholders in the enterprise—important people, people who were going to inherit this facility and use it. They've been pushed progressively down in the organization to where the average scientist probably doesn't come in contact with [the director] more than two or three times a year. It's actually fairly rare for them to have a one-on-one conversation with [the director]. . . . And so the scientists have seen

---

13. This was for technical reasons having to do with their lack of experience in the digital electronics that the new regime wanted to see installed (see below). Nevertheless, it was experienced as a loss of control by the scientists who were previously supervising them.

themselves, I think, pushed way down in the organization.... There is a significant undercurrent within the scientific group of dissatisfaction—a feeling that they're cogs.

And so—Robbie's administration was not perfect. I think many of the engineers felt under Robbie somewhat oppressed, that they weren't given empowerment—authority to do the jobs that they wanted—the appreciation and so on—but Robbie was very good with the scientists. I think it's flipped the other way in many ways.

A scientist who did leave told me that where Vogt had kept scientists on his side by telling them they were having more fun (doing more-exciting science) than the engineers, he had a different experience under the new regime.

[1997] There is no attempt to allow scientists to have fun under the current structure—just the opposite. If I were ever to try to model [the] management philosophy, I'd say [they measure] success by how miserable people are. The more misery the better.

Another scientist who left was bitter about the changed work practices.

[1997] It was like an atmosphere where you had the feeling that they expect results according to a schedule. And that was on a background of a prototype 40-meter, and nobody had ever attempted to get a machine of that complexity running.

... I remember one day when [the project leader] said something like, "I do not really care what you guys do, I just want to make sure that at any given point there are at least three people in the lab. I don't care—you can sweep the floor—I want three people in the lab." And that's about his approach.

... I worked two-and-a-half years on the 40-meter.... In the beginning and in-between, I would say I spent a good part of my life in that lab. Like there were numerous evenings when I did not get out before midnight...just because it was necessary to understand what was wrong if something was wrong. One day I got an email from the project manager that he is putting a schedule in: from 4 to 8, whatever. "The 40-meter has to be used properly, and you guys have to start working." So I went to him and I said, "You know what..., in case it interests you, that schedule you just distributed, should I adhere to it, I would be very happy, because it is less than I currently work, so you just cut three hours out of my work."

A third member who left told me, as quoted earlier, [1997] "[T]here are times for doing things and times for seriously thinking about what you are doing." The new project manager, on the other hand, saw it rather differently.

> [1997] When I came to LIGO, I saw a group of people, and I frequently said to people—"This looks like a battered wives' shelter.". . . .
>
> What I saw, then, was people coming in at 9:00 or 9:30—turning on the laser, which then took an hour and a half to warm up, and then working for an hour or two to try to tune something, and then breaking for a one- to two-hour lunch, during which time some people went off to the gym and swam for an hour and then sat out on the veranda at "The Greasy" [Caltech's refectory] for an hour, greeting people, and then maybe back in the lab at two o'clock, and maybe work for a few more hours and left around 5:30, and turned everything off, so that it cooled down and lost its state. And then the whole thing began again. So in a 168-hour week, the number of hours in which real progress could be made—real head-scratching—was incredibly small. You don't do that on an experiment. All my life I've worked on experiments that when you got the thing on, there was an imperative in this apparatus that said, "Keep me in a state that's known, go in a sensible rational way from one state to another, study me." So [when] I began to press for lengthening the days, I also noticed that between nine and four in the afternoon, there would often be six or seven people, but at 8:30 there was no one.
>
> So I had a couple of meetings. And I said, "You, [named person], you're the senior person here. What I want you to do is, take your people and spread them out. Some people should come in at seven in the morning and work till three or four, and some should come in and work till nine or ten at night; and if we have enough people, we could work round the clock. Let's turn the laser on and never turn it off. Let's overlap by half an hour, these shifts. Let's have a 'handoff'—an intellectual handoff [at shift changes]—'What have you been doing,' and 'What are the hard problems,' and 'I can't get it to do this,' 'Let's see what you can do in the next eight hours.'"
>
> This is the way experiments work, if you've visited nuclear physics and particle physics experiments, a lot of astronomy experiments. There's a team—there's a relay race with a handoff, and the machine is on, and it's babied, and people understand its state, and people who are on shift know that something is going awry, and they understand the health of the thing, and they transmit the information on the health of the thing from shift to shift—OK? None of that was happening; it was allowed to go fallow every night—very inefficient.

I have to tell you, I have no regrets—none! none! none!—about what has happened [the eventual loss of many of the staff]. I tried to coach these people by presenting to them the way you would use six or eight physicists to extract the scientific best out of an instrument. I gave them models of how to put together a week, scheduling in times when you turn things off and do some maintenance; scheduling blocks of time when you do quiet experiments; these are very simple examples.

This procedure was seen by an ex-member of the 40-meter team in a different way.

[1997] [W]hile some of us were working hard in the lab trying to solve some of these technical problems, he was insisting on a rigid structure of schedules—seemingly arbitrary rules such as "No, we couldn't eat lunch together, because somebody had to be in the lab at all times"—somewhat anathema to the spirit of encouraging research. And these things left me puzzled, and I'm sure I left him puzzled as well.

Routinization of the Science

The industrial relations dimension of this dispute is intimately related with a much more complex set of cultural changes that the high-energy physics approach brought with it. The new leadership had a different conception of the science of gravitational wave detection and, in particular, a different conception of the science that lies at its heart—interferometry.

At the most general level, as a result of their experience, high-energy physicists, even more than other scientists, tend to a belief in the repro-ducibility of the world and the amenability of science to routinization. LIGO Director Barish's strong claim, quoted earlier, sums it up:

[1999] [R]unning these interferometers is not magic. It is amenable to analysis and systematic studies.

Project Manager Sanders, also from high-energy physics, said,

[2000] The thing that the high-energy community does is take a very complicated problem and bash it until it is under control, and it's reproducible, and then go ahead and make a million of them.... I even said to myself—I'm glad to be out of high-energy physics, because much of what you do experimentally is this kind of mass production. And with LIGO it is more a kind of one-off. But the level at which you have to take something complex and move it to something that's routinized is, in fact, the same.

On the other hand, a scientist who had been with the project a long time, and is still with it, said,

[1996] [M]uch of the history of this has tended to be a bit more artisanry than engineering, and perhaps more artisanry than science.... There is a lot of it that can be operated in that very schedule-driven, cost-driven, engineering kind of mold. The actual construction of the interferometers themselves has to just be a nice engineering problem. But there have to be some pieces of it that are...more artisanry, and it's hard to know exactly how to balance those things.

And Vogt believed the science was still immature.

[1996] [I]n other fields where I have worked before—in radio astronomy and high-energy astrophysics—when I introduced changes in the system, they worked. Because the system was mature; there was a complete mature discipline in place with a lot of experts, and if I didn't know the answer, there were hundreds of people I could have a conversation with and they put me on the right track.
...[But], for example, in the Caltech 40-meter, when we recombined the beam, we found that the noise is a factor of 2 above the predicted noise. You know, we have been working on that for 8 months, and it's still the same. You know, I believe it's easier to work on the 40-meter than the 4-kilometer. And so if we run into these problems in the 40-meter and we can't turn it on, there can be very severe repercussions.

Before returning to the details of this disagreement, I will set it in a larger context.

CHAPTER 35

MECHANISM VERSUS MAGIC

The 40-meter team's row with the new management, I have argued, can be seen as a symptom of the growth of LIGO from small to big science, but it was also an argument about the role of models in physics.

> [2000] *LIGO Physicist:* I was remembering an epiphany that I had. While I was a postdoc at MIT, I audited a class in the engineering department on modeling techniques. And the class was kind of a weird class, and I kept feeling like something of an outsider because I realized that every time the professor wanted an example of something—and if you were a physicist he would have said, "Let's treat the simple case of a harmonic oscillator this way"—the guy would always say, "Let's treat an internal combustion engine this way."
>
> Nevertheless, I had one of my deepest revelations about the nature of science listening in his class one day when the professor said, "No model is complete." And when I first heard it, I thought this is just outrageous—[I thought] this is just an argumentative thing. It was only later that I realized that not only was it correct

but it was deep, and it was the kind of deep statement that I'd never heard any physicist say. Though good physicists know it, it's never taught. And the kind of opposite thing is taught by implication in the way our physics courses are set up. And it is taught by saying, "Assume you have an X," and it lays out the model for you, and you solve the equations. And we seldom ask, in the real world—in this thing we care about—Is the model right?

I think experimenters in general are good at this, and at least good experimenters are able to pose a bunch of questions, and list a bunch of worries, and start to ask, "OK, what should I include and what not?" And so a lot of people in our community do this unconsciously, but I was still astounded by hearing it said for the first time and realizing that I disagreed with it.

And then I recall you [Collins] at the table [he refers to an earlier conversation, where similar points were discussed] making a very apt remark referring to the biggest controversy in your management paper [a draft of these chapters], when you put on this impish grin and said, "Interferometers aren't magic." And, of course, you were meaning to [bring up the argument between Barry Barish and Bob Spero (a key member of the 40-meter team) which bears on the point]. And there's a sense in which that's right. And there's a sense in which each of them in their quotes that you got from them staked out too extreme a position that I don't think either of them would really defend calmly if they weren't angry at each other.

When Barry says that interferometers are not magic, he does not mean that the first model you are going to write down will contain everything, but he does mean that you should try modeling and keep beating on the model until it agrees. And Bob similarly, when he gave the impression that only deep intuition could do it, didn't mean that models are of no use, though he actually kind of lived his life without paying much attention to modeling, though I'm sure he wouldn't defend that.

Collins: So this is a kind of megatension. Sometimes physicists talk as though the world can be modeled, and sometimes not.

Respondent: That's right, and sometimes conversations don't make sense unless everyone shares the knowledge of "Are we talking about a model or are we talking about the real world?" And I had my epiphany about that, or at least about how people tend to live more in one thing than the other, after a physics colloquium at Syracuse some years ago on black holes [coincidentally, one of the things discussed in chapter 24]....And I walked out of the room after the colloquium with my friend..., and I asked him, "Do black holes have such-and-such a property?"—I don't remember what it was.

And he turned to me and said, "Do you mean in classical relativity, in string theory, in quantum gravity, in super gravity?"

And I said [respondent's heavy emphasis], "*No, in the real world do they have this property?*"

And then he laughed and he said, "I have no idea!" And at first I wanted to strangle him, and then I realized there was a learning moment there.

The argument about the role of models is either a big one or a small one. The small argument is that while we will one day be able to model the world well enough to build devices like gravitational-wave detecting interferometers from plans and schedules, we have not yet reached that point. This, I believe, is what the 40-meter team believed.

The big argument is about the nature of our relationship with the world. As I put it in the introduction, it is a tension between the world seen as an exact, calculable, plannable sort of place, just waiting for us to get the sums right, and the world seen as dark and amorphous, bits of which, from time to time, we are lucky enough to catch in our speculatively thrown nets of understanding, if we throw them with sufficient skill. The invention of modern science can be said to be the invention of the idea of phenomena—underlying relationships that stay constant and drive the world like a clockwork mechanism, hidden only by the welter of particular circumstances in which they occur. The job of the scientist is to strip away the particular, leaving the general exposed. In due course, the phenomenal nature of even what once seemed particular will become clear—the noise will become signal, and science will advance another step toward understanding the mechanism of the world. We might think of gravitational-wave detection science as a material metaphor or model of this idea: the whole idea of LIGO is to isolate the detecting device from the world so that only the pure phenomena—the waves—remain to be seen. Where this cannot be accomplished, the job is to understand the remaining "noise" well enough to subtract it from the phenomena.[1] Furthermore, what was once noise will become phenomena in turn, until the whole machine is understood.

Science's success in this project has been enormous, and yet, every time it seems to be on the point of accomplishing the dream, something weird happens. The quantum refuses to be caught in the mesh of the net;

---

1. James McAllister (1996) expresses the point very clearly, describing the thought experiment as used by Galileo as the ultimate stripping away of circumstance and an approach which showed the way to modern science. The "Princess and the Pea" metaphor used in chapter 14 is apt when science is seen this way.

mathematics turns out to have irredeemable gaps; complexity threatens to sweep the whole net away; and who knows what surprises are to come.

Though it may seem odd to say it, these are small troubles, easily handled, compared to the big issue of the relationship of the net of understanding to human social life. It is we who make the nets and throw them, and it may be that we fool ourselves about how much of what is out there is waiting to be caught. Sociologists have revealed over and over again that an appearance of orderliness is often an illusion: it is maintained by overlooked but continual human "repair" of the unraveling net of order. Bureaucracy runs smoothly only when its operatives continually break the rules to smooth out the operation; replace the operators with nonhumans, or with humans trained only superficially, and inefficiency results.[2] Supposedly "intelligent" machines, meant to replace humans in various skilled roles, work only to a certain point and do not work at all unless the humans put right their continuing and potential disastrous errors. Devices such as spell-checkers or automated translation programs and speech transcribers are striking illustrations of the reluctance of human language to be caught in the net. Closer inspection shows that even arithmetical calculators—arithmetic being a paradigm of order—rely on us to smooth the link between the abstract world and the world we live in.[3] Perhaps the mechanical world exists only in our dreams.

Turning to science, sociology of scientific knowledge prefers the backward-running arrow of causality, from waves in social space-time to the phenomena they are taken to represent. Could it be that there is more to this picture than a metaphor? Could it be that order is invented and imposed rather than discovered? Is there something we could investigate that would tell us? No—because whatever investigation we dream up will leave ontology untouched. Repeated uncoverings of even severe disorder may be taken to reveal either something deep about the world or a mere shortfall in brilliance and temporary setback. When order is reluctant to emerge in interferometry, is it a matter of science's place in the universe, or is it that LIGO had not yet reached the stage of being amenable to forward planning? Is it that such order as there appears to be is fallibly imposed by us, or that the order in the particular bit of the world called hypersensitive interferometry had yet to be sufficiently uncovered? Is it that no model is complete, or that the model of the interferometer was

2. See, for example, Gouldner 1954 and Garfinkel 1967, chap. 6, which is entitled "Good Organizational Reasons for 'Bad' Clinic Records."

3. For a detailed and extended working out of some of these themes, see Collins 1990 and Collins and Kusch 1998.

incomplete?[4] Whichever, the backward-causality way of looking at things does highlight different aspects of scientific life, and these aspects show themselves in the practice of science; the sociological view redirects our gaze to things that might otherwise have remained below the horizon of attention. Such things are to be seen in the disagreement between the 40-meter team and the new LIGO management.

THE MESH OF THE NET AND THE TRANSFER OF KNOWLEDGE

Over the years, the sociologists' backward way of looking at things has thrown up such differences in other places. One is in models of the transfer of scientific knowledge. If the net and the world were matched to each other, then knowledge of the world would be easy to transfer. It could be captured in accurate, shared, symbolic ways of representing reality that could move freely among scientists, carrying all there was to know. Anything that was incomplete about a model would be unimportant. If, on the other hand, gaps in the models were a serious matter, then attempts to transfer knowledge symbolically would tend to fail. The bits that fell through the holes in the net of the symbolic representation would turn out to be vital.

Now, it happens that in the transfer of certain experimental abilities, the gaps in the model *are* important. Certainly in those cases of new and difficult sciences that have been studied in detail, knowledge seems to have to be transferred by the movement of people who know how to apply the models and the recipes *and how to repair the gaps in the net,* not just by the transfer of the information, models, or recipes themselves. Only when the typically human net-repairing capacities have been learned—the model here being the learning of a new spoken language—does the technology transfer. One of the early illustrations of this was the transfer of the skills required to build a new kind of laser. A study showed that in the early 1970s, successful builders of transversely excited atmospheric pressure (TEA) lasers were those who developed social relationships with other successful TEA laser builders, whereas those who had no such relationships failed.[5] To build a

---

4. I do not claim that the members of the 40-meter team themselves had anything philosophically this "big" in mind.

5. Collins 1974, 1985/1992. In that case study, all those who failed had no social relationship, but of course there could always be counterinstances; after all, the laser was invented by someone who had no one else to learn from. What we are dealing with here, however, is transfer of abilities, not reinvention.

TEA laser successfully, it seemed you had to learn a culture or language of laser building, not just follow a set of instructions. In sociological jargon, the model that fit knowledge transfer in the case of the TEA laser was an "enculturational model" as opposed to an "algorithmical model."

Much of what was to follow in the argument between the 40-meter team and the new management can be seen in terms of the extent to which each group believed, at that time, that knowledge about interferometry could be adequately captured in symbolic form. The 40-meter team did not write much down, but they could claim that this did not show that they had not developed a great deal of knowledge and skill pertaining to interferometery. The knowledge and skill, however, were embodied in their practices and discourse and could not be explicated. Their value to the LIGO project could not, then, be "read off" from what could be found in print, but was located in the uniqueness and extent of their experience—something that they would still be able to apply directly to work on the LIGO interferometers and/or transfer to the rest of the team through social interaction. The issue has been described before as a matter of the transfer of tacit knowledge.[6]

What Is Tacit Knowledge?

The idea that scientists have "tacit knowledge" was first introduced by the physical chemist Michael Polanyi. Tacit knowledge has been shown to have an influence in, among other things, the laser building we have just described, the development of nuclear weapons, biological procedures, and veterinary surgery. There is also a burgeoning literature on tacit knowledge and expert systems and other "intelligent machines," while the philosophy of tacit knowledge and the notion of its practice in general is substantial.[7]

---

6. I am not sure if the 40-meter team really would claim this. In a sense, I am inventing a kind of ideal-type 40-meter team and then inventing the arguments that such a group might use. Thus it is I putting this matter in terms of transfer of tacit knowledge, not the 40-meter team.

7. Michael Polanyi (1958) invented the term *tacit knowledge*. Collins (1974, 1985/1992) discusses the TEA laser. MacKenzie and Spinardi (1995) consider tacit knowledge and atomic bomb design. Jordan and Lynch (1992) and Cambrosio and Keating (1988) talk about biological procedures. For veterinary procedures, see Pinch, Collins, and Carbone 1996. For the idea applied to experts systems and machines in general, see Collins 1990, Collins and Kusch 1998, and, for example, the articles in Goranzon and Josefson 1988. For a critique of the idea, see Turner 1994. For a discussion of this critique and an analysis of the way the idea of tacit knowledge has been used in sociology of science, see Collins 2001a; and for more of this, and the roots of the idea of the importance of practice in Wittgenstein (1953) and the phenomenological tradition, see the rest of the articles in Schatzki et al. 2001.

Oddly, Jordan and Lynch do not use the term *tacit knowledge* in their discussion of continuing variation of craft practices in biological preparations, nor do they refer to the previously existing

For the purposes of this book, the idea of tacit knowledge can be developed by starting from the practice of scientists rather than from the usual philosophical arguments. It is the case that some scientists can do certain experiments or measurements, while others cannot. This might be because the ones who cannot succeed are bad at hand-eye coordination or other experimental skills; it might be because the unsuccessful scientists do not have the right equipment or specimens on hand; or it might be because they lack tacit knowledge. To make life easier and to avoid some philosophical deep water, I will *define* tacit knowledge in a simple way: knowledge or abilities that can be passed between scientists by personal contact but cannot be, or has not been, set out or passed on in formulas, diagrams, or verbal descriptions and instructions for action. Where transfer of tacit knowledge is a problem, it can sometimes be solved by an exchange of visits: experimenter B, who cannot accomplish a measurement or make a piece of apparatus work, will often succeed after spending time in the laboratory of already accomplished experimenter A, or after having A work for a period in B's laboratory.

Referring to the 40-meter team, it could be argued that their tacit knowledge of interferometry could only be preserved by keeping them together in the team, where others could learn by observing and talking with them informally about their work. Novices would learn the language of interferometer building by living in the community of fluent "interferometry speakers," and they could learn interferometry only in this way (barring reinvention), just as fluency in a spoken language can be attained only by living among fluent speakers of that language. To try to break this sentiment down, we can say that at least five kinds of scientific knowledge *can* be passed between an imagined accomplished scientist A and an imagined novice B by personal contact and not in written or other symbolic form. In the sections that follow, I am not making the case for the 40-meter team; rather, I am merely systematizing a case that could be made whether or

---

and well-known literature on the same topics as explored in the physical sciences. Cambrosio and Keating stress the way that scientists themselves use similar categories to describe the "artistic" and "magical" aspects of their work as those developed here. They claim that scientists' own categories are sharper and more useful than those developed by sociologists. Certainly one criterion of success in the description of tacit knowledge is whether the outcome seems plausible to scientists—that is, whether it matches their world as they already understand it. In this chapter, I try to provide a more systematic exploration of that world than is common in the physical sciences and then to draw out some implications which certainly vary from scientists' current practice. Should all this be a failure in the eyes of scientists, it is still a worthwhile exercise, I believe, for the object of much sociology of scientific knowledge is to explore the world of scientific knowledge for the sake of nonscientists.

not it has any foundation in the reality of a high and unique level of skills among the team.

1. Concealed Knowledge

def.: expert A does not want to tell "the tricks of the trade" to others, or journals provide insufficient space to include such details. A laboratory visit reveals these things. Concealed Knowledge is not a very interesting or subtle category and should not perhaps even be described as a form of tacit knowledge. The limitations on knowledge transfer concern logistics or deliberate concealment, so failure to transfer concealed knowledge does not bear at all on our level of understanding of a science such as interferometry. Indeed, the very fact that scientists know how to conceal knowledge shows that it is not really tacit—they already know how to represent it in some kind of symbolic form, or they would not be in a position deliberately to conceal it. This category does not bear on the argument between the new LIGO management and the 40-meter team in that the management believed not that the 40-meter team were deliberately concealing things but that they had nothing to conceal.

The next four categories of tacit knowledge are much more interesting. They reveal some of the reasons that, however much we write, we may not be able to express what we know in ways other than by demonstrating the corresponding actions or engaging in the discourse in the broadest sense.

2. Mismatched Salience

def.: There are an indefinite number of potentially important variables in a new and difficult experiment, and different experimenters focus on different ones. Consequently, A does not realize that B needs to be told to do things in certain ways, and B does not know the right questions to ask. The problem is resolved when A and B watch each other work.

The 40-meter team knew an indefinite number of things about interferometry, so by definition, they could not write them all of them down. They may have made choices about what to commit to print and what not to commit to print that did not mesh with the gaps in the potential 4-kilometer builder's knowledge, but given the size of the problem, there was nothing they could do in the way of writing to resolve the matter. Nevertheless, novices could, perhaps, have learned these things by working with their more experienced colleagues.

3. Ostensive Knowledge

def.: Words, diagrams, or photographs cannot convey the kind of information that can be understood by direct pointing, demonstrating, or feeling.

The difference between categories 2 and 3 is that in the former case, everything could be committed to symbols if there was an indefinite amount of paper, whereas in the latter case one would not know how to write down what one knew even if one had an indefinite amount of paper, because we do not know how to express it other than through the joint manipulation of material objects. This category could have applied to the 40-meter team's tacit knowledge.

4. Unrecognized Knowledge

def.: A performs aspects of an experiment a certain way without realizing their importance; B will pick up the same habit during a visit, while *neither* party realizes that anything important has been passed on. Much unrecognized knowledge becomes recognized and explained as a field of science becomes better understood.

Here the experimenter does not know that he or she knows certain things about how to manipulate the apparatus, so it would never occur to him or her to write them down. This kind of problem was referred to in the case of those trying to repeat Joe Weber's experiments (chapter 6), the most striking description being that of the scientist who said, "[I]t's very difficult to make a carbon copy. You can make a near one, but if it turns out that what's critical is the way he glued his transducers, and he forgets to tell you that the technician always puts a copy of *Physical Review* on top of them for weight, well, it could make all the difference."

Of course, talking in terms of "forgetting" may be to miss the point that if you do not know that it is important, then there is nothing to forget. In the case of the building of the TEA laser, one such example was the length of the "top lead." Only lasers with a top lead shorter than eight inches would work, but no one knew that this was a crucial variable. Copying the layout of someone else's physically implemented laser, however, would mean that the top lead would be short enough and there would be a chance that the laser would function—working from a circuit diagram would not. So, scientists had a much better chance of making a working laser if they first spent time in successful scientists' laboratories, but they could not say why. We do not know how much unrecognized knowledge the 40-meter team had. In any case, the new leadership of LIGO wanted more knowledge to be explicated and recognized faster. Their worry was that the science—the formal capture of the knowledge—was not going fast enough.

5. Uncognized/Uncognizable Knowledge

def.: Humans do things such as speak acceptably formed phrases in their native language without knowing how they do it. Such abilities can be passed on *only* through apprenticeship and unconscious emulation, and aspects of their experimental practice are similar.

Uncognizable knowledge is the most philosophically contentious case. Reductionists would want to say that all our abilities will one day be understood at the level of the physics and chemistry of the body and brain, so that category 5 will collapse into 4; others believe that abilities such as language are irreducibly social accomplishments, which means they will never be understood at the level of brain functioning.[8] These two ways of looking at the world bear upon what I have called the big question: is the world ready-made to fit our nets of reason, or is it that we will never catch more than a bit of it? Does catching a bit of it from time to time mislead us into thinking we can catch it all?

As indicated, for our purposes we do not need to decide about the big question. Whether some or all uncognized knowledge is *uncognizable* need not concern us when we study the way tacit knowledge works, for two reasons.

If we take knowledge as a whole, the fact that natural language and similar human accomplishments are currently not fully understood means that now and for the foreseeable future, even that which can be articulated in language rests on a foundation of uncognized abilities even if they are not forever uncogniz*able*.

Turning to scientific experimentation, the reasoning is still less contentious. So long as science continues to develop, new experiments will be continually passing through a stage in which they are not fully understood, and certain aspects of the skill required to do them will be passed between experimenters only tacitly. For the purposes of the argument between the 40-meter team and the new management, category 5 can be collapsed into category 4, so we do not need to worry about the reductionist problem.

THE "Q" OF SAPPHIRE

Before turning back to the 40-meter team, we can explore these categories further by describing an episode of gravitational wave science from the late 1990s. This was the measurement of the quality factor, or "Q," of sapphire, the material proposed for the substrate of the mirrors of Advanced LIGO.

---

8. The classic text here is Wittgenstein 1953. This category of tacit knowledge is crucial in the debates such as that over whether computers will ever fully mimic the achievements of social beings.

For about twenty years prior to the turn of the twentieth century, the team led by Vladimir Braginsky at Moscow State University as part of a larger program on low dissipation systems had been claiming to have measured quality factors ("Q"s) in sapphire up to 4×10^8 at room temperature.[9] Despite years of effort, these measurements were successfully repeated in the West only in the summer of 1999. The failure to transfer the tacit knowledge of how to make the measurements had been responsible for at least some of the delay.

The quality factor of a material indicates the rate of decay of its resonances—how long it will "ring" if struck. (A bell that rings for a long time has a high "Q" and vice versa.) The number 4×10^8 means 400 million, and relates to the time in seconds that the ringing in an object takes to die to half its original amplitude. A high "Q" therefore indicates a long "ringdown" time. A long ringdown, in turn, implies that the object must ring with a very pure tone—to use the jargon, it has a "narrow resonance peak." The mirrors for Advanced LIGO will have to be made from a material with the highest possible "Q" so that the tails of the resonance band are narrow. This means that these tails are less likely to spread into the frequency range of gravitational waves and become mixed up with the signals that the interferometer is designed to detect. The higher the "Q," that is, the purer the tone of the mirror materials, the more sensitive will the interferometers be, because the noise levels will be lower. The Russian measurements suggested that sapphire would be the best material.

Because of sapphire's promise, efforts were made at universities including Caltech, Stanford, Perth, and Glasgow to repeat the Russian measurements.[10] But until the summer of 1999, no one outside Moscow State University had succeeded in measuring a "Q" higher than about 5×10^7 in sapphire. One American scientist told me that "there had been a certain amount of doubt in the [Western] community, because the only really high 'Qs' that had been measured above 10^8 at room temperature had been in Moscow," while a scientist from Moscow State told me that certain Western universities had indicated that they did not trust the Russian findings.

In the summer of 1998, after a series of failed efforts to measure "Qs" comparable with the Russian claims, members of a Glasgow University group visited Moscow State University for a week to learn the Russian

9. Braginsky, Mitrofanov, and Panov 1985.

10. The Russians more recently discovered that sapphire has some less desirable properties that may make it less suitable as mirror material. Among the community, this revelation is known as "the Braginsky bombshell" and will be mentioned later in the book, but it does not affect the argument presented here.

technique. Shortly thereafter, a member of the Moscow team—whom I will refer to as "Checkhov"—worked in the Glasgow laboratory for a week. In neither case was a high "Q" measurement achieved.

Checkhov had left a piece of Russian sapphire with the Glasgow laboratory (after running experiments on other crystals with them for a week), but the highest "Q" they could obtain with this specimen was about 2×10^7. And this was after attempting to match the Russian measurement over three weeks, during which they tried twenty different suspension combinations, each with a number of ringdowns at different vacuum pressures (see below for experimental details). When they finally emailed Checkhov to explain their problems, he reported that he had checked back in the Moscow laboratory notebooks and discovered that the "Q" of that particular piece of sapphire was not as good as he had said!

In the summer of 1999, Checkhov again visited the Glasgow group, bringing another piece of sapphire with him. After another week of effort, in mid-June 1999, a "Q" of over 10^8 was measured in the West for the first time; a similar result was achieved for a sample of American-grown sapphire. Subsequently, the measurements have been repeated with no Russian present by a member of the Glasgow team (who was present during Checkhov's visits to Glasgow), working at Stanford on an American-grown sample.

Components of Tacit Knowledge in "Q" Measurement

The method of measuring "Q" is to suspend the crystal—which might be a cylinder five to ten centimeters long and one to ten centimeters in diameter—in a sling around its midpoint. The sling is a single thread or wire which is looped around the crystal, the ends being held by compressing them in a clamp above the crystal. The crystal is thus balanced at the end of a pendulum, which helps isolate it from vibrations transmitted from the apparatus. The suspended crystal is loaded into a vacuum chamber, which is pumped down. One end of the crystal is painted with a dot of aluminum so that it acts as a mirror for the laser interferometer used to measure the vibrations which shines through a porthole. The crystal is driven up (set ringing) by an electrostatic end plate generating an AC field at the crystal's natural frequency. The field is switched off and the decay of the vibration, measured by the interferometer system (which can compensate for gross movements of the whole crystal), can be seen on a chart recorder or fed directly to a computer for analysis. The rate of decay can be converted to the "Q" of the sapphire. For a high-"Q" crystal, it might take twenty minutes or so to register sufficient decay to provide a good measurement. A lower "Q" crystal requires only a minute or so to give an easily measurable result.

Sapphire crystals have no perfect modes, so even if they are suspended exactly around the midpoint, some movement will be transmitted to the suspension fibers; therefore it is effectively the "Q" of the crystal-pendulum system that is being measured. A false low reading will result from losses of energy in the system. Significant energy can be transferred from crystal to suspension if one of the pendulum's natural frequencies of vibration is similar to that of the crystal—that is, to make the system work properly, the pendulum length must be "antimatched" to the crystal frequency. Friction losses between fibers and clamp must be avoided by making the clamp contact the fibers sharply where they first enter the clamp area—but not so sharply that the fibers are severed. Energy can also be lost in friction between crystal and fiber, and there are potential friction losses within the fiber itself—thus the choice of fiber and its preparation are both important. There are also thermodynamic losses between the vibrating elements and the residual air in the vacuum chamber to consider. The art of the experiment is to minimize all these losses.

By watching Checkhov work, the Glasgow group learned that good measurements had to be accomplished by trial and error over many repeated runs—they learned that the experiment remained difficult even *after* a first success had been achieved. As Donald (a pseudonym) put it,

> [1999] I think the thing that we learned most of all was patience. [We] would experiment away for a morning, perhaps, and after several runs we would end up with the same "Q"; in the past we would have been tempted to say that was the "Q." What we learned from [Checkhov] was that he was much more patient than that. He would go for days before he would believe [such a result]. He would keep varying the parameters by tiny amounts, because he knew to do that from the work he had done previously. And there would be enormous time put into it. And we would be sitting watching....
>
> And once you know to do that, [you can succeed]—but until you know that, it's hard.

Checkhov's approach, however, also revealed ways in which each of the many runs could be done more efficiently. To pump a vacuum chamber down to a very low pressure takes a long time. The Glasgow group had been pumping down for about 2.5 hours prior to each measurement, while Checkhov's practice cut the time in half, sacrificing an order of magnitude or two in vacuum. His practice showed that most of what needed to be learned could be learned at a higher pressure, reserving the lowest pressure runs for a final measurement only. Checkhov also used very short suspensions. The Glasgow group had used suspensions comparable in length with those that would eventually be employed in full-scale laser interferometers, but Checkhov used as short a length as possible so as to make frequency matching with the crystal less likely (the modal frequencies are further apart in short strings). Thus, with Checkhov's approach, fewer setups were wasted, and less time and care had to be spent on getting the length of the pendulum right so as to make sure it was antimatched to the crystal.

Let me try to describe what was going on in terms of the fivefold classification of tacit knowledge. In this case there does not seem to have been any category 5 (Uncognized/able Knowledge) transferred between Moscow and Glasgow, because both groups already shared the same broad "language" of science. Differences in this kind of knowledge show themselves only where very big differences in scientific worldview are juxtaposed.

Knowledge about the degree of vacuum and the length of the suspension belongs to categories 1 and/or 2 (Hidden Knowledge/Mismatched Salience). This is because degree of vacuum in exploratory runs is not likely to be noted in a published report; likewise, gross pendulum length seems like a choice that would be made on grounds other than experimental efficiency. Yet with trial-and-error experiments, efficiency is very important if enough runs are to be carried out to press the measurements to the limit. Certainly, the most appropriate choices became clear to the Glasgow group only through watching the Moscow practices. It is true that a diagram showing a crystal supported by a very short pendulum is shown in Braginsky's book (fig. 35.1), but it could easily be read as a schematic representation giving no information about length.[11]

Though the importance of the clamping could be described, and has been described, it was Checkhov's way of working that revealed the possible importance of repeated minute adjustments to the clamp should high "Q" not be achieved at first. To describe the principle of clamping and to mention its importance is not the same as revealing its importance through the care that is taken in practice; we do not have an exact language for describing "degree of care that needs to be taken," so coming to understand it is a matter of ostensive knowledge (category 3).

Something similar applies to the material of the suspension fibers. Checkhov used very fine Chinese-silk thread, which he supplied to the Glasgow group (who had earlier used steel piano wire). Trial and error

---

11. Braginsky, Mitrofanov, and Panov 1985, p. 27; I present a close impression of the original.

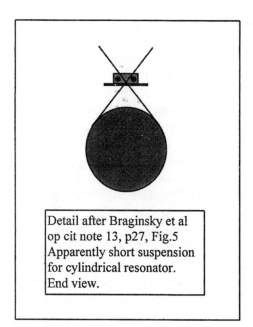

Detail after Braginsky et al
op cit note 13, p27, Fig.5
Apparently short suspension
for cylindrical resonator.
End view.

Fig. 35.1 Crystal hanging from a thread as represented in Braginsky et al.'s book.

had shown the Russians that other kinds of silk thread gave lower "Q"s. It was also known that fine tungsten wire gave still better results but that it had to be polished carefully to just the right (indescribable) degree, and that the clamping problem was particularly acute with tungsten. Donald believed it was the hardness of the tungsten that made the clamping so critical—the compressibility of silk allowed a certain leeway in the design of the clamp. Thus silk was used for most runs, with tungsten (which might improve the "Q" by a factor of 2), being preserved for a final measurement once the general area of the expected result had been defined by the easier method. The nature of suspension materials and clamping seem to belong in categories 2 (Mismatched Salience), 3 (Ostensive Knowledge), and 4 (Unrecognized Knowledge): they are matters whose salience became clear for the Glasgow group only after working with Checkhov. For both parties, the science was slowly emerging, turning knowledge that no one knew they could or should express into something that could be articulated as the importance of previously unnoticed parts of the procedure became revealed.

Polishing of tungsten and greasing of both tungsten and silk had been found to be vital. In Braginsky, Mitrofanov, and Panov's book, we find (at p. 29), "The presence of a fatty film (e.g., pork fat) at the points of contact between the suspension fiber and the resonator is important." It was believed that grease between fiber and crystal prevented frictional losses. Greasing turned out to be critical, but there is no vocabulary to describe exact amounts of pork fat (the Glasgow group used commercially available lard after watching Checkhov, whereas they had used Apiezon brand grease previously).

Working with Checkhov revealed two methods of greasing a fine silk thread. A thicker Italian-silk thread was first greased with a "daud" (a Scots dialect word) of lard and wiped with a cloth until most of the lard had been absorbed or rubbed off. The crystal was then mounted and balanced in a loop of this thread. The greased Italian thread would leave a thin track on the crystal. The crystal was then dismounted and rehung on fine, Russian-supplied Chinese thread, which would now be sitting in the thin ring of grease left by the thicker Italian thread. The run I witnessed produced a slightly lower "Q" than had been expected, and the reasons described to me indicate a nice case of ostensive knowledge.

> [1999] *Ericson* [pseudonym]: It's very difficult to be precise about the amount of grease you apply, because you're just applying grease to the thread. If you apply too much, the "Q" tends to fall off, because it's too loose and it will wobble and you will get an erratic ringdown. But if you have too little grease, then the thread may stick and slip rather than sit smoothly on the mass. In this case, I think there probably wasn't quite enough grease, which is why it [the "Q"] is slightly lower than what I thought it might be. But if you get it spot on, you can usually get a very high result.... I think there's not quite enough.
>
> *Collins:* And that's just from your looking at it.
>
> *Ericson:* Yeah—that's just empirical—from my experience of doing this before, I can sort of tell. When you take off the greased thread and you see this band of grease, there's a feel for what's enough and what's too much. And that looked less—but not too far off.

The second method of greasing thread demonstrated by Checkhov, and used interchangeably with the first method, was direct greasing of the fine thread with human body grease. Checkhov would run the fine Chinese thread briefly across the bridge of his nose or behind his ear. The ear method was adopted by the Glasgow group, though it turned out that only some people had the right kind of skin. Some, it turned out, had very effective and reliable grease; others' grease worked only sporadically; and some experimenters' skins were too dry to work at all. All this was discovered by trial and error and made for unusual laboratory notebook entries, such as

"Suspension 3: Fred-greased Russian thread; Suspension 12: switched from George-grease back to Fred-grease" and so forth. As with James Joule's famous measurement of the mechanical equivalent of heat, it turns out that the experimenter's body could be a crucial variable.[12] Knowledge of how to apply the right amount of grease to the system has aspects that belong in categories 2, 3, and 4.

It may be that the measurement of the "Q" of sapphire involves more craft skill than usual and therefore is untypical in terms of the importance of interpersonal contact. Most gravitational wave scientists whom I spoke with were unhappy with the amount of what they called "black magic" involved in the Russian methods of "Q" measurement and were looking for some simpler way of doing things. It seems one American group managed to measure high "Q"s in glass by an entirely different method, though it has to be said that it was not one that turned out to be unproblematic; and in July 1999, an Australian group briefly mentioned an independent replication of the Russian results using a tungsten wire support that appeared to be relatively straightforward. I have not been able to investigate either case in the kind of detail described above. But even if the need for tacit skills is especially marked in this case, the same broad analysis will be applicable wherever a new kind of apparatus is to be built or a new procedure carried out.

TACIT KNOWLEDGE AND INTERFEROMETRY

What the new management at LIGO wanted from the 40-meter group was progress toward the routinization of their tacit knowledge, if they had any. I can summarize the four ways in which procedures that were once esoteric and difficult because of their tacit component become routine. *First,* as we interact socially, that which was not obvious becomes obvious; this is what happens in the case of concealed knowledge, mismatched salience, and ostensive knowledge. *Second,* as we understand more science we learn to make more tacit knowledge explicit: we come to know in a conscious way elements of our knowledge which we did not know we knew. Unrecognized knowledge becomes recognized and can subsequently be passed on without personal contact. *Third,* social contact between scientists spreads knowledge that is still tacit throughout the community; that is, more scientists learn the new experimental language even though no one can set it out, so that much more can be transferred by explicit symbols (just as spouses

12. See Sibum 1995 for the Joule experiments.

or teammates can transfer a great deal with only a word or a gesture, relying on a common background of tacit understanding). This mechanism applies to unrecognized knowledge so long as it remains unrecognized, and to uncognized/able knowledge. *Fourth,* mechanical, or "turnkey," methods for packaging the experiment are worked out, replacing the need for tacit knowledge.[13]

The question was, Had enough time elapsed for the 40-meter team to have succeeded in turning tacit knowledge into symbols or reproducible physical objects? The new management believed that the era of the formula should either be here already or approaching much faster than the 40-meter team were accomplishing. What they wanted was more of the first and fourth mechanisms described above, and they were either prepared to sacrifice the second and third mechanisms, or they did not believe there was anything there to be sacrificed. This point forms an important background for the discussion to come in chapter 41.

Two Identical Machines?

Another way to approach the problem of tacit knowledge, or the extent to which we have or can capture the world in symbols, is to ask whether enough is known about interferometry to make it possible to build two identical interferometers. If it was possible to set out the recipe for interferometer building in sufficient explicit detail, then it ought to be possible to build any number of identical copies. Wait! What do we mean by "identical"? As the Heraclitus said, one cannot step into the same river twice, so no two things can be identical. The point, however, is this: if one understands interferometry sufficiently well, one will know which subset of the indefinitely large set of descriptors of an interferometer one needs to know to make one that is functionally identical—one that locks and produces steady interference fringes with a predictably low level of noise.

I applied this test of the tacitness of knowledge to the TEA laser, spending some time with a laser builder who was trying to build a second copy of one he had successfully completed. Here the tacitness of the knowledge that bears on the test can belong only to category 4—Unrecognized Knowledge—because experimenters who are one and the same person (1) cannot keep secrets from themselves, (2) must share saliences, (3) must have identical ostensive knowledge, and (5) must share a scientific language with

---

13. For an extended discussion of the transformation of tacit knowledge into "turnkey" knowledge, see Collins 2001c, passim; and Collins and Kusch 1998, chap. 9.

themselves. The only thing they might not be able to share with their doppelganger-selves is bits of unconsidered behavior, the importance of which they have not recognized. But even under these circumstances, it took the laser builder I worked with two whole days from first assembly to successful firing, and here we are talking about a device approximately a meter in length with about ten large discrete parts.[14]

LIGO is running an experiment of the comparison type—it is building three interferometers with the same team, all meant to be identical, except that one is half the length of the other two. And LIGO took it as a matter of deliberate and considered policy to make these devices as like each other as possible in order to eliminate unwanted and troublesome differences. The key, especially on a device as large and complex as LIGO, is what is called configuration control. Configuration control means keeping records of every procedure and every change made to each interferometer so that every team member ought to be able to read the notes and understand exactly what every component is and why every setting stands where it does. But even where all the material is explicit, the recordkeeping required for configuration control is huge. Dennis Coyne, LIGO's senior engineer, explained the problems of maintaining configuration control at the Hanford and Louisiana sites.

> [2000] There are . . . aspects of configuration control . . . which we have not done terribly well, partly because it's cultural. If you are an engineer, I think you're predisposed maybe; you understand the need for configuration control better than if you're a physicist. But if you come from an environment like JPL [the Jet Propulsion Laboratory], or the industrial-military complex, configuration control is maintained at all times. In the university environment, it's just an anathema—it's culturally strange to them. Though I've seen some few converts who become zealous about it.
>
> Anyway, the main thing you want to do is . . . have a drawing that you build a part to. You want to make sure you don't lose that drawing—that it's filed away—and if you change it—if you find you need to change it—that you document that change, or otherwise you get the kind of situation we have currently in the 40-meter lab. It's built by a bunch of graduate students or postdocs. So they come in, they do their postgraduate work over a few years. They make changes in the middle of the night. No one knows but themselves. They can even forget. Several months later they say, "I remember, I did something to that board, but . . ." They can't say, "Well, here's the

14. Collins and Harrison 1975.

sketch, and I changed these values" and, you know. And that's obviously important here, because if you do it in Hanford, you probably want to repeat it down here [Livingston], and when you go to embellish upon that, you decide you need, say, a feed-forward loop around what you already have, you want to pull out the board and say, "How am I going to change that?" And if your drawing doesn't match what's been done out in the field, it's a problem.

The other problem that they have, again, in the 40-meter lab is that as you commission, there's always a need to reconfigure. To do kind of "what ifs." "What if we take this signal, drop that other one down by a factor of 10, and instead feed this around and so pull up some cables and stretch them across the floor and connect up some temporary instrumentation and say, 'That works great, [even though] that may not be the way we want to keep it finally...'"

Or maybe it didn't work and you just leave it. You [need to] fly back to your institution..., so you just leave it and rush out to the airport. And nobody knows, and it gets buried in the cable tray. That's the kind of fundamental configuration control things.

The problem of the sheer amount of material involved when it comes to describing procedures is also well expressed by Coyne. Here he uses the term *tacit knowledge* himself:[15]

And then there are tacit knowledge things...that touch on configuration control to some extent. But it's a much more difficult thing to do, obviously. It's much more ethereal.... For example, when we came down here and did the instrumentation, and beginning commissioning. We learned from what we did up at Hanford. Some of it was straightforward—we learned that that circuit didn't work and we had to change this circuit value—change this resistor—no problem—so long as you made the change on the document, you can repeat that, and you're already ahead of the game when you come down here. But there's an awful lot to learn, and for some of the relatively simple stuff, like installing the big dumb structures, like the seismic stuff, we had to figure out what the procedures were and the how tools for installing would work, and we had to work that in real time in Hanford. But then we wrote that down in procedures, but even then, you know, the procedures probably were 40 to 50 pages. It's at the limit of what people will actually read and remember and into all the detailed steps. But it doesn't capture everything.

15. Though I am not sure whether it was Coyne or myself who first introduced the term into the conversation.

So you really have to have people, if you want to do it fast and efficiently and not make mistakes; you send people down who've done it before, because a person can remember a lot better if they've done it than if they read it. So some of the tacit knowledge captured in the procedures.

Commissioning is a lot more difficult. It's much less prone to writing down the procedure. Eventually it will be. It's like TV repair. Debugging a TV—a TV is a fairly complex apparatus—and it takes a fair amount of skill and knowledge. And there are some rote procedures you can go through to isolate the problem, but those TVs are commodity items—you know they've been worked on for years and years. In a brand-new machine, where even the designers are not sure how its going to operate, it's hard to convey the tacit knowledge other than by education—by one-on-one, or by one-on-a-few, training.

Coyne also provided an example of how things were done when difficulties were encountered. At Hanford, some of the mirror-actuators, which are glued onto the mirrors, fell off. No one really understood why.

> *Coyne:* In the case of optics installation, there, because we had so many problems in the art of bonding these magnets onto the mirror, and it caused so much schedule delay, we had [named persons]—the three principal people who did it at Hanford, and struggled through it all. We had them come down here and do it all [laughter]. . . . and the others kind of helped a little, but they did it.
>
> *Collins:* So you didn't even make any attempt to transfer it in that case; you just transferred the people.
>
> *Coyne:* We were so afraid in that case, because frankly we just don't understand what went wrong at Hanford. We have theories, there were a bunch of—we just kind of scrubbed the whole procedure and went through every detail. There was Kapton [?] tape applied to one of the fixtures that might have introduced some silicon contamination that might have compromised the bond—so rather than risk any of that, we just said, "Look, we'll just get the same people, the same thing."
>
> And we were very careful about when we did the optics installation, too. Because things at Hanford weren't quite right; the fixtures weren't quite right, we didn't have the right procedures, and we ended up knocking off perfectly good magnets bonded on mirrors—knocking them off. So we had [named person], who was the principal person up at Hanford, [other named person's] counterpart, come down here and work them and direct most of that installation and do most of the hands-on.

And the second installation, I was here and did some of it hands-on, and just very carefully, making sure that procedure was transferred well and that knowledge. And I think it's paid off, it's worked.

Of course, people are learning all the time. As Coyne explained,

> The mode cleaner's a good example. The mode cleaner we aligned . . . on Wednesday. And the effort that took—I can't remember exactly—but on the order of quite a few weeks—say three or four weeks up at Hanford—we did in three days. Because we knew the procedure, we knew the tools, and we had the tools in hand.
>
> Stan came Friday night. Stan and I and Sanni and Peter Saulson started working on locking the mode cleaner Friday night, and discovered one little problem after another—no major things—things like pins swapped on cables and a few other problems like that—some software problems, and we got it locked on Sunday in a low-finesse polarization. Had it robustly locked in a high-finesse polarization, and everything seemed to be good—at least as good as Hanford—by Wednesday. One week—and that locking up at Hanford took easily a month or two.
>
> So the time compression is enormous, and it's partly because some of the electronic fixes they found up there were already incorporated. It's not all solved—we found some electronics glitches here that were puzzled as to why it works at Hanford, but we had to make a change here. And we don't understand quite yet. And we're going to take a measurement of the residual loop error and compare it to Hanford to try to understand what's going on. It could just be that the environment is noisier—there are a number of possible explanations.

As we see, some things are done faster because explicit changes have been made to improve the device at one site, and these modifications can be transferred; some things are done faster because experienced personnel are transferred from one site to another; and some things remain obstinately different—a fix that works at one place does not work at the other.

Barry Barish, the champion of the orderly world ("interferometers are not magic"), noting my interest in the difference between sites, assured me there was no magic there either.

> [2000] *Barish:* I don't think it [technology transfer between the sites is] a meaty subject for you in LIGO. And the reason is, we've been so concerned about it for so long because we don't want to invent the wheel twice that we basically

build in enough so that it's just natural and it's not an issue. You're not going to find an issue on how we transfer the technology from Hanford to Livingston.

Collins: I'd be surprised if that wasn't interesting.

Barish: OK, we'll see.

Collins: [T]here's a problem about changed circuits in one place haven't been transferred to another.

Barish: They have been. That's what I'm telling you, is when you look at it—I brought up all those issues. When you look at it, what you'll find is—which I don't think is very interesting—is that you've thought about things and you ran them well, is that stuff is all transferred. We've thought about all that and it is all transferred, and you'll find that they have the modified circuits in Livingston because we've built that into the system.

Collins: [unconvinced] Alright.

Sanders: Do you know of a specific example when it didn't happen?

Barish: The only place where we fall down, and I just don't think it's very interesting, is the psychological one. Some people just—and you can imagine, there's ego—you don't want to go feel that the other guy's the expert that you're learning from. That's the only place that we have trouble. We worry a lot about the problem. I go to Livingston a lot and I'm very sensitive to my part of it, which is that we shouldn't put in any extra effort—any more than we really need to do a smooth technology transfer. I can't identify anything that we're not doing basically right except the psychological issue, which is a hard one that we have to struggle with all the time, which is that people want to have their own identity as individuals and they want to invent how to do something better. . . . The mechanisms to do the transfer, enough overlap of people, people going from [Hanford] to Louisiana, are well designed. We don't do everything well, but we do that well.

Of course, Barish is probably right that the procedures were as well designed as they could be, and the problems were solved insofar as was possible, but the last phrase is germane. The problem had to be solved by the transfer of people, not symbolic representations, and this speaks to the problem of tacit knowledge implicitly just as Coyne speaks to it explicitly.

The Ultimate in Symbolic Representation: Computer Modeling

Just as a glass is either half full or half empty, the same facts can be seen as either revealing the amenability of the world to symbolic representation or revealing the world's recalcitrance. The culture of high-energy physics, and its immense success over five decades, encourages those brought up in the tradition to see the glass as half full—the world can be modeled, and any backwardness in modeling is a result of pigheadedness, a penchant for mystery, or laziness. This is the implication of Barish's "interferometers are not magic." Those brought up in other experimental fields and traditions, however, might be more conscious of scientific artisanship and the salience of tacit knowledge and the like. This tension runs throughout this book.

Another way in which Barish's view was expressed was in his determination that the interferometers be represented with computer models. One of the major innovations introduced by the new management team was a stress on computer representation and the start of the construction of an "end-to-end model" of the LIGO devices. It is now hard to find anyone in the LIGO team who does not think that this approach has been a success, and other national groups use the same techniques, proudly displaying the "fit" between their computer models' predicted noise patterns and the noise found in the real machine. Nevertheless, the meaning of the success of the computer modeling needs to be examined carefully before drawing superficial conclusions.

The question about models is whether they represent the world well enough to suit the purposes in the minds of their developers. For example, scientists have spent a long time trying to model the world's atmosphere with computer models, but it is unclear, to say the least, that they have succeeded. If their success were obvious, then there would be fewer arguments about the nature and direction of future climate change. If the models were better, then, chaotic regions aside, long-range weather forecasting would be better. Perhaps long-range weather forecasting will improve as the power of computers increases along with the number of points on Earth's surface at which we can take measurements. Perhaps the workings of Earth's atmosphere will one day be understood as a machine, but for now it is not. Worse, to understand fully, we will have to understand people fully, because people contribute to climate change, and that makes the challenge of atmospheric modeling seem like a seriously intractable problem.

On the other hand, computer models seem to have been a great success when it comes to modeling the processes that take place inside an exploding hydrogen bomb. At the time it was first done, working out what happened inside an exploding H-bomb was the most difficult calculation that had ever been made in science. But in another sense it is an easy problem, because the world of the exploding H-bomb is self-contained. These shifts in perspective—what was once thought of as a hard problem coming to be seen as an easy problem, and what was once seen as an easy problem coming to be seen as a hard problem—keep happening as the domain of the computer increases in size. For example, once upon a time ability at "mental arithmetic" was taken to be the true sign of cleverness, but the existence of pocket calculators means that arithmetic is now considered easy. On the other hand, speaking and understanding speech was once thought to be easy because anyone could do it—even those who could not do mental arithmetic—but continuing attempts to reproduce humanlike speech transcription with computers have shown that, at least as far as our current knowledge is concerned, it is an impossible problem. What is difficult and what is easy in this sense have to do with the extent to which the problem touches first, on an open-ended or complex physical world, and second, on the way it touches the social world.

Setting aside some subtleties about the social nature of science, modeling the explosion of a nuclear weapon has turned out to be easy on both counts.[16] Fluid flow, which is the crucial mechanism of H-bomb detonation, is, of course, complex. And even in the confined space of an H-bomb there is so much going on that the modeling has to be based on a small sample of the fluids.[17] Nevertheless, it has turned out that none of the complexities of hydrodynamics are fatal for the computer models and that the sampled elements can be treated as representative. Furthermore, when an H-bomb is exploding, it is about as isolated from the rest of the world as any physical entity can be. When the atomic chain reaction within the bomb starts, things move so fast that whatever is happening in the outside world is, in comparison, standing still. Likewise, the forces and energy releases generated inside the bomb are so great that anything that impinges upon them from outside will be invisible. Peter Medawar said the success of science depends on picking problems that are amenable to what he called *The Art of the Soluble*.[18] The nuclear bomb is an example par excellence, however difficult the sums might be. This is why, since the banning of nuclear testing, the development of nuclear warheads has taken place entirely within the realm of computer models. Will these newly developed devices in actually work? Let us hope we never find out.

In terms of isolation from the world, an exploding nuclear weapon is almost the opposite kind of thing to a sensitive interferometer. An exploding

16. But cf. MacKenzie and Spinardi 1995. Also, this argument should not be taken to mean that controlling nuclear weapons is easy. I refer only to the modeling of what happens from ignition to the completion of most of the nuclear processes.

17. Galison (1997, chap. 8) discusses how the modeling process developed.

18. Medawar (1967) talks of science as the art of the soluble. In contrast, sociology concentrates on problems that represent, par excellence, the art of the insoluble. Of course this is not to say that modeling the hydrodynamic processes in the heart of an exploding bomb is easy.

nuclear weapon has tremendous influence on the world, but the world has no influence on it. In contrast, the workings of an interferometer have no influence on the world, while almost any outside disturbance at almost any rate of change can have an enormous influence on it. The success of models for nuclear weapons and, perhaps, in high-energy particle accelerators, whose closed nature may approximate that of the bomb, do not necessarily mean that modeling can ever cope with all the problems of an interferometer.

It remains, however, that interactions with the outside world are not a big issue in terms of model building. There have been some unexpected influences on the inteferometers, such as the effect of acoustic noise (see chapter 41), but these touch on nothing deep; nor would the 40-meter team stake their claim on the problem of understanding outside influences. Rather, the deep problems, if there are any, are going to turn on the interrelationship of the interferometer's components. The inside of an interferometer is in many ways more complex than the inside of a nuclear bomb. There are multiple mirrors; there are electronic feedback circuits, each of which might interact with the other; there are vestigial gases; there are wisps of stray scattered light; there are fluctuations in the laser; there are thermal inhomogeneities in the materials; there is the gravitational pull of the operators—any of these can have feedback effects on all the others. There is, in sum, every kind of metamorphosed state of at least two of the four fundamental forces of nature trying to confound the model in every way possible, and the whole point of an interferometer means that everything that might affect it is amplified by the acute sensitivity of the device itself. The question, big or small, is whether a real interferometer is going to turn out to be more like an H-bomb or more like the weather.

However it turns out, for the time being, the interferometer is not like an H-bomb. As Matt Evans, one of the contributors to the model, put it,

> [2000] For an interferometer simulation, I would be worried if someone came up to me and claimed that they could understand an interferometer with just a computer model. There's too many details that may or may not be important and are easily forgotten if you're not in there in front of it trying to work out why it's broken. It would never occur to you, the reason why the thing is broken, until you have to diagnose it.
>
> You see, with nuclear warheads, they've been doing it long enough for them to believe they have all the important details, and they know what the important details are and they can include them in the model and this it works out, but this field is still pretty new.

Whichever way it does eventually turn out, and irrespective of deep questions about the nature of the physical universe, the *usefulness* of modeling is not to be straightforwardly equated with the extent to which a model captures the world completely. Modeling, even where it is markedly unsuccessful in terms of its grandest aims, is a useful investigative tool. For example, weather forecasting may never be an exact science, but it has certainly been improved by attempts to model it.

Another place where modeling simply does not work in terms of producing an accurate outcome is in the economies of nations. Robert Evans studied the practices of a panel of econometricians brought in by the British government to predict, from the beginning of the year, that year's inflation and unemployment. He found the modelers could, and did, make huge differences in their forecasts by making slight changes in the historical periods from which they drew their time series of input data. He also found the econometricians' forecasts varied hugely from one to another, but that the behavior of the economy was different from any of their forecasts and a long way from anything that would have been predicted by taking an average of the economists' models. Yet the econometricians were still the best people from whom to take advice, inaccurate though their models were. This was because the discipline of modeling had made them much more knowledgeable of the way one variable in the economy related to another.[19]

Donald Mackenzie's study of the interaction between computing and mathematics demonstrates the same broad point. One goal of mathematicians is to find rigorous proofs for the reliability of computer programs, especially those involved in safety-critical applications. But the goal has not been achieved where the programs are long and complicated. Nevertheless, as Mackenzie shows, the effort put into attempts to produce proofs of the validity of programs has led to the elimination of far more program "bugs" than were eliminated by other methods. Thus have programs been improved by mathematical procedures even if the procedures have never delivered the products intended by the mathematicians.[20]

In the same way, computer modeling has already proved its worth in interferometry without demonstrating that it is the decisive answer to either the big or the small question of whether interferometers are magic. It seems widely agreed that the attempts to lock the first and subsequent LIGO interferometers have been helped by the use of a "stripped-down" computer model. "Lock" is the state of an interferometer needed for steady observation. To accomplish lock, the laser, mirrors, and output must be in a state

19. Evans 1999.
20. MacKenzie 2001.

of balanced and steady mutual feedback, which calls for the solution of a problem in "twenty-dimensional space," as one scientist described it to me. To look at the overall plan of LIGO is to realize that the big interferometers are but the outer shell of a nest of controlling optical and electronic apparatus, most components of which themselves consist of suspended mirrors with feedback systems. A LIGO system is, in fact, rather like a fractal, with the big interferometers being made up of ever smaller models of themselves. At the center we have first, a reference cavity, which tells the laser how to configure itself; second, a mode cleaner, which cleans the output of the laser, making the light coherent and monochromatic; and third, the inner stage of the interferometer proper, the power-recycled Michelson interferometer, which is a reduced-scale model of the whole interferometer that uses the inner mirrors of the device's multiply reflecting arms without switching in the distant mirrors. The power-recycled Michelson is the heart of the large interferometer.

Before there can be any hope of making the big interferometers work, the embedded units must be made to work, and big or small, the process is similar. It is a matter of setting each of dozens of feedback control loops so that the mirrors are moved just the right amount when they go out of line to bring them back into alignment and keep the whole optical-mechanical spiderweb still. Indeed, that analogy is probably worth pursuing: think of an interferometer as a spiderweb, with each strand having a little controller to push or pull it ever so slightly this way or that. When a fly bangs into the web, the sensors have to move the controllers just the right amount to counteract the effect so that all the strands of the web stay still. As one can imagine, just one sensor-controller moving too much or too little will vibrate the whole web and cause the other sensors and controllers to act in their turn, with a great risk that the vibrations will be amplified rather than stilled. (Think about how you would build a device to keep a spiderweb still!)

The level of difficulty involved in achieving lock increases as the sensitivity of the device increases. This is a matter of scale. On a human scale, the occasional earthquake aside, Earth is a metaphor for all that is solid: What can be more solid than the earth beneath our feet? But as the scale decreases—and an interferometer is so sensitive that it is like a huge microscope magnifying the most microscopic movements of Earth to an enormous size—so the solid ground becomes a restless, quaking bog. The job of locking, then, is like holding still a spiderweb mounted on a jelly trampled by elephants.

In the case of the locking of the first interferometers, the model developer and user was Matthew Evans, a graduate student at Caltech working under Barish's supervision. As Evans explained, the use of the model was not a one-way process. It was not that one built a complete model of an interferometer and then "applied" it. On the contrary, for its application to locking, the crucial feature of the model was its difference from the real interferometer. Certain parts of the noise spectrum that were known not to have much effect on the locking could be eliminated in the model, and other bits of noise could be temporarily held stationary while the interaction of some subset of the apparatus was examined. In effect, instead of simultaneously trying to control twenty dimensions, one could hold eighteen of them constant and study the interaction of two.

The other feature of this model, as in all modeling, was back and forth iteration. If the device itself did not behave as the model predicted, one would change the model, circling ever closer to a match that would work for all practical purposes. In other words, the model was being used like an extremely complicated paper and pencil rather than a complete representation of the world. This is very nicely expressed in the words of a respondent.

> [2000] *Respondent:* Well—look, Harry, there's always this funny thing about modeling. I think modeling is like going swimming in very cold water: you just tip your toes in as much as you can. You can talk about building a very, very all-inclusive thing that's got everything modeled in it, but if it's really all inclusive, and you don't make any approximations, that's a nice ideal, because it sounds good and you don't have to think too hard about whether your approximation's good—I think in the long run, things like that are totally useless. And the reason you want a model is that you don't understand some complicated system. If you make the model as complicated as the system you don't understand, you probably don't understand the model either, so it's not a thinking tool.
>
> So somehow it's got to be a simplification of the real world, and the thing you always worry about is what you're throwing away when you make approximations. And so you never get what you want in modeling. So comments like "Is it ready?" or "Is it useful?"—I think they can be useful long before they're, quote, "ready." And they might actually lose usefulness by the time they are ready and they get overdeveloped.
>
> *Collins:* So it's another way of helping you think through the problem?
>
> *Respondent:* Yeah—it's really a discipline. The way I look at it, it's a thinking tool, and that will really evolve as the problems change. For instance, eventually I think that when we're trying to improve the sensitivity of the instrument, and we can manipulate it on command, and it does more or less what we tell

it, then I think you really start to get persnickety and worry about the details of all these little things that might create subtle amounts of noise, modeled exactly the way they're built. I think right now we're at a stage where we can ask simpler questions, but they're really quite important, and I think what we're trying to do is get intuition from it, and intuition comes from doing something repetitively enough—you know, near-like things repetitively enough—so that you recognize patterns; and I think right now we don't recognize patterns, we just see stuff go by on this data viewer program, which acts like an oscilloscope, on whichever of many channels we want to look at.

So right now what we're trying to do is get some intuition for how to recognize different patterns. And so what we were just talking about was what kind of interferometer scenarios, with things going wrong with these different configurations, and so one of the first questions is, we sort of mentally divide things into the sort of [things the mirrors can do—e.g., changes in alignment versus changes in separation]. The question is, how does one step through these different cases of what's not right? And then there's a whole set of knobs that get set.

As it was, the first interferometer to which the model was applied, the Hanford 2K, took many months to be made to lock.[21] There is, of course, far more than modeling involved in making an apparatus work. Here is what one afternoon of these months looked like according to my field notes from September 2000:

> I spend about 3 hrs just sitting by the 2-km laser with David Shoemaker and Nergis [Mavalvala] as they try to solve a 60-Hz hum problem on one of the leads in preparation for the big switch-on. This was extraordinary because they simply could not get the "opto-isolator" (?), which they were trying to fix on the end of this line, to work at all, and they could not understand why. A lot of the time was spent trying to work out whether the connections at the ends of all the cables were compatible. So they had a fistful of converters and gender changers which they kept trying in different combinations—like me trying to install a printer in the old days, with all the different possible switch combinations. This whole problem was ludicrously trivial for two such competent persons to be dealing with, yet utterly familiar for anyone who has ever done any practical work. I suggested several times that it ought to be

---

21. There are widely varying definitions of how long it took, depending on how reliable a lock with how much power is taken as the end point.

dealt with by a technician, not them, but this is not the way things are set up here.

By the time I had to leave, they had made no progress in solving the problem in spite of bringing in a third party and stringing a completely new cable from one console to another, with new kinds of special boosters and so forth.[22] This was an endless round of using testing equipment on combinations of wires, with the two of them just squatting on the floor, trying things out, like any two kids. Progress on this huge, hundreds-of-millions-of-dollars project was being held up for a whole afternoon by a simple can't-find-the-source-of-the-fault-in-the-wiring problem of the sort one might encounter in the home.

This process, I noted, kept Shoemaker at the site until 10:00 p.m. and Mavalvala until 2:00 A.M. These were not untypical work hours nor untypical work practices. The actual bringing of this huge, complex machine to an operational state was the work of no more than half a dozen heroes, with the model being just one tool to aid their *craft*.

Matt Evans described the model's contribution to the process as follows:

> [2002] I think it was sort of an indispensable tool along the way, and sort of for developing the initial idea for how to make the acquisition code work it was a very important tool. In hindsight, I think if I'd known everything before I started, I could have just written the code and said, "It's obvious—it should work this way." But I didn't, and no one else did. And there are a few periods of time that I remember fondly, sitting there at 2 a.m., and running back and forth between the control room and the computer room, where I would run something in my model and see how it turned out, and decide that I had fixed some bug that was in the code and then I would haul the code over, and recompile it, and load it on the interferometer computers that were actually running in the physical system, and play with it for an hour or so, and then go back to the computer room—bounce back and forth on an hour-to-hour basis.

Evans explained that the model had hardly changed since it was used to help lock the 2K at Hanford and subsequently had been used to aid the switch-on of the 4K interferometers at the other two sites. Locking the latter devices had gone faster as a result, but were not without their

---

22. At this time I was suffering from an undiagnosed cracked tibia (I thought it was a twisted ankle), and the intense pain, relieved by huge doses of painkillers, meant I had to cease observations by about 5:00 p.m. each day and lie down in my hotel room.

problems due to extra seismic noise and the like. Evans, like Barish (his thesis supervisor), divides problems into those which are matters of real physics and those which are merely technical.

> [2002] [I]n some sense, we understand all of the physics that goes into interferometers. There's little there that we don't know how to write up the equations for, but knowing exactly what's important—there's a number of effects that we know are there but we don't include in the model because we don't think they're important—
>
> *Collins:* Such as what?
>
> *Evans:* There's a lot of just sort of "technical dirt" that may or may not be important, and you don't really know until you get there and you find that the interferometer is not really behaving in the same way as the simulated thing, and you have to start asking, "What bit of dirt did I forget to put in my model?"

Referring to the remaining difficulties of locking the 4K devices even after the lock-acquisition code was completed, Evans said,

> [2002] [E]ssentially, the acquisition work was done well over a year ago, and all the challenges in making the 4-km lock have been technical problems. Being a software guy I'd say [laughing] hardware problems.

Once the interferometers are working and it comes to modeling the noise, things get more complicated, because here the object is to model something closer to the complete interferometer, not a stripped-down idealization of it. Once more, iteration is the key. No one claims that the whole interferometer can be modeled without repeatedly referring to the real object; but the question remains whether all the features of the device have been included in the components of the model: is interferometry well-enough understood to make it possible for repeated iterations between model and reality to result in good convergence between model and world? That is not to say that the model is simply used in its complete form, but the nearer the noise predicted by the model is to the noise produced by the interferometer, the more will the modelers feel that they can switch off bits of the noise at will and still be confident that the interactions between the remaining noises in the model are telling them something about the world.

We will not know for a while the answer to the big questions about the representativeness of models of the big interferometers, but in the meantime, to repeat, the high-energy physicists' tendency in this regard is to see the glass as half full. Their "model of the model" is something that is capturing the world—at least to the extent that we can see that there is no "magic". Under this model of the model, lack of correspondence between model and world will be at worst a temporary matter—a matter of technological "dirt" and a bit more development. Yet people like those in the 40-meter team at LIGO are more likely to see this glass as half empty—as exhibiting the lack of fit between the model and reality. Though, in terms of the big question, they may subscribe to the overall "deterministic machine" view of the world, they will see the model's failure to correspond as at least showing that the science of interferometry has not yet reached the stage where modeling is going to be as revealing as it could be with more understanding. For the 40-meter team, at the very least, much more experimentation and theorization would be needed before the models would do the job.

CHAPTER 36

THE 40-METER TEAM VERSUS THE NEW

MANAGEMENT, CONTINUED

These alternate views of science and the world—magic or not magic—laden with tacit knowledge or amenable to systemization and representation in symbols—are the larger context in which the continuing argument between the 40-meter team and the new management at LIGO can be read. We left Project Manager Gary Sanders saying that he was glad to be out of the mass production business: [2000] "I'm glad to be out of high-energy physics, because much of what you do experimentally is this kind of mass production. And with LIGO it is more a kind of one-off. But the level at which you have to take something complex and move it to something that's routinized is, in fact, the same." A member of the old 40-meter team agreed with the description of high-energy physics given by Sanders but disagreed wholeheartedly about what it meant for interferometry.

> [1997] [W]orking together to accomplish something great is
> opposite to the atmosphere that the new management would like
> to see. They would rather see people working together, or not, to do
> something that's routine. And that is partly because a lot of the

success of this high-energy physics work is done by this transformation—of turning a difficult project into a lot of small pieces that are routine. So there are pieces of LIGO that can be handled that way—including this matter of computers and data analysis—but there at the heart of it is this piece that you can only ignore for so long at your peril if you want to detect gravity waves. And that is the fundamental interferometer research that is not routine—it just can't be, because nobody's done it before, and it's very difficult.

How did the new management see what they found?

[1997] Within the first few days that I looked at the 40-meter, I saw this clinking-clanking, rickety system. And yet we were making great demands of it. And so on one of my first visits to the 40-meter I met with the people who were responsible for the electronics on it—and I had led an electronics group . . . and so I looked at the electronics they had, and I couldn't believe how badly implemented it was. [It was] not documented—modules that were supposed to be the same were by appearance different, with different components—the differences weren't documented; the way cables and grounds were done—there were a whole bunch of technical things that were incredibly sloppy. So I spoke to the electronic engineer and the junior engineer who worked on these things and found out that they were very, very analytical—the senior guy—and understood the details of the input circuit, and so on, but paid no attention to producing an intellectual work product into an object that was consistently reproducible—which in many ways is the basis of scientific research.

The new management felt they had to transform this approach. They sacked electronics engineers and replaced their designs with the kind of digital electronics that would eventually prove vital on the full-scale interferometers, and placed their own people in charge.

The 40-meter team thought that though digital electronics would be useful in the long term, it was not a pressing matter given scarcity of time and resources. The electronics had been initially constructed to suit the demands of a more exploratory approach, and they thought that it was good enough to do the job it was needed for. They saw their chief contribution as the deep study of noise sources such as would form a foundation for interferometry science for LIGO I and succeeding generations. The 40-meter team, we might say, were not ready for either their device or themselves to be digitized.

The new leadership believed that the 40-meter was no longer producing interesting results and that in any case, as technically realized, its time as an instrument for fundamental research had passed. They said that unknown noise sources, "a bit of the slogan of those you interviewed," would be specific to the 40-meter and would probably have to do with the lack of standardization—"I saw this clinking-clanking rickety system"—with which it had been put together over the years. They said the device's suspension was too crude to go further in researching seismic noise. Moreover, high levels of seismic noise prevented it from being used for understanding high-frequency thermal noise in the mirrors. Also, the laser power was too low for it to be used for studies of noise in the light beam. To study these sources of noise required either special test rigs or the much better insulated and sensitive full-scale LIGO. And it certainly is the case that in a project of this sort, which is pushing the frontiers of sensitivity to the limits, prototyping can go only so far, because building on a smaller scale means that one will not encounter the sources of noise that might confound the much more sensitive experiment itself. (For this kind of reason, the French-Italian VIRGO project decided not to build a small-scale prototype at all.)

The two sides also disagreed about the significance of the 40-meter's "locking" problems. Lock seemed to be achievable only for short periods on the 40-meter prototype interferometer, and the conditions under which it could be achieved were ill understood. Lock certainly could not be achieved at will, and this was something that would most definitely be needed if the duty cycle of the 4-kilometer devices was to stand scrutiny from its public and its paymasters. Yet, other things being equal, locking a 4-kilometer interferometer would be at least 100 times as hard as locking a 40-meter interferometer. Thus, during an NSF review committee meeting I attended in 1997, there was much talk about whether the large LIGO interferometers would ever be made to lock. But members of the 40-meter team claimed that the locking problem had been exaggerated and that once lock had been achieved on the 40-meter, it was easy to sustain. The problem was, rather, that as soon as locking had been achieved for a short time, they would find out what they needed to know and move on to another interferometer configuration; that meant they had to start lock acquisition all over again. The team also claimed that lock would be easier to achieve on the larger interferometers because of their relatively favorable seismic isolation.[1] They believed the difficult thing would be to combine lock with high sensitivity.

---

1. First fleeting locks were achieved on LIGO's 2-kilometer interferometer in September 2000.

About the only thing that the two sides of this dispute agreed upon was that the electronics in the 40-meter was badly engineered—though they disagreed about how important this was—and that the 40-meter work, insofar as it was achieving deeper understanding, was badly documented. Nevertheless, both sides believed that a careful study of the documentation would support them: on the one hand, it would show that the 40-meter team produced nothing after the beginning of the 1990s that affected the design of LIGO, while new people who have worked on the 40-meter more recently have done valuable science on problems such as new interferometer configurations; on the other hand, it is claimed that the documentation would show that nothing deep had been learned about interferometry since the 40-meter team left.

PSYCHOLOGY

Differing conceptions of the nature of the science also appeared when it came to ideas about the personal competencies needed by scientists who were to work successfully on interferometry research. A scientist who left the project said,

> [1997] What I found disappointing was that after two years the project manager still didn't really know what it meant to do interferometric detection of gravitational waves.

In contrast, another member of the management team explained the personal qualities needed for those working on what I have called a project characterized by a high degree of coordination as opposed to autonomy.

> [1997] Once you professionalize, the guys who are very good in the lab where you control everything no longer have their arms around it all. Other people can work very well in that environment. They interface with the experts who built the electronics and understand what they need to of that; they interface with the computer people and do very well at that; and some people can work in this broader environment technically. Some people make the mistake of saying that as soon as you are in this broader environment, it's a management problem—it's not a management problem! The technical part is actually more technical and more sophisticated. So if we talk about making the 40-meter, or look at LIGO—it's technically much more sophisticated than a smaller lab, and one person can't have their arms around it all.

One of the new management team spoke critically about the cognitive style of a leading member of the 40-meter team thus:

> [1997] He was available as the Oracle of Delphi—he was available as a guru. One of the things I've noticed in scientists is that there are people who are doers, and then there are people who somehow fall into a mode of observer-critic, or gadfly, where they use how smart they are to point out problems, but never to resolve them or solve them, or to move past them.
>
> ...He was venerated—he was referred to as "the guru." And so people could ask him questions, and sometimes he would expound an answer.

The "guru" tag, as it happens, was used by those from both sides of the argument to describe this team member, but the old team members used it as a term of veneration—for one who could see more deeply into the problems than others and was always ready with technical insights.

> [1997] [He] was a really bright, deep thinker.... He was a person who did see the forest for the trees. But he saw *all* the trees. And he was a wonderful guru for the project.

Temporality

The decision to build a planned and reliable LIGO I on schedule fed backward into other aspects of the project. I have argued that the strategic choices were necessary, but not every member of the old 40-meter team saw it this way.

> [1997] [I]t's conceivable that I might have stayed at LIGO, given the physical constraint of weak signals and difficulty of building the detectors, if I sensed that the people in charge of LIGO were focusing on that—which I saw as the central problem. And I didn't [sense that]. My take was that the inclination was to reduce it to an engineering problem as much as possible; the management is not technically cognizant—I don't think they ever claimed to be.... But the direction in terms of who's hired—who you try very hard to keep—to stay—how you shape the research, what kind of emphasis you give to the experiments as opposed to the engineering—the people making those decisions didn't know the technical details; I don't sense that they appreciated how hard and how important the physics was.

Here he referred to the promise of the calm after "the sandstorm":

> [1999] The organizational strategy called the LIGO Laboratory did not address my concerns that the LIGO effort was heading in the wrong direction— namely that of emphasizing relatively easy problems in engineering, rather than focusing efforts on improving detector sensitivity. . . . Yes, LIGO could fail due to inadequate engineering. But I believed then [at the time he left], as now, that LIGO is more likely to fail from inadequate physics research. I hope it succeeds in detecting signals, but the longer an experimental research program focused on demonstrating fundamentally predicted sensitivity levels is deferred, the fainter that hope becomes. The LIGO Laboratory may well be the best organizational scheme for restoring that focus. To date, however, I've seen no evidence of it generating the needed research, or even a clear vision of how to accomplish it.

In other words, this scientist wanted resources shifted out of the immediate project of building a LIGO to a specified set of sensitivity targets and into what he saw as intrinsically less controllable work concerned with much higher levels of sensitivity. In November 1999, he told me,

> [1999] Extending your [Collins's] offer to fantasize a bit [about what members of the 40-meter team might have done had they still been working on the project], I would commission a separate test interferometer—perhaps with 10-meter arms—to study the recycled optical configuration. And to really indulge in the fantasy of doing things right, I'd work at an intermediate scale—a few hundred meters—to extend the presumed 40-meter success before jumping to 4 km. Of course, not all of this could have been accomplished in the two-plus years since I left. At a modest level of effort, I believe we could have retired much of the risk associated with displacement noise.

The project manager saw it from the opposite angle.

> [1997] [W]hat came out toward the end was that he thought what should have been done was to work longer with the 40-meter, or a slightly longer instrument, for maybe years or a decade, then you'd know enough. Which was kind of a childlike misunderstanding of what's going on in the world. The fact is, someone, his mentor—Robbie Vogt, who he admires—got the thing funded. We're building it. And it is a discovery instrument. So in some sense, "Get on the bandwagon." But he really wanted to stay down there and

study these things to his satisfaction. And, you know, at a university, I guess, one should be able to do that…but this is the LIGO project!

In April 2000, Bob Spero, one of the principal scientists in the 40-meter team, predicted that LIGO would not reach a strain sensitivity of 10^{-20} until 2002. This has turned out to be about right. At the same time, he predicted that the interferometer would not improve by a further factor of 10, to the design sensitivity, until three years later—2005.[2] A still clearer prediction would be made in 2003 (see chapter 41).

The Spero prediction contrasts with a promise by LIGO to have a full year's accumulated data at its design sensitivity by the end of 2004.[3] We will soon know how it works out, though it has to be stressed that even if LIGO does not meet its promises on time, this does not prove that the intermediate-prototype route would have achieved the goal in less time.

CULTURE

To allow the 40-meter team to do what they wanted the way they wanted to do it would have required not only a shift of resources but also a loosening up of the consensual work style that goes with a coordinated project. Members of the old 40-meter team experienced what they saw as pressure to conform.

> [1997] From the first day they singled the old scientists out, verbally and actually. They came up with names like "the old guard" or "the old critical guard."…My feeling, from the beginning, was that they didn't pay any attention to us.

But of course the 40-meter team's vocal distrust of the new work style was always going to prevent "new LIGO" from running smoothly, because it would always represent a body of discontent at the heart of the project.[4] In

2. Emails to me April 25, 2000, and February 14, 2003 (the latter giving permission to go "on the record" in response to my specific request).

3. This promise has slipped somewhat, and the very fact that it existed has been somewhat muddied; more recent statements talk of completing the year's science run by 2006. Barish tells me that the full-year's data by the end of 2004 was a "plan," not a firm NSF "milestone."

4. Members of the 40-meter team tell me that they tried to follow the new work style but found it made them much more inefficient. They claim their appearance of recalcitrance stemmed from their readiness to confront the management and *argue* for the inefficiency of the new way of working rather than from principled objection to it.

other words, the group presented a problem of internal team politics for the new leadership irrespective of scientific considerations, and it would stand in the way of their changing the cultural environment to something like that of high-energy physics. Any change in intellectual and work style has to involve a social as well as an organizational change.

CONCRETE SUCCESS

In the story of the 40-meter prototype, then, we find disagreements on all three cognitive dimensions: psychological, temporal, and cultural. The new management considered that the 40-meter team were incompetent or uncooperative, whereas the 40-meter team thought the new management were insensitive, starved their effort of the resources needed to meet the demands made on them, and failed to appreciate the good, intuitive-style work that had been done on very difficult problems. The new leaders thought the 40-meter team did not know how to do complex science of this kind in an efficient and orderly way, whereas the 40-meter team believed that their new leaders failed to understand the nature of the problems and the "three steps forward, two steps back" character of interferometer science. The new management considered the 40-meter team had refused to face up to the need to get LIGO I up and running, whereas the 40-meter team thought that the new leadership were too ready to neglect difficult scientific problems in order to concentrate on solving easy but uninteresting technological problems.[5]

The difference in view is clear enough in the way the new management treated not just the 40-meter team but the whole 40-meter prototype. They did not feel that they could learn anything from the prototype that would be useful for LIGO. Now they had to move straight on to LIGO proper and to escape from the paralysis of being continually in a state of preparation for the move. Field notes I made in 1997 describe an NSF review meeting held at Caltech where the 40-meter had just achieved a considerable success by locking in a new configuration.

> [Caltech, October 1997] [T]he 40-meter project had been robbed of personnel
> whenever LIGO needed more person power. People had been taken from the

---

5. Members of the 40-meter team point out that in their new location, JPL (the Jet Propulsion Laboratory in Pasadena), they are even more under the yoke of budgets and timetables but working happily, suggesting, they say, that it was not their unwillingness to accept big-science working practices that lay at the heart of the problem. The fact remains, for whatever reason, that that was not the way they succeeded in working on the 40-meter.

40-meter to work on the problems of distance and alignment control for LIGO. In the small group meeting arranged to discuss the role of the 40-meter, Gary Sanders at one point made it clear that he would sacrifice the goals of the small interferometer for the sake of improving the chances of LIGO; he did not see the 40-meter as the route to the design of LIGO. It was as though the 40-meter represented the aspirations of the "old guard"—their false god, whereas LIGO was the true god. The 40-meter (I am overdramatizing somewhat here) was to be sacrificed not only literally but, and more important, symbolically, for the machine that represented the new management philosophy.

Thus, though the postdocs who now ran the device seemed remarkably competent, if it were to succeed it would be in spite of management decisions rather than because of them. And the postdocs were doing a great job.[6]

It is clear that a cultural and social transformation had been accomplished. In the words of one of the new leadership team,

> [2000] *Respondent:* By thought, familiarization—
> *Collins:* Socialization?
> *Respondent:* Socialization—I guess that's a word you might know—something's happened.

I have argued that irrespective of scientific considerations, the new management had no choice but to build a full-scale LIGO and build it quickly and effectively. I have argued that the legacy of Vogt's regime left no choice but to take the money and make with it something that outside agencies could see and touch. At the very least, this would mean that management virtuosity could be offered as a proxy for positive scientific results. And first indications are that the new management succeeded. I'll use a passage from my field notes from a meeting at Hanford in 1997 to evoke the success.

> [1997] [T]he vans [from the hotels] drove along metaled roads and drew up in front of a glazed entrance with a small, paved forecourt. Between the

---

6. The small group meeting on the 40-meter was a remarkable social occasion. About 30 men sat in an office in the basement of Caltech, while side by side on a sofa at the side of the room sat the three young women who were now taking the scientific responsibility for the device. They were cross-examined and answered competently in a matter-of-fact tone, giving a very good account of themselves. For those interested in gender roles in science, it would have been a fascinating moment.

concrete paths the desert dust lay as dirty as ever, and though the building was not exactly elegant, it did speak of civilization.

Stepping through the entrance, the members of the LIGO Program Advisory Committee found themselves in a small, warm lecture hall with spotlights, plastered walls, and rows of blue seats. About fifty people filled the room and the meeting began.

Barry Barish opened proceedings with a telling remark. He said it felt to him just like a meeting of the Superconducting Super Collider team. Somebody in the audience, remembering the fate of the SSC, shouted out, "You don't mean that, do you?"...

Barish introduces the principal members of the team who have brought LIGO to its present state of near readiness. Everyone is smiling. Otto Matherney, who had been in charge of the construction project, says, "A year ago there was a lot of sagebrush and a lot of tumbleweed blowing about, and you will see that we have built a facility and we are in here." We are told that the toilets flushed for the first time just two days ago, but that the tap water was not yet cleared for drinking. Again, what we see is the team's ability to get things done when they have to be done. Fifty important people were to come to the site for a meeting, and a week ago the site was not fit to house them; but long ago, Fred Raab had agreed that the PAC meeting would be held here, and here it was—functioning as it should—at least, nearly so.

LIGO Hanford was now a place fit for professionals. Where before there had been trucks, builders, dust, chemical toilets, dust, and more dust, now there was warmth, cleanliness, carpets, offices, email terminals, flush toilets, and a kitchen that could make the coffee to go with the snacks that would satisfy fifty people in suits. The steel tubes that made up the arms of the vacuum system were complete and covered from end to end with concrete covers. The five buildings, one huge one at the corner, one midstation on each arm, and one end station on each arm, were complete. The number of people gathered, the finish on the structures, and, for me, the transformation that had taken place in a mere six months, gave the search for gravitational waves a very real feeling of solidity, and energy. These people had shown "can do," and it was hard not to believe that if they could do all this, they could detect gravity waves.

There are very few who are ready to criticize new leadership's approach to building LIGO's facility. And it may well be that the success of this construction has been responsible not only for keeping the project alive but for attracting many more creative physicists to it. The building of the facility, though it may seem little more than concrete foundations and steel tubes, is far from a trivial thing. In many ways it is a stirring accomplishment. I have seen theoretical physicists go away from a tour of the partially completed facility filled with new optimism for the future of gravitational wave astronomy, even though, as cannot be too strongly stressed, it is not concrete and steel that will detect gravitational waves.

When the subject is interferometry, however, the matter is less obvious. That the new management has not neglected the longer term future is clear from the way they have nurtured plans for Advanced LIGO and prepared the way for subsequent generations of interferometers in the way LIGO I's facility has been constructed. But there were still choices to be made about priorities. We will see how the future unfolds.

CHAPTER 37

REGIME 4 (AND 5): THE COLLABORATION

REGIME 4

Whichever view one takes of the amenability of interferometry to routinization, the need for skilled interferometer scientists is as great as ever. Furthermore, in this field there is, or certainly was, no body of trained postdocs who can be brought in to fill a gap when someone leaves. Robbie Vogt put it this way:

> [1996] Barry's [Barish's] philosophy works perfectly well in high-energy physics, because for every person in his team, if he loses that person, there are probably a hundred outside who are equally qualified. But for every person who is knowledgeable in the art of gravitational wave interferometry, there is nobody out there.

How, then, is LIGO coping with the loss of interferometer scientists from the 40-meter team now that the facility is complete and it is time for interferometry to take center stage?

The answer is that they seem to be coping well.[1] By the turn of the millennium, the LIGO Laboratory was growing fast and training up new talent, and MIT was installing special facilities featuring full-size vacuum chambers in a short interferometer to do fundamental research on noise. Even more marked, however, was the contribution being made to LIGO by another organization, the LIGO Scientific Collaboration.[2] The LSC allows LIGO to avail itself of the scientific skills of the worldwide interferometer community. It could be said that the loss of the 40-meter team has been less damaging than it might have been because it has been replaced by new talent drawn from a far wider pool. If the change in management style between the early 1990s under Vogt and the late 1990s when the LSC began to contribute significantly were to be represented graphically, the first diagram would show a narrow, vertical, exclusive organization with Vogt at the top, while the second diagram would include a wide horizontal structure drawing in a still relatively uninitiated pool of talent from distant locations. This change has required an open approach to both management ideology and management practice, perhaps another feature drawn from the international collaborations of high-energy physics. Unlike earlier transitions, there has been no change of leadership personnel as "Regime 4" supplanted Regime 3, and no great upheaval—it is, then, entirely my fancy to describe matters in terms of four rather than three management regimes.

The origin of the LSC can be conveniently associated with the McDaniel Report, commissioned by NSF to recommend how best to organize LIGO to produce science.[3] The model produced by the McDaniel Committee was again drawn from high-energy physics: the idea was that there would be a facility on the one hand—equivalent to an accelerator producing a stream of particles—and a body of users on the other who would use the facility to do their own experiments with their own detectors placed within the facility.[4]

1. Let me stress again that, now that they are gone, the new leadership does not consider that the loss of the 40-meter team scientists represents a problem; as far as they are concerned, the problem was the lack of productivity of the 40-meter team even when they were in post. At the time, they did try hard to hold onto the leading figures.

2. I asked Barish how much he thought the changes from LIGO Project to LIGO Laboratory and then LSC had been planned and how much they had been a response to events. He said he thought that it was a mixture of both. I suspect the importance of the contribution of the LSC has come as the biggest surprise, and this was echoed in a remark of Sanders. Bear in mind, however, that the importance of an LSC-like organization is anticipated in the comments of NSF review committees that met in 1992 and 1993.

3. "Report of the NSF Panel on the Use of the Laser Interferometer Gravitational-Wave Observatory (LIGO)." The panel met on June 24–25, 1996.

4. Boyce McDaniel was himself a high-energy physicist. I have been told that what the McDaniel Committee would recommend was well understood before the conclusions were reached officially.

But it is much harder to separate the facility from the experiment in LIGO as compared to particle accelerators. There is nothing for LIGO to produce that is equivalent to a stream of particles and would allow us to say it is "working" prior to the installation of sensitive interferometers. There are plenty of ways in which it could fail to work—vacuum leaks, outgassing of metals or seals, poor seismic characteristics, and so forth—but it is still only the success of the interferometers that can provide the criterion for the proper working of the whole thing. Thus the distinction between teams of users and teams of device operators that works well in high-energy physics has no real meaning for LIGO and its non-US equivalents, at least not in the near term.[5]

But, because LIGO has kept to its schedules, it promises to be the first large interferometric installation to gather scientific data. And because LIGO has more than one interferometer, they can check on each other, and it is likely that for some time it will be producing the best data. Scientists can be attracted into LIGO's orbit by the promise of a share in the credit for producing this data, and a share in a project that commands far greater financial resources than any of the other similar projects in the world. Thus, members of the relatively poverty-stricken Scottish-German team have been constructing high-quality interferometers for longer than the Americans, but they have been forced by their budget to keep the length of the arms down to 600 meters, limiting their long-term sensitivity. The existence of the LSC has provided a means by which they can now lend their ideas and talents to the much larger LIGO in exchange for a share of the glory. Many US and Australian universities are also providing specialist expertise in lasers, suspensions, mirror materials, and so forth. Consequently, for at least the first half decade of LIGO's life, members of the LSC have been making contributions to LIGO I and are well in the lead of the LIGO Laboratory in pressing forward with Advanced LIGO. Indeed, Advanced LIGO has been described as the creation of the LSC.[6]

I will take as my principal example of this relationship the contribution of the European GEO600 project to LIGO, because it represents the cleanest and least ambiguous case for the argument being mounted here.[7]

5. I am assured that the separation works well as an accounting device for the distribution of funds, but there is no clear separation of personnel.

6. Up to 2003, the LSC was led by Rai Weiss, who proudly advanced its successes at public meetings, praising the creative scientists generously, irrespective of their rank and nationality.

7. I am sorry to say that this will involve me stressing the contributions of the Scots-German group, which, as a British citizen myself, trying to be impartial, I find somewhat embarrassing. It is, however, by far the most striking case, and I am strictly *following* the lead of LIGO spokespersons, such as Rai Weiss, in highlighting the GEO contribution.

The Glasgow group, it will be recalled, were among the first to start designing large-scale interferometers but, they could not obtain funding for their plans. Eventually they teamed up with the German group to build a much smaller joint interferometer than either team wanted. The joint Italian-French VIRGO group, though entering the field much later, did obtain funds in the $100-million range to build a kilometer-scale device that had a chance of competing with LIGO, while GEO was scraping up funds amounting to less than $10 million. GEO, by doing much of the building themselves, as in a traditional laboratory project, and by keeping everything small, were able to proceed.

Heartbreaking though this stunting of their ambitions has been for GEO, in some ways it has turned out to be an advantage. GEO has always been able to work as a small, coherent team with none of the management upheavals or tensions of LIGO. Nor has there been anything like the management overhead in terms of transparency and external scrutiny that LIGO has experienced. GEO has been able to do its work within the traditional privacy of a laboratory project precisely because it is cheap, whereas LIGO has been able to do nothing without scrutiny because it is expensive. In GEO it is still interferometer scientists who make the choices, so it does not feel like highly coordinated science. LIGO, on the other hand, because of the necessary degree of oversight, has had to be relatively conservative in its technical choices in order to be certain that things would get built on time and work to specification. GEO has experienced the opposite pressure: to have any chance of competing in sensitivity with the much larger LIGO, it has had to work continually at the far edge of technology. That is why GEO has taken the lead in developing features that will go into Advanced LIGO, such as the multistage mirror suspension pendulums and advanced detector configurations. GEO's contribution to Advanced LIGO could be said to be another triumph for relatively autonomous science.[8]

Nothing is being claimed here about the *psychological* advantages accruing to GEO—GEO personnel do not seem to have *invented* anything major in the period since they started building their device—the special cognitive

8. Weiss, in characteristically colorful terminology, told a meeting of the LSC that the Glasgow group had given the whole community a "kick up the tuchus" (Yiddish slang for backside) with its advice, based on experience, to press ahead with signal recycling—an advanced configuration—at the earliest opportunity. To spoil the picture a bit, LIGO's Fabry-Perot arm configuration is more advanced than GEO's simple reflection, but this is left over from the early Drever days.

I want to avoid the impression that LIGO I is somehow routine and boring, while GEO is exciting. This is very far from the case: to make decisions and spend huge sums before the scientific facts are in is far from boring. The sheer size of LIGO is a matter of massive scientific bravado on the part of the agencies that fund it, the politicians who support it, and the managers who manage it. The GEO team would have been delighted to have been funded to the tune of hundreds of millions of dollars even if it did bring more management oversight with it.

feature comes under what I have called the *temporal* and *cultural* dimensions: more adventurous technologies could be championed sooner and longer under this less scrutinized regime. Thus, we can say that small, disorganized science was successful at the very outset of the LIGO project when Ron Drever was inventing many of its major features, and that small, unscrutinized science has again been important at the very outset of Advanced LIGO. Ironically, GEO would have been an ideal setting for a scientist like Drever (personality factors aside). We could reinforce this view by looking at other LSC contributions—for example, the almost completely independent Moscow State University group, with no money to build devices on any significant scale, has made deep and vital contributions to the whole scientific outlook of gravitational wave science as well as crucial technical advances.[9] Regime 4 has, then, seen an injection of the small-science approach at just the point when Advanced LIGO was a *developing science* rather than a *mature science*. If the 40-meter team was the major representative of autonomous science within LIGO, then I would argue that the much larger LSC is their replacement.

Of course, it is the combination of the LSC's autonomous science and LIGO's coordinated science that will bring about gravitational astronomy, should it be Advanced LIGO that does it. The argument is simply that both kinds of science will have made their necessary, but individually insufficient, contributions.

"REGIME 5"

I had finished writing up the main themes of this analysis by the end of 1999, and the following two paragraphs are taken, unaltered (except for the excision of some irrelevant material between them), from a draft written at the time:

> At the turn of the millennium, LIGO is going through a peaceful and happy stage. Prospects for the funding of LIGO II look good, while the facility is almost completed and installation of LIGO I has been going smoothly.[10] If

9. I refer to Braginsky's quantum nondemolition ideas, the Moscow group's careful experimentation and analysis of "systems of low dissipation" such as pendulums, their advancing of the cause of sapphire mirrors, their very delicate measurement of the quality factor of sapphire, and their alerting of the community to the problems of thermo-elastic noise in sapphire.

10. "LIGO II" was the original name for "Advanced LIGO." The usage was changed about 2001, when it was thought that asking for funds for a "LIGO II" might be taken by politicians to be giving a hostage to fortune—the funding of a LIGO III, a LIGO IV, and so forth.

there are to be difficult times for LIGO I, they are still a year or so in the future. But the potential for strain between big science and little science has not gone away. The members of the LSC do not have the same concrete commitment to LIGO as the LIGO leadership. That is to say, any member of the LSC can forget about LIGO at any time and go back to doing whatever work they please. Members of US universities can decide to do a different kind of science whenever their current grants run out, or cease to put all their efforts into the big project; members of, say, GEO, can abandon LIGO and go back to concentrate fulltime on their European work. There are three things to keep the members of the wider community working for LIGO and meeting the schedules they have promised. One is the written agreements they have signed as members of the LSC—the so-called MOUs (Memorandums of Understanding); but these are unenforceable. There is the prospect of having their names on the first discovery papers—the right of any member of the team who makes a substantial contribution. And there is the sense of membership of the scientific team founding gravitational astronomy and the more generalized peer recognition they hope to gain for their contributions.[11]

. . . But these ways of controlling scientists were not found to be adequate in LIGO's earlier stages. Barish and Sanders found they could not rely on teamwork and scientific self-interest alone to pull things through according to their model, and therefore the same question remains today: To what extent will the leadership feel it necessary to organize and control the work of the members of LSC, each making only a small financial contribution to the project, in order to guarantee responsible stewardship of the $450 million? And yet some of the key contributions have been made by the LSC to the spending of the $450 million only because they were made under relatively autonomous, small-science-type conditions. This problem must be resolved.

What happened very shortly after this passage was written was another change in management style in the direction of coordinated science, this time applied to the LSC. I learned about it in January 2000.[12] The prospect of LIGO II/Advanced LIGO gaining significant funding meant that it was felt that additional LIGO I–type control was needed. Thus, in 2000, LSC

11. A fourth, rather less grand reason is that membership of a successful project will lead to jobs in the future. For substantial discussions on "the reward system of science," see, for example, Hagstrom 1965, Zuckerman 1969, Merton 1973, and Crane 1976.

12. In retrospect, I could have seen it coming if I were a more skillful observer. At the NSF meeting in Washington in October 1999, which I attended, the straws were blowing in the wind, but at the time I did not spot them. (At that meeting I was concentrating on the way management virtuosity was being put forward as the basis of the case for funding LIGO II.)

work specific to Advanced LIGO was brought under the control of the LIGO Laboratory—the Caltech-based group. I asked the project manager, Gary Sanders, why this was happening, and he explained as follows:

> [2000] Why are we doing this? Because our sponsor, to say nothing of ourselves, need assurance that we're going to manage this successfully. There needs to be someone held accountable, there needs to be someone organizing things and measuring progress and able to handle the scrutiny.... Any project this large will get scrutiny from the upper reaches of the NSF and from Congress.
>
> Now that's a kind of cynical view: the LSC is subjecting itself to the management of the LIGO Lab so that it will survive Congressional scrutiny—a cynic would say it that way—a realistic person would say a loose federation of collaborators needs to have leadership and management so that the project can be carried out successfully. Decisions have to be made. Processes have to be put in place.
>
> What we're doing now is spreading—promulgating—the management technique that we've used inside the LIGO Laboratory across the LSC. And we have to have a balance between minimization of risk and assurance of success to whatever extent that's possible without distorting or abusing all these researchers. They have to be able to feel the same intellectual stimulation, but they also have to understand that what we're taking on is a mutual responsibility.

Henceforward, then, LSC Advanced LIGO projects will come under the kind of tight control typical of the inside of the LIGO Project rather than the much looser control that NSF exercises over its grant-winning scientists. For example, the LIGO Laboratory demands weekly progress reports from all its scientists, whereas NSF leaves principal investigators to run their own projects, the accountability happening on a project-by-project timescale rather than a week-by-week reporting in. One LSC scientist explained his fears about the potential intrusiveness of LSC control, feeling that he would no longer be able to decide changes in such small things as choice of equipment without passing through the bureaucratic mill. Furthermore, the plan is that in due course, as the current NSF-funded projects expire, direct financial control of these LSC projects will pass to the LIGO Laboratory. As I was told by an NSF spokesperson,

> [2000] [S]tarting in 2002, the plan is that part of the money is going to be centralized.... The supervision of graduate students and all should be

handled locally. [But] money for engineering and large-scale equipment will
be handled through the laboratory and tightly coordinated.

According to the model put forward above, this kind of change is inevitable as the LSC plans for Advanced LIGO to come under the kind of
external scrutiny that is associated with the spending of $100 million or
more. This is what fits and backs up our general analysis of the evolution of LIGO management. In the early period of Advanced LIGO, scientists
have been able to be more adventurous in choosing their lines of research
and their technological options than they could have been under the con-
current heavily scrutinized and timetabled LIGO I organization. They have
been able to do this because the LSC members have been in different institutions, separate from the LIGO Project or Laboratory and away from
temporal and cultural constraints. But now that Advanced LIGO's design
choices are being frozen, and the science is maturing and beginning to
draw the scrutiny that is bound to be associated with big funding, the
LSC institutions are being gathered into the established organization. The
question remains, however, about the need for pockets of unaccountable
inventiveness for any of the future generations of interferometers, the successors to Advanced LIGO if there are any.[13] One would imagine that the
same sequence will be repeated if research directed at still more advanced
generations of interferometers begins to command significant resources.
This will be another test of the thesis.

CONCLUSION

I have traced and explained the four, or perhaps five, management regimes
in terms of autonomous and coordinated models. The very early days of
interferometry were a period of developing science that flourished and produced appropriate creativity with almost no organization. I have suggested
that an important transition in a developing science is the moment when
control over the direction of scientific work is taken from the inventors and
passes to those with less of a commitment to any particular line of development. In the case of LIGO, that moment happened when Robbie Vogt took
over the project. The Troika period was the first attempt at a transitional
step and it failed, being neither one thing nor the other, and providing only

---

13. Cryogenics and diffractive interferometry, potential technologies for a successor to Advanced LIGO, are already being funded by NSF.

the conditions for conflict without the means of resolution. In contrast, the Vogt period moved interferometry away from the autonomous model and into the coordinated science era by removing control over ideas from the scientists. This allowed design features to be frozen, engineering designs and cost estimates to be worked out, and the project to be funded.

The Vogt regime, however, failed to recognize the need for transparency, and the internal organization and timetables that go with it. It failed to allow an element of control to pass to the political sphere, and it did not meet the need to draw on a wider range of talent than can be found in a small, strongly led team. The Barish-Sanders regime finally delivered the timetabling and transparency that have brilliantly produced LIGO I's facility and promise to give rise to interferometers with a specified, relatively conservative level of sensitivity and an ambitious duty cycle. One might look at this next stage as passing an element of control still further away from the inventors and into the sphere of public accountability.

I have argued that LIGO is an unusual big-science project because it cannot work at all without taking a very big step. This means it is essentially trying to solve the kind of very high-risk, ill-researched problems usually associated with a *developing science,* though with big-science spending, facilities, and associated scrutiny. This, I have argued, means that there is a strain between autonomous and coordinated science and technology at the very heart of the LIGO project. The intimate relationship between ideas about the nature of the scientific and technological knowledge, the way the work should be organized, and the strains that this can produce was seen in the transitional stage of the 40-meter laboratory. The dramatic nature of the management transitions were a consequence of the need for the two kinds of science to be successful along with disagreement about the timing and nature of the transitions.

In LIGO I, we have seen a steady increase in the degree of coordination. It is probable that, irrespective of the technical and philosophical arguments, this change was necessary for LIGO's survival. The display of management efficiency has also made it at least conceivable that funding for Advanced LIGO will be agreed before LIGO I has given rise to any significant scientific results.

It could be argued that the project would have benefited still more had the transition from autonomous to coordinated science been less gradual, since the in-between stages were not a great success. On the other hand, a sudden transition (on the lines of the Manhattan Project), with all its associated upheavals, may only be possible or desirable if funds are assured (as they were with the Manhattan Project); in LIGO's case, funds were not assured until the transitional Vogt regime had won them. If this argument is correct, it is the Troika alone that represents an avoidable organizational mistake.[14]

The transition to "Regime Five" reiterates the story of the earlier changes. A period of relatively unsupervised work provides the creativity, and then a period of coordination satisfies the demands of a wider constituency once large money is involved and a degree of maturity has been achieved.

In chapter 31, we asked whether there was a kind of science—developing science—that could be done well only under one kind of organization, while only mature science could be done under another. Throughout the analysis, we have seen the way that political necessities and exigencies have confounded the two questions. Nevertheless, it seems that there is enough in this story to suggest that there are indeed two kinds of science best suited to two kinds of organization. The intriguing thing about LIGO, as opposed to other big sciences, is that it is small-science-like in that it is taking such a large step forward in one go. The future may tell us which kind of science interferometry is.

14. Though even here Rich Isaacson argues that the Troika experience was a necessary step on the learning curve for the scientists involved.

PART V

BECOMING A NEW SCIENCE

CHAPTER 38

POOLING DATA: PROSPECTS AND PROBLEMS

[1998] [M]ost of us have been independent entrepreneurs for years, and it's not an easy experience—I speak mainly for myself—but I think there are a few others of you—where it is that for the first time you really have to collaborate, not only in the technology, and you can't always hold that little idea that you had—one's personal idea that you got credit for—but now you find it's being applied and it's universal and it's become something that is part of the common domain. And one has to look at the joy of doing the common thing rather than the individual thing. And I think that's now going to extend clearly into the data analysis, because we're dealing with . . . a fairly big project. . . . [The problem is,] how do you actually deal, in concert, together, where everybody has been fiercely independent up to now.

—Senior experimenter addressing a gravitational wave
data analysis workshop in 1998

[2001] It is obvious on ethical grounds that you should not keep the data, but it is also obvious to me that you've worked so hard for this data that you consider it yours. Maybe wrongly so, but you want to be the first to look at it, and to analyze it, and to get credit for any results.

—Interview with junior analyzer of data

[1998] Now, the question is . . . when is the end of the competition, when is the
beginning of the collaboration?

—Leader of the Italian-French VIRGO interferometer group

TECHNICAL INTEGRATION AND SOCIAL COHERENCE

The whole worldwide enterprise of gravitational wave detection is like the
relationship between the LSU bar group and the Frascati team that I discussed in chapter 22, but writ large. In other words, all the detectors in the
world need one another, but they do not always share a scientific culture.

Sources and Strategies of Detection

The laboratories need one another because, for most gravitational wave signals, the more independent confirmations of coincident pulses, the better.
This does not apply to every class of signal. As explained when I described
the contents of the "Blue Book" produced by the MIT team in the 1970s,
there are four classes of gravitational wave signal. What are referred to as
burst sources emerge from supernovas and other astrophysical catastrophes
whose waveform cannot be calculated. Burst sources include the "wild card"
of unknown sources, in which some gravitational wave scientists invest a
lot of hope.

Inspiraling binary neutron stars have a calculable waveform, which
makes detection easier because the signal can be matched to the waveform template. (Inspiraling binary black holes should emit a strong signal
but have a waveform that as yet cannot be fully calculated.)

Continuous sources include such things as pulsars, which are rapidly
spinning asymmetrical neutron stars. Even though the signal from a pulsar
is weak, because the frequency is known and constant it can be integrated
over a long period. This can make the detection of a continuous wave convincing enough to announce even if only one detector is involved.

The fourth potential source of signals is the stochastic background. This
is a combination of cosmic background gravitational radiation and the
aggregate of many faint examples of the first two sources described above.
Such a signal is best detected by correlating the output of two detectors
placed relatively near each other.

The first "discovery" of gravitational waves will be couched in terms of a
confidence level: "There is only one chance in N that this coincidental pulse
of energy could have been caused by chance alone." Because of the field's

early history, and the accumulated opposition by LIGO's enemies, the statistical confidence level associated with an announcement will need to be especially high if the results are to be accepted as a "discovery."[1] In the case of the first two kinds of sources described above—bursts and inspirals—more detectors seeing the same event in coincidence translates into more statistical significance. There is a good chance that the first confirmed sighting will be of one of the first two kinds of sources.

For the bars, the skepticism of the outside world is magnified because it matches the agenda established by the dominating interferometry community. In consequence, the pressure on the bar groups to pool their data and to be conservative is overwhelming. Hence the bar groups are setting their thresholds for what might count as a signal so high that, the Frascati group aside, they no longer expect to see any bursts. Interferometer groups, on the other hand, have to plan for the day when they will announce a discovery.

There is an arithmetical corollary to the increase in statistical confidence that comes with coincidences between more detectors that could be important when the interferometers begin to announce results. At the level of sensitivity attainable by the first gravitational wave detectors, visible gravitational events are likely to be very rare—somewhere between three or four per century to three or four per year. Rare events, even when associated with a high level of probability, are less convincing than repeated events. Single events can more easily be glossed to be the result of some kind of systematic error or fluke and probably not be believed until the corresponding large population of events is discovered (see chapter 41).

Now one can calculate the likelihood of an accidental coincidence between two detectors at various signal strengths. If the signal is small compared with the noise, we say we have a low signal-to-noise ratio (SNR). If the SNR is low, there are lots of noise peaks around the level of the signal, and accidental coincidences will be comparatively frequent. If the SNR is high, on the other hand, there will be less chance of an "accidental." Given the rarity of the expected real signals, one wants to make the chance of an accidental still more rare—let us say one wants the chance to be no more than once per thousand years. Suppose one has two detectors and that the SNR needed in each single detector to make an accidental this rare is 10 to 1. If one has three detectors, one would not need such a high SNR in each to

---

1. We may also expect that once several gravitational events have been agreed to have been seen, the statistical standards needed for an observational claim will fall. This, too, is a matter of sociology, or a more mathematical counterpart of it known as Bayes' theorem.

be equally convincing, because the likelihood of an accidental coincidence in three detectors is so much lower.

Let us suppose (these numbers are a fiction) that with three detectors one needs an SNR of only 5 to 1 in each instead of 10 to 1 to gain the same level of confidence.[2] Effectively, this doubles the sensitivity of the detectors—three detectors can be used to look for signals of half the strength needed by two detectors while still giving the same confidence level. Now, the displacement caused by gravitational waves diminishes linearly as the distance of the source, so doubling the sensitivity means one can look twice as far into space for similar events. It follows from simple geometry that doubling the distance to which you can see means an eightfold increase in the number of potential sources.[3] So, if increasing the number of detectors by one does decrease the needed SNR by something like 50%, the potential source frequency will be increased eightfold, and this is a huge improvement given the small number of sources that are likely to be within range when the detectors first go on air. It would be far more convincing if scientists could say that they were seeing, say, two or three events a year rather than one every three or four years.[4] This is another way of explaining why collaboration between laboratories is very attractive for that class of gravitational wave events whose statistics are improved with coincidences among many detectors. When full-scale gravitational-wave *astronomy* (as opposed to detection) begins, of course there will be no choice but to combine results from national detector projects. This follows from the need for long baselines if the directions of sources of gravitational radiation are to be pinpointed.

The technical advantage of data sharing is only part of the story, however. Think in terms of scientific credit. Sharing data is like joining a betting syndicate: there is a better chance of a share in the prize—a credible discovery claim—but one can no longer win the whole prize for oneself. How does one rank the technical imperatives against the gamble of joining a syndicate? To understand the choices that laboratories make, we need to analyze these broad options in more detail. We can start by considering different kinds of *technical integration* among laboratories on the one hand and different kinds of *social coherence* on the other.

2. I learned this argument from Barry Barish, director of the LIGO project, when we spoke at the Gravitational-Wave International Committee meeting in Paris in December 1998.

3. Assuming that the distribution of sources in three-dimensional space is roughly uniform, which we will take it to be for the sake of the argument.

4. I should add that these events would be roughly in line with theoretical expectations; Joe Weber's early claims were *unconvincing* because he saw far too many events to be compatible with the theory.

Technical Integration

Three points on a scale of technical integration, running from low to high, can be described as corroboration, coincidence analysis, and correlation.

1. *Corroboration:* A laboratory processes and analyzes its data to the point at which it can present them as independent *findings,* for confirmation or otherwise by other independent laboratories.
2. *Coincidence analysis:* A team's data are reduced to an *event list,* which can be transformed into "findings" by coincidence analysis with event lists from other laboratories. This is how data are handled by the resonant-bar collective, IGEC (International Gravitational Event Collaboration).
3. *Correlation:* Raw data streams from different laboratories are pooled prior to processing.[5]

In the case of corroboration, the data and any findings they contain always remain identified with a named laboratory—the origin is always distinct. In the case of coincidence analysis, a finding cannot be identified with the contributing laboratories, since findings emerge only after data have been combined (chapter 22). But there can be any amount of local and independent processing of the data stream before it is submitted for combination with the data from other laboratories. In the case of correlation, contributing laboratories lose their identities and local, independent data processing is minimized.

Social Coherence

Corroboration, coincidence analysis, and correlation can be accomplished more or less easily under different kinds of social relations among the laboratories. Thus, corroboration, though it *may* be an entirely cooperative exercise, is also possible where there is intense competition between institutions. Correlation, on the other hand, implies that the identity of the contributing groups is lost in the process of combining data and extracting

---

5. Krige (2001) describes what amounts to creative crossing of the boundaries between these categories by Carlo Rubbia when his team "discovered" the "W-boson." Rubbia managed to extract statistical confirmation for his insecure findings from another CERN team *before announcing them,* without allowing the other team to attain any status higher than corroborator. Whether corroboration or "replication" should be thought of as a form of data sharing is a moot point, and it is not thus treated in the award of Nobel Prizes, etc. For analytic convenience, we will treat it here as a form of data sharing. We might also think of these three types of data sharing as turning on the degree of "externality" that experimenters give to their data before passing them to others (Pinch 1986). The less local processing, the less externality.

a finding; this means that groups contributing their data to a correlation lose the ability to compete with each other.[6]

Social coherence describes the social relations among laboratories. There are two dimensions to social coherence: system integration and moral integration.[7]

System integration involves a bureaucratically organized combination of groups under some formal set of agreements or common institutional umbrella. All formal arrangements of this sort, as analysis of bureaucracies shows, depend on trust, and on sufficient mutual understanding for the parties to be willing to "repair" rules in mutually acceptable ways whenever they fail to meet unforeseen eventualities. But that said, it is not hard to recognize distinctively formal arrangements of this sort. We might think of such agreements in terms of Ferdinand Toennies' concept of association. Sociologists, drawing on the ideas of Toennies, would refer to the character of such an arrangement as *gesellschaftliche*.[8]

Moral integration means adopting common attitudes, habits, norms, evidential cultures, and so forth and is likely to involve much higher levels of trust and colleagueship than is necessary for system integration. Turning to the sociological literature again, this kind of integration corresponds with Toennies' concept of community. It is a *gemeinschaftliche* arrangement.

An example of a very high level of social coherence would be between two detector laboratories belonging to the same group—such as Joe Weber's Maryland and Argonne labs, or the Guido Pizzella group's Frascati and CERN labs. Argonne cannot compete with Maryland to be first to detect gravitational waves, and Frascati cannot compete with Geneva, not only because they need each other to establish findings but also because the groups are not separate institutions, neither organizationally nor culturally.[9]

---

6. Correlation includes coherent analysis, in which the combination of data takes into account the differential sensitivities of the detectors caused by their different orientations to a source and the distance between them. In other words, the sky is divided up into "cells," and data from separate detectors are combined differently according to which cell (i.e., direction) is being probed.

7. David Lockwood (1964), in an early and influential paper, made the distinction between "system integration," which was to do with the interlocking of institutions within a society, and "social integration," which was to do with the commonality of social norms. I have more or less taken over this idea but made the terminology a little more transparent. Lockwood's "social integration" is replaced with the more unambiguous "moral integration," while I invent "social coherence" as a generic term covering any combination of social/moral and system integration.

8. Gouldner (1954) discusses bureaucracies. Toennies (1987) initially wrote at the turn of the nineteenth century.

9. Which is not to say that competitions over technical excellence or speed and style of work may not arise.

I could redescribe chapter 22's analysis of relations between Louisiana and Frascati as a tension between technical integration and social coherence. Frascati needed a level of technical integration sufficient to conduct coincidence analysis, but the two laboratories had a very low level of social coherence; they were different social organizations, they inhabited a different institutional context, and they drew on different evidential cultures and cleaved to different sets of norms about the way science should be reported.

The history of international collaboration among the interferometers has been a story of the increasing need for technical integration bringing about a subtle but steady decrease in competition and increase in social coherence. This slow change has come about as the difficulty of the search for gravitational waves has become more and more apparent, so that the ambitions of any one laboratory to make the first breakthrough have faded. In this respect, the interferometers are now repeating the history of the bars. The bar collective formed when social outliers discovered that their hopes of making an independent, or semindependent, discovery were fruitless. Unless there is an unexpected finding, this is the broad direction in which the international interferometer community can be expected to go.[10]

High technical integration is most easily managed in the context of high social coherence, because at worst, system integration provides a bureaucratic framework for collaboration; and at best, moral integration engenders trust and facilitates the transfer of tacit knowledge.

LEVELS OF TENSION

Are Data Public Property?

The world of gravitational wave detection does not consist only of laboratories. There are a range of actors with an interest in the way the groups cooperate. At the highest level are the funding agencies, who have an interest in what I will call public property rights. At the lowest level are the interests of individual scientists, young and old, powerful and not so powerful, inside and outside the main teams; they are concerned with private

---

10. There are some cultural differences between the Hanford and Livingston interferometer groups, but these differences are a luxury that can be afforded because the two groups are institutionally bound together, and in any case the differences described are not deep—having more to do with style than views about data analysis.

property rights.[11] As I said in the introduction, in modern physics the train of inference from output to input is enormously long. To transform outputs into something that can count as findings is a complex, collective activity. But who should be included in the collective, and when, where, and how should they contribute? Where should what we might call the locus of collectivization be? Unsurprisingly, different actors have different views.

To start with the funding agencies, these tend to be national bodies with national interests at heart.[12] When we turn to the examination of the strategies of individual laboratories, we find that, say, the French funding agency is concerned to make sure that a large financial input will result in a commensurate gain in national prestige, and this implies a degree of national independence. The British agency, on the other hand, is more concerned to see its investment guaranteed by international cooperation.[13]

Who owns data generated by taxpayers' money? The question may not be pressing for small science, but with a big, publicly visible project, such as LIGO, the rights of scientists are more salient. Should any scientist of any nationality be given access to the data stream on demand, or should the data be treated as a national treasure, the property of American scientists alone? Or is it that the data stream belongs not even to any American scientist but to the LIGO scientists alone—the people who are creating the data? Outsiders have not shared in the half-century-long process of conceiving the ideas, generating the political credibility, and building and refining the detectors. If the data were public, an outsider could win the race even if the only thing they discovered was some ingenious method of data analysis, and this does not seem fair.

To make the data widely available also increases the scientific risk. The balance of risk and reward is different for insiders and outsiders. An outsider has no investment in the future of the project and has relatively little to lose by making a spurious claim.

There is also what we might call a data analyst's regress, which is a subspecies of the experimenter's regress: The only way to tell if one's data analysis is correct is to have it discover real effects, but the only way to find out if effects are real is to analyze data in a correct way. As in the case

11. Like outsiders, young scientists will have invested comparatively little of their professional "capital" into a project.

12. The European Space Agency will share the funding of LISA, a gravitational wave detector in space. French and Italian agencies jointly fund VIRGO, and British and German agencies jointly fund GEO600.

13. The Australian Research Council will not spend large sums until the science has been proved by other countries.

of the experimenter's regress, one way in which analysts try to break out of the circle is with a species of calibration, which in this field is known as a mock data challenge. False events are inserted into a data stream for blinded analysts to extract. But like calibration in general, mock data challenges have the flaw that one can never be sure that the form of the mock data will be the same as the form of the yet-unanalyzed real data.

On the other hand, there is one technical argument for *widening* access to data. A huge amount of data will be generated, and it can be analyzed in an open-ended number of ways; perhaps as many analysts as possible *should* be encouraged to engage with the interferometers' output. As Bernard Schutz, the leading theorist, told me,

> [2001] *Schutz:* The point is that we don't know what gravitational waves we're going to see. It's like taking a picture—you don't know what's going to be in the picture—and you can't just hide that away and say, "I've looked at that picture and there's nothing there." You've to be able to make that data available to all the people who have data rights, but in the long run, both LIGO and GEO are going to make their data public. And it has to be done in a way that people can analyze—that they can look at.
>
> *Collins:* If you had a "full" set of templates, this would be unnecessary, because you could check all possibilities and there would be no need to farm it out. You could then produce an "event list"? [What I am suggesting here is that if they could anticipate all possible astrophysical models, there would be no problem.]
>
> *Schutz:* That's right. That's what particle physicists do. So at CERN they have a big detector; the experiment is designed to detect a certain kind of event, they throw away everything else that could be fantastically interesting physics because it's not triggered, and they only tell you what the result is. You never see the original data.—*They* never see the original data!
>
> *Collins:* But you're saying that in the case of interferometers, because we're probably never going to have a full set of templates, because we're always exploring, you've got to make the data public so that new people can look for new things.
>
> *Schutz:* ...I think controlling access to make sure that people know how to handle the data, how to look at vetoes and so on—that's fine. But I think in the long run we're going to have to open up access in a controlled way, to the scientific community.

This, as we will see, was the conclusion of the LIGO group, that the "home team" would be given a first run at the data but that eventually they would have to become public.

Experimenters and Theorists

So much for the interests of the agencies and the wider scientific community, who stand "above" the groups. What about groups such as "experimenters" on the one hand and "theorists" on the other?

One of the most vituperative exchanges I have ever witnessed at a physics conference was between Rai Weiss, the leading experimentalist, and Bernie Schutz, the leading theorist. One of Schutz's students had presented a paper showing how a data analyst could remove at least some instrument noise from a data stream without referring to the experimenter. Weiss felt this was inappropriate and highhanded. Worse, Schutz had upset the community on a previous occasion by taking data from a prototype interferometer at Glasgow and analyzing it without referring to the experimental team or visiting their laboratory.[14] The experimenters insisted that no theorist do anything with their data until they were satisfied that they had eliminated instrumental artifacts. Another data analyst explaining the depth of the experimentalists' anger described Schutz to me as "someone who has never seen a soldering iron in his life."

Schutz told me that he thought neither experimenter had realized how modest were the data analysts' ambitions. In the first case, his student had intended only to show one way to remove interference from the well-known 50-Hertz mains frequency, but he now realized that the row reflected

> [2001]...sensitivities that I see in most experimenters at some level. That is, that the only person who really understands the data is the experimenter, and a theorist like me, coming in and saying, "This is how you treat the data—modify the data—before you start drawing conclusions," that's something that makes [an experimenter] very suspicious.

In the case of the Glasgow results, Schutz explained that he considered he was doing a demonstration on what might equally well have been randomly generated data; he had used the data stream for convenience, whereas the experimenters had taken it that he was showing that their experiment was badly contaminated with noise.

> [2001] So that was a lesson to me in the fact that experimenters are sensitive from the point of view of personal property in allowing their raw data to be published.

14. This is known as data from the "100-hour run" on the 10-meter Glasgow interferometer.

Schutz acknowledged that he had come to appreciate the position of experimentalists better as a result of these experiences.

> [2001] I do have a feeling of sensitivity toward the experimenters that I didn't have a long time ago. They are reluctant to show their data, they are reluctant to give it away until they're sure it's qualified, because there's a matter of pride, among other things—that people can see what mistakes they've made, or how this or that does not work properly.

Experimenters, we might say, are reluctant for anyone outside their laboratory to see "naked" data.[15]

On the other hand, a young data analyst was convinced that these complaints had more to do with experimentalists' proprietorial feelings about their data—private property rights—than expertise or pride.

> [2001] And my very personal opinion is that anybody who says that it [withholding data] is because you're not sure that other people will analyze it right—I mean, that's rubbish—it's just because you want to have control of your data. This is what I think—I'm convinced of this.... I'm sure that [the experimenter] believes what he says, but I also think that people—It sounds much better, even to oneself, if one says it's because "if you don't have the deep understanding of the instrument, blah, blah, blah, you don't do it right." I think that's partly true, but I'm convinced that's simply not the real reason.

The sentiments expressed here probably go back to the old tensions in science between theory and experiment, which themselves once reflected the difference in social class between those who work with the head and those who work with the hand. Working with the hand is hard and sweaty, whereas working with the head is to share in God's grand design. In modern science, the cultural baggage remains; the great icons are mostly thinkers such as Newton, Einstein, Bohr, and Hawking; the great experimenters, such as Faraday, are revered in a rather different way. But quite aside from the culture of theory and experiment, the fact remains that theorists, as we have seen earlier in the book, can do their work with less investment and therefore in a more playful, adventurous spirit, whereas experiment is always deadly serious or the funding agency will want to know why.

---

15. Just as historians and sociologists are treated as prurient when they look at naked data such as those from Millikan's oil drop experiment (Holton 1978) and all the case studies that have followed.

One cannot blame experimenters for not wanting theorists to come in from the outside and use their hard-won data with any less of a sense of responsibility than goes with their job.[16]

Raw or Processed Data?

Since the lowest level of technical integration on the corroboration–coincidence analysis–correlation scale affords the most opportunities for competitive gain, why would anyone opt for a higher level? The answer is that a higher level of technical integration is technically more efficient. A significant event is more likely to be found within combined data streams when the joint data are raw, because all the information in one data stream can be compared with all the information in the other; no information is lost before data is compared.[17] This notion is clearly in tension with the claim that only experimenters who have worked on the specific machine know how to remove the noise. In Weiss's words,

> [1999]...I think the obligation of the experimenters—in this case when you do start comparing data—is actually for each to take its best cut in making what is a reduced data set that gives you as much as you can establish free of instrument signature....That's where all the cunning, all the crafting, and everything goes....I think that for the first time—at least that's the way I would figure to go—is [for each group] to make its own instrument's idiosyncratic-thing-free data set.

At an earlier meeting, Bill Hamilton of the LSU group had made the same point, drawing on his experience with bars.

> [1998] I[I]f the air springs are low, we see a bus hitting a certain pothole; that means our vibration isolation is not just exactly good. No—each detector is different. VIRGO will be different from LIGO. LIGO Livingston will be different from LIGO Hanford. Hopefully it will be much closer because you

16. Sibum (2003) discusses this distinction and points out (private correspondence) that the degree of responsibility felt by experimenters may be historically specific. In earlier days, when experiment was cheap, experimentalists, too, may have taken a light-hearted approach toward their work.

17. This idea may have been invented by theorist Sam Finn and for a long time seems to have been most vigorously supported by him in the gravitational wave data analysis field. It is now widely accepted. A more recent development is coherent analysis, in which the raw streams are standardized for the relative positions and orientations of the detectors on Earth's surface in relation to a putative source or series of hypothesized sources.

have a common way of taking data . . . you can take that as a prediction, if you want, but I'll bet you I'm right.[18]

It is possible that the interferometers will be less afflicted with these problems than the bars. The interferometers have to be understood better than the bars if they are to realize their advantage as broad-band instruments—instruments that can recognize the pulse shape of gravitational waves as well as register a burst of energy; this means understanding the noise better. And, as we have seen, in the spirit of high-energy physics, the interferometer scientists are using computer simulations to model their instruments—an effort that requires resources well beyond those available to the bars.[19] How complete the models will become, in the light of what we know about tacit knowledge, remains to be seen.

There are two other opposed arguments: on the one hand, local processing of data offers too many opportunities for unwitting statistical massage; on the other, common statistical analysis allows for common statistical artifacts.[20]

The Problem of Recruitment and the Problem of Control

To pool data, one must find others willing to collaborate. Let us call this the problem of recruitment. Once recruited, desertion is always a possibility; recruits must not be tempted to break away and do their own preemptive analysis. Let us call this the problem of control. In the ideal case, high technical integration seems to resolve the problem of control, for there is no data left which can be associated with a single laboratory; but this does not stop individuals from breaking away and analyzing the *shared* data to which they have gained access.[21] Of course, high social coherence solves both problems—by definition.

18. For a discussion of variation between apparently identical paper-handling machines, see Kusterer 1978.

19. Knorr-Cetina (1999) explains that when things go wrong in physics, physicists try to solve the problem by understanding the world more thoroughly, whereas in the life sciences, they just throw the preparation away and start again.

20. An "analytic theory of replication" (Collins 1985/1992, chap. 2, pp. 34–38) holds that corroboration of results is more powerful when the confirming group is sociometrically distant from the originator up to the point where the confirming group still shares enough common scientific background to be a credible replicator—something that requires a degree of sociometric proximity. The exact choice of sociometric distance for suitable confirmers is likely to be a compromise based on the historical and social background of each science and each discovery claim within it. Here we see the social and the technical so intimately interwoven that it may make little sense to separate them conceptually.

21. As seems to have happened in the COBE project—see chapter 39.

Source- and Project-Specific Arguments

These general considerations are qualified by the context of each specific detector project. Both the detector design—the kinds of sources it is best able to see—and the style and structure of project management bear on the decisions. Detector design makes a difference because some sources, such as continuous sources, can be reasonably confidently detected with only one device and need only corroboration from others; correlation confers little or no technical advantage. Or it may be that there are wavebands to which only certain detectors are sensitive, so that collaboration with others in respect of sources that emit in these wavebands is ruled out.

Management style and organization also make a difference because if a group is leading the race for the scientific prize, its members will be happier. Skilled interferometer scientists are in short supply and work in an international marketplace. Only if a loyal staff can be retained can a noncollaborative strategy be maintained.

THE DATA-POOLING DECISION TREE

As we see, the problem of data pooling is complex, with many cross-cutting arguments on a number of dimensions. In practice, scientists tend to solve problems like this by referring to experience and precedent, and we will look at some cases shortly. Here, however, is a summary of the analytic points that contribute to the data-pooling "decision tree."[22]

Evidential Threshold and History: Scientific fields demand different levels of statistical confidence. In gravitational waves, the level is high, which encourages collaboration.

Technical Integration: There are three levels of increasing technical integration, with corresponding decreasing levels of local processing.

1. Corroboration
2. Coincidence analysis
3. Correlation

Social Coherence: Social coherence has two dimensions that make technical integration easier, the second still more than the first.

---

22. One may think of this exercise as analogous to causal modeling in sociology, but without the numbers.

1. System integration
2. Moral integration

Funding Agencies and Public Property Rights: Funding agencies might prefer national independence or international collaboration. Data are gathered with taxpayers' money, and agencies differ in the extent to which they will encourage these data to be made public.

Private Property Rights: There are two problems with getting others to share data.

1. The problem of recruitment
2. The problem of control

Arguments for High Technical Integration: There are two.

1. The more technical integration, the more efficient the analysis.
2. More analysis close to home—that is, less-technical integration—creates more opportunities for statistical massage.

Arguments against High Technical Integration: There are four.

1. Those who know their own experiments are best able to remove artifacts.
2. The more independent analysis there is, the less opportunity is there for common mistakes.
3. Home teams have a moral right to be the first to extract findings from hard-won data.
4. The release of "undressed" data may unfairly expose unrefined experimental technique.

Source- and Project-Specific Arguments: The way a project is managed and the kind of source it can see best will affect the way the decision tree "plays out."

* * *

With this framework in mind, we can visit each of the projects in turn and use it to illuminate their attitudes toward data pooling.

CHAPTER 39

INTERNATIONAL COLLABORATION AMONG THE
INTERFEROMETER GROUPS

The Resonant Bars: IGEC

By 1997, the chance that any one resonant bar group, or even pair of groups, might make a positive discovery had diminished while the threat of the interferometers had increased. At that time, the four bar groups created the International Gravitational Event Collaboration (IGEC). Massimo Cerdonio, the chair, wanted the individual groups to develop joint data-processing methods with a high degree of technical integration. Looking back in 2001, he remarked to a meeting of the IGEC group, "[M]any years ago [we said t]hat we will try to make the different kinds of analyses converge on a single kind of analysis, so we have something much more comparable. So it has not happened. And I want just to let you know it has not happened."

A high level of technical integration in the bar groups has not come about for two reasons. First, the bar experimenters insist that only they know how to conduct the analysis satisfactorily for their own devices. Experimenters feel that they barely understand the sources of noise within their own apparatuses and that there is little chance that an outsider would be able to separate interesting events from noise. In the light of our discussion of the role of tacit knowledge in experimentation, this is very reasonable, and we have already seen the point made by Bill Hamilton of LSU as a warning to the interferometer groups. At the same 1998 meeting of the Gravitational-Wave International Committee where Hamilton spoke, members of the bar groups were asked if it would be possible to analyze the raw data coming from other bar groups. There was general agreement that to understand another detector would involve a site visit of many months at the very least.

The second reason that IGEC maintains only a low level of technical integration is that the groups still cannot agree about what their joint protocol should comprise. In particular, the Frascati Group, reflecting their open evidential culture, want to dig deeper into the noise than the other groups. In our terms, IGEC has imposed system integration on a foundation of low moral integration; a high level of technical integration is difficult without moral integration.

A practical compromise has been worked out which allows each group to process their data to the point where they can extract events—that is, potential gravitational wave signals—and submit 100 such events per day for joint analysis. (I understand that the Frascati Group, in line with their open evidential culture, would prefer that each group submit 1000 events per day.) This is, of course, the middle level of technical integration—coincidence analysis. It allows a great deal of autonomy to each constituent group to decide how to make the "first cut" of the data. Thus at a meeting of IGEC in 2001, a spokesman for one group said,

> [2001] I think that we understand that every group uses different criteria for defining events.... Now, I think that we should still give to every group full responsibility for the event lists—how [events are] defined.... discussion is very useful at the cultural level, especially if it turns out that one of us can change their method or can influence the other to change their method—their criteria.... [But] I think that any discussion should start from the point that every group has their own criteria and takes their own responsibility for defining the event list. Otherwise, I think that the discussion will be too long, and we'll never arrive to the next step in the analysis of our data.

The *symptoms* of lack of moral integration among the bar groups have decreased to the extent that the problems of control discussed in chapter 22, when the Louisianans worried that the Frascatians would analyze *joint* data

independently, are no longer with us even though the joint Louisiana Protocol has been abandoned. But as we saw when I first discussed the founding
of IGEC (chapter 25), this could be a technical matter; the larger number of
contributing groups provides a powerful opposition to maverick interpretations. Second, in IGEC, data analysis is done by a task force drawn from
across the groups, and the majority of its members share a conservative
data analysis philosophy. This, again, is an increase in system integration,
not moral integration. Up to 2003, differences of opinion and approach
among the bar groups were still marked. Since all the collective data analysis done to date by IGEC has yielded no positive results, however, there
has been nothing to cause the cultural differences to be brought into the
open, and cooperation continues in a spirit of friendly disagreement.[1]

There seems little point nor likelihood of a further increase in technical
or social coherence among the bar groups. In any case, their attention is
being directed elsewhere. The Louisiana bar group is cooperating more and
more with LIGO, which means they will rotate their bar out of alignment
with the other IGEC bars at least for short periods (which may get longer
as LIGO stays on air longer). The Frascati group is beginning a continuous
wave search for the pulsar remnant of SN1987A. For this they need to make
a small reduction in the length of their bar, making it marginally less suitable for joint searches with other bar groups. A continuous signal, of course,
can be verified by one group alone. Were they to find indications of a signal,
Frascati would need to seek only corroboration, the lowest level of technical
integration, which puts no demands on either system or moral integration.
And the Frascatians are tending toward independence in other ways, too.

Since IGEC has brought the joint output of the resonant bars under
control, and since the interferometers have secured their funding (at least
in the short and medium term), relations between the two detector communities are improving (with the possible exception of Frascati). In May of
2002, at a workshop on the island of Elba, resonant-mass groups and interferometer groups were spoken of as equals. My field notes made during
the afternoon of May 20 reveal that the bars *as a group* were not making
positive claims but putting forward new upper-limit records. That is, they
were stating with confidence that they had *not* seen gravitational waves of
more than a certain strength during a certain observational period. They
were also setting records for continuous running in coincidence (without

1. Except that in late 2002, the Frascati group would publish a paper with a positive claim
based on an independent analysis of their own data. This was to cause some distress in IGEC,
though it broke only the spirit of the agreement, not letter (Astone et al., 2002; see chapter 41).

seeing anything). This kind of claim in no way invades the interferometers' territory and makes it easier for the resonant bar scientists to be greeted as colleagues. When the proceedings were drawing to a close, the summarizer, Peter Saulson, told us that the meeting had revealed more collegiality between the bars and the interferometers than ever before.

LIGO

LIGO, in effect, leads the worldwide gravitational wave detection community, both in the perceived importance of its spokespersons' opinions and in the institutional roles they fill.[2] Because of this we will spend much more time on LIGO than on the other groups.

Having more than one interferometer, LIGO is in a much stronger position than any other group to make a credible independent discovery announcement in respect of the largest class of gravitational wave phenomena—burst sources. Confirmation of burst sources usually turns on finding coincident signals between detectors either through coincidence analysis or correlation. The LIGO group is unaffected by the moral arguments against technical integration so far as internal data exchanges between their "two and a half" detectors are concerned, because they are maximally socially coherent. LIGO, then, can choose whichever level of technical integration best suits the search, and base the choice purely on technical considerations.

In spite of these advantages, LIGO has been at the forefront of proposals for *international* collaboration. Why is this, given that LIGO seems to have more to lose and less to gain through collaboration than any other group?

Keeping in mind the Antiforensic Principle, a cynic might say that LIGO has less to lose through collaboration than appears at first sight. The reason is that since LIGO is the clear leader of the international community, the bulk of the reward for the first discovery, in whatever form it comes, is bound to accrue to LIGO. The cynic's argument gains in force from the fact that so far, the major institution for bringing about international collaboration among interferometers has turned out to be the LIGO Scientific

---

2. For example, Barry Barish, as well as being the director of LIGO, was also the founding chair of the Gravitational Wave International Committee (GWIC), which was set up to foster collaboration among the detector groups and continues in the role up to the time of this writing. The importance of informal leadership is shown by the extent to which other groups around the world seek positive opinions from the LIGO leadership when they press for funding for their own experiments, and the way the British-German project is now justified by its role within LIGO (see below).

Collaboration. If the first discovery is made as a result of data pooled under the aegis of the LSC, a great deal of credit is bound to go to the group represented by the *L* in the acronym.

Less cynically, there is good reason to think that the LIGO leadership have been assiduous in fostering collaboration because of their background and experience. Both the director, Barry Barish, and the influential project manager, Gary Sanders, coming from high-energy physics, are used to international collaboration (for instance, Barish has worked for long periods in Italy). In a field such as high-energy physics, where machines are hugely expensive, there is no alternative but to cooperate. Given the technological difficulties of interferometers, collaboration would seem completely natural to high-energy physicists drawn to the field.

A strong push in the direction of collaboration has also come from Rai Weiss, the leader of the MIT group. Weiss, too, has important experience of collaboration, having played a leading role in the Cosmic Background Explorer (COBE) project. Still more marked in his statements is his sense of the harm done to the field by what are now accepted to be the spurious claims of the early years. Weiss wants to make sure that the science of gravitational waves does not damage its credibility further with ill-founded discovery claims, and therefore he wants to minimize risk. The same fears that drove the interferometer scientists to confront Weber's later claims, the SN1987A results, and the Perth-Rome coincidences are making them cautious about the analysis of data from their own interferometers—at least in the first instance. Weiss is also very concerned about the problem of control—the problem of people breaking away from agreements and striking out on their own; he experienced this problem at first hand during his time with COBE.

Invoking Precedent

The LIGO management set up the LSC to bring in outside groups of scientists. As the LSC grew, it drew in more and more separate universities and research laboratories, both in the USA and abroad. By May 2002, it included 21 separate institutions comprising 26 research groups and 281 individuals; by early 2003, the figures were about 40 and 400. Thus, potentially, there is a much greater range of interests represented within LIGO than within any of the other groups. So LIGO has the biggest problem to solve in establishing a data analysis strategy.

With complex problems such as these, formal rules cannot provide a path through the thicket of overlapping principles. In practice, the solution is reached by leaning on precedent and experience as much as by calculation.[3] Precedents that seem to have been influential in guiding the LSC are the COBE project, in which Weiss was a leading player; international neutrino search collaborations, in which Barish continues to be a major figure; and the traditions in high-energy physics, which form the background of many of the new leadership. At a 1998 workshop, where the rules for handling data were being explored, all three of these were invoked.[4]

The COBE Precedent

COBE gave rise to the maps, published a few years back, of the unevenness in the three-degree cosmic background radiation. These figures were somewhat nauseatingly described by some publicity-seeking scientists as "The Face of God." The COBE collaboration was formed in 1973 and stayed together for 23 years until 1996. The NASA-driven philosophy that informed the COBE project was that the satellite, and the rest of the experiment, were paid for by the taxpayer, so the data must be treated as public property. NASA insisted that the COBE scientists prepare data in a form that could be made available to any interested scientist who would like to analyze them. Nevertheless, the "home-team" scientists were given two years to clean and prepare the data, and complete such analysis as they could, before the wider release. As Rai Weiss, who held a senior position in the COBE collaboration, said,

> [1999] [T]hey [the COBE scientists] had an obligation to deliver this data and deposit it in a public place.... The idea was that you deposited that data there, and your obligation was that you had to write manuals for how to handle that data. It was a hands-off thing—very different from [high energy physics], and very different from the way we intend to be [in gravitational wave physics], because we can't make our instrument so perfect that we've taken out all the instrument signature [noise]. And, by the way, I might as well tell you that COBE didn't succeed with that either. It's just that we had to take our very best guesses fairly early.... You had some sort of manual, but I don't think it was any good [laughter].

3. Orr (1990) explains the importance of shoptalk and "war stories" to photocopier repair engineers. The technology is very different, but the way good practice is encapsulated in descriptions of experience is similar.

4. The Third Gravitational Wave Data Analysis Workshop (GWDAW 3) held at Pennsylvania State University in November 1998.

In this statement we see the strain between making data public and the need for local processing to remove artifacts. Experimenters believe it is their task and responsibility to clean data before they are studied by analysts who do not know the experiment. The work, we are warned, is likely to be done badly if it is done speedily.

We have seen that the technical necessity argument and the private property argument are difficult to untangle; the need to clean data effectively also provides time for the experimentalist to scour the data in private. Weiss referred wistfully to the problem of controlling the COBE data.

> [1999] Because of the ground rules that this system had, it got itself into some interesting troubles.... We had to give the data away, effectively, before it was all analyzed. And Mark Davis, who was at Berkeley, managed to use the ... data to get [one of the] result[s] before the people who actually worked on the experiment did. It happens—it's part of the deal—OK?

The problem of common data-handling mistakes was also discussed by Weiss in terms of the COBE precedent. COBE tried to solve it by making itself into a microcosm of a competitive scientific community. To quote Weiss again,

> [1999] You had a prescription for how papers would be written, and also an internal review of these papers.... And there were elaborate rules for public release.
>
> ... The DMR—Differential Microwave Radiometer—which is the one that made that very stunning discovery [that viewed through the temperature of the cosmic background radiation, the sky looked blotchy], but on a marginal set of data. If you ever look at that data, you'll see the signal-to-noise was probably not much better than our first attempt. Most of the noise [was what] you saw in that wonderful map that people have. Where you see little hotspots in the universe over here and cold spots over there [represented by red and blue color coding in the "map" that was widely published], most of that's noise. Most of that had nothing to do with the universe at all. [Only] about 30% of it was signal in the very first pictures.
>
> ... That was so tricky and so worrisome for everybody on the team that we actually had competing groups, with them deliberately not talking to each other, having their own algorithms, their own process, their own group of software people working for them. And then there had to be a shootout within the team, so that this thing could or could not be published, and that may yet happen to us also. So it isn't that every team is completely

orthogonal. You look at the thing and think where are you very prone to make error.[5]

This kind of approach works so long as the problem of control does not arise. In the case of COBE, one of the competitive subgroups did try to break and produce an early publication. Weiss explained that he had to tell them, [1999] "I won't let that happen. How do you know that's right? There are all these checks that have to be done, and they have not been done yet." The group in question tried to circumvent Weiss by going straight to NASA, but "we calmed everything down." Nevertheless, there was a breakaway publication by another subgroup. As Weiss said, "it was never clean and never simple."

Neutrino Searches and High-Energy Physics as Precedents

Neutrino detection has some of the characteristics of gravitational wave detection.[6] A network of neutrino detectors awaits the pulse that will be emitted by an exploding star, or supernova, in our galaxy. The theory says that the neutrinos are emitted early in the explosive process and should arrive at Earth an hour or so before the visible light. The hope is that if the network of detectors agrees that it has seen a neutrino pulse, it will be able to inform the astronomical community in time for them to point their optical telescopes in the right direction and observe the explosion from the moment the light arrives. Gravitational waves, if they are emitted, should arrive on Earth at roughly the same time as the neutrinos, so the network of gravitational wave observatories could, in principle, team up with the neutrino detectors and run in coincidence with them.[7]

Theoretical models of supernova explosions are significantly less reliable at predicting outputs of gravitational waves than neutrinos, however. There is a great deal of uncertainty about how many supernovas will emit detectable gravitational waves, and how strong they will be. Neutrino detection, by contrast, is an established activity (though this does not make it completely reliable).

As with neutrino detection, the worldwide network of gravitational wave detectors will be built and run by independent teams. COBE was different; there, a purpose-built team was assembled from the outset, and a

5. Weiss went on to say, "The [first] announcement was made on the first year and a half of data, but when we got to four years it was very solid."

6. At the 1998 meeting it was Barry Barish who referred to neutrino detection and high-energy physics as precedents.

7. Preparations for such a linkup are being made (Barish, email, August 2001).

degree of independence in the component parts had to be specially en-gineered. Thus the problems of social control are greater in gravitational interferometry and neutrino detection than in the COBE collaboration.

In automating the interactions, a "technical fix" has been found for the neutrino network in the absence of the opportunity for good social control. All neutrino groups "subscribe" to a central computer, keeping it continuously informed of the state of their instrument. Each group is allowed to alert the central computer when it thinks its detector has seen an interesting flux of neutrinos during any ten-second time period. If such a positive report is made, the computer will reply with a report on the status of the other detectors in the network during that same ten seconds; it will say which of them was switched on during that period, what the level of sensitivity of each active detector was reported to be, and whether it reported a positive result in the same period. If all switched-on and sensitive sites report a strong positive result simultaneously, they will each know about the result of the others, and this will trigger a report to the astronomical community telling them to be alert for the appearance of a supernova within the next hour at a specific point in the sky.

Certain features of this arrangement are worth noting. "Seeing" a flux of neutrinos is not a yes/no phenomenon. As Barish explained, [1999] "The discussions inside of an experiment about what we send, when we send it, what reliability we have, are incredibly—I don't know—there's a lot of controversy about that. Inside of our experiment, we have actually two different tracks that analyze it and have to agree before a signal's sent . . . so there's a big internal thing."

The choice of what to report is constrained by the fact that groups are allowed to make only a limited number of positive reports. A level of reporting is chosen so that there ought to be no more than one loud, collective false alarm per century, and this allows individual detectors to report "gold-plated" events roughly once per month. The choice of how many events to report has the potential to divide the groups in the same way as the members of IGEC are divided. In this case, the number of reports allowed to be submitted by any one detector is based on the assumption that there will be a supernova in our galaxy roughly three times a century on average, so a really strong candidate for detection will be a flux of neutrinos of a size seen only once every thirty years or so—well above the normal, day-to-day noise fluctuations of the devices. This determines the evidential threshold for any one device.

Actually, that criterion applied rigidly would make for too severe a re-striction on reporting, so it has been relaxed. A really loud alarm will be sounded in the astronomical fire station only if most of the detectors see a similar signal, so it does not matter if individual machines make relatively frequent alarms; these are ignored when they are not confirmed by the others. Also, there are plans in place to allow a second level of "silver-plated" events to be reported.[8]

Another vital feature of the neutrino network is that no group is allowed to interrogate the computer to find out what other groups are doing unless it itself is currently reporting a positive result. In other words, when a positive result is reported, the group doing the reporting is "blind" to what the other groups are seeing. This prevents post-hoc reanalysis of the output of an individual device based on what the other groups are seeing. If you failed to report the event that the others saw, you missed it! The counterpart of this provision is that each group must report immediately when it is interrogated via computer by a team with a positive sighting. It must make an immediate report even if it has *not* seen an effect. This, it seems, was one of the most difficult agreements to bring about, as Barish explained: [1999] "Some people don't like that, of course, because they don't like the fact that somebody knows whether they thought, online, they didn't have it."

The reluctance on the part of scientists to make their *non*-sightings instantly available for public scrutiny is easily understood. If there was a supernova and you did not see it even though your instrument was reportedly running in a sensitive state, then it makes it evident that you were doing something wrong. Note that the incompetence is not so subject to the experimenter's regress as it might be, because barring a major upset in physics, the criterion of whether your instrument should have seen neutrinos is not only the agreement of all the other groups but also the unmistakable light signature of a supernova roughly one hour later. The declaration that must be made is more like gambling on a horse race, where you win or lose according to what the horse does, than gambling on the stock exchange, where, in principle, you can win by getting everyone else to agree with you about how stocks are behaving.[9] According to Barish, scientists in this field are uncomfortable with the instant reporting back, because they believe they need time to check that their instrument really was seeing nothing at a point in time. In other words, the neutrino scientists, just like

8. Barish, email, August 2001.
9. Barry Barnes (1983) has tried to make this a difference of principle in the analysis of types of entities found in the world. Barnes's work is insightful, but here we must bear in mind that there was a time when the existence of the other correlates of supernovas, the bursts of light and neutrino fluxes, were themselves matters of stock-exchange-like agreement.

the gravitational wave experimenters, are unhappy about releasing naked data to the prurient gaze of the world at large (though they have agreed to do it).

A final interesting feature of this protocol is that each group retains the right to report positive results on its own irrespective of what the other groups are reporting. In other words, they retain the right to act outside the agreement as well as inside it. Thus, even if a general alarm is not sounded when a supernova occurs, because many devices were actually running less sensitively than their teams believed, a competent team still has a record of its report for the subsequent "told-you-so" session. Thus, they do not try to resolve fully the problem of control: the independence of each group is preserved even as they collaborate. As Barish put it,

> [1999] Now, what do the experimenters want? This is where the sociology is similar or different than here [gravitational wave detection]....The individual experiments don't want to be bound by anything they see from other experiments. Because the fact that somebody sees that the other experiment says it's sensitive but didn't see anything—people don't want that to prevent them from announcing a result—they just don't want it. Why?—they just feel the other guy is probably incompetent—his experiment must not be working even if he thought it was.

That the neutrino groups are ready to live with this feature of their protocol, whereas most gravitational wave groups are concerned with finding a way to eliminate independent reporting, almost certainly reflects the fields' differing histories.

The neutrino groups feel their data belong to them, but a collaborative checking system has still been worked out. In terms of technical integration, this arrangement draws on both the medium and the lowest level: the coincidence analysis has a medium level of technical integration, but the groups retain the right to make an independent announcement irrespective of the coincidence analysis results; the independent announcement could be corroborated by telescopes that see light or by post-hoc analysis of the other groups' data.

Collaboration in Practice: The LSC

The precedents discussed above reveal different ways of organizing data sharing with different levels of system and moral integration. COBE, as one team, should have had high moral integration, though the breakaways show that it was far from perfect. Nevertheless, a degree of system "disin-tegration" was artificially engineered to safeguard against common biases. The neutrino groups have good system integration but low moral integra-tion. Their collaboration is managed according to a signed agreement to follow a set of rules, as with IGEC. With these in mind, let us turn back to gravitational waves.

In April 1998, Weiss circulated a letter to "leaders of the gravitational wave project" proposing a new kind of system integration.[10] It proposed what was the human equivalent of the computer-controlled arrangement of the neutrino groups, a strategy that, Weiss said, would enable the field to be "opened responsibly." He called for the results of *all* the instruments that were observing—prototypes, resonant masses, and so forth—to be in-cluded in the analysis. After local analysis, the results would be brought to a council composed of representatives from each group. After due consid-eration, the individual groups would submit publications, and an ensuing paper issued by the council would reconcile all the observations, includ-ing the reasons for there being no data from some detectors. The council would maintain an inventory of the schedules of all the worldwide detec-tors and should be formed within a year. But Weiss's council has not yet been formed. The first international collaboration with real substance was to be within the LSC.

The LSC was not conceived as a mechanism for international collabo-ration. In 1998, at a meeting of the Gravitational-Wave International Com-mittee, Barish said, "The LSC is part of LIGO. It's something separate. It's our science arm . . . it's not invented to be the international collaboration for all the detectors—it's not what it's for." Later at the same meeting he told me, [1998] "We don't consider it [the LSC] something to analyze VIRGO data, or do GEO or TAMA data.—I don't want to bastardize it."

But the crucial feature of the LSC, which was to allow it to transform itself into something far more powerful, was that it was open to all comers. This arrangement was favored by the National Science Foundation, because it went a long way toward "squaring the circle" of public property rights. It meant that although the data produced with public money that supported LIGO remained, for a limited time anyway, in the private domain of LIGO scientists, "LIGO scientists" came to be defined as anyone who joined the LSC—and that meant anyone, wherever they came from, who was willing

---

10. The complete letter can be found at WEBQUOTE under: "Ray Weiss's Letter Proposing Integration among Interferometer Groups."

to "pay the entry fee" of making a significant contribution to the LIGO project. This was, again, a model drawn from high-energy physics. As Barish explained to a 1998 data analysis workshop,

> In both cases [high energy and COBE], I contend, the data is very much available, but under a different model. The model in NASA [the COBE model] is that the data's made available by putting the data into a form that anyone can analyze—it's very well documented and so on. The high-energy-physics model is quite different than that. It's that in a given experiment, the data's available so long as someone joins the collaboration for that experiment, and the collaborations are open; they're not closed.

Again, speaking to me at the same workshop: [1998] "Anyone who wants to analyze LIGO data can, they just have to do it by joining the LSC."

From the beginning, the LSC was never to fit the model drawn from high-energy physics, at least not in every respect. As we have seen, in the developing years of gravitational wave science it is not possible to make a clean distinction between the developers of a detector and its users. It will not be possible to make such a distinction until gravitational wave detectors become robust and reliable, and this is unlikely to come about for decades. In the meantime, gravitational wave scientists have to innovate and theorize about device-building even more than they have to innovate and theorize about the meaning of data. This has become ever clearer as the LSC has evolved to take responsibility for designing the next generation of much more sensitive LIGO detectors—Advanced LIGO.

There are other reasons for the success of the LSC. One of them is that it has engendered a degree of moral integration alongside the system integration even though this has involved a loss of autonomy for the members. In 2000, Gary Sanders, the LIGO project manager, described the process to me as he had experienced it.

> [2000] At the first Aspen meeting in January 1995, I first floated the idea that there should be a users' group. I was attacked for this. People were very suspicious.... There were very spirited discussions during which I was attacked.... It took three Aspen meetings to get to the stage of the LSC.... The resistance was strong at first and then subsided.... The people were afraid—you know. These are people who've self-selected themselves to lead an academic lifestyle with small groups, and suddenly there's this specter of a different way of doing things and less of academic freedom.... X declared the words *high-energy physics* to be pejorative. It got that blatant in

the discussion.... But here we are—it looks as though people are pulling together.... Now people don't seem to want to run from this.... People realize it's a collective enterprise, and they do have to pay a little to get this.

Thus the LSC, though it has quite intrusive formal reporting rules for its members, has developed aspects of a morally integrated community.

VIRGO

From the beginning, the French-Italian VIRGO has had the reputation of being the most independent and competitive group. As one of the leaders put it to me, European unity was the "only answer" to the power of the American project to suck up prestige and personnel. Of course, as the second biggest interferometer in the world (after the two LIGO 4-kilometer devices), if competition is possible at all, VIRGO is in the strongest position.

Of those groups who have gained funding to build a large or medium-sized interferometer, VIRGO was the last to enter the field of gravitational wave detection. Both LIGO and GEO have their roots in the early and mid-1970s, a time when the traumas of the Weber dispute convinced scientists that making a convincing discovery was going to be difficult. Furthermore, by the 1990s, funding cuts had forced the British and Germans to pool their efforts to build just one small interferometer. In contrast, many of VIRGO's personnel were newcomers with no understanding of the field's bitter adolescence.[11] Furthermore, VIRGO did not build a prototype, so the team's first experience of full-blown interferometry was later still.

From the viewpoint of other groups, then, the 3-kilometer VIRGO detector seems to have had a charmed birth (though a spokesperson explained to me that it seemed less easy from the inside), and when I began fieldwork in the middle 1990s it was resented by other groups. VIRGO seemed to have "sprung from nowhere" with funds to build a large device even as the long-standing German-British collaboration was being squeezed almost to death.

Members of other groups thought that VIRGO personnel were arrogant. Many came from the elite of high-energy physics and thought of

11. Not everyone in the VIRGO group was a newcomer, however. The exceptions included Alain Brillet, the leader of the French side of the project, who began work on interferometers for gravitational wave detection as early as 1979. The French side of the project also included a number of scientists having fifteen or so years of experience and a track record for developing important features in advanced lasers, mirror coating, optics measurement, theory, and computer modeling. Some of these have been subsequently adopted by other groups. The novel idea for a very low frequency suspension for VIRGO (see below) was also well established.

gravitational wave detection as a race between competing groups to be first to build the most powerful machine. They did not grasp the subtle problems of this new science so painfully learned by others; they believed that a little cleverness would lead them straight to the prize. In later years, a leader of the VIRGO project was to admit to me that the impression of arrogance may have had a basis in fact; VIRGO's early years were characterized by a competitiveness learned in high-energy physics, and it had taken years for personnel to understand that real collaboration would be needed.

Spokespersons for VIRGO's French funding agency (the Centre National de la Recherche Scientifique) seemed to place a great value on making the discovery. One very senior CNRS official explained to me, though I am not quite sure how serious she was, that it was the "Latin temperament" that allowed the Italians and French to be funded so generously, while the British and Germans were starved of money in spite of their much longer pedigree and experience. VIRGO, she said, had been supported because it was exciting—a great adventure—even though there was no detailed cost-benefit analysis of the sort that characterized the Anglo-Saxon and German approach. Likewise, at a meeting in 1998 of the Gravitational-Wave International Committee, an organization whose sole purpose was to foster collaboration, another senior French funding official, to whom I will give the pseudonym "Giscard," shocked the non-VIRGO delegates by saying that gravitational wave detection was "not entirely a matter of collaboration." I initially heard this as a few innocent words, but it was interpreted by the Americans, British, and Germans as a deeply significant statement, especially as it came from such a highly placed source. A delegate explained,

> [1998] *Delegate:* To me it was the most out-of-context and revealing thing that anyone said yesterday.
>
> *Collins:* The remark about competition?
>
> *Delegate:* It was more than competition. It was competition—it was something—[Giscard] didn't use the word *undermining*, but it was some inference to too much collaboration somehow undermines—
>
> ... [I]f you take it out of context, it just sounds like an observation. I think it has much more significance than that. I mean, why did [Giscard] get up and say that? [Giscard] also said something about low frequencies and discovery. And there's more to it than that. It's *who* said it. I mean, this is [Giscard], not somebody in the trenches!
>
> And I've heard a bunch of other comments that tell me that for whatever reason that I can't really pull out that there's some theme or story,

> or something, in this VIRGO sociology that's got to do with the glories of a
> first discovery. And I think that it's pretty blatant when it starts to become
> said in a form like this.... [W]here's [Giscard] got that? It's different from
> some scientist "down the hall," with his ambitions.
>
> ...I can only guess, and I could be wrong what lies behind it.... [But
> t]hey have to make people like [Giscard] happy on their side—behind
> them—supporting them—whatever. And there might be more inferred
> promises of glory, and dangers of losing that glory, that are coming out.

Falling back on the Antiforensic Principle, we cannot be certain what was in the mind of Giscard or the French-Italian team, but we can be certain what was in the mind of my respondent and the feelings that were abroad in the community. This interpretation of the remark did not spring from nowhere.

So much for the desire to be first; but VIRGO had funds to build only one interferometer, big though it was. I have explained the importance of coincidences in this field, so how can VIRGO compete with a LIGO that has two and a half interferometers of its own?

VIRGO has established for itself a potential window of opportunity by designing a detector that can see to lower frequencies than the other interferometers. The mirrors are supported by a superattenuator, which is a very tall cascade of springs supported by an inverted pendulum. A suspended mirror stays still, because so long as the movement of the pendulum support is rapid, it will be back where it started before the mirror "has had time to notice" that anything has changed. In the vertical plane, the same effect can be achieved by using soft springs. It is not hard to see that these isolation techniques will work less well at low frequencies: if the pendulum support moves very slowly, the mirror will have time to "catch up" with the support before the support moves back again. The VIRGO design team put a huge effort into developing the superattenuator with many stages of springs, designed to isolate their mirrors from much lower frequencies than the other first-generation interferometers could handle; if it works, it will be a major technical accomplishment.[12] LIGO will be able to search for gravitational waves only down to frequencies of about 40 Hertz (cycles per second), since any lower frequency signal will be mixed up with unwanted noise; VIRGO, on the other hand, should be able to see signals of only 10 Hertz or so.

---

12. I understand that University of Western Australia personnel contributed significantly to the design of the superattenuator, especially the "inverted pendulum" component.

The many stages of springs in the superattenuator, each suspended from the one above, give the appearance of a segmented tower, and the VIRGO team have made capital out of the resemblance the Tower of Pisa. As well as being witty, and "branding" the device by association with Italy, it means that the *essential difference* between VIRGO and the other designs, which gives it a possible competitive advantage, cannot be forgotten by its competitors and its funders.[13]

The advantage of low frequency is that heavy inspiraling binary star systems can be more easily seen. As the components of these systems get closer to one another the circling get faster and a large burst of gravitational waves with a characteristic pattern is emitted in the last few moments as the stars approach. The more cycles of the inspiral that can be seen, the better the statistical significance; low frequency allows the pattern to be detected from an earlier stage so long as the stars are heavy enough to be emitting strongly while they are still a little way apart. There may be inspiraling black-hole systems, then, that only VIRGO will be able to see with confidence. To some extent, this offsets the fact that it is a single detector.

Of course, even a single detector can see events that are correlated with bursts of light, neutrinos, or gamma rays, and VIRGO could look for continuous sources on its own, again going to lower frequency. Such results as these would need to be corroborated by other laboratories, but corroboration leaves the initial discoverer clearly identifiable for allocation of credit. As one VIRGO scientist with a reputation for championing independence put it, [1997] "You need a very good signature to make a claim, like a continuous source or like a burst with good signal-to-noise ratio. If you have such a signal, that would be convincing in any of the detectors. There are windows that are not open to all detectors, so you could conceive of such a situation." He admitted, however, that such a circumstance was unlikely to arise.

VIRGO suffered disadvantages in addition to its being a lone detector. The organization is a loose consortium of laboratories in France and Italy exhibiting less social coherence than any other single detector group. Indeed, the individual laboratories may have competed with each other over resources and project philosophy.

As we have seen, VIRGO's schedule was delayed by problems of land purchase, flooding of buildings, and, perhaps, this loose management structure. An independent strategy is all very well when a project is ahead of the

---

13. The analogy may turn against VIRGO, as the supports of its beam tubes are sinking uncomfortably fast into the unsuitable soil (not necessarily a fatal problem, unless it affects more major components or gets out of hand).

field, but if it falls behind it is hard to maintain the loyalty of the team. Everyone wants to be on the team that makes the first detection, and groups or collaborations with more than one detector will start to look more attractive if they are pulling ahead. The problem is exacerbated if the home team is a loose consortium, more characterized by system integration than moral integration, because this is not the kind of institution that engenders loyalty. For VIRGO, things were made still worse by the comparatively low salaries of European scientists compared with their American counterparts. As a result VIRGO, began to lose important staff to LIGO. Working for LIGO they could earn more, possibly feel more valued, and have the better chance of being on the team that made the first detection.

The innovative engineer Riccardo De Salvo, a onetime colleague of Barish's in high-energy physics, did leave VIRGO for LIGO about this time. Another VIRGO member, attracted by the open-to-anyone policy of the LSC, suggested that VIRGO scientists (and even whole VIRGO groups) be allowed to join the LSC even if their first allegiance remained with VIRGO. The LSC allowed members of other projects to join, and most members of the GEO team had taken advantage of the arrangement. The VIRGO member had already started cooperating with the LIGO team over data analysis. However, this degree of cooperation was considered unacceptable to the VIRGO management, and they put a stop to it. They made it clear that they were not prepared to allow members of their project team to become LSC members. In late 2001, a senior VIRGO spokesperson told me,

> [2001] I said many times, to [the named leaders of LIGO], that we would join immediately a new, LSC-based international collaboration if its name was more neutral... and if the purpose was to support not only LIGO but also the other projects. It is obvious that LIGO would still have a louder voice than any other group, but it [must] not be understood as colonization—that, of course, we [could] not accept. It is [now up to LIGO] to make a step in the right direction.

But the LIGO leadership, in spite of the success of the GEO collaboration, still has an ambivalent attitude toward the LSC, insisting that it is not a vehicle for analyzing VIRGO data. I'll invoke the Antiforensic Principle and decline trying to pin down the intentions further.

As of this writing in early 2002, some of the pressure on VIRGO seems to have been lifted. For a number of years, VIRGO's date of commissioning seemed to be a victim of "von Neumann's constant." That is, the time to completion remained the same however much the calendar advanced. But it now looks as though the beginning of 2003 really will see a working VIRGO. If in 2003, when it finally goes on air, VIRGO were in better shape than LIGO was when it first locked—which it might be given that it has used full-sized components as the prototype—then it would not be so far behind LIGO.

In part, the speeding up and the relief of pressure on VIRGO has come as a result of a management transformation and the formation of a new management umbrella, EGO, which stands for *European* Gravitational Observatory. EGO, however, is strangely named; it is meant to work in something of the same way as the LSC, allowing outsiders to bid to work on projects with VIRGO. Unfortunately GEO, is also in Europe, but it was not much consulted in the setting up of EGO, nor does it figure in the discussion of where EGO is going. EGO was set up to solve VIRGO's problems and made little reference to GEO (though spokespersons told me this was an oversight caused by the speed with which things had to be done, not a deliberate snub). It looks as though the *E* in *EGO* is going to cause as much trouble as the *O* in *LIGO*, not to mention that to European gravitational wave scientists outside the EGO inner circle, EGO seems an all-too-appropriate acronym.

For VIRGO, one advantage of EGO is that it is set up as a private company under Italian law, which means that it can create new positions with good salaries. The previous bureaucratic umbrella was the Italian high-energy physics agency Istituto Nazionale di Fisica Nucleare, which, on civil service principles, had a fixed number of posts, with fixed and rather low salaries. Under EGO, I was told, people who had left VIRGO were returning to the fold.

This is not all there is to EGO, however. Plans for a second large European interferometer are being canvassed. Independence from America is still a goal. One idea is to build a second detector relatively close to the existing VIRGO installation. Detectors that are close to each other can search for cosmic background radiation by correlating their outputs; should they find it, all they would need is corroboration by another pair of detectors. The search for cosmic background is, then, something that is naturally carried out by a single national team, and if the cosmic background were made the goal, EGO could retain its independence in a natural way while still contributing to the international search for, and location of, other sources.

At the same time, in 2001, VIRGO's leadership, at least in the person of Adalberto Giazotto, began to talk in terms of the international community becoming "one machine."[14] Mid-2002 would see Giazotto's plan for a new

14. I first heard Giazotto say this at the Amaldi meeting in Perth, Australia, in July 2001, but was told that it had been said at other conferences and workshops earlier in the year.

kind of LSC-like organization that would make this possible. Just what level of social coherence would be planned or accomplished remains to be seen. It is worth noting that at a meeting in May of 2002 it was Giazotto who set out the idea of coherent analysis, a highly technically integrated data-pooling scheme.

GEO

The British-German team running GEO is a paradigm of the growth of social coherence among what were once disparate groups. The two competing groups were forced together by financial exigency, and they are now one institution with unified scientific and work values and no visible internal friction.

With interferometers, size matters. Arm length sets an upper limit to broadband sensitivity, and as time goes by, GEO's limited length and narrow beam tubes (which cannot be lengthened, because the light would diffuse so much over a longer distance that it would hit the walls of the narrow tubes), will prevent it from competing with larger devices. Any chance for GEO comes from its pioneering use of another technical innovation, which enables it to make its detector especially sensitive in a narrow bandwidth—signal recycling. GEO600 could try to achieve the first detection by conducting a narrowband search from some promising source of continuous waves, such as a pulsar, the frequency of which is known; or it could increase its sensitivity across a slightly broader range and run in coincidence with LIGO. It is the latter strategy that will be pursued first.

GEO600 has grown ever closer to the American project. A high proportion of the GEO600 team are now members of the LSC. In turn, the LSC has made major contributions to the future of LIGO by becoming the main design team for the next generation of US detectors, and members of the GEO group have been in forefront of this development; they have brought with them their experience of the advanced techniques that they have used in their attempt to make GEO sensitive in spite of its small size.[15] Ironically, the GEO group has received a very substantial increment in funds to carry through their contributions to the LIGO project—or joint LIGO-GEO

---

15. In addition to signal recycling, they, along with Stanford University, have pioneered an advanced way of hanging mirrors from pendulum wires known as monolithic suspensions. They have also developed three- and four-stage pendulum supports, which have been directly imported into the Advanced LIGO design.

project, as it might now be called—so membership in the LSC has brought financial as well as scientific gains to the GEO team.[16] Since there is such a large overlap between membership of GEO and membership of the LSC, the question of trust in regard to data pooling does not arise; LIGO's data belong to GEO personnel and the arrangement is reciprocal, at least in principle. GEO scientists' membership in the LSC defines their relationship with LIGO as one of high social coherence.

There are signed agreements between GEO and LIGO binding them to data sharing, but there is more than system integration going on. The GEO personnel get on well with their American counterparts and vice versa, and they see the world largely in the same way.[17] Thus a high level of technical integration will be easy to achieve once everyone agrees how much local processing should be done.

TAMA

The interferometer built by the TAMA group in Tokyo has 300-meter arms and is known as TAMA300. It is located in a seismically noisy environment, which may limit its sensitivity significantly, especially during work hours. Given this and its limited arm length, TAMA is likely to be the least sensitive of the new generation of interferometers in the long term. TAMA, however, has led the field in installation and commissioning. It was the first to demonstrate that a "big" interferometer could be made to work at all—something that was not entirely clear until it had been demonstrated—and it has been very successful in putting design into practice, in increasing the reliability and duty cycle of the device, and in the long process of "driving down the noise."

16. The British contribution is more than $11 million, and the German contribution is likely to match it. The attitude of these agencies is in marked contrast with that of the French. The British and German agencies seem to prefer to give money to projects involving transatlantic collaboration, whereas the French agency favors, or at least favored, transatlantic competition.

17. Social coherence is also fostered by exchange of personnel and developing personal relations. In particular, graduate students tend to take postdoctoral positions in one of the other institutions, and more-senior personnel take fellowships or move for short periods to the other group. Glasgow students have gone to Caltech and to Stanford (also a member of the LSC), and relations between Stanford and Glasgow groups are cordial. Another element not discussed here are the preexisting social relationships, which foster collaboration between groups. The groups of gravitational radiation researchers are made up of scientists, many of whom have previously worked in other fields where they would have formed social bonds and trust relationships that they brought with them to the new science. (For example, the Italian scientist discussed earlier who transferred from VIRGO to LIGO.)

Now that TAMA is no longer the only working large interferometer, plans to run in coincidence with LIGO seem to be going smoothly, though an earlier agreement to compare data with the 40-meter device did not work out. This sequence of events, and the problem as perceived on the American side, fits the overall sociological argument being put here—namely that real cooperation only begins when individual groups come to understand that finding a believable gravitational wave signal is so hard that no individual group or small subset is likely to be able to do it alone, so that joining the "betting syndicate" is the most sensible option. I can say with equal confidence that spokespersons for the Japanese group assure me that earlier failures to exchange data were due to contingencies not related to any potential competitive advantage.

TWO MORE FACTORS IN THE DECISION TREE

In the previous chapter, I set out a "data-pooling decision tree," which concluded with a paragraph entitled "Source- and Project-Specific Arguments." We can now add to this tree the factors affecting individual groups:

- *Detector design*, which affects the chances that a group has of making an independent discovery. This factor will interact with theoretical estimates of source distribution.
- *Organization*, which determines whether building schedules will be met, and determines whether a loyal team can be held together long enough to realize the chance of a "long shot."

The future will be interesting. In the long term, the need for a long baseline will surely bring all the detectors together in one way or another. In the interim, the groups still have competitive options open to them.

CHAPTER 40

WHEN IS SCIENCE? THE MEANING OF UPPER LIMITS

(A gravitational wave laboratory in Louisiana, January 12, 2002. Scientists are discussing the prospects for their current gravitational wave detection run.)
Pessimistic scientist: With this sensitivity, we won't see shit!
Optimistic scientist: But if we don't see shit well enough, it still might be write-uppable.

WHAT IS AN UPPER LIMIT?

Most of this book has been either about claims to have detected gravitational waves or about how to best detect them. But that is not the only thing you can do with a gravitational wave detector. You can also use it *not* to see gravitational waves. That is exactly what Richard Garwin and James Levine spent most of their time doing with their resonant bar detector, and that is what Tony Tyson, Dave Douglas, Peter Kafka, Vladimir Braginsky, Ron Drever, and others finished up doing with theirs. All these people published results in scientific journals that said that there were no gravitational waves on the order of Joe Weber's claims. Effectively, what they did was to set "upper limits" on the flux of gravitational radiation. Their work was valued by the scientific community, because it contrasted with what Weber had claimed—it represented new information about what was in the heavens.[1]

The new generations of interferometers now coming on air are almost certainly not going to see gravitational waves for some time. In the meantime, they are going to need to present themselves as doing something. What they are going to do is describe new upper limits to the flux of gravitational waves. They are going to say, "Though we have not seen any gravitational waves, our new detectors are sensitive enough to show that there is even less gravitational radiation out there than any other detector has been able to demonstrate."

What is happening here? Is this steady progress in science, or does it amount to the transmutation of elements? Is the lead of scientific failure being turned into the gold of scientific success? Under what circumstances do sciences publish negative results and describe them as "findings"? In many sciences, when the scientists look for something and do not find it, they slink away. The journals reflect this prejudice. Notoriously, in most sciences it is extremely difficult to publish a negative result. So what is different here?

The "Logic" of Upper Limits

Why is the meaning of a negative result so ambivalent? Consider what I will call the "logic" of upper limits. Almost certainly you cannot feel vibrations in your copy of *Gravity's Shadow,* but according to physicists the volume is vibrating to some degree due to the effect of gravitational waves passing through it. That you can't sense these vibrations means that there is no pair of binary neutron stars about to spiral into each other within about 100 kilometers of Earth. If there were, you would sense the gravitational waves emitted by the binary system in the form of repeated distortions in the shape of this book at twice the frequency of the rotation of the system; the vibration would be a millimeter or two in amplitude.[2] Thus, you have just

1. An upper limit states, "The flux of gravitational wave is less than n bursts, of strength y, within frequency band x, per day/week/year," or "The strength of continuous gravitational waves within frequency band z is less than p."

2. Thanks to Peter Saulson for the "back-of-the-envelope" calculation. Neutron stars are only about 10 kilometers in diameter, so there is room for a pair of them to spin around each other at a distance of 100 kilometers and affect the size of your book by a couple of millimeters. At close ranges, however, gravitational waves are much weaker than "Newtonian" gravitational forces. Neutron stars tend to have a mass of about 1.4 times that of our Sun, and the "ordinary"

made an observation belonging to the field of gravitational wave detection. Furthermore, no twenty-first-century scientist would dispute your result. This is odd, because gravitational wave science, as we have seen over and over again, is so hard that almost any result is likely to be difficult to obtain and depends on the utmost brilliance and perseverance founded on huge stores of tacit knowledge.

Let us call the combination of you and *Gravity's Shadow* the BOok Gravitational WAve Detector, or BOGWAD for short. To begin to understand what is going on, we can note that although BOGWAD's quiescence would be readily taken to show the *absence* of strong gravitational waves, any vibration in BOGWAD would *not* be accepted as revealing the *presence* of gravitational waves. Should you claim to have detected gravitational waves with BOGWAD, no one would believe you. At that point your qualifications, experience, skills, and so forth would be called into question; you would, in short, fall foul of the experimenter's regress.

There are three interlinked features of the "logic" of setting an upper limit that make it different from making a positive claim. The first feature is that *to set an upper limit is to see nothing,* and one can see nothing however insensitive the apparatus.

The second feature is that while the sensitivity of a detector needed to make a positive observation of gravitational waves is set by nature—it depends on how strong the gravitational waves are—*the observer can choose the level of an upper limit.* Observers can make upper-limit claims that are as high and undemanding (and uninteresting) as they like and check them with detectors of arbitrarily less sensitivity.[3]

The third feature is that in setting an upper limit, *signal and noise do not have to be separated.* One can set an upper limit by claiming only that the size of the signal is no greater than the *sum* of the signal and the noise. If there is a lot of noise, the upper limit thus set will constrain the world less than it otherwise would; but since the signal cannot possibly be larger

---

gravitational forces associated with two such stars cavorting violently 100 kilometers from Earth would have ripped you and Earth to pieces long before you had a chance to feel the vibrations from the gravitational waves. At long distances, the effect of gravitational waves is greater than the effect of ordinary gravity, which is why, potentially, gravitational astronomy can be done with gravitational waves, but it cannot be done by sensing ordinary gravitational forces. If one does not like the fact that BOGWAD can only be a thought experiment because of the devastating effect of ordinary gravity, one can place the source a bit further away and use a crude amplifier for the signals—a slightly more complicated apparatus than BOGWAD but no different in principle.

3. I am referring to nature in the naive way, which the sociology of scientific knowledge questions. But all we need to concern ourselves with at this point is the freedom to set an upper limit as high as you like compared with the lack of freedom in the case of positive signals. Where these constraints come from need not concern us right now.

than the signal plus the noise, the claim cannot be wrong. Of course, the more noise that can be eliminated, the stronger the upper limit and the more interesting the result in terms of science. One of the most interesting scientific results ever, the apparent absence of variations in the speed of light as Earth moves in its orbit, discovered in the famous Michelson-Morley experiment of 1887, was an upper limit.[4] But the *logic* of upper limits does not require the result to be interesting.

To sum up, the reasons we can make an observation with BOGWAD that will be widely accepted are that we have not been required to separate signal from noise, so we do not need the corresponding skills and resources; we have chosen a level for our upper limit that is so high that no one would want to contradict it; and our result consists of not seeing anything. With this logic in mind, we can examine real gravitational wave experiments.

UPPER LIMITS IN THE PRACTICE OF GRAVITATIONAL WAVE SCIENCE

An instrument is a device for detecting a signal, and making it work better means enhancing the signal-to-noise ratio (SNR). To make a gravitational wave detector more sensitive, one must eliminate noise and/or understand it well enough to be able to subtract it from the signal. It is only in logic that an upper limit can be as high as the experimenter wishes; in practice, to be worthy of scientific note, it has to be low. The question is, When does an upper limit become low enough to be scientifically interesting?

We can quickly supply the solution for two extreme cases. At one extreme, an upper limit to gravitational radiation flux set by BOGWAD is not interesting to anybody. At the other extreme, an upper limit set by LISA, the gravitational wave detector in space due to fly about 2011, could be very interesting to anyone interested in the constitution of the universe and the nature of science. LISA will consist of three satellites which bounce light between them, forming an interferometer with arm lengths measured in millions of kilometers. If LISA can be made to work to its full potential, it will be so sensitive to gravitational waves (of low frequency) that if it fails to see them, the upper limit it sets will imply some fundamental mistake in our understanding of the physical world: either the theory of relativity

---

4. The Michelson-Morley experiment is often said to have been a "proof" of the theory of relativity. It was less of a proof than it might have been, since it did not show that the variation in the speed of light was zero (which is what relativity needs), only that it was not of the order of the speed of Earth in its orbit. The dispute was to continue for another 40 years (see Collins and Pinch 1993/1998).

or our models of the evolution of the universe will be wrong, and it may be possible to tell which.

All being well, the same will apply in the case of Earth-based detectors two generations beyond those currently coming on air. For the sake of convenience, I'll refer to the generations of detectors as LIGO I, II, and III even though this nomenclature is no longer in use and not all detectors are American. What we are saying is that if LIGO III (or its non-American equivalents) reaches the levels of sensitivity now envisaged for it, yet fails to detect gravitational waves and can set only an upper limit, that upper limit, like a LISA upper limit, will give rise to widespread interest.

What about the more difficult in-between cases of upper-limit setting? In rough order of likelihood, to see gravitational waves (which for simplicity I'll refer to as sensitivity, though LISA will be looking for different sources), we have the following sequence of detectors: BOGWAD, room-temperature bars, cryogenic bars, the 40-meter, LIGO I, LIGO II, LIGO III, LISA. Our problem is to understand when the setting of an upper limit with these devices is interesting as science and why it should be more interesting than just failing to see something.

Something can only be interesting to *someone*. To begin to answer the question about what is interesting, we have to consider different audiences for upper limits, because different audiences are interested in different things. Once more, the general idea of the target diagram is useful. We can divide the world of those interested in science into five: the general public, or "taxpayers," for short, physicists in general, astronomers and astrophysicists, funding agencies, and the core group of scientists who actually make the detectors and analyze the results. In this model the funding agencies are much nearer the scientists than they are in our initial representation, which placed "funders and policymakers" beyond the rings of the "scientific community." This is because we are now thinking of the specific parts of the funding agencies charged with running gravitational physics; we are thinking, in other words, of people like Rich Isaacson, who himself made contributions to the theory of gravitational waves, and during his term of office was effectively an integral part of the community.[5] These different audiences' interests in upper limits are set out schematically in figure 40.1.

In this figure the horizontal axis shows the five different audiences. Sensitivity of detection increases up the vertical axis with the generations of detectors indicated. The vertical direction also indicates the passage of time, with the room-temperature bars now being forgotten in terms of

---

5. In 2001, Isaacson retired and his role was taken by Beverley Berger.

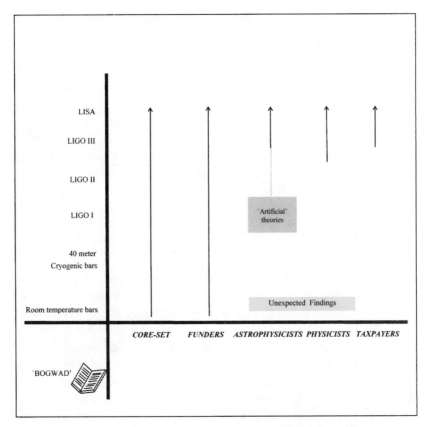

Fig. 40.1 Upper limits produced by detectors of varying sensitivity and their audiences.

scientific significance and the current focus of attention being at the LIGO I level. The presence of a line or other mark on the graph indicates that the corresponding audience has an interest in an upper limit set by the detector of corresponding sensitivity. At a first approximation, audiences become more and more demanding of upper limits as we move from left to right—that is, as we move away from the core of the science. This I will now explain, starting at the right and moving left, though I do not have much to say about the two rightmost audiences.

Taxpayers

GRAVITATIONAL WAVES NOT DETECTED WITH INSENSITIVE DETECTOR—such a newspaper headline would not convince taxpayers that their science dollars were being well spent. For most of us, what is exciting about gravitational wave science, and what was used to persuade the senators to fund LIGO and made the French funders act in accordance with their Latin temperament, is the discovery of gravitational waves, not their nondiscovery in the form of setting upper limits. No one said to those senators and funders, "Just give us the money, and we'll show you how little gravitational radiation there is!" It may be that a very low upper limit, such as might be set by LIGO III or LISA, could be of wide interest because it would overturn our understanding of the universe; in such a case even Einstein's theory of gravitation could be put in question, and there seems to be a wide interest in anything to do with Einstein; but we are a long way from being in a position to set this kind of upper limit.[6]

Physicists

Evidence on the lack of interest of physicists as a whole in high upper limits comes from remarks about the behavior of tenure review committees, though it is rather sketchy. The most heartfelt and colorful comment comes from a scientist who believed that his graduate students were unfairly treated by examining committees at his prestigious university when all they had to present were technically sound results. The results, though they demonstrated scientific and technical virtuosity, were seen as lacking in positive scientific significance. He explained to me that this was why he had become convinced that gravitational wave detectors had to be large (and expensive) enough to engage the interest of physicists by producing positive results.

> [2000] And the reason for why it became big, at [my institution], and in my head, was fundamentally because of an incident that happened with two students who had done a beautiful job, and they got this shit from my own colleagues. And I said, "I'm never gonna put students through this again." If we are going to continue with this, we are going to have to do it on a scale such that even if we don't see anything, no [expletives deleted] theorist, OK, can confront one of my students and say, "What did you discover?" and give him a sneering, [expletive deleted] ride, OK? So it has to be something where the upper limit is good enough. And you say, "Yeah—we have made a

---

6. The evidence we have for what taxpayers are interested in comes only from our competence as native members of Western society. This is the kind of evidence to which anthropologists aspire in respect of the societies they study.

scientific statement." And it was then, back in the middle '70s, that I decided, OK, you have to have a big structure.

A very similar attitude was reported to me in respect of another leading institution. The writer is one of the leading younger analysts in the gravitational wave field.

> [1997] I have been told that one of the reasons that junior faculty in experimental gravitational physics at [named institution] have not been successful in getting tenure is that there are—after the probationary period of six years—there are no positive science results to show for their work. Instrument development that does not lead to a completed, interesting science experiment does not count; uninteresting upper limits don't count. I have also been told that one of the issues raised at both [this and another institution], whenever new faculty appointments in this area are considered, is that no junior faculty member can be expected to have demonstrated the level of accomplishment [i.e., positive results in a completed experiment] expected for a promotion to a tenured appointment. This is another pressure that plays in the system, but is not generally remarked on.[7]

This quotation, along with at least one other story concerning failure to obtain tenure in another field of physics as an apparent consequence of failure to achieve a positive result, indicates the tenor of some of the discourse in the field, but it does not necessarily reflect institutional behavior. Furthermore, I was assured by a senior figure that the quotation did not match his experience, and that while a positive result would help a tenure decision, tenure could still be granted for good science in the absence of new findings. A statistically representative analysis of past tenure decisions could reveal the importance of positive findings and the way this factor varies from institution to institution and the way it varies with time, but no such study has been completed. I can only say, with the Antiforensic Principle in mind, that awareness of the possibility of such an effect is present in the discourse of physicists, and that this indicates that physicists think about these things—they think that negative results are less impressive than positive ones—irrespective of the statistics of tenure decisions.

Of course, this is not all there is to it. We know, on the basis of our broad understanding of the academic culture, that physicists in general would not concern themselves with high upper limits—those that merely reflect the

7. Email, November 1997.

absence of phenomena that are not expected to be there anyway—if they
had no administrative reason to be concerned. (Maybe someone should
count the number of Nobel Prizes awarded for setting an upper limit!)

Astronomers and Astrophysicists

The broad view of astronomers toward LIGO has already been discussed.
Astronomers—as we know from the damage caused by the fact that the
O in *LIGO* stands for "Observatory"—are not impressed by nonobservations
unless they are strikingly in conflict with expectations. On the other hand,
as I have said, the very high upper limits set by the room-temperature
bars of Weber's rivals were astrophysically interesting in showing Weber
to be wrong. But it was only Weber's unexpected claims that made room-
temperature bar results important. In figure 40.1, the interest in upper
limits engendered by Weber's claims is indicated by a gray box.

Most of Weber's initial claims were published in *Physical Review Letters*
(PRL). This prestigious journal is reserved for rapid publication of important
results, and for many physicists in the gravitational radiation "business" it
is the preferred outlet (with *Nature* as an alternative). The early 1970s also
saw the succession of counterclaims, effectively setting upper limits, published in *PRL* and *Nature*. We will take publication in high-prestige, rapid-
publication journals such as these as a rough and ready indicator of the
importance of a result, though here again a statistically representative analysis would enrich the argument.[8]

After the 1975 "watershed," when Weber's claims lost nearly all their
credibility, not only new positive claims of the Weber kind but also new
upper limits set by room-temperature bars ceased to be specially scientifically important. Thereafter, no further room-temperature bar results, either

8. The papers in *PRL* are Levine and Garwin 1973, 1974; Garwin and Levine 1973; Tyson 1973b;
Drever et al. 1973; and Douglas et al. 1975. Negative papers were also published in the rapid-
publication journals of other countries. As early as 1972, a Russian group published in the Russian-
language *JETP Letters*, which is routinely translated into English (Braginsky et al. 1972), and in 1973
an Italian group published in *Lettere al Nuovo Cimento* (Bramanti, Maischberger, and Parkinson 1973).
A Japanese group (Hirakawa and Nariha 1975) also published an upper limit in *PRL*. This upper limit
is of a rather different sort that did not bear directly on the Weber claims. It looked for continuous
waves emitted from the pulsar in the Crab Nebula. It's special interest appears to lie in its novelty
and ingenuity, a reason for publication that I will discuss when we look in depth at the 40-meter
result.

There are scientists who consider that the importance of outlets such as *PRL* has faded with
the growth of the electronic transmission of preprints and that to try to publish in such a journal
is to play a status game without scientific significance. This, again, would be an interesting topic
for research. Perhaps it would be found that their attitude toward rapid publication journals has
changed in recent years.

positive or negative, were reported in *PRL* or *Nature*.[9] One can see, then, that the interest in negative results from the room-temperature bars was, in a sense, "artificial." The measurement was not interesting because of the way it bore on the nature of the universe in some absolute way but because it contradicted what would have been a very interesting claim had it been true.

As the years have gone by, astrophysical theory has been filled out in greater detail, and astrophysicists now consider that they understand better the distribution of well-known sources of gravitational waves in the heavens. Though the possibility of unknown sources is invoked from time to time, as we have seen, a strong consensus has formed around the idea that the sky is relatively dark in terms of gravitational radiation. With nothing for them to disprove, the upper limits set by the much more sensitive cryogenic bars, even though they were lower than those set by the earlier room-temperature bars, are less scientifically important. It is probably safe to assume that this is why results published in 1989 and 1999 appeared in less prestigious journals.[10]

We now turn to the upper limit published in *PRL* in 1999. This paper emerged from the 40-meter prototype interferometer located at Caltech.[11] As we will see, this result can be said to be "on the cusp" of astrophysical meaningfulness, making it a particularly interesting case. As we will also see, it was also on the cusp of publishability by *PRL* about the same time other not too dissimilar upper-limit papers were being rejected by the same journal; indeed, the paper itself was initially rejected, and the decision to submit to *PRL* was the subject of a minor but sociologically

9. Positive claims included Weber et al. 1976 and Ferrari et al. 1982. Negative claims included Allen and Christodoulides 1975 and Brown, Mills, and Tyson 1982.

10. Amaldi et al. 1989; Astone, Bassan, Blair, et al. 1999; and Astone, Bassan, Bonifazi, et al. 1999. The same is true of the early, and not very demanding, interferometer result, Nicholson et al. 1996. In the year 2000, a cryogenic bar result was published in *PRL* (Allen et al. 2000), but this represented a substantial technological advance, since it involved the combining the results of five separate resonant detectors. The importance of such technological advances is described below. Confusingly, there are two Allens involved in this story. We will spend most of our time on Allen et al. 1999, where the principal author was Bruce Allen. The author in the case of Allen et al. 2000, the cryogenic bar upper-limit paper, is Z. A. Allen.

11. Allen et al. 1999. Most of the interviews and discussions and the main framework of the analysis for this chapter were completed and set out in draft before I knew that the file on the 40-meter results existed; my interview with Bruce Allen on this topic was the last interview I conducted that concerned upper limits. Thus the Allen interview and the 40-meter file have been used to flesh out and enliven an analysis that was already largely in place. The Allen et al. 1999 result was not produced by what has been referred to throughout as the "40-meter team." Indeed, one of the complaints about the 40-meter team is that they did not produce any "science" of this kind out of their own results. The Allen et al. result was produced by a much younger group of analysts, not experimentalists.

revealing disagreement among the contributors.[12] Furthermore, due to the generosity of the authors, we have a great deal of data about this case; team members analyzing the data were spread across a number of institutions, and there was a lengthy email interchange among them and other commentators. This email archive was preserved, and I was given access to it.

Astrophysics and the 40-Meter Upper Limit

The 40-meter interferometer was limited in sensitivity by its arm length and was also noisy. Its primary purpose, of course, was not to detect gravity waves but to explore the technology needed for the 4-kilometer LIGO I. The person who assembled the team who used the 40-meter data to set an upper limit was Bruce Allen. It is widely agreed that he and the team demonstrated initiative and ingenuity as they extracted the upper-limit result. Furthermore, it was the best such result from an interferometer, and the first to concentrate on a particular kind of gravitational wave source—inspiraling binary neutron stars. The team used knowledge of the typical waveform that would be produced by such a source to disregard some of the noise in the detector—that which did not follow the waveform. They were, then, helping to pioneer the template matching or "matched filtering" that would be a fundamental part of the data analysis for interferometric gravitational wave detection of burst sources.

On the other hand, the upper limit the team succeeded in setting on these events—binary inspirals in our own galaxy—was about one event every two hours, or about 4400 events every year. They were able to say that if there were such events taking place in our galaxy, they were 90% confident that there were no more than about this number per year. To understand the significance of this—how much of a surprise it was, or how valuable it was in terms of *information*—it has to be compared with the predictions of theorists. Theorists calculated that in our galaxy, there should be roughly one such event every million years! The 40-meter upper limit was, then, between nine and ten orders of magnitude higher than the flux predicted by the astrophysicists, a result that initially seems little more impressive

---

12. Disagreements among actors serve the same function for sociologist-observers as the deliberately engineered breaching experiments described by Garfinkel (1967). In terms of sociological substance, not too much should be read into the fact that the paper was initially rejected. *PRL* turns down quite a high proportion of first submissions that are accepted subsequently. One of the authors told me that the group were quite encouraged by the initial response even though it was a rejection.

than the result produced by BOGWAD. So why was the result published by
PRL?[13]

A first possibility is that the paper had significant astrophysical meaning
because it was an empirical result—even if it engaged with no seriously
held hypothesis, it was important just by virtue of the fact that it was
empirical. In our correspondence, a member of the 40-meter upper-limit
team emphasized this point. In an email to me dated December 31, 2001,
he wrote,

> [2001] [W]e are ~10 orders of magnitude below the interesting rate. But do
> note that ours is the first published limit based on gravitational wave
> observations. This is significant in itself.

This argument would be stressed again and again in the debate over the
meaning of the paper that was to come.

The importance of the empirical result would be enhanced if the theoretical estimate were not very confidently held, as was the case in the
early days of Joe Weber's claims. The estimate that the 40-meter upper
limit touched on—the frequency of inspiraling binary neutron stars in our
galaxy—is based on models of the evolution of stars, matching this with
surveys of the heavens that attempt to map the population of binary systems ripe for collapse. The "one per million years" figure is taken from a
paper by Sterl Phinney published in 1991. In 1998 another theorist who
knew the field well expressed his view of Phinney's estimate in an email
to the 60-meter upper limit group. His comment gives the flavor of the
current theoretical astrophysics.[14] In the passage below, "NS-NS" stands for
binary system comprising two neutron stars, while "BH" stands for "black
hole."

> [1998] Current estimates based on stellar evolution theory all come in around
> $3 \ 10^{(-5)}/(\text{yr Gal})$ for NS-NS [that is, roughly one per 30,000 years in the
> galaxy], and are all over the map for NS-BH, BH-BH. The story (which I've
> discussed with Sterl) for NS-NS rates based on radio observations is that
> (a) we've surveyed 15 times more of the galaxy without finding new binaries

13. BOGWAD's result, by the way, is 20 orders of magnitude away from the astrophysical
predictions. Should the point of the introductory discussion of BOGWAD not have been entirely
clear, it can now be seen that it begs the question of why the 40-meter result had any scientific
meaning at all.

14. Quotation taken from the email exchange among the contributors to the paper.

[since his 1991 publication]; (b) when re-doing the estimates, some of those
surveys aren't as relevant as others, so they must be folded in very carefully;
(c) the astrophysics community is somewhat more comfortable with a few of
the "fudge factors"...discussed in his 1991 ApJ (eg, the angular size of the
radio beam and its consequences for what fraction of pulsars are visible).
(a) + (b) + (c) give a rate that's roughly the same as Sterl's 1991 number.

We can see from this that astrophysical estimates were not strong, and
some varied from the original Phinney estimate by as much as a factor of
30 [1 per 1 million years : 1 per 30,000 years], but even if one of the higher
astrophysical estimates were the one with which the 40-meter result should
be compared, the gap between observation and theory would still be huge—
eight or nine orders of magnitude rather than nine or ten.

When I chatted with Barry Barish at a conference in March 2002, five
different ways for theory and an upper-limit observation to interact became
clear.

1.   There could be no engagement, because it was inconceivable that there could
     be as much effect as the upper limit rules out. For example, an upper limit
     might rule out only the possibility that more of the energy of a system is
     emitted as gravitational waves than is emitted by the system in all forms.
2.   There could be a measurement that set an upper limit well above what any
     existing theory predicts and that, therefore, does not constrain current
     theory.
3.   A measurement could refer to a domain that is so ill constrained by theory
     that any measurement is worthwhile in case it constrains future theories.
     Barish told me that any measurement of the stochastic background of
     gravitational waves set by the current generation of interferometers would be
     like this. The measurements would not amount to much, but the theory was
     so ill developed that any measurement may turn out to be useful as theories
     emerged.
4.   A real physical measurement constrains theories that are out there but are
     not believed very seriously. (Weber's claims supported theories of this kind
     that were current at the time, and I will discuss another case shortly.)
5.   A measurement sets a constraint on widely believed theories. This is the
     LISA/LIGOIII scenario, which may even extend to LIGO II.

Thinking in these terms, what the email contribution referring to the
state of theory made clear is that the 40-meter result was not a case like
category 3. Theory, though it had some slack, was not so forgiving that

any measurement at all would be astrophysically valuable. Even though the prediction of the number of inspiraling systems has large error bars, they are not so large that the 40-meter measurement is a valuable constrain upon them. If the choice lies between the 40-meter result being an instance of category 2 or category 3, it must belong to category 2.

That things were seen this way by at least some of the contributors or potential contributors to the 40-meter analysis was clear from the email interchange. The distance between the upper limit and the theoretical prediction worried some of the contributors enough to cause them to question whether the result could be said to have any astrophysical significance at all. One contributor (not an author) wrote,

> [1998] Overall it is a good paper as it is written. Still, my position is that it is good that the data processing schemes for the grav. wave detectors are being explored and tried out. However, I do not see what astrophysical impact the result of < 0.9/hour has [this was a response to an earlier draft in which the measured rate had not yet been finalized]. So I stick to my belie[f] that PRL might not be the right journal for it.

Another contributor, who gave the authors a great deal of help in working the paper into its final form, refused to be a coauthor of the paper for similar reasons even though he had been one of the team who developed the 40-meter device. On April 17, 1998, he wrote to the team,

> [1998] This paper is not about astrophysics or instrumentation, it's about extracting signals from noise. Its value is in the development of statistical tools that can be applied to LIGO datasets when they become available.

To this he received a reply as follows:

> [1998] I don't quite agree. We do put a new type of physical, observational limit on a process.

On May 18, the second commentator wrote,

> [1998] It's not even necessary that the result weigh in on some specific model that's been published. But I'd feel better about sending the paper to PRL if I knew it said something specific about some *imagined scenario for events—even if the scenario were somewhat speculative.* Can you think of anything along those lines? If not, perhaps we should consider an alternative venue to PRL. [My emphasis]

He eventually concluded (August 14),

> [1998] On the off chance that there's any uncertainty in where I stand on the
> paper and that uncertainty is delaying progress, here it is:...I decline
> authorship on the draft as it now stands. I'm inappropriate as an author,
> since its value is based on analysis work I didn't contribute to. Also, the
> "main goal" [of the paper] point just bothers me too much.

As it was, the paper was submitted to *Physical Review Letters,* but it was
initially rejected for reasons related to the views of these two commentators. The editor said, "[W]e judge that while the work probably warrants
publication in some form, it does not meet the special criteria of importance and broad interest required for Physical Review Letters." The editor
recommended submission to *Physical Review D,* a prestigious journal but one
that publishes longer papers whose results are not considered worthy of
urgent publication.

Two referees commented on the original submission. The first wrote a
couple lines recommending publication, but the editor had drawn on the
remarks of "Referee B," who said, among other things,

> I recommend that this manuscript not be published in *Physical Review Letters.*
> While the gravitational wave data analysis reported is certainly of interest to
> researchers in that field, I do not feel it satisfies *PRL's* criteria of importance
> and broad interest.... The fact that the upper limit for the event rate found
> is many orders of magnitude worse than those of stellar population studies
> makes this work seem more like a test of analysis techniques than an actual
> search for gravity waves. The bulk of the new material in the paper
> concentrates on the analysis methods used, and I find that very little new
> physics is learned from this search. This work appears to be a good step
> forward in interferometer data analysis, therefore, I encourage the authors to
> resubmit the paper in a longer form to *Physical Review D.*[15]

15. As Trevor Pinch pointed out to me, a similar line was taken by a referee reporting on an
early upper-limit submission on a flux of neutrinos. In 1955, Ray Davis submitted a paper setting
an upper limit on neutrinos emitted from a nuclear reactor. A referee commented as follows:

> Any experiment such as this, which does not have the requisite sensitivity, really has no
> bearing on the question of the existence of neutrinos. To illustrate my point, one would
> not write a scientific paper describing an experiment in which an experimenter stood on
> a mountain and reached for the moon, and concluded that the moon was more than
> eight feet from the top of the mountain! (Bahcall and Davis 1982, p. 245, quoted in Pinch
> 1986, p. 56)

Davis eventually measured the flux of neutrinos emitted from the Sun. This early paper was eventually published in *Physical Review* (Davis 1955).

The team, over the signature of Bruce Allen, asked the editor to reconsider.

> Referee B [said] that the upper rate limit found was many orders of
> magnitude worse than would be expected from stellar population studies.
> This is true, and is stated in the manuscript (same paragraph). But it ignores
> an important distinction: in contrast to these other bounds, ours is the first
> bound that is based on direct gravitational-wave observations. It does not rely
> on stellar evolution models, or on electromagnetic observations.

PRL then sent the manuscript to two other referees. Referee C thought it should be published, though he/she remarked as follows:

> Interest: this is a substantial advance in a subfield of physics. *Not astrophysics,*
> *certainly,* as Ref B points out, but rather the burgeoning field of
> interferometric detection of GWs. [my emphasis]

Referee D also thought the paper should be published after some changes. We see, then, that at least two of the contributors to the interchange and at least two out of four referees (the other two did not comment on the specific point) did not think the paper constituted astrophysics.

In the paper as finally published, the value of an empirical observation is stressed once more:

> Using stellar population models one can forecast an expected inspiral rate
> of [one per million years] . . . far below our limit. However, unlike these
> model-based forecasts, our inspiral limit is based on direct observations of
> inspirals. (p. 1501)

The room for debate about the meaning of the result in terms of astrophysics is illustrated further in the following interview-conversation I conducted in 2000 with two senior physicists (who shared no close institutional ties with any member of the 40-meter upper-limit team). In this conversation, I played devil's advocate.[16]

> *Collins:* [S]etting fairly high upper limits can be described in two ways, either as
> doing science or not doing science. See, let's take the thing that Bruce Allen

---

16. This conversation took place long before I had seen the 40-meter Web site or interviewed Bruce Allen about the upper limit.

has done with the 40-meter. [He used a] single interferometer. A lot of people want to make a big deal out of this.... They want to say, "Look, this thing has produced science." I have to say this is where I start—even though I am supposed to be neutral on this—this is where I start to be a bit cynical. Because really it isn't really science . . . at all.

Senior Physicist 1: I disagree totally.

Collins: Alright—go on.

SP1: You see, you believe the theorists too much. You believe we know that there's not a source there. What Bruce did was look at actual data and analyze it. And, from observations, say something about nature. If you believe the theory, then you conclude, "But that's not an interesting statement, because from pure theory we have no expectation that there's anything there."

Collins: But the statement that you just made could be made about any detector, however insensitive. It's exactly true of the bar detectors and so on.

SP1: When I first heard that the supernova 87A had produced detectable neutrinos, I said, "So what, why are they making a big deal of it?" I'm just going from pure theory. There are these theories of cellular structure, and processes, and nuclear reactions, and they all say there are "gonna be neutrinos, blah, blah, blah." But they've actually seen the neutrinos. That means this whole intellectual construct is correct.

Collins: No—that's not the right analogy for the 40-meter results, because the 40-meter hasn't seen anything. [The neutrino detectors actually saw an expected flux of neutrinos, not nothing.]

SP1: They [the 40-meter analysts] did not see something in perfect agreement with theory!

Collins: Well, you know, this cup [holding up coffee cup] doesn't see gravity waves either, and it's in perfect agreement with theory. Look, no gravity waves!

Senior Physicist 2: I see what you're saying, but there is a dividing line. There is a class of experiments that are not really doing science at all. And then there is a class that are clearly doing science. Then there's that dividing line, and I would say the 40-meter experiment comes at the dividing line. It's true that it was known from other investigations and other evidence that the upper limit on seeing gravity waves, coming from whatever, could be set well above [below] what they were going to see. That was already known. However, what [SP1] says is also true. There could have been some surprises; there could have been a surprise. And so this is sort of a borderline. But this is a real instrument.

SP1: How do you know there are not collapsing black holes in globular clusters around our galaxy? If they were there, they would have shown up. . . .

> SP2: There are two aspects. One is looking for real gravity waves—which are on the border line. Then there's the instrumental science of how you make an instrument like this work.

The very last phrase of this conversation will take us into the discussion of the meaning of high upper limits for funders and for the core group, but first I will draw out other aspects of the conversation.

Negative and Positive Claims Revisited

It is claimed that the 40-meter instrument could have seen collapsing black holes in globular clusters around our galaxy or some such—by which is meant inspiraling binary black holes. We know from the earlier quotation in the exchange about the 40-meter that estimates "are all over the map for NS-BH, BH-BH," and this might leave room for a BH-BH surprise. But this possibility was discussed within the 40-meter email exchange. It will be recalled that one of the correspondents suggested that someone might at the very least come up with an "imagined scenario for events" even if it was "somewhat speculative." But this correspondent answered his own question.

> (May 19 1998) Here's my feeble attempt at coming up with astrophysical significance to the R90 [the measured result]: it might set a limit on the density of black hole binaries in the galactic halo. That would be worthwhile to publish, on the basis of previously published papers estimating such populations. Alas, this attempt fails because the galactic halo was out of reach of the 40 m.

If, however, we ignore this "out-of-range" problem, we can draw out a deeper point first referred to in the discussion of BOGWAD. Both senior scientists are claiming that the 40-meter analysis was real astrophysics because it could have produced a surprise (effectively shifting the finding from a category 2 to a category 3). But it could not have produced a surprise, because if it had found a piece of data consistent with a BH-BH inspiral, it would not have been believed to be a BH-BH inspiral.

As has been argued in the opening section, to make a positive sighting of gravitational waves, it would be necessary to distinguish signal from noise. The 40-meter setup cannot do this, because to separate signal and noise in this field it is almost always vital to cross-check findings on two detectors sufficiently far apart for their respective noise to have no common source. This is the crucial first step in separating signal and noise in nearly all gravitational wave research. As the paper itself makes clear,

> Without operating two or more detectors in coincidence, it is impossible to characterize the non-Gaussian and non-stationary background well enough to state with confidence that an event has been detected. (p. 1500)

We can add that to establish a positive result, much more care would also have to be taken over "environmental monitoring"—which is making sure, by measuring every conceivable external influence on the detector, that nothing else is causing the effect. Thus the difference in the logic of upper limits shows itself in differences in the way the observations have to be done. As explained, to set an upper limit in the face of a putative signal, one merely raises the upper limit rate, and this does not require separation of signal and noise. Lest this is still not clear, one need only think of how carefully the physical and environmental monitoring (PEM) channels of two detectors would have to be examined before any positive claim could be made. The PEM channels monitor changes in seismic, electrical, and cosmic ray activity and so forth so that these uninteresting causes of disturbance in the detector can be eliminated before a positive claim is made. The PEM channels were looked at in the case of the 40-meter result, but they were not examined in the same fine-grained way that would be necessary to mount a positive claim. In short, if the detector had been disturbed by the signals from BH-BH inspirals, no one would have known; it would merely have made the upper limit still higher. This would not have caused any surprise, however, as the upper limit was so high anyway that raising it further would take almost nothing from its value as astrophysical information.

What the senior physicists might have said (and this is, perhaps, what they meant), is that since (if) the 40-meter was capable of seeing these events in the galactic halo, and since some astrophysicists had suggested that there were such events, in setting an upper limit the team may have been engaging with a seriously held astrophysical scenario, and it was this that might have made their upper limit interesting to astrophysicists. But the paper did not rest its case for publication on its engagement with this phenomenon. I will return to this point shortly, after dealing with some other aspects of the conversation.[17]

Use of the Term *Science*

"Science" is one of the more obviously "socially constructed" ideas in gravitational wave research. In spite of its inability to separate signal and noise,

---

17. My interpretation had better be right, or the senior physicists were effectively offering a license to speculate about unknown sources and the ability of insensitive detectors to see them— something they were adamant should not be offered to the Frascati group.

the senior physicists were adamant that the 40-meter instrument had done some "science" even though it was an isolated device. A senior leader of the LIGO project also told me that the 40-meter result constituted "a significant search for binary inspiral sources in our galaxy," and criticized the old 40-meter team for not doing the "science" that was there to be done. On the other hand, on other occasions, physicists have insisted that only two or more gravitational wave detectors together are capable of doing "science." That is why LIGO has two detectors in spite of the fact that at the 1986 review meeting to decide whether the LIGO ideas should proceed (chapter 27), Richard Garwin, arguably the most influential person on the panel, argued that there should be only one.[18] On another occasion, here is how "Senior Physicist 1" argued the case for building two detectors, an argument that most other LIGO scientists agreed with wholeheartedly. Once more, I take the role of *provocateur*:

[2000] *Collins:* Then there's Garwin's argument that if you aren't going to detect gravity waves, why have two detectors?

SP1: You couldn't tell if it was working if you just had one.

Collins: I don't understand—please explain.

SP1: Well, there's a huge background noise, so how can you tell if there's a signal or not?

Collins: Oh no—you can't. But the point is, you don't expect to see a signal at this stage—not really.

SP1: Oh—OK. So if you're not expecting a signal, you can just build something and measure its noise.—Sounds right—sounds reasonable.

Collins: And, of course, that's what GEO has done.

SP1: Well, once you have something else you can run in coincidence with [GEO has LIGO], you can do science. People are not going to spend ten years of their lives just building technology. I think that's the real reason we don't just build one.

Here the respondent describes working with a single detector as "just building technology" rather than doing science.[19]

18. Garwin told me this in an interview.

19. Still other scientists argued that it was necessary to have two detectors even if no positive signal was going to be seen. They argued that because one can separate signal from noise only if one has two detectors, it is necessary to have two detectors to mount a sensible noise reduction program even if all one was trying to do was improve the technology without doing astrophysics. They argued that with only one detector, one might be trying to "improve" the instrument by eliminating what appeared to be non-Gaussian noise events that were really signals! This argument seems to make sense only if there is reason to expect to see a fairly regular stream of signals.

Bear in mind that it is I alone who have had the luxury of retrospectively poring over these coffee-bar conversations. The object is not to show how smart I am, nor to treat such conversations as though they were evidence in a court of law, but to use them to represent the science "language game" as it was being played out in this temporal and social space. We can use our native competence to be sure that in this space there would be little flexibility about the notion of science, were it BOGWAD or LISA that had given rise to the upper limit findings—there is no way BOGWAD would be counted as doing science and no way a sound LISA upper limit would not be counted as doing science; the social consensus concerning these instruments and their powers is too strong. But the notion of "science" is more flexible when it comes to the 40-meter, and the use of the term can be aligned with the argument being pursued at the time. When trying to justify the building of two detectors rather than one, then "science" involves separating signal from noise; when trying to justify the worth of the 40-meter upper limit, then "science" need not involve separating signal from noise, and "science" can be done with one detector alone. In sum, depending on who you were, where you were, and what you were arguing, the 40-meter results were either just inside the boundary of science or just outside it. Now we turn to the future and LIGO.

Astrophysics, LIGO I, and the Uses of Speculation

> [1996] You will have a scientific result . . . and it will be a real astrophysics result, for LIGO in the first six months. . . . [The result] will most likely be an upper limit, but it will be an upper limit that will be an order of magnitude or two smaller than the existing one, and that means it will constrain theories, because it's now much more constraining. It will be a real astrophysics result. And that means the astronomers for the first time will have to take us seriously as astronomers, because we have determined something about the energy density of the universe—a number they want—and they will respect us—and they will be more tolerant if it takes us another year to find the coalescing binaries. You see, you feed them something right up front. You say, "We are producing science"—We don't just build a beautiful detector, we have done science with LIGO.

These are the enthusiastic words of an ex-leader of the LIGO project. The upper limit to which he referred is on the stochastic background radiation. He considered that even LIGO I can produce an upper limit on the stochastic background that will interest astrophysicists. The stochastic background number, he believed, will constrain theories and justify LIGO to its old enemies, the astronomers. This is reasonable, since, according to Barish, the cosmic background lies in category 3 for the relationship between findings and theories; that is, the theories are so ill specified that any empirical number is a useful one. The ex-leader implied, however, that LIGO's improved upper limit on the number of inspiraling binary neutron stars will not interest astrophysicists—it will still be too far above theoretical expectations. Thus, for this respondent, setting an upper limit on the stochastic background will be astrophysics, while setting a better upper limit than the 40-meter on inspiraling binaries will not be astrophysics—he agrees it is category 2.[20]

But astrophysics, as we have seen, can be born of speculation. In the 40-meter debate, it was suggested that significance could be gained if the upper limit bore on even an *imagined* astrophysical scenario. Because of LIGO I's greater sensitivity, it should be less hard to think up some theory for it to encounter. LIGO I's range, assuming it can be made to work according the specification, *does* stretch out to the Galactic Halo. Speculation does not have to turn out right to be of value; its value lies in the opportunity it provides for an instrument to prove it wrong. It is another way in which seeing nothing can transmuted into science. Perhaps setting an upper limit on BH-BH inspirals in the Halo will "do the trick" for LIGO I.

There is also a more detailed theory that could serve the same function for first-generation LIGO. In 1998, Hans Bethe and Gerry Brown published "Evolution of Binary Compact Objects That Merge." This paper claimed that a high population of black hole–neutron star merging binaries probably exists, which would consequently give the first generation of LIGO a ten-times better chance of seeing such events than had been thought. According to this paper, LIGO I would need to improve its sensitivity by only a factor of 2 to see one such event per year.

Hans Bethe is a much-respected physicist and Nobel Laureate. Gerry Brown also has won many prizes for physics and is a highly respected theorist. Nevertheless, the shoptalk among the astrophysics-knowledgeable community suggests that the paper is markedly speculative. It is said that it is only taken seriously at all because of the fame of its authors. The joke going around is that if the same paper had been written by Gerry Bethe and Hans Brown, no one would have heard of it. Still, the paper *was* written

---

20. The reference to binary neutron stars *in the quotation* is to a positive sighting, not an upper limit. Upper limits on binary neutron stars are not mentioned. That is why I say the meaning of upper limits on BN-BN inspirals for this respondent is *implied*. According to some physicists, this sentiment—about the importance of a potential LIGO I upper limit on background radiation—is somewhat overenthusiastic.

by two famous authors, and it may lend some interest to upper limits set by LIGO, or some credibility to arguments suggesting that money spent on only a minor upgrade could bring the devices into the realm of potentially observable gravitational waves. We can call the theories which have reputations like Bethe-Brown artificial theories, and they are represented in a gray box in figure 40.1. In respect of the astrophysical significance of upper limits, they serve the same function as Weber's early positive claims.

The Funders and the Core Group

We now turn our attention to the next two audiences for upper limits. Above, I quoted an ex-director of the LIGO project as suggesting that a useful upper limit on the cosmic background radiation would be one of the first of its results. He went on to say, [1996] "And NSF will be pleased and Congress will be pleased.... It will be a tremendous morale boost if we do new science, and we will also keep the wolf from the door. But at the same time, what will be another criterion of respect for LIGO will be if we achieve what we predict very quickly, even if it sees nothing."

This last sentence expresses the reason for setting upper limits that will always appeal to funders and to the core group of scientists irrespective of the astrophysical significance of the results. The same sentiment was expressed by "Senior Physicist 2" quote above, who said that apart from doing astrophysics, the importance of the 40-meter result was developing "the instrumental science of how you make an instrument like this work." The point is again set out in the 1999 40-meter paper itself: "Our study also demonstrates methods being developed to analyze data from the next generation of instruments" (p. 1501).

And, of course, the same point once held true for the much more insensitive room-temperature bar detectors. When they were built, they, too, were remarkable technological achievements that could justify confidence in the funding organizations by demonstrating their technical virtuosity. Here, for example, a scientist recorded in 1972 referred to one of the least-sensitive early bar detectors: [1972] "...has analyzed his data and came with up a tremendous upper limit on the flux—really worthless—but he needed to do that because he needed to continue. He needed justification for continuing his work there..."

What setting an upper limit does is demonstrate control over the apparatus and the chain of inference from detector to claimed signal even if the apparatus is too insensitive to see a signal. On LIGO's Web site, one can find such a sentiment expressed by another spokesperson for the project, Rai Weiss, who apparently did not think of it as a significant search for astrophysical sources: "While the [40-meter] instrument's sensitivity is too low to detect any astrophysical sources of gravitational radiation, it does provide insight into the nature of a real data stream, and the properties of instrument noise"—that is, work done on it is seen here not as significant new astrophysics but useful practice for analyzing LIGO's data when it comes "on stream."[21]

Could such a claim be made for BOGWAD? Could BOGWAD be of interest to funding agencies because of its technological achievements if not its astrophysical findings? The answer has to be no. The reason is that there is no significant chain of inference from BOGWAD to final claim. Setting an upper limit with the 40-meter is different from setting an upper limit with BOGWAD—not because the 40-meter's upper limit is ten orders of magnitude better than BOGWAD's but because the 40-meter is a broadly similar device to the large interferometers. LIGO has grown incrementally from the 40-meter in a sense that does not apply to BOGWAD. Prolonged periods of practice at sensing vibrations, and increases in the format of *Gravity's Shadow*, however large, are not going to make you and your book into a gravitational wave detector of a kind that might detect gravitational waves, whereas the 40-meter result required practice and proficiency in the use of lasers, suspended mirrors, amplifiers, computers, and statistics. To build, operate, and *analyze the data* emerging from the 40-meter or any one of its successors *is* to take a step in the direction of detecting gravitational waves.[22] Thus, even though setting an upper limit with one of these devices does not accomplish much of direct impact on astrophysics, it is a step toward doing gravitational astrophysics, because it demonstrates that a whole system has been completed and integrated. In this way, every such device brought to completion will have assured the funding agencies supporting it that their money was being spent on a project with a potential payoff. Seeing nothing with any old machine does not do it; seeing nothing with

21. <http://www.ligo.caltech.edu/LIGO_web/lsc/lsc.html>, accessed July 2000.

22. It is very easy to make a mistake with this kind of inference. Those trying to build intelligent machines make the mistake over and over again. They note the performance of the current generation of so-called intelligent machines, note the rate of improvement in chip density and speed, and extrapolate from there to assured success. But there is every reason to think that progress of an entirely different and unforeseeable kind is needed to go from where we are now to humanlike performance in a machine. By contrast, in the case of gravitational wave detection there is every reason to think that continued development in the same general direction will eventually achieve the goal even though it will be very difficult.

the immensely complex set of technologies and inferences that constitute a "gravitational wave detector" can do it.

Upper limits set by LIGO are going to be interesting to the funding agencies for the same reason. They will show that something that can be called "science" is being done even in the absence of a positive detection. LIGO is going to want another slice of funds to spend on Advanced LIGO before they have confirmed the existence of gravitational waves. To be in a position to say that they are already doing science will help their cause. Thus, in the year 2000, a senior LIGO spokesperson told me,

> *Respondent:* It's hard for me to imagine them giving us the money [for construction of LIGO II] at the beginning of the science run unless we haven't already got early scientific results. . . . To me it's a couple of papers. I think in the middle of 2002, when we're having our final big reviews on the construction of LIGO II, if we hold the construction schedule that I'm touting right now, we ought to be able to show a few papers—a couple of papers setting limits. Not discoveries, unless that drops into our lap—unless we're really lucky—but a paper setting a limit on the rate of binary inspirals out to some distance—a limit on the rate of stochastic or asymmetric supernovae— some credible papers already submitted—it may not be submitted to *Physical Review*—it may be some electronic preprint service—but they [will be] out there in the community. We've published!
>
> That's a standard that says we should focus now not just on delivering the code and being able to ingest the first bits of data, but people should position themselves—in effect writing the paper they think they are going to publish, leaving out the numbers and arranging the people and the software so that they are ready to show that they are an astrophysicist, not just an instrument builder.
>
> *Collins:* But these upper limits are not going to be fantastically interesting science.
>
> *Respondent:* Right, but to a member of a review panel of this thing, they'll say, "Look, these guys have made this thing work, they've got the community organized, they're putting out the first scientific papers, and just looking at the derivative [i.e., the rate of increase] of what they've done, these guys look like they're doing what they should be doing—they're making a good show of it." If you don't have that science, then there's this great mystery: "Are these guys scientists or are they just tinkerers?" Because we're not in this ultimately—all of us—as tinkerers, we're in this to answer a bunch of scientific questions.

The pressure is coming from above as well as below. As two spokespersons with responsibility for LIGO in the NSF explained to me in 2000,

> *Spokesperson 1:* ...I attended the [Review Committee meeting of LIGO] at Caltech in May [2000]....And the presentation by the lab was actually focusing on delaying the science run....And the Review Panel said: "Don't look at it that way. Don't say you're delaying. As soon as you're ready to turn these instruments on and begin taking data, you're into science. It's an engineering run, but you're into science as well. And you're not going to stop your ... collaborators from looking at data and trying to find sources, are you?— No!—OK, then that's science. And so don't say you're delaying for a year, because you're still doing engineering. The engineering run is still doing science." That was the philosophy that was introduced at that meeting, and they adopted it.
>
> *Spokesperson 2:* OK, that's telling them how to describe what they're doing in a clearer way, and more accurate way. But the actual pressure to take all of this data, and exercise all the computing algorithms and all of that—that's coming from the LSC [LIGO Scientific Collaboration]. They're driving it. They say, "You're sitting on all of this data, and it not only will help you in diagnosing the performance but we can expect science and want to expect science—give it to us. Let's start setting up procedures."

The members of the core group, as we might now expect, experience as successes what to the most distant audiences look like failures. Everything that has been said about the funders applies to the core group: the extraction of upper limits from prototype machines, or devices still too insensitive to see positive signals, shows that science and technology are being done even if it is not yet astrophysics.[23]

---

23. Actually, this is to oversimplify: there are even a class of *astrophysicists* to whom these results are interesting—those who take a special interest in the use of gravitational waves as an astronomical tool and who therefore want to learn how to frame the kind of estimates that will enable experimenters to make a comparison with their observations. Bruce Allen drew the attention of the editor of *PRL* to this group when he wrote back insisting the paper was worth publication. He said, "They [the results] show astrophysicists the range of issues that must be considered to obtain physical limits from gravitational wave observations." The 40-meter analysis showed this group of astrophysicists what manner of estimates they would need to produce so as to maximize the astrophysical usefulness of the upper-limit findings of new generations of detectors. These astrophysicists, insofar as they are doing this kind of work, can, however, be said to belong to the core group for the purposes of this analysis rather than to the astrophysical community. (I am grateful to Jolien Creighton for pointing this out to me. It suggests that the 40-meter paper was, as some physicists would put it, more than an "MPU"—Minimum Publishable Unit).

CONCLUSION

Setting an upper limit is definitely discovering nothing in one sense—it is the discovery of no signal. If the upper limit is high enough, it is also discovering nothing in another sense—nothing of scientific importance. If the upper limit is really high, as with BOGWAD, discovering an upper limit may not be doing science at all.

Here we have explored the "meaning" of upper limits and the meaning of "science." One can set very high upper limits with complete confidence even though one is not doing science. Upper limits start to be scientifically meaningful as they get lower. We have looked first at the way gravitational wave upper limits gain astrophysical importance and then at the way they gain technological importance. We have shown the way these two sources of meaning appeal to different audiences.

There can be disagreement about both the astrophysical significance and the technological significance of a result. We have explored the disagreement about the astrophysical meaning of the 40-meter upper limit and noted that it could be argued that the 40-meter paper was published in *PRL* largely because of its technological significance. A couple of other cases document the potential disagreements about technological significance. In 1996, a paper reporting an upper-limit result on the stochastic background radiation produced by *two* resonant bar detectors working in coincidence was submitted to *PRL* and turned down. The authors believed that though the result represented no advance in terms of astrophysics, the analysis represented a very important technological innovation: for the first time in such an observation, the noise was measured using two detectors, allowing much more noise to be eliminated. Referee "A" did not think this innovation merited publication.

> I do not recommend this paper for publication in the *Physical Review Letters*. The paper reports on an improvement of previous measurements on an upper limit to the gravitational wave stochastic background in a narrow bandwidth near one kilohertz. No physics of importance is extracted from that result to justify rapid publication in the *Physical Review Letters*. The fact that the experiment involves two instead of one antenna permits an improvement in the upper limit but, in itself, is not a surprising new approach.

The disappointment was to be repeated when in 2000 a paper was submitted to *Physical Review D*, a journal with less-stringent criteria of publication. In that case, the crucial referee comment was as follows:

The paper does not introduce significant new physics, and the statistical treatment is not clearly presented. If there were significant physical content to the upper limits presented in Figure 2, then the statistical treatment might be of interest. Under that circumstance, though, the interpretation of the upper limit estimation would require a clearer exposition.

As we see, (astro)physical significance and technological significance are not necessarily treated as independent by physicists. This leaves a great deal of space for judgment about the overall importance of an upper limit. Here a lack of physical content is taken to disqualify publication of a technological advance, though it did not do so for the 40-meter paper.[24]

The significance of technological advance is difficult to measure. Perhaps a little more can be said about astrophysical or "scientific" progress. It seems likely that when gravitational wave detectors become much more sensitive, and assuming no positive sightings are achieved, what will count as a significant advance in setting an upper limit will need to be less and less of an improvement on what went before. We ought to see the distance between each new upper limit and the previous one published in a high-prestige journal steadily narrowing. The technological reason for this is that it gets harder and harder to improve sensitivity as the sensitivity improves (as it gets harder to climb the higher you get up a mountain), but, more important, the results do not have to be much better in order to constrain more and more theories as the realm of theoretical prediction is approached. Applying this thought "in reverse," as it were, a measure of the astrophysical significance of a result might be the extent to which an upper limit has to improve, in the absence of significant technological advance, to be worth publishing in a top journal. Bruce Allen has suggested to me that nowadays a publishable improvement would be a couple of orders of magnitude.[25] What will count as a publishable upper-limit result a decade from now?

What is our overall conclusion? It has been prefigured in the epigraph to this chapter. Setting an upper limit may amount to seeing nothing, but

24. As we have seen, it did not rule out publication in *PRL* of Allen (B.) et al. 1999, nor did it rule out Allen (Z. A.) et al. 2000. It may be appropriate to understand the varying judgments in terms of the argument between different groups of gravitational researchers, which is discussed above under the heading of evidential cultures (chapter 22), or it may simply be "the luck of the draw." Some of the referees volunteered their identity to me (in confidence), and thus I know that the author of the second referee's comment above would have drawn the same conclusion in respect of the 40-meter paper had he been a referee, which he was not. I do not, however, know the reasoning behind the whole chain of rejection for these papers.

25. Private correspondence.

even if this is "nothing" in the strong sense of the word, such that there is no engagement with theory, if the nothing is seen well enough the result can be "write-uppable"; it can still be science for scientists. The possibility of the transmutation of the lead of failure into the gold of science is in the eye of the beholder.

PART VI

SCIENCE, SCIENTISTS, AND SOCIOLOGY

CHAPTER 41

COMING ON AIR: THE STUDY AND SCIENCE

WHY DOES LIGO NEED TWO INTERFEROMETERS IF IT IS REALLY A PROTOTYPE?

LIGO I will start, and possibly finish, by setting upper limits. With luck it might see an event or two, but unless something very strange occurs, it will never become a gravitational "observatory." The LIGO project, however, was justified to scientists by its potential for gravitational astronomy, and that is why a series of more sensitive detectors was planned from the outset. The astrophysical scenario on which the proposal was based—a relatively dark gravitational-wave sky—implies that the best LIGO I can ever be is a "detector." So why is it resourced to the level of a finished instrument rather than a prototype?

We know the answer in terms of the way historical events unfolded. What might seem like rational decision-making in the core of the science is not necessarily seen as rational decision-making in the outer layers of the target diagram. Tony Tyson said that when he attacked LIGO in Congress, his mistake was to argue that his favored trajectory, involving more prototyping before commencing on the full-scale project, would involve a delay of only a year or two and therefore a relatively small saving. At the level of Congress it was hard to see what the fuss was about; if the thing was ever going to be built, one might as well get on with it. But is there any rationale other than the coarse grain of congressional thinking?

Think about what could have been done instead. The components of the first potential *observatory*—Advanced LIGO—could all have been developed on a smaller prototype or a single device rather than two. These components include the multistage pendulums, the sapphire mirrors, the advanced seismic isolators, and the powerful laser. Vital techniques of data analysis could have been honed by comparing streams of real-time and delayed data, perhaps with some artificial signals inserted. The argument that in the absence of two detectors one might be confused by real gravitational wave signals does not hold up: if there are enough gravitational wave signals to confuse LIGO, then astrophysics as we know it is wrong. Of course, astrophysics may be wrong, and that would be very exciting; but one cannot plan a science on the assumption that everything you know about the sky is incorrect. Were you to plan that way, there would have been no justification for building LIGO in the first place, because things might equally well be wrong in the other direction, and there would be no sources of waves within range of even an Advanced LIGO.

Look at it this way: if the argument for building two 4-kilometer interferometers is the possibility of seeing an unknown source, then why did it not apply equally to the 40-meter device? If we know so little about the sky, then the 40-meter's noise elimination program could also have been confounded by real but undetected signals. There *was* a reasonable justification for building more than one detector when Weber was going on air because so little was know about gravitational wave sources, but now we know much more about what is in the sky. And it is that "much more" which justified the building of LIGO and was a crucial weapon deployed in the battle with the resonant mass program.

Building only one LIGO detector, on the other hand, would have put off the day of big spending and strong scientific coordination until much more of the risk of failing to achieve astronomy had been eliminated. Furthermore, there would have been none of the seismic troubles and additional costs of the Livingston interferometer. It is this kind of argument that gave rise to the 40-meter team's idea that they would like to have built a 400-meter device to "retire much of the risk" before full-scale interferometry began.

I have tried this argument on many gravitational wave scientists and have heard only one technical counterargument that remained standing after the point had been clarified.[1] It was said that only now have the detectors become sensitive enough to enter the realm where there might reasonably be a surprise that would justify building two. But even if this is scientifically reasonable, it is not the kind of argument on which you can base scientific expenditure: "Let's spend a lot because a big surprise is a little less unlikely."

All that said, I want to argue that the decision to build two interferometers was the correct one. The reasons have to do with human beings and social context rather than science and technology. Building only one detector, or a succession of smaller prototypes, would the right thing to do if science were done in a social, psychological, and political vacuum; if there were a big pot of money into which scientists could reach ad lib; if scientists were robots without families who could be put into cold storage for indefinite lengths of time without damage; and if the only constraint were to do the most science for the least money.

As things are, gravitational wave detection, from instigation to first detection, is a rough match for an adult lifespan or professional career (including that of this author), and it would be hard to persuade scientists to devote their lives to a scientific enterprise with a significantly longer time span.[2] As Rich Isaacson of the NSF put it to me, [1995] "Once you . . . can run in coincidence, you can do science. People are not going to spend ten years of their lives just building technology. I think that's the real reason we don't just build one." Reinforcing the point, at least two senior scientists (I did not ask the question systematically) assured me that they would have abandoned the project early on if only one detector had been built; what could be achieved with one detector only would not have satisfied their criteria for the project of a lifetime (and I guess the same goes for me).[3]

Again, as explained above, at the time of this writing, the first results are being reported by the various small groups who are analyzing the data. All these results are upper limits, yet one can sense the electricity in the

---

1. For an earlier example of the deployment of the argument, see chapter 40.

2. Are there sciences that run to longer time scales without the chance of a significant midway result?

3. Critics, such as members of the old 40-meter group, do not accept this argument. They suggest that international collaboration could resolve the problem of coincidences. But international collaboration could not be relied on at the time the decision to build two detectors was made and is still not assured.

air which is present only because if there was a really suggestive signal on one of the detectors, one knows one could check it against the other; this makes the huge task of sifting through the data far more enticing.

Finally, if LIGO I sees an event or two, though this will not make it an observatory, the financial floodgates are likely to open, ensuring the full implementation of gravitational astronomy. This means that, given luck, the decision to try to see *anything* in the way of gravitational waves as soon as possible, as opposed to going more directly toward gravitational astronomy in a systematic and less expensive way, is likely to have been a good one. After all, if the less expensive way reduces the assurance of eventual success, it would be a false economy.

In sum, having reached the point where the funding decision was ex-posed to the politicians—where the locus of accountability had moved to the outer circle of the target diagram—it made sense to go for a complete in-strument rather than a prototype. One can imagine a world where it might have been different, but we are not dealing with that imaginary world. The conclusion is that, given the world we are in, LIGO I is no bigger than it should be.

LIGO AND THE FIRST UPPER LIMITS

In mid-2002, LIGO brought all three of its interferometers into a condi-tion to run in coincidence. Now the noise floor is being lowered toward the promised level and the duty cycle is building up. But LIGO has fallen behind schedule. It has to be said that though the individual problems were not anticipated, that there would be unanticipated problems *was* an-ticipated by almost everyone; rightly, there is not a lot of blaming going on.

Some embarrassment comes from the effect of seismic noise on the detector at the Livingston site; passing trains throw the device out of lock, shutting it down for more than an hour each night. Worse, during the day, logging activity in the surrounding woods does the same. It seems that something about the sequence of logging was misunderstood at the time the site was being selected. Apparently, the LIGO management was told that trees were harvested every ten years or so, seemingly an acceptable incidence for logging-induced downtime. What wasn't understood was that close patches of forest would be harvested for ten-year-old trees on a much more regular basis, making the instrument useless on workdays.

The net result is that the Livingston interferometer can collect data only during the night and between trains, consequently reducing the duty cycle. But this reduction is not immediately fatal; other work can be done during the fallow periods, and since during this shake-down period there is much work to be done, Livingston is still operating at about 60% of the duty cycle of the Hanford detector (as of March 2003). What is more problematic is the effect of the unsocial running hours on the domestic life of the scientists. Working such hours can be managed for short and intense periods, such as during the initial locking phase, but it cannot be kept up indefinitely. Furthermore, certain kinds of family arrangements (such as single parenthood) rule out even a rotating-shift pattern of work. LIGO Livingston has to be able to work 24 hours a day if the duty cycle is to be enhanced and if the social problem is to be solved. The solution is to fit more seismic isolation beneath the vacuum chambers, and this upgrade is being designed and implemented. But it means unanticipated costs, so it will not be ready until the beginning of 2004.

All the LIGO instruments are also having trouble with unexpected noise in the electronics. Partly, the noise stems from the mounting of digital electronics alongside analogue electronics in the same frames, allowing electrical cross talk between the systems. There is also a problem of radio frequency noise, which will mean rebuilding some of the electronics to the substantially higher specification typical of radio telescopes—again a matter of cost and delay.

Another source of trouble is acoustic noise. Sounds disturb the exposed parts of the device in many ways. Many noisy fans and the like have had to be removed from the proximity of the interferometer, and it may be that parts of the device not enclosed in vacuum tanks will have to be fitted with acoustic shields. Robert Schofield, who is working on this noise problem, believes that it will have to be reduced by a factor of a hundred to a thousand before it ceases to be an obstacle to LIGO's reaching design sensitivity.

During the first coincidence run (E7), Hanford was still more trouble-some than Livingston. Hanford turns out to be highly vulnerable to seismic noise caused by the high winds which are prevalent in that part of the country. Unacceptable winds blow for an average of 10% of the time, and it may turn out that even more time is lost to these effects.

Another extraordinary event, in June 2002, was the parting of a mirror suspension wire on the Hanford 2-kilometer device with the subsequent collapse of a mirror, causing the detector's control magnets to tear off. There are stops designed to prevent a mirror from falling too far and causing such damage, but in this instance they failed. The cause of the fall was an earthquake on the Chinese border (minor earthquakes of this magnitude happen twice a month or so), which shook the interferometer enough to make a servo oscillate. This in turn moved a mirror so far out of alignment that it directed the infrared laser beam onto a suspension wire and softened it to breaking point. A new mirror, which had already been prepared, was hung in place and realigned without too much difficulty. The trouble is that the vacuum system had to be opened and pumped down again, and this takes time. Still more worrisome is that most of the other suspended mirrors in all three interferometers could suffer a similar accident. A "simple" solution, already executed in the case of the mirror that fell, is to fit baffles in front of the suspension wires. Unfortunately, fitting the baffles to all the mirrors means opening many vacuum systems, which has the potential to cause more contamination and delay.

In May of 2002, a strange noise, related to the inverse of the cube of the frequency, became apparent in the Hanford 2-km interferometer. That this was probably nothing fundamental was shown by the fact that it did not appear on the Livingston device. I was given conflicting accounts as to whether the Hanford 4-kilometer had yet reached the levels of sensitivity that would reveal the new noise if it was there. If the existence of the "one-over-f-cubed" noise showed anything, it showed that we still do not know enough about interferometers to guarantee that two devices that look "the same" to us really are the same, thereby giving us just a glimpse of what the "kingdom of magic" might look like. Caltech graduate student Matt Evans, who contributed to the computer model of the detector, told me he could not reproduce the noise in his computer simulation, and he had concluded by the middle of 2002 that it was nothing significant. "I think it may be unique to the 2K, which means it probably means it's, you know, an insect flew into a coil driver somewhere, which is hard to see— literally a bug somewhere in the hardware." By the middle of 2002, people seemed to have forgotten the one-over-f-cubed noise, and it must have lost its definition as other noises were eliminated. At that date, the noise curves of all three interferometers were making steady progress downward toward the design specification, though everyone knew that it would get harder as the promised land approached.

As of the early part of 2003, the 40-meter team are still arguing the case that LIGO needed more prototype work if the noise was going to be properly understood. In May of 2002 I met a member of the old team who, though not wishing LIGO ill, pointed out to me that the long interferometers that had been built were proportionately more noisy than the short ones, intimating that there was still space for something deep and unknown to arise that affects only long devices. He also pointed out that one important source of unanticipated trouble was the very digital electronic controls that the 40-meter team were reluctant to install. This respondent still believed that the 40-meter team's view of things could be maintained.

What will it mean if the 40-meter team turn out to be more right than the new management? Consider the aforementioned acoustic noise. Would LIGO be better off if more research had led to a better understanding of acoustic noise before the full-scale device was built? Barry Barish told me that it was not entirely a bad thing that LIGO was suffering from teething troubles of this kind. The art of managing such a project, he suggested, was *not* to spend so much on the design and prototyping that the machine worked as soon as it was switched on; it was cheaper and quicker to spend less on development and fix problems as they came along. There are other technologies, such as space missions, which have to work first time, but they are much more expensive for that very reason. LIGO is not a project of this kind. So for the 40-meter team the acoustic noise problem could, perhaps, be seen as a risk that might have been retired; for Barish it is normal life within a responsible spending regime.

Barish's argument seems reasonable so long as we are not in a special hurry to go straight to gravitational wave astronomy. If Advanced LIGO is funded, then LIGO I will have done its job of building a platform of political, technological, and scientific credibility from which the leap to the next level of sensitivity can be made. There is, however, one "horror scenario" that could unfold (though one would have to bet against it): a close-by supernova or other loud gravitational wave event happens when one or more of the LIGO detectors is out of commission or swamped by noise that could have been anticipated by a more NASA-like approach to the project. This would be as embarrassing as the SN1987A event, when all the cryogenic bars were off air. Of course, if the detectors were still unbuilt and we were still at the prototype stage, then the risk would be still higher.

Setbacks and skeptical predictions notwithstanding, in February 2003 LIGO was given a plenary session at the American Association for the Advancement of Science meeting in Denver. For the first time, the device was presented to the public as a working scientific instrument and the first tentative results were offered for public consumption. I noted that the atmosphere was positive and that there was only one hostile question from the audience; it was about the cost of Advanced LIGO.[4]

In March 2003, I was present at the LIGO Scientific Collaboration meeting in Livingston, Louisiana. The scientists were happy, bringing travelers' tales of praise in the community: "LIGO is working!" was the reported sentiment. "What a magnificent achievement." At the same time, the scientists were presenting their findings to their colleagues and putting the final touches to the first papers that would express scientific results. There would be four upper-limits papers on bursts, inspirals, continuous sources, and stochastic background respectively.

It seemed that LIGO was now benefiting from the fierce criticism that had beset it in the past. The project was a year or two behind its planned schedule, experienced some unpredicted problems, and certainly had not seen any gravitational waves. On the other hand, because so many critics had said that the vacuum system would not work, the laser would never be made to lock, the whole idea of making measurements of that degree of sensitivity was ridiculous, and so forth, the fact that it worked at all was a cause for congratulation. Still, absolutely nothing had happened to make the *O* in *LIGO* any less of a hostage to fortune. The astronomers are still in a position to ask, "What have you observed?" But, for the moment, LIGO's glass is seen as half full rather than half empty.

Note that it is almost inevitable that LIGO will have to make a success out of the astrophysically meaningless. In an ideal world, the first detector would complete its search of the heavens, and all the lessons that could be learned would be learned, before the design of the next stage, Advanced LIGO, was finalized. In the world as we know it, however, there is a large team of trained and experienced human scientists who have to be kept busy if they are not to disperse. There are long lead times for gaining political and financial authorization, for the decisions to pass through the appropriate committees, and for the budget to be finalized and signed. There are long lead times for constructing parts of the device, such as the sapphire test masses. LIGO I has to justify the next slice of spending in the absence of any positive scientific results, and all it has to offer are management virtuosity and upper limits.

In sum, LIGO is like the dog walking on its hind legs, of which Dr. Johnson remarked that the wonder is not how well it does it but that it does it at all. If Advanced LIGO is funded before the novelty wears off and skeptical astronomers rediscover their voices and demand that the dog

---

4. Positive reports were also to come back from the American Physical Society meeting in April.

talk as well as walk, then the upper limits set in the early part of 2003 will have been a resounding success, however meaningless they are in terms of astrophysics.[5]

THE FRASCATI PAPER

In November 2002, the Frascati group shocked the rest of the gravitational wave community by publishing a paper in the respected journal *Classical and Quantum Gravity (CQG)*, suggesting that they had seen coincidences between their NAUTILUS and EXPLORER bars. They were confronted at the Gravitational-Wave International Committee (GWIC) and the Gravitational Wave Data Analysis Workshop (GWDAW), which took place sequentially in Kyoto, Japan, in December 2002.

A paper roundly criticizing their statistical procedures was published in *CQG* by theorist Sam Finn. In it, Finn said that the statistical analysis was too weak to merit consideration even as a preliminary indication that there was a signal to investigate.[6] The story of the tension between the Frascati team and the Americans continues, but I can draw out a more interesting consequence that I am going to claim as a potential minor triumph for sociological analysis. At a meeting in Cambridge, Massachusetts, in November 2002 and at the December meeting in Kyoto where the Frascati results were discussed, I found the LIGO community angry that the Italians had, once again, broken ranks. They told me either that the results were not worth talking about or that having to confront them empirically, which they felt they had to, was a nuisance and a waste of time.[7] I argued that they should be delighted—I think I said, "They should be paying the Frascati group for giving them this early opportunity to do some real science." I was arguing that even if the Frascati group's results turned out to be wrong, they would fulfill the function of one of the gray boxes in figure 40.1; if LIGO I can refute the Frascati results, it will be much more of a "scientific" result than setting an upper limit. At the LSC meeting in March 2003, it was decided that the LIGO group responsible for analyzing burst sources with unknown

5. I can't resist another metaphor, this one drawn from American football: LIGO has gained good field position without scoring a touchdown; it must not fail to score the field goal.

6. Their reply criticizing Finn, placed on an electronic preprint exchange, was submitted to *CQG*, though I did not manage to find it up to June 2003. (See Astone et al. 2002; Finn 2003; Astone et al. 2003.)

7. The leadership said that being published in an archival journal necessitated an empirical confrontation.

waveforms should try to confront the Frascati results as speedily as possible. I await a change of heart among the LIGO group, and should they produce a refutation, I await with interest the reception of the results by the scientific community. If the reception benefits LIGO, I want it on record that I said it first.

LIGO and the First Positive Results

Now let us look forward to a time when the first credible intimations of *positive* results make themselves felt. Some of the most uncompromising statements about how to act in these circumstances were expressed by LIGO scientists when the "evidential collectivists" among the bar community were claiming to have seen puzzling signals that might be consistent with gravitational waves. Disdain was expressed for those who made a claim in such a way that if it turned out to be right, they could gain credit, but if it turned out to be wrong, they could say they never meant it as anything but provisional. The gravitational wave standard should shine brightly or not at all, it was said; there were to be no drab "ifs" and "buts," only the glitter of certainty.

The effect of LIGO scientists' criticism of the bar community was marked. When the International Gravitational Event Collaboration (IGEC) was formed to coordinate the results gathered by the bars, it marked a further step in the attempt to "globalize" gravitational wave physics by imposing the dark image of the sky and the conventional view of detector sensitivity on the group. IGEC adopted a standard for signals that would allow their detectors to see no more than one event every ten years or so— and this meant looking only for signals that would stand out well above the noise, and would produce an accidental result on more than two detectors hardly ever. This effectively killed the bar scientists' last hope of a collective positive claim. At the same time, it amounted to the imposition of evidential individualism.

It will be interesting to see if LIGO maintains this standard regarding its own data. Even as the year 2000 came toward its end, the standard was beginning to rock. At least some LIGO scientists were saying IGEC's thresholds were too demanding.[8] LIGO's birth as a real detector is at least initially giving rise to the kind of optimism that once characterized the

---

8. A LIGO scientist told me that he had told Bill Hamilton he was "crazy" for maintaining such a high standard for potential data points (see chapter 26).

bar community. When I remarked to LIGO scientists that "LIGO I has virtually no chance of seeing anything," I might be told, "The trouble with you, Harry, is you believe what the astrophysicists tell you." Or I would be told that there is always something unexpected to be found whenever we look at the sky with a new instrument that is orders of magnitude more sensitive than those that went before. Or I was told that only now were we reaching the level of sensitivity that made observations astrophysically reasonable. LIGO scientists, then, including some of those who had been most scornful of the evidential collectivism of the bar groups, were now allowing themselves the kind of license to speculate about a brighter sky that they had earlier foresworn.

Will something like evidential collectivism reemerge among the interferometer groups? Will they play a little with the concept of what it means to broadcast a scientific result? I hope so; it always seemed to me that the standards imposed on the bar groups were too harsh and remain unrealistic for such a brand-new science. Let me set out my prejudice for what it is worth (not much): while I agree entirely with the strategy of the "new management" in going for the early and efficient completion of a full-scale instrument capable of making observations, I am not entirely convinced by the scientific rationale for that view. I think the approach was entirely justified by the political, financial, and human context of the overall project, which made the 40-meter team's recalcitrance ill advised. Nevertheless, I do not think that large-scale interferometry is likely to be as free of "magic" as, say, high-energy physics. I think the first results are going to be tentative and exploratory and that to wait for a level of confidence that is as secure as that routinely obtained in high-energy physics, before making positive indications public, would be a mistake. This may require some backtracking in terms of "the gold standard" for public announcements, but that would be less damaging, both scientifically and politically, than what would probably turn out to be a futile attempt to keep every speculation hidden away from the public gaze.[9]

Image, Logic, and Duty Cycle

There are two possibilities for a first signal—a buildup of a population of signals to the point of statistical significance or a single, loud event

---

9. Of course, judicious "leaking" of information could "square the circle." Scientists could find ways of constructively using the status hierarchy of different forums for announcements and quasi announcements, indicating the presence of something while not publishing (see chapter 42, the section "Presenting and Legitimating Knowledge").

that stands out on all the detectors that are on air. Peter Galison divides high-energy physics into two kinds, which map onto these two possibilities. There is evidence that comes in the form of a unique, perfect image such as is generated by a device like a bubble chamber or the track of a particle through an emulsion; and there is proof in the form of statistical general-izations drawn from masses of events such as are recorded by scintillation counters, spark chamber, and the like.[10] Gravitational wave detection can also be thought of in these two ways, but because the events to be detected are not under the control of the experimenter, the consequences are more marked.

If one is looking for a buildup of a population of signals, one can be relatively relaxed about duty cycles and delays in reaching full oper-ational capacity, because such problems merely slow down the process of discovery; a reduction of 20% in the duty cycle means only that it will take 20% longer to gather a statistically significant number of data points. If, on the other hand, the first result is likely to be a unique event, then any period of downtime is a terrible risk, potentially giving rise to the "horror scenario" described on page 737: the unique occurrence may happen during a period of downtime. This would be a great scientific disaster after all that money had been spent; in gravitational wave physics if a supernova is missed today, it is gone forever.

This returns us to evidential cultures. High-energy physics, at least of the statistical variant, is a race to gather enough events (scintillations, sparks, etc.) to reach a level of statistical significance that justifies a discovery claim. Given agreement about what this level should be, evidential individualism can be maintained. But if the first sighting of a gravitational wave is a single event, then, unless the event is huge or correlated with other kinds of radiation, the standard will be hard to maintain. A single event will be there and then it will be gone, leaving lots of scope for argument. Will the event be suppressed if it does not meet the standards of evidential individualism?

THE CHALLENGE

In March of 2003 I showed Bob Spero, a key member of the original 40-meter team, a diagram of LIGO's steady progress toward its sensitivity target. Fig-ure 41.1 shows the sensitivity of the Livingston interferometer (at the time the best one when it was working) as it lay between January and June

---

10. Galison 1997.

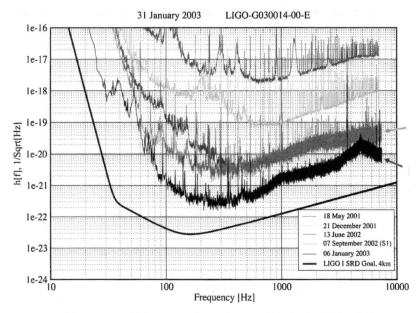

Fig. 41.1 LIGO's increasing sensitivity. Source unknown. Courtesy of Gary Sanders, LIGO project, California Institute of Technology.

2003.[11] The heavy line in the figure is the sensitivity target, and the sequence of spiky lines is the measured performance. Frequency is indicated on the horizontal scale on the diagram, while the vertical scale is a logarithmic plot of sensitivity. The graph shows LIGO I within a factor of 10 or so of its planned sensitivity over most of the frequency range, though there is a four-orders-of-magnitude problem at the low end of the scale. In response to my badgering, Spero wrote a prediction about the future of the devices and signed it. He said that, assuming LIGO's design sensitivity would enable it to see neutron star coalescences to a distance of 20 megaparsecs (which it does), by end of 2005 it would still be able to see no further than seven megaparsecs.[12]

In June of 2003, the LIGO group made a contrasting claim at a review committee meeting. They claimed that at least one interferometer would have reached its sensitivity goal by the end of 2004. They based this estimate

11. Not much in the way of improvements had been made between January and June because data gathering and other work was being done.

12. A megaparsec is 3.26 million light-years. The "signing ceremony" took place at Mijare's Mexican Restaurant on Palmetto Drive in Pasadena on March 13, 2003. This way of indicating sensitivity is a good one, because it allows the measured line to slip below the solid line in some places while still not achieving the intended reach for inspiraling systems.

on an explanation of each of the remaining noise sources and how they were to be resolved. In short, in June 2003, the LIGO team members in question were sure that the kingdom of magic had been banished as far as noise in LIGO I was concerned.

There is clear water between the two predictions—a year and a factor of 3—and we shall soon see who was right.

CHAPTER 42

METHODOLOGY AS THE MEETING OF TWO CULTURES:

THE STUDY, SCIENTISTS, AND THE PUBLIC

DANCING WITH SCIENTISTS

The philosopher and critic of sociology Peter Winch wrote in 1958 of the idea of the germ theory of disease. He drew a contrast between the discovery of a new germ within the germ theory and the first development of the theory itself. Winch remarked that the establishment of the germ theory, unlike the discovery of a new germ, meant that doctors and surgeons had to learn new ways of doing things. For example, understanding the germ theory meant that the action of cleaning was transformed. Or, to look at it another way, that surgeons began painstakingly cleaning their outer garments before operating is what gave germs their meaning. Winch was pointing out that that what we think and what we do are, in a deep sense, the same thing. Taken together, ways of thinking and doing are what Wittgenstein called a "form of life."

Winch used this Wittgensteinian identification of concepts and actions to argue that the deep problems of sociology were really "misbegotten epistemology"–that is, that understanding actions amounts to the understanding of the interrelationship of concepts.[1] Subsequent generations of sociologists have realized that this cannot be a critique of sociology, because "Winch can be turned on his head"; the argument can just as easily be read as implying that epistemology is really misbegotten sociology. It is the debate that is misbegotten—the point is, concepts and actions are part of each other.

Epistemology and sociology are indistinguishable! If there is one key idea in the sociology of scientific knowledge, this is it. We must learn to grasp the sense of what scientists do if we want to think as they think, and understand what they think if we want to grasp the sense of what they do. In this chapter, we will reflect on what this means for the relationship between the analyst and the scientist. But first, what does it mean for scientists themselves, the quintessential establishers of new meanings in the world?

One thing that follows is that one can never fully capture the world in a formula or description, because the meanings of formulae and descriptions cannot be separated from what is done with them. And people learn what is to be done with them by being immersed in social groups as well as being taught explicitly.[2] Hence the conduct of scientific life, including the carrying out of experiments, is redolent with the unspoken and unspeakable—tacit knowledge. This is the "idea" behind much of this book. The irreducibility of actions to explicit rules is what makes the second set of ripples—the ripples in social space-time—so important, because it is this set that established meaning in the world. It is why science progresses "funeral by funeral." It is what gives rise to the experimenter's regress and the longevity of scientific disputes; to establish the identity of experiments is a historical not a logical process. It is what makes it so interesting to study the demise of the idea of the new cross section and center-crystal instrumentation and worth noting that such notions live on more easily in the actions of theorists than experimenters. It is why the technical arguments for the demise of a technology such as the spheres are never as decisive as the political fact of their demise. It is what gives rise to questions of

---

1. Epistemology is the theory of the foundations of knowledge. It should be clear that Winch's discussion of the germ theory (p. 121) anticipates Kuhn's (1962) distinction between normal and revolutionary science. The works referred to are Winch 1958 and Wittgenstein 1953.

2. An immediate consequence is that we should not expect "intelligent" machines to be able to mimic human abilities faithfully until they are capable of living out our social lives; the continuing failures of intelligent machines are easily understood once we see this (Collins 1990; Collins and Kusch 1998).

magic versus mechanism discussed in chapter 35, including that new experimental techniques (such as the measurement of the "Q" of sapphire) are transmitted by personal interaction, not the exchange of documents, and that computer modeling of physical domains is fraught with ambivalences. It is what makes it hard to make two identical machines, and necessary that people as well as papers travel from a working interferometer to a nascent device. And it is why we need a "test" to find out whether the 40-meter team of the LIGO project or the new management are right about the extent to which noise in interferometry is fully understood.

Does the last sentence not defeat the sense of this entire passage? If noise can be "fully understood," then cannot ways of acting be reduced to marks on paper? No—that is to mistake the generality of the argument. What is happening when understandings move from the obviously tacit to the not-so-obviously tacit is that the foundations of knowledge are becoming ubiquitous, so that less has to be said in order for the appropriate action to follow. It is rather as though we can say that the rules of English are captured in a book of grammar plus the dictionary; this is true, but only for English speakers.[3] Most of what is unspoken most of the time is so deeply embedded in ordinary life as to be invisible. The argument between Barry Barish and Bob Spero is only about the stage that has been reached in the transformation of the esoteric but unspoken to the ubiquitous but unspoken, not about the extent to which any such apparent transformation rests on a foundation of the unspeakable—we already know it does.[4]

Now let us turn back to the analyst and the scientist. Part of the job of the social analyst is to render the invisible component of social-epistemological life visible. One way to do this is to study those moments when what seems strange or new is being turned into what seems ordinary. Turning the strange into the taken-for-granted is what science does in its laboratories, its conferences, its publications, and its publicity. A sociological understanding of science, then, means mastering science after the fashion of a scientist, but it also means being able to step back far enough from that world to enable new "forms of life" to be topics as well as accomplishments. The sociologist has to engender a degree of "estrangement" from a form of life so that it can be described even as it is achieved.

3. For an extended study of the transformation of the unspoken to the explicit, see Collins and Kusch 1998.

4. Just to drive home the point, learning the unspeakable is the major purpose of the apprenticeship element of a scientific training and a good part of the function of the endless conferences.

Estrangement is not easy. A form of life, that is, "a way of being in the world," implies a "taken-for-granted-reality"; as has just been argued, when a form of life is mastered, things just seem to follow so naturally that it seems impossible that they could be any other way.[5] One trick used throughout this book for escaping the tyranny of the taken-for-granted is to seek out controversy. The very goal of argument is to make problematic what an opponent thinks is natural or obvious; the sociologist can "piggyback" on scientists' arguments using maverick scientists as allies in the process of estrangement. One can immediately see a problem for the researcher: on the one hand, the sociologist wants to interact with scientists intensively enough to grasp their world in a thoroughgoing way; on the other, the sociologist is going to treat the troublemaking arguments of respondents' scientific enemies with seriousness. Estrangement, then, is likely to result in a bumpy ride.

Let us return to the germ theory and imagine the scene in the anteroom of an operating theater. You are a sociologist investigating surgery. The surgeons are preparing to go to work. They wash and scrub their hands. They use a nail brush and disinfectant. The cleaning is repetitive and ritualistic—almost ostentatious. The nurses hold out latex gloves into which the surgeons insert their hands in a strangely awkward manner. The nurses drape gowns over the surgeons' clothes, place masks over their mouths and bonnets over their hair. All the movements are stylized and contorted. You get the idea: this is what *germs* mean for surgeons. Germs are very small and very pervasive. To remove them requires the application of lots of special fluids and lots of scrubbing. The surfaces from which the germs have been removed must not touch the surfaces that have been scrubbed clean. Even the surfaces that have been cleaned by scrubbing must not touch the surfaces that have been, as the surgeons say, *sterilized* by heating. Germs are invisible, but their invisibility does not prevent the whole surgical team from accepting their mysterious influence. What you are watching when you watch these balletic movements is "the social construction of germs." Each member of the team, as they dance the dance of the germs, is making the germs real for each other member. The choreography is like mime, conspiring to make the invisible visible.

You do not take sides. You do not remonstrate. You do not say, "Don't be silly, there's nothing there," nor do you say, "The emperor has no clothes."

5. *Taken-for-granted-reality* is the term is used by philosophical phenomenologists such as Alfred Schutz (1964), whereas *form of life* is Wittgensteinian.

You remain neutral, but you do not immerse yourself too thoroughly. You stay in your street clothes, because to fall entirely into the ritual would take your attention from it *as* a ritual. To see things for what they are, you must nurture your estrangement. As the surgical team move toward the door of the operating theater you make to accompany them—to observe them as they work. They block the way. You have not learned the dance. These people are entranced! Only those who can dance are allowed into the holy of holies.[6]

What I have done in this book is to describe the dance, not of germs but of gravitational waves. Now, rather than continuing to describe the dance, I am going to reflect upon my observations of it. I want to ask questions about my dancing with gravitational wave scientists. How did I compare with the fictional sociologist in the surgical anteroom? There are four questions:

1. To what extent *could* I take part in the "dance" of gravitational wave detection if I wanted? To what extent did I master the taken-for-granted world of the scientists? Or, to "domesticate" the question: how much gravitational wave science expertise do I have?
2. To what extent *did* I take part in the dance during the study?
3. To what extent is a refusal to dance intrinsically damaging to the scientists' project in the way that the failure to "scrub up" would be fatal to the surgeons' project?
4. What did the scientists make of my performance?

Of these, question 3 is the most intellectually interesting, but the set of questions taken as a whole bears on much more than the methodology of the study. Ordinarily, the discussion of methodology provides a warrant for a study's findings; it explains why the findings ought to be believed. Here, however, to answer questions 1 to 4 I have to discuss the extent of my own scientific expertise as I see it and as the scientists see it. Who is an expert? How is expertise mastered? How are levels of expertise ranked? The answers are as much a discussion of the nature of knowledge as of method. The answers also bear on the increasingly important question of the relationship between expert and nonexpert. Let me try to disentangle some of this.

---

6. The example of the relationship between the idea of germs and the actions of surgeons is taken from Winch 1958, p. 121, but the description and the metaphors are my own.

PRESENTING AND LEGITIMATING KNOWLEDGE

Once I had a friendly but heated argument with Barry Barish concerning the claims of the 40-meter team. I tried to act as devil's advocate for the leader of the team, Bob Spero, arguing that there could still be unknown sources of noise that would be uncovered as LIGO tried to reach its sensitivity target.[7] Barish told me I had not understood the science. He told me I should read the papers that the 40-meter team generated, whereupon I would find there was nothing in them.

> [2000] If this is a science and Spero is really making scientific statements, show me the documents so I can refute them, instead of anecdotal discussions and comments that come down the hall and this or that. If somebody wants to tell me that this is not subject to analysis, I want to see the argument. And if you can tell me that all he does is make a claim, then bullshit—I don't care. I want to see why. I challenge him through you right now, because I don't see the documents. I've read everything—OK—that he's ever written. What is it [the unknown noise source]? Because maybe it isn't subject [to analysis], maybe this has some tricks in it, something I don't see, but to tell me that he just knows in his heart somewhere that this is not subject to analysis like the rest of science—show me something!

Barish found graphs on his computer comparing actual noise with the model of noise generated by theory; they matched quite closely, indicating that all sources of noise were understood and that the 40-meter team must therefore be wrong.[8] I said there was still room to fit something new between the model and the actuality and suggested that even though the 40-meter team had written nothing down, they might still know things tacitly; their understanding of interferometers might be "embodied," not explicit. A lot was going on in this conversation, and we will return to it again and again. But let me start with Barish's stress on written information.

All scientists, and that includes social scientists, consider that the status of their knowledge claims is in part a function of how and where the

---

7. I recorded the whole argument. It was originally to be part of this book but had to be cut on grounds of length. It can now be found on WEBQUOTE under "Barish and Collins Debate Noise." Bob Spero, the representative of the 40-meter team whose role I tried to take, considers that I somewhat overplayed his hand.

8. This was long before anyone had suggested, as they did in June 2003, that they knew how to eliminate the remaining noise.

claims are made. As a *first-order* approximation, the status of knowledge claims increases as we work our way up from laboratory talk, through conference presentations and preprints, to a publication in a high–prestige, peer-reviewed journal. This status hierarchy was much discussed in the traditional sociology of science that was dominant before the 1970s. Peer review in the published journals was one way to instantiate the Mertonian norm of "organized skepticism," which was taken to ensure the validity of scientific findings.

The sociology and history of scientific knowledge looked more closely at the practices of peer review and publication in general. It engaged in a *second-order* analysis. Rather than simply accepting the differences between types of sources, it looked at the way these practices were established and how publications and other sources were molded by context. Historians described the development of the "literary technology" that first made readers of papers into "virtual witnesses" and showed that it rested on social understandings: the experimenters themselves had to occupy the right social location—that of gentlemen, a status that is now to some extent occupied by the institutional affiliation of the author. Sociologists studied the processes through which laboratory findings were converted to published works, sometimes treating the publication as establishing the knowledge rather than the knowledge as giving rise to the publication.[9] Here we have seen that it is more complex even than this: publications, even in the same journals, do not carry identical values. Thus Weber and Radak's 1996 paper on the correlation between gravitational waves and gamma-ray bursts has been hardly read at all. Papers published in 1976 and in 1982, both suggesting the existence of gravitational waves, and both in the prestigious *Physical Review,* made no impact. We know, then, that the author and the context of a paper deeply affect the way it is read and received.

Again, we have seen that the meaning of a publication is different for evidential individualists and evidential collectivists. Evidential collectivists' papers often include a disavowal of their status as knowledge claims, saying only, "We've found these data, they seem impossible, make what you will of them." Evidential individualists believe, "It should not be possible (as has been the case in this field before) to equivocate and claim validity for an observation at one moment and then deny it later when challenged."[10]

9. For the Mertonian tradition, see, e.g., Merton 1957. Shapin and Schaffer (1987) and Shapin (1994) discuss the development of literary technology. For the displacement of social class by institutional affiliation, see Collins 2001c. Knorr-Cetina (1981) and Latour and Woolgar (1979) treat the publication as the locus of scientific knowledge.

10. The quoted sentence is from a personal email sent by a scientist to me in 1997.

We can see, then, that though the second-order analysis of publication is an improvement upon the over-abstracted treatment of the traditional sociology of science, the role of publication as the means of constructing scientific knowledge has been exaggerated still. To understand the meaning of a publication, one must know how it is received in the community. One cannot read the meaning of a paper from its contents or its placement because, in a sense, physics is an oral culture.[11]

In the contemporary world, physicists rarely read published papers, and when they do they rarely read them to obtain knowledge.[12] They become apprised of colleagues' work either through electronic or other preprints or, in the case of the core set or core group, through talk, the email network, and conference corridors and cafes. When it comes to the transmission of knowledge, and that includes knowledge about how a published claim is to be interpreted, it is this talk that is crucial.[13] Thus it is that Kip Thorne's verbal refutation of Weber's new cross section (chapter 21) could be more in- fluential than Leonid Grischuk's published refutation. It is because physics is an oral culture that an extraterrestrial, using the published literature alone, would not come to know terrestrial science.

This leads to the possibility of a *third order* of analysis, the use of place- ment of knowledge claims as a symbolic resource. We might say that a published paper serves the same function in physics as does a miracle in the Catholic Church. One does not have to have witnessed three miracles to make a saint, one has only to know that they have taken place. Likewise, once the oral culture has established the meaning of a knowledge claim, one does not have to read the paper to accept it, only know that it has been published. The *third order*, then, concerns the way the first order is

11. Thanks to Dan Kennefick this succinct phrase. I believe Dan is writing a paper on the wider significance of physics being an oral culture.

12. Given this, it might seem that Weber and Radak's 1996 paper not being read meant little, because it would have been known about anyway. But this is not the case (though a few more people may have known about it than had read it).

13. Here is a fact that has become such a taken-for-granted part of my life with these physicists that I almost forgot to write it down anywhere. The physicists spend a lot of time at conferences and meetings giving presentations to one another. But they generally do not write these presentations out in words; they use a few "viewgraphs," or, nowadays, the almost ubiquitous PowerPoint. The PowerPoint presentations are then placed on a Web site, and as far as the physicists are concerned, their talks have been recorded and are accessible to all. But what they actually said about the PowerPoint slides at the meetings remains inaccessible, along with their meaning! We came across a fine example of the problem when I discussed Kip Thorne's 1994 presentation to the National Science Board (chapter 27). We have his slides, which include optimistic guesses about what LIGO I would be able to see as well as better-grounded estimates for binary-neutron star sources. With this as our documentary evidence, we struggle to know for certain how the relationship of the guesses to the grounded estimates was presented and perceived by the audience.

reflexively used by scientists.[14] A very recent example is the response to the Frascati group's November 2002 publication of the marginally positive result in the refereed journal *Classical and Quantum Gravity*. In chapter 41 it was explained that it was intended that LIGO would confront this paper empirically. At the time of the publication, the LIGO leadership said they did not want to be diverted into confronting the paper but felt impelled to do so because, having been published in a refereed journal, the paper counted as "archival." The archive has to be pristine, so an archival paper that is empirical has to be empirically confronted. It is certainly widely agreed that if the paper had been circulated only on the electronic preprint server, it would not have merited more than oral confrontation even though its contents would have been just as widely known.[15]

Turning back to earlier passages of this book, I have argued that Joe Weber's understanding of science exhibited a kind of "scientism" in the way he would treat the *publication* of a scientific result as the *accomplishment* of a scientific result. In chapter 20, a respondent is quoted making just this complaint: "unfortunately, the thing I'd missed is that he subsequently used that, because when he gave a talk and people said they didn't believe it, he then said to them, 'well, it's been published in a refereed journal.' So he was using that to advance his case."

A nice third-order distinction can be found in Kip Thorne's and others' predictions about the potential success of LIGO. For example, in 1987 Thorne allowed to be published the details of a bet he made with Jerry Ostriker that gravitational waves would be detected before the year 2000.[16] And yet Thorne can quite reasonably state that he has never made such claims in any official setting where the future of funding turned on the likelihood of their being true. In informal settings, many LIGO scientists will express firm optimistic *beliefs* about LIGO's potential: "It will see many unexpected sources"; "It will see inspiraling black holes which have lots of energy"; and so forth, though they would not commit themselves to such predictions in any place where they would have to be called to scientific account. There is nothing sneaky or dishonest about this: it is quite

14. Collins and Pinch (1979) pointed out that a publication in a major journal could be rendered as "not quite a publication" by the addition of an editorial disclaimer. A nice example is *Nature* of July 1988, when the editor both published certain homeopathic claims and simultaneously disowned them in an editorial (E. Davenas et al., *Nature*, 333, 816, 1988; *Nature*, 334, 287, 1988). The editor was, in effect, both publishing and not publishing at the same time; he published while simultaneously removing *Nature*'s imprimatur.

15. The discussions took place at meetings in Kyoto in December 2002 which I attended. The published paper is Astone et al. 2002.

16. Thorne 1987.

reasonable to believe one thing even when one only calculates something less. But it is a third-order phenomenon that provides the license for making an optimistic estimate while, as it were, not making it at the same time; to outsiders this can be seen as equivocation, just as evidential individualists see evidential collectivists' publication practices as equivocal.

It seems likely, as I have argued in chapter 27, that the astronomers' distrust of LIGO spokespersons turned on their view that the fine distinction between a measured formal claim and an optimistic hunch might not be appreciated by nonscientists. The situation was not helped because, being a big project, nonscientists were crucial in the funding decisions; the understanding of the symbolic meaning of different placement of knowledge claims that is common to the inner core of the physics community might not be shared by those outside it. If this is the case, then ambiguity between what is formal and what is informal seems to allow equivocation.

We might also say that Thorne also made use of the knowledge-constituting hierarchy of different forums in the way he chose to handle Weber's new cross-section claims. By only *talking* about the cross section, rather than *publishing,* Thorne resolved a moral dilemma: "I did not want to make a big deal of this in public. I felt that Joe had pioneered this field. He had pioneered it in a wise direction. . . . And I just did not want to be attacking him in public." In this way Thorne could defeat Weber's claims without administering the symbolic hammer-blow of refutation in a refereed journal publication that his colleagues wanted to see.[17]

Writing and the Sociologist

For all the reasons discussed above, the sociologist has to reflect carefully when scientists say, "Read the papers if you want to understand the science!" There is a lot more known in science than is represented in the papers. The meaning of the papers is impenetrable in the absence of the kind of nuanced understanding that scientists obtain from their own embeddedness in the oral culture. It may be that sciences such as high-energy physics are so well organized that nothing that counts as a knowledge claim has no counterpart in a publication, but if this is so it is exceptional.[18] Finally, "read the papers" can be a reflexive, third-order demand: "Science

17. Thorne's analysis was eventually published in an edited volume, but this still does not have the same symbolic impact as a paper published in a refereed journal.

18. For example, Rai Weiss has published little of the vast amount he knows about the practice of large-scale interferometry.

makes knowledge by publishing papers, and if you want to understand what counts as knowledge, that is where you will find it."

Much of this has long been understood at the concrete level. In 1963, Peter Medawar, himself a scientist, wrote a piece called "Is the Scientific Paper a Fraud?" in which he explained that written papers falsify the chronology and process of discovery. The scientific paper, if it is to be useful in its symbolic role as a knowledge claim, must work, as we might say, as a "virtual replication device." Outsiders must be made to feel that anyone anywhere with access to the resources could repeat the work, and that in reading the paper they have replicated it in a virtual sense. What follows is that all the craft skills associated with any particular scientist, group of scientists, or specific location has to be hidden. Certainly nothing of which there could be a sociology must be allowed to be seen to have been taking place. Hence the passive voice: in the scientific paper things are not made to happen, they just happen.

So should the sociologist read the papers? Of course. But should the sociologist read them as a description of either the development of the science, the current state of the science, or the nature of the science? No! We can sum up this whole passage by saying that it is important not to give too much weight to the apparent *reliability* of written documentation at the expense of the *validity* of the findings. That is, the apparent openness of the contents of documents to verification by anyone should not be overvalued, since their meaning can be understood only through the medium of talk.[19]

THE BOUNDARIES OF SCIENCE

If only things were so simple! Unfortunately, to study science, one has not only to internalize it, estrange oneself from it, and reflect upon it, one has also to make judgments within it. For example, to be a sociologist of gravitational wave science, one has to decide what is gravitational wave science and what is not. Such judgments are equivalent to those one would make if one were to treat the widely read and accepted scientific paper in the way scientists treat it: as representing the truth. This takes us, then, to question 2: to what extent is the scientific dance going on in this book?

19. If one actually examines the way documents are used by historians in seriously contested situations (such as Evans 2002), one discovers that they establish their meaning by reference to, for example, less-reliable but more-valid sources. (In the Evans case, official documents were interpreted by reference to diaries recording what other people said. The diaries are there for anyone to look at, but their content is a lot less verifiable than my tape recordings!)

Clearly, there is a lot of science in it. For example, I have tried to explain the science of gravitational wave detection. In writing these sections I have benefited from the advice and guidance of respondents, but there is nothing in the book that I do not understand in one way or another, so there is a lot of dancing going on. In general, the science which I accept and endorse is that which is thoroughly bedded down in the way of life that constitutes gravitational wave detection, while the science that I estrange myself from is that which is still being formed.

Where to treat scientific findings as a matter of fact and where to treat them as a matter of analysis is my choice—a matter of my judgment. Even where I refer to something which seems to be unquestioned fact, such as "Inspiraling binary neutron stars are the most well-understood source of gravitational radiation," I *could* have chosen to treat the claim as a subject of analysis. I could have developed a series of historical theses about how consensus was reached that those glittering things in the night sky are distant suns, that some are neutron stars, that there are binary neutron star systems that spiral into each other near the end of their lives, that these emit gravitational waves, that we know the form of the waves they emit, and that these waves can be detected. If I did this assiduously, every "scientific" claim which is treated here as a matter of fact without any estrangement could have been turned into a "sociological" claim. This was what I meant when I said in the introduction that the causal arrow could be seen as traveling in either direction, from the stars to the *Physical Review* or from the *Physical Review* to the stars. In this book I treat the causal arrow as emanating from the stars most of the time and emanating from social space-time some of the time. Let me give an example of the choice at work.

The Model Interferometer and Realism

Let us consider an element of gravitational wave science that is well embedded both in the world of my respondents and in my world—the basic idea of interferometry. Interferometry seems to offer little in the way of choice: it is so much taken for granted that I would have difficulty relativizing it. Something that makes it still harder to treat as a topic for sociological analysis is that I have built a crude interferometer myself. One evening in early 2001, I sat in my Cardiff office in a nearly deserted building, fiddling with a laser pointer, some bits of wood and wire, a clamp stand, some old mirrors, a beam splitter from a scrap comparison microscope borrowed from Bath University, and a lot of adhesive putty, when suddenly, blobs of black appeared within the reddish glow of light on my makeshift screen.[20] Almost immediately I guessed I was seeing interference, and I was sure of this once I had reproduced the effect a few more times and discovered how sensitive it was to the smallest disturbance of the setup. It was sensitive, furthermore, in the right kind of way—a lateral movement of the dark shapes corresponding to whichever mirror I nudged with my wisp of wool; touch the back of one mirror and the movement was in one direction; touch the back of the other and the movement of the dark shape went the other way. I had not really believed I could build an interferometer—an instrument always talked of in terms of its delicacy and recalcitrance—so when I saw the fringes I rushed into the corridor whooping with joy (now I know what that cliché means), looking for someone to tell.

At that moment there was a force upon me—the force of *realism*. I knew I had made an interferometer, crude though it was, not because there was a network of scientists agreeing that I had done it but because I could see the dark fringe of interference moving around under my control! So much for the second set of ripples; this was just me and the world—me and the first set of ripples. Here was a perfect example of the lone scientist in communion with nature!

For the social scientist, the trick is not to pretend there is no force of realism. The force is everywhere in everyday life as well as in moments such as those I have described. The trick is not to deny it but to deal with it. So—suppose I wanted to analyze that moment when the force was upon me, not as a realist but as a methodological relativist—or even a philosophical relativist. How could I have done it?[21]

By assiduous use of the archives I could, perhaps, have transported myself back to the historical moment when interferometers were first being developed, and there I may have been able to rediscover doubts and arguments about the meaning of the fringes. Failing that, I could stay in the present and merely note that at best, the dark blob of interference was a case of what was referred to in the introduction as calm-weather, cup-and-saucer seeing. The blob was vulnerable to all the problems of ordinary seeing—mass hallucination, "I must be dreaming," "It's a fake," and so forth.

20. Many thanks to Bob Draper of Bath University for helping me out with a loan of equipment on this occasion as on so many others.

21. Crude critics of sociology of scientific knowledge, such as those who suggest we might jump out of 30-story windows to demonstrate our arguments, seem to have no conception of this. They seem to assume that the sociologists are such fools that they have not noticed the way the force of realism impacts on our everyday lives (as much in drinking a cup of coffee as in not jumping out of windows or making interferometers). For the distinction between philosophical and methodological relativism, see Collins 2001b and chapter 43.

Less subtly, the whole chain of inference from the blob sighting to "interference" between rays of light is a long one and depends on my accepting a whole raft of theories and other taken-for-granted assumptions about how light works and travels through this strange device. Furthermore, since I doubt that I could have made the device without resorting to the peculiar light source I used—a laser—the effect depends on my accepting the whole business of lasers on trust. Thus, if some well-qualified scientist were to look at my device and tell me that what I was seeing was not an interference fringe pattern at all but some mundane effect, I would not have much in the way of a counterargument.[22] And so with a bit of effort I could "deconstruct" my interference fringes and banish "the force."

For the argument of the book, however, there is no need to avoid the force of realism as it exhibits itself in the case of the interferometer. We can simply use the idea of the interferometer and interference fringes as the supporting framework for other kinds of arguments. On the whole, the exact choice about where to relativize and where not to relativize is not very carefully worked out; most of the time it does not need to be. Most of the time, some things are treated as scientific facts and some as facts-in-the-making depending on the dynamics of the story. Most of the time, the principle of methodological relativism, when it is applied to facts-in-the-making, needs be seen as no more than a version of a methodological guideline found in every science: concentrate on the explanatory variable. In this case, it implies that the science be "held constant," as it were. For facts-in-the-making, the science must not be taken to explain itself on pain of circularity and/or the dimming of the sociological gaze.

Letters in Green Ink

Setting the dividing line between science and nonscience has two consequences for what lies on the taken-for-granted side of science; some bits of science, such as the interferometer, are treated as unproblematic in a positive way, while some bits of science are treated as unproblematic in a negative way. Let me give an example of the latter.

Well-known scientists often receive letters from people who believe they have discovered some new "secret of the universe." Occasionally, I receive

---

22. A couple of paragraphs back, I said there was no network of scientists agreeing with me that fringes had been seen, but there was a network. It was the virtual network of scientists from the past.

them, too. These letters are widely referred to as having been "written in green ink," because there is often something quirky about their form as well as their content. I have received such missives handwritten on lined paper with no indication of where they come from, typed on both sides of the paper from edge to edge, and so forth. The use of green ink is another such characteristic (the physical form of a letter can confer legitimacy or otherwise as much as any other social convention). One of the things I do when I get these letters is turn off my relativism, symmetry, estrangement, and all the rest, just as I did with the interferometry. But in this case I dismiss the contents without thinking much rather than endorse it without thinking much. In deciding that these letters lie beyond my remit, I am making a judgment *within* science.

A slightly more difficult case was two technical papers, published in physics journals, that described experiments to measure a certain gravitational effect. These were sent to me along with a cover letter informing me that they bore upon the topic of gravitational radiation. I read the papers and concluded that the experiments described, though easy to confuse with gravitational wave work, were not trying to measure the right kind of delicate forces. Without conferring with respondents, and trusting entirely to my own scientific judgment, I wrote as much to the sender of the papers and ignored them thenceforward.

In these cases, I make the same kind of decision as the gravitational wave physicists when they decide to ignore, say, Weber's claims to have found correlations between his pulses and gamma ray bursts. The difference is that they exclude more than I exclude: my "scientificness zone" is wider than the physicists' scientificness zone. The physicists' job is to draw a well-defined zone and make it as narrow as possible. My job is to study the way they define their space, so their space has to lie within my (slightly, see chapter 43) larger space. Thus, I must take seriously pieces of protoscience that the scientists reject out of hand so that I can see how they go about making these items unreasonable.

What follows from my approach is a new understanding of the moments when new science is being made. Those periods look far more deeply embedded in the social than they once did. In the long term, it may or may not be that some kind of transcendental scientific logic overrides all the mundane forces that I describe, but in the short term methodological relativism reveals such mundane forces as there are. Furthermore, methodological relativism shows that the short term is somewhat longer than conventional histories of science would suggest.

METHODOLOGY AS LIVED EXPERIENCE: THE GROUCHO PARADOX

What are the consequences for my interactions with scientists of my wide zone? This is question 3: In what ways does the story of the surgeons and the social scientist with which I introduced this section of the book represent my relations with my respondents? Is there something in the very approach of the methodological relativist that is antagonistic to the project of gravitational wave detection such as would make the metaphorical surgeons want to bar me from the door of their metaphorical operating theater?

Compare the problem of science with more typical anthropological field sites. Imagine I was studying the conflict between Catholics and Protestants in Northern Ireland. There could be no conflict between them if there were no way of identifying them as different groups, and that identification turns, ultimately, on a difference of religious practice, including what happens at services of worship. In the Catholic Mass, it is said that the bread and the wine *become* the body and blood of Christ, whereas in the Protestant church these elements merely *symbolize* the body and the blood. In analyzing this conflict, the last thing the sociologist is going to take into account is whether transubstantiation really occurs; in terms of the actual transformations of the bread and wine, the sociologist will be a methodological relativist. His or her analysis will be about how the beliefs are maintained and reinforced, and this will include the socioeconomic forces that tend to drive groups further apart once the first dividing line is established.

The sociologists' kind of analysis is fine so long as it is broadcast well away from the scene of the conflict. But if the analyst lived in the society, at least in certain times and places, he or she would be forced to identify with one group or another on pain of death or injury; neutral analysis is not an acceptable social identity at the point where a battle between competing beliefs is being fought out (as, in a less tragic way, it was not an option for John Hasted and John Taylor—see note 37).

Now transfer the predicament to the study of a modern scientific controversy. Science is what we learn at school: if you "understand" science, you pass your science exams; if you understand science really well, you can *become* a "scientist." To understand science really well means to "be on the side" of those who do not make scientific mistakes—who know where the border is located. This is understanding as identification with the science. Analytic understanding of science—a refusal to accept, at least for the time being, that what has become defined as a mistake is a mistake—is hard to maintain in our society, because *it is our society*. All of us, including social analysts, live here, where faith in science is one of the things that makes the society work. All of us live science; it is a cornerstone of our values.[23]

But let me insist on my good intentions in respect of science until I am blue in the face, and the tension will still not go away. It is the job of science to turn disagreement into agreement, or at least a decision—that is what scientists do. It is the job of sociology, on the other hand, to hold the boundary open or reopen it once it has been closed. It is the job of sociology, then, to make salient even nearly forgotten disagreement—this is what concentrating on the process rather than the outcome implies—and this threatens closure. In our story, we saw how closure was managed by the scientific community for the cases of Joe Weber, Giuliano Preparata, and in a less dramatic form, Guido Pizzella and the 40-meter team who departed from the LIGO project. Yet here they are again, alive and kicking in the pages of this book, long after the scientists who are making gravitational wave science have forgotten about them. For instance, the 40-meter team had left the LIGO project by the mid-1990s, yet they take a starring role in chapter 36, even being given, in chapter 41, a public platform to express their view about the future of LIGO! This is what having a wider boundary implies.

Worse still, if closure is to work in science, it has to be seen to work according to scientific norms—a matter of wisdom-free calculation based on the evidence or logic represented in the constitutive journals. The very form of life of science demands that the process be assimilated to the outcome. But what we see in this book is that closure is a matter of wisdom and judgment. To insist that a closure (at least in the short term) was something other than calculative or "rational" is, given the canonical model of science, to invite its reopening. Yet this is the job of the analyst. Can this be other than irritating at best, and dangerous at worst? The sociologist can readily agree that scientific consensus formation is vital, and the analyst who wrote this book has not the slightest interest in challenging any of the closures that have been described; but, willy-nilly, the very act of describing, from a sociologist's point of view, the way they came about is to challenge them. Neutrality on matters of science conflicts, then, with the whole project of science. It is no surprise that the analyst's prized impartiality is often mistaken for scientific incompetence or mischievousness.

We might say that there is an "expert's regress" that is the equivalent of the experimenter's regress. Scientific expertise is judged according to the conclusions reached by the expert. Refuse to accede to scientific judgments

---

23. And as we explain in Collins and Evans 2002, this is the kind of society we prefer to live in.

and you make yourself out to be a scientific fool. As one of my respondents warned me in a friendly way,

> [W]ith your present emphasis, you're about to become a lightning rod again.... You need to be careful about your credibility... it's the same issue that the experimentalists face. A scientist stakes his reputation on his results. If they are silly, half-baked, or incorrect, people stop listening, and it's almost impossible to get back one's good name ever again.

The dilemma is like that of Groucho Marx, who famously said that he would not join a club that would have him as a member. The analyst's version of the "Groucho Paradox" is that the value of being a member of the science club depends on being able to declare views that are grounds for disqualification.

To return to the debate about unknown noise sources in LIGO, one of the points that Barish made was that the debate in which I had chosen to be devil's advocate was not an interesting one. At one point he said, "[I]f you're going to do sociology of science, you have to deal with science.... You have to deal with it enough to know whether the debate has any sense." And that I cannot avoid making decisions about what makes scientific sense is exactly what I admitted when I described my treatment of the interferometer on the one hand and the green-ink letters on the other. Fortunately, the Spero case is not like the green-ink letters, at least from where I stand. In the debate about unknown sources of noise, I am not trying to be a scientist but to represent a scientific position which, though it is dead as far as the powerful boundary makers are concerned, is still not dead to one who is lagging a little way behind. In such cases, one must be assiduous in one's symmetry. Were I to be otherwise, I would be telling Bob Spero, "My scientific judgments on matters of interferometry are better than your scientific judgments," yet Bob Spero works on interferometry at the prestigious Jet Propulsion Laboratory in Pasadena, and I am merely a sociologist.

The Lived "Tu Quoque" Argument and Academic Values

Given the unavoidable tension between this approach to sociology and science, why should the physicists put up with me?[24] In its early years, sociology of scientific knowledge was attacked by philosophers using what

---

24. There may be all kinds of mundane reasons, but I want to make a more general point.

became know as the *tu quoque,* or "you, too" argument. The claim is that if you believe that what can be believed follows from something other than good reasons, then there cannot be good reasons for believing that very claim. Therefore, the claim rests on a paradox and must be false. The *tu quoque* argument is too cute to be taken seriously; if we abandoned all the grounds for belief that fail when turned upon themselves, we would have very little left in the way of beliefs. Take the view that "the only reliable beliefs about the natural world rest on experiment or systematic obser-vation." Should that claim be rejected because it itself does not rest on experiment or systematic observation? Anyway, there is an equal and oppo-site paradox—what we might call the paradox of confirmation—that cancels out the *tu quoque* argument. It is simply that there is no positive claim that does not depend on some premise or other, so that each proof depends on some previous proof and so forth. In the words of the funny story about what it is that supports the world, it is "turtles all the way down." This makes realism at least as incoherent as relativism.

The old *tu quoque* argument can, however, be replaced with a naturalized version. It comes in two variants—the general and the short term. To save philosophical debate of the kind I want to avoid, in the description below I include, in square brackets, phrases that cover the less contentious short-term version.

The lived *tu quoque* argument goes as follows: We can show by systematic observation that contrary to the liberal academic view beloved of scientists, scientific arguments are rarely settled [in the short term] by reason. That is to say, it is rare for a determined arguer to be convinced [in the short term] by reason. Therefore, scientific conflicts are usually settled [in the short term] by something other than reason—let us say, for argument's sake, they are settled by socially marginalizing opponents. But the sociologist's war-rant for gaining access to scientists' physical and interpretative space is the liberal academic view, namely that science (including social science) can proceed best only if everyone has equal access to data and the right to present them and interpret them as they see fit. Thus the sociologist demands the right to be proved wrong by reason rather than marginaliza-tion. But if the sociologist is right, then the only way he or she can be proved wrong [in the short term] is by marginalization. Hence to be proved wrong, the sociologist must be marginalized—which would prove that the sociologist was right.

What follows? Compartmentalization! The way to set out to do science is as though it works according to the liberal academic view—whether it does or not is another matter. Even if arguments are rarely settled by reason [in the short term], we must continue to act as though they were, because only by going on in that way can we preserve the idea of science.[25]

Luckily, I work in an area where my respondents and I share liberal academic values. Only because of this can I continue to interact with them even while espousing a view that, in spite of myself, is antagonistic to their project in all the ways outlined above. In other fieldwork situations, such as a deep ethnic division underpinned by religious bigotry, matters might be very different. Here, however, even those who scorn my work feel obligated to give me access to theirs because of their adherence to the very liberal academic values that my work exposes to danger. As consolation, let me point out that I live with the same paradox: I, too, try to live according to liberal scientific values even as I show that they must often be honored in the breach.[26]

ANECDOTES ABOUT INTERACTION

For the above reasons of principle, my relations with my respondents *ought* to be good and on the whole they are, in spite of the tensions. Let me, however, briefly depart from the main thrust of the analysis to describe how these relations were realized in practice.

---

25. Repeat the argument in another context. Let us say we are dealing with the sociology of the criminal justice system, and the sociologist discovers that according to the way the courts make their decisions, the overwhelming correlate of innocence is wealth. Does it follow that judges should be replaced by accountants? No! This is definitely a case where one cannot get "ought" from "is." One must still cleave to the principles of justice even if one knows they are not working.

26. Interestingly, sometime after I wrote this passage, one of the senior scientists on the project put this point to me quite explicitly. He said that he was willing to work with me even though he could not understand what I was trying to do or where it was leading, because he felt an obligation to me as a fellow academic.

Some "postmodernists" take the opposite tack and consider that the kind of analysis of science done here and elsewhere makes adherence to scientific principles otiose; unfortunately, that turns social studies of science into more of a political movement than an academic discipline. There are also those whose prime motivation is not to attack science but to intervene on the side of the scientifically powerless. It has been argued that treating two sides of a scientific argument symmetrically, as in any kind of relativism, favors the powerless; and since this cannot be avoided, one might as well get on with supporting the powerless in a self-conscious manner. But no such thing follows. One might just as well argue that since publishing this book will involve the felling of trees for paper, one might as well endorse the destruction of forests. (Incidentally, some of the fiercest criticism of my early work came from Joe Weber and his colleagues, because I did not come down firmly in favor of Weber's ideas.) For a remark to the effect that symmetry inadvertently favors the powerless, see Collins and Pinch 1979. For a discussion about whether or not this precipitates self-conscious support of the powerless, see Scott, Richards, and Martin (1990); the response by Collins (1996); and the important comment of Ashmore (1996). For an attempt to revive the demarcation between science and political movements, see Collins and Evans 2002.

A Nice Irony regarding Systematic Analysis

For sociology of scientific knowledge (SSK), part IV of the book, which deals with the management transformations undergone by LIGO, exemplifies a nice irony. SSK, as can be seen in earlier parts of the book, usually reveals the untidiness of the lived world that lies behind the neat accounts of investigation and discovery found in scientific papers and textbooks. As we have seen, scientists often object to SSK, saying that what is crucial is the science itself, not details of laboratory activity or the agreement and disagreement among scientists; as I have been admonished, "there is science and there is gossip." In part IV, however, the roles of scientist and sociologist are reversed. Here, the sociologist is trying to do some social science—that is to say, to make some social-scientific categories, with neat divisions between one type of entity and another. So here, the social scientist is simplifying and schematizing in order to extract something general from the untidiness of everyday life. And here it has been respondents who have been complaining about the superficiality that such a piece of structure-making entails; it is the scientists who want to see the details of persons and personalities drawn out and the sociologist who wants to minimize their role.[27] There is, of course, no right and wrong of the matter, there is just one kind of project or another.

The Range of Individual Responses

Turning from the general to the particular, as it happens, my respondents include some who sympathize with the project to the extent that they are able to locate themselves within my worldview and read my work not as scientists but as fellow social analysts. On occasion they even correct my sociology. Others understand my position intellectually and help me develop it, but warn me that it is eccentric and liable to endanger my reputation. Still others give me whatever I ask for in the way of information without any particular enthusiasm or interest in what I do with it. Others misunderstand the project and dislike what they see coming out of it, but still help me out of a sense of academic obligation. A few others seem to dislike what comes out of the project so deeply that any help they give

---

27. Jon Agar (private communication) points out that it can be difficult to separate structure and personality when a person is used to symbolize a project in the public mind—for instance, Bernard Lovell came to represent not only Jodrell Bank but also British radio astronomy in the public mind in the UK (even at a time when it might be argued that the less spectacular Cambridge arrays of small telescopes were making more discoveries [Agar 1998]).

is grudging and set about with warnings, even threats. This last group believe they know the answer to question 3 and that my project, if it has any potency at all, is likely to be harmful. Because of the usual tendency of scientists to see things in individualistic terms, this last small class of respondents assume my motives are suspect.

To give an example of how this works, I sent a confidential draft of a paper about the management of LIGO to twelve respondents. One responded with a congratulatory endorsement and a number of helpful corrections. One told me the draft was "chock-full of insights" but suggested numerous ways in which his role might be more favorably interpreted, some of which resulted in changes. One told me the draft was "very interesting" but did not comment further. One told me the draft was so seriously flawed as to require more in the way of commentary than was written in the draft itself, but did not, in the end, provide any specific criticisms except to say that I had overlooked the much more interesting psychological problems of those involved. One was critical but arranged an extremely helpful ninety-minute teleconference with me, during which there seemed to be some meeting of minds and which resulted in alterations to the paper and added sections on methodology. One said he had been unfairly treated and responded with a list of specific points, nearly all of which have resulted in changes to the paper in the direction requested. One said that the draft reflected a lack of understanding of the science and was journalistic and anecdotal, and provided technical information to set right some of the mistakes; this resulted in my handling the particular points of disagreement more carefully. One told me he would be commenting in a major way, but has not yet done so. One came back with an extremely helpful two-and-a-half-hour transatlantic phone call correcting a number of mistakes, offering a variety of alternative interpretations, winning some points and conceding others in the debate, and suggesting that I had nearly all of it right if some it wrong (and that it was infuriating to physicists not to have the physics dealt with as physics—though also conceding that there might be value in having a different way of looking at things). Another made a series of corrections which have been incorporated, and two made no response. I also had a useful response from a scientist working closely with one of the respondents to whom the paper was sent.

The generalizations about the principles and practice of my relationships with my respondents as a group should be set against the background of this wide range of individual responses. But it is the generalizations that remain the most important for the arguments presented here.

SOCIOLOGY OF SCIENTIFIC KNOWLEDGE, SCIENTIFIC JUDGMENT, AND TYPES OF EXPERTISE

The reasoning in this chapter follows a sinuous course; at some times I find myself arguing that I must master the science, or at least elements of it, in order to do the sociology; at other times it is the essential difference between my expertise and that of the scientists that seems to be crucial. This section is about more of the differences between my respondents and me.

Consider my analysis of the extinction of the Italian group's claims to have seen evidence consistent with gravitational waves (chapter 22). My analysis reveals that the micropolitics of science played a significant role in the process. Does this analysis show, then, that the science was "impure" or "distorted"? The answer is yes if what happened is compared to the canonical model of science. To be a scientist in the process of doing science is to endorse the canonical model—and quite rightly, too. So for the unreflexive scientist at the bench, that part of the outcome of the debate with the Italian group has to be a cause of concern. To the social analyst, however, the canonical model of science is simply wrong.

To express this in terms of a new metaphor, the canonical model is rather like a child's painting: there is the sky, there is the earth, there is the sun, a person, a nose, a mouth, and so on. A child's painting is a graphic transliteration of the world represented by names for objects. We know that this kind of graphic transliteration does not look much like the world as we know it. A mature "realist" painting, on the other hand, which at first glance captures the world as we ordinarily see it, is much more confusing than the child's painting when examined closely—it is all is indistinct daubs. It is the indistinct daubs that make the world of science that we as analysts are trying to describe. Paint the canonical version, with its discrete and well-defined objects, and what you get does not look anything like science. Or again, we might say that the analysis presented here is like Newton's experiment with the prism. Newton passed white light through a prism and showed it was made up of many colors. This did not show that the white light he used was not "true" light, or contaminated light, just that white light was not a simple thing. To the analyst, the science we see in these pages is not contaminated or distorted, it is just science. There is no purer science. That is one way in which the analyst's judgments are, and should be, different from those of the working scientist, who must value scientific integrity above all things.

The same point, that what is mere description to the analyst can sometimes seem like judgment to the working scientist, can be made with a thought experiment. Let us project ourselves into the future and imagine that the current generation of interferometers has seen no signals and Advanced LIGO, with its sapphire mirrors, has just been "switched on." Now suppose that Advanced LIGO is flooded with clear coincident signals consistent with a flux of gravitational radiation similar in amplitude and frequency to those found by Joe Weber and, as a corollary, inconsistent with the null results of LIGO I. Now suppose that it slowly dawns on the scientific community that Advanced LIGO is not working in the way it was designed to work but that instead the sapphire mirrors are acting as Weber-type resonant masses and responding to gravitational waves according to Weber's long-dismissed quantum analysis of the cross section, which would render them highly responsive. The interferometer in this thought experiment is merely measuring displacements of the end of the sapphire-crystal mirrors (as with the measurements of the "Q" of sapphire described in chapter 35). To complete the fantasy, let it be that it turns out that sapphire and aluminum alloy (at room temperature) share some rare property that makes Weber's enhanced cross section for gravitational waves apply to them but not to frozen aluminum alloy or the fused quartz used for the mirrors of LIGO I. Finally, suppose that everyone comes to agree to all of this. In such an eventuality, should I celebrate? Should I adopt the mantle of the outside observer who saw the defensibility of Weber's claims more clearly than the scientists themselves? Should I celebrate the fact that I alone among the commentators had revealed the flaws in the scientific process that led to the demise of Weber's claims?

Let me say that were this to come about, it would make an irresistible story. It would also help to prove the general thesis insofar as it can be expressed in the form "Scientific conclusions are underdetermined by theory and experiment." But there would be nothing touching on matters of scientific knowledge per se for the analyst to celebrate. As I have stressed, I have made no scientific judgments, and therefore it cannot be the case that any of my scientific judgments could be proved right or wrong by the way the science turns out. Indeed, if anything, the massive change of scientific consensus that would comprise Weber's views coming to be accepted would cause trouble for my thesis. How would I explain that all these "vested interests" were overturned by mere experimental findings? Far from celebrating, I would have a new project on my hands: to investigate this sociologically astonishing turnabout.

Yet it could be even worse! Should Weber's findings be revived, it could be argued that the whole of Collins's study of the social nature of gravitational wave science is vitiated by the fact that he examined a case of science where the entire core community was doing a bad job. For the sake of the integrity of my study, I have to insist that what we see in these pages is not flawed science but exemplary science.

My Scientific Expertise

What has been established so far is (i) I am not making scientific judgments; (ii) I do not want to make scientific judgments about matters at the core of my study; and (iii) I have to make scientific judgments about matters outside the core of my study. But do I have enough expertise to make these choices? This is question 1: To what extent can I choose to dance with the scientists? How much of the appearance of choice is not really a choice but something forced upon me by my scientific ignorance? Let us start by looking at some more differences between the social analyst and the scientist.

Disdain

There is one obvious way in which I am less than an adequate participant in the life of the scientists: it is a corollary of the necessity of taking seriously both viewpoints that make up a scientific dispute. I cannot generate the same levels of disdain for core physicists that one core physicist can generate for another. This may not be damaging to the project, but it must, once more, be at least irritating. I can get a sense of what is going on by reversing the case and imagining the physicists reacting to my claims in the same way as I respond to theirs.

For example, certain philosophers seem to take it to be their role to be apologists for science, and they appear to have little more comprehension of the sociology of scientific knowledge than the worst of the "science warriors."[28] Suppose one of my respondents were to tell me, "Well, Harry, you came in for some pretty heavy criticism by that philosopher in that [named] edited book I was reading the other day." I would be disgusted. I would try to explain that my meticulously researched case studies, based on years of development of new philosophical and sociological traditions, should not be compared with a couple of hours of witless dredging through tired old

---

28. For example, Koertge 2000 contains some examples of this kind of thing. This is not to say that every contributor to the volume is equally mindless.

orthodoxies. I would explain that I treat such criticism with disdain and that my respondent should treat it with equal disdain.

And yet that is exactly the kind of disdain that I "deconstruct" when physicists express it of other physicists. When they say to me that "X has gone off the rails" or "Y was never a careful observer anyway," I look at them smugly and say, "Oh, no; it's just that X and Y have lost their credibility, but their ideas could still be shown to be quite reasonable from their point of view." Thinking about this, I discover anew what saints my respondents can sometimes be. I would certainly have difficulty being patronized to the extent of being told that the comparison of my arguments with those of the despicable class of philosophers was just a matter of point of view!

Conflicting Ways of Being in the World and the Art of the Insoluble

Another way in which I am less than perfect as a participant-observer is more a matter of cognitive style than the logic of viewpoints. Scientists are natural optimists, while sociologists are natural cynics. Scientists' entire manner of speaking and body language reflect their optimism. Even those scientists who believe that large-scale interferometry is still magic know that one day all will become clear. The idea that the world can be ordered and controlled, in the long term if not the short, is expressed from moment to moment in the way scientists go about solving their technical problems. They are calm and certain in their actions; they know that mistakes will be made and that they might have to spend many, many hours chasing false leads; they know that their goals might take many times as long to accomplish as they expect; they know, mostly, that the interferometer will not fit comfortably inside the skin of its symbolic representations. But these things are seen as "troubles," not fundamental features of the world.

Sometimes it would be nice to live in this secure world—the world I inhabited when I was a science student at school. The way science is taught is a series of solvable puzzles—in Peter Medawar's phrase, it is "the art of the soluble." Continual exposure to solved puzzles leads one to believe that with enough brilliance and effort, all problems can be solved.[29] In contrast, the social scientist is taught "the art of the insoluble"—nothing is ever really clear. For the social scientist, there is no equivalent of the instant and unambivalent reward one gets from solving problems—such as the tremendous "high" I got from making my first interference fringes.

29. After an exclusive diet of the art of the soluble (Medawar 1967), an encounter with the world of frontier science can come as a shock.

This difference in training is, perhaps, one of the less obvious reasons the sociologists' primary focus is the "magic"—the disorder. To the social analyst, the idea of disorder comes more naturally; order is what social actors create, not discover. The symbolic net which scientists throw across the world is seen by the sociologist as a way of trying to tame a deep disorder. The sociologist, then, is always looking for the places where bits of the world escape from the net—like a balloon bursting out of a parcel. This can seem salacious as well as profane. To the scientists, burstings-out of the balloon indicate moments of failure of understanding, design, or experimental skill. So the sociologist should not be surprised that the most well-intentioned acts of analysis are sometimes seen by scientists as accusations of scientific incompetence. In a difficult and delicate project such as LIGO, where a team has to be held together by the internal attraction of its own optimism, sociological analysis can even be a threat.[30]

Types of Expertise

All these reservations in mind, let us consider my scientific expertise in a more positive light. I justify my approach toward the analysis of science by calling it "participant comprehension." In this method, one interacts with one's respondents as much as possible with a view to internalizing their world as far as one is able—that is, to try to attain some approximation to their form of life (never forgetting the need to step back occasionally). If some reasonable level of internalization can be accomplished, it is argued, then one can describe that world from the inside as well as the outside. The subjective becomes the objective.[31] But I have just explained the many ways in which I do not internalize the scientists' "form of life." I don't

30. Rai Weiss once told me that he distrusted my project, because I did not have the same commitment to the detection of gravitational waves as the scientists. He suggested that I would be just as happy, and have just as much to write about, if the project of ground-based gravitational wave detection turned out to be a failure. It is true that I will still have something to write about if the project is a failure. But I would prefer the search to be a success, partly because I want to see my new friends and colleagues succeed, and partly because success will round out my project: the most exciting stage of the research to come will be watching the scientists decide that they have enough data to make a positive announcement. Then I will have been able to compare the making of a scientific success with the making of scientific failure. Furthermore, should the science be a resounding success, my study will be more salient.

31. Thus it is with my understanding of my native language, English. It is because I am a representative of the social collective of native English speakers that I know nearly everything about ordinary spoken English that there is to know. I do not have to check with others whether my sentence construction is correct; I just know, reliably. (Naturally, there are exceptional sentences, but that does not affect the point.) For more discussion of participant comprehension as a method, and its compromises, see Collins 1984.

share their disdain for other physicists, I don't share their visceral optimism, I don't always share their view of what is correct science and what
is false science, and so forth. So I want to eat my cake and have it, too.
What is going on? One thing that is going on is compromise. All social
scientific method involves compromise. The important thing to know is
when the compromise is fatal and when it is acceptable. How can we judge
this?

Let us start from the participant comprehension end of things: this is
the perspective that tells one to gain as much understanding as possible. I
have much less understanding of the science than the scientists. What can
I say about my level of expertise?

There are a whole range of ways of "knowing something about science."
For example, I know (I think) that the rest mass of the neutrino is "4eV."
But in knowing that, I know almost nothing. For example, I also know that
my dining table is 4 feet long, and that is quite a different way of knowing
something that includes the number 4. But my understanding of the way
dining tables fit into the world enables me to do something further with
the knowledge that my table is 4 feet long. For example, I can work out
whether or not I can seat eight for dinner. I cannot do anything like this
in the world of neutrinos in consequence of my knowledge that their rest
mass is 4eV. "Trivial Pursuit" aside, for me "40,000" would carry just as
much information as "4" when it comes to the rest masses of neutrinos. To
go back to the germ theory of disease, knowing that the rest mass of the
neutrino is 4eV affects my actions in almost no way at all, and therefore,
for me, it is almost devoid of meaning.

Here is another way of quasi knowing. I found it on a beer mat made
for the Babycham company in 1985. The beer mat gives an answer to the
question, "What Is a Hologram?" It says,

> A hologram is like a 3 dimensional photograph—one you can look right into.
> In an ordinary snapshot, the picture you see is of an object viewed from one
> position by a camera in normal light.
>
> The difference with a hologram is that the object has been photographed
> in laser light, split to go all *around* the object. The result—a truly
> 3 dimensional picture!

This explanation is capable, presumably, of making at least some people
feel that they now know more about holograms. The words on the beer mat
are not simply nonsense, nor would we be likely to mistake them for, say, a
riddle or a joke. Presumably, there are people now alive who have studied

the beer mat and, if asked, "Do you know how a hologram works?" would reply, "Yes," whereas immediately before they had read the beer mat they would have answered, "No," to the same question. Yet the explanation on the beer mat would not enable me to make a hologram or even to argue about holograms.[32] Let us call these kinds of ways of knowing science "beer-mat knowledge."

I consider that in respect of gravitational wave science, I have more than beer-mat knowledge. I have what I will call "'interactional' expertise." This is not as much expertise as scientists have; scientists have "contributory expertise," that is, expertise sufficient to enable them to contribute to the science in question, to get a job in the science, and so forth.[33] Interactional expertise will not get you a job. But interactional expertise is a big step up from no expertise at all and a big step up from beer-mat knowledge. Interactional expertise enables one to carry on conversations with scientists about their science—conversations that will hold the scientist's interest and not be too much of a chore for either party. In such a conversation, the party having interactional expertise may supply a lead, anticipate a response in order to shorten and speed the interchange, supply a word or an idea when the expert pauses. These are things that make a conversation flow and reveal that both parties are fully engaged.

With enough interactional expertise, one can even convey useful information to the expert. For example, I was once able to tell Gary Sanders that the Japanese TAMA interferometer was not suffering from light scattering from the inside walls of its beam tubes. I could do this because I knew enough about interferometry to know that it was an interesting question and had asked it when I was touring the TAMA interferometer. My host, Seiji Kawamura, realizing from our interchanges that I knew enough to understand, volunteered to me the method they had used to test for this reflection—shaking the tubes and looking for an effect on the output. I immediately understood the logic of the test and knew it was important to tell Gary Sanders that it had been done. This additional information conveyed to him that I knew what I was talking about when I told him there was no trouble from the walls of the tubes. I transfer little bits of technical knowledge quite often in my fieldwork, sometimes because I am asked, sometimes because the gaps in scientists' awareness of what is going on make themselves evident in conversation.

---

32. The examples of the rest mass of the neutrino and the hologram are taken from Collins 1990.

33. For a first discussion of this classification of expertises, see Collins and Evans 2002.

Interactional expertise can even allow one to take a devil's advocate position if one is brave enough. While one could never invent a counter-argument in respect of a technical matter—that would need contributory expertise—one can sometimes *represent* such a position, with more or less success. This is what I was trying to do in my debate with Barish mentioned earlier in this chapter.[34]

For a long time, I thought the dividing line between interactional and contributory expertise was the mathematics. One cannot become a practicing physicist without a degree of mathematical ability, because one would not pass the university examinations that would lift one onto the first rungs of the professional ladder. But mathematics is not the whole answer. In other sciences, mathematics is not an issue and cannot be the dividing line between interactional and contributory expertise. Even in physics, as I learned, there are brilliant experimenters whose mathematics is not very good; Ron Drever is an example. So it is not just my mathematical limitations that define my lack of contributory expertise, it is more. I just don't know how to design, build, and diagnose the electronic circuits, computer programs, mirror suspensions, beam tubes, optical cavities, or anything much else that goes into an interferometer. On the occasions when I get a real insight into what my respondents can do with the technical tools at their disposal, I generally come away feeling humbled. Yet it remains that to a surprising extent, I can talk about their work; this is not a point of principle but an empirical discovery that could have gone the other way. I have tried to study another field of science—the theory of amorphous semiconductors—where it did go the other way, and I failed to master even a vestige of interactional expertise.

---

34. On occasions, I think I stray slightly beyond interactional expertise. After a bit of a struggle I have, after all, managed to build a crude interferometer. Furthermore, as a sociologist, I have also learned the elements of statistics. On four or five occasions over a period of nearly a decade, I have even won arguments with physicists about physics, usually when the scientist in question wasn't quite as current as I was with the particular detail at stake. The most recent example was when I was told that gravitational waves did not have a shadow and I was able to point out that they must have at least a vestigial shadow, or one would not be able to detect them at all: detection requires that energy be taken from the waves, and therefore the waves must be marginally less energetic on the far side of a detector (or any material they pass through). On one remarkable occasion I was assured that my critical comment on the proposed design of a prototype interferometer was so apt that the team would now rethink an important feature of the instrument in a major way, and this did actually happen. But these little bits of competence barely raise me above the interactional expertise level when it comes to gravitational wave detection science and technology proper. My claim remains that I can manage my end of conversation about many aspects of gravitational wave detection so long as the topic is broad principles of design or data analysis without needing mathematical or algebraic analysis. (Such mathematical ability as I had at school I have allowed to atrophy.)

It is also an empirical discovery that interactional expertise in a field is sufficient to do the sociology of scientific knowledge in that field; contributory expertise is not necessary. The results of the many studies have shown this.[35]

That contributory expertise adds relatively little to the ability to analyze the sociology of a field is less surprising when one remembers that scientists have contributory expertise only in narrow areas. To the outsider, the community of physicists presents what sometimes seems like an impenetrable wall of expertise; to the insider it can look very different. Thus in 2003, one respondent, knowing the approach I was taking, wrote to me as follows:

> When we look around at one of these [physics] meetings, we don't see a mass
> of scientists, but a bunch of people with different skills, personalities,
> histories, weaknesses, technical accomplishments, roles, ambitions....
> Included in that bunch are a certain fraction for whom we feel a fair amount
> of disdain [he is referring to my usage].... But we are all a part of a
> community, as measured by the fact that we've seen these folks many times
> before at meetings where we interact, and expect to see them many times
> later.... As you know by now ... we have widely varying ranges of skill at
> "math" and at other ideas that constitute the ostensible subject of our work,
> but there is so much to making things go that key contributors can often be
> weak in what is supposedly basic.... "Interactional expertise" is all that any
> of us have in much of the subject of our work.

INTERACTIONAL EXPERTISE AND DECISION MAKING
IN THE WIDER WORLD

Now let us move further from the main purpose of the book and investigate a "spin-off": the role of these different kinds of expertise, first in the world of science and then in the wider world. To continue to exist, the sociology of scientific knowledge has to establish the right of certain nonscientists to comment on aspects of science that have traditionally been taken to be the

---

35. I can compare the studies I have done myself where I had expertise that ranged from none, through interactional, to contributory. Having no expertise is fatal; having interactional or contributory expertise is equivalent in terms of the sociology that is produced (but would not be good enough for "technical history of science"—see chapter 43). The same lesson can be learned by examining the sociological output of other researchers whose types of expertise range across these levels. See also the generous comment of ex-physicist Alan Franklin (1994).

business of scientists alone. Once scientists were allowed to do their work with a degree of privacy more appropriate to a religious cult, occasionally issuing a report of a revelation for public consumption. The sociology of scientific knowledge has to invade the inner sanctum. The issue is pressing because the doors can no longer be kept closed. Even if we wanted to leave technological decision making to the scientists alone, we could not, because the pace of scientific and technological consensus formation is slower than the pace of politics. This is becoming more obvious as science and technology press ever more upon the public domain.[36] The study of the meeting point of experts and nonexperts becomes, then, ever more important.

Looked at in this way, my "methodology" is itself a substantive research project. With tongue in cheek, I'll call my own project an exercise in "high-energy sociology." In high-energy physics, particles are projected at targets, and their subsequent trajectories tell the observer something about the forces involved in the interaction. I am, as it were, a "particle" projected at the "target" of the gravitational wave community.[37] In this section of the

36. What Richard Garwin was doing (chapter 9) was trying to make scientific consensus formation, in the case of Weber's claims, catch up with the micropolitics of funding agencies. As we have seen, he did not entirely succeed. Though he accelerated the scientific process within the core set, he did not produce consensus in the outer rings of the target, nor even uniform agreement among scientists. Joe Weber was able to obtain grants from taxpayer-funded sources almost up to his death, and his scientific allies continued to rally to him for about a quarter of a century after Garwin's 1975 "coup." Such a pattern is not untypical of scientific controversy (Collins and Pinch 1993/1998).

37. The analogy of the trajectory of a particle was used by Trevor Pinch and me in 1981. As we put it, two physicists—Professors John Hasted and John Taylor—"fired themselves" at the scientific community when they tried to investigate aspects of parapsychology. Though the physicists claimed repeatedly that they had no preconceptions about the truth of the matter, and were merely investigating in an open-minded way, the scientific community insisted on deflecting them into one of two mutually exclusive positions: they must be either "believers" or "skeptics." There was simply no social role for neutral investigators; the resistance of the social target to this identity was too great for the particles to be able to penetrate. As a result, Taylor's trajectory became that of a skeptic and Hasted's that of believer (Taylor 1980, Hasted 1981).

The same kind of idea underlies the method of what I have elsewhere called the "proxy stranger." The "stranger" part of "proxy stranger" refers to the method of anthropologists, who consider that they are especially sensitive to the mores of societies when they first enter them and before familiarity has done its damaging work. In the method of the *proxy* stranger, however, a third party is "fired off" to interact with a social group or process with which he or she is unfamiliar, while the investigator, who is relatively familiar with the society under investigation, looks on. The interaction between the proxy stranger and the society is revealing to the observer. In that kind of investigation, there is no need to worry about the proxy stranger's lack of expertise; the more strange the projectile, the more interesting the trajectory. (Collins and Kusch 1998, chap. 9; Hartland 1996).

Garfinkel's (1967) famous "breaching experiment" is another example of something close to the proxy stranger, except that in this case the "projectile" *pretends* to be a stranger, while in fact he or she understands the society. My favorite example was invented by Peter Halfpenny, who persuaded

book, it is my trajectory that we are looking at. For these purposes it is good that there are gaps in my scientific expertise. If my expertise were perfect, I'd be no good as a "particle" in the high-energy sociology experiment: instead of bouncing around in the target and yielding information, I'd just slide straight through.

High-Energy Sociology and the Core of an Esoteric Science

Referring to the target diagram in the introduction, we can consider my interactions in the different rings of the community and its publics. Consider first decision making within the core group of specialist scientists and my debate with Barish about potential unknown noise sources in LIGO. I thought I put up a fairly creditable performance acting as devil's advocate, championing the role of the 40-meter group against Barish's claim that all LIGO's potential noise sources were well understood. But let us not worry about whether I actually did perform well; let us just suppose I did. Let us suppose, then, that in this argument, or some other argument, I managed to use my interactional expertise to defend a position in physics against a physicist. Would such a capacity entitle the outsider to a share in making scientific judgments within the core of the physics itself?

The claim might seem absurd, but think of it this way: when Barish took over the LIGO project, he, too, had no contributory expertise in large-scale interferometry, and the fact remains that he has invented nothing that one can name that has contributed to that science. It might even be that Barish's insistence on the reading of the literature and the modeling of the system (the nonmagical element of things) were symptoms of his *not* being well enough embedded in the community to understand the esoterica through the oral culture. A manager coming in from outside has to use more-formal methods of "getting up to speed." A bone of contention between the new management and the 40-meter team was this very issue, with Barish claiming that he had the kind of technical expertise needed. But Barish is a successful manager of the project, so this means that lack of hands-on experience and absence of deep embedding in the oral culture of

his students to board buses and ask for two tickets, one for themselves and one to reserve the seat next to them.

One might also think of the controversy studies in sociology of science as examples of high-energy sociology. A controversy study is an analysis of interacting human particles that reveals much about the social structure of the medium in which they interact (e.g., Collins 1983b).

a science cannot be an obstacle to successful management of science.[38] So if I'm so smart and possess all this interactional expertise, why don't I take over the management of LIGO? (Don't worry, I intend this as a reductio ad absurdum.)

There are many obvious answers. A manager like Barish commands great respect in the wider scientific community for successful work in other sciences, and this provides good political potential. Looking inward, there are personal qualities such as the authority that comes with a high scientific reputation in other fields. And there are formal managerial skills—the knowledge of how to handle large numbers of people and large sums of money. But let us set all these aside to concentrate on the scientific capacities which Barish himself insisted were the crux of the matter. Barish possesses two kinds of scientific credentials for the job of manager of this otherwise unfamiliar science. First, he has a reservoir of formal abilities that accompany scientific accomplishment in other fields, and a good proportion of this will be directly transferable. Second, he has long experience of making scientific judgments in other fields; Barish knows what it is to make a high-level scientific judgment. The first of these is simply a body of knowledge common to many kinds of scientists at all levels, including novices. The second, which is crucial for a manager, we will refer to as referred expertise; it is a body of experience gained elsewhere that has a reference here, because the judgments to be made are of the same general type even if the science is different.

What does this mean for the argument between Barish and me about noise in LIGO if we allow, for argument's sake, that I did a reasonable job? Setting aside all the political nous, the authority, and the technical accomplishments that Barish has and I don't, the difference between him and me is the following: I was just having fun, because nothing turned on my views; yet for Barish, to choose a position in a debate amounts to making a decision about how to act next. For me there is nothing at stake in these arguments; if I turn out to be hopelessly wrong, I can say, "Well, what would you expect from a sociologist?" If I turn out to score an occasional point, it is like ragging the schoolmaster. The difference is a difference in the forms of life in which we are embedded, he as scientist, me as analyst. In nearly every case (we will look for an exception at the end of the chapter), interactional expertise is not a sufficient basis to make

38. For what it is worth (perhaps slightly more than usual, as I am judging a social skill and have managed a large school of social sciences myself), Barish seems to me to be an excellent manager of the quiet, authoritarian type: the Clint Eastwood of gravitational waves, as it were.

judgments of this sort. Thus, irrespective of any fleabites I might inflict, Barish's scientific judgment—the judgment he will act on—remains that there are no unknown fundamental noise sources to be found in LIGO, and he has to take responsibility for the judgment. In spite of (what we will allow to be, for argument's sake) my fairly creditable performance in the argument, at this level I am not *contradicting* him, because nothing turns on what I think.

This is why, in spite of the uncertainties in the heart of science that are made more salient by sociological analysis, nothing has or should change in the decision-making chain in the core of an esoteric science (barring the possible exception to be discussed below). The scientists have already discounted their uncertainties even though they may not have articulated them. Like any other decision makers, they live with the chance of being wrong but must still act. This is the expertise of the core set, and this is where decision-making rights in matters of esoteric science should remain. If this book makes any difference to esoteric science, it should affect only the way scientific judgment is thought about, not the way it is done.

If it were Bob Spero making the arguments rather than his advocate, it would have been different. Had Spero been in charge he would have made a different set of choices: he would have made choices predicated on the idea that noise in large-scale interferometry was not sufficiently well understood to justify the shutting down of work on prototypes. But Bob Spero has contributory expertise, so what we would be seeing would be a clash between two higher level sets of expertise.

Barish argued that if I read the papers carefully, I would be forced to reach the same scientific conclusions, and to make the same scientific judgments, as he. On this he is wrong. I am not making scientific judgments but describing the grounds of the disagreement in terms that enable us to pick out the judgmental aspects of the decisions made by those whose job it is to judge. Where he is right is in his response to my claim, "You see, I'm being devil's advocate because I'm trying to put their viewpoint." He replied, "But it's science, not viewpoints." A manager's job is not to see everyone's viewpoint but to establish the correct viewpoint and make the appropriate decisions. When I try to take Spero's position, I am acting out a role, not living the life of physics.

But we are not finished yet. Let us return to the slightly more tricky case of the Perth-Rome coincidences. I argued that the scientific outcome was driven in part by the power relations between bars and interferometers. Does this not open up the debate to all potential power brokers? I would say no. The judgments, "rational" or otherwise, are still the business of the core group of scientists and no one else.[39] Again, the only difference my analysis might make is to the way scientists *think* about the way they make decisions, not about the way they make decisions. Scientific decisions, right or wrong, are best when they are based on the judgment of the most skilled and experienced persons—the scientists. What I show is only that these decisions are based on experience and judgment rather than exact calculation.

Review Panels

Moving out from the core, let us turn to decision-making forums where a heterogeneous distribution of expertise is the norm. An implicit judgment of the value of expertise other than contributory expertise is made when committees of so-called scientific peers meet to make judgments on the work of other scientists. Often the peers have no contributory expertise in the sciences they are judging.

Consider the NSF Review Panel that met in 1996 and was discussed extensively in chapter 23. In that chapter, I offered some scientific assessments of the proceedings. None of the other scientists were themselves contributors to the gravitational wave detection *experiments*. Some were theorists and some were experimenters in other fields. As a result, in terms of the science and the assessment of the potential success of the sphere program, the future of which was being discussed, I felt I could *(not should!)* have made a nontrivial contribution.

What the other panel members brought with them that I did not possess was the kind of referred expertise that I have discussed above. They had experience running other scientific projects; this was experience that could be applied to this case even though the science was different.

Much more striking, however, the committee members brought with them political sensitivities that, as I argued in chapter 23, were the crux of the matter. The whole committee meeting was really a mutual exercise in sniffing out (or, perhaps, "socially constructing") the direction of the political wind. What the meeting amounted to was the granting of official recognition to the direction of the wind; this was done in the guise of scientific judgment. If I had been in a position to make a contribution to that panel decision, even if my contribution was reasonable in terms of the science, it would have been politically naive. Ironically, the trouble with my contribution would have been that it would have been inadequate in being too

---

39. For this argument worked out more technically, using the example of Shapin's (1979) analysis of the Edinburgh phrenology debates, see Collins and Evans 2002.

narrowly technical! For example, I would have put the case for the spheres too strongly, not being sensitive enough to understand that the technicalities were not the heart of the matter. The committee almost certainly made the right decision in terms of the political realities—to have funded the spheres would have severely reduced the viability of the whole gravitational wave detection program—but I would have been so busy worrying about my technical competence and input, and being fair to the scientific claims of sphere proponents, that I would have missed the bigger picture.

As we move to review committees at higher and higher levels we discover less and less contributory or even directly referable expertise. Consider the decision to fund LIGO: The congressional representatives asked the most naive questions during the hearing (chapter 33), displaying a degree of scientific ignorance that would have shamed any sociologist. And they were almost certainly more gullible than I would have been when it comes to "the wonders of science." So the technicalities are not the issue in the majority of high-level decision-making committees; the issue is political sensitivity and political representativeness.

Currently, the choice about the range of expertise assembled to make these kinds of decisions is a matter of custom and practice; the bigger the decision, the more nonexperts are brought into the matter. This is the irony to which I have pointed earlier in the book: the more money you want to spend on science, the less expert will the decision makers be.

The Political Sphere

The whole analysis so far has been conducted for an esoteric science, gravitational wave physics. But can this analysis speak to the case of sciences that are exposed more to the public domain? Would it also work if the science were to do with global warming, the safety of nuclear power stations, the human food chain, health, and so forth? Here the case might differ. To explore the difference, we can use our imaginations to convert gravitational wave science into a public domain science. Imagine—and let me stress that this is entirely fantasy—that it was discovered that immersion in a certain (largish) flux of gravitational radiation, combined with living near an overhead electrical power line, caused cancer. And imagine that the discovery that Earth was bathed in gravitational radiation with a strength that would be just detectable by the Perth-Rome detectors would mean that those near power lines were at risk. In such a circumstance, who should be involved in decision making about whether the Perth-Rome coincidences were to be taken seriously? What kinds of levels of expertise and political sensitivities should be brought to the table in these circumstances? It does not do to say that in such a case the decision becomes the prerogative of politicians working according to the precautionary principle, or we would spend our entire national budgets on scares dreamed up by cranks (as may well be the case with the existing debate over the deleterious effects of power lines on health). The political decisions have to stand on a foundation of science. Yet science cannot work at the pace of political decision making.

I do not know the answer to the question about who should make the scientific decisions in circumstances such as these. I want to suggest only that the classification of types of scientific expertise, including contributory expertise, interactional expertise, and referred expertise, might be useful in the discussion and that this book provides some materials for thinking about them.[40] I suspect that in these circumstances the relative value of interactional expertise might increase as compared with its weight within core science. In these circumstances, a study of the way the trajectory of the Perth-Rome coincidences (or the like) came to an end might be a legitimate input to the scientific judgment as to the existence of high fluxes of gravitational waves. Perhaps or perhaps not—I offer the idea for debate.

METHODOLOGICAL CONCLUSION

Throughout this chapter, I have traveled a complex course between the need for more or less expertise and the significance of a variety of kinds of expertise for social studies of science and for scientific judgment. This is the beginning of a study, not the end.

There is one firm conclusion that I want to draw, in the absence of which the value of this whole study would be enormously reduced. We have seen—and in this chapter especially I have tried to be as honest about it as possible—that there are inevitable tensions between the social analysis of science under the rubric of methodological relativism and science itself. This is the conclusion I want to draw: we should not try to avoid the minor inconvenience caused by social analysis of science by barring the doors of science's inner sanctums to the social analyst. To do this would damage science, would damage the idea of freedom of discourse that underlies science, and would damage liberal democracy; and it might produce poorer decisions than would otherwise be made where science and technology enter the public domain.

40. See Collins and Evans 2002.

CHAPTER 43

FINAL REFLECTIONS: THE STUDY AND SOCIOLOGY

WHAT HAS BEEN ARGUED IN THIS BOOK?

The history of gravitational wave detection has been traced from the very first glitch in 1962 as described in a Ph.D. thesis to the point when instruments costing hundreds of millions of dollars are coming on air. Unusually for such a project, the science has led the analysis throughout: the scientists' flurries of excitement have given rise to the sociological flurries of excitement; my grant applications have expressed the same hopes and fears as the scientists' grant applications; my project is worth funding because the search for gravitational waves is worth funding; if the founders of the scientific field die before gravitational waves are detected, then, probably, I will die before my project is completed.

Furthermore, the whole second episode of sociological investigation, starting in the mid-1990s, was inspired by a *scientist*. In 1993 (see chapter 44), Joe Weber convinced me that gravitational wave research was about to blaze up once more, and I concluded that a new and fascinating sociological conflagration would result. Of course, though the sociology has been paced and colored by the science, the driving force of my project has been not the detection of gravitational waves but the meaning of knowledge. The science, the sequence of events, and the achievements and failures of the individuals have been described, but the framework has been a series of questions about knowledge.

Over the course of four decades, claims to have detected gravitational waves directly using purpose-built antennae have been made in about fifteen papers published in the scientific journals. These fifteen or so papers report the findings of six distinct sets of observations or analyses. The original Weber group published approximately ten papers between the later 1960s and the middle 1970s claiming positive observations with pairs of room-temperature detectors; these I count as one set of observations. Then, in 1982, there was a positive report by the Frascati group together with Weber. There were a set of positive reports concerning waves from Supernova 1987A. There was the claim of coincidences between the Frascati group and the Perth group (though it did not appear in the refereed journals). There was Weber's 1996 publication concerning correlations with gamma ray bursts. Finally, there was the 2002 positive report by the Frascati group. These reports were supported by commentaries and discussions which tried to make the observations compatible with widely accepted theories and astrophysical models. The fate of the 2002 claim is still to be decided, but as regards the other five claims, the fourteen or so "results" papers, and the commentaries associated with them, very few scientists can be found who still think they are worth further investigation. The first portion of *Gravity's Shadow*, up to the middle of part III, is devoted to an examination of the trajectory in social space-time of these first five claims and their associated publications.

In this first major passage of explanation, we can see the working out of all three stages of what has been called "The Empirical Programme of Relativism" (EPOR).[1] The first stage is the establishment of the extent to which experimental results allow "interpretative flexibility." A "result" is not a "result" until the community gives it meaning. There is a big black hole in social space-time into which science sinks if no one responds to it. Some of the papers went straight into the black hole, while some escaped as a result of a lot of hard work and others escaped because of the contingencies of time and place in which they were born. In the case of those that were not immediately swallowed up, we watched the ripples first expand and then expire. A scientist wanting to support maverick ideas can burrow deep into

---

1. Collins 1981b.

the existing "dam" of theories and findings and loosen the "stones" of assumptions that hold the edifice together. (The dam is always a little crumbly anyway because of the role of tacit knowledge and the experimenter's and the theoretician's regresses.) The "logic" of this kind of scientific debate is supported, then, by a social fabric whose warp is the unwillingness to question everything that is logically questionable and whose weft is trust. Hardly any scientists ever actually make *direct* observations of anything that bears exactly on a point of dispute, so in the end, nearly all must be agreement—sometimes explicit, sometimes unspoken or unnoticed.

Agreement is a social matter, and different communities can reach different kinds of agreement. One community exemplifying one scientific way of being survives while another dies, and scientific history unfolds. As with all history, multiple causes bring about the future. Some causes spring from experimenters and theorists at the center of the debate; some spring from the actors in the further reaches of the community.

This takes us to the second stage of EPOR: analysis of the way the potentially ever-ramifying tree of scientific potential is closed down in particular cases by forces outside those that are normally considered "scientific." (Here the term *scientific* is used as an "actor's category"—what I have elsewhere called the canonical view of science.) Given interpretative flexibility and its consequences, *judgments* have to be made about when a scientific adventure has run its course. In the absence of such judgments, science would choke on its own potential; the task of moving on would be endlessly hampered by the task of closing every loophole on the past. The role of funding agencies in shutting down certain lines of advance is, then, vital. Science is, on the one hand, a quintessentially open-ended activity where individual heroes can fight against the most powerful authority; on the other hand, it cannot work in the absence of social control. The second stage of EPOR explores the ways in which social control is exercised.

The third stage of EPOR looks at the way that wider social forces influence the mechanisms of closure. The clearest example in the book is the influence of the funding of LIGO on the attempts by the bar groups to maintain their credibility. The implication is the following counterfactual statement: without LIGO and its funding battles in the late 1980s, the data from the bar groups would have been handled differently. How events would have turned out in the end is hard to say, but in the meantime the analyses would have been carried out, for example, more flexibly, with lower thresholds for what should be brought into the signal-noise analysis, and with less stress on the draconian compulsory blinding. In sum, one way or another, there would have been more freedom to treat the observations as explorations rather than definitive claims, and the bar program as a whole would have continued to "tune" its procedures for longer rather than been pressured to affirm or deny the existence of gravitational waves in every publication. In short, "evidential collectivism" could have lasted longer had LIGO not been the principal influence on the framing of the debate.[2]

Only some of the science covered in this book is about discovery claims, and perforce the same is true of the sociology. From part II onward, the analysis increasingly turns toward the change in technology from resonant masses to interferometry. Nevertheless, questions about knowledge remain central. The way that interferometry came to be *seen* as the superior technology is the subject of much of part III of the book. The relative weight given to theory over empirical claims is crucial to this transformation, and this in turn influenced what was to count as the right way to analyze data. There were disputes about the relative sensitivity of the two technologies, the absolute potential of interferometers, and the strengths and incidence of sources of gravitational waves. As always, more than one interpretation of the same evidence was available, and these interpretations played out differently in the different rings of the target diagram. In the core, the difference between an informed guess and a calculated estimate was clear, but it may not have been so clear in the outer rings.

Part IV moves on to interferometry itself and its transformation from a small to a big science. LIGO and the other interferometers were still building in the 1990s, and they gave rise to no directly induced ripples of protoknowledge. Nevertheless, the organizational and political traumas associated with the birth of the new big science turned on different views of our knowledge of large-scale interferometry. Was interferometry ready for the degree of routinization made necessary by its exposure in the political arena? Was it a science like high-energy physics, which would fit comfortably into a symbolic description of itself, or was there still "magic" to be dispelled through less-organized research? Was, and is, LIGO a small science spending big-science money? We will soon know.

In the first section of part V, the costs and benefits of international cooperation are examined, and once more the politics and the science are shown to be intimately related. Build an innovatory design on schedule and an independent strategy is viable, at least in the short term. Whether you think such a strategy has promise also depends on how you interpret

---

2. Some readers will be disappointed that Stage Three of EPOR has not been delivered in a more red-blooded fashion, but attempts to explain changes in technology as a result of the pork barrel or the influence of the military-industrial complex were not convincing (chapter 29).

the evidence of what is in the heavens, and on the relative value of false negatives and false positives in your view of the world.

The second section of part V takes us back to knowledge claim trajectories, but of a peculiar kind. Now we see the first results coming from the interferometers essentially saying, "There are no gravitational waves stronger than this." Why, we ask ourselves, should anyone be interested? How does such a ripple ever get going, and why does it not fall immediately into the black hole of obscurity? This, again, is a question about what counts as knowledge in the world, and once more it turns out to depend on which sector of scientific society—which ring of the target diagram—you are looking at.

Knowledge is also center stage in this, the final part (part VI) of the book. Here, however, the topic is my knowledge and the knowledge I am generating. The question is how these knowledges relate to the world of scientists. How similar and how different is my worldview from that of my respondents? The analysis becomes a study of interaction between experts and nonexperts in the core of science and in the realm of science policy and science politics.

In the introduction to *Gravity's Shadow*, I set out the main themes of the study as follows: the construction of truth and falsity; the triumph of one technology over another; the necessity for choice without complete justification; the pruning of the potentially ever-ramifying branches of scientific and technological possibility if science is ever to move forward; the different role of the rings in the target diagram; the growth of science from small to big; conflicting styles and cultures of science; and the tension between the world seen as an exact, calculable, plannable sort of place, just waiting for us to get the sums right, and the world seen as dark and amorphous, bits of which, from time to time, we are lucky enough to catch in our speculatively thrown nets of understanding. The brief resume of the book given above shows where these themes have been explored.

Is Gravitational Wave Science Representative?

Though every stage of EPOR is exemplified in the first half of the book, gravitational wave science seems not to be a fruitful site for looking at the influence on science of "big-P" political interests or of the military-industrial complex. Gravitational wave science cannot be taken to *represent* other sciences in this respect. Are there other places in which the topic is too specialized to allow the findings to be applied to science as a whole?

A case study is not a statistically representative sample; each case study is special in one way or another. As in any other science, worries about statistical representativeness are reduced as the number of case studies increases, and as it happens, many of the conclusions of *Gravity's Shadow* are backed up by other studies. In particular, the broad outlines of the way scientific knowledge travels through social space-time is as well confirmed as any qualitative social-science finding.[3] Most of the existing work concerns small science, however, and it has been suggested that the gravitational wave science still to come will be less subject to the experimenter's regress than the earlier periods described here.[4] Is this so?

The Experimenter's Regress and Big Science

The experimenter's regress, it will be recalled, draws on the fact that the results of a second experiment cannot bear on the results of a first experiment unless it is agreed that both experiments were competently done; where there is deep conflict about what the conclusions should be, such agreement is unlikely. The normal criterion for competence—that the experiment produces the correct outcome—is not available where what counts as correct outcome is the very subject of the controversy.

The experimenter's regress is easy to witness where there are many competing experiments emerging from independent laboratories. Where institutions merge, however, as is the trend in large-scale interferometry, one might expect the regress to become less evident. Thus the International Gravitational Event Collaboration (IGEC—chapter 25), had it managed to draw the competing bar laboratories into a morally integrated group, would have closed down much of the potential for such controversy. As it happens, IGEC has not been effective in preventing breakaway claims, but interferometry is doing better. If VIRGO, the French-Italian group, finally joins the "LIGO club" or some variant of it, will it mean that the experimenter's regress will disappear?

Social coherence on an international scale does not eliminate the potential for disagreement; it means only that any disagreements will take

---

3. Early works on the broad topic include Fleck 1935/1979; Kuhn 1961, 1962; Collins 1975, 1981c, 1985/1992; Holton 1978; Latour and Woolgar 1979; Shapin 1979; Knorr-Cetina 1981; MacKenzie 1981; Pickering 1981a, 1981b, 1984; Pinch 1981, 1986; Travis 1981; Collins and Pinch 1982; Gieryn 1983; Lynch 1985; Galison 1987; and Shapin and Schaffer, 1987. For summaries of the field, see articles in the *Annual Review of Sociology* by Collins (1983b) and Shapin (1995).

4. Previous studies of big science that do bear on the question to some extent include Pickering 1984 or Knorr-Cetina 1999.

place within one large group rather than between many smaller groups. What may change, however, is the extent to which the symptoms of such disagreement become visible to the outside world. As we saw in chapter 39, Rai Weiss would prefer to handle all such disagreements "in-house," and the project leaders share Weiss's view that the LIGO Scientific Collaboration (LSC) should set up a system of internal checks, mounted by competing analysis groups, before any results emerge from the project; Barry Barish explained to me that this was the only way to do the science properly, because only "internal referees" know enough of the details of the science to do an adequate job of criticism.

There are two different ways of looking at what is going on here: one can either talk of taking the maximum care inside the project to ensure no mistakes are being made, or one can say that one is making sure that disagreements take place inside the project rather than outside. Of course, any project should make sure that trivial and consensually agreed mistakes are eliminated before papers are published, but we are concerned with deeper disagreements. One either conceals deep disagreements or exposes them to the wider community. Evidential individualists try to keep scientific disagreement hidden; evidential collectivists are happy to have them play out in public. It could be that if evidential individualism becomes the ethos of the integrated international interferometer community, we will no longer see any disagreements and no longer witness the experimenter's regress in action. I hope this will not happen, because it is healthier for the relationship between science and the public if the salience of conflict creates the conditions for the public's expectations of science to become more realistic. And science itself might be damaged if it becomes the norm that visible disagreement is a symptom of failure. I hope, therefore, that I will be in a position to record any disagreements within the future LSC even if the leaders manage to keep them within the project. I hope I will be able to document the extent to which the experimenter's regress still has a role. I promise, however, that my analysis will lag behind their analysis. The scientists will have had plenty of time to handle things according to their preferences before I discuss them in public.

More Similarities and Differences between Gravitational Wave Science and Other Sciences

Gravitational wave science is a high-risk, long-term project, and as such it does not represent the day-to-day world of most scientists. Does this make it possible to generalize from this project to the wider world of science? No, if the object is to understand the timbre of ordinary life in the laboratory; yes, if the object is to understand the role of science as our most reliable source of knowledge.

Neither physical nor social science is necessarily done best by looking at statistically representative cases.[5] For example, in some styles of high-energy physics, truth is established by finding the *best* image for showing what happens when particles are exposed to enormous forces. As argued in the introduction, the same may apply to social groups: it could be that only when the right kind of group is exposed to the right kind of extreme conditions that we see the processes we want to see. Gravitational wave detection science is exposed to extreme conditions. The science is very hard, the step in the step function is very large, the science is insecure and subject to external attack. These forces ensure that the work has to be vocally defended and that its deep assumptions and its internal and external rationales and political maneuvers are more easily seen. Again, the stresses of growth from small to big are made visible in a peculiarly dramatic way, because LIGO is making this change for the first time and over a short timescale. Features of a special science, once they are revealed as a result of the special circumstances, can be applied to other sciences, just as a bubble chamber photograph of a single particle subjected to huge forces can be generalized to the properties of all other similar particles.

On the other hand, perhaps LIGO's special features as an organization make it a poor model of institutional change and control, the subject of much of the latter part of the book. Certainly it would not be sensible to look to LIGO for a description of how a transition from small to big is managed smoothly. On the other hand, features of LIGO that put it under special strain as an organization may, once more, reveal the deeper forces at play. My claim is that the description of the transition presented here could be useful to the next scientific project that is to undergo a similar transition; LIGO shows the worst that can happen short of complete failure, and it may show how to avoid the pitfalls in the future.

Gravitational wave detection is also unusual in that international collaboration is more than a matter of financial exigency, it is integral to the science. Nevertheless, though the particular choices made may be affected by the specifically scientific imperative, the broad analysis of costs and benefits is unaffected. Indeed, it could be argued, as above, that the

---

5. Sociologists, with their tradition of large-scale surveys, have a misplaced confidence in representativeness.

scientific importance of collaboration highlights the dilemmas and choices in interesting ways.

There is nothing special about the "philosophical" part of the analysis of upper limits or the different meanings of upper limits for different audiences. Upper limits can show that an experiment is working, in a sense, even when it is finding nothing. Because of this, upper limits, even where they have little astrophysical significance, may have special importance to LIGO when it asks for funds ahead of the generation of positive results. This, however, does not affect the general upper-limit analysis.

Most of the scientific claims described in this book have been controversial. Every finding offered for publication in this field is offered in the context of a history of dispute, and we have seen the way this has affected refereeing and publishing. In terms of publication practices, gravitational wave science is, then, typical of controversial sciences, not of the ordinary run of routine sciences. On the other hand, if we want to understand the role of peer review and the like where difficult problems emerge from deep and divided interests—the appropriate terrain for any study of science that springs from a concern with the nature of knowledge—then controversial science is the right place to look.

Finally, gravitational wave science is quintessentially esoteric, whereas the majority of sciences encountered by the majority of us are much less obviously the business of core-group experts alone. This makes the relationship between this science and the general public a special one. Nevertheless, I argue that the "purity" and "hardness" of the case makes it ideal for exploring the philosophical and moral dilemmas of the relationship between experts, nonexperts, and quasi experts.

THE PLACE OF *GRAVITY'S SHADOW* IN THE UNFOLDING OF SOCIOLOGY OF SCIENTIFIC KNOWLEDGE

In the early part of the 1970s, I was lucky enough to contribute to what would become a kind of revolution in our understanding of the sciences. As we know, there is nothing new under the sun, and like all revolutions, this one had many precursors.[6] Nevertheless, during the 1970s it felt as though new ways of thinking about and analyzing science were being established. The predominant existing philosophies of science took their subject to be a kind of logic: there were fixed data points generated by experiments, and

6. The most noteworthy being Ludwik Fleck.

the job of philosophy of science was to show how they related to theories. The predominant sociology was concerned with discovering the set of
"norms" explaining how scientists ensured that their output was subject
to rigid quality control so that the data points would be reliable. What
happened in the 1970s was that analysts began to look much more closely
at what actually happened in the laboratories and networks of scientists
where the data points were being established. The analysts were informed
by new(ish) philosophical ideas that made us think about science as a kind
of language—one among many—rather than a kind of logic that stood outside and above human discourse. We found that the laboratory practice did
not support either the predominant philosophical or sociological models
of science; scientists did not act according to the models that the philosophers and early sociologists had constructed for them—scientists could not
act in the way the philosophers and sociologists thought they *must* act for
science to make sense.

As the revolution gained momentum new kinds of sociological methods
grew up. Instead of looking at science as an idea, or scientists as a uniform
group, case studies of the establishment of particular pieces of scientific
knowledge came to seem more and more attractive. The boundaries of a
field of study in what became known as the sociology of scientific knowledge (SSK) became defined by a scientific question rather than a question
about populations of scientists. The former subject of inquiry, "scientists,"
was replaced by another: "this or that discovery or nondiscovery."

In those days, each new paper or book was a clarion call: "Look at
science this new way—it's much more interesting and exciting!" One of
the features of the "new way" was that scientific conclusions were to be
explained, and this meant they could not figure as *explanations.* The analyst
had to ignore the scientific facts of the matter on pain of producing a
circular argument: "This truth came to be established because it was true."
Scientific truth had to drop out of the explanatory equation if the new way
was to make sense.

The success of this way of looking at science was sufficient to inspire
some quite radical claims, including some made by me. It turned out that
with enough ingenuity, you could explain any scientific truth as the outcome of unfolding social life within specific scientific communities; different communities of scientists, using the same theoretical and experimental
methods, could support different truths. In retrospect, it is not surprising
that the new way could support very radical worldviews because that is
how it is in science, as we have seen in the pages of this book—different
truths do survive side by side in science, at least in the short term. As the

revolution developed, new books and papers reflected the unfolding realization of what could be accomplished in this way. Authors would include
a concluding section that revealed their ever more radical credentials; we
learned how little it was possible to believe while leaving the world relatively unchanged. All this was vital in bringing about the revolution. We
had to shock ourselves out of a complacent state of mind and make ourselves and our colleagues look at science as a much richer activity than
had been thought.

Time has passed, and *Gravity's Shadow* is written as though this is simply
the right way to do things. Of course, the "Science Wars"—the unpleasant
war of words between a group of self-appointed spokespersons for science
and the social scientists—are not long over (if they are over), so perhaps a
book like this serves more of a purpose in establishing a potential change
than ought to be the case 30 years after the revolution was declared.[7] Or
perhaps the revolution is being seen through rose-tinted spectacles—it may
have taken place only among the analysts of science. In that case, *Gravity's
Shadow* may have more than a consolidating role to play.

THREE KINDS OF APPROACHES TOWARD HISTORY AND SOCIOLOGY OF SCIENCE: TECHNICAL HISTORY, THE MIDDLE RANGE, AND CULTURAL STUDIES AND THE LIKE

Gravity's Shadow represents a positive choice of a position in social studies
of science: methodological relativism. As we have seen, methodological relativism involves deliberately averting the gaze from scientific arguments so
as to investigate the social relations of the science more assiduously. This
involves no claims that are philosophical per se. The position, in a sense,
takes the middle ground, but it is not a compromise. It has been chosen
not *because* it takes the middle ground but because it does the right kind
of work.

In chapter 42 we looked at the scientific judgments that have to be
made *by the analyst* when carrying out a study such as this in spite of the
principle of methodological relativism; the gaze cannot always be averted
from the science. In this study, the theory of inspiraling binary neutron
stars and the idea of interferometry are treated as unproblematic; they are
a fixed background—the "scenery" against which the action of methodological relativism is played out. What is to count as scenery and what is to

---

7. I will discuss the science wars in greater detail below. See also Labinger and Collins 2001.

count as action is often up to the analyst. I received two unsolicited papers supposedly on gravitational wave science, and I treated them as scenery (in this case belonging in the "wings" of gravitational wave science along with the letters in "green ink" that I receive).

These choices are methodological, not logical or philosophical. As argued above, if I were doing a more radical sociology of science, I could choose to treat interferometry or the binary neutron stars as a subject for analysis—as action, not scenery. I could do the same with the unsolicited papers, treating them as within the boundary of my study; I would then examine the process through which they were excluded from gravitational wave physics by physicists. If my intention were to do something still more radical—let us say my topic was something like "the meaning of science in the West"—even the process by which "green ink" letters were rejected could be subject to analysis. As it happens, in this study I lag only a little way behind the scientists in their boundary-defining work. In a more radical study, the analyst would be further behind, and the boundary would be broader.

We can use this idea to review the different approaches toward the social analysis of science that we find today. Note that the picture has much in common with the choices that scientists themselves make, as can be seen by thinking back to the idea of the dam of scientific assumptions that I used as a metaphor in chapter 5. Recall that Joe Weber was described as removing stones from the dam in order to stop the waters of scientific impossibility from closing over his findings. To shift between our two metaphors, where his critics wanted to treat a range of assumptions as part of the scenery, Weber wanted them to remain part of the action. Looking in more detail, we noted that there was a choice about the depth or profundity of the stones that could be removed. Excavate deep into the dam wall and science's reservoir would empty, allowing weeds to flourish on the lakebed of science—everything would be action, nothing would be still.

The methodological choice that faces social analysts of science is like Weber's choice. How much should be still, how much should be action? The difference is that the analyst is not trying to effect a permanent remodeling of the dam, merely a temporary shifting of stones so as better to examine the process by which the dam is assembled. In *Gravity's Shadow,* the anchorings of stones near the top of the dam have been examined; other analysts might choose to go deeper. The sociologist Malcolm Ashmore has done us all a favor by burrowing more deeply into the dam and showing what such an exercise looks like. I suspect that he would find it natural to use the "green ink" letters as a sociological probe, indicating the boundaries that we erect around the very notion of science in our society. My argument with Ashmore, insofar as I have one, turns not upon his logic but his choice of problem; it is a matter of sociological strategy.

Perhaps one can use this approach to categorize social studies of science more generally. For example, the narrowest boundaries with the most scenery and the least action are set by what we might call "technical historians of science," a term I learned from Jed Buchwald. A technical historian of science reproduces the thought of scientists, following their path through the world and closing down the ramifying tree of possibilities in the much the same way as the scientists closed them down. A feature of this kind of history is that it needs as much technical expertise as the original science. (Of course, the original scientists also had to set the new questions and find the new answers.) Another feature of this kind of analysis is that it is likely to find approval among scientists. Scientists will recognize and feel at home with the style of reasoning, because it is very close to their own. A consequence is that this kind of study is not likely to reveal science (as opposed to history) in an unusual or surprising light, because this kind of study is very close to science itself. It also follows that this kind of account can be as esoteric as the science itself and tends to have a small and specialized audience of readers.[8]

I move a little further back from this intimate embrace with science in the current work. Studies like this, which have wider boundaries; have less scenery and more action than technical history. Consequently, these studies depend less on technical understanding. I have argued that interactional expertise is sufficient for these kinds of analyses. Almost as a corollary, scientists are not likely to feel immediately comfortable with this work for all the reasons outlined in the previous chapter. Here the analyst reaches toward a kind of reason or cause for what comes to count as scientific truth which is somewhat different from that found in the world of the scientist. In the case of studies like this one, it is hard work to get scientists to appreciate the strange perspective. It follows, however, that valuably counter-commonsensical results are more easily generated. There is also the possibility that this perspective can engage more readily with science and especially with science policy. Another consequence of the lower level of expertise required is that the work can be accessible by a wider readership than in the case of technical history, though some scientific literacy is still needed.

Still further back from the scientific frontier are academic analyses, which can be done in the absence of even specialist interactional expertise.

8. It can, of course, be popularized, and gain an enormous audience, but I am talking about academic studies.

Typical are analyses that treat science as a form of culture, often taking their lead from other kinds of cultural studies. Semiotics, the study of signs, applies equally well to works of science and works of art; under this approach, science is just discourse. The most well-known exemplars of this approach are Bruno Latour and Michel Callon. By its nature, this approach requires still less in the way of scientific understanding from its practitioners or its consumers, and it can attract an audience across the academic disciplines. Callon and Latour have succeeded brilliantly in co-opting an audience drawn from humanities in general. Alan Sokal's famous hoax was executed on a journal that publishes within this tradition, the aptly titled *Social Text*.

Thus within this schema, in terms of proximity to the technicalities of science, this book is behind the technical historians but in front of the cultural analysts. I have suggested that technical historians need contributory expertise in the sciences they study and methodological relativists need only interactional expertise, while semioticians and cultural studies experts need little or no scientific expertise.[9]

The middle one of these three possibilities has been chosen here because it can provide new perspectives on science, but perspectives that are not so radical that they are unlikely to engage with the world of scientists. To use a scientific metaphor, it is a matter of impedance matching, or the general principle of which impedance matching is an example: there must be some relationship between one thing and another in substance and dimension if one is to affect the other. Radical philosophies open our eyes to new problems for analysis, but that cannot be the end of the matter. Consider: one would never approach a scientist at the bench and say, "The trouble with your conclusions lies in the problem of induction." Too abstract a philosophical critique either destroys the subject in its entirety or leaves it untouched. This is not to say that something like the problem of induction cannot be the way into a new set of inquiries, but the art is to find the set of inquiries that engages with the world; the philosophical problem, of itself, will not do so. Methodological relativism, I am arguing, as opposed to less radical or more radical approaches, has the right kind of substance and scale to make for a particular kind of interesting engagement with science.[10]

9. See Latour and Woolgar 1979, where scientific ignorance is proclaimed as a virtue. Which is not to say that there are not cultural analysts of science who are more scientifically literate than I, it is just that such literacy is not integral to the exercise; the same applies to the other levels. We are talking of what level of expertise is necessary, not what is found in practice. Furthermore, most analysts have a tendency to display as much scientific skill as they can muster in order to fend off criticism, even when it is unnecessary for their type of analysis.

10. See also Collins and Yearley 1992.

Many other analysts of science who would not describe themselves as methodological relativists, for example those associated with the
Edinburgh "Strong Programme" and even some "realists," are effectively
methodological relativists in their practice and fall into the same middle
way as I am advocating in spite of themselves. Indeed, my own early
work fell under this description even when I thought I was a much more
radical kind of philosophical relativist. If I look back at my own earlier
publications, I find there is no need to change anything in the light of
my switch from philosophical to methodological relativism except for a
few rhetorical flourishes. This is because the message of methodological
relativism is simply to concentrate on social causes, and this is what all
good social history and sociology of science already does—so long as the
underlying philosophy is not such as to discourage such an approach.

This should be unsurprising, since methodological relativism, being a
methodology, ought to be compatible with many philosophical positions.[11]
To go to the other end of the spectrum, consider Martin Rudwick's *Great
Devonian Controversy*, a historical study which attempts to explore the detailed working-out of dispute with a view to establishing a realist viewpoint.
The only thing the methodological relativist would want to change is the
amount of attention granted to two marginal mavericks in the story.[12] This
aside, Rudwick, in trying to write as though contemporaneously, effectively
analyzes in the fashion of a methodological relativist (though he does not
use the term). If Rudwick had switched his philosophical claims around
by 180 degrees, it would have made little difference to his book. Or consider a couple of beautiful historical studies contributing to our modern
understanding of science, Gerald Holton's analysis of Millikan's oil-drop experiment and John Earman and Clark Glymour's study of Eddington's 1919
eclipse observations. Both of these were written by historians so violently
opposed to philosophical relativism that they verge on being classed as

11. In passing, let me say that philosophical relativism, in spite of the many arguments to the
contrary, can, I believe, be maintained if not proved. Likewise for realism. But even if philosophical
relativism could be demolished, methodological relativism would remain unaffected. For more on
this argument, see the contributions of Bricmont and Sokal and of Collins in Labinger and Collins
2001. For a recent philosophical discussion of these positions, see Bloor, unpublished. For symmetry,
see Bloor 1973, 1976. David Bloor has criticized methodological relativism even though it seems to
me to be a corollary of the principle of symmetry in the so-called strong program in the sociology
of knowledge. Bloor, a philosopher, considers the idea too "idealist." For this argument, see Barnes,
Bloor, and Henry 1996. Nevertheless, in face-to-face discussion Bloor and I have established that,
though we may disagree on philosophy, the consequences for the practice of studies of science of
our respective positions are almost identical. Bloor and I also agree that our positions are very
similar when compared with the more radical forms of analysis that are sometimes given the label
"postmodernism."

12. Rudwick 1985. In the 1980s, Trevor Pinch and I wrote reviews of his book along these lines.

"science warriors."[13] Yet their integrity as historians still results in studies informed by what is, in effect, methodological relativism (I am sure they would reject the term); they show the way the scientists, as one might say, "constructed" their results out of ambivalent data and how the social context of the times resulted in the acceptance of the findings. They are among the finest studies of this kind.

SCIENCE WARS

Gravitational waves cause ripples in space-time that in turn cause ripples in social space-time that, under the right circumstances, cause us to believe there are gravitational waves—or the other way around. I am trying to argue that which way you see it has consequences for nothing other than the methodology of social analyses of science. But this is disingenuous. *Beliefs* about such things as the direction of causality are among the most powerful forces on the planet, because they are widely taken to be powerful. In this they are like beliefs in gods, transubstantiation, and witches. People die as a result of their expressed beliefs or nonbeliefs in such things even when their only physical correlate is the words in which they are expressed. Arthur Miller's play *The Crucible,* about the Salem witch hunts, stands for any such life-and-death struggle over things whose only consequences follow from the expression of beliefs in them. Here is a quotation from one of the witch hunter's speeches in the play:

> But you must understand, sir, that a person is either with this court or he be counted against it, there be no road between. This is a sharp time, now, a precise time—we live no longer in the dusky afternoon when evil mixed itself with good and befuddled the world. Now, by God's grace, the shining sun is up, and them that fear not light will surely praise it.[14]

Those who experienced the period of bitter argument between certain natural scientists and social analysts of science will understand Miller's play well. I myself have been made to stand in the dock, as it were, at the British Association for the *Advancement* (Lord help us) of Science while a Fellow of the Royal Society stabbed a finger in my face and demanded,

13. More than "verge" in the case of Holton. The works are Holton 1978; Earman and Glymour 1980.

14. Miller 1952, p. 84; from a speech of Deputy Governor Danforth.

"Are you or are you not a relativist?" We have to get away from this kind of thing and see that when social analyses are done with scientific integrity, they are simply a part of the rich world of systematic academic inquiry. What counts or what *ought* to count in such studies is not the philosophy but the methodological approach. This still allows many kinds of research, more or less radical, depending on how broadly one draws the boundaries around what counts as "science in the making."

Gravity's Shadow has adopted a position a little way behind the scientists: far enough to have troubled them when they found that their taken-for-granted world was not my taken-for-granted world, yet near enough, I hope, for the analysis to seem neither bizarre nor totally out of touch with their concerns. But if my respondents had been determined science warriors, they would have found excuse enough to eject me from their company. Their tolerance and, when this failed, their academic values, have enabled me to maintain my balance on the tightrope. It has been a stimulating walk, and I can only hope that they will not find the view from my precarious position entirely without interest.

CHAPTER 44

JOE WEBER: A PERSONAL AND METHODOLOGICAL NOTE

First Contact

It is the late fall of 1972. I am a graduate student embarking on my first visit to America since I was sixteen. I have written to all the scientists I want to interview, explaining who I am and what I am doing, but I have not asked them to write back. I have said that I will be in the USA in a couple of weeks and that I will contact them to arrange an interview when I arrive. My theory is that if I ask for a written reply to my letters, the easiest response is no, and it will be hard to retrieve the situation. I think I have a better chance if I speak on the telephone.

To this project I bring more than sociological experience; for example, a friend and I have run our own little business bagging horse, cow, and pig manure and selling it from door to door. We carry the bags on our shoulders, knocking on people's doors as we make our way down the street. The bags leak, and on a wet day we can clear a cafe or a pub in a couple of minutes. Customers' first reaction to a stinking salesman with a filthy product is not always good, but we make a lot of money. So I am used to giving people the opportunity to interact with me under less than ideal circumstances.

Joe Weber is the most important person for me. Though my Ph.D. will not stand or fall on talking to him, it is a close-run thing. Finally, I telephone him from Canada. He picks up the phone.

"Hello—Professor Weber?"

"This is he."

"My name's Harry Collins. I wrote to you a couple of weeks ago to explain that I would be coming to America and hoped to interview you in connection with my research project on gravity wave science. Can we fix up an appointment?"

"No, we cannot. You are asking for the most precious commodity in the world, Joe Weber's time. It's more precious than gold, more precious than rubies, more precious than diamonds. I don't see why I should waste it on you. If you were a physicist I would consider it, but you are not a physicist. The important thing is to do physics, and I have no time for anything else. This is a very important experiment, and I am fully committed to it."

"Professor Weber, I realize how important your experiment is, and that is why I have chosen to do research on it for my Ph.D.—because it is so important. I think that the experiment and the debate surrounding it should be described for a wider audience. I have already spoken to many of the scientists involved in the field, but, of course, you are the key to the whole project. It is you who are the central scientist in the area, and without your contribution my project will be very incomplete. In any case, I think your viewpoint should be properly represented in what I write, because I would not want to get a one-sided picture."

"Everything I have to say is already published. You can read what I have to say in the papers I have written. There is nothing I have to add that you cannot read about already."

"It may be true that from a physicist's point of view you have written down everything you want to say in your papers, but as a nonphysicist there are questions I want to ask that are not already answered in the papers. For example, one of the things I want to know about is how often, and how, you make contact with others in the field. This is the kind of question I am asking everyone else. I want to check that the answers they are giving me are accurate. Also, I want to ask you about the source of your ideas—whether you thought up this way of detecting gravity waves yourself, or whether you are building on the ideas of someone else. As you say, this may all be a waste of time, but I am certainly prepared to drive down from Quebec to Maryland to talk to you briefly—say, for half an hour—at my own risk. If you are right and there is nothing new to learn, all you will have wasted will be half an hour, but it will make an enormous difference to my project. Because I have to drive from Quebec it will take me a couple of days to get to Maryland, but I could come on any of the following days . . . at a time that will suit you. If we make an appointment for just half an hour, I would be quite happy and will not need to bother you again."

"Alright, if it is at your own risk. I think it will be a waste of time because I have nothing to say that is not in the publications, but if you want to do this at your own risk I will spare half an hour, but no more. And I will not talk about anything that is not a matter of science. We could meet up at 12:30 on Thursday, October 12, or later if it would suit you better. Since you are coming all this way, I will try to make it as easy for you as possible."

"Thanks, Professor Weber—that is really generous of you and will help my project enormously. I will meet you at your department in a couple of days, and I am looking forward to it."[1]

WEBER AT IRVINE

It is the spring of 1993. The publications that emerged from my Ph.D. research have been a great success, and I am a well-known academic in my field of sociology of scientific knowledge. I have interviewed Weber once again, in 1975; but since 1976, when I conducted interviews with the German gravitational group, I have not taken much interest in what

---

1. This dialogue is somewhere between a paraphrase and a reconstruction of the conversation—unfortunately, I did not tape-record it, but I made some notes in my notebook which record the "more precious than diamonds, rubies, etc." sentiment. My memory of the conversation's general form is quite clear, however. It was an important moment in my life: without Weber's cooperation, my whole project would look thin. I also remember the "diamonds, gold, rubies" statement quite clearly, because the colorful remark took me quite by surprise (I may have the order of the objects wrong). I have used my subsequent experience of Weber's speaking style to fill out some of the other details that I cannot remember with equal clarity. Thus, while this is far from a verbatim transcript, the content, the rhythm and order, and the style of the conversation are, I believe, pretty close to the conversation that took place on that day. What I do have recorded is a comment I made in 1975 to one of my interviewees who was an ex-colleague of Weber's. We were talking over the fact that Weber had become a much more mild-mannered and cooperative person in the last six months or so. I remarked that setting up my forthcoming interview with Weber had been a pleasure compared to the last time. I described the 1972 encounter thus: "I have engraved on my mind the phone call that I made to him in 1972 to ask him for an interview. And he said quite terrible things to me. I did actually see him in the end, and I was really quite frightened of talking to him."

has been happening in gravitational wave research. I have been publishing material based on the earlier work, but the focus of my new research has been elsewhere.

In the spring of 1993, I am teaching a semester at the University of California at San Diego. Somehow I know that Weber spends some of his time at the university's Irvine campus, which is not an unreasonable drive north from San Diego. It is eighteen years since I have had any contact with Weber, but I think it would be interesting to meet him again, just for "old times' sake." I am just curious to find out what happened. I telephone Irvine, and to my delight I discover him in his office. He seems mellower than I remember him, and he is quite happy with the idea of seeing me. I drive up to see him on Friday, May 21.

Weber has an office at Irvine because his second wife, Virginia Trimble, has a post there. To my surprise and relief, he meets me there in a friendly manner. Perhaps he is pleased that someone he last saw eighteen years ago should want to speak with him again. I soon discover that my initial assumption, that his old work would no longer be central to his life, is quite wrong. Weber is still deep into gravitational waves and seems more certain than ever that his ideas are not only correct but revolutionary. He has invented a new way to analyze the sensitivity of a resonant bar—its "cross section"—and now believes it to be a million times more sensitive than he thought it was when he did his much-criticized earlier work. Weber now has allies, as he relates.

> [1993] Amaldi asked the most brilliant theorist in Italy—Giuliano Preparata—to go over my cross-section calculation. He's professor of theoretical physics at University of Milan. And Preparata went over my analysis and told Amaldi he thought it was wrong. And he proceeded to publish a paper suggesting it was wrong. But about a month later he had second thoughts, and he did a completely independent analysis that, remarkably, got the same cross section that I had published in '84 and '86. And so, since then, he has, as it were, come over and become a supporter.

The Italian experimental team led by Guido Pizzella is also giving him support, though, to use Weber's term, they "oscillate" somewhat. These are exciting times, I am given to understand, and I have reentered the field just as a major revolution is about to take place in our understanding of the interaction of gravitational waves with matter. The big laser interferometers, I learn, are a waste of taxpayers' money and are essentially a confidence trick. The new cross-section calculation renders them obsolete. Yet Weber has a battle on his hands against powerful and entrenched forces.

> [1993] [O]f course, if the 1984 cross section is right, it will discourage the
> United States Congress from giving Caltech their $211,000,000.

The new theory invented by Weber and reformulated by Preparata works not only for gravitational waves but also for the detection of neutrinos. Neutrinos are very hard to detect and can pass through huge quantities of matter with no effect on their trajectory. Traditionally, to detect neutrinos, very large tanks of liquid are used in the hope that just a few tens of atoms in the tank will be transmuted by their passage. Weber can detect neutrinos, he believes, with crystals small enough to be held in the hand.

Weber has become older and smaller since I last saw him. He is still as lively as a cricket, but he had turned 74 two days earlier. As I recall, he wears a baggy navy suit, a white shirt, and a tie. His trousers are a little short, his shoes look like they are on the end of sticks, and this, with his bush of hair, adds to the impression of Einstein-like brilliance mingled with eccentricity.

The Irvine campus—which is made largely from concrete—is relieved with masses of red bougainvillea. It is a bright, hot, sunny day. We stroll across the campus toward the cafeteria to get our lunch. My prevailing memory is of Weber standing in his Chaplinesque suit in a concrete archway in the sun, surrounded by an archway of red flowers, holding in his hand a cylindrical crystal of silicon that he says can be used to detect neutrinos in place of those ridiculously massive underground tanks full of liquid. I thought I had a photograph of this scene, but it turns out to be an imaginary composite of two photographs that I actually did take: one is of Weber in his office holding the crystal; the other is of Weber in the archway with the bougainvillea, but not the crystal. And the shirt isn't white but striped, but my memory of the trousers is about right.

Weber and I spend the day together and have a great talk. He seems happy and relaxed; he is very proud of his second wife. He tells me all about his family—how his first wife died and how that side of the family, including his sons, are afflicted with a gene that makes them produce high levels of cholesterol and die young. He tells me how his second wife was much sought after, but she, though not Jewish herself, had chosen a Jewish man; he was lucky enough to be the one, though he was 24 years older than she.

I had gone to see Weber out of curiosity—just to revisit a time that had long passed. Yet I finished the day determined to follow up all that he had told me. The research project I would apply for—"The Life after Death of Scientific Ideas: Gravity Waves and Networks"—just fell into place. I'd apply for a grant that would last a year to enable me to explore what Weber had told me; to get to Italy and interview Preparata and Pizzella; and to talk with a few of the skeptics and find out what they made of the latest revolution in the making. I thought I would probably get the grant—it was a "natural." I would be able to complete the research in a year. I had no idea I would still be doing work on gravitational waves ten years later.

Weber's early work, and that of his critics, was the inspiration for my best-known journal paper and the source of a good proportion of my professional reputation. I had done lots of other things in the meantime, but now Weber's conversation with me was to give rise to another phase of my professional life.

THE LAST INTERVIEWS

The revolution that Weber promised in 1993 did not happen, but he kept up his spirits and applied for grant after grant. In the middle of a conversation about other things, he would often turn aside to assure me that he had no intention of committing suicide, and would provide me with an account of other scientists who had done so. His beautiful wife, he explained, was sustaining him.

But things were not going well, and Weber was getting older. He was no longer so friendly when I met him. Once I arrived at 9:15 a.m. when my appointment was for 10:00, and I bumped into him in the corridor while I was merely checking out the exact location of his Maryland office with a view to returning at the appointed time. He flared up at me—"You don't even know how to keep time." Then he shouted that if I was going to record anything, he was not going to say anything. I had to renegotiate the use of the tape recorder all over again. This anger was not directed at me alone—he was losing the sympathy of others of his erstwhile colleagues and acquaintances in the gravitational wave field because of the anger and scorn he would offer when they didn't agree with him.

And his conversation was becoming more predictable: he would not so much converse as reiterate his theories over and over. At that stage I was wanting to look at early documents and correspondence, but he would insist on my spending an hour or so listening to the theories which I already knew inside out before I could spend a few minutes with the documents. I think the cause of these frustrating encounters was that Weber could not understand how anyone who understood his theories could fail to be convinced by them. All academics are like that. If we do not believe that what we hold to be true is self-evident once it is properly explained, we would have no reason to believe it ourselves. But I could not persuade Weber that it was not my job to be convinced or not convinced—it was just my job to record the convictions of others.

During one of these interviews, it must have been the penultimate, Weber once more explained to me his theory of center-crystal instrumentation. He explained it with the aid of the blackboard in a corner of his extraordinarily cluttered Maryland office with just room for the two of us to sit.[2] Weber drew a diagram of a bar detector on the board with arrows to show the way the phonons bounced backward and forward, energizing the piezocrystals on each pass.

About six months later, I visited the Maryland office again. Again Weber insisted on explaining his theory. Again he got up to draw on the board. But, to his surprise, the diagram he needed was already there. It was the one he had drawn for me six months previously. Had I been the only visitor?

JOE WEBER IS DEAD, AND MY LIFE AS A SOCIOLOGIST IS EASIER

It's about 10 past 8 on a Saturday evening—February 3, 2001. I have to cook dinner in a short while, but I'm doing a bit of work on this section of the book in a spare half hour. I'm actually working through Weber's 1976 paper and setting out its contents. And I suddenly realize how easy this is, because Joe won't be reading it. I don't have to think about whether Joe will think I have given a fair description of the paper, or whether Joe will agree that I have all the technicalities exactly right, or whether Joe will actually want me to say some of the things I will say about it. Would I have said that he admitted his computer error "gracelessly" had he still been alive? I won't know whether Joe would have been hurt by my pointing out that this paper, in spite of the huge effort that went into it, was forgotten the instant it was published. I do know that my earlier work was attacked by both Weber and certain of his acquaintances, because they thought it

---

2. Incidentally, the only office that I have seen more cluttered belongs to Ron Drever. Is there a thesis there?

revealed a bias against him; the only "bias" I could see was that it didn't forthrightly endorse his ideas.

I can honestly say—and I hope this does not sound too mawkish—that I wish the work were not quite so easy as a result of Weber's death. He didn't like what I wrote about him because he could never see further than science and scientific method, and he thought sociology, or any other comment that took time away from the laboratory, was a waste of time. He was often distant and difficult at the beginning of an interview and always very formal in correspondence, but by the time I'd been with him an hour he would be cracking jokes and making wry comments about himself and his critics; so like many others, I couldn't help developing a strange kind of affection for the man. This was someone whose academic raison d'etre had been crushed to death but who refused to have his spirit broken—that kind of obstinacy is infectious.

In spite of the influence of Joe Weber on my life, when you add it all up I didn't spend very long in his company—perhaps 24 hours in all. For a significant proportion of that time, he was yelling at me; and for another, still larger proportion, he was going through formulaic repetitions of things I already knew. But it was his relentless enthusiasm for his ideas that encouraged me to look further into the gravitational wave business in the mid-1990s. Whenever I spoke with Weber, I felt I had a great sociology project on my hands. In 1993 I came away wanting to find out what was happening with him and his resonant-bar-building colleagues, not to study the interferometers. The interferometers became interesting only later.

But now that Joe Weber is dead, I also feel something else: it is all so much easier. For the sociologist working on a contemporary case, every word is judged by living, breathing people. I am, in my writing, remaking the world of those who made it in the first place. That means I have a bigger team of judges on my back than other writers. The usual team includes the sociologists and all the other readers whom I have to satisfy, my alter ego telling me that what I am writing is rubbish, and my funding agency. But my judges also include my respondents in two forms: those who are virtual and anticipated, and those who are really waiting out there, sometimes to give me a very hard time.

Actually, this is one of the reasons I like contemporary sociology as opposed to history; there is something satisfying, agonizing though it is at the time, in going through the fire and coming out reasonably intact. Joe Weber's fire has been put out, and though it is easier, I feel a little guilty about enjoying the cool breeze.

There was a fashion in French postmodernist literary criticism to say, "The author is dead." It means that once a novel or any other work of art has been written, it is passed into the hands of its interpreters and critics, the author's intentions being of no relevance. In contemporaneous sociology, the author is not dead, because the health of an extended field study turns on the willingness of the multiple authors (the respondents) to tolerate the interpretations of the interpreter. If integrity is to be preserved, this can mean that the interpreter sometimes has to change the authors' understanding of their world as well as his or her own understanding. But now, in the case of Joe Weber, the author is dead in the literal sense, and this means he is dead in the metaphorical sense, too.

JOE WEBER IS DEAD, AND MY LIFE AS A SOCIOLOGIST IS HARDER

It is February 19, 2001. I have spent a lot of time revisiting the whole first decade or so of the Weber business. The first time I visited, in the 1970s, I read such papers as I could get my hands on, but the bulk of my understanding and analysis came from interviews with scientists—the interviews from 1972 and 1975 that I have quoted extensively in this book. My final visit to this episode is much more a matter of reading published papers, conference proceedings, letters, and any other documents I have gathered over the years. Three things have changed between first and last visit.

The Times Have Changed

The great social change that I am trying to describe and document has taken place. When I was writing in the early 1970s, it was still possible that there were lots of gravitational waves that had just been momentously discovered by Weber; that was what we were still in the process of finding out. One could talk of Weber's gravitational waves without being ridiculed or feeling silly—without feeling that one had committed some terrible scientific faux pas. Toward the end of the period, the tide of scientific opinion was turning quickly against Weber, but it hadn't yet crystallized. Weber's failure would be finally realized in the utter lack of impact that his latest papers would have on his fellow scientists, but utter lack of impact, like any other null result, takes time to reveal itself; it is as hard to reveal a null in terms of social impact as to prove a null in terms of the existence of gravitational waves—indeed, they are the same process.

Joe Weber Is Dead

Upholding a "symmetrical" view as a sociologist of science is not easy. By 1975, nearly every scientist I spoke with would tell me how wrong Joe Weber was, even if they weren't quite ready to commit themselves in print. To maintain my "estrangement" from this growing consensus was becoming increasingly hard, but I had the great advantage of being able to talk with Joe Weber. Joe Weber was a great talker. He never expressed anything but unqualified confidence in his results when talking with me and he was very good at pulling out one argument after another, like rabbits out of a hat, to show both why he was right and why critics were wrong. So, if I was struggling to maintain my estrangement and symmetrical stance when I had just interviewed a bunch of Weber's critics, I came away from a chat with Joe Weber with a tape recorder full of confidence. When I remembered our conversation, or if I listened to that tape, it was as though I had taken a symmetry pill or a sip of estrangement medicine. As I have described, when I started my second episode of serious fieldwork in the mid-1990s, I started it as a result of getting a big dose of symmetry from Weber himself, and discovering that there were others, such as Guido Pizzella and Giuliano Preparata, who could also fill my prescription for symmetry medicine. But now both Weber and Preparata are dead, so my supply of medicine is drying up, and maintaining symmetry is getting harder. In 2001, there are no Weber-type fluxes of gravitational waves to be found in social space-time. If I want to enjoy Weber-type gravitational waves, I have to resurrect them all by myself.[3]

What Is Left Are Writings Refracted through a Dense Medium

This very day I have attended a lecture about how we perceive. The lecturer exposed us to a bit of Led Zeppelin's music played backward. We were supposed to hear the secret message that the band were said to be putting out to the true adherents of their Satanic cult. All one could hear was wavery sound with the word *Satan* enunciated twice. Then the lecturer projected onto the screen the full text of the words that were said to be audible in the backward version and played the sounds again; having seen the words, one could hear them far more readily than one might expect.

---

3. This would change once more in 2002 with the new Frascati Group claims.

When I now read, as I have just read, the correspondence between Joe Weber, Dick Garwin, and others, such as Dave Douglass, I read it knowing how things turned out. I read this correspondence as through a template that allows me to focus on where Weber went wrong and hides all those places where he went right. The pattern of the template is Weber desperately struggling to hide his mistakes and shift his position; he wriggles and struggles to maintain his foothold on the island. Knowing that he's going to drown, I see his feet and hands grasping and slipping where once I saw them clinging and climbing. The difference between grasping and slipping and clinging and climbing is almost nothing—it is just what you are primed to see.

And reading the published papers is the same. I know that the arguments Weber puts forward are going to be ignored. Knowing that, they read as futile strugglings rather than decisive refutations of his critics. "The author is dead" and the meaning of those papers is now entirely in the hands of his readers. The dominating source of my interpretation is the interpretation of all those living readers who now surround me.

After I had written this, I stumbled across the following comments at the end of my transcript of my 1975 interview with Weber (chapter 8). After the interview, as I see from my transcript, I went to the cafeteria in the University of Maryland to reflect. My notebook reads:

> I now make some comments onto my tape recorder while sitting in the Maryland cafeteria.
>
> I note the totally different impression I have after talking to Weber as compared to talking to Douglass, Levine, and Garwin.

Now I switch from making notes on paper to speaking into the recorder. I transcribe this talk later:

> After talking to Weber, I come away with the impression that there is a real, genuine, Northeast Coast conspiracy between Douglass, Garwin, and so on, to shoot Weber down, and they have succeeded to the extent of *Physical Review* rejecting his last paper and to the extent of his whole project nearly collapsing from lack of funds.
>
> Weber himself is totally confident that nature will speak for him. . . . His latest results demonstrate the effect with four and one-half standard deviations of confidence, and the reason he doesn't think anybody else has got it is because they are using the wrong algorithm or one or two other

things. And the German groups only used the right algorithm for sixteen days of their work.[4]

Here is the point. I am bouncing around between different peoples' "constructions of reality." Given the way history has turned out, it is almost impossible for a scientist, reading the little bit of transcript above, to think other than "But Weber was wrong. If Collins can believe that what he said has any value, it is because he is so determined to be open-minded that his brains have fallen out." And quite rightly, too, now that we have the science of 2003.

Worse, even I am struggling to let it make sense to me. I am devastated by Garwin and Levine's point made in chapter 9 that when Weber lowers his threshold, he gets huge numbers—800—of zero-delay events and a high level of statistical significance. I am also deeply troubled by the "second law of thermodynamics" argument in chapter 11, note 2. For me, these arguments are "killers." But what I can't now do is go back to Weber and ask him for his interpretations. And I cannot invent what he would say—he knew his science better than I ever will, both because he lived it and because he was a physicist and I am a sociologist. That's the sociological problem about Joe Weber no longer being here.

Postscript

I didn't spend long in Weber's company, and what I had at stake was something different from what the physicists had at stake. That may be one of the reasons that others of his erstwhile colleagues in the business lost patience and came to see him as a danger to them and to science. Weber absolutely refused to give up his ideas and become an honored "elder statesman." David Blair explained it to me once.

> [1996] If Weber had done the smart thing for him—politically, socially, and everything else—he would have—but maybe not the true, not really the correct thing to have done given his beliefs—would have been to say, "Well, yes, I admit I was wrong, and all this," and then he would be the venerable leader of the field. And, er, we would all acknowledge "Uncle Joe" in every lecture. And people have tried and tried and tried to get him to accept that role.

4. Weber says that *Physical Review* rejected his last paper. I presume this is reflected in the long delay between submission and publication of what became Lee et al. 1976.

> I was actually—I was hosting the conference in 1988 in Perth when Preparata attacked Weber, which was before he came up with this theory that agreed with Weber, and I was completely intimidated by Weber after that. I had about—I was stood up against the wall for about two hours and shouted at and I've experienced many episodes of that since then, and before then; but I was organizer of that conference and I arranged a special presentation to Weber as the, sort of, founder of the field to try to get him to, you know, sort of shift his stance. Unfortunately, the presentation was er—this presentation occurred like thirty minutes after he'd finished his standup screaming at me for allowing Preparata to speak without informing him so he could prepare a defense or something.

But Blair, along with others who knew him well, still seemed to have the same kind of feelings as I; this was a man who was immensely frustrating to deal with but whose positive features of character came through in the end.

For the new generation of scientists—those who know the man only through his work—he was either someone who didn't know how to do statistics and didn't know that he didn't know, or someone who willfully distorted the record to support his ideas. The more reflexively aware of these critics understand the way perceptions of a person take on a life of their own once the world has been reshaped. As one of my respondents, quoting a famous literary spat, wrote to me recently,

> [2002] I can't help it, when I read Weber every word seems to cry "Crank!", including "and" and "the."[5]

---

5. Originally Mary McCarthy speaking of Lillian Hellman.

CODA

This coda reports developments since the main manuscript was finished in the summer of 2003. One such development is that, all being well, a permanent paper copy of the material from WEB-QUOTE will be deposited in the archives of the Niels Bohr Library of the American Institute of Physics, College Park, Maryland, USA.

STOP PRESS, MARCH 2004

The Frascati Results

The Frascati group have continued to support their maverick result published in *Classical and Quantum Gravity* (*CQG*). While admitting it did no more than indicate something interesting—"watch this space"—they insisted on their right to publish. Sam Finn, it will be recalled, had written a paper, published in *CQG* in early 2003, which claimed that their result amounted to nothing, since the odds against its having arisen due to chance were only about 1 in 3. Finn's calculation depended on the assumption that the particular boxes in the histogram singled out for examination by the Frascati group had been chosen only after their data were collected rather than selected in advance. The Frascati group, led by Pia Astone, began loosening a new stone in the dam of assumptions that supported the orthodox view, the so-called classical approach to statistics. Astone, citing papers and a new book by D'Agostini, declared that classical statistics were deeply flawed; a Bayesian approach was necessary; and one could make no sense of a physics result unless one was comparing it with a model.[1] The model could be selected at any time—the crucial feature was that it was selected *independently* of the analysis; it did not have to be selected *prior to* the analysis.[2]

At the GWIC meeting in Tirrennia, near Pisa, in July 2003, the strains showed in the response of the Rome group to suggestions emerging from the GWIC leadership that there should be worldwide agreements about statistical methods and the proprieties of publishing. The Rome group felt it was being put into a straitjacket and voted against all the proposals. Nevertheless, the proposals passed on a show of hands. Since the agreements were, in the nature of things, voluntary, there was no diplomatic crisis.

Just to illustrate that genuine disagreement exists, at the Amaldi conference which immediately preceded the GWIC meeting a young theorist who had visited the Frascati group told me that he was excited about their result, and that he was sure the Frascati group had merely stumbled upon the galactic disk as the potential source, not invented it to fit the data. He also said that the direction of the galactic disk made sense astrophysically, since the disk was well supplied with neutron stars that could produce high-frequency gravitational waves when they suffered "earthquakes" and the like. We awaited the next raft of Frascati results, which were promised for the end of the year, but, disappointingly, have not emerged in time for inclusion in this coda. I will report further confirmed developments on the Web site www.cardiff.ac.uk/socsi/gravwave.

International Alliances

By July of 2003 the rate of international alliance formation and dissolution had accelerated. The catalyst for some of this movement was the Frascati group. What with NIOBE closing down, and the Frascati group acting independently for all practical purposes, this left only AURIGA from Padua and

---

1. See, for example, D'Agostini, Giulio, 2003, *Bayesian Reasoning in Data Analysis: A Critical Introduction* (Singapore: World Scientific). D'Agostini is, as it happens, Astone's husband.

2. Personal communication and see Astone, P., G. D'Agostini, and S. D'Antonio, 2003, "Bayesian Model Comparison Applied to the Explorer-Nautilus Coincidence 2001 Data," *Classical and Quantum Gravity* 20, no. 17: S769–S784 (September 7, 2003).

the Louisiana bar, ALLEGRO, as contributing members of IGEC, the bar alliance. AURIGA had decided to share data with the LSC after the fashion of Louisiana's ALLEGRO, so the remaining four bars had split clearly into two groups—those working with the interferometers and those working independently. The difference in evidential cultures had been realized in stark political form.[3]

As for the interferometers, TAMA, the Japanese group, had signed a data analysis agreement with LIGO, but had also signed an agreement with the Frascati group, thereby opening the door to strained relationships ("my enemy's enemy is my friend"). GEO was ever more tightly integrated with LIGO, though even here there was a new destabilizing factor. The Frascati group let it be known that Bernard Schutz, GEO's overall boss, had formed an alliance with them for the purpose of statistical analysis; he would act as their statistical consultant. The international web of cooperation and competition was also becoming ever more complex as a result of personal friendships and other relationships (including at least two marriages) that develop in an international community in which members meet one another often. Personal alliances increasingly cut across the official networks represented by signed agreements, creating unexpected areas of moral integration that can cause embarrassment to groups for whom it would be easier to stay at arm's length.

In July 2003, VIRGO's long-running dilemma about which way to jump in the matter of cooperation was becoming still more pressing. As the interferometers, including VIRGO, come on line, it is becoming clear how many people are needed to run an interferometer and analyze the stream of data.[4] LIGO solves this problem with its still-growing LSC, but VIRGO cannot do the same; it is a loose collaboration of laboratories from different nations with no strong central leadership that can enforce rules associated with a "flat" organization—the individuals who have built VIRGO are not always prepared to sacrifice their property rights in the data, and newcomers can be unwelcome. In addition, there is not enough money in Europe to fund the satellite labs of an LSC-like organization: VIRGO was funded to be built

---

3. Some initial overtures made by Frascati to LIGO had not borne fruit.

4. VIRGO, incidentally, was not startlingly more rapid in reaching its promised sensitivity as a result of its "no-prototype" policy, but it might be that the lessons that could have been learned from the use of the center station as a prototype had not been learned assiduously enough; for example, I understand the center station mode cleaner never worked well enough to allow everything else to be tested as well as it might have been.

GEO, in the meantime, became the first large interferometer to prove the signal recycling principle.

and to find gravitational waves, and the funders want results, not R&D. At the time of writing, VIRGO-LIGO relationships are markedly strengthening.

Sensitivity

As for the argument about sensitivity, it is now easier to follow, as LIGO has begun to give figures for the distance at which it can see an inspiraling binary neutron star event (with stars of 1.4 solar masses). A range of about 20 megaparsecs will be achieved if the noise curve for the two 4-kilometer detectors (the spiky line on figure 41.1) reaches down to the Science Requirements Document (SRD) curve (the solid line) along a good stretch of its length (with the low frequency end being the least important). As the positions of those lines draw together, the range of a single detector will be approaching 14 megaparsecs, giving 20 megaparsecs for two 4-kilometer interferometers working in coincidence. As described, according to earlier plans, this level was to have been reached by the end of 2002, but the latest ambition is to reach it by the end of 2004, so that a full year's worth of data (at 50% running time) can be accumulated by the end of 2006. The counterclaim that I badgered out of Bob Spero was that LIGO would still be at only one-third of design sensitivity (7 megaparsecs) by the end of 2005. Spero's claim translates into a single-detector sensitivity of about 4.5 megaparsecs by the end of 2005. At the beginning of 2004, the Hanford 4-kilometer by itself has seen as far as 6.5 megaparsecs from time to time, so it looks as though Spero was wrong. We wait to see if LIGO will hit its 2004 target, though the team is confident.

There are two important things to stress about the way I have put this. First, Spero was very reluctant to produce a number to fit my bet, and he should not be held responsible for my framing of this argument. Second, the LIGO management consider that my framing trivializes the science. This is my responsibility, and I may not have been wise to take what I thought to be a revered tradition in physics (see, for example, the stream of bets made by Kip Thorne, including his published bet with Jerry Ostriker that LIGO would detect gravitational waves before the year 2000 [chapter 27]) and try to apply it within a sociological analysis. In chapter 42 above, we find the following:

> It is the job of science to turn disagreement into agreement, or at least a decision—that is what scientists do. It is the job of sociology, on the other hand, to hold the boundary open or reopen it once it has been closed. It is the job of sociology, then, to make salient even nearly forgotten

disagreements—this is what concentrating on the process rather than the outcome implies, and this threatens closure. In our story we saw how closure was managed by the community for the cases of Joe Weber, Giuliano Preparata, and in a less dramatic form, Guido Pizzella and the 40-meter team who departed from the LIGO project. Yet here they are again, alive and kicking in the pages of this book long after the scientists who are making gravitational wave science have forgotten about them. For instance, the 40-meter team had left the LIGO project by the mid-1990s, yet they take a starring role in chapter 36, even being given, in chapter 41, a public platform to express their view about the future of LIGO!

I wrote this passage in a self-congratulatory way, thinking that I had engendered in my respondents an understanding of the nature of my project that would enable them to appreciate why I was providing a platform for defeated scientists, even though it reopened old tensions. It turned out that in at least some cases, I had done a much worse job than I imagined. When I explained that the disagreement between the 40-meter group and LIGO would be revisited in this Coda and on my Web site in the form of a runoff between the sensitivity predictions of Spero and the LIGO team, I was violently criticized for turning serious science into a game show. It became clear, once more, that the social space-time through which gravitational waves travel—the second set of ripples—is generated by forces just as strong in their own way as those that give rise to the first. I believe that my analysis, quoted above, correctly identifies the causes; but my insensitivity to their centrality in the scientific culture I was analyzing, and my failure to bridge the gap between the scientific and sociological worldview as thoroughly as I had imagined, is nothing to be proud of. Fortunately, the damage has been largely repaired, leaving only a small area of agreement to disagree.[5]

An Extra Source of Noise

What is the sociological analysis that I (wisely or unwisely) tried to sharpen and focus by introducing a bet? It was the debate about whether the fundamental noise sources in LIGO were as thoroughly understood as had been claimed. Members of the 40-meter team believed that LIGO had left the prototyping stage too early, before everything was sufficiently understood; leaders of the LIGO team believed that there was nothing left to discover aside from "technical noise," and that this was better and more efficiently

5. I should add that the disagreement was with only a couple of respondents.

beaten down on the full-scale device, since the prototype's technical-noise profile was likely to be too different from that of the final device. I think I can say that all parties would agree that if there was nothing to find but mundane technical noise, then the strategy of the LIGO leadership would have been correct. I have also explained (above) that in my view the LIGO leadership's strategy of building big when they could was the only option that made political sense. Still, irrespective of the political strategy, we are left with the sociological/philosophical question of whether LIGO, at the time it was built, still held fundamental surprises—was it a mature science? In the nature of things, I am in a small to vanishing minority when I consider this possibility, which opens me up to the accusation that I am trying *to do* science from a position of incompetence and exposes further my use of the views of the 40-meter team—now long absent from the heart of the debate—as my license for giving life to a viewpoint long absent from the "official" position.

Interestingly, there *has* been a surprise in terms of sources of noise, and it has to do with the way light bounces between mirrors at the heart of the devices. A scientist working in the field of microscopy, with no previous institutional affiliation with LIGO, read the literature on interferometers and concluded that light in LIGO's arms could twist the mirrors on their axes if the beam is off-center, and the subsequent feedback could cause an instability to grow.[6] This finding has required an adjustment to the feedback algorithm for the mirrors in LIGO I and will mean some rethinking of the design of Advanced LIGO.[7]

Whether this is just something "unexpected in an unexpected way" or something "fundamental" is a matter of interpretation, with the tide flowing overwhelmingly in the nonfundamental direction (for example, even members of the 40-meter team were unwilling to count this new source of noise as a vindication for their view).[8]

That an outsider can discover, and be allowed to discover, an important new noise at the heart of an instrument of this size and importance is sociologically extraordinary. It also reflects very well on the new management's willingness to open up the LIGO project. Anyone talented can offer

6. Sidles, John, and Daniel Sigg, "Optical Torques in Suspended Fabry-Perot Interferometers." This is an electronic preprint that may be found by entering title or authors in a search engine. My understanding is that it is also to be published in *Physical Review A*.

7. My thanks to David Shoemaker for steering me through these technicalities.

8. Furthermore, it seems the problem had been looked at many years before by an Italian team, but an incorrect conclusion had been drawn—for a reference and discussion, see the Sidles and Sigg paper mentioned in footnote 6.

a contribution, and thus LIGO can harvest technological experience from a much richer field than under any of the previous management regimes. Indeed, the discoverers of the new noise, the flamboyant John Sidles and his University of Washington team, have been made members of the LSC.

We await LIGO's final accomplishment of its sensitivity target to be sure that there really are no other big surprises; many scientists agree that beating down the last factor of 2 in sensitivity will be the crucial test, though others believe they can predict and accomplish all the adjustments necessary for reaching the SRD curve and know how to exceed that target. Still others say that the original target was set so conservatively that the achievement is no great surprise, but that is an argument that is indefeasible.

Evidential Individualism in Practice

At the time of writing, many groups and individuals within the LSC are analyzing data. Several of the initial analyses could have been interpreted as indicating a positive result, but any such suggestion was soon withdrawn: there was too little evidence; the inferences were astrophysically unreasonable; or they had various other problems and inconsistencies. Of more concern was what to do with the stream of upper limits that were being generated. It was turning out that data analysis was an immensely time- and labor-consuming task that was going much slower than the data collection. In November 2003, when the third science run, "S3," was about to commence, only one of the papers describing the results of S1 had been accepted for publication. The deeper problem in the winter of 2003 was the S2 results. These were discussed at length at the November LSC meeting in Hanford, but the anxiety was over whether there would be anything ready to report to the GWDAW meeting scheduled for mid-December.

The LSC—at least, the data analysis sections of it—is a closed meeting (Collins is allowed in), but GWDAW is an open meeting. The problem was that the S2 results discussed at the LSC meeting had not yet passed through the stringent self-review procedures designed to ensure that no claim made in the public domain has to be withdrawn later—the practical face of LIGO's strict evidential individualism. Now, however, to some delegates the procedure seemed too clumsy and slow to make sense in light of the timetable of meetings. GWDAW is an annual gravitational wave data analysis workshop, and it seemed wrong that LIGO's latest data should be withheld from such a meeting. As one of the delegates put it to widespread, self-deprecatory chuckles, LIGO people had complained about the Frascati group publishing their data without providing an opportunity for wide consultation in the community, yet they themselves were about to miss an opportunity for just such a wider discussion. To put the problem more analytically, the more rigid the evidential individualism, the longer are data withheld from public scrutiny and, therefore, public criticism.

What we are seeing here is the complexity of the territory between extreme evidential collectivism and extreme evidential individualism. What exactly counts as a release of data to the wider world? Where does, say, ArXiv, the electronic preprint exchange, stand on the evidential culture scale? This is a nice tension that we hope to watch unfold when putative positive data come in. We could be said to be engaged here in exploring the detailed structure of what I called "seeing's space-time" in the introduction.

What Next?

Before this book is published, LIGO Livingston should have long been back on air with its improved seismic isolation in place; not long after publication the spiky lines of the updated figure 41.1s for the two big interferometers should have been brought nearly into coincidence with the SRD represented by the solid lines, and LIGO should be able to see inspiraling binary neutron stars at a distance of 20 megaparsecs. Something more, presumably, will have happened with the Frascati results: new data will back them up, or not back them up; if new data do back them up, there will or will not be a continuing argument about the propriety of the way they were generated. VIRGO should be gathering data, and perhaps it will be fed into the LIGO-GEO data analysis scheme. With luck, LIGO will be the sweetheart of a good proportion of the scientific world, and Advanced LIGO will be funded.

The Next Generation

Only when compiling the index for this book did I realize that the vast bulk of the hardworking scientists from just below the top rank to those at the bottom, who actually make the large interferometers function, are mentioned only rarely if at all. I'm sorry about this—it is a misrepresentation of what actually takes place in a big scientific project. Furthermore, the new generation will have to navigate their way through a society much less ready than that of their forebears to grant, without reflection, that science is the quintessence of truth. It would be nice if this book were to make the new world a bit easier to understand and to cope with, because without the kind of heroic science described in these pages, our lives would be impoverished in every sense.

LIGO, by the way, has rediscovered Weber as a pioneer; one of his original bars is exhibited in the foyer of the Hanford installation along with a series of explanatory posters on the wall. And a computer cluster at Cardiff University dedicated to gravitational wave data analysis has been named for him.

STOP PRESS AGAIN, APRIL 2004

Joe Weber and Giuliano Preparata developed the quantum coherence ideas that gave rise to the enhanced cross sections for gravitational waves and neutrinos in crystals. In chapter 20, footnote 23 (p. 347), I argued that these ideas might survive, but only as theories, and I gave some examples of recent papers that continued the tradition. It seems, however, that the new cross section is not entirely dead experimentally. An article in *New Scientist* explains that the Milan-based tire company, Pirelli, is investing $150,000 per year in an experimental project to test the neutrino theory, with a view to using neutrinos to communicate through the Earth.[9] The scientists involved learned of the work from Preparata, who was at the University of Milan and also worked with Pirelli on cold fusion; they have salvaged some of Weber's old experimental apparatus. This fits our model of Pascalian funding—it would be hard to imagine the work being funded from public sources.

9. See Durrani, Martin, 2004, "From Tyres to Neutrinos," *New Scientist* 17 (April): 36–39.

APPENDICES

What is small?

What is "a strain of 10^{-16}," or 10^{-21}? The strain of 10^{-16} reported by Weber in *Physics Today* is a change in length of less than the diameter of an atomic nucleus in an aluminum bar a meter or so in length, a couple of feet in diameter, and weighing a couple of tons.

What is "less than the diameter of an atomic nucleus"? The thickness of a human hair is about 1/100 of a millimeter (there are about 25 mm to the inch). An atom is 5 orders of magnitude smaller than that. An "order of magnitude" is what you have when you multiply or divide by 10. To get to "5 orders of magnitude" smaller, you must divide by 10, five times. So an atom is 1/100,000 of the size of a hair; that is, you could lay about 100,000 atoms side by side across a hair. To put this into the physicists' terminology, the diameter of an atom equals the thickness of a human hair multiplied by "ten to the minus five" (10^{-5}). The minus sign before the five implies *divide;* if there is no minus sign, then one must multiply so:

$$\text{diameter of atom} = \text{thickness of human hair} \times 10^{-5}$$
$$\text{thickness of human hair} = \text{diameter of atom} \times 10^{5}$$

A molecule of water is three atoms bound together. Every cup of wine that passed through Julius Caesar's body contained so many molecules that if it flowed into the Tiber and mixed evenly with all the seas in the world, and thence turned into rain in the normal way of things, and thence found its way to the tap in your kitchen, then the chances are you drank some of these atoms when you last had a cup of coffee. If the oceans, rivers, and resulting rain were all mixed evenly, a cupful of water from the tap would contain about eleven of the molecules that flowed through Julius Caesar when he drank a particular cup of wine. If Caesar drank 100,000 cups of wine in his lifetime, then, assuming all the water in the world is now mixed, each cup of coffee you drink contains about a million molecules that have passed through Julius Caesar's bladder (and Adolph Hitler's and everyone else's).

An atom is sometimes thought of in terms of the solar system, with the nucleus in the middle and the electrons circling around it like the planets. This image is useful in that the nucleus of an atom is very small compared with the atom itself. In fact, the nucleus of an atom is to the atom as the atom is to a human hair—if it were physically possible, you could lay 100,000 nuclei side by side across an atom, and therefore you could lay 10,000,000,000 nuclei across a human hair (width of an atomic nucleus = width of human hair $\times$ 10^{-10}). Putting some of this together:

$$\text{diameter of atom} = \text{thickness of human hair} \times 10^{-5}$$
$$= \text{millimeter} \times 10^{-7}$$
$$= \text{meter} \times 10^{-10}$$

$$\text{diameter of nucleus} = \text{meter} \times 10^{-15}$$
(because there are 100,000 nucleus widths to the atom)

Sinsky's calibration experiment seemed to show that Weber's detector could see changes in length of about one-tenth of this—one-tenth of the diameter of an atomic nucleus ($= \text{meter} \times 10^{-16}$).

Cardiff Bay

As I write these words I am looking at Cardiff Bay from the window of my study. It is an enclosed body of water with a surface area of about a square mile. Imagine that it were confined within vertical banks. If some water were poured into the lake, the level would rise. How much water would you have to put into the lake to make the level rise by 10^{-21}—the change to be measured by LIGO's interferometers, the latest instruments to try to detect gravitational waves? The answer is 1/100,000 of a drop!

Now imagine you had to measure this rise in the level of the waters. One would build a roof to eliminate wind-driven waves, but what if someone walks past a mile away—even that will cause ripples on the surface of the bay that will be huge compared with the effect we are trying to measure. So let us suspend the whole bay so it is no longer in contact with the ground. The gravitational force of the Moon and the Sun will pull the waters into hills and valleys far larger than the effect we want to see. And what about the tiniest temperature change in the water? That is one way to think about how hard it is to isolate a gravitational wave detector so that it can see the tiny forces to be measured.

In the early days of this field, it seems to have been the fashion to refer to
the object under investigation as "gravitational radiation." More recently,
the term more frequently used is "gravitational waves." I am told that the
change came about when scientists were beginning to look for sites to build
big detectors. *Waves* sounds less threatening than *radiation* and no one wants
more radiation in their state, so nowadays everyone talks of "waves." But
both terms mean the same and can be used interchangeably when political
sensibilities are not at stake. If there is a technical difference, it might
be that gravitational radiation is more naturally used of a diffuse, long-
term source such as the background radiation left over from the big bang,
whereas *waves* implies something more sharply defined—the gravitational
waves from a binary star.

The early preference for the term *gravitational radiation* was also useful
to distinguish the phenomenon from gravity waves, which are a different
matter entirely. Gravity waves are waves in Earth's atmosphere caused by
changes in air pressure.

An issue of the scientific journal *Nature* dated November 3, 1972, reveals
a strange coincidence. The first "letter," which is written by Joe Weber and
which I discuss in some detail in chapter 10, is entitled "Computer Analyses
of Gravitational Radiation Detector Coincidences." Immediately following
it is a letter entitled "Atmospheric Gravity Waves to be Expected from the
Solar Eclipse of June 30, 1973." It concludes: "We would also like to draw
attention to the possibility that the daily heating and cooling at sunrise
and sunset could give rise to effects of this kind.... This is likely to produce
a continuous daily spectrum of gravity waves and could provide an important source of the gravity waves observed in the ionosphere under normal
conditions" (p. 32).[1] This discussion has nothing to do with gravitational
waves.

Unsurprisingly, the term *gravitational waves* is often carelessly foreshort-
ened to *gravity waves* even when it is the former that is under discussion.

---

1. The article is Beer and May 1972.

Fortunately, context always prevents any real confusion, as there is almost no connection between the fields. (Though atmospheric *gravity waves* could have a disturbing effect on an interferometric *gravitational wave* detector—they could be a source of noise. In turn, changes in the density of the air surrounding a detector will cause changes in the gravitational pull of the air on the mirrors. This is called gravity-gradient noise.)

APPENDIX INTRO.3: ROGER BABSON'S ESSAY, "GRAVITY—OUR ENEMY NUMBER ONE"

When I was a boy my oldest sister was drowned while bathing in Annisquam River, Gloucester, Mass. Yes, they say she was "drowned", but the fact is that, through temporary paralysis, or some other cause (she was a good swimmer) she was unable to fight Gravity which came up and seized her like a dragon and brought her to the bottom. There she smothered and died from lack of oxygen.

What T.B. taught me

When I was a young man I was ill for a year with tuberculosis. I had had a very confining indoors job in a room with no ventilation and, when getting a severe cold at Buffalo, I had not the resistance to throw it off. These were the days before the oxygen tents and I was taken West to high altitudes. When asking the doctors, "Why the high altitude?" they replied, "Because there is less moisture in the air at a high altitude and hence *relatively* more oxygen." They explained how Gravity pulls the moist and humid air down into the valleys and on the low lands in the vicinity of the seacoast. Therefore, to fight this effect of Gravity, I convalesced in the West, but finally settled in Wellesley Hills, Mass. The modern system of rain-making is another illustration of the constructive use of Gravity.

At one time in my career I was indirectly the largest stock-holder in Holtzer-Cabot Electric Company. An important part of this company's work was to make motors used in ventilating systems, especially factories and public buildings. I hence was interested in ventilation. Even before my connection with the Holtzer-Cabot Company, I became acquainted with the owner of the Sturtevant Blower Company of Jamaica Plain, Boston, Ex-Governor Eugene Foss. He was an enthusiast on fresh air. This was before the days of air-conditioning. Although he would now strongly recommend air-conditioning for its ability to dehydrate the air, yet even air-conditioning does not give the *air-circulation* which the old blower systems gave. Mr. Foss said to me once: "One of the greatest causes of illness is *bad air* in which people are being 'drowned' as if they were thrown into the ocean. Old man Gravity is responsible." Gradually I found that "old man Gravity" is not only directly responsible for millions of deaths each year, but also for millions of accidents which Mr. Foss probably had in mind. Broken hips and other broken bones as well as numerous circulatory, intestinal and other internal troubles are directly due to the people's inability to counter act Gravity at a critical moment.

My grandson's death

The above thoughts had been more or less dormant in my brain until the summer of 1947 when my grandson, Michael, a splendid swimmer of 17 years of age, was drowned in Lake Winnipesaukee, N. H. He was in a motorboat with a party of people, one of whom fell overboard while the boat was quickly turning. Michael immediately pulled off his coat and shoes and jumped in after her. The boat, however, had left the woman several hundred feet behind; hence my grandson was obliged to swim back to find her. He succeeded in getting the woman back into the boat, she is healthy and happy today, but that "dragon" Gravity came up and snatched Michael! He was so exhausted he couldn't fight this force which pulled him to the bottom. In fact, it was five days before his body was found.

There are thousands of such accidents every summer, notwithstanding the fact that most boats carry life preservers which are practical anti-gravity aids. If these would be more freely used, deaths from drowning would greatly be reduced, but most people—especially good swimmers—think it is a sign of weakness or is sissified to use these aids for fighting Gravity. This is a great mistake and should be corrected by all swimming teachers.

Since Michael's death I have become more and more interested in the subject of Gravity. It seems as if there must be discovered some partial in-sulator of Gravity which could be used to save millions of lives and prevent accidents. In addition it is my belief that someday Gravity will be harnessed to provide a free source of power. For these and other reasons, I am greatly interested in the Gravity Research Foundation.

Importance of air currents

In closing let me return to the original purpose of this article, namely the factor which Gravity performs in connection with improper ventilation. As Gravity pulls the water-ladened and impure air down into the valleys and low spots of the ground, so it holds the stagnant impure air down in the lower portion of rooms, offices, stores and factories. People who are obliged to spend most of their time in such contaminated quarters, ultimately suffer therefrom. In fact, millions of people finally are "drowned" in this contaminated air through lack of a proper supply of oxygen, due to the fact that carbon-dioxide, a poisonous gas, is heavier than oxygen. Such people do not die all at once as did my Michael when the oxygen was shut off immediately and completely. They die gradually due to an insufficient supply of oxygen over a period of years. Perhaps it would be fairer to say that their health is impaired and their life is shortened by an insufficient supply of oxygen.

We as yet have discovered no partial insulator of Gravity. Therefore we must overcome Gravity by indirect ways like using elevators, life preservers and oxygen tents. For the purpose I have in mind, the quiet electric fan is most useful, especially where air conditioning is unavailable. The electric fan constantly "pushes" the heavy foul air away from our faces. Hence instead of breathing and re-breathing this same impure air in which we are submerged of which the oxygen is constantly being reduced, we are constantly getting a fresh supply.

I am the largest stockholder of the Gamewell Company which has installed more fire alarm systems than any company in the world. From nearly 75 years devoted to the study of fires we learn that most of the "deaths by fire" come not through burning but through suffocation, a form of "drowning." Furthermore, according to J. Wendell Sether of the National Board of Underwriters, a majority of our dwelling house deaths occur upstairs in bedrooms from downstairs fires. When a fire gets underway, super-heated combustion gases ranging from 800 to 1,000 degrees in temperature defy Gravity and quickly flood the upper halls of a house, hotel or office building. These lethal gases enter bed- and other upper rooms through open doors and transoms and asphyxiate the occupants. This means that Gravity can work against us in two opposite ways. Moral: Always sleep with your door and transom closed and depend upon an open window and an electric fan for ventilation.

It is best to live on a hill in a fairly elevated rural section*; but a supply of fresh air is available in any modern city if it is allowed to enter the rooms. The air in every room therefore can gradually be changed but an

---

*As a footnote let me pass along one closing thought: Not only is the rural air purer; but if you are so located that it blows through a pine forest before you breathe it you are very fortunate. The air of the woods is not only pure but medicated. It possesses chemical and/or electrical qualities which are of great value. The ideal sanitarium would be so constructed that the patients would sleep in the forest, but be wheeled out into the sunshine during the day. Piney-woods air, shaded sunlight and sparkling spring water with proper food and rest should cure almost anyone.

electric fan greatly aids this change. Hence the air in the room is less stagnant than it would otherwise be. The great advantage, however, of a quiet, moderate breeze blowing on one's face is that it makes it impossible for us to breathe again the contaminated air that we have just breathed out or exhumed. Many people seem to dread, by instinct, air blowing on their faces; but we should train young children to ask for it. This is one of the aims of the Gravity Research Foundation.

APPENDIX III.1: COLONIAL CRINGE

The special geographical position of the Perth group explains their sensitivity to what they perceive as bullying. The University of Western Australia is at the western edge of the Australian continent, and the next large cities are far away; from Perth it is about 1500 miles across the Nullarbor Plain to Adelaide, which itself feels isolated and looked down on by the establishment of Melbourne, Canberra, and Sydney (Melbourne is a further 500 miles east). Canberra is the government seat, and the three eastern cities between them house most of the population. Perth, then, is cut off even from the rest of Australia, and isolated twice over from the rest of the world. David Blair and his group, therefore, have to work hard to be noticed. As Blair explained,

> [1998] Australia's got quite a few things called Australian National Scientific Research Facilities. And every single Australian National Research Facility is in New South Wales [the state which includes Sydney]. If it's called a National Facility, or Australian this or that, it's New South Wales, actually, and they give it that name to justify using all the taxpayers' money in New South Wales. So Western Australia has always had a slight secessionist mentality— we get the thin end of the stick, and we contribute so much to Australia's exports, et cetera, and we don't get anything back. So I saw this [AIGO, the Australian International Gravitational Observatory] as something where Western Australia could have a national facility. We'd have the first national facility in Western Australia.

He went on:

> [A]lways there has been a sort of "colonial cringe" mentality here—that all this important stuff has to be done over there. And I saw this as being culturally important, because here was something that we could really do ourselves. So we got this collaboration going within Australia.

The problem gets more acute, because Perth's relationship to Australia is like Australia's relationship to the rest of the world:

Americans believe that nothing exists beyond their borders. Australia doesn't really exist. It's just a theoretical concept. So we don't get referenced by them . . . they don't read our papers; they are so much into the—you know—get on the phone and ring people, and people moving around, and you refer to the people who you had a conversation with in the last six months.

But these people that occasionally materialize out of some other universe once a year are almost forgotten. And if you don't want to be forgotten, you have to hammer, hammer, hammer the fact that we exist by going to meetings and things—even though we feel there are too many of them.

Some Americans who have paid visits to Perth have gone away impressed by the level of activity that they found, but many Americans, as I can report from a number of conversations, do not like this "hammer, hammer, hammer." For these and other reasons, Blair's active campaign on behalf of UWA has been to some extent counterproductive, adding to the tensions over the Perth-Rome coincidence claims.

APPENDIX V.1: THE METHOD

The archive

As the project has gone forward I have been more and more conscious that I may have been collecting materials that could one day be of deep historical interest. Let us suppose that gravitational wave astronomy becomes a reality and results in many important discoveries. Now imagine that we had the equivalent archive pertaining to something like the debates about the meaning of the Michelson-Morley experiments. Imagine that we had Michelson's and Miller's voices on tape, and Einstein's and Eddington's view of the experiments. For that kind of reason I have been preserving all the material very carefully in a fireproof safe. There are copies of all the interviews and a copy of this manuscript with every quotation and other remark attributed to its author.

All this material has been collected under an assurance of confidentiality and I do not expect it to be released in my lifetime, but I hope to make arrangements for it to be preserved just in case it is of interest to generations hence.

Fieldwork

A summary of fieldwork completed up to August 2003 (including fieldwork associated with earlier phases of the extended project on gravitational wave research) is given in the next paragraphs (a figure in brackets indicates multiple visits to that location).

Visits to the universities of Bristol (3+), Reading, Glasgow (4+), Sussex, Cardiff (many visits, as I now work in Cardiff), Hannover (2), Leiden, Maryland (5+), MIT (4+), Morehouse, Irvine, Stanford (3), Caltech (9+), Rochester (3), Louisiana State (6+), Tor Vergata-Rome, Padua, Milan, Perth, Adelaide, Canberra, Amsterdam, Princeton, Princeton Institute for Advanced Studies (2).

I have also been to industrial laboratories, including IBM (3), Bell Labs (3), Hughes Aircraft, CSIRO Sydney, NASA-Goddard Space Flight Center, Marshall Space Flight Center.

834

Other locations visited include the Max Planck Institute-Munich; Sophia-Antipolis; the LIGO site at Hanford Nuclear Reservation (5); the LIGO site at Livingston, Louisiana (7); the GEO600 site near Hannover (3); the VIRGO site near Pisa (4); the Frascati labs (3+); the National Science Foundation in Washington, DC (9+); The Albert Einstein Institute in Potsdam; the Japanese National Astronomical Observatory; the offices of April Burke and Kevin Kelly in Washington; the homes of Joel Sinsky in Baltimore, David Zipoy in Punta Gorda, Florida, Bob Forward near Inverness, Frank Schutz in Cayucos Beach, and Peter Saulson in Baton Rouge. Peter Saulson has also visited me in my own home.

I have attended conferences or committee meetings in Washington, DC ([NSF] 2); Pisa (3); Orsay; Jerusalem; Geneva; Boston; Hanford, Washington (2); Livingston, Louisiana (4); the California Institute of Technology (Caltech) (4); Stanford University; Pennsylvania State University; University of Florida, Gainesville; Paris; Perth, Australia; Trento; Elba; Kyoto; and Hannover.

I have also spoken with gravitational wave physicists in cafes, restaurants, corridors, cars, planes, and even a boat, and am in regular touch using email, the electronic Web, and, occasionally, the telephone. All of these methods of communication make increasingly less sense to set out in list form as my interactions with the community have grown in depth.

Other sources for my research include published and unpublished papers, newspaper reports, papers stored at the Smithsonian Institution, and, as explained, confidential materials at the National Science Foundation.

REFERENCES

Abramovici, Alex, William E. Althouse, Ronald W. P. Drever, Yekta Gursel, Seiji Kawamura, Fred J. Raab, David Shoemaker et al. 1992. "LIGO: The Laser Interferometer Gravitational-Wave Observatory." *Science* 256 (17 April): 325–33.

Agar, Jon. 1998. *Science and Spectacle.* London: Harwood Academic Publishers.

Aglietta, M., G. Badion, G. Bologna, C. Castagnoli, A. Castellina, W. Fulgione, P. Galeotti, et al. 1989. "Analysis of the Data Recorded by the Mont Blanc Neutrino Detector and by the Maryland and Rome Gravitational-Wave Detectors during SN 1987A." *Il Nuovo Cimento* 12C (1): 75–103.

Allen, B., J. K. Blackburn, P. R. Brady, J. D. E. Creighton, T. Creighton, S. Droz, A. D. Gillespie, et al. 1999. "Observational Limit on Gravitational Waves from Binary Neutron Stars in the Galaxy." *Physical Review Letters* 83 (8): 1498–1501.

Allen, W. D., and C. Christodoulides. 1975. "Gravitational Radiation Experiments at the University of Reading and the Rutherford Laboratory." *Journal Of Physics A* 8: 1726–33.

Allen, Z. A., P. Astone, L. Baggio, D. Busby, M. Bassan, D. G. Blair, M. Bonaldi, et al. 2000. "First Search for Gravitational Wave Bursts with a Network of Detectors." *Physical Review Letters* 85 (24): 5046–50. (December 11.)

Amaldi, E., O. Aguiar, M. Bassan, P. Bonifazi, P. Carelli, M. G. Castellano, G. Cavallari, et al. 1989. "First Gravity Wave Coincidence Experiment between Resonant Cryogenic Detectors—Louisiana-Rome-Stanford." *Astronomy and Astrophysics* 216 (June 7): 325–32.

Anon. 1969. "Gravitational Waves Detected." *Science News* 95 (June 21): 593.

Aplin, P. S. 1976. "Pinning Down Gravity Waves." *Physics Bulletin* (December): 538–40.

Ashmore, Malcolm. 1989. *The Reflexive Thesis: Wrighting Sociology of Scientific Knowledge.* Chicago: Univ. of Chicago Press.

Ashmore, Malcolm. 1993. "The Theatre of the Blind: Starring a Promethean Prankster, a Phoney Phenomenon, a Prism, a Pocket, and a Piece of Wood." *Social Studies of Science* 23 (1): 67–106.

Ashmore, Malcolm. 1996. "Ending Up on the Wrong Side." *Social Studies of Science* 26 (2): 305–22.

Astone, P., D. Babusci, M. Bassan, P. Bonifazi, P. Carelli, G. Cavallari, E. Coccia, et al. 2002. "Study of the Coincidences between the Gravitational Wave Detectors EXPLORER and NAUTILUS in 2001." *Classical and Quantum Gravity* 19 (7): 5449–65.

Astone, P., D. Babusci, M. Bassan, P. Bonifazi, P. Carelli, G. Cavallari, E. Coccia, et al. 2003. "On the Coincidence Excess Observed by the Explorer and Nautilus Gravitational Wave Detectors in the Year 2001." http://arxiv.org/archive/gr-qc/0304004.

Astone, P., M. Bassan, D. G. Blair, P. Bonifazi, P. Carelli, E. Coccia, C. Cosmelli, et al. 1999. "Search for Coincident Excitation of the Widely Spaced Resonant Gravitational Wave Detectors EXPLORER, NAUTILUS, NIOBE." *Astroparticle Physics* 10 (January 1): 83–92.

Astone, P., M. Bassan, P. Bonifazi, P. Carelli, E. Coccia, C. Comeli, V. Fafone, et al. 1999. "Search for Gravitational Radiation with the Allegro and Explorer Detectors." *Phys. Rev. D* 59 (12): 2001–7.

Astronomy Survey Committee, National Academy of Sciences, National Research Council. 1973. *Astronomy and Astrophysics for the 1970s.* Vol. 2, *Reports of the Panels.* Washington, DC: National Academy Press.

Bahcall, J., and R. Davis Jr. 1982. "An Account of the Development of the Solar Neutrino Problem." In *Essays in Nuclear Physics,* ed. C. Barnes, D. Clayton, and D. Schramm, 243–86. Cambridge: Cambridge Univ. Press.

Baldi, Stephane. 1998. "Normative versus Social Constructivist Processes in the Allocation of Citations: A Network Analytic Model." *American Sociological Review* 63 (6): 829–46. (December.)

Barnes, Barry S. 1983. "Social Life as Bootstrapped Induction." *Sociology* 4:524–45.

Barnes, Barry, David Bloor, and John Henry. 1996. *Scientific Knowledge: A Sociological Analysis.* London: Athlone Press.

Bartusiak, Marcia. 2000. *Einstein's Unfinished Symphony.* Washington, DC: Joseph Henry.

Bassan, M. 1994. "Resonant Gravitational Wave Detectors: A Progress Report." *Classical and Quantum Gravity* 11:A39–A59.

Battersby, Stephen. 2002. "On the Quest of a Wave." *New Scientist* 7:26. (September.)

Beer, Tom, and A. N. May. 1972. "Atmospheric Gravity Waves to be Expected from the Solar Eclipse of June 30, 1973." *Nature* 240 (5375): 30–32. (November 3.)

Bel, Luis. 1996. "Static Elastic Deformations in General Relativity." http://arxiv.org/archive/gr-qc/9609045.

Bell, Colin. 1978. "Studying the Locally Powerful." In *Inside the Whale: Ten Personal Accounts of Social Research,* ed. C. Bell and S. Encel, 14–40. Oxford: Pergamon.

Bertotti, B., and A. Cavaliere. 1972. "Gravitational Waves and Cosmology." *Il Nuovo Cimento* 2B (2): 22.

Bethe, Hans A., and G. E. Brown. 1998. "Evolution of Binary Compact Objects that Merge." *Astrophysical Journal* 505:780–89, http://arxiv.org/archive/astro-ph/9802084.

Bijker, Wiebe E. 1995. *Of Bicycles, Bakelites, and Bulbs: Toward a Theory of Sociotechnical Change.* Cambridge, MA: MIT Press.

Bijker, Wiebe, Tom Hughes, and Trevor Pinch. 1987. *The Social Construction of Technological Systems.* Cambridge, MA: MIT Press.

Billing, H., P. Kafka, K. Maischberger, F. Meyer, and W. Winkler. 1975. "Results of the Munich-Frascati Gravitational-Wave Experiment." *Lettere al Nuovo Cimento* 12 (4): 111–16. (June 25.)

Blair, David, ed. 1991. *The Detection of Gravitational Waves.* Cambridge: Cambridge Univ. Press.

Blair, David, and Geoff McNamara. 1997. *Ripples in a Cosmic Sea: The Search for Gravitational Waves,* with a foreword by Paul Davies. St. Leonard's, NSW: Allen and Unwin.

Bloor, David. 1973. "Wittgenstein and Mannheim on the Sociology of Mathematics." *Studies in the History and Philosophy of Science* 4:173–91.

Bloor, David. 1976. *Knowledge and Social Imagery.* London: Routledge and Kegan Paul.

Bloor, David. Unpublished. "Relativism at 30,000 Feet." Unpublished manuscript.

Bogdanov, Grichka, and Igor Bodanov. 2001. "Topological Field Theory of the Initial Singularity of Spacetime." *Classical and Quantum Gravity* 18 (21): 4341–73.

Bonazzola, S., M. Chevreton, and J. Thierry-Mieg. 1974. "Meudon Gravitational Radiation Detection Experiment." In *Gravitational Radiation and Gravitational Collapse,* ed. Dewitt-Morrette Cecile Dordrecht, 39. Boston: Reidel.

Boughn, S. P., M. Bassan, W. M. Fairbank, R. P. Giffard, P. F. Michelson, J. C. Price, and R. C. Taber. 1990. "Method for Calibrating Resonant-Mass Gravitational Wave Detectors." *Review of Scientific Instruments* 61:1–6.

Boughn, S. P., W. M. Fairbank, M. S. McAshan, H. J. Paik, R. C. Taber, T. P. Bernat, D. G. Blair, and W. O. Hamilton. 1974. "The Use of Cryogenic Techniques to Achieve High Sensitivity in Gravitational Wave Detectors." In *Gravitational Radiation and Gravitational Collapse,* ed. Dewitt-Morrette Cecile Dordrecht, 40–51. Boston: Reidel.

Bourdieu, R. 1997. "The Forms of Capital." In *Education: Culture, Economy and Society,* ed. A. H. Halsey, Hugh Lauder, Phillip Brown, and Amy Stuart Wells. Oxford: Oxford Univ. Press.

Braginsky, V. B., A. B. Manukin, E. I. Popov, and V. N. Rudenko. 1973. "The Search for Gravitational Radiation of Non-Terrestrial Origin." *Physics Letters* 45A (4): 271–72.

Braginsky, V. B., A. B. Manukin, E. I. Popov, V. N. Rudenko, and A. A. Khorev. 1972. "Search for Gravitational Radiation of Extra-Terrestrial Origin." *JETP Letters* 16:108–12. (Trans. from Russian letters to *Journal of Experimental and Theoretical Physics,* August 5.)

Braginsky, V. B., A. B. Manukin, E. I. Popov, V. N. Rudenko, and A. A. Khorev. 1974. "An Upper Limit on the Density of Gravitational Radiation of Extraterrestrial Origin." *JETP Letters* 39:387–92. (Trans. from Russian letters to *Journal of Experimental and Theoretical Physics* 66:801–12).

Braginsky, V. B., V. P. Mitrofanov, and V. I. Panov. 1985. *Systems with Small Dissipation,* trans. Erast Gliner with an introduction by Kip S. Thorne. Chicago: Univ. of Chicago Press.

Bramanti, D., and K. Maischberger. 1972. "Construction and Operation of a Weber-Type Gravitational-Wave Detector and of a Divided-Bar Prototype." *Lettere al Nuovo Cimento* 4 (17): 1007–13.

Bramanti, D., K. Maischberger, and D. Parkinson. 1973. "Optimization and Data Analysis of the Frascati Gravitational-Wave Detector." *Lettere al Nuovo Cimento* 7 (14): 665–70. (August 4.)

Brautti, G., and D. Picca. 2002a. "The Sensitivity of Antennas for Gravitational Wave Detection." *International Journal of Modern Physics A,* 17 (3): 327–34. (January 30.)

Brautti G., and D. Picca. 2002b. "Thermal Stimulation of Gravitational Wave Antennas." *International Journal of Modern Physics A* 17 (8): 1111–16. (March 30.)

Bressani, T., E. Del-Giudice, and G. Preparata. 1992. "What Makes a Crystal 'Stiff' Enough for the Mossbauer Effect?" *Il Nuovo Cimento* 14D (3): 345–49. (March.)

Bressani, T., B. Minettie, and A. Zenoni, eds. 1992. *Common Problems and Ideas of Modern Physics.* Singapore: World Publishing Company.

Broad, William. 1992. "Big Science Squeezes Small-Scale Researchers." *New York Times,* December 29, pp. C1, C9.

Brown, B. L., A. P. Mills, and J. A. Tyson. 1982. "Results of a 440-Day Search for Gravitational Radiation." *Physical Review D* 26 (6): 1209–18.

Brown, Malcolm W. 1991. "Experts Clash over Project to Detect Gravity Waves." *New York Times,* April 30, C1, C5.

Callan, C., Dyson, F., and Treiman, S. 1987. "Neutrino Detection Primer" JASON, The MITRE Corporation, 7525 Coshire Drive, Mclean, VA 22102-3481 [JSR-84-105 July 1987].

Cambrosio, Alberto, and Peter Keating. 1988. "'Going Monoclonal': Art, Science, and Magic in the Day-to-Day Use of Hybridoma Technology." *Social Problems* 35 (3): 244–60.

Capshew, J. H., and K. A. Rader. 1992. "Big Science: Price to Present." *Osiris* 7:3–25.

Christodoulides, C., and W. D. Allen. 1975. "Gravitational Radiation Experiments at the University of Reading and the Rutherford Laboratory." *Journal of Physics A* 8:1726.

Christodoulou, Demetrios. 1991. "Nonlinear Nature of Gravitation and Gravitational-Wave Experiments." *Physical Review Letters* 67 (12): 1486–89. (September.)

Coleman, J. 1988. "Social Capital in the Creation of Human Capital." *American Journal of Sociology* 94:95–120. Suppl. no. ss.

Collins, Harry. "Harry Collins's Gravitational Research Project." http://www.cf.ac.uk/socsi/gravwave/index.html.

Collins, H. (Harry) M. 1974. "The TEA Set: Tacit Knowledge and Scientific Networks." *Science Studies* 4:165–86.

Collins, H. M. 1975. "The Seven Sexes: A Study in the Sociology of a Phenomenon, or The Replication of Experiments in Physics." *Sociology* 9 (2): 205–24.

Collins, H. M. 1981a. "The Role of the Core-Set in Modern Science: Social Contingency with Methodological Propriety in Science." *History of Science* 19:6–19.

Collins, H. M. 1981b. "Stages in the Empirical Programme of Relativism." *Social Studies of Science* 11:3–10.

Collins, H. M. 1981c. "Son of Seven Sexes: The Social Destruction of a Physical Phenomenon." *Social Studies of Science* 11:33–62.

Collins, H. M. 1983a. "The Meaning of Lies: Accounts of Action and Participatory Research." In *Accounts and Action,* ed. G. N. Gilbert and P. Abel, pp. 69–78. London: Gower.

Collins, H. M. 1983b. "The Sociology of Scientific Knowledge: Studies of Contemporary Science." *Annual Review of Sociology* 9:265–85.

Collins, H. M. 1984. "Concepts and Methods of Participatory Fieldwork." In *Social Researching*, edited by C. Bell and H. Roberts, pp. 54–69. London: Routledge and Kegan Paul.

Collins, H. M. 1985/1992. *Changing Order: Replication and Induction in Scientific Practice.* Beverly Hills and London: Sage. 2nd ed., Chicago: Univ. of Chicago Press.

Collins, H. M. 1987. "Pumps, Rock and Reality." *Sociological Review* 35:819–28.

Collins, H. M. 1988. "Public Experiments and Displays of Virtuosity: The Core-Set Revisited." *Social Studies of Science* 18:725–48.

Collins, H. M. 1990. *Artificial Experts: Social Knowledge and Intelligent Machines.* Cambridge, MA: MIT Press.

Collins, H. M. 1994. "A Strong Confirmation of the Experimenters' Regress." *Studies in History and Philosophy of Science* 25 (3): 493–503.

Collins, H. M. 1996. "In Praise of Futile Gestures: How Scientific Is the Sociology of Scientific Knowledge." *Social Studies of Science* 26 (2): 229–44.

Collins, H. M. 1998. "The Meaning of Data: Open and Closed Evidential Cultures in the Search for Gravitational Waves." *American Journal of Sociology* 104 (2): 293–337.

Collins, H. M. 1999. "Tantalus and the Aliens: Publications, Audiences and the Search for Gravitational Waves." *Social Studies of Science* 29 (2): 163–197.

Collins, H. M. 2000. "Surviving Closure: Post-Rejection Adaptation and Plurality in Science." *American Sociological Review* 65 (6): 824–45.

Collins, H. M. 2001a. "What Is Tacit Knowledge." In *The Practice Turn in Contemporary Theory*, edited by Theodore R. Schatzki, Karin Knorr-Cetina, and Eike von-Savigny. London: Routledge.

Collins, H. M. 2001b. "One More Round with Relativism." In *The One Culture?: A Conversation about Science*, edited by Jay Labinger and Harry Collins, pp. 184–95. Chicago: Univ. of Chicago Press.

Collins, H. M. 2001c. "Tacit Knowledge, Trust, and the Q of Sapphire." *Social Studies of Science* 31 (1): 71–85.

Collins, H. M. 2002. "The Experimenter's Regress as Philosophical Sociology." *Studies in History and Philosophy of Science* 33:153–60.

Collins, H. M., and Robert Evans. 2002. "The Third Wave of Science Studies: Studies of Expertise and Experience." *Social Studies of Science* 32 (2): 235–96.

Collins, H. M., and R. Harrison. 1975. "Building a TEA Laser: The Caprices of Communication." *Social Studies of Science* 5:441–50.

Collins, H. M., and Martin Kusch. 1998. *The Shape of Actions: What Humans and Machines Can Do.* Cambridge, MA: MIT Press.

Collins, H. M., and T. J. Pinch. 1979. "The Construction of the Paranormal: Nothing Unscientific Is Happening." In Sociological Review Monograph. no. 27, *On the Margins of Science: The Social Construction of Rejected Knowledge*, ed. Roy Wallis, pp. 237–70. Keele, UK: Keele Univ. Press.

Collins, H. M., and Trevor J. Pinch. 1981. "Rationality and Paradigm Allegiance in Extraordinary Science" [In German.] In *The Scientist and the Irrational*, ed. Hans Peter Duerr, pp. 284–306. Frankfurt: Syndikat.

Collins, H. M., and Trevor J. Pinch. 1982. *Frames of Meaning: The Social Construction of Extraordinary Science.* London: Routledge and Kegan Paul.

Collins, Harry, and Trevor Pinch. 1993/1998. *The Golem: What You Should Know about Science*. Cambridge: Cambridge Univ. Press. 2nd ed., Canto.

Collins, Harry, and Trevor Pinch. 1998/2002. *The Golem at Large: What You Should Know about Technology*. Cambridge: Cambridge Univ. Press. Canto paperback ed.

Collins, H. M., and Steven Yearley. 1992. "Epistemological Chicken." In *Science as Practice and Culture*, ed. A. Pickering, pp. 301–26. Chicago: Univ. of Chicago Press.

Cooperstock, F. I. 1992. "Energy Localization in General Relativity: A New Hypothesis." *Foundations of Physics* 22 (8): 1011–24.

Cooperstock, F. I., V. Faraoni, and G. P. Perry. 1995. "Can a Gravitational Geon Exist in General Relativity." *Modern Physics Letters A* 10 (5): 359–65.

Crane, Diana. 1976. "Reward Systems in Art, Science, and Religion." *American Behavioural Scientist* 19 (6): 719–34.

Davis Jr., R. 1955. "Attempt to Detect the Anti-Neutrinos from a Nuclear Reactor by the $Cl^{37}(v,e-)Ar^{37}$ Reaction." *Physical Review* 97:766–69.

Dear, Peter. 1990. "Miracles, Experiments, and the Ordinary Course of Nature." *ISIS* 81 (309): 663–93.

Dear, Peter. 2002. "Science Studies as Epistemography." In *The One Culture: A Conversation About Science*, ed. Jay Labinger and Harry Collins, pp. 128–41. Chicago: Univ. of Chicago Press.

de-Solla-Price, D. 1963. *Little Science, Big Science*. New York: Oxford Univ. Press.

Dewitt-Morrette, Cecile, ed. 1974. *Gravitational Radiation and Gravitational Collapse*. Dordrecht: Reidel.

Dickson, C. A., and B. F. Schutz. 1995. "Reassessment of the Reported Correlations between Gravitational Waves and Neutrinos associated with SN 1987A." *Physical Review D* 51:2644–68.

Dietrich, Jane. 1998. "Realizing LIGO: How a Huge Instrument to Detect Gravitational Waves Survived Its Growing Pains and Now Nears Completion." *Engineering and Science* 61 (2): 8–17.

Douglass, D. H., R. Q. Gram, J. A. Tyson, and R. W. Lee. 1975. "Two-Detector-Coincidence Search for Bursts of Gravitational Radiation." *Physical Review Letters* 35 (8): 480–83. (August 25.)

Drever, R. W. P., J. Hough, R. Bland, and G. W. Lesnoff. 1973. "Search for Short Bursts of Gravitational Radiation." *Nature* 246 (December 7): 340–44.

Duhem, Pierre. 1981. *The Aim and Structure of Physical Theory*. Trans. P. P. Wiener. New York: Athenaeum.

Dyson, Freeman J. 1963. "Gravitational Machines." In *Interstellar Communication: A Collection of Reprints and Original Contributions*, ed. A. G. W. Cameron. New York: W. A. Benjamin, Inc.

Eardley, D. M. 1983. "Theoretical Models for Sources of Gravitational Waves." In *Rayonnement Gravitationnel*, ed. Nathalie Deruelle and Tsvi Piran, pp. 257–94. Amsterdam: North-Holland.

Earman, John, and Clark Glymour. 1980. "Relativity and Eclipses: The British Eclipse Expeditions of 1919 and Their Predecessors." *Historical Studies in the Physical Sciences* 11 (1): 49–85.

Edge, David O., and Michael J. Mulkay. 1976. *Astronomy Transformed: The Emergence of Radio Astronomy in Britain*. London: John Wiley and Sons Inc.

Einstein, Albert, and Nathan Rosen. 1937. "On Gravitational Waves." *Journal of the Franklin Institute* 223:43–56.

Epstein, Steven. 1996. *Impure Science: AIDS, Activism and the Politics of Knowledge.* Berkeley and Los Angeles: Univ. of California Press.

Evans, Lawrence. 1996. "Should We Care about Science 'Studies'?" *The Faculty Forum* 8 (1): 1–2. (October.)

Evans, Richard J. 2002. *Telling Lies about Hitler: The Holocaust, History and the David Irving Trial.* London: Verso.

Evans, Robert. 1999. *Macroeconomic Forecasting: A Sociological Appraisal.* London: Routledge.

Faulkner, William. 1929. *The Sound and the Fury.* London: Jonathan Cape.

Ferrari, V., G. Pizzella, M. Lee, and J. Weber. 1982. "Search for Correlations between the University of Maryland and the University of Rome Gravitational Radiation Antennas." *Physical Review D* 25 (10): 2471–86.

Festinger, L., H. W. Riecken, and S. Schachter. 1956. *When Prophecy Fails.* New York: Harper.

Field, G. B., M. J. Rees, and D. W. Sciama. 1969. "The Astronomical Significance of Mass Loss by Gravitational Radiation." *Comments on Astrophysics and Space Science* 1:187.

Finn, L. S. 2003. "No Statistical Excess in EXPLORER/NAUTILUS Observations in the Year 2001." *Classical and Quantum Gravity* 20: L37–L44.

Fleck, Ludwick. 1935/1979. *Genesis and Development of a Scientific Fact.* English ed., Chicago: Univ. of Chicago Press.

Forward, Robert Lull. 1965. "Detectors for Dynamic Gravitational Fields." Ph.D. diss., Univ. of Maryland.

Forward, Robert Lull. 1971. "Multidirectional, Multipolarization Antennas for Scalar and Tensor Gravitational Radiation." *General Relativity and Gravitation* 2 (2): 149–59.

Forward, Robert Lull. 1978. "Wideband Laser-Interferometer Gravitational-Radiation Experiment." *Physical Review D* 17 (2): 379–90. (January.)

Franklin, Alan. 1994. "How to Avoid the Experimenters' Regress." *Studies in the History and Philosophy of Modern Physics* 25 (3): 463–91.

Franson, J. D., and B. C. Jacobs. 1992. "Null Result for Enhanced Neutrino Scattering in Crystals." *Physical Review A* 46 (5): 2235–39.

Frasca, S., and M. A. Papa. 1995. "Networks of Resonant Gravitational-Wave Antennas." *International Journal of Modern Physics D* 4 (1): 1–50.

Galison, Peter. 1987. *How Experiments End.* Chicago: Univ. of Chicago Press.

Galison, P. 1992. "The Many Faces of Big Science." In *Big Science: The Growth of Large Scale Research,* ed. P. Galison and B. Hevly. Stanford, CA: Stanford Univ. Press.

Galison, P. 1997. *Image and Logic: A Material Culture of Microphysics.* Chicago: Univ. of Chicago Press.

Galison, Peter, and B. Hevly, eds. 1992. *Big Science: The Growth of Large Scale Research.* Stanford, CA: Stanford Univ. Press.

Garfinkel, H. 1967. *Studies in Ethnomethodology.* Englewood Cliffs, NJ: Prentice Hall.

Garwin, Richard L. 1974a. "The Evidence for Detection of Kilohertz Gravitational Radiation." Paper presented at the "Fifth Cambridge" Conference on Relativity, Massachusetts Institute of Technology, June 10.

Garwin, Richard L. 1974b. "Detection of Gravity Waves Challenged." *Physics Today* 27 (12): 9–11. (December.)

Garwin, Richard L. 1975. "More on Gravity Waves." *Physics Today* 28 (11): 13. (November.)

Garwin, Richard L., and James L. Levine. 1973. "Single Gravity-Wave Detector Results Contrasted with Previous Coincidence Detections." *Physical Review Letters* 31 (3): 176–80. (July 16.)

Geison, Jerry. 1995. *The Private Life of Louis Pasteur*. Princeton, NJ: Princeton Univ. Press.

Gerstenshtein, M. E., and V. I. Pustovoit. 1963. "On the Detection of Low Frequency Gravitational Waves." *Soviet Physics-JETP* 16:433–35 (Orig. pub. in Russian 1962.)

Gibbons, G. W., and Stephen William Hawking. 1971. "Theory of the Detection of Short Bursts of Gravitational Radiation." *Physical Review D* 4 (8): 2191–97.

Gibbs, W. Wayt. 2002. "Ripples in Spacetime." *Scientific American* (April): 62–71.

Gieryn, Thomas. 1983. "Boundary-Work and the Demarcation of Science from Non-Science: Strains and Interests in Professional Ideologies of Scientists." *American Sociological Review* 48:781–95.

Gieryn, Thomas. 1999. *Cultural Boundaries of Science: Credibility on the Line*. Chicago: Univ. of Chicago Press.

Godin, Benoit, and Yves Gingras. 2002. "The Experimenter's Regress: From Skepticism to Argumentation." *Studies in the History and Philosophy of Science* 30A (1): 137–52.

Goffee, Robert, and Richard Scase. 1995. *Corporate Realities: The Dynamics of Large and Small Organizations*. London: Routledge.

Gooding, David. 1985. "In Nature's School: Faraday as an Experimentalist." In *Faraday Rediscovered: Essays on the Life and Work of Michael Faraday, 1791–1876*, ed. David Gooding and Frank A. L. James, pp. 105–35. London: MacMillan.

Gooding, David. 1990. *Experiment and the Making of Meaning*. Dordrecht: Kluwer.

Goranzon, Bo, and Ingela Josefson, eds. 1988. *Knowledge, Skill and Artificial Intelligence*. London: Springer-Verlag.

Gordy, Walter. 1988. "The Nature of the Man, William Martin Fairbank." In *Near Zero: New Frontiers of Physics*, ed. J. D. Fairbank, B. S. Deaver Jr., C. W. F. Everitt, and P. F. Michelson, pp. 7–18. New York: W. H. Freeman and Co.

Gouldner, Alvin W. 1954. *Patterns of Industrial Bureaucracy*. New York: Free Press.

Greiner, Larry. 1998. "Evolution and Revolution as Organizations Grow." *Harvard Business Review* (May–June): 55–68.

Gretz, D., M. Lee, and Joseph Weber. 1974. "Gravitational Radiation Experiments in 1973–1974." University of Maryland Technical Report no. 75-017 (August).

Grischuk, L. P. 1992. "Quantum Mechanics of a Solid-State Bar Gravitational Antenna." *Physical Review D* 45:2601–8.

Grischuk, L. P. 1994. "Density Perturbations of Quantum-Mechanical Origin and Anisotropy of the Microwave Background." *Physical Review D* 50 (12): 7154–72.

Guston, David H. 2000. *Between Politics and Science: Assuring the Integrity and Productivity of Research*. Cambridge: Cambridge Univ. Press.

Hacking, Ian. 1983. *Representing and Intervening*. Cambridge: Cambridge Univ. Press.

Hagstrom, Warren. 1965. *The Scientific Community*. New York: Basic Books.

Hamilton, W. O. 1988. "Near Zero Force, Force Gradient and Temperature: Cryogenic Bar Detectors of Gravitational Radiation." In *Near Zero: New Frontiers of Physics*, ed. J. D. Fairbank, B. S. Deaver Jr., C. W. F. Everitt, and P. F. Michelson. New York: W. H. Freeman and Co.

Harry, Gregory M., Janet L. Houser, and Kenneth A. Strain. 2002. "Comparison of Advanced Gravitational-Wave Detectors." *Physical Review D* 65 (8): 082001, 1–15.

Hartland, Joanne. 1996. "Automating Blood Pressure Measurements: The Division of Labour and the Transformation of Method." *Social Studies of Science* 26 (1): 71–94. (February.)

Harvey, B. 1981. "Plausibility and the Evaluation of Knowledge: A Case Study in Experimental Quantum Mechanics." *Social Studies of Science* 11:95–130.

Harwit, Martin. 1981. *Cosmic Discovery*. New York: Basic Books.

Hasted, John. 1981. *The Metal Benders*. London: Routledge and Kegan Paul.

Hawking, Stephen William. 1971. "Gravitational Radiation from Colliding Black Holes." *Physical Review Letters* 26 (21): 1344–46.

Heilbron, John L. 1992. "Creativity and Big Science." *Physics Today* 45 (11): 42–47.

Hesse, Mary B. 1974. *The Structure of Scientific Inference*. London: Macmillan.

Hevly, B. 1992. "Reflections on Big Science and Big History." In *Big Science: The Growth of Large Scale Research*, ed. P. Galison and B. Hevly, pp. 355–63. Stanford, CA: Stanford Univ. Press.

Hirakawa, Hiromasa, and Kazumichi Nariha. 1975. "Search for Gravitational Radiation at 145Hz." *Physical Review Letters* 35 (6): 330–34.

Holton, Gerald. 1978. *The Scientific Imagination*. Cambridge: Cambridge Univ. Press.

Hu, Enke, Tongren Guan, Bo Yu, Mengxi Tang, Shusen Chen, Qinzhang Zheng, P. F. Michelson et al. 1986. "A Recent Coincident Experiment of Gravitational Waves with a Long Baseline." *Chinese Physics Letters* 3 (12): 529–32.

Hu, Renan, Nanfeng Jiang, Yicheng Liu, Rongxian Qin, Dajun Tan, Jingfa Tian, Guozong Wang et al. 1982. "Progress Report on the Gravitational Wave Experiment in Beijing-Guangzhou." In *Proceedings of the Second Marcel Grossman Meeting on General Relativity*, ed. R. Ruffini, pp. 1133–44. Amsterdam: North-Holland.

Hulse, R. A., and J. H. Taylor. 1975. "Discovery of a Pulsar in a Binary System." *Astrophysical Journal* 195 (2): L51–L53.

Infeld, Leopold. 1941. *The Evolution of a Scientist*. New York: Doubleday.

Jensen, O. G. 1979. "Seismic Detection of Gravitational Radiation." *Rev Geophysics* 17 (8): 2057–69.

Johnston, R., J. Currie, L. Grigg, B. Martin, D. Hicks, E. N. Ling, and J. Skea. 1993. *National Board of Employment, Education and Training, Commissioned Report no 25*. Canberra: Australian Government Publishing Service.

Jordan, Kathleen, and Michael Lynch. 1992. "The Sociology of a Genetic Engineering Technique: Ritual and Rationality in the Performance of the 'Plasmid Prep.'" In *The Right Tools for the Job: At Work in 20th Century Life Sciences*, ed. Adele Clark and Joan Fujimura, pp. 77–114. Princeton, NJ: Princeton Univ. Press.

Kafka, P. 1972. "Are Weber's Pulses Illegal?" Essay Submitted to Gravity Research Foundation.

Kafka, P. 1973. "On the Evaluation of the Munich-Frascati Weber-Type Experiment." In *Colloques Internationaux du Centre National de la Recherche Scientifique, no. 220: Ondes et Radiations Gravitationelles*, pp. 181–201.

Kafka, P. 1975. "Optimal Detector of Signals through Linear Devices with Thermal Noise Sources, and Application to the Munich-Frascati Weber-Type Gravitational Wave Detectors." Lectures presented at the International School of Cosmology and Gravitation, Erice, Sicily, March 13–25; MPI-PAE/Astro 65 (July 1975).

Kennefick, Daniel. 1999. "Controversies in the History of the Radiation Reaction Problem in General Relativity." In *The Expanding Worlds of General Relativity*, ed. H. Goenner, J. Renn, J. Ritter, and T. Sauer. Einstein Studies, vol. 7. Boston: Birkhauser.

Kennefick, Daniel. 2000. "The Star Crushers: Theoretical Practice and the Theoretician's Regress." *Social Studies of Science* 30 (1): 5–40.

Kevles, Daniel K., and Leroy Hood. 1992. *The Code of Codes: Scientific and Social Issues in the Genome Project.* Cambridge, MA: Harvard Univ. Press.

Knorr-Cetina, Karin. 1981. *The Manufacture of Knowledge.* Oxford: Pergamon.

Knorr-Cetina, Karin. 1999. *Epistemic Cultures: How the Sciences Make Knowledge.* Cambridge, MA: Harvard Univ. Press.

Koertge, Noretta, ed. 2000. *A House Built on Sand: Exposing Postmodernist Myths about Science.* Oxford: Oxford Univ. Press.

Krige, John. 2001. "Distrust and Discovery: The Case of the Heavy Bosons at CERN." *ISIS* 95:517–40.

Kuhn, Thomas S. 1961. "The Function of Measurement in Modern Physical Science." *ISIS* 52:162–76.

Kuhn, Thomas S. 1962. *The Structure of Scientific Revolutions.* Chicago: Univ. of Chicago Press.

Kundu, P. K. 1990. "Gravitational Radiation Field of an Isolated System at Null Infinity." *Proceedings of the Royal Society of London* A 431:337–44.

Kusterer, K. C. 1978. *Know-How on the Job: The Important Working Knowledge of "Unskilled" Workers.* Boulder, CO: Westview Press.

La Sapienza. 1997. Internal Report no. 1088, 20 May 1997. Rome: Univ. of Rome, La Sapienza.

Labinger, Jay, and Harry Collins, eds. 2001. *The One Culture? A Conversation about Science.* Chicago: Univ. of Chicago Press.

Lakatos, I. 1970. "Falsification and the Methodology of Scientific Research Programmes." In *Criticism and the Growth of Knowledge,* ed. I. Lakatos, and A. Musgrave, pp. 91–195. Cambridge: Cambridge Univ. Press.

Lakatos, I. 1976. *Proofs and Refutations.* Cambridge: Cambridge Univ. Press. (Orig. pub. in *British Journal for the Philosophy of Science* 14 (1963): 1–25, 120–39, 221–45, 296–342.)

Langmuir, I. 1989. "Pathological Science." *Physics Today* 42 (10): 36–48 (October. Orig. pub. 1968 as Langmuir, Irving, "Pathological Science," ed. R. N. Hall. General Electric R and D Centre Report, no. 68-C-035, New York. [Edited version of December 18, 1953, colloquium.])

Latour, Bruno, and Steve Woolgar. 1979. *Laboratory Life: The Social Construction of Scientific Facts.* London and Beverly Hills: Sage.

Lee, M., D. Gretz, S. Stepple, and J. Weber. 1976. "Gravitational-Radiation-Detector Observations in 1973 and 1974." *Physical Review D* 14 (4): 893–906.

Levine, James L. 1974. "Comment on a Publication by Bramanti, Maischberger and Parkinson on Gravity Wave Detection." *Letter al Nuovo Cimento* 11 (4): 280–82.

Levine, James L., and Richard L. Garwin. 1973. "Absence of Gravity-Wave Signals in a Bar at 1695 Hz." *Physical Review Letters* 31 (3): 173–76. (July 16.)

Levine, James L., and Richard L. Garwin. 1974. "New Negative Result for Gravitational Wave Detection, and Comparison with Reported Detection." *Physical Review Letters* 33 (13): 794–97.

Linsay, Paul, Peter Saulson, and Rainer Weiss. 1983. *A Study of a Long Baseline Gravitational Wave Antenna System* [aka the "Blue Book"]. Prepared for the National Science Foundation under NSF Grant PHY-8109581.

Lockwood, David. 1964. "Social Integration and System Integration." In *Explorations in Social Change*, ed. George K. Zollshan and Walter Hirsch, pp. 244–57. London: Routledge and Kegan Paul.

Logan, Jonathan L. 1973. "Gravitational Waves—A Progress Report." *Physics Today* 26 (3): 44–52. (March.)

Lynch, Michael. 1985. *Art and Artifact in Laboratory Science: A Study of Shop Work and Shop Talk in a Research Laboratory*. London: Routledge and Kegan Paul.

Lynch, Michael. 1995. "Springs of Action or Vocabularies of Motive." In *Wellsprings of Achievement: Cultural and Economic Dynamics in Early Modern England and Japan*, ed. Penelope Gouk, pp. 94–113. Aldershot, UK: Variorum.

Lynch, Michael, and David Bogen. 1996. *The Spectacle of History: Speech, Text and Memory in the Iran-Contra Hearings*. Durham, NC: Duke Univ. Press.

Lynch, M., E. Livingstone, and H. Garfinkel. 1983. "Temporal Order in Laboratory Work." In *Science Observed: Perspectives on the Social Study of Science*, ed. K. D. Knorr-Cetina and M. Mulkay, pp. 205–38. London: Sage.

MacKenzie, Donald. 1981. *Statistics in Britain 1865–1930*. Edinburgh: Edinburgh Univ. Press.

MacKenzie, Donald. 1998. "The Certainty Trough." In *Exploring Expertise: Issues and Perspectives*, ed. R. Williams, W. Faulkner, and J. Fleck, pp. 325–29. Basingstoke, UK: MacMillan.

MacKenzie, Donald. 2001. *Mechanizing Proof: Computing, Risk, and Trust*. Cambridge, MA: MIT Press.

MacKenzie, Donald, and G. Spinardi. 1995. "Tacit Knowledge, Weapons Design and the Uninvention of Nuclear Weapons." *American Journal of Sociology* 101 (1): 44–99.

Mannheim, Karl. 1936. *Ideology and Utopia: An Introduction to the Sociology of Knowledge*. Chicago: Univ. of Chicago Press.

Martin, B. R., J. E. F. Skea, and E. N. Ling. 1992. *Report to the Advisory Board for the Research Councils and the Economic and Social Research Council*. Swindon, UK: ESRC.

Mather, John C., and John Boslough. 1998/1996. *The Very First Light: A Scientific Journey Back to the Dawn of the Universe*. Harmondworth, UK: Penguin (1st ed. Basic Books.)

McAllister, James. 1996. "The Evidential Significance of Thought Experiments." *Studies in History and Philosophy of Science* 27 (2): 233–50.

McCurdy, Howard E. 1993. *Inside NASA: High Technology and Organizational Change in the U.S. Space Program*. Baltimore: Johns Hopkins Univ. Press.

Medawar, Peter B. 1963/1990. "Is the Scientific Paper a Fraud?" *Listener* 70:377–78. (12 September. Repri. in Pyke, D., ed., *The Threat and the Glory, Reflections on Science and Scientists*. Oxford: Oxford Univ. Press.)

Medawar, Peter B. 1967. *The Art of the Soluble*. London: Methuen & Co. Ltd.

Merton, Robert K. 1942. "Science and Technology in a Democratic Order." *Journal of Legal and Political Sociology* 1:115–26.

Merton, Robert K. 1957. *Social Theory and Social Structure*. Rev. ed. New York: Free Press.

Merton, Robert K. 1973. *The Sociology of Science: Theoretical and Empirical Investigations*. Chicago: Univ. of Chicago Press.

Mervis, Jeffrey. 1991. "Funding of Two Science Labs Revives Pork Barrel vs. Peer-Review Debate." *The Scientist: The Newspaper for the Science Professional* 5 (23): 1, 11. (November 25, 1991.)

Michaels, Anne. 1997. *Fugitive Pieces*. London: Bloomsbury Publishing.

Miller, Arthur. 1952. *The Crucible*. London: Heinemann.

Miller, Arthur. 1987. *Timebends: A Life*. London: Methuen.

Mills, C. W. 1940. "Situated Actions and Vocabularies of Motive." *American Sociological Review* 5:904–13.

Misner, Charles W. 1972. "Interpretation of Gravitational-Wave Observations." *Physical Review Letters* 28 (15): 994–97.

Misner, Charles W., Kip S. Thorne, and John Archibald Wheeler. 1970. *Gravitation*. New York: W. H. Freeman and Co.

Moss, G. E., L. R. Miller, and R. L. Forward. 1971. "Photon-Noise-Limited Laser Transducer for Gravitational Antenna." *Applied Optics* 10:2495–98. (November.)

Mullis, Kary. 2000. *Dancing Naked in the Mindfield*. New York: Vintage Books.

Nader, Laura. 1969. "Up the Anthropologist—Perspectives Gained from Studying Up." In *Reinventing Anthropology*, ed. D. Hymes, pp. 284–311. New York: Random House.

Nicholson, D., C. A. Dickson, W. J. Watkins, B. F. Schutz, J. Shuttleworth, G. S. Jones, D. I. Robertson et al. 1996. "Results of the First Coincident Observations by Two Laser Interferometric Gravitational Wave Detectors." *Physics Letters A* 218:175–80.

Orr, Julian. 1990. "Sharing Knowledge, Celebrating Identity: War Stories and Community, Memory in a Service Culture." In *Collective Remembering: Memory in Society*, ed. David S. Middleton and Derek Edwards, pp. 169–89. Newbury Park, Calif.: Sage Publications Limited.

Paik, H. J. 1974. "Analysis and Development of a Very Sensitive Low Temperature Gravitational Radiation Detector." Ph.D. diss., Stanford Univ.

Paik, H. J. 1976. "Superconducting Tunable-Diaphragm Transducer for Sensitive Acceleration Measurements." *Journal of Applied Physics* 47 (3): 1168–78.

Pamplin, B. R., and H. M. Collins. 1975. "Spoon Bending: An Experimental Approach." *Nature* 257:8. (4 September.)

Parsons, Talcott. 1969. *Political and Social Structure*. New York: Free Press.

Perutz, Max. 1995. "The Pioneer Defended: Review of 'The Private Life of Louis Pasteur,' by Gerald L Geison." *New York Review of Books*, 21 December.

Phinney, E. S. 1991. "The Rate of Neutron Star Binary Mergers in the Universe: Minimal Predictions for Gravity Wave Detectors." *Astrophys. J. Lett* 380:L17–L21.

Pickering, Andrew. 1981a. "Constraints on Controversy: The Case of the Magnetic Monopole." *Social Studies of Science* 11:63–93.

Pickering, Andrew. 1981b. "The Hunting of the Quark." *ISIS* 72:216–36.

Pickering, Andrew. 1984. *Constructing Quarks: A Sociological History of Particle Physics*. Edinburgh: Edinburgh Univ. Press.

Pinch, Trevor J. 1981. "The Sun-Set: The Presentation of Certainty in Scientific Life." *Social Studies of Science* 11:131–58.

Pinch, Trevor J. 1986. *Confronting Nature: The Sociology of Solar-Neutrino Detection*. Dordrecht: Reidel.

Pinch, Trevor, H. M. Collins, and Larry Carbone. 1996. "Inside Knowledge: Second Order Measures of Skill." *Sociological Review* 44 (2): 163–86.

Polanyi, Michael 1958. *Personal Knowledge*. London: Routledge and Kegan Paul.

Polanyi, Michael. 1962. "The Republic of Science, Its Political and Economic Theory." *Minerva* 1:54–73.

Popper, Karl R. 1959. *The Logic of Scientific Discovery*. New York: Harper & Row.

Preparata, G. 1988. "Quantum Mechanics of a Gravitational Antenna." *Il Nuovo Cimento B* 101 (5): 625–35.

Preparata, G. 1990. "'Superradiance' Effect in a Gravitational Antenna." *Modern Physics Letters A* 5 (1): 1–5.

Preparata, G. 1992. "Coherence in QCD and QED." In *Common Problems and Ideas of Modern Physics*, ed. T. Bressani, B. Minettie, and A. Zenoni, pp. 3–35. Singapore: World Publishing Company.

Press, William H., and Kip S. Thorne. 1972. "Gravitational-Wave Astronomy." *Annual Review of Astronomy and Astrophysics* 10:335–74.

Price, R. H., J. Pullin, and P. K. Kundu. 1993. "The Escape of Gravitational Radiation from the Field of Massive Bodies." *Physical Review Letters* 70 (11): 1572–75.

Quine, W. V. O. 1953. *From a Logical Point of View*. Cambridge, MA: Harvard Univ. Press.

Rich, Ben R., and Leo Janos. 1994. *Skunk Works*. New York: Little Brown and Co.

Richards, Evelleen. 1991. *Vitamin C and Cancer: Medicine or Politics*. Basingstoke, UK: MacMillan.

Riordan, Michael. 2001. "A Tale of Two Cultures: Building the Superconducting Super Collider, 1988–1993." *Historical Studies in the Physical Sciences* 32 (1): 125–44.

Rudwick, Martin. 1985. *The Great Devonian Controversy: The Shaping of Scientific Knowledge among Gentlemanly Specialists*. Chicago: Univ. of Chicago Press.

Runciman, W. G. 1966. *Relative Deprivation and Social Justice: A Study of Attitudes to Social Inequality in Twentieth Century England*. London: Routledge and Kegan Paul.

Sapolsky, Harvey. 1972. *The Polaris System Development: Bureaucratic and Programmatic Success in Government*. Cambridge, MA: Harvard Univ. Press.

Saulson, Peter R. 1994. *Fundamentals of Interferometric Gravitational Wave Detectors*. Singapore: World Scientific.

Saulson, Peter. 1997a. "How an Interferometer Extracts and Amplifies Power from a Gravitational Wave." *Classical and Quantum Gravity* 14:2435–54.

Saulson, Peter. 1997b. "If Light Waves Are Stretched by Gravitational Waves, How Can We Use Light as a Ruler to Detect Gravitational Waves?" *Am. J. Phys* 65:501–5.

Saulson, Peter. 1998. "Physics of Gravitational Wave Detection: Resonant and Interferometric Detectors." In *Proceedings of the XXVIth SLAC Summer Institute*, ed. Lance Dixon, pp. 113–62. Stanford, CA: SLAC-R-538. [Reports a meeting of 1998.]

Schatzki, Theodore R., Karin Knorr-Cetina, and Eike von-Savigny, eds. 2001. *The Practice Turn in Contemporary Theory*. London: Routledge.

Schulman, L. S. 1972. "Gravitational Shockwaves from Tachyons." *Il Nuovo Cimento* 2B:38–44.

Schutz, Alfred. 1964. *Collected Papers II: Studies in Social Theory*. The Hague: Martinus Nijhoff.

Sciama, Dennis W. 1969. "Is the Galaxy Losing Mass on a Time Scale of a Billion Years?" *Nature, Physical Science* 224:1263–67.

Sciama, Dennis. 1972. "Cutting the Galaxy's Losses." *New Scientist* 53:373–74. (February 17.)

Sciama, D. W., G. B. Field, and M. J. Rees. 1969. "Upper Limit to Radiation of Mass Energy Derived from Expansion of Galaxy." *Physical Review Letters* 23 (26): 1514–15.

Scott, P., E. Richards, and B. Martin. 1990. "Captives of Controversy: The Myth of the Neutral Social Researcher in Contemporary Scientific Controversies." *Science Technology and Human Values* 15:474–94.

Shaham, Jacob. 1973. "Sidereal Period of Weber Events." *Nature, Physical Science* 246:25–26.

Shapin, Steven. 1979. "The Politics of Observation: Cerebral Anatomy and Social Interests in the Edinburgh Phrenology Disputes." In *On the Margins of Science: The Social Construction of Rejected Knowledge*. Sociological Review Monograph no. 27, ed. R. Wallis. Keele, UK: Keele Univ. Press.

Shapin, Steven. 1994. *A Social History of Truth: Civility and Science in Seventeenth-Century England*. Chicago: Univ. of Chicago Press.

Shapin, Steven. 1995. "Here and Everywhere: Sociology of Scientific Knowledge." *Annual Review of Sociology* 21:289–321.

Shapin, Steven, and Simon Schaffer. 1987. *Leviathan and the Air Pump: Hobbes, Boyle and the Experimental Life*. Princeton, NJ: Princeton Univ. Press.

Shaviv, G., and J. Rosen, eds. 1975. *General Relativity and Gravitation: Proceedings of the Seventh International Conference (GR7), Tel Aviv University, 23–28 June, 1974*. New York: John Wiley.

Sibum, H. Otto. 1995. "Reworking the Mechanical Value of Heat: Instruments of Precision and Gestures of Accuracy in Early Victorian England." *Studies in History and Philosophy of Science* 26 (1): 73–106.

Sibum, H. Otto. 2003. "Experimentalists in the Republic of Letters." *Science in Context* 16 (1–2): 89–120.

Simon, Bart. 1999. "Undead Science: Making Sense of Cold Fusion after the (Arti)fact." *Social Studies of Science* 29 (1): 61–85.

Sinsky, Joel Abram. 1967. "A Gravitational Induction Field Communications Experiment at 1660 Cycles per Second." Ph.D. diss., Univ. of Maryland Dept. of Physics and Astronomy, Technical Report no. 662.

Sinsky, Joel Abram. 1968. "Generation and Detection of Dynamic Newtonian Gravitational Fields at 1660 cps." *Physical Review* 167 (5): 1145–51.

Sinsky, Joel Abram, and Joseph Weber. 1967. "New Source for Dynamical Gravitational Fields." *Physical Review Letters* 18 (19): 795–97.

Smith, R. W. 1992. "The Biggest Kind of Big Science: Astronomy and the Space Telescope." In *Big Science: The Growth of Large Scale Research*, ed. P. Galison and B. Hevly, pp. 184–211. Stanford, CA: Stanford Univ. Press.

Sonnert, G., and G. Holton. 1995. *Gender Differences in Science Careers*. New Brunswick, NJ: Rutgers Univ. Press.

Srivastava, Y. N., A. Widom, and G. Pizzella. 2003. "Electronic Enhancements in the Detection of Gravitational Waves by Metallic Antennae." http://arxiv.org/archive/gr-qc/0302024 v1 (15 pp., February 7.)

Stayer, D. M., and G. Papini. 1982. "Detecting Gravitational Radiation with Quartz Crystals." In *Proceedings of the Second Marcel Grossman Meeting on General Relativity*, ed. R. Ruffini. Amsterdam: North-Holland.

Tart, Charles T. 1972. "States of Consciousness and State-Specific Sciences." *Science* 176:1203–10.

Taylor, J. H., and J. M. Weisberg. 1982. "A New Test of General Relativity: Gravitational Radiation and the Binary Pulsar PSR 1913+16." *Astrophysical Journal* 1 (253): 908–20. (February 15.)

Taylor, John. 1980. *Science and the Supernatural*. London: Temple Smith.

Thomsen, Dietrick E. 1968. "Searching for Gravity Waves." *Science News* 93:408–9 (April 27.)

Thomsen, Dietrick. 1970. "Gravity Waves May Come from Black Holes." *Science News* 98 (26): 480. (December 26.)

Thomsen, Dietrick E. 1978. "Does Gravity Wave?" *Science News* 113 (11): 169–74. (March 18.)

Thorne, Kip S. 1983. "The Theory of Gravitational Radiation: An Introductory Review." In *Rayonnement Gravitationnel*, ed. Nathalie Deruelle and Tsvi Piran, pp. 1–57. Amsterdam: North-Holland.

Thorne, Kip S. 1987. "Gravitational Radiation." In *300 Years of Gravitation*, ed. Stephen Hawking and Werner Israel, pp. 350–458. Cambridge: Cambridge Univ. Press.

Thorne, Kip S. 1989. "Gravitational Radiation: A New Window onto the Universe." Unpublished manuscript.

Thorne, Kip S. 1992a. "Gravitational-Wave Bursts with Memory: The Christodoulou Effect." *Physical Review D* 45 (2): 520–24. (January.)

Thorne, Kip S. 1992b. "On Joseph Weber's New Cross-Section for Resonant-Bar Gravitational-Wave Detectors." In *Recent Advances in General Relativity: Proceedings of a Conference in Honour of E. T. Newman*, ed. A. Janis and J. Porter, pp. 241–50. Boston: Birkhauser.

Thorne, Kip S. 1992c. "Sources of Gravitational Waves and Prospects for Their Detection." In *Recent Advances in General Relativity: Proceedings of a Conference in Honour of E. T. Newman*, ed. A. I. Janis and J. R. Porter, pp. 196–229. Boston: Birkhauser.

Toennies, Ferdinand. 1987. *Community and Association*. East Lansing: Michigan State Univ. Press. (Orig. pub. late 1800s.)

Travis, George David. 1980. "On the Importance of Being Earnest." In *The Social Process of Scientific Investigation: Sociology of the Sciences Yearbook, 4*, ed. Karin Knorr, R. Krohn, and R. Whitley, pp. 165–93. Dordrecht: Reidel.

Travis, George David. 1981. "Replicating Replication? Aspects of the Social Construction of Learning in Planarian Worms." *Social Studies of Science* 11:11–32.

Travis, John. 1993. "LIGO: A $250 Million Gamble." *Science* 260:612–14. (April 30.)

Traweek, Sharon. 1988. *Beamtimes and Lifetimes: The World of High-Energy Physicists*. Cambridge, MA: Harvard Univ. Press.

Tricarico, P., A. Ortolan, A. Solaroli, G. Vedovato, L. Baggio, M. Cerdonio, L. Taffarello et al. 2001. "Correlation between Gamma-Ray Bursts and Gravitational Waves." *Phys. Rev. D* 63 (8): 082002, 1–7.

Turner, Stephen P. 1990. "Forms of Patronage." In *Theories of Science in Society*, ed. Susan E. Cozzens and Thomas F. Gieryn, pp. 185–211. Bloomington: Indiana Univ. Press.

Turner, Stephen P. 1994. *The Social Theory of Practices: Tradition, Tacit Knowledge and Presuppositions*. Oxford: Polity Press.

Tyson, J. A. 1973a. "Gravitational Radiation." *Annals of the New York Academy of Sciences* 224:74–91.

Tyson, J. A. 1973b. "Null Search for Bursts of Gravitational Radiation." *Physical Review Letters* 31 (5): 326–29. (July 30).

Tyson, J. A. 1991. Testimony before the Subcommittee on Science of the Committee on Science, Space and Technology, United States House of Representatives, March 13.

Tyson, J. A., C. G. Maclennan, and L. J. Lanzerotti. 1973. "Correlation of Reported Gravitational Radiation Effects with Terrestrial Phenomena." *Physical Review Letters* 30 (20): 1006–9.

Waldrop, M. Mitchell. 1990. "Of Politics, Pulsars, Death Spirals—and LIGO." *Science* 249:1106–8. (September 7.)

Wallis, Roy, ed. 1979. *On the Margins of Science: The Social Construction of Rejected Knowledge.* Sociological Review Monograph no. 27. Keele, UK: Univ. of Keele Press.

Weber, Joseph. 1959. "Gravitational Waves." New Boston, NH: Gravity Research Foundation. (1st Prize Essay.)

Weber, Joseph. 1960. "Detection and Generation of Gravitational Waves." *Physical Review* 117 (1): 306–13.

Weber, Joseph. 1961. *General Relativity and Gravitational Waves.* New York: Wiley Interscience.

Weber, Joseph. 1962. "On the Possibility of Detection and Generation of Gravitational Waves." *Colloques Internationaux du Centre National de la Recherche Scientifique: XCI Les Theories Realtivistes de la Gravitation, Royaumont, 21–27 Juin 1959.* pp. 441–50. Paris: Editions due Centre National de la Recherche Scientifique. (Content first enunciated at a 1959 conference.)

Weber, Joseph. 1966. "Observation of the Thermal Fluctuations of a Gravitational-Wave Detector." *Physical Review Letters* 17 (24): 1228–30.

Weber, Joseph. 1967. "Gravitational Radiation." *Physical Review Letters* 18 (13): 498–501. (March 27.)

Weber, Joseph. 1968a. "Gravitational Waves." *Physics Today* 21 (4): 34–39.

Weber, Joseph. 1968b. "Gravitational Radiation from the Pulsars." *Physical Review Letters* 21 (6): 395–96.

Weber, Joseph. 1968c. "Gravitational-Wave-Detector Events." *Physical Review Letters* 20 (23): 1307–8. (June 3.)

Weber, Joseph. 1969. "Evidence for the Discovery of Gravitational Radiation." *Physical Review Letters* 22 (24): 1320–24. (June 16.)

Weber, Joseph. 1970a. "Gravitational Radiation Experiments." *Physical Review Letters* 24 (6): 276–79.

Weber, Joseph. 1970b. "Anisotropy and Polarization in the Gravitational-Radiation Experiments." *Physical Review Letters* 25 (3): 180–84. (July 20.)

Weber, Joseph. 1971a. "Gravitational Radiation Experiments." In *Relativity and Gravitation,* ed. C. G. Kuper and A. Peres, pp. 309–22. New York: Gordon and Breach.

Weber, Joseph. 1971b. "The Detection of Gravitational Waves." *Scientific American* 224 (5): 22–29. (May.)

Weber, Joseph. 1972. "Computer Analyses of Gravitational Radiation Detector Coincidences." *Nature* 240:28. (November 3.)

Weber, Joseph. 1974. "Weber Replies." *Physics Today* 27 (12): 11–13. (December.)

Weber, Joseph. 1975. "Weber Responds." *Physics Today* 28 (11): 13, 15, 99. (December.)

Weber, Joseph. 1981. "New Method for Increase of Interaction of Gravitational Radiation with an Antenna." *Physics Letters A* 81 (9): 542–44. (February 23.)

Weber, Joseph. 1984a. "Gravitons, Neutrinos and Antineutrinos." *Foundations of Physics* 14 (12): 1185–1209.

Weber, Joseph. 1984b. "Gravitational Antennas and the Search for Gravitational Radiation." In *Proceedings of Sir Arthur Eddington Symposium, Nagpur, India, January 21–27,* vol. 3, ed. Joseph Weber and T. M. Karade.

Weber, Joseph. 1985. "Method for Observation of Neutrinos and Antineutrinos." *Physical Review C* 31 (4): 1468–1475.

Weber, Joseph. 1986. "Gravitational Antennas and the Search for Gravitational Radiation." In *Proceedings of the Sir Arthur Eddington Centenary Symposium, Nagpur, India, January 21–27*, ed. Joseph Weber and T. M. Karade, 3:1–77. Singapore: World Scientific.

Weber, Joseph. 1988. "Apparent Observation of Abnormally Large Coherent Scattering Cross-Section Using KeV and MeV Range Antineutrinos, and Solar Neutrinos." *Physical Review D* 38 (1): 32–39.

Weber, Joseph. 1991. "Gravitational Radiation Antenna Observations, Theory of Sensitivity of Bar and Interferometer Systems and Resolution of Past Controversy." Lecture presented at the International School of Cosmology and Gravitation, Erice, Sicily.

Weber, Joseph. 1992a. "Supernova 1987A Gravitational Wave Antenna Observations, Cross-Sections, Correlations with Six Elementary Particle Detectors, and Resolution of Past Controversies." In *Recent Advances in General Relativity*, ed. A. I. Janis and J. R. Porter, pp. 230–40. Boston: Birkhauser.

Weber, Joseph. 1992b. "Gravitational Radiation Antenna Observations, Theory of Sensitivity of Bar and Interferometer Systems, and Resolution of Past Controversy." In *Current Topics in Astrofundamental Physics*, ed. N. Sanchez and A. Zichichi, pp. 508–34. Singapore: World Scientific.

Weber, Joseph, D. Gretz, M. Lee, and S. Steppel. 1976. "Gravitational Radiation Detector Observations in 1973 and 1974." *Phys. Rev. D* 14 (4): 893–907.

Weber, Joseph, and J. V. Larson. 1966. "Operation of Lacoste and Romberg Gravimeter at Sensitivity approaching Thermal Fluctuation Limits." *J Geophys Res* 71 (24): 6005–29.

Weber, Joseph, M. Lee, D. J. Gretz, G. Rydbeck, V. L. Trimble, and S. Steppel. 1973. "New Gravitational Radiation Experiments." *Physical Review Letters* 31 (12): 779–83. (September 17.)

Weber, Joseph, and B. Radak. 1996. "Search for Correlations of Gamma-Ray Bursts with Gravitational-Radiation Antenna pulses." *Nuovo Cimento B* 111 (6): 687–92.

Weber, Max. 1930. *The Protestant Ethic and the Spirit of Capitalism*. London: Unwin Hyman.

Weinberg, A. 1967. *Reflections on Big Science*. Cambridge, MA: MIT Press.

Weisberg, Joel M., and Joseph H. Taylor. 1981. "Gravitational Radiation from an Orbiting Pulsar." *General Relativity and Gravitation* 13:1–6.

Weiss, R. 1972. RLE Quarterly Progress Report, no. 105, 54–76. (April; RLE = Lincoln Research Laboratory of Electronics.)

Weiss, R. 1979. "Gravitational Radiation—The Status of the Experiments and Prospects for the Future." In *Sources of Gravitation Radiation*, ed. L. Smarr, pp. 7–35. Cambridge: Cambridge Univ. Press.

Whitley, Richard. 1984. *The Intellectual and Social Organization of the Sciences*. Oxford: Oxford Univ. Press.

Whyte Jr., William H. 1957. *The Organization Man*. New York: Doubleday.

Wiggins, Ralph A., and Frank Press. 1969. "Search for Seismic Signals at Pulsar Frequencies." *Journal of Geophysical Research* 74:22. (October 15.)

Wilson, Brian, ed. 1970. *Rationality*. Oxford: Blackwell.

Winch, Peter G. 1958. *The Idea of a Social Science*. London: Routledge and Kegan Paul.

Wittgenstein, Ludwig. 1953. *Philosophical Investigations*. Oxford: Blackwell.

Zipoy, D. M. 1966. "Light Fluctuations due to an Intergalactic Flux of Gravitational Waves." *Physical Review* 142:825–38.

Zuckerman, Harriet. 1969. "Patterns of Name-Ordering among Authors of Scientific Papers: A Study of Social Symbolism and Its Ambiguity." *American Journal of Sociology* 74:276–91.

Zuckerman, Harriet, and Robert K. Merton. 1971. "Patterns of Evaluation in Science: Institutionalization, Structure and Functions of the Referee System." *Minerva* 9:66–100.

INDEX

Abramovici, Alex, 506n17
academic freedom and management, 688
Academic Freedom and Tenure Committee (Caltech), 583
acoustic noise, 735, 741
acoustic stacks, 44, 63
acronyms, 262
adaptations to rejection, 352ff
Adelberger, Eric G., 447n17
adult lifespan, significance of, 733
Advanced LIGO, 493, 554, 556, 564, 565, 590, 591, 595n7, 612, 613, 646, 649–55, 688, 695n15, 722, 732, 737, 738, 768. *See also* LIGO II
Agamemnon, 588n5
Agar, John, 546–47n1, 548n4, 765n27
Aglietta, M., 365n7, 373n15
AIDS causation heresy, 351
AIGO, 467, 527, 528, 529, 828
algorithm, for extracting GW signals from bars, 157, 168, 182, 186, 188, 191, 192, 201, 204
algorithmic and enculturational models, of knowledge transfer, 388, 608

ALLEGRO, 224, 231, 232, 413, 417, 458, 459, 475, 488, 815, 858, 860; and continuous sources, 462, 474
Allen, Bruce, 467, 468, 488n6, 707nn10–11, 708, 713, 723n23, 725n24
Allen, W. D., 87, 202, 202n5
Allen, Z. A., 707n10, 725n24
Alvarez, Luis, 88, 394
Amaldi, Edoardo, 218, 219, 224n13, 226, 298, 319, 359, 382, 407, 803
amateur publications and networks, 356
American Institute of Physics, 813
anisotropy. *See* sidereal anisotropy
Antiforensic Principle, 412, 577, 578, 578n4, 582, 583, 679, 691, 693
anti-LIGO group, 501n11
Aplin, Peter, 139, 140, 140n11, 193, 194, 228
Archer, Edward, 340
Argonne National Laboratory, 47n17, 85
Aristotle, 327
arrogance, in the scientific community, 418, 419, 689, 690
art of the soluble (and insoluble), 627, 627n18, 774